HUMAN PHYSIOLOGY
The Mechanisms of Body Function

As an additional learning tool, McGraw-Hill also publishes a study guide to supplement your understanding of this textbook. Here is the information your bookstore manager will need to order it for you: 66967-8   STUDY GUIDE TO ACCOMPANY HUMAN PHYSIOLOGY, Fourth Edition

HUMAN PHYSIOLOGY
The Mechanisms of Body Function

Fourth Edition

Arthur J. Vander, M.D.
Professor of Physiology
University of Michigan

James H. Sherman, Ph.D.
Associate Professor of Physiology
University of Michigan

Dorothy S. Luciano, Ph.D.
Formerly of the Department of Physiology
University of Michigan

McGraw-Hill Book Company
New York   St. Louis   San Francisco   Auckland   Bogotá   Hamburg
London   Madrid   Mexico   Montreal   New Delhi
Panama   Paris   São Paulo   Singapore   Sydney   Tokyo   Toronto

HUMAN PHYSIOLOGY: THE MECHANISMS OF BODY FUNCTION

567890 HALHAL 898

ISBN 0-07-066966-X

This book was set in Times Roman by Ruttle, Shaw & Wetherill, Inc.
The editors were Mary Jane Martin and Susan Hazlett;
the designer was Charles A. Carson;
the production supervisor was Charles Hess.
The drawings were done by Fine Line Illustrations, Inc.
Halliday Lithograph Corporation was printer and binder.

Library of Congress Cataloging in Publication Data

Vander, Arthur J., date
   Human physiology.

   Bibliography: p.
   Includes index.
   1. Human physiology.   I. Sherman, James H., date
II. Luciano, Dorothy S.   III. Title.   [DNLM: 1. Phys-
iology.   QT 104 V228h]
QP34.5.V36   1985   612   84-21828
ISBN 0-07-066966-X

TO OUR PARENTS
AND TO JUDY, PEGGY, AND JOE
WITHOUT WHOSE UNDERSTANDING
IT WOULD HAVE BEEN IMPOSSIBLE.

CONTENTS

PREFACE

The primary purpose of this book remains what it was in the first three editions: to present the fundamental mechanisms of human physiology. Our aim has been to tell a story, not to write an encyclopedia. The book is intended for undergraduate students, regardless of their scientific background. The physics and chemistry requisite for an under standing of the physiology are presented where relevant in the text. Students with little or no scientific training will find this material essential while others, more sophisticated in the physical sciences, should profit from the review of basic science oriented toward specifically biological applications. Thus this book is suitable for most introductory courses in human physiology, including those taken by students in the health professions.

The overall organization and approach of the book is based upon a group of themes which are developed in Chapter 1 and which form the framework for our descriptions: (1) all phenomena of life, no matter how complex, are ultimately describable in terms of physical and chemical laws; (2) certain fundamental features of cell function are shared by virtually all cells and, in addition, constitute the foundation upon which specialization is built; (3) the body's various coordinated functions—circulation, respiration, etc.—result from the precise control and integration of specialized cellular activities, serve to maintain relatively constant the internal composition of the body, and can be described in terms of control systems similar to those designed by engineers.

In keeping with these themes, the book progresses from the cell to the total body, utilizing at each level of increasing complexity the information and principles developed previously. Part 1 is devoted to an analysis of basic cellular physiology and the essential physics and chemistry required for its understanding. Part 2 analyzes the concept of the body's internal environment, the nature of biological control systems, and the properties of the major specialized cell types—nerve, muscle, and gland—which constitute these systems. Part 3 then analyzes the coordinated body functions in terms of the basic concepts and information developed in Parts 1 and 2. In this way we have tried to emphasize the underlying unity of biological processes.

This approach has resulted in several characteristics of this book: (1) Cell physiology has received extensive coverage. (2) We have been willing to spend a considerable number of pages logically developing a single cellular concept (such as the origin of membrane potentials) required for the understanding of total-body physiological processes, and which we have found to offer considerable difficulties for the student. (3) We have made every effort not to mention facts simply because they happen to be known, but rather to use facts as building blocks for general principles and concepts. In this last regard, we confess that even an introductory course in physiology must teach a rather frightening number of facts, and our book is no exception. We have tried to keep in mind, however, the words of John Hunter, the eighteenth-century British anatomist and surgeon: ''Too much attention can not be paid to facts; yet too many facts crowd the memory without advantage, any further than they lead us to establish principles.'' (4) We have employed a very large number of figures as an aid to developing concepts and explanations. These figures also provide an excellent summary of the most important material in each chapter. (5) We have not shied away from pointing out the considerable gaps in current understanding. In summary, our book may be long in areas where other texts are brief, and vice versa; moreover, our approach requires the student to think rather than simply to memorize.

In this fourth edition, the level, scope, and emphasis remain unchanged. The text has been completely updated, a process involving changes too numerous to list here. We have also done considerable rewriting in order to improve clarity of organization and presentation.

Arthur J. Vander
James H. Sherman
Dorothy S. Luciano

A FRAMEWORK FOR HUMAN PHYSIOLOGY

I

One cannot meaningfully analyze the complex activities of the human body without a framework upon which to build, a set of viewpoints to guide one's thinking. It is the purpose of this chapter to provide such an orientation to the subject of human physiology—the mechanisms by which the body functions.

Mechanism and Causality

The **mechanist** view of life holds that all phenomena, no matter how complex, are ultimately describable in terms of physical and chemical laws and that no "vital force" distinct from matter and energy is required to explain life. The human being is a machine—an enormously complex machine, but a machine nevertheless. This view has predominated in the twentieth century because virtually all information gathered from observation and experiment has agreed with it. But **vitalism**, the view that some force beyond physics and chemistry is required by living organisms, is not completely dead, nor is it surprising that it lingers specifically in brain physiology, where scientists are almost entirely lacking in hypotheses to explain such phenomena as thought and consciousness in physicochemical terms. We believe that even this area will ultimately yield to physicochemical analysis, but we also feel that it would be unscientific, on the basis of present knowledge, to dismiss the problem out of hand.

A common denominator of physiological processes is their contribution to survival. Unfortunately, it is easy to misunderstand the nature of this relationship. Consider, for example, the statement "During exercise a person sweats because the body needs to get rid of the excess heat generated." This type of statement is an example of **teleology**, the explanation of events in terms of purpose. But it is not an explanation at all in the scientific sense of the word. It is somewhat like saying, "The furnace is on because the house needs to be heated." Clearly, the furnace is on, not because it senses in some mystical manner the house's "needs," but because the temperature has fallen below the thermostat's set point, and the electric current in the connecting wires has turned on the heater.

Is it not true that sweating actually serves a useful purpose because the excess heat, if not eliminated, might have caused sickness or even death? Yes, it is, but this is totally different from stating that a "need" to avoid injury caused the sweating to occur. The cause of the sweating, in reality, was an automatically occurring sequence of events initiated by the increased heat generation: Increased heat generation → increased blood temperature →

increased activity of specific nerve cells in the brain → increased activity, in turn, of a series of nerve cells, the last of which stimulate the sweat glands → increased production of sweat by the sweat-gland cells. Each of these steps occurs by means of physicochemical changes in the cells involved. In science, to explain a phenomenon is to reduce it to a causally linked sequence of physicochemical events. This is the scientific meaning of causality, of the word ''because.''

On the other hand, that a phenomenon is beneficial to the person is of considerable interest and importance. It is attributable to evolutionary processes which result in the selecting of those responses having survival value. Evolution is the key to understanding why most bodily activities do indeed appear to be purposeful. Throughout this book we emphasize how a particular process contributes to survival, but the reader must never confuse this survival value of a process with the explanation of the mechanisms by which the process occurs.

A Society of Cells

Cells: The basic units

Cells are the simplest structural units into which a complex multicellular organism can be divided which still retain the functions characteristic of life. One of the unifying generalizations of biology is that certain fundamental activities are common to almost all cells and represent the minimal requirements for maintaining cell integrity and life. Thus, a human liver cell and an amoeba are remarkably similar in their means of exchanging materials with their immediate environments, of obtaining energy from organic nutrients, of synthesizing complex molecules, and of duplicating themselves.

Each human organism begins as a single cell, the fertilized ovum, which divides to form two cells, each of which divides in turn, resulting in four cells, and so on. If cell multiplication were the only event occurring, the end result would be a spherical mass of identical cells. However, during development, cells begin to exhibit a variety of specialized functions, each cell becoming specialized in the performance of a particular function, such as developing force and movement (in muscle cells) or generating electric signals (in nerve cells). The process of transforming an unspecialized cell into a specialized cell is known as **cell differentiation**. In addition to differentiating, cells migrate to new locations during development and form selective adhesions with other cells to form multicellular structures. In this manner, the cells of the body are arranged in various combinations to form a hierarchy of organized structures. Differentiated cells with similar properties aggregate to form **tissues** (nerve tissue, muscle tissue, etc.). These combine with other types of tissues to form **organs** (the heart, lungs, kidneys, etc.) which are linked together to form **organ systems** (Fig. 1-1).

About 200 distinct kinds of cells can be identified in the body in terms of detailed differences in structure and function. However, when cells are classified according to the general types of functions they perform, four broad categories emerge: (1) **muscle cells**, specialized for the production of mechanical forces which produce movement; (2) **nerve cells**, specialized for initiation and conduction of electric signals over long distances; (3) **epithelial cells**, specialized for the selective secretion and absorption of organic molecules and ions; and (4) **connective-tissue cells**, specialized for the formation and secretion of various types of extracellular connecting and supporting elements. In each of these functional categories there are several cell types that perform variations of the general class of specialized function. For example, there are three different types of muscle cells—skeletal, cardiac, and smooth muscle cells—all of which generate forces and produce movements but which differ from each other in shape, mechanisms controlling their contractile activity, and location in the various organs of the body.

Muscle cells may be attached to bones and produce movements of the limbs or trunk. They may be attached to skin (as, for example, the muscles producing facial expressions) or they may enclose hollow cavities so that their contraction expels the contents of the cavity, as in the case of the pumping of the heart or the expulsion of urine by the bladder.

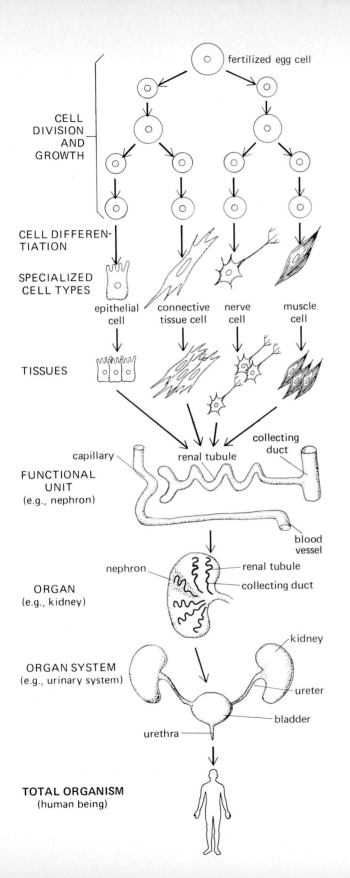

fertilized egg cell

CELL DIVISION AND GROWTH

CELL DIFFERENTIATION

SPECIALIZED CELL TYPES

epithelial cell   connective tissue cell   nerve cell   muscle cell

TISSUES

FUNCTIONAL UNIT (e.g., nephron)

capillary   renal tubule   collecting duct   blood vessel

ORGAN (e.g., kidney)

nephron   renal tubule   collecting duct

ORGAN SYSTEM (e.g., urinary system)

kidney   ureter   bladder   urethra

TOTAL ORGANISM (human being)

FIGURE 1-1. Levels of cellular organization.

Muscle cells also surround many of the hollow tubes in the body; their contraction changes the diameter of these tubes, as in the case of the muscle cells surrounding blood vessels.

Nerve cells have the ability to generate electric signals and the ability to propagate these signals rapidly from one area of the body to another. The signal may influence the initiation of new electric signals in other nerve cells, or it may influence secretion by a gland cell or contraction of a muscle cell. Thus, nerve cells provide a major means of controlling the activities of other cells.

Epithelial cells are located mainly at the surfaces which cover the body (or individual organs) or which line the walls of various tubular and hollow structures within the body. These cells form the boundaries between compartments and function as selective barriers regulating the exchange of molecules across them. For example, the epithelial cells at the surface of the skin form a barrier which prevents most substances in the external environment from entering the body through the skin; the epithelial linings of the lungs, gastrointestinal tract, and kidneys regulate the exchange of molecules between the blood and the external environment, i.e., across an epithelial barrier. Epithelial cells are also found in glands which form from the invagination of the epithelial surfaces (see Chap. 6 for classification of glands).

Connective-tissue cells, as their name implies, have as their major function connecting, anchoring, and supporting the structures of the body. Connective-tissue cells typically have a large amount of extracellular material between them. The cells themselves include those in the loose meshwork of cells and fibers underlying most epithelial layers and also include other cell types as diverse as fat-storing (adipose) cells, bone cells, and red and white blood cells. Many connective-tissue cells secrete into the fluid surrounding them molecules which form a matrix consisting of various types of protein **fibers** embedded in a **ground substance** made of polysaccharides and protein. This matrix may vary in consistency from a semifluid gel (in loose connective tissue) to the solid crystalline structure of bone. The extracellular fibers formed by these cells include ropelike **collagen** fibers (which have a high tensile strength and resist stretching), rubber-band-like **elastic** fibers (which provide the elastic qualities found in various organs), and fine, highly branched **reticular** fibers.

Tissues

Most specialized cells are associated with other cells of a similar kind, forming aggregates known as **tissues**. Corresponding to the four general categories of differentiated cells, there are four general classes of tissues: (1) **muscle tissue**, (2) **nerve tissue**, (3) **epithelial tissue**, and (4) **connective tissue**. It should be noted that the term ''tissue'' is frequently used in several different ways. It is formally defined as just described, i.e., as an aggregate of a single type of specialized cell. However, it is also commonly used to denote the general cellular fabric of any given organ or structure, e.g., kidney tissue or lung tissue, each of which in fact usually contains all four classes of tissue.

Organs and organ systems

Organs are composed of the four kinds of tissues arranged in various proportions and patterns—sheets, tubes, layers, bundles, strips, etc. For example, the kidney consists largely of (1) a series of small tubes, each composed of a single layer of epithelial cells; (2) blood vessels, whose walls consist of an epithelial lining and varying quantities of smooth muscle and connective tissue; (3) nerve fibers with endings near the muscle and epithelial cells; and (4) a loose network of connective-tissue elements that are interspersed throughout the kidney and that also form an enclosing capsule. The structure of many organs is organized into small, similar subunits often referred to as **functional units**, each performing the function of the organ. For example, the kidneys' 2 million functional units, the nephrons, are the tubules with their closely associated blood vessels. The total production of urine by the kidneys is the sum of the amounts formed by the individual nephrons.

Finally, the last order in classification is that of

the **organ system**, a collection of organs which together perform an overall function. For example, the kidneys, the urinary bladder, the tubes leading from the kidneys to the bladder, and the tube leading from the bladder to the exterior constitute the urinary system. There are ten organ systems in the body. Their components and functions are given in Table 1-1.

To sum up, the human body can be viewed as a complex society of cells of many different types which are structurally and functionally combined and interrelated in a variety of ways to carry on the functions essential to the survival of the organism as a whole. Yet the fact remains that individual cells constitute the basic units of this society, and almost all of these cells individually exhibit the fundamental activities common to all forms of life. Indeed, many of the body's different cells can be removed and maintained in test tubes as free-living cells.

There is a definite paradox in this analysis. If each individual cell performs the fundamental activities required for its own survival, what contributions do the different organ systems make? How is it that the functions of the organ systems are essential to the survival of the organism as a whole when each individual cell of the organism seems to be capable of performing its own fundamental activities? The resolution of this paradox is found in the isolation of most of the cells of a multicellular organism from the environment surrounding the

---

TABLE 1-1. Organ systems of the body

System	Major organs or tissues	Primary functions
Circulatory	Heart, blood vessels, blood	Rapid flow of blood throughout the body's tissues
Respiratory	Nose, pharynx, larynx, trachea, bronchi, bronchioles, lungs	Exchange of carbon dioxide and oxygen; regulation of hydrogen-ion concentration
Digestive	Mouth, pharynx, esophagus, stomach, intestines, salivary glands, pancreas, liver, gallbladder	Digestion and absorption of organic nutrients, salts, and water
Urinary	Kidneys, ureters, bladder, urethra	Regulation of plasma composition through controlled excretion of organic wastes, salts, and water
Musculo-skeletal	Cartilage, bone, ligaments, tendons, joints, skeletal muscle	Support, protection, and movement of the body
Immune	White blood cells, lymph vessels and nodes, spleen, thymus, and other lymphatic tissues	Defense against foreign invaders; return of extracellular fluid to blood; formation of white blood cells
Nervous	Brain, spinal cord, peripheral nerves and ganglia, special sense organs	Regulation and coordination of many activities in the body; detection of changes in the internal and external environments; states of consciousness; learning
Endocrine	All glands secreting hormones: Pancreas, testes, ovaries, hypothalamus, kidneys, pituitary, thyroid, parathyroid, adrenal, intestinal, thymus, and pineal	Regulation and coordination of many activities in the body
Reproductive	Male: Testes, penis, and associated ducts and glands	Production of sperm; transfer of sperm to female
	Female: Ovaries, uterine tubes, uterus, vagina, mammary glands	Production of eggs; provision of a nutritive environment for the developing embryo and fetus
Integumentary	Skin	Protection against injury and dehydration; defense against foreign invaders; regulation of temperature

body (the external environment), and the presence of an internal environment.

The Internal Environment

An amoeba and a human liver cell both obtain their energy by the breakdown of certain organic nutrients; the chemical reactions involved in this intracellular process are remarkably similar in the two types of cells and involve the utilization of oxygen and the production of carbon dioxide. The amoeba picks up oxygen directly from the fluid surrounding it (its external environment) and eliminates carbon dioxide into the same fluid. But how can the liver cell obtain its oxygen and eliminate the carbon dioxide when, unlike the amoeba, it is not in direct contact with the external environment (the air surrounding the body)? Supplying oxygen to the liver is the function both of the respiratory system (comprising the lungs and the airways leading to them), which takes up oxygen from the external environ-

ment, and of the circulatory system, which distributes oxygen to all parts of the body. In addition, the circulatory system carries the carbon dioxide generated by the liver cells and all the other cells of the body to the lungs, which eliminate it to the exterior. Similarly, the digestive and circulatory systems, working together, make nutrients from the external environment available to all the body's cells. Wastes other than carbon dioxide are carried by the circulatory system from the cells which produced them to the kidneys and liver, which excrete them from the body (Fig. 1-2). The kidneys also regulate the concentrations of water and many essential minerals in the plasma. Thus the overall effect of the activities of organ systems is to create within the body the environment required for all cells to function. This is called the **internal environment**.

The internal environment is not merely a theoretical physiological concept. It can be identified quite specifically in anatomical terms: The body's

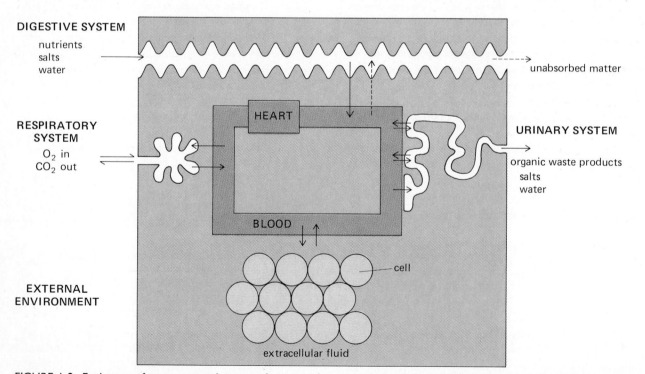

FIGURE I-2. Exchanges of matter occur between the external environment and the circulatory system via the digestive system, respiratory system, and urinary system.

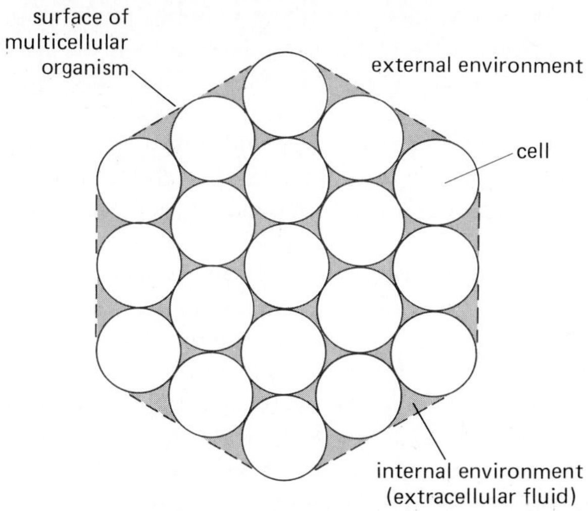

FIGURE 1-3. Extracellular fluid is the internal environment of the body.

internal environment is the **extracellular fluid** (literally, fluid outside the cells), which bathes each of the body's cells (Figs. 1-2 and 1-3). In other words, the environment in which each cell lives is not the "external environment" surrounding the entire body but the local extracellular fluid surrounding that cell. It is from this fluid that the cells receive oxygen and nutrients and into which they excrete wastes. A multicellular organism can survive only as long as it is able to maintain the composition of its internal environment in a state compatible with the survival of its individual cells. The French physiologist Claude Bernard first clearly described in 1857 the central importance of the extracellular fluid of the body: "It is the fixity of the internal environment which is the condition of free and independent life. . . . All the vital mechanisms, however varied they may be, have only one object, that of preserving constant the conditions of life in the internal environment." This concept of an internal environment and the necessity of maintaining its composition relatively constant is the single most important unifying idea to be kept in mind while attempting to unravel and understand the functions of the body's organ systems and their interrelationships.

To summarize, the total activities of every individual cell in the body fall into two categories: (1)

Each cell performs for itself all those fundamental basic cellular processes (movement of materials across its membrane, extraction of energy, protein synthesis, etc.) which represent the minimal requirements for maintaining its individual integrity and life; and (2) each cell simultaneously performs one or more specialized activities which, in concert with the activities performed by the other cells of its tissue or organ system, contribute to the survival of the total organism by helping maintain the stable internal environment (extracellular fluid) required by all cells. These specialized activities together constitute the coordinated body processes (circulation, respiration, digestion, etc.) typical of multicellular organisms like human beings.

Clearly, the society of cells which constitutes a human body bears many striking similarities to a society of persons (although the analogy must not be pushed too far). Each person in a complex society must perform individually a set of fundamental activities (eating, excreting, sleeping, etc.), which is virtually the same for all persons. In addition, because the complex organization of a society makes it virtually impossible for any individual within the society to raise his or her own food, arrange for the disposal of wastes, and so on, each individual participates in the performance of one of these supply-and-disposal operations required for the survival of all. A specialized activity, therefore, becomes an additional part of one's daily routine, but it never allows the individual to cease or to reduce the fundamental activities required for survival.

Body-fluid compartments

Extracellular fluid is located in two so-called compartments. Approximately 80 percent of the extracellular fluid surrounds the body's cells (excluding the blood cells). Because it lies between cells, it is also known as **intercellular** or, more often, **interstitial fluid**. The remaining 20 percent of the extracellular fluid is the fluid portion of the blood, the **plasma**. The blood (plasma and blood cells suspended in the plasma) is continuously circulated by the action of the heart to all parts of the body. As will be seen in Chap. 11, the plasma exchanges oxygen, nutrients, wastes, and other metabolic products with the interstitial fluid as the blood passes through the capillaries of the body. Because of the transcapillary exchanges between plasma and interstitial fluid, solute concentrations are virtually identical in the two fluids, except for protein concentration. With this major exception, the entire extracellular fluid may be considered to have a homogeneous composition which is very different from that of the **intracellular fluid** (the fluid inside the cells). These differences will be described in Chap. 6.

To generalize one step further, the human body can be viewed as containing three fluid compartments: (1) blood plasma, (2) interstitial fluid, and (3) intracellular fluid (Fig. 1-4). The major molecular component of all three compartments is water, which accounts for about 60 percent of the normal body weight, and amounts to about 42 liters (L) in a standard man. Two-thirds of the total body water (28 L) is intracellular fluid. The remaining one-third of the total body water (14 L) is the extracellular fluids, 80 percent as interstitial fluid (11 L) and 20 percent as plasma (3 L).

The above figures for the volumes of fluids in the various compartments of the body are those of a 70-kg (154-lb) man. Obviously, body measurements vary between different individuals. Some measurements vary directly with body weight and can be expressed as a percentage of the total body weight, whereas others depend on age, sex, and state of health as well as body weight. The typical values for a healthy, 70-kg, 21-year-old male have been used for years as representative measurements for an individual. This standard was chosen because most of the data collected over the years on normal, healthy individuals has come from measurements made on male medical students. Measurements on females of the same age, corrected for differences in total body weight, give values of a similar magnitude, although slight sex-specific differences do exist. Thus, a female has a slightly lower total body-water content than a male of the same weight because her body is composed of a higher percentage of fat-storing (adipose) tissue, which contains less water than other tissues. Unless otherwise stated, quantitative values given in this text refer to a healthy, 70-kg, 21-year-old male.

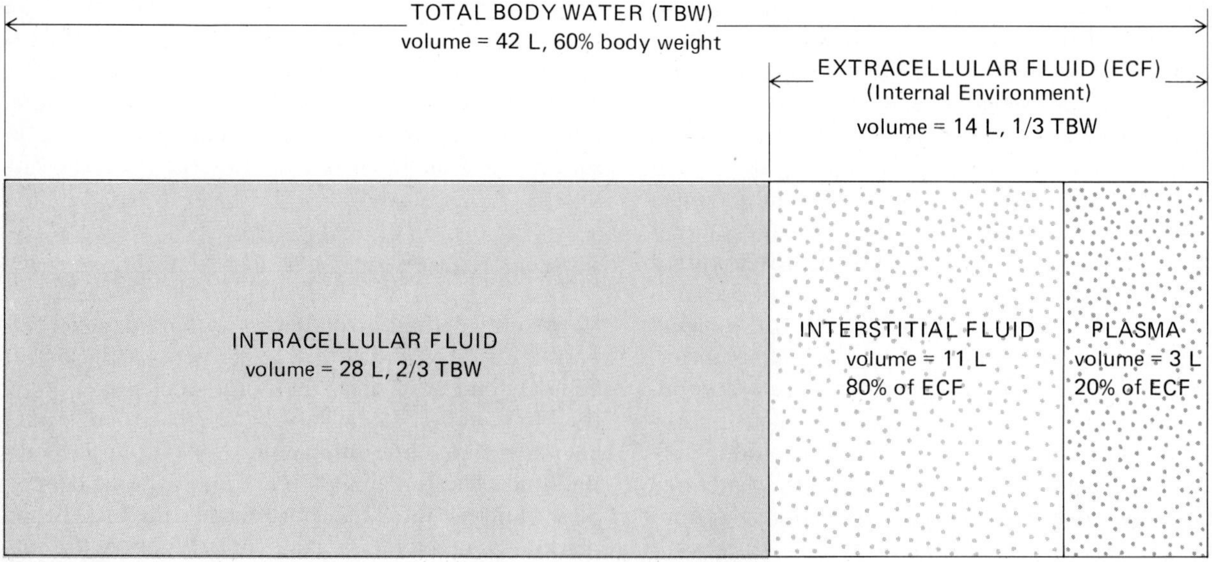

FIGURE I-4 Fluid compartments of the body. Volumes are for a "standard" 70-kg man.

The Importance of Control

Implicit in life is control. Regardless of its level of organizational complexity, no living system can exist without precise mechanisms for controlling its various activities.

Every one of the fundamental processes performed by any single cell (an amoeba or a liver cell) must be carefully regulated. What determines how much sugar enters a cell? Once inside the cell, what proportion of this sugar is to be utilized for energy or transformed into fat or protein? How much protein of each type is to be synthesized, and when? How large is the cell to grow, and when is it to divide? The list is almost endless. Thus, an understanding of cellular physiology requires knowledge not only of the basic processes but also of the mechanisms which control them. Indeed, the two are inseparable.

In any multicellular organism like a human being, these basic intracellular regulators remain, but the existence of a multitude of different cells organized into specialized tissues and organs obviously imposes the requirement for overall regulatory mechanisms. Information about all important aspects of the external and internal environments must be monitored continuously; on the basis of the content of this information, ''instructions'' must be sent to the various tissue and organ cells (particularly muscle and gland cells) directing them to increase or decrease their activities.

This processing of information is performed primarily by the nervous and hormonal systems. Thus these two systems contribute to the survival of the organism by controlling the activities of the various bodily components so that any change (or impending change) in the body's internal environment automatically initiates a chain of events wiping out the change or leading to a state of readiness should the impending change actually occur. For example, when (for any reason) the concentration of oxygen in the blood significantly decreases below normal, the nervous system detects the change and increases its output to the skeletal muscles respon-sible for breathing movements. The result is a compensatory increase in oxygen uptake by the body and a restoration of normal internal oxygen concentration.

A Rationale

With this introductory framework in mind, the overall organization and approach of this book should easily be understood. Because the fundamental features of cell function are shared by virtually all cells and, in addition, constitute the foundation upon which specialization develops, we devote the first part of this book to an analysis of basic cellular physiology. Much of this cell biology is now referred to as **molecular biology** in recognition of the ultimate goal of explaining cellular processes in terms of interactions between molecules.

At the other end of the organizational spectrum, the third part of the book describes how the body's various coordinated functions (circulation, respiration, etc.) result from precisely controlled and integrated activities of specialized cells grouped together in tissues and organs. The theme of these descriptions is that each of these coordinated functions (with the obvious exception of reproduction) serves to keep some important aspect of the body's internal environment relatively constant.

The second part of the book provides the principles and information required to bridge the gap between these two organizational levels, the cell and the body. Control systems are analyzed first in general terms to emphasize that the basic principles governing virtually all control systems are the same, but the bulk of Part 2 is concerned with the major components of the body's control systems (nerve cells, muscle cells, and gland cells). Once acquainted with the cast of major characters—nerve, muscle, and gland cells—and the theme—the maintenance of a stable internal environment through the interactions of these components—the reader will be free to follow the specific plot lines—circulation, respiration, etc.—of Part 3.

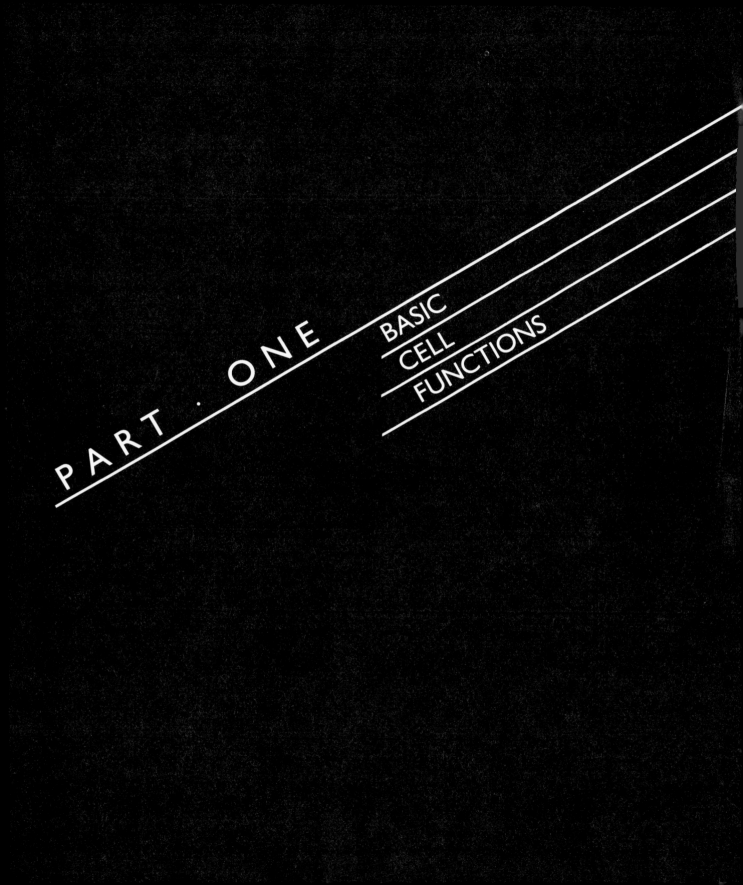

PART . ONE

BASIC
CELL
FUNCTIONS

CELL STRUCTURE

The human body is composed of trillions of cells, each of which is a microscopic compartment. In fact, the word "cell" means a small chamber (like a jail cell). Within each cell, numerous chemical reactions, collectively referred to as **metabolism,** take place. Cell survival depends upon this metabolic activity, which in turn is dependent upon a continuous supply of chemicals (nutrients) from the surrounding extracellular fluid. These nutrients provide the chemical energy and raw materials necessary to build and maintain cell structure and to carry out cell functions.

Cells are very small and must be so for the following reason. After crossing a barrier (the plasma membrane) at the cell surface, nutrients are distributed throughout the cell interior by diffusion—a physical process involving the random motion of individual molecules, as will be described in Chap. 6. Diffusion over long distances is a very slow process. If a cell were to grow larger and larger, it would eventually reach a size where the rate at which nutrients could be distributed throughout the cell by diffusion would not be sufficiently rapid to meet the metabolic demands of the cell for energy and raw materials, and the cell would die.

Thus, there are functional (physiological) limitations on the size of cells. The cells of a mouse, a human being, and an elephant are all approximately the same size. An elephant is large because it has many, many cells, not because it has larger cells. A majority of the cells in a human being have diameters in the range of 10 to 20 μm,[1] although cells as small as 2 μm and as large as 120 μm are present. A cell 10 μm in diameter is about one-tenth the size of the smallest object that can be seen with the naked eye; therefore, microscopes must be used to examine the structure of cells.

Microscopic Observations of Cells

The size of the smallest object that can be clearly seen through a microscope depends upon the wavelength of the radiation used to illuminate the specimen—the smaller the wavelength, the smaller the object that can be resolved. With a light microscope, the smallest object that can be seen clearly is about 0.2 μm in diameter, and this instrument can be used to visualize individual cells as well as some aspects of intracellular structure. However, an electron microscope is necessary to make visible the fine details of cellular structure that exist inside of most cells.

Instead of light rays, an electron microscope uses electrons. A beam of electrons can be focused on a specimen by electromagnetic lenses, just as glass lenses focus a beam of light in a light microscope. A greater resolution is achieved by an electron microscope than by a light microscope, because electrons behave as waves with much smaller

[1] See Appendix I for a listing of metric units of measurement and their English equivalents.

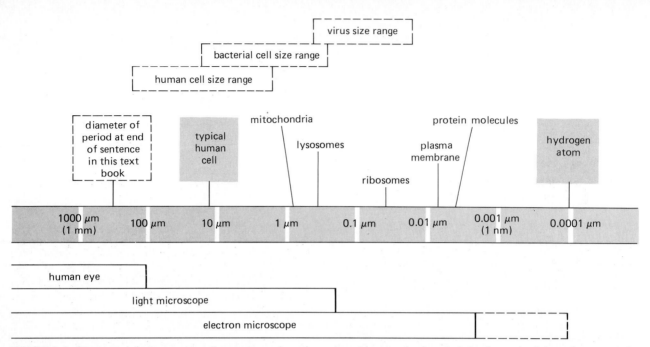

FIGURE 2-1. Size range of microscopic cell structures, plotted on a logarithmic scale. A typical size for each of the cell components is given; each has a range of sizes around this typical value.

wavelengths than those of visible light. Cell structures as small as 0.002 μm (2 nm) can be resolved with an electron microscope, and resolutions approaching 0.2 nm can be achieved under very special conditions with nonbiological materials. This degree of resolution approaches the size of individual atoms, the smallest of which, hydrogen, is about 0.1 nm in diameter. In fact, very large molecules, such as the molecules of proteins and nucleic acids, can be seen, although the individual atoms that make up these large biological molecules are still beyond the limits of effective resolution. The distribution of microscopic structures lying in the size range from atoms to cells is illustrated in Fig. 2-1.

Although living cells can be observed with a light microscope, this is not possible with an electron microscope. To form an image with an electron beam, most of the electrons must pass through the specimen, just as light passes through a specimen in a light microscope. However, electrons can penetrate only a very short distance through matter; therefore, only very thin specimens can be exam-

ined. Cells must be specially prepared for observation in an electron microscope by cutting them into thin sections, like sections of a salami. Each section is on the order of 100 nm thick, which is about one-hundredth of the thickness of a typical cell.

Because electron micrographs (such as Fig. 2-2) are the images of very thin sections of a cell rather than of the whole cell, they can often be misleading. Structures that appear as separate objects in the electron micrograph may actually be continuous structures that are connected through a region lying outside of the plane of the thin section. For example, a thin section through a ball of string would appear as a collection of separate lines and disconnected dots, even though it was originally one continuous piece of string.

The microscopic exploration of the human body has revealed that cells do not all have the same general shape. Very few cells are smooth spheres, which is the image of a cell most likely to come to mind. The free-floating white blood cells most closely approximate this generalized image. Most cells are surrounded by and anchored to neighboring cells, and such immobilized cells usually have a polyhedral or approximately boxlike shape. Many cells have shapes that are related to specific functions that they perform. For example, the disklike shape of red blood cells shortens the diffusion distance between extracellular and intracellular compartments, thereby allowing the rapid exchange of oxygen and carbon dioxide (the transport of which in the blood is the major function of red blood cells). Muscle cells have an essentially cylindrical shape. Nerve cells have, extending from a central region of the cell, one or more cylindrical processes which are used to receive and transmit electrical signals. Many of the relations between cell shape and cell function will be described in more specific terms in later chapters. For the remainder of this chapter we will be concerned with the structures that are found inside most cells (but not every cell) in the human body.

Cell Organelles

Compare an electron micrograph of a section through a real cell, Fig. 2-2, with a diagrammatic illustration of a typical cell, Fig. 2-3. What is immediately obvious, both from the diagram and the electron micrograph, is the extensive amount of structure inside a cell. A human cell is not simply a chemical solution surrounded by a limiting barrier, the **plasma membrane**. The cell interior is divided into a number of compartments surrounded by intracellular membranes. These membrane-bound compartments, along with some nonmembranous particles and filamentous structures, are known as **cell organelles** (little organs). Each cell organelle performs specific functions which contribute to the functioning and survival of the cell as a whole. The cell structures found in most cells are listed in Table 2-1. In this chapter we briefly describe each of them, while later chapters will elaborate upon the mechanisms by which they perform their functions.

The interior of a cell is composed of two large regions: (1) the **nucleus**, a spherical or oval organelle located near the center of a cell; and (2) the **cytoplasm**, the region located outside the nucleus. Most cell organelles are in the cytoplasm, suspended in a fluid known as the **cytosol** (Fig. 2-4). Molecules entering a cell across the plasma membrane enter the cytosol, from which they may then pass into one of the cell organelles. The **intracellular fluid**, which is the total fluid (water plus dissolved molecules) inside a cell, comprises not only the cytosolic fluid but also the fluid inside the organelles including the nucleus. Thus, intracellular fluid is not a fluid of uniform chemical composition.

Membranes form the major structural element in cells and are found surrounding the entire cell (the plasma membrane) and enclosing most of the cell organelles. Membranes are very flexible structures, much like a piece of cloth; they can be bent and folded but cannot be stretched without being torn. This flexiblity, along with the fact that cells are filled with fluid, allows cells to undergo considerable change in shape without disruption of their structural and functional integrity.

Although membranes perform a variety of functions, as will be described in Chap. 6, their primary function is to act as a selective barrier to the passage of molecules, allowing some molecules to cross while excluding others. Thus, the plasma membrane regulates the passage of substances into and out of the cytosol, whereas the membranes surrounding cell organelles allow selective movement of substances between the organelles and the cytosol. Because this selectivity can be altered, the rates and types of molecular passage can be controlled. One of the consequences of the restricted movements of molecules across various membranes is that the products of chemical reactions can often be confined to specific cell organelles.

Almost all cells contain a single **nucleus**, generally located near the center of the cell. The overall function of the nucleus is the transmission and expression of genetic information. Surrounding the nucleus is a barrier, the **nuclear envelope**, composed of two membranes separated by a small space. At regular intervals along the surface of the nuclear envelope the two membranes become joined, form-

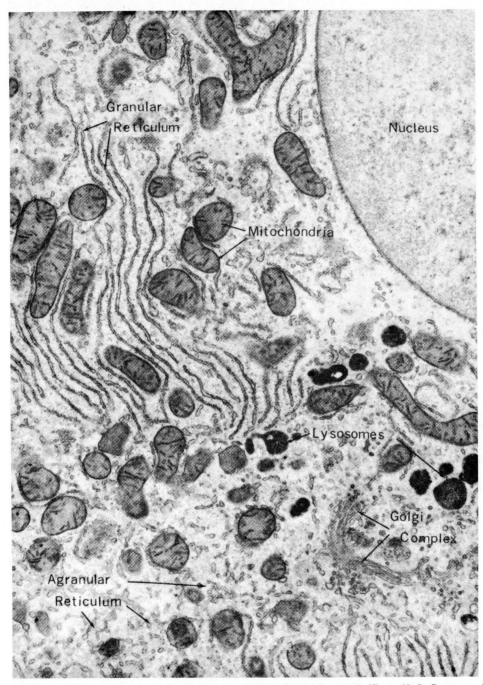

FIGURE 2-2. Electron micrograph of a thin section through a portion of a rat liver cell. [From K. R. Porter, in T. W. Goodwin and O. Lindberg (eds), "Biological Structure and Function," vol. 1, Academic Press, Inc., New York, 1961.]

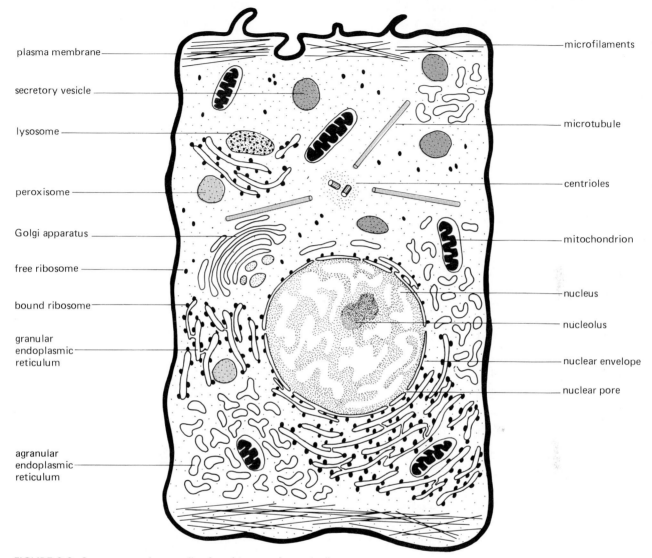

plasma membrane

secretory vesicle

lysosome

peroxisome

Golgi apparatus

free ribosome

bound ribosome

granular endoplasmic reticulum

agranular endoplasmic reticulum

microfilaments

microtubule

centrioles

mitochondrion

nucleus

nucleolus

nuclear envelope

nuclear pore

FIGURE 2-3. Structures and organelles found in most human cells.

ing the rims of circular openings known as **nuclear pores** (Fig. 2-3). Through these pores molecular messages move into and out of the nucleus.

The most prominent structure within the nucleus is the densely staining **nucleolus**, a highly coiled, filamentous structure associated with numerous granules but not surrounded by a membrane. The nucleolus is the site at which ribosomes, major cytoplasmic organelles, are assembled. Some

of the molecules that are used to make a ribosome are synthesized in the region of the nucleolus, whereas others are synthesized in the cytoplasm and then moved into the nucleus. The completed ribosomal particles move into the cytoplasm, where they become the sites at which protein molecules are synthesized.

The nucleus contains, in addition to the nucleolus, a fine network of threads known as **chro-**

TABLE 2-1. Cell structures found in most cells

Plasma membrane

Nucleus

    Nuclear envelope

    Chromatin

    Nucleolus

Cytoplasm—membrane-bound structures

    Endoplasmic reticulum

        Granular

        Agranular

    Golgi apparatus

    Mitochondria

    Lysosomes

    Peroxisomes

Cytoplasm—nonmembranous structures

    Ribosomes

    Microfilaments

    Intermediate filaments

    Microtubules

    Centrioles

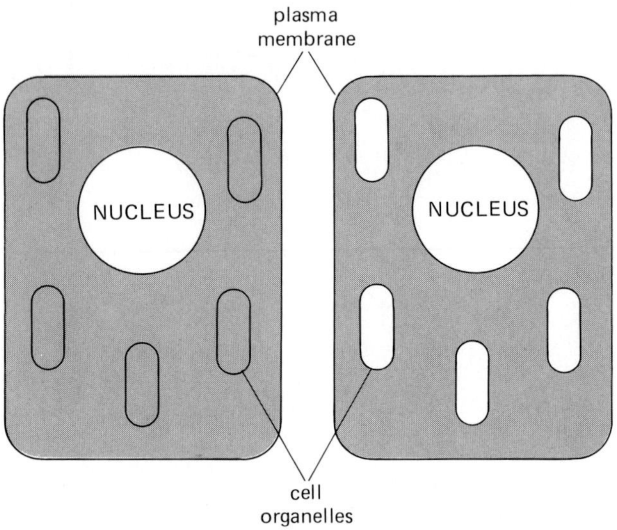

CYTOPLASM (in color)   CYTOSOL (in color)

FIGURE 2-4. Comparison of the locations of cytoplasm and cytosol.

matin, which are coiled to a greater or lesser degree, producing variations in the granular density of the nucleus. These threads (which become highly condensed to form **chromosomes** at the time of cell division) contain molecules of DNA (see Chaps. 3 and 4), which store genetic information and pass it on from cell to cell each time a cell divides. This genetic information directs the synthesis of protein molecules on the cytoplasmic ribosomes by way of messenger molecules that pass from the nucleus to the cytoplasm.

The most extensive cytoplasmic cell organelle is the system of membranes which forms the **endoplasmic reticulum** (Fig. 2-5). It extends throughout the cytoplasm in the form of relatively flattened sacs or as a branching tubular network. The membranes enclose a space that is continuous through-

out the membranous network as well as with the space between the two membranes of the nuclear envelope (Fig. 2-3). Two types of endoplasmic reticulum can be distinguished: **granular** (rough-surfaced) and **agranular** (smooth-surfaced) **endoplasmic reticulum**. The granular endoplasmic reticulum has ribosomes, which appear as small particles, embedded in its cytosolic surface and has a flattened sac appearance when seen in an electron micrograph. The agranular endoplasmic reticulum has no ribosomal particles on its surface and has a branched, tubular structure. Both granular and agranular endoplasmic reticula exist in the same cell, and the lumen enclosed by both these sets of membranes appears to be continuous. The relative amounts of the two types of reticulum vary in different types of cells and even within the same cell with changes in cell activity.

Ribosomes, in addition to being bound to the granular endoplasmic reticulum, are also found free in the cytoplasm. Each ribosome is about 200 nm in diameter and is composed of a large number of protein molecules and several nucleic acid molecules (see Chap. 4). The ribosomes, as noted above, are assembled in the nucleolus and then pass into the cytoplasm. The function of the ribosomes is to

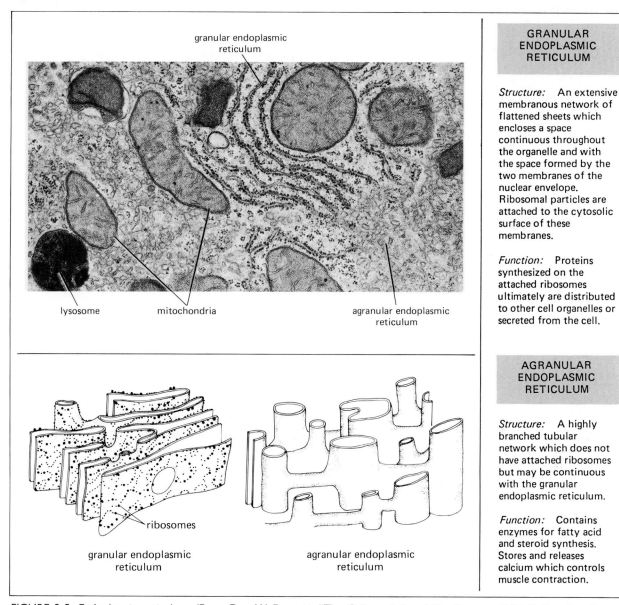

granular endoplasmic
reticulum

lysosome   mitochondria   agranular endoplasmic
reticulum

ribosomes

granular endoplasmic
reticulum

agranular endoplasmic
reticulum

GRANULAR ENDOPLASMIC RETICULUM

Structure: An extensive membranous network of flattened sheets which encloses a space continuous throughout the organelle and with the space formed by the two membranes of the nuclear envelope. Ribosomal particles are attached to the cytosolic surface of these membranes.

Function: Proteins synthesized on the attached ribosomes ultimately are distributed to other cell organelles or secreted from the cell.

AGRANULAR ENDOPLASMIC RETICULUM

Structure: A highly branched tubular network which does not have attached ribosomes but may be continuous with the granular endoplasmic reticulum.

Function: Contains enzymes for fatty acid and steroid synthesis. Stores and releases calcium which controls muscle contraction.

FIGURE 2-5. Endoplasmic reticulum. (From Don W. Fawcett, "The Cell, an Atlas of Fine Structure," W. B. Saunders Company, Philadelphia, 1966.)

synthesize protein molecules from amino acid subunits, (see Chaps. 3 and 4) using genetic information sent by messenger molecules from the genes in the nucleus.

The proteins synthesized by those ribosomes which are attached to the granular endoplasmic reticulum pass into the lumen of the reticulum as they are synthesized. These proteins are destined to be transferred to another organelle, the Golgi apparatus. In contrast, the proteins synthesized on the free ribosomes are released into the cytosol, where they perform their functions.

The agranular reticulum performs several functions, including the synthesis of various types of

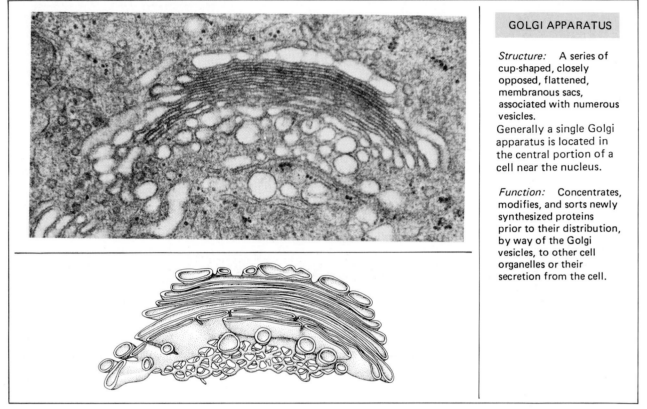

Structure: A series of cup-shaped, closely opposed, flattened, membranous sacs, associated with numerous vesicles. Generally a single Golgi apparatus is located in the central portion of a cell near the nucleus.

Function: Concentrates, modifies, and sorts newly synthesized proteins prior to their distribution, by way of the Golgi vesicles, to other cell organelles or their secretion from the cell.

FIGURE 2-6. Golgi apparatus. *(From W. Bloom and D. W. Fawcett, "Textbook of Histology," 9th ed., W. B. Saunders Company, Philadelphia, 1968.)*

"fats," some of which are incorporated into the basic structure of cell membranes (see Chaps. 5 and 6).

The **Golgi apparatus** (Fig. 2-6), named in honor of Camillio Golgi, who first described this cell organelle, consists of a series of closely opposed, flattened membranous sacs which are slightly curved, forming a cup-shaped structure. Most cells have a single Golgi apparatus located near the nucleus, although some cells may have several. Associated with this organelle, particularly near its concave surface, are a number of membrane-enclosed vesicles. As mentioned above, proteins that are synthesized on the ribosomes attached to the granular endoplasmic reticulum are transferred to the Golgi apparatus. The Golgi apparatus, during the passage of various proteins through its lumen, selectively sorts them into vesicles and in some manner determines the address to which each vesicle will be delivered. When proteins are secreted from cells, the vesicles produced by the Golgi apparatus are known as **secretory vesicles.**

Mitochondria are rod- or oval-shaped organelles that are surrounded by two membranes, an inner and an outer membrane (Fig. 2-7). The outer membrane is smooth, whereas the inner membrane is folded into sheets or tubules known as **cristae,** which extend into the internal space (**matrix**) of the mitochondrion. Mitochondria are found scattered throughout the cytoplasm. Large numbers of them are present in cells that use large amounts of energy; for example, a single liver cell may contain 1000 mitochondria. Less active cells contain fewer mitochondria. These organelles are concerned with

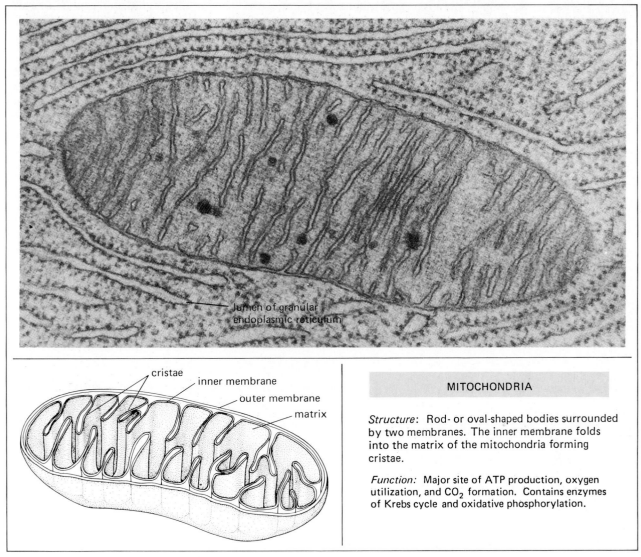

Lumen of granular
endoplasmic reticulum

cristae
inner membrane
outer membrane
matrix

MITOCHONDRIA

Structure: Rod- or oval-shaped bodies surrounded by two membranes. The inner membrane folds into the matrix of the mitochondria forming cristae.

Function: Major site of ATP production, oxygen utilization, and CO_2 formation. Contains enzymes of Krebs cycle and oxidative phosphorylation.

FIGURE 2-7. Mitochondria. *(Courtesy of K. R. Porter.)*

the chemical processes by which energy is made available to cells in the form of molecules of adenosine triphosphate (ATP) (see Chap. 5). Most of the ATP used by cells is formed in the mitochondria. The chemical reactions of this process consume oxygen and produce carbon dioxide.

Lysosomes are spherical or oval bodies surrounded by a single membrane which encloses a densely staining, granular matrix (Fig. 2-3). They

are difficult to distinguish in electron micrographs from other organelles of similar size and shape but of quite different chemical composition, such as secretory vesicles and peroxisomes (see the following). A typical cell may contain several hundred lysosomes. Lysosomes digest (break down) various complex substances, such as bacteria and cellular debris, that have been engulfed by the cell. They may also digest other cell organelles that have been damaged and are no longer functioning normally. The lysosomes are thus a highly specialized intracellular digestive system. They are an especially important part of the various types of specialized cells that make up the defense systems of the body (see Chap. 17).

Peroxisomes are similar in structure to lysosomes, being oval bodies enclosed by a single membrane. The chemical composition of the peroxisome matrix (lumen) is, however, quite different from that of the lysosomes. Like mitochondria, peroxisomes consume oxygen, although in much smaller amounts, and this oxygen is used in various chemical reactions that are not associated with ATP formation (see Chap. 5). This organelle can also destroy certain products of oxygen reactions which can be quite toxic to cells, specifically hydrogen peroxide—hence the organelle's name.

In addition to all the membrane-enclosed organelles we have described, the cytoplasm of most cells contains a variety of rodlike filaments of various sizes. This cytoplasmic filamentous network is referred to as the cell's **cytoskeleton** and, like the bony skeleton of the body, is associated with processes that maintain and change cell shape and produce cell movements.

Four classes of filaments (Fig. 2-8) are recognized on the basis of their diameter and the types of protein they contain. In order of size, starting with the thinnest, they are (1) **microfilaments**, (2) **intermediate filaments**, (3) **muscle thick filaments**, and (4) **microtubules** (Table 2-1). Microfilaments, along with microtubules, can be rapidly assembled and disassembled, allowing a cell to change its cytoskeletal framework according to changing requirements. The other two types of filaments, once assembled, are less readily disassembled.

Microfilaments, which are composed of the contractile protein **actin,** make up a major portion of the fibrous cytoskeleton in all cells. Intermediate filaments are most extensively developed in cells, or in regions of cells, that are subject to mechanical stress. Muscle thick filaments, composed of the contractile protein **myosin,** are found only in muscle

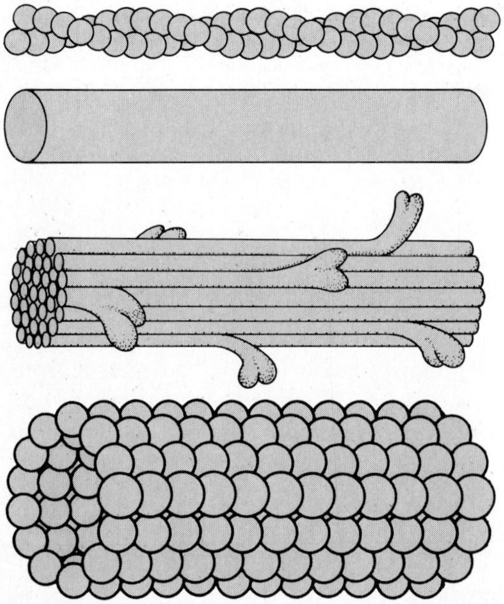

CYTOSKELETAL FILAMENTS	DIAMETER, nm	PROTEIN SUBUNIT
microfilaments	7	actin
intermediate filaments	10	several proteins
muscle thick filaments	15	myosin
microtubules	25	tubulin

FIGURE 2-8. Cytoskeletal filaments that are associated with cell shape and motility.

cells. However, myosin molecules not in filamentous form are also present in many if not most nonmuscle cells, where they interact with microfilaments to produce local forces and movements.

Microtubules, which are composed of the protein **tubulin**, are hollow tubes. They are the most rigid of the cytoskeletal filaments and are frequently found in long, cylindrical processes, such as nerve cells, where they provide the framework which maintains the cylindrical shape. Microtubules are also associated with movements of parts of cells. For example, when a cell divides, it duplicates its chromosomes and sends one copy to each new cell in the process known as **mitosis** (see Chap. 4). The structure immediately responsible for the separation of the chromosomes, known as the **mitotic spindle**, is composed of microtubules, as are the organelles which direct the mitotic process, the **centrioles**. Another example of the role of microtubules in cell motility are the hairlike extensions—**cilia**—on the surface of some epithelial cells. In tubes that are lined with ciliated epithelium, the cilia wave back and forth, propelling the luminal contents along the tubes. Microtubules have also been implicated in the movements of organelles within the cytoplasm, such as the directed movement of secretory vesicles from the Golgi apparatus to the plasma membrane.

3

CHEMICAL COMPOSITION
OF THE BODY

Atoms and molecules are the chemical units of cell structure and function. In this chapter we describe the structures and some characteristics of the major classes of chemicals in the body. The specific functions and reactions of these various classes of molecules will then be discussed in subsequent chapters. This chapter is, in essence, an expanded glossary of chemical terms and structures; like a glossary, it should be consulted according to need.

Atoms and Molecules

Atoms

Chemical elements. Atoms are the smallest units of matter that, singly or in combination with other atoms, form all substances. Ninety different types of atoms are known to exist naturally in the universe, and over a dozen additional atoms have been created artificially in atomic reactors. Matter composed only of identical atoms is known as a **chem-** **ical element**. For example, oxygen, carbon, and iron, which are pure chemical elements, are composed only of oxygen atoms, carbon atoms, and iron atoms, respectively. A one- or two-letter symbol is used as a shorthand identification of each element, as will be seen later in Table 3-3.

Atoms are composed of even smaller units of matter, **subatomic particles**, which differ in mass, electric charge, and location in an atom (Table 3-1). The chemical properties of atoms can be de-

TABLE 3-1. Characteristics of major subatomic particles

Subatomic particle	Mass relative to mass of electron	Electric charge	Location in atom
Electron	1	Negative (−1)	Orbiting the nucleus
Proton	1836	Positive (+1)	Nucleus
Neutron	1839	Neutral (0)	Nucleus

scribed in terms of three subatomic particles—**protons**, **neutrons**, and **electrons**—which are themselves composed of even smaller subatomic particles. The protons and neutrons are confined to a very small volume at the center of an atom, the **atomic nucleus**, whereas electrons revolve in orbits around the nucleus. The miniature-solar-system model of an atom is an oversimplification, but it is sufficient to provide a conceptual framework for understanding the basic chemical properties of atoms. Protons and neutrons are approximately equal in mass and are about 2000 times heavier than an electron. Thus, almost all of the mass of an atom is located in its nucleus. Each of the three particles has a different electric charge: Protons have one unit of positive charge, electrons have one unit of negative charge, and neutrons are electrically neutral (i.e., they have no electric charge). Since the protons are located in the nucleus, the atomic nucleus has a net positive charge equal to the number of protons it contains. However, the entire atom has no net electric charge because the number of negatively charged electrons orbiting the nucleus is equal to the number of positively charged protons in the nucleus.

Atomic number. The atoms of each chemical element contain a characteristic number of protons which distinguishes them from the atoms of other elements. This number is known as the **atomic number**. For example, hydrogen, the simplest atom, has an atomic number of 1, corresponding to its single proton (Table 3-2), whereas calcium has an atomic number of 20, corresponding to its 20 protons. (Since an atom is electrically neutral, the atomic number is also equal to the number of electrons in an atom of a given element).

Atomic weight. The **atomic weight** of an atom is a relative value that indicates its weight relative to that of other types of atoms. The atomic weight of carbon is 12, which indicates that a carbon atom is 12 times heavier than the smallest atom, hydrogen, which has an atomic weight of 1. But a carbon atom is only half as heavy as a magnesium atom, which has an atomic weight of 24. Since almost all of the weight of an atom is due to its neutrons and protons, the atomic weight of each atom is approximately equal to the sum of these two types of particles. It is not exactly equal to this sum because the electrons do have some weight, and the weights of the proton and neutron are not exactly equal. The atomic weights in Table 3-2 have been rounded off to the nearest whole number.

One **gram-atomic weight** of a chemical element is an amount of the element in grams equal to the numerical value of its atomic weight. Thus, 12 grams of carbon is 1 gram-atomic weight of carbon. Since the atomic weight scale is a relative scale, 1 gram-atomic weight of any chemical element contains the same number of atoms—6×10^{23}—which

TABLE 3-2. Atomic composition of the most abundant elements in the body

Element	Atomic number	Number of electrons	Number of protons	Number of neutrons	Atomic weight
Hydrogen	1	1	1	0	1
Carbon	6	6	6	6	12
Nitrogen	7	7	7	7	14
Oxygen	8	8	8	8	16
Sodium	11	11	11	12	23
Magnesium	12	12	12	12	24
Phosphorus	15	15	15	16	31
Sulfur	16	16	16	16	32
Chlorine	17	17	17	18	35
Potassium	19	19	19	20	39
Calcium	20	20	20	20	40

is known as 1 **mole**. Thus, 12 grams of carbon, 1 gram of hydrogen, and 24 grams of magnesium—1 mole, respectively, of each element—each contain the same number of atoms—6×10^{23}.

Atomic composition of the body. Of the 90 natural chemical elements, only 24 appear to be essential for maintaining the structure and function of the body (Table 3-3). In fact, just four of these elements—hydrogen, oxygen, carbon, and nitrogen—account for about 99 percent of the atoms in the body. These four elements, along with small amounts of sulfur and phosphorus, combine to form carbohydrates, lipids, proteins, and nucleic acids as well as water, a substance which makes up 60 percent of body weight.

Most of the seven elements listed in Table 3-3 as "major minerals" are present as electrically charged atoms, known as ions (to be discussed below). These elements play important roles in the electrical properties of cells and in the transfer and utilization of chemical energy. Some of them also constitute a major part of the extracellular matrix formed by connective tissues, including the solid matrix of calcium and phosphorus atoms in bone tissue.

The 13 **trace elements** are present in extremely small quantities, but they are nonetheless essential for normal growth and function. For example, iodine plays a role in the functioning of the thyroid hormone, and iron is important in the transport of oxygen by the blood. It is likely that additional trace elements will be added to this list as the chemistry of the body becomes better understood.

Many other elements can be detected in the body in addition to the 24 elements listed in Table 3-3. These elements enter through the foods we eat and the air we breathe but do not have any known essential chemical function. Some elements, e.g., mercury, not only are not required for normal function but can be extremely toxic.

Molecules

Molecules are formed by linking atoms together. The formula of a chemical compound indicates the number and types of atoms that the molecule contains. Thus, the formula for a molecule of water, which contains two hydrogen atoms and one oxygen, is H_2O, and that for glucose, a sugar, is $C_6H_{12}O_6$. This formula indicates that glucose contains six carbon atoms, twelve hydrogen atoms, and six oxygen atoms. Such formulas, however, do not indicate which atoms are linked to which in the molecule.

Covalent chemical bonds. Atoms are held together by electric forces between the charged subatomic particles of the interacting atoms. The outermost electrons in atoms, known as the **valence electrons,**

TABLE 3-3. Essential elements in the body

Symbol	Element
Major elements: 99.3% total atoms	
H (63%)	Hydrogen
O (26%)	Oxygen
C (9%)	Carbon
N (1%)	Nitrogen
Major minerals: 0.7% total atoms	
Ca	Calcium
P	Phosphorus
K (Latin *kalium*)	Potassium
S	Sulfur
Na (Latin *natrium*)	Sodium
Cl	Chlorine
Mg	Magnesium
Trace elements: less than 0.01% total atoms	
Fe (Latin *ferrum*)	Iron
I	Iodine
Cu (Latin *cuprum*)	Copper
Zn	Zinc
Mn	Manganese
Co	Cobalt
Cr	Chromium
Se	Selenium
Mo	Molybdenum
F	Fluorine
Sn (Latin *stannum*)	Tin
Si	Silicon
V	Vanadium

play the major role in establishing a bond between two atoms. Several types of bonds are recognized, on the basis of the strength with which they hold atoms and molecules together. The strongest chemical bond, a **covalent bond**, is formed between two atoms when each atom shares one of its valence electrons with the other atom. The atoms in most molecules in the body are linked by covalent bonds.

The atoms of some elements can form more than one covalent bond and thus become linked simultaneously to two or more other atoms. Each element has a characteristic number of covalent bonds it can form. The number of chemical bonds formed by the four major elements are hydrogen, 1; oxygen, 2; nitrogen, 3; and carbon 4. When the structure of a molecule is diagramed, each covalent bond is represented by a line. The chemical bonds of the four major elements can thus be represented as

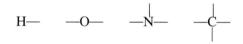

A molecule of water containing two hydrogen atoms linked to a single oxygen atom can be illustrated as

H—O—H

In some cases, two covalent bonds—a **double bond**—are formed between the same atoms; an example of this is carbon dioxide (CO_2):

O=C=O

Note that in this molecule the carbon atom still forms four covalent bonds and each oxygen atom only two.

Molecular shape. When atoms are linked together, molecules with various shapes can be formed (Fig. 3-1). Although we draw diagrammatic structures of molecules on flat sheets of paper, these molecules have a three-dimensional shape. When more than one covalent bond is formed with a given atom, the bonds are distributed about the atom in a pattern

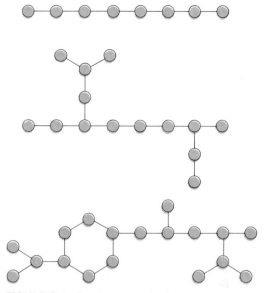

FIGURE 3-1. Combination of spherical atoms to form molecules of various shapes.

which may or may not be symmetrical (Fig. 3-2). For example, the four bonds of a carbon atom occur at the four corners of a tetrahedron with the carbon atom at the center. The three bonds of a nitrogen atom form the three legs of a tripod, and the two bonds of an oxygen atom form an angle of 105°.

Molecules are not rigid, inflexible structures. Within certain limits, the shape of a molecule can be changed without breaking the covalent bonds linking its atoms together. A covalent bond is like an axle around which the atoms so joined can rotate. As illustrated in Fig. 3-3, a sequence of six carbon atoms can assume a number of different shapes as a result of rotations around various bonds in the sequence. However, in contrast to a single covalent bond, a double bond between two atoms is rigid, preventing rotation. As we shall see, the three-dimensional shape of molecules is one of the major factors governing molecular interactions in the body.

Ions

An atom is electrically neutral since it contains equal numbers of electrons (which are negative) and protons (which are positive). If, however, an atom

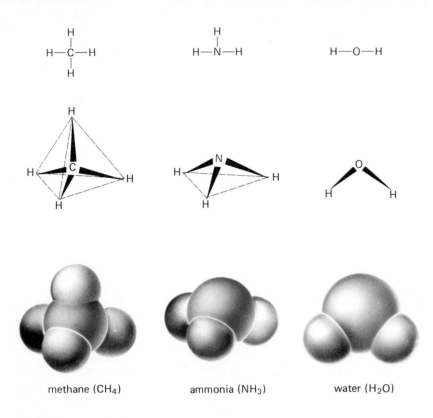

methane (CH₄)   ammonia (NH₃)   water (H₂O)

FIGURE 3-2. Geometric configuration of chemical bonds around the carbon, nitrogen, and oxygen atoms.

gains or loses one or more electrons, it acquires a net electric charge and becomes an **ion**. For example, when a sodium atom loses one electron, it becomes a sodium ion (Na^+) with a net positive charge; i.e., it now has only 10 electrons but 11 protons. On the other hand, a chlorine atom can gain an electron and become a chloride ion (Cl^-) with a net negative charge (it has 18 electrons but only 17 protons). Some atoms can gain or lose more than one electron to become ions with two or even three units of electric charge, e.g., calcium ions (Ca^{2+}). Just as the number of valence electrons that an atom has determines the number of covalent bonds it can form, these valence electrons also determine the ability of an atom to form an ion, and the charge on the ion. Hydrogen and the major mineral and trace elements readily form ions, whereas carbon, oxygen, and nitrogen do not. Table 3-4 lists the ionic forms of the major mineral elements. Ions that have a net positive charge are called **cations**; negatively charged ions are called **anions**.

The process of removing electrons from or adding them to an atom to form an ion is known as **ionization**. Since ions are formed by removing one or more electrons from one atom and transferring them to another atom, the process of ionization leads to the formation of two ions, a cation and an anion.

Ionization can also take place in atoms that are linked to other atoms in molecules, giving rise to ionized molecules. Two commonly encountered groups of atoms within molecules which undergo ionization are the **carboxyl group** (—COOH) and the **amino group** (—NH₂). The formulas for molecules containing such groups would be written as R—COOH and R—NH₂, where R signifies the remaining portion of the molecule. The carboxyl group ionizes when the oxygen atom linked to the hydrogen atom captures the hydrogen atom's only electron to form a carboxyl ion (R—COO⁻) and a hydrogen ion (H^+) (Fig. 3-4). In contrast, the amino group can bind a hydrogen ion, forming an ionized amino group (R—NH₃⁺). Ions and molecules with

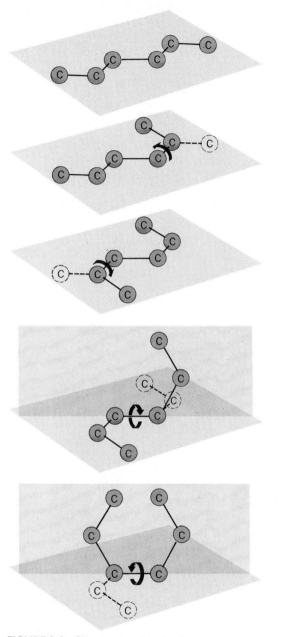

FIGURE 3-3. Changes in molecular shape occur as the molecule rotates around different carbon-to-carbon bonds, transforming its shape from a relatively straight chain into a ring.

a number of ionized groups, when dissolved in water, can conduct electricity and are often referred to as **electrolytes**.

TABLE 3-4. Ions formed from mineral elements

Atom		Ion		Electrons gained or lost
Sodium	Na	Sodium ion	Na^+	1 lost
Potassium	K	Potassium ion	K^+	1 lost
Chlorine	Cl	Chloride ion	Cl^-	1 gained
Magnesium	Mg	Magnesium ion	Mg^{2+}	2 lost
Calcium	Ca	Calcium ion	Ca^{2+}	2 lost

Polar molecules

As we have seen, when the electrons of two atoms interact, the two atoms may share electrons equally, forming an electrically neutral covalent bond, or one of the atoms may completely capture an electron from the other atom, forming ions. Between these two extremes there exist covalent bonds in which the electrons are not shared equally between two atoms but instead reside closer to one atom of the pair, which thus acquires a slight negative charge, while the opposite atom, having partly lost an electron, becomes slightly positive. Such bonds are known as **polar covalent bonds**. Molecules containing polar bonds do not have a net electric charge, as do ions; instead they have a polarized separation of charges around their polar bonds. For example, the bond between hydrogen and oxygen in a **hydroxyl group** (R—OH) is a polar covalent bond in which the oxygen is slightly negative and the hydrogen slightly positive. The electric charge associated with the ends of a polar bond is considerably less than the charge of a fully ionized atom

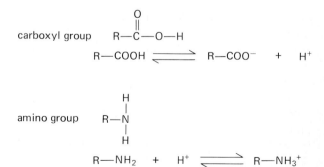

FIGURE 3-4. Ionization of carboxyl and amino groups. (R represents the remainder of the molecule.)

(for example, the oxygen in the polarized hydroxyl group has only about 13 percent of the negative charge associated with the oxygen in an ionized carboxyl group, R—COO$^-$).

Atoms of nitrogen and oxygen tend to form polar bonds with hydrogen atoms, whereas the bonds between carbon and hydrogen, or between two carbon atoms, are electrically neutral. Thus, molecules may have regions which are electrically neutral, polarized, or ionized. Molecules containing significant numbers of polar bonds or ionized groups are known as **polar molecules**, whereas molecules composed predominantly of electrically neutral bonds are known as **nonpolar molecules**. As we shall see, the physical characteristics of these two classes of molecules, especially their solubility in water, are quite different.

Water. Hydrogen is the most numerous atom in the body, and water is the most numerous molecule; 99 out of every 100 molecules are water molecules. The bonds linking the two hydrogen atoms to oxygen are polar bonds; therefore, the oxygen in water is slightly negative and each of the two hydrogen atoms has a slightly positive charge (Fig. 3-5). The positively polarized regions near the hydrogen atoms of one water molecule are electrically attracted to the negatively polarized regions of the oxygen atoms in adjacent water molecules (Fig. 3-5) forming a **hydrogen bond**. A hydrogen bond is a very weak link, having only about 4 percent of the strength of the covalent bonds linking the hydrogen and oxygen within a single water molecule. Therefore, clusters of hydrogen-bonded water molecules are continuously formed and broken, accounting for the fact that water is in a liquid state over a wide range of temperatures. As the temperature is lowered, the weak hydrogen bonds are less frequently broken, and larger and larger clusters of water molecules are formed. At 0°C, the water freezes into a continuous crystalline matrix—ice. Hydrogen bonds can form not only between water molecules but between water and other polar molecules, between two other types of polar molecules, or even between polar regions within a single molecule. These weak hydrogen bonds between and within

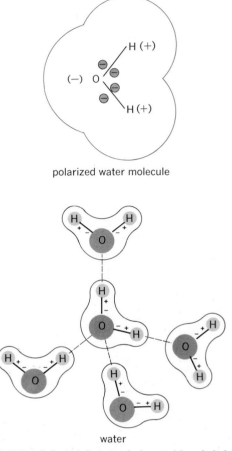

polarized water molecule

water

FIGURE 3-5. (A) Polarized chemical bonds link the hydrogen and oxygen atoms in a water molecule. (B) Hydrogen bonds between adjacent polarized water molecules.

molecules play an important role in the interactions between molecules and in determining the overall shape of large molecules, especially molecules of proteins and nucleic acids.

Water molecules take part in numerous chemical reactions, of the following general type:

$$R_1—R_2 + H_2O \rightleftharpoons R_1—OH + H—R_2$$

In this reaction the covalent bond between R_1 and R_2 is broken in the process of adding a hydroxyl group and a hydrogen atom to R_1 and R_2, respectively. Reactions of this type are known as **hydro-**

lytic reactions (or simply hydrolysis). The breakdown of many large molecules in the body occurs by hydrolysis. Note that the above reaction can also occur in the opposite direction, synthesizing R_1—R_2 from R_1—OH and H—R_2, in which case a molecule of water is formed during the reaction.

Solutions

Substances dissoved in a liquid are known as **solutes**, and the liquid in which they are dissolved is the **solvent**. Solutes dissolved in a solvent form a **solution**. Water is the most abundant solvent in the body, and a major portion of cellular metabolism takes place in the intracellular fluid, which is composed of water and dissolved solutes. However, not all molecules will dissolve in water.

Molecular solubility. In order for a substance to dissolve in water it must be electrically attracted to water molecules. For example, table salt (NaCl) is a solid crystalline substance because of the strong electrical attraction between the sodium ions (which are positive) and the chloride ions (which

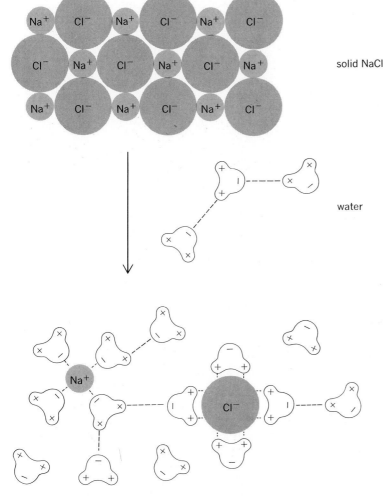

solid NaCl

water

solution of sodium and chloride ions

FIGURE 3-6. The ability of water to dissolve sodium chloride crystals depends upon the electrical attraction between the polar water molecules and the charged sodium and chloride ions.

are negative). This strong attraction between two oppositely charged ions is known as an **ionic bond**. When a crystal of sodium chloride is placed in water, the polar water molecules are attracted to the charged sodium and chloride ions (Fig. 3-6); the ions become surrounded by clusters of water molecules which decrease the electric force of the ionic bonds, allowing the sodium and chloride ions to separate and go into solution, i.e., to dissolve.

Molecules dissolve in water provided they contain a sufficient number of polar bonds or ionized groups that are electrically attracted to the polar water molecules. Thus, the presence of ionized carboxyl and amino groups or polar hydroxyl groups in a molecule promotes solubility in water. In contrast, molecules composed predominantly of carbon and hydrogen are insoluble in water since their electrically neutral covalent bonds are not attracted to the water molecules.

When nonpolar molecules are mixed with water, the solution separates into two phases, as occurs when oil is mixed with water. The strong attraction between polar water molecules "squeezes" the nonpolar molecules out of the water phase. (However, such a separation is never 100 percent complete, and very small amounts of nonpolar solutes will remain dissolved in the water.) Although nonpolar molecules do not dissolve to any great extent in polar solvents, they do dissolve in nonpolar solvents such as oil.

Some molecules contain polar or ionized groups at one end of the molecule and a nonpolar region at the opposite end; such molecules are called **amphipathic**. When mixed with water, amphipathic molecules form spherical clusters known as **micelles** in which the polar regions of the molecules are located at the surface of the micelle, where they are attracted to the surrounding water molecules, and the nonpolar ends are oriented toward the center of the micelle (Fig. 3-7). Other nonpolar molecules can dissolve in the central nonpolar regions of a micelle and thus occur in aqueous solutions at far higher concentrations than would be possible in the absence of micelle formation. As we shall see, the orientation of amphipathic mole-

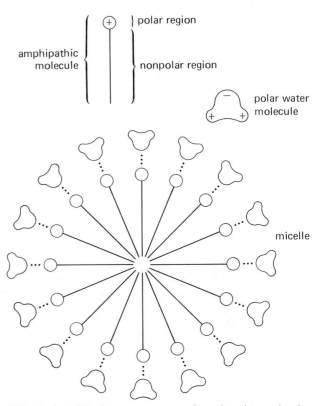

FIGURE 3-7. Micellar organization of amphipathic molecules dissolved in water.

cules in water plays an important role in the structure of cell membranes and in both the absorption of nonpolar molecules from the gastrointestinal tract and their transport in the blood.

Concentration. **Concentration** is defined as the amount of material present per unit of volume. The standard unit of volume in the metric system is a **liter** (L). (One liter equals 1.06 quarts, or approximately three 12-ounce cans of soda pop.) Smaller units are the milliliter (mL, or 0.001 L) and the microliter (μL, or 0.001 mL) (see Appendix I). One way of expressing the amount of a chemical compound is by its mass in grams; its concentration then is expressed as the number of grams of compound present in a liter of solution.

A comparison of the concentrations of two different types of compounds on the basis of number of grams per liter of solution does not directly indicate how many molecules of each compound are present. For example, 10 g of compound X, whose molecules are heavier than those of compound Y, will contain fewer molecules than 10 g of compound Y. Concentrations in units of grams per liter are most often used when the chemical structure of the solute is unknown. When the chemical structure of a molecule is known, concentrations are usually expressed as moles per liter, which gives the number of molecules of a given substance that are present in the solution. One mole of any substance contains the same number of molecules—6×10^{23}. The definition of a mole of a compound is similar to that of a mole of a chemical element; it is the weight of a compound in grams equal to its molecular weight. Therefore

$$\text{Number of moles} = \frac{\text{weight in grams}}{\text{molecular weight}}$$

The **molecular weight** of a molecule is equal to the sum of the atomic weights of all the atoms in the molecule. For example, glucose ($C_6H_{12}O_6$) has a molecular weight of 180 ($6 \times 12 + 12 \times 1 + 6 \times 16 = 180$), thus 180 g of glucose equals one mole of glucose. A solution containing 180 g of glucose in 1 L of solution is said to be a **one molar** solution of glucose. If 90 g of glucose were dissolved in enough water to produce 1 L of solution, the solution would have a concentration of 0.5 moles per liter (mol/L). The concentrations of solutes dissolved in the body fluids are much less than 1 molar; many have concentrations in the range of millimoles per liter (mmol/L, or 0.001 mol/L), while others are present at even smaller concentrations—micromoles (μmol, or 0.000001 mol) or nanomoles (nmol, or 0.000000001 mol) per liter.

Hydrogen ions and acidity. As mentioned earlier, the hydrogen atom consists of a single proton orbited by a single electron. The hydrogen ion (H^+), formed by the loss of the electron, is thus a single free proton. Hydrogen ions are formed by the ionization of molecules which liberate a hydrogen ion. Such molecules are called **acids**. For example,

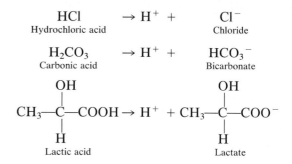

Conversely, any substance which can accept a hydrogen ion is termed a **base**. In the reactions above, bicarbonate and lactate are bases, since they can bind hydrogen ions. When hydrochloric acid is dissolved in water, it completely dissociates into hydrogen and chloride ions, which do not recombine in solution. However, in the case of lactic acid, only a fraction of the total number of lactic acid molecules in solution are ionized at any instant. Therefore, if the same number of hydrochloric and lactic acid molecules are dissolved in the same volume of water, the hydrogen ion concentration will be lower in the lactic acid solution than in the hydrochloric acid solution. Hydrochloric acid and similar mineral acids, which are 100 percent ionized in solution, are known as **strong acids**, while molecules like carbonic and lactic acids, which do not completely ionize in solution, are **weak acids**.

The **acidity** of a solution refers to the hydrogen-ion concentration in the solution; the higher the hydrogen-ion concentration, the greater the acidity. [It must be understood that the hydrogen-ion concentration refers only to the hydrogen ions that are free in solution and not to those that are bound, for example, to amino groups ($R—NH_3^+$).] The hydrogen-ion concentration is frequently expressed in terms of the **pH** of a solution, which is defined as the negative logarithm to the base 10 of the hydrogen-ion concentration:

$$pH = -\log H^+$$

Thus, a solution with a hydrogen-ion concentration of 10^{-7} mol/L will have a pH of 7, whereas a *more* acid solution with a concentration of 10^{-6} mol/L will have a pH of 6. Note that as the acidity *increases,* the pH *decreases* and that a change in pH from 7 to 6 represents a tenfold increase in the hydrogen-ion concentration.

Pure water has a hydrogen-ion concentration of 10^{-7} mol/L (pH $= 7.0$). Solutions with a lower hydrogen-ion concentration are said to be **alkaline**, while those with a higher hydrogen-ion concentration are said to be **acid**. The extracellular fluids of the body have a hydrogen-ion concentration of about 4×10^{-8} mol/L (pH $= 7.4$) and are thus slightly alkaline. Intracellular fluids tend to be slightly more acid than extracellular fluid. In the extracellular fluid, hydrogen ion concentrations beyond the tenfold pH range of 7.8 to 6.8 are incompatible with human life if maintained for more than a brief period of time. But even small changes in the hydrogen-ion concentration can greatly change the physiological functioning of the body.

As we have seen earlier, the ionization of carboxyl and amino groups involves hydrogen ions. Thus, the electric charge on a molecule that results from the ionization of these groups is dependent on the hydrogen-ion concentration. Increasing the hydrogen-ion concentration will decrease the ionization of carboxyl groups and increase the ionization of amino groups. Lowering the hydrogen-ion concentration will have the opposite effect. If the electric charge on a molecule is altered, its interaction with other molecules or with other regions within the same molecule will be altered, and thus its functional characteristics will be altered. All of these effects can result from changes in the concentration of hydrogen ions.

Classes of Organic Molecules

Because most naturally occurring carbon molecules are found in living organisms, the study of these compounds became known as **organic chemistry**, and molecules which contain carbon are known as **organic molecules**. (**Inorganic chemistry** is the study of the chemistry of noncarbon molecules.) Although organic chemistry grew out of the study of molecules found in living organisms, the chemistry of living organisms, **biochemistry**, now forms only a portion of the broad field of organic chemistry.

The one property of the carbon atom which makes life possible is its ability to form four separate covalent bonds with other atoms, in particular with other carbon atoms. There is almost no limit to the number of carbon atoms that can be combined in this way. Since carbon atoms can also combine with other atoms such as hydrogen, oxygen, nitrogen, and sulfur atoms, a vast number of molecular structures can be formed with relatively few chemical elements. Moreover, very large organic molecules can be made by linking together smaller molecular subunits to form long chains known as **polymers**.

Most of the thousands of organic molecules in the body can be classified into four groups: **carbohydrates**, **lipids**, **proteins**, and **nucleic acids** (Table 3-5). In addition, some molecules are composed of portions of molecules from several of these classes.

Carbohydrates

Although carbohydrates account for only 3 percent of the organic matter in the body, they play a central role in the chemical reactions which provide cells with energy. The major carbohydrates are composed of atoms of carbon, hydrogen, and oxygen, with a hydrogen atom and a hydroxyl group linked to most of the carbon atoms. The proportions of carbon, hydrogen, and oxygen in most carbohydrates are represented by the general formula $C_n(H_2O)_n$, where n is any whole number. The presence of the polar hydroxyl groups makes carbohydrates quite polar and, hence, readily soluble in water.

The simplest carbohydrates are the **monosaccharides**, the most important of which is **glucose**, a 6-carbon molecule ($C_6H_{12}O_6$) often called "blood sugar." There are two ways of representing the structure of glucose (Fig. 3-8), and these apply to monosaccharides in general. The first is the conventional representation for organic molecules, but the second gives a better idea of the three-dimensional shape of monosaccharides such as glucose. Five carbon atoms are linked with an oxygen atom

TABLE 3-5. Major categories of organic molecules in the body

Category	Percent of body weight	Majority of atoms	Subclass	Subunits	Characteristics
Carbohydrates	1	C, H, O	Monosaccharides (sugars)		Glucose is major fuel used by cells to provide energy
			Polysaccharides	Monosaccharides	Small amounts of glucose stored as glycogen
Lipids	15	C, H	Triacylglycerols	3 fatty acids + glycerol	Insoluble in water Major store of reserve fuel which can be used by most cells to provide energy
			Phospholipids	2 fatty acids + glycerol + phosphate + small charged nitrogen molecule	Molecule has polar and nonpolar ends Major component of membrane structure
			Steroids		Cholesterol and steroid hormones
Proteins	17	C, H, O, N		Amino acids	Large polymers of amino acids Major source of nitrogen Many functions
Nucleic acids	2	C, H, O, N	DNA	Nucleotides containing the bases adenine, cytosine, guanine, thymine	Store genetic information
			RNA	Nucleotides containing the bases adenine, cytosine, guanine, uracil	Translate genetic information into protein synthesis

in a ring which lies in an essentially flat plane with the hydrogen and the hydroxyl groups lying above and below the ring. If one of the hydroxyl groups below the ring is shifted to a position above the ring, as shown in Fig. 3-9, a different monosaccharide results. Most monosaccharides in the body contain five or six carbon atoms.

Larger carbohydrate molecules can be formed by linking a number of monosaccharides together.

"Table sugar," or **sucrose** (Fig. 3-10), is composed of two monosaccharides, glucose and fructose. (Note that a water molecule results from the formation of the covalent bond between glucose and fructose.) Carbohydrate molecules composed of two monosaccharides are known as **disaccharides**. When many monosaccharides are linked together to form large polymers, the molecules are known as **polysaccharides**. Starch and cellulose, both found

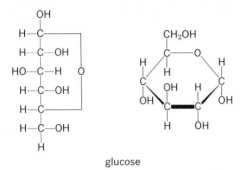

FIGURE 3-8. Two ways of diagraming the structure of the carbohydrate glucose.

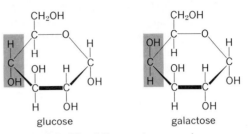

FIGURE 3-9. The difference between the monosaccharides glucose and galactose depends upon the position of the hydroxyl group on the fourth carbon atom.

in plant cells, and **glycogen**, present in animal cells and often called "animal starch," are examples of polysaccharides. All three of these polysaccharides are composed of thousands of glucose molecules linked together in long chains (Fig. 3-11). The difference between them depends on where the chemical bonds are located that link the glucose molecules together and on how frequently branches appear along the molecule. Cellulose is a straight-chain polysaccharide with no branches; glycogen is highly branched, and starch less so.

Lipids

Lipids are molecules composed predominantly of hydrogen and carbon atoms. Since these atoms are linked by neutral covalent bonds, lipids are non-polar molecules which are insoluble in water. It is this physical property of insolubility in water (and solubility in nonpolar solvents) which characterizes this class of organic molecules. The lipids, which account for about 40 percent of the organic matter in the body, can be divided into three subclasses: **triacylglycerols**, **phospholipids**, and **steroids**.

Triacylglycerols. The triacylglycerols, or neutral fats, constitute the majority of the lipids of the body, and it is these molecules which are generally referred to simply as "fat." Neutral fats are formed by linking together two types of molecules, **glycerol** and **fatty acids** (Fig. 3-12). Glycerol is a 3-carbon molecule which by itself is a carbohydrate. A fatty acid consists of a chain of carbon atoms with a carboxyl group at one end (Fig. 3-12). The carboxyl groups of three fatty acids are attached to the three hydroxyl groups of glycerol to form a molecule of triacylglycerol.

Because fatty acids are synthesized in the body by linking together 2-carbon fragments, most fatty acids have an even number of carbon atoms, 16-

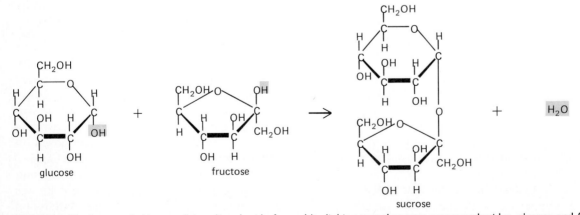

FIGURE 3-10. Sucrose (table sugar) is a disaccharide formed by linking together two monosaccharides, glucose and fructose.

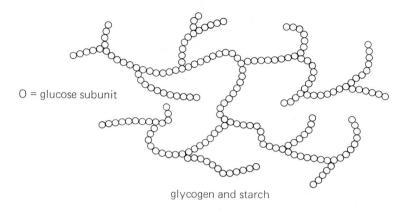

O = glucose subunit

glycogen and starch

FIGURE 3-11. Branched chain structure of two polysaccharides—glycogen and starch. Although both are composed of glucose subunits, glycogen is more highly branched than is starch.

and 18-carbon fatty acids being the most common. When all of the carbons in a fatty acid chain are linked to single bonds, the fatty acid is said to be **saturated**. Some fatty acids contain double bonds, and these are known as **unsaturated** fatty acids. If more than one double bond is present, the fatty acid is said to be **polyunsaturated** (Fig. 3-12). Animal fats generally contain a high proportion of saturated fatty acids, whereas vegetable fats contain more polyunsaturated fatty acids. The three fatty acids in a molecule of triacylglycerol need not be identical; therefore, a variety of neutral fats can be formed with fatty acids of different chain lengths and degrees of saturation.

Phospholipids. Phospholipids are similar in overall structure to neutral fats with one important difference. The third hydroxyl group of glycerol, rather

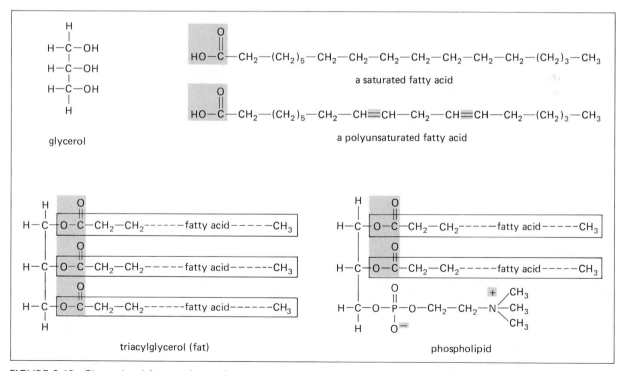

FIGURE 3-12. Glycerol and fatty acids are the major subunits that combine to form triacylglycerol (fat) and phospholipids.

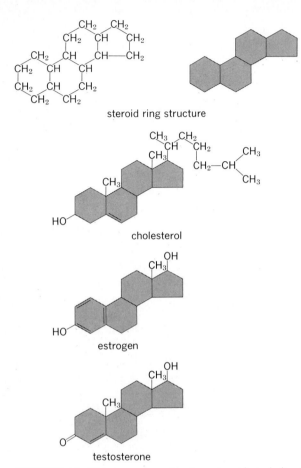

steroid ring structure

cholesterol

estrogen

testosterone

FIGURE 3-13. Ring structures of various steroids, including the female and male sex hormones estrogen and testosterone.

than being attached to a molecule of fatty acid, is linked to a phosphate group. A smaller polar or ionized nitrogen-containing molecule is usually attached to this phosphate group (Fig. 3-12). Both the phosphate and nitrogen groups are usually electrically charged. These groups constitute a polar region at one end of the phospholipid molecule, and the fatty acid chains provide a nonpolar region at the opposite end; therefore, these molecules are amphipathic. When mixed in water, they become organized into micelles, with the polar ends of the molecules associating with the water molecules.

Steroids. Steroids have a distinctly different structure from that of the other two subclasses of lipid

molecules. Four interconnected rings of carbon atoms form the basic skeleton of all steroids (Fig. 3-13). A few polar groups may be attached to this ring structure, but they are not numerous enough to make a steroid water-soluble. The steroid family includes, among others, cholesterol and the male and female sex hormones, testosterone and estrogen, respectively.

Proteins

The term "protein" comes from the Greek *proteios* ("of the first rank"), which aptly describes their importance. These molecules, which account for about 50 percent of the organic material in the body, play critical roles in almost every physiological process. Proteins are composed of carbon, hydrogen, oxygen, nitrogen, and small amounts of other elements, notably sulfur. Most proteins are very large molecules, containing thousands of atoms, and like most very large molecules, they are formed by linking together a large number of small subunits. The subunit of protein structure is an **amino acid**; thus, proteins are polymers of amino acids.

All amino acids have an amino (—NH₂) and a carboxyl (—COOH) group, both attached to the terminal carbon of the molecule (Fig. 3-14). A third bond of this terminal carbon is to hydrogen (with one exception) and the fourth to the remainder of the molecule, which is known as the **amino acid side chain**. The proteins of living organisms, including those of humans, are composed of 20 different amino acids, corresponding to 20 different side chains; Fig. 3-14 illustrates the structure of only seven of them. Some of these side chains are nonpolar, others are polar but not ionized, and a few contain ionized groups.

Protein molecules are sequences of amino acids in which the carboxyl group of one amino acid is linked to the amino group of the next amino acid in the sequence; in the process a molecule of water is formed (Fig. 3-15). The bond between an amino group and a carboxyl group is called a **peptide bond**. It belongs to the general class of polar covalent chemical bonds. Note that when two amino acids

CHARGE ON SIDE CHAIN   SIDE CHAIN   AMINO ACID

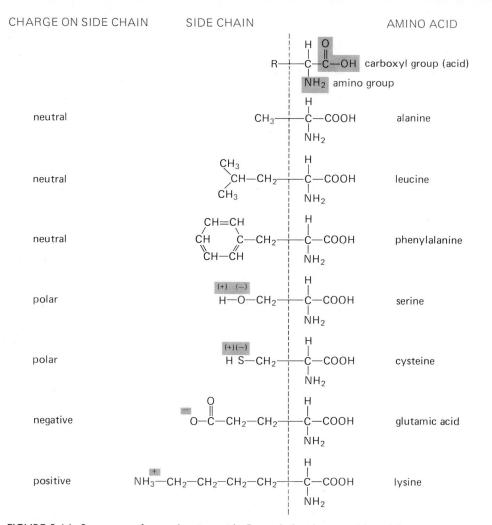

FIGURE 3-14. Structures of several amino acids. R stands for the remainder of the amino acid side chain.

are linked together, one end of the resulting molecule has a free amino group and the other has a free carboxyl group. Each of these free groups can become linked to other amino acids, and so on, thereby forming a **polypeptide chain**. The peptide bonds form the backbone of the polypeptide chain while the remainder of each amino acid sticks out to the side of the chain.

The specific sequence of amino acids along a polypeptide chain constitutes the **primary structure** of the protein molecule. There are two variables in this primary structure: (1) the total number of amino

acids in the sequence, and (2) the specific type of amino acids at each position along the sequence. If the number of amino acids in a polypeptide chain is less than about 50, the molecule is known as a **peptide**; if the sequence is more than 50 amino acid units long, the polypeptide is known as a **protein**.

Each position along a polypeptide chain may be occupied by any one of the 20 different amino acids. Let us consider the number of different peptides that can be formed that are only three amino acids long. Any one of the 20 different amino acids may occupy the first position in the sequence, any

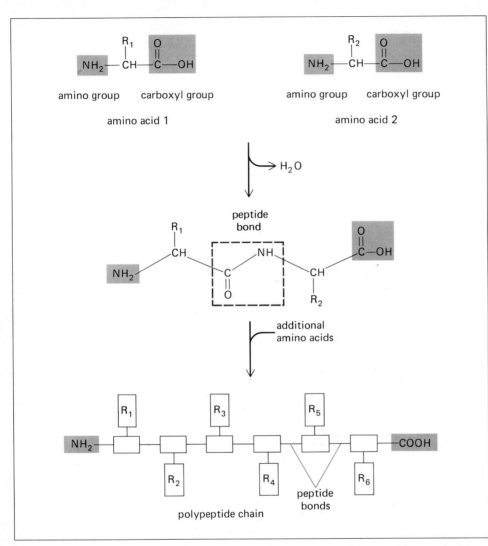

FIGURE 3-15. Primary structure of a protein molecule: A sequence of amino acids are linked together by peptide bonds between the carboxyl group of one amino acid and the amino group of the next amino acid in the sequence to form a polypeptide chain.

one of 20 the second position, and any one of 20 the third position, for a total of $20 \times 20 \times 20 = 8000$ possible sequences. If the peptide is six amino acids in length, $20^6 = 64,000,000$ possible combinations can be formed. Thus, with 20 different amino acids an almost unlimited variety of polypeptide chains can be formed by altering both the amino acid sequence and the total number of amino acids in the sequence.

The primary structure of a polypeptide chain is analogous to a string of beads, each bead representing a single amino acid (Fig. 3-16). Since amino acids can rotate around their peptide bonds, a polypeptide chain is flexible and can be bent into a number of three-dimensional shapes (**conformations**) just as a long piece of flexible rope can be twisted into many types of configurations. This phenomenon provides the key to understanding the

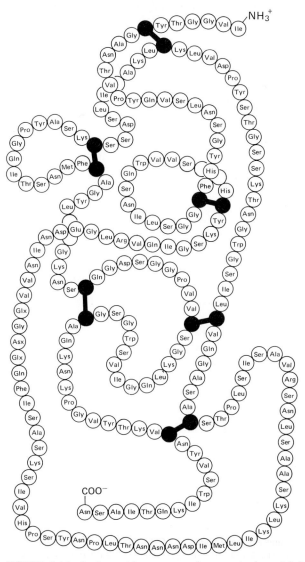

FIGURE 3-16. Amino acid sequence of a protein (trypsin) illustrating the folding of the linear sequence of amino acids (circles). The bars cross-linking various segments of the protein chain represent covalent bonds between the side chains of cysteine amino acids. (*Adapted from Sidney Bernhard.*)

functions of proteins, as we shall see in the next chapter.

Three factors determine the three-dimensional shape of the polypeptide chain once the amino acid sequence has been formed: (1) electrical attractions between various polar and ionized regions along the polypeptide chain, (2) electrical attractions between these regions and surrounding water molecules, and (3) covalent bonds linking the side chains of two amino acids. Let us look at each of these in turn.

An important example of the first factor is the interaction between peptide bonds. Because individual peptide bonds are polar, they interact with each other through hydrogen bonds (Fig. 3-17). The maximal interaction of these hydrogen bonds, which are located at regular intervals along a polypeptide chain, produces a coiling of the chain into a helical conformation known as an **alpha helix**. The peptide bonds are, of course, not the only polar or ionized sites in a protein; some of the amino acid side chains also contain polar bonds or ionized groups which are attracted to oppositely charged regions of the polypeptide chain. These electrical interactions tend to disrupt portions of the regular alpha-helix coil, producing bends and irregular, random conformations.

When proteins are dissolved in water, the various polarized and ionized regions of the molecule are attracted to the surrounding polar water molecules. Because of these forces of attraction, the polar side chains come to be located along the surface of the molecule, i.e., in contact with water. In contrast, the nonpolar side chains are not attracted to water, and they become located in the interior of the molecule, removed from contact with water. These nonpolar side chains, if they come close to each other, are attracted to each other by a very weak force—**van der Waals force**—resulting from the weak electromagnetic field which surrounds atoms.

Finally, the third factor—covalent bonding between side chains—is provided by the amino acid **cysteine**. The side chain of this amino acid (Fig. 3-14) contains a **sulfhydryl group** (R—SH), which can undergo a chemical reaction with a sulfhydryl group in another cysteine side chain to produce a **disulfide bond** (R—S—S—R), linking the two amino acids together (Fig. 3-18). Disulfide bonds form strong links between portions of a polypeptide chain (or between two polypeptide chains), in contrast to the

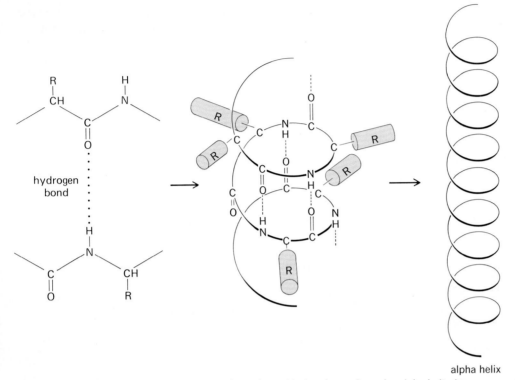

FIGURE 3-17. Hydrogen bonds between polarized peptide bonds produce the alpha-helical structure of polypeptide chains.

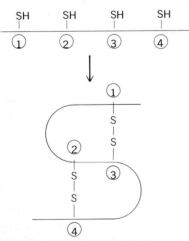

FIGURE 3-18. The sulfhydryl groups of cysteine side chains form disulfide bonds cross-linking portions of a polypeptide chain.

weaker hydrogen and ionic bonds which are more easily broken.

To recapitulate, a protein consists of a hierarchy of structure (Fig. 3-19) beginning with its linear sequence (primary structure). This is formed into regions of alpha-helical and random-coil configurations (**secondary structure**). Additional bends and folds due to other bonds (both disulfide and hydrogen) within the molecule, as well as to interactions with the surrounding water molecules, complete the protein's total three-dimensional shape (**tertiary structure**). Moreover, yet another level of structural organization (**quartenary structure**) exists in **oligomeric proteins** (Fig. 3-20), i.e., proteins which contain more than one polypeptide chain (an example is the hemoglobin molecule, which consists of four polypeptide chains). Each chain is known as a **protomer**. Protomers are sometimes held together by hydrogen and ionic bonds, as in hemoglobin. In some cases the chains may be

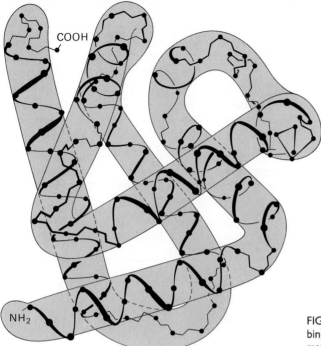

COOH

NH₂

FIGURE 3-19. Tertiary structure of a protein molecule (myoglobin) containing regions of alpha-helical and random-coil conformations. *(Adapted from Albert L. Lehninger.)*

linked together by disulfide bonds, as in the hormone insulin, which consists of two polypeptide chains.

Table 3-6 provides a summary of the types of bonding forces which contribute to the conformation of protein molecules. These same bonds are also involved in other interactions between molecules, which will be described in later chapters. In the next chapter we shall consider how the three-dimensional shape of proteins is related to their functions and how a cell goes about assembling these large molecules from amino acid subunits.

Nucleic acids

The nucleic acids contribute very little to the body's weight, but these molecules are the largest and most specialized of all. They are responsible for the storage of genetic information and its passage from parent to offspring and from cell to cell. It is the nucleic acids which determine whether one is a human being or a mouse, or whether a cell is a muscle cell or a nerve cell. There are two types of

single polypeptide chain

oligomeric protein

FIGURE 3-20. Oligomeric protein composed of four polypeptide chains (protomers). *(Adapted from Albert L. Lehninger.)*

TABLE 3-6. Bonding forces between atoms and molecules

Bond	Strength	Characteristics	Examples
Covalent	Strong	Shared electrons between atoms	Most bonds linking atoms together to form molecules
Ionic	Strong	Electrical attraction between oppositely charged ionized groups	Attractions between ionized groups in amino acid side chains contributing to tertiary structure; attractions between ions in a salt
Hydrogen	Weak	Electrical attraction between polarized bonds, usually hydrogen and oxygen	Attractions between peptide bonds forming the alpha helix secondary structure of proteins and between polar amino acid side chains to form tertiary protein structures; attractions between water molecules
Van der Waals	Very weak	Attraction between nonpolar molecules and groups when very close to each other	Attractions between nonpolar amino acids contributing to tertiary protein structure; attractions between nonpolar fatty acid chains in micelles

nucleic acids, **deoxyribonucleic acid (DNA)** and **ribonucleic acid (RNA)**. The DNA molecules store genetic information coded in terms of their repeating subunit structure, whereas RNA molecules are involved in the decoding of this information into instructions for linking together a specific sequence of amino acids to form a specific protein. Both types of nucleic acids are polymers composed of linear sequences of repeating subunits. Each subunit, known as a **nucleotide**, has three components: a phosphate group, a sugar, and a ring of carbon and nitrogen atoms known as a **base**. The phosphate group of one nucleotide is linked to the sugar of the adjacent nucleotide to form a chain with the bases sticking out to the side of the phosphate-sugar backbone (Fig. 3-21).

DNA. The nucleotides in DNA contain the sugar **deoxyribose**—hence, the name deoxyribonucleic acid. Four different nucleotides are present in DNA, corresponding to the four different bases that may be attached to deoxyribose. The four bases are divided into two classes: (1) the **purine** bases, **adenine** and **guanine**, which have two rings of nitrogen and carbon atoms, and (2) the **pyrimidine** bases, **cytosine** and **thymine**, which have only a single ring (Fig. 3-22).

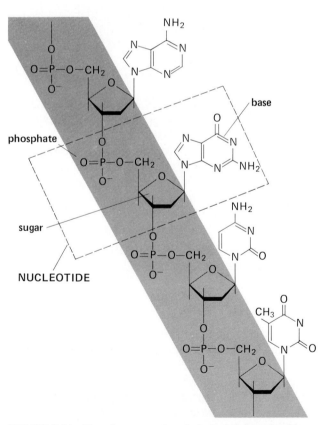

FIGURE 3-21. Phosphate-sugar bonds (color) link nucleotides in sequence to form nucleic acids.

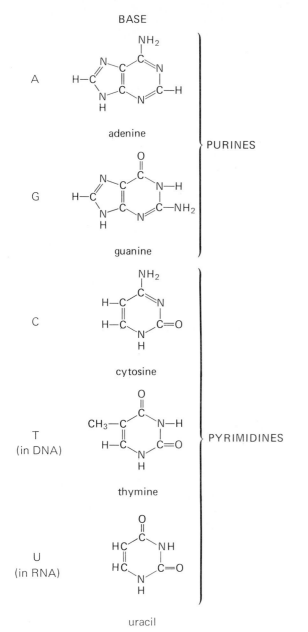

FIGURE 3-22. Structures of the major bases found in nucleic acids.

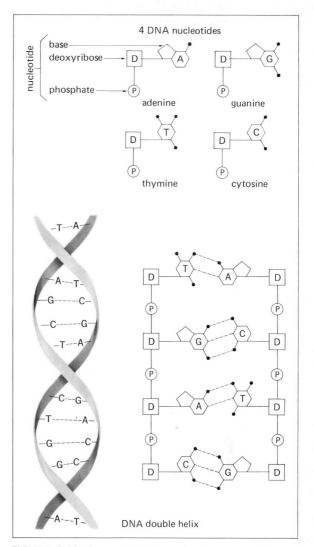

FIGURE 3-23. Base pairings between two nucleotide chains form the double-helical structure of DNA.

A DNA molecule consists of not one, but two chains of nucleotides coiled around each other in the form of a double helix (Fig. 3-23). Polar groups along the ring of a base on one chain form hydrogen bonds with those of a base on the second chain to link the two chains of DNA together. The location of the hydrogen-bonding groups in the four bases is such that adenine (A) can only pair with thymine (T), and guanine (G) with cytosine (C). As we shall see in the next chapter, this specificity in base-pairing provides the mechanism for duplicating and transferring genetic information.

RNA. The general structure of RNA molecules differs in only a few respects from that of DNA (Fig. 3-24): (1) RNA consists of a single (rather than a double) chain of nucleotides, (2) in RNA, the sugar in each nucleotide is **ribose** rather than deoxyribose, and (3) the pyrimidine base, thymine, in DNA is replaced by the base **uracil** (U) in RNA (Fig. 3-22). Although RNA contains only a single chain of nucleotides, portions of this chain may bend back upon itself to undergo base-pairings with other nucleotides in the same chain.

Mixed classes of organic molecules

In addition to the molecules which fall clearly into one of the four major classes just described, there are a number of molecules which are formed by linking together molecules from these different classes. In some cases the linkage involves strong covalent bonds; in others, the association may be quite weak. Some examples of such molecules that will be encountered in subsequent chapters are given here.

Glycolipids are amphipathic molecules that are similar in structure to phospholipids. The lipid portion consists of two long nonpolar chains to which are attached, at one end, one or more monosaccharides. Glycolipids are located in the plasma membrane and appear to be involved in the recognition of cells by defense cells and by viruses.

After certain proteins have been synthesized, one or more monosaccharides are covalently attached to specific side chains to form **glycoproteins**.

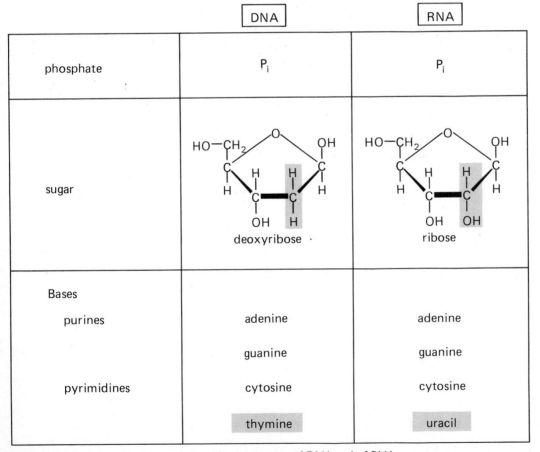

	DNA	RNA
phosphate	P_i	P_i
sugar	deoxyribose	ribose
Bases		
purines	adenine	adenine
	guanine	guanine
pyrimidines	cytosine	cytosine
	thymine	uracil

FIGURE 3-24. Comparison of the nucleotide composition of DNA and of RNA.

Most of the proteins in the extracellular fluid as well as most secreted proteins appear to be glycoproteins.

Although the name **lipoprotein** implies a structure analogous to that of the glycoproteins, with lipid molecules replacing the monosaccharides, the actual structure of these substances is quite different. The lipoproteins consist of micellar aggregates of lipid molecules which are partially coated by a layer of protein. The lipoproteins are involved in the transport of lipids in the blood.

Nuclear DNA is associated, by ionic bonds, with several positively charged proteins. This complex of protein and nucleic acid is known as **nucleoprotein**. A similar complex between RNA and protein occurs in ribosomes. The function of nucleoproteins will be discussed in the next chapter.

4 MOLECULAR CONTROL MECHANISMS
DNA AND PROTEIN

The outstanding accomplishment of twentieth-century biology has been the elucidation of the chemical basis of heredity and its relationship to protein synthesis. Whether an organism is a human being or a mouse, has blue eyes or black, has light skin or dark is determined by the proteins it possesses. Moreover, within an individual organism, muscle cells differ from nerve cells or epithelial cells or any other type of cell because of proteins. Proteins not only form the major structural elements in these cells but also are essential components of practically every cell function, different functions being performed by different sets of proteins in different cells.

Crucial for an understanding of protein function is the fact that each type of protein consists of a specific sequence of amino acids which determines its three-dimensional shape (conformation). In the first section of this chapter the relation between protein conformation and protein function is described.

So preeminent is the role of proteins in cellular function that the primary hereditary information passed from cell to cell is a set of instructions specifying the amino acid sequences of the proteins that can be synthesized by that cell. This information is coded into the molecular structure of DNA molecules; the middle section of this chapter describes the nature of this code and the decoding mechanism that leads to protein synthesis.

Given that different cell types have different proteins and that the specifications for these proteins are coded in DNA molecules, one might be led to conclude that different cell types have different sets of DNA molecules. However, such is not the case; all cells in the body (with the exception of sperm or ova) receive the same genetic information when DNA molecules are replicated and passed on to daughter cells at the time of cell division. Cells differ in structure and function because only a portion of the total genetic information common to all cells is translated in the process of synthesizing specific proteins. In other words, different portions of the genetic information are translated in different types of cells. The last section of this chapter describes some of the factors that govern the selective expression of genetic information, as well as the process of cell division.

SECTION A.
PROTEIN BINDING SITES

Characteristics of Protein Binding Sites

Proteins differ from other classes of molecules in their ability to bind organic molecules and ions selectively. This binding may be so specific that one type of protein can bind only one type of organic molecule or ion and no other. Such selectivity allows a protein to "identify" (by binding) the presence of one particular type of molecule in a solution containing hundreds of different molecules.

A **ligand** is any molecule or ion which is bound to the surface of a protein by forces other than covalent chemical bonds. These forces are either electrical attractions between oppositely charged, ionized (or polarized) groups on the ligand and protein or the weaker attractions between adjacent nonpolar regions of the two molecules due to their electric fields (van der Waals forces). The chemical groups in only a small region on the protein surface, known as the **binding site**, are responsible for holding the ligand on the protein surface. The reversible binding of a ligand (L) to a protein binding site (P_b) to form a bound complex ($P_b \cdot L$) is written:

$$L + P_b \rightleftharpoons P_b \cdot L$$

The dot between P_b and L in the bound complex ($P_b \cdot L$) indicates the absence of a covalent bond between P_b and L. A protein may contain several binding sites, each specific for a different ligand.

Chemical specificity

The force of electrical attraction between opposite charges or polarized regions of molecules decreases markedly as the distance between the charges increases, and the even weaker van der Waals forces act only between nonpolar groups that are very close to each other. Therefore, in order for a ligand to bind to the surface of a protein, its atoms must be close to those of the protein. This happens only when the shape of the surface of the ligand is complementary to the shape of the surface of the protein binding site, such that the two molecules fit together like pieces of a jigsaw puzzle (Fig. 4-1). Therefore, a protein binding site can bind only those ligands having a complementary shape. This property of a protein binding site is known as its **chemical specificity**. Thus, the chemical specificity is determined by the shape of the protein surface in the region of the binding site. In the previous chapter we described how the three-dimensional shape of a protein is determined by the location of the different amino acids along its polypeptide chain. Accordingly, proteins with different amino acid sequences

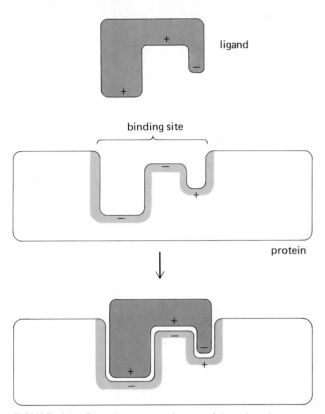

FIGURE 4-1. Complementary shapes of ligand and protein binding site determine the chemical specificity of binding.

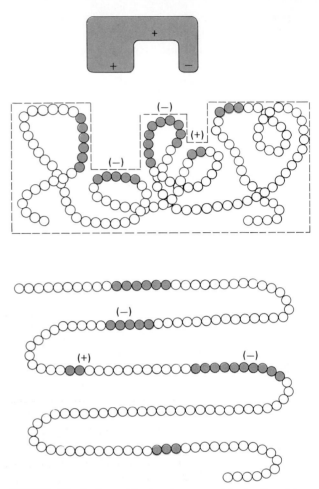

FIGURE 4-2. Amino acids that interact with the ligand at a binding site may be at far-removed sites along the polypeptide chain, as indicated in this model showing the three-dimensional folding of a protein, with the unfolded polypeptide chain shown below it.

have different shapes and, therefore, different binding sites, with different chemical specificities.

As illustrated in Fig. 4-2, the amino acids that interact with a ligand need not be adjacent to each other in the primary protein structure, since the folding of the polypeptide chain may bring various segments of the molecule into juxtaposition. Thus, amino acids from widely separated regions of a protein may participate in the binding of a ligand to the surface of the protein.

Some binding sites have a chemical specificity that allows them to bind only one type of ligand, whereas other binding sites may have a broader range of specificity, enabling them to bind a number of different ligands. For example, protein X in Fig. 4-3 has a binding site that can combine with three different ligands, whereas protein Y has a greater (i.e., more limited) specificity and can combine only

with a single ligand. Clearly, the degree of chemical specificity depends on the shape of the binding site; the more exactly its shape conforms to the shape of the ligand, the greater its chemical specificity.

Affinity

How strongly a ligand is held to the binding site is known as the **affinity** of the binding site for the ligand. The affinity of a binding site determines how likely it is that a bound ligand will leave the protein

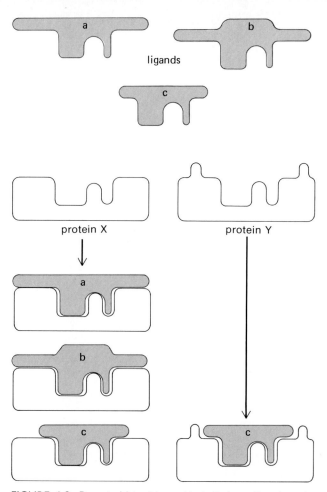

FIGURE 4-3. Protein X is able to bind all three ligands, a, b, and c, which have similar chemical structures; protein Y can only bind ligand c.

chemical specificity, but they may have different affinities for it. For example (Fig. 4-4), a ligand may have an ionized group with a negative charge which would bind strongly to a binding site that contained a positively charged amino acid side chain but would bind less strongly to a binding site of the same shape which did not have a positive charge. In addition the closer the surface of the ligand and binding site, the stronger the electrical interactions of both polar and nonpolar groups; hence, the more closely the shape of the ligand matches the shape of the binding site, the stronger the binding.

Saturation

At any time, a binding site on a protein is either occupied by a ligand or it is not. In a solution containing a number of identical protein molecules (each with a single binding site, for example) and complementary ligands, some of the proteins may have their binding site occupied by ligands and others not. When all the binding sites are occupied, the population of binding sites is said to be fully (100 percent) **saturated**—there are no unoccupied sites available to bind additional ligands. When half of the available sites are occupied, the system is 50 percent saturated, and so on.

The probability that a given binding site will be occupied by a ligand depends upon two factors: (1) the concentration of free ligand in the solution and (2) the affinity of the binding site for the ligand. The greater the ligand concentration, the greater the probability of a ligand encountering an unoccupied binding site and becoming bound (Fig. 4-5). The physiological implication of this dependence on concentration can be shown by an example. If the ligand were a molecule which exerts its biological effect by binding to a specific protein, the effect of the substance would increase with increasing concentration until all the binding sites become occupied. Further increases in ligand concentration would produce no further increase in the magnitude of the effect since there would be no additional sites to be occupied. To generalize, a continuous increase in the magnitude of a chemical stimulus (ligand concentration) which exerts its effects by

surface and return to its unbound state. Binding sites which bind a ligand tightly are said to be **high-affinity** sites, whereas those that bind weakly are **low-affinity** sites.

Affinity and chemical specificity are two distinct properties of binding sites. Chemical specificity depends only on the shape of the binding site, whereas the affinity (strength of binding) depends on the types of electrical interactions between ligand and binding site, as well as on the shape of the binding site. Two different proteins may be able to bind the same ligand, i.e., may have the same

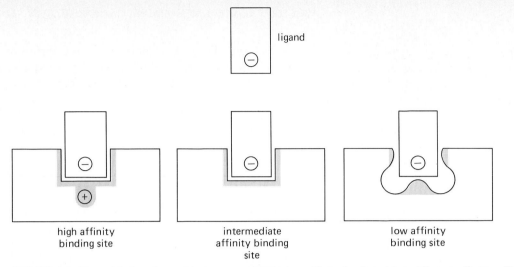

FIGURE 4-4. Three binding sites with the same chemical specificity for ligand but different affinities. The high-affinity site closely matches the shape of the ligand and has a positive charge that will interact with the negative charge on the ligand. The surface of the intermediate-affinity binding site closely matches the surface of the ligand but does not have an opposite electric charge. The low-affinity site has no charge, and its surface comes near the surface of the ligand at only a few points.

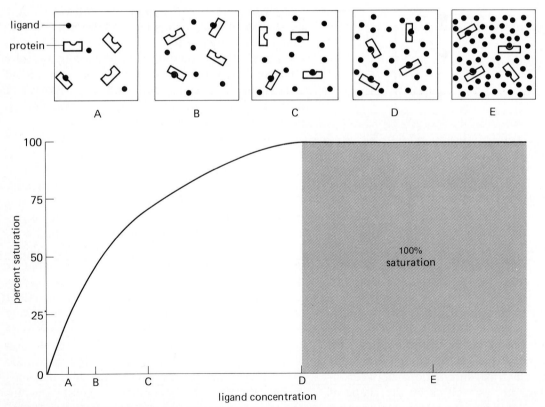

FIGURE 4-5. Increasing ligand concentration increases the number of binding sites that are occupied (increases percent saturation). At 100 percent saturation, all the binding sites have been occupied, and further increases in concentration of ligand will not increase the amount bound.

binding to specific sites on proteins does not produce a continuous increase in the biological response to the stimulus.

The second factor which determines the occupancy of a binding site is the affinity of the binding site for the ligand. Collisions between molecules in a solution and those on a binding site can dislodge a loosely bound ligand, much as tackling a football player may cause a fumble if the ball is not held in a tight grip. If a binding site has a high affinity for a ligand, even a low concentration of ligand will result in a large proportion of the binding sites being occupied, since once bound to the site, the ligand will remain bound for long periods of time. A low-affinity site, on the other hand, requires a much higher concentration of ligand to achieve the same degree of saturation, since with the shorter duration of occupancy of low-affinity sites, more frequent encounters are required to maintain a given degree of saturation. One measure of the affinity of a binding site is the ligand concentration necessary to produce 50 percent saturation of the binding sites; the lower the ligand concentration at 50 percent saturation, the greater the affinity of the binding site (Fig. 4-6).

Competition

As we have seen, more than one ligand may be able to bind to certain binding sites. Consider the situation in which two ligands, A and B, are able to bind to a particular site. Obviously only one ligand, either A or B, can occupy a given site at one time, and therefore when both are present in solution they will compete with each other for the available sites. **Competition** is the effect that one ligand has

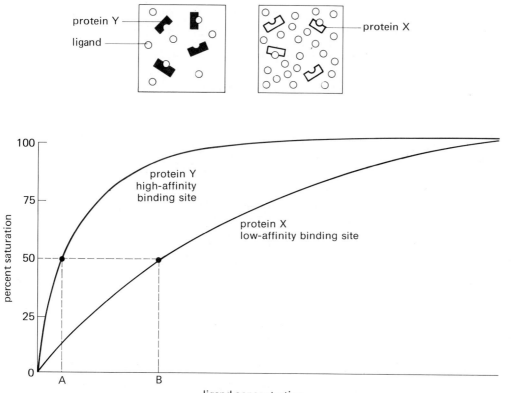

FIGURE 4-6. When two different proteins, X and Y, are able to bind the same ligand, the protein with the higher-affinity binding site (protein Y) is 50 percent saturated at a lower ligand concentration than is required to saturate 50 percent of the lower-affinity binding sites on protein X.

upon the binding of another ligand when both are able to bind to the same site. The number of sites occupied by each type of ligand will depend on the relative affinities of the binding site for the different ligands and on the concentrations of the ligands. If the concentration of A remains constant and the concentration of B is increased, the number of sites occupied by A will decrease because some of them will become occupied by B.

As a result of competition, the biological effects of one ligand may be markedly diminished by the presence of a competing ligand. For example, many drugs produce their effects by competing with the body's natural ligands for binding sites; by occupying the binding site, the drug prevents the natural ligand from binding and producing its response.

Functions of Protein Binding Sites

Having defined the physical characteristics of ligand binding to protein binding sites—chemical specificity, affinity, saturation, and competition—we now ask what roles such binding plays in cell function. A detailed answer to this question is given in the remaining chapters of this book, since practically every cell function, and its regulation, is mediated by specific interactions between ligands and their corresponding binding sites on proteins. In this section, a few examples will suffice to illustrate general principles.

Aggregation of proteins into organized structures

An obvious result of binding is the holding together of two molecules—the ligand and the protein containing the binding site. Multimolecular aggregates arranged in specific three-dimensional patterns (for example, ribosomes and protein filaments) can be constructed by this type of binding of the individual component molecules. The proteins in these structures may contain one or more binding sites for nonprotein ligands, but in addition, they have binding sites specific for other protein molecules in the multimolecular aggregate. In other words, the proteins in such cases act both as binding sites and ligands. When the various proteins of a multimo-

lecular structure are placed in a solution, the structure will actually assemble itself, since the location of each molecule in the structure is determined by its binding to other proteins. If one were to take the component pieces of a clock and shake them up in a basket, one would not expect them to fall into the positions that they occupy in the functioning clock, but this is, in large part, the way a cell goes about forming its structures. It synthesizes a number of different proteins and mixes them together in solution. Because of the specificity of their binding-site–ligand interactions, the proteins aggregate into specific structures. Thus, the only information required by the cell is that which specifies the sequence of amino acids in each protein, since these sequences determine the shape of the binding site; the cell does not require additional instructions on how to combine the proteins into more complex structures.

The assembly during embryonic development of multicellular tissues and organs from individual cells provides another example of selective binding. Different types of cells have in their plasma membranes proteins which contain specific binding sites for a class of proteins known as **aggregation factors**, each of which itself contains multiple sites. Therefore, one site on the aggregation factor binds to the surface of one cell while another binds to the surface of an adjacent cell, thereby linking the two cells together. Given a variety of surface binding sites on different cells and corresponding aggregation factors, cells can be arranged in various patterns. However, this is only one of several processes which contribute to the anatomical organization of tissues and organs.

Enzymes

Substances which increase the rates of chemical reactions but do not themselves undergo any net chemical change during the reaction are known as **catalysts**. Certain proteins, called **enzymes**, are the catalysts for the chemical reactions in the body. The property of enzymes which enables them to catalyze certain chemical reactions is the chemical specificity of their binding sites. The ligands which bind to the enzymes' binding sites are known as

substrates, and the molecules that are ultimately formed from these ligands in the chemical reaction catalyzed by the enzyme are known as **products**. The mechanism by which the binding of a substrate to the enzyme increases the rate of a chemical reaction will be discussed in the next chapter.

Practically every chemical reaction in the body is catalyzed by a specific enzyme. In the absence of enzymes, most of the reactions that occur in living organisms would proceed at a negligible rate. Therefore, the types of chemical reactions that occur within a particular cell depend upon which enzymes are present in that cell. Furthermore, particular types of reactions are localized to particular regions of a cell because specific sets of enzymes occur in those regions; for example, the chemical reactions that occur in the mitochondria are different from those that occur in the nucleus or in the endoplasmic reticulum because each of these cell organelles contains a different set of enzymes.

Certain enzymes are present in almost all cells, whether they be muscle cells, nerve cells, etc., and to that extent these cells are all similar (although they are not identical, since the concentrations of these common enzymes may differ from cell to cell). In contrast, other enzymes may be absent from certain cells and present in others, which results in certain cells being able to carry out some reactions that do not occur in other cells. The crucial point is that cell chemistry differs somewhat from cell to cell because different types of cells contain different amounts and types of enzymes. Moreover, a given cell may undergo changes in its activity as a result of changes in the types of enzymes it synthesizes and the amounts of enzyme synthesized. Again we see that a cell is what it is because of the instructions it receives concerning the types of proteins (enzymes, in this case) it can synthesize and the rates at which they are synthesized.

Detection of chemical messengers

In addition to holding molecules and cells together and catalyzing chemical reactions, protein binding sites play a key role in the communications systems of the body. Chemical signals that are transmitted between cells or between one area of a cell and another are detected by specific binding sites for the chemical messengers. For example, most cells are exposed to the same hormones (chemical messengers), which reach them by way of the circulatory system, but only those cells that contain binding sites for a particular hormone will respond to it. Thus, hormone-binding sites provide the basis for the selective action of hormones.

There are many other types of chemical signals that depend on binding to specific binding sites in order to produce a response. For example, the stimulation of a cell by a nerve cell depends on a chemical agent which is released from the nerve cell and which binds to specific binding sites on the plasma membrane of the responding cell. Within cells, chemical signals can influence the functioning of certain enzyme molecules by binding to specific sites on the enzyme. Many other examples of the binding of chemical messengers to protein binding sites will be described in later chapters.

Regulation of Binding-Site Characteristics

Because proteins are associated with practically every cell function, the mechanisms for controlling these functions involve the control of protein activity. The regulation of protein synthesis determines the types and amounts of proteins in a cell, whereas modulation of existing proteins alters their shape and, thus, their functional activity. The mechanisms for altering the shapes of protein binding sites will be discussed in this section, whereas later sections of this chapter will describe the regulation of protein synthesis.

As described in the previous chapter, the shape of a protein depends on electrical interactions between charged or polarized groups that are distributed along the polypeptide chain as determined by its amino acid sequence. These electrical interactions take place between different portions of the polypeptide chain as well as between the polypeptide chain and other molecules in the medium surrounding the protein. Any change in the distribution of electric charge along the polypeptide chain or in the polarity of the molecules immediately surrounding a protein will alter the distribution of electrical attractions and thereby alter the shape of the protein.

A number of nonspecific factors in the environment of a protein have the potential for altering protein shape. Two of the most important are temperature and acidity, but normally these factors are not used to control protein shape because they are maintained relatively constant in the body. However, should large changes in temperature or acidity occur, as in disease, they can produce marked alterations in the shape of proteins and thus in protein function.

The two mechanisms that are normally used by cells to alter protein shape are allosteric and covalent modulation.

Allosteric modulation

Whenever a ligand binds to a protein binding site, the attracting forces between ligand and protein (which hold the ligand on the binding site) also alter the distribution of forces within the protein, thereby producing a change in protein shape. In many cases the attraction between binding site and ligand causes the shape of the binding site itself to change so that it more closely matches the shape of the ligand. However, the binding of ligand can also change the shape of the protein in other regions that are not part of the binding site.

If a protein contains two separate binding sites, the binding of a ligand to one site can alter the shape of the second binding site and, hence, its ability to bind the second ligand. This is the molecular basis for the **allosteric modulation** of protein function (Fig. 4-7), and such proteins are known as allosteric (other shape) proteins. One binding site on an allosteric protein, known as the **functional site**, carries out the specific function associated with the protein; for example, if the allosteric protein were an enzyme, then the functional binding site would bind substrate molecules and catalyze their conversion to product molecules. The other binding site is the **regulatory site**, and the ligand that binds to this site is known as a **modulator molecule**, since its binding to the regulatory site will regulate the functional site on the protein (Fig. 4-7). The modulator molecule is a chemical substance entirely different from the ligand that binds to the functional site.

Modulator molecules belong to the general class of chemical messengers, which carry a signal from one site to another; the regulatory site to which they bind is the equivalent of a molecular switch. In some allosteric proteins the binding of the modulator molecule to the regulatory site turns *on* the functional site by producing a change in the shape of the functional site which allows it to bind its specific ligand; in other allosteric proteins the binding of a modulator turns *off* the functional site by producing a change in shape that prevents the functional site from binding a ligand. Binding of the modulator molecule may, in some cases, alter the

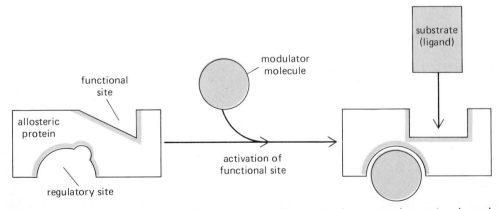

FIGURE 4-7. Allosteric modulation. Binding of a modulator molecule at a regulatory site alters the shape of the functional site (here an enzymatic site) which allows the site to bind its substrate.

affinity of the functional site for its ligand rather than the ability of the site to bind a ligand. In any case, the function performed by an allosteric protein is regulated by the presence or absence of a modulator molecule bound to the regulatory site.

In a population of identical allosteric proteins (as would exist in a cell), the number of proteins that are turned *on* or *off* depends on how many of the regulatory sites are occupied by modulator molecules. Therefore, the magnitude of the cell function (as determined by the number of functional sites that are operating) can be regulated by varying the concentration of the modulator molecule. As we shall see in later chapters, many types of chemical messengers modify cell functions by acting as modulator molecules that alter the activities of specific proteins.

It should be emphasized that most proteins are not subject to allosteric modulation. Only certain proteins associated with each type of cell function need to be regulated in order to control the function. It is these key proteins that are subject to either allosteric or covalent modulation.

Covalent modulation

A second way to alter the shape, and therefore the function, of protein molecules is to attach various chemical groups, through covalent bonds, to the side chains of certain amino acids in the proteins. This is known as **covalent modulation**. In most cases (but not all) phosphate is the chemical group that is attached, and the process is known as **phosphorylation**. Phosphate has a net negative charge, and so when phosphate is attached to one of the amino acid side chains in a protein it introduces a negative charge into this region of the protein where previously there had been none. This alters the distribution of forces between this region and other regions of the protein, producing a change in protein conformation (Fig. 4-8). If the change in protein shape occurs in the region of a binding site, it will change the binding characteristics and, hence, control the activity of the binding site. Phosphorylation of a protein may turn *on* a binding site or turn it *off*, depending on the specific protein and the type of shape change that occurs at the binding site.

Unlike allosteric modulation, in which there are no covalent bonds formed between the modulator molecule and the protein, covalent modulation involves chemical reactions in which covalent bonds are formed. These reactions are mediated by specific enzymes. (In contrast, no enzymes are required for allosteric modulation.) When the reaction involves phosphorylation, the enzyme is a **protein kinase**, which catalyzes the transfer of phosphate

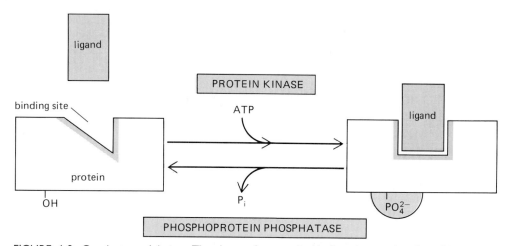

FIGURE 4-8. Covalent modulation. The shape of a protein binding site can be altered by the attachment, through a covalent chemical bond, of an ionized phosphate group to one of the amino acid side chains. The enzyme protein kinase catalyzes the phosphorylation of the protein while a second enzyme, phosphoprotein phosphatase, removes the phosphate group, returning the protein to its original shape.

from a molecule of adenosine triphosphate (ATP) (see Chap. 5) to a hydroxyl group present on the side chain of certain types of amino acids.

$$\text{Protein—OH} + \text{ATP} \xrightarrow{\text{protein kinase}} \text{protein—PO}_4^{2-} + \text{ADP}$$

If such a reaction is to be used to control protein activity there must also be a mechanism for removing the phosphate group and returning the protein to its original shape. This dephosphorylation is accomplished by a second enzyme known as a **phosphoprotein phosphatase**

$$\text{Protein—PO}_4^{2-} + \text{H}_2\text{O} \xrightarrow{\text{phosphoprotein phosphatase}} \text{protein—OH} + \text{HPO}_4^{2-}$$

Thus, two enzymes are required to control protein activity by covalent modulation, one to add the modulating group and one to remove it.

Although most proteins contain amino acids capable of being phosphorylated, only those proteins that are substrates for specific protein kinases actually are phosphorylated. There are many different protein kinases with specificities for different proteins. Several different kinases may be present in the same cell. In general, the chemical specificities of the phosphoprotein phosphatases are broader, and a single enzyme can dephosphorylate many different phosphorylated proteins.

Different modulator molecules activate different protein kinases and thus control different sets of phosphorylated proteins. The phosphoprotein phosphatase is usually not regulated and remains active at all times. Therefore, the activity of the phosphorylated protein depends only on the activity of the protein kinase: As long as the kinase is active the proteins that are dephosphorylated by the phosphoprotein phosphatase are rephosphorylated by the kinase, and so most of the protein is maintained in its phosphorylated state. Turning off such a protein requires only stopping the kinase reaction, since the phosphoprotein phosphatase will remove the phosphate groups and keep the protein in its dephosphorylated state.

SECTION B.
GENETIC INFORMATION AND PROTEIN SYNTHESIS

Genetic Information

Molecules of deoxyribonucleic acid (DNA) contain instructions coded into their molecular structure for the synthesis of cell proteins. As described in the previous sections, the protein composition of a cell determines the cell's structure and functional activity. The portion of a DNA molecule which contains the information determining the amino acid sequence of a single polypeptide chain is known as a **gene**. A single DNA molecule contains many genes—the units of hereditary information.

As an example of the expression of genetic information, consider eye color in people. The color is due to the presence of pigment molecules in certain cells in the eye. A sequence of enzyme-mediated reactions is required to synthesize these pigments, and the genes determining eye color control the synthesis of specific enzymes in this pathway. Note that the genes for eye color do not contain information about the chemical structure of the eye pigments themselves; rather they contain the information required to synthesize the enzymes which mediate the formation of the pigments.

Although DNA contains the information necessary for the synthesis of specific proteins, DNA does not itself participate *directly* in the assembly of a protein molecule. Most of the DNA in a cell is in the nucleus (a small amount is in the mitochondria), whereas most protein synthesis occurs in the cytoplasm. The transfer of information from DNA to the site of protein synthesis is the function of molecules of RNA (ribonucleic acid) whose synthesis is governed by the information in the genes in

DNA. This mechanism of expressing genetic information occurs in all living organisms and has led to what is now called the "central dogma" of molecular biology: Genetic information flows from DNA to RNA and then to protein:

$$DNA \rightarrow RNA \rightarrow protein$$

Figure 4-9 summarizes the general pathway by which the information stored in DNA influences cell activity via protein synthesis.

DNA and the genetic code

As described in the previous chapter, DNA consists of two polynucleotide chains coiled around each other to form a double helix. Each chain is a sequence of nucleotides (see Fig. 3-21) joined together by phosphate-sugar linkages. Each nucleotide in DNA contains one of four different bases (adenine [A], guanine [G], cytosine [C], or thymine [T]—see Fig. 3-22), and each of these bases is specifically paired, A to T and G to C, with a base on the corresponding polynucleotide chain of the double

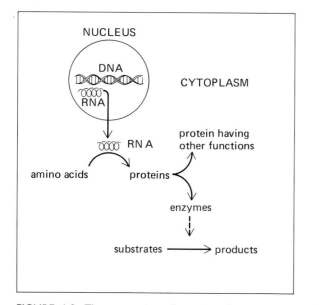

FIGURE 4-9. The expression of genetic information in a cell occurs through its transcription from DNA to RNA in the nucleus followed by the translation of the RNA information into protein synthesis in the cytoplasm.

helix (see Fig. 3-23). Thus, both chains contain a precisely ordered sequence of bases, one chain being complementary to the other. It is the sequence of these bases along the polynucleotide chains which provides the code ultimately specifying the sequence of amino acids in proteins.

The genetic language is similar in principle to a written language, such as English, which consists of a set of symbols forming the letters of an alphabet. The letters are arranged in specific sequences to form words, and the words are arranged in linear sequences to form sentences. The genetic language contains only four letters, corresponding to the four bases A, G, C, and T. The words are three-base sequences, which specify particular amino acids (i.e., each word in the genetic language is only three letters long). These words are arranged in a linear sequence along the chains of DNA, and the sequence of words specifying the structure of a single protein makes up a gene (Fig. 4-10); thus a gene is equivalent to a sentence, and the entire collection of genes in a cell is equivalent to a book.

How are the four "letters" of the DNA alphabet, A, G, C, and T, arranged to form at least 20 different three-letter code words? Since the four bases can be arranged in 64 different combinations ($4 \times 4 \times 4 = 64$), a triplet code actually provides more than enough code words. It turns out that not just 20, but 61 out of the 64 possible triplet combinations are used to specify amino acids. This means that a given amino acid is usually specified by more than one code word (of the 20 different amino acids, 18 are represented by more than one code word). The three code words which do not specify amino acids are known as **termination** code words; they perform the same function as does a period at the end of a sentence—they indicate that the end of a genetic message has been reached.

The genetic code is a universal language used by all living cells; for example, the code words for the amino acid tryptophan is the same in the DNA of a bacterium, an amoeba, a plant, and a human being. Although the same code words are used by all living cells, the messages they spell out—the sequences of code words (the "sentences") which determine the amino acid sequences in proteins—

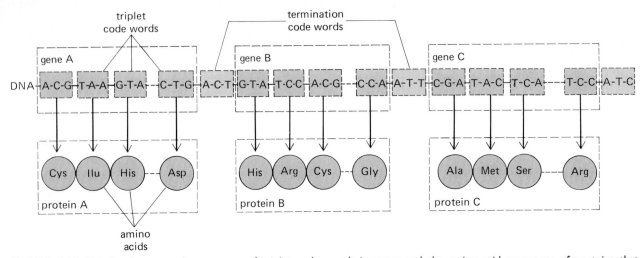

FIGURE 4-10. Relation between the sequence of triplet code words in genes and the amino acid sequences of proteins that correspond to them. The names of the individual amino acids are abbreviated. Note that all three genes are part of the same DNA molecule, which may contain hundreds of genes in a linear sequence.

are, of course, different in each organism. The universal nature of the genetic code is one piece of evidence which supports the concept that all forms of life on earth evolved from a common ancestor.

The giant molecules of DNA are in the cell nucleus and are unable to pass through the nuclear membrane into the cytoplasm; yet it is in the cytoplasm, on ribosomes, that proteins are synthesized. Since DNA cannot leave the nucleus, a message carrying genetic information must pass from the nucleus to the cytoplasm. This message is in the form of RNA molecules known as **messenger RNA (mRNA)**. The transfer of genetic information from DNA to protein thus occurs in two stages: First, the genetic message must be passed from DNA to mRNA (**transcription**); second, the message in mRNA is used to direct the assembly of the specific sequence of amino acids to form a protein (**translation**).

Protein Synthesis

Transcription: Synthesis of messenger RNA

Transcription is the process by which the information represented by a sequence of nucleotides in DNA is used to determine the sequence of nucleotides in mRNA. Recall from Chap. 3 that ribonu-

cleic acids are single-chain polynucleotides, whose nucleotides differ from DNA in that they contain the sugar ribose (rather than deoxyribose) and the base uracil (rather than thymine). The other three bases, adenine, guanine, and cytosine, occur in both DNA and RNA.

Free (uncombined) nucleotides containing a sequence of three phosphate groups (nucleotide triphosphates) form the pool of subunits that are used to form molecules of RNA. To form messenger RNA the four types of free ribonucleotides (represented by the four different bases) must be linked together in a sequence determined by the sequence of deoxynucleotides in a DNA gene.

Recall (from Chap. 3) that, in DNA, the two polynucleotide chains are linked together by hydrogen bonds between specific pairs of bases—A pairing with T, and G with C. Transcription begins with the breakage of these weak hydrogen bonds so that a portion of the DNA double helix uncoils. The bases in the free ribonucleotide triphosphates are then able to pair with their exposed DNA counterparts. Free RNA nucleotides containing adenine will pair with any exposed thymine base in DNA; likewise, RNA nucleotides containing G, C, and U will pair with exposed C, G, and A bases respectively, in DNA, (uracil, which replaces thymine in

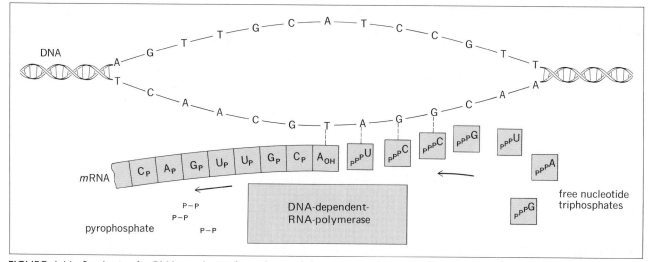

FIGURE 4-11. Synthesis of mRNA on the surface of one of the two strands of a DNA molecule through base pairing between free ribonucleotides and the bases in DNA.

RNA, pairs with adenine). In this way, the nucleotide sequence in DNA (the genetic code) determines the sequence of nucleotides along a molecule of mRNA.

Once the appropriate nucleotides have been lined up properly by pairing with bases in DNA, they are joined together by an enzyme known as **RNA polymerase**; this enzyme catalyzes the splitting-off of two of the three phosphate groups from each nucleotide and the covalent linkage of the RNA nucleotide to the next one in sequence (Fig. 4-11). RNA polymerase is active only when bound to DNA and will not link free nucleotides together

(in what would be a random sequence) when they are not base-paired with DNA. DNA acts as a modulator molecule which allosterically activates the RNA polymerase.

Since DNA consists of two strands of polynucleotides, both of which are exposed during transcription, it should be possible to form two different mRNA molecules, one from each strand (Fig. 4-12). These two mRNA molecules would have different nucleotide sequences because the two chains of the DNA helix are not identical and, thus, would code for two entirely different proteins. However, only one of the two potential molecules of mRNA

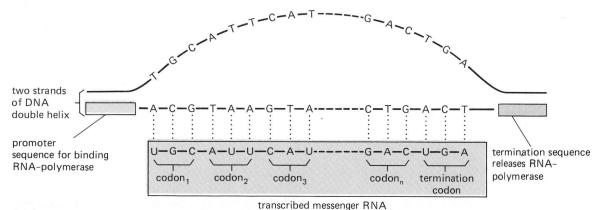

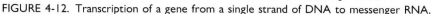

FIGURE 4-12. Transcription of a gene from a single strand of DNA to messenger RNA.

is ever formed. Which mRNA is formed depends on a specific sequence of nucleotides, called the **promoter**, that is located at the beginning of the gene; the promoter, which combines with a binding site on RNA polymerase, is present in only one of the two DNA strands. Since this binding is required for the initiation of transcription, only the strand of DNA containing the promoter sequence will be transcribed. Because the promoter for any given gene may be located in either strand of the DNA double helix, some genes are transcribed from one strand while other genes in the same DNA molecule are transcribed from the opposite strand. Beginning at the promoter end of a gene, the RNA polymerase moves along the DNA strand, joining one ribonucleotide at a time to the growing mRNA chain until it reaches a **termination sequence** at the end of the gene which causes it to release the newly formed molecule of mRNA.

In a given cell, the information in only a few of the thousands of genes present in DNA is transcribed to molecules of mRNA at any given time. Which genes are transcribed depends on the ability of RNA polymerase to bind to the promoter site on a particular gene. Various mechanisms are used by cells to block selectively or make accessible the promoter region of a specific gene. Such regulation of gene transcription provides a means of controlling the synthesis of specific proteins and thereby the activities of cells. It will be considered later in this chapter.

It must be emphasized that the nucleotide sequence in a molecule of mRNA is not identical to that in the corresponding strand of DNA, since its formation depends on the pairing between complementary, not identical, bases (Fig. 4-12). Thus, during transcription, the RNA code words which are formed are complementary to the code words in DNA. These triplet base sequences in mRNA corresponding to code words in DNA are called **codons**. It is the sequence of these codons in mRNA that constitutes the message carried to the cytoplasm where it will be translated into a specific sequence of amino acids. A **termination codon** in mRNA is formed from the complementary termination code word at the end of the gene in DNA.

Although the entire sequence of nucleotides in a DNA gene is transcribed into a corresponding sequence of nucleotides in mRNA, only a portion of this sequence actually specifies the sequence of amino acids in a protein. Sequences of DNA nucleotides within a gene which are the code words for specific amino acid sequences in a protein are known as **exons**. Within a single gene, exon regions are separated from each other by noncoding sequences of nucleotides known as **introns**. Both exons and introns are transcribed into the nucleotide sequence of mRNA as it is formed.

Before passing to the cytoplasm, a newly formed molecule of mRNA must undergo **RNA processing** (Fig. 4-13) to remove the intron sequences. Nuclear enzymes which identify specific nucleotide sequences at the beginning and end of an intron remove these sequences and splice the end of one exon to the beginning of another exon to form a molecule of mRNA with no intron segments. As would be expected, most RNA splicing occurs between one exon and the next exon in sequence; however, in some cases the exons of a single gene can be spliced together in several different sequences, resulting in the formation of different mRNA molecules from the same gene and giving rise, in turn, to several proteins with different amino acid sequences.

The mRNA molecule formed as a result of this RNA processing is 75 to 90 percent shorter than the originally transcribed mRNA, which means that 75 to 90 percent of the nucleotide sequences in DNA structural genes are introns (noncoding sequences). What role, if any, such large amounts of "nonsense" DNA may perform is unclear.

Once transcribed and processed, a molecule of mRNA must move into the cytoplasm, where its message can be used to direct the synthesis of a specific protein. Although molecules of mRNA are considerably smaller than DNA molecules, they are still very large. (Since a typical protein may consist of a sequence of 100 amino acids, the mRNA molecules coding for it must contain 100 codons, i.e., 300 nucleotides.) Such large molecules do not readily pass through biological membranes (see Chap. 6). The nucleus is enclosed by two membranes (Fig. 4-14) traversed at intervals by porelike structures which do allow passage of mRNA molecules. These

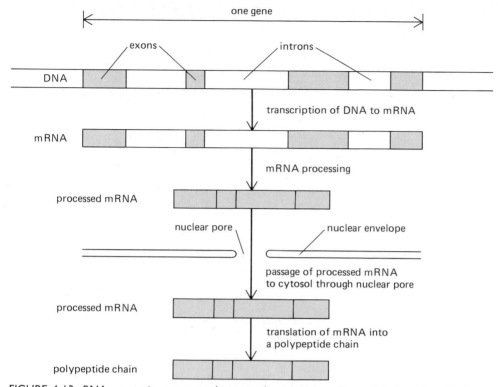

FIGURE 4-13. RNA processing removes the noncoding sequences (introns) before the mRNA passes through the nuclear pores to the cytosol.

apparent openings are not simply holes through which these mRNA molecules can move but are filled with proteins which selectively mediate the passage of large molecules. The nuclear pores also seem to allow the selective entry into the nucleus of certain proteins (RNA polymerase for example) that are synthesized in the cytoplasm. The molecular mechanism responsible for the selective passage of large molecules through these pores is unknown.

Translation: Synthesis of polypeptides

Having reached the cytoplasm, a mRNA molecule binds to a ribosome, the cell organelle which mediates protein assembly. Before describing this assembly process we must first describe the structure of a ribosome as well as the characteristics of two additional types of RNA that are involved in protein synthesis.

Ribosomes. Ribosomes are small granules (about 23 nm in diameter) located in the cytoplasm, either suspended in the cytosol (free ribosomes) or attached to the surface of the endoplasmic reticulum (bound ribosomes). Protein synthesis takes place on the ribosome. When the synthesis is complete, the protein is released from the ribosome. In the case of free ribosomes the protein is released into the cytosol, whereas in the case of bound ribosomes the protein is released into the lumen of the endoplasmic reticulum where it will undergo some modifications in structure and eventually be secreted from the cell or incorporated into various cell organelles by processes that will be described in Chap. 6. The number of ribosomes in a cell, and their distribution between free and bound forms, varies between different cells and during different periods of cell activity, such as during cell growth or when the cell is secreting large amounts of protein.

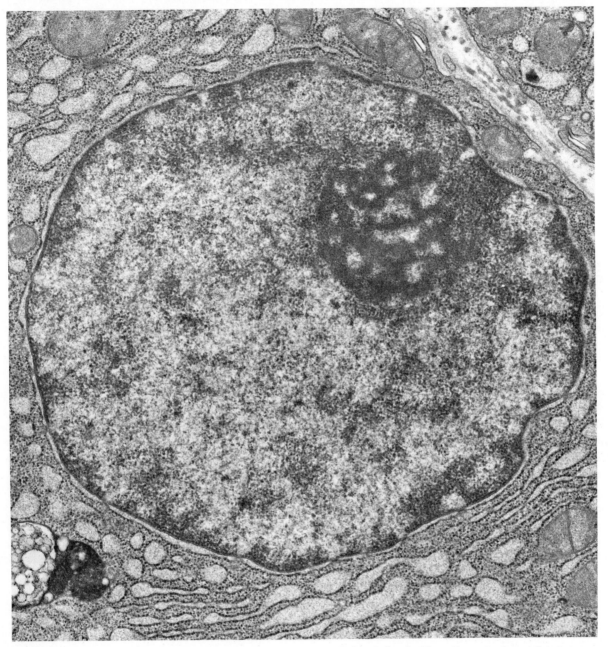

FIGURE 4-14. Electron micrograph of the nucleus in a pancreas cell. Note the double nuclear membrane that is interrupted at intervals by nuclear pores. The densely stained nucleolus is located in the upper right portion of the nucleus. (Courtesy of K. R. Porter.)

Each ribosome consists of over 70 protein molecules in association with a type of RNA known as **ribosomal RNA (rRNA)**. Ribosomal RNA is synthesized in the nucleus, DNA serving as a template for the positioning of its nucleotides in the same manner as just described for synthesis of mRNA. (Transcription of certain genes, known as **structural genes**, yields mRNA which will be used to form proteins, whereas transcription of other genes yields forms of RNA which assist in protein synthesis but do not carry coded information determining protein structure.) The genes which code for rRNA are associated with the **nucleolus**, a densely staining nuclear organelle consisting of filamentous and granular elements (Fig. 4-14); moreover, the nucleolus is the site at which the subunits of ribosomes are assembled. The ribosomal granules are formed by a combination of rRNA (transcribed from rRNA genes in DNA) and certain proteins that have been synthesized in the cytoplasm and then transferred to the nucleus. The ribosomal granules then move into the cytoplasm (probably through the nuclear pores) where they combine with still other proteins to form the two large subunits of ribosome structure—a larger subunit known as the 60S subunit and a smaller 40S subunit.

When an mRNA molecule arrives in the cytoplasm, one end of it binds to the 40S subunit, which then binds to the 60S subunit to form a fully functional ribosome (Fig. 4-15). A portion of the messenger RNA molecule appears to lie in a groove between the two subunits of the ribosome. The proteins of the ribosomal particle provide the enzymes and other proteins required for the translation of mRNA's coded message into protein.

Transfer RNA. In the translation process, how do individual amino acids become arranged in the proper sequence to form a protein molecule? By themselves, free amino acids do not have the ability to bind to the bases in mRNA. Rather, orientation of the amino acids on the mRNA molecules involves yet a third type of RNA known as **transfer RNA (tRNA)**. Transfer RNA molecules are the smallest (about 80 nucleotides long) of the three types of RNA. Like mRNA and rRNA, tRNA is synthesized by base pairing with DNA nucleotides

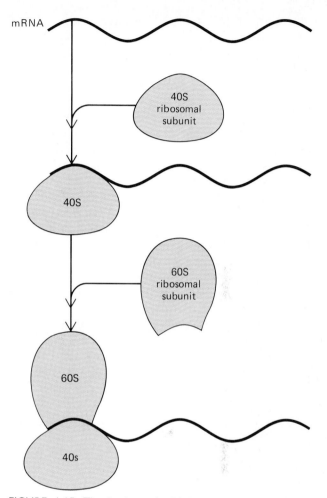

FIGURE 4-15. The binding of mRNA to a ribosome occurs first to the 40S subunit, to which the 60S subunit then binds to form a functional ribosome.

at specific tRNA genes in the nucleus; it then moves to the cytoplasm. The key to the functioning of tRNA is that each of these molecules can combine both with a specific amino acid and with a codon in mRNA specific for that amino acid. This permits tRNA to act as the identifying link between the amino acid and the mRNA codon.

Transfer RNA is covalently linked to its amino acid by an enzyme known as **aminoacyl-tRNA synthetase**. There are 20 different aminoacyl-tRNA synthetase enzymes, each of which catalyzes the linkage of only one type of amino acid to its particular type of tRNA. The next step is to link the tRNA molecule (bearing its attached amino acid) to the mRNA codon for that amino acid in mRNA. As one might have predicted, this is achieved by base pairing between tRNA and mRNA. One three-nucleotide sequence in a given tRNA molecule is exposed and can base-pair with a complementary codon in mRNA; this three-nucleotide sequence is appropriately termed an **anticodon**. Since there are 61 different mRNA codons for the 20 different amino acids, one might expect to find 61 different molecules of tRNA, each with a different anticodon. There are, however, fewer than 61 types of tRNA molecules because some of the anticodons are able to pair with several of the different codons specifying the same type of amino acid.

Figure 4-16 illustrates the binding to mRNA of a tRNA molecule specific for the amino acid alanine. Note that it is covalently linked to alanine at one end and base-paired with G-C-C in mRNA at the other end. Its anticodon for this base pair is C-G-I (tRNA has several unusual bases, the I representing **inosine**, which base-pairs with cytosine).

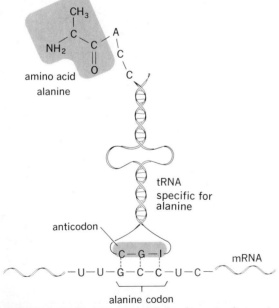

FIGURE 4-16. Base pairing between the anticodon region of a tRNA molecule with the corresponding codon region of an mRNA molecule. The exposed loop in the central region of tRNA is the site of interaction with aminoacyl-tRNA synthetase.

Protein assembly. The interactions between mRNA, rRNA, tRNA, and ribosomal proteins take place on the surface of the ribosomal particles. The initiation of mRNA translation is a complex process involving several specific ribosomal proteins and rRNA that interact to identify the initial codon sequence in mRNA and to catalyze the formation of a peptide bond between the first and second amino acids in the peptide chain that is to be synthesized. This initial step appears to be the slowest step in the ribosomal process of translation, and the rate of protein synthesis can be regulated by factors that influence this initiation process. The coded message in mRNA is read in only one direction, always beginning at the mRNA molecule's end which corresponds to the amino acid located at the free-amino end of the polypeptide chain.

The large 60S subunit of the ribosome has two binding sites for tRNA; one holds the tRNA molecule that is attached to the most recently added amino acid in the progressively lengthening chain of amino acids, and the second binding site holds the tRNA containing the next amino acid to be added to the chain. Ribosomal enzymes catalyze the formation of a peptide bond between these two amino acids. Following the formation of the peptide bond, the tRNA at the first binding site is released from the ribosome and the tRNA at the second site (now linked to the peptide chain) is transferred to the first binding site as the ribosome moves one codon space along the mRNA, making room for the binding of the next amino acid–tRNA molecule (Fig. 4-17). This process is repeated over and over as each amino acid is added in succession to the growing peptide chain (at a rate of two to three amino acids per second). When the ribosome reaches the termination codon specifying the end of the polypeptide chain, the link between the polypeptide chain and the last tRNA is broken and the completed protein is released from the ribosome.

The synthesis of protein molecules is an energetically expensive process. Energy equivalent to four molecules of ATP (the energy currency of cells, as will be discussed in Chap. 5) is used in the process of forming each peptide bond in the polypeptide chain. Some of this energy is used to form the amino acid–tRNA link and some to move the

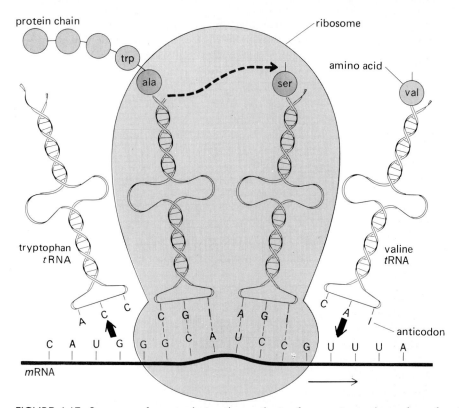

FIGURE 4-17. Sequence of events during the synthesis of a protein on the surface of a ribosome. The three-base anticodon of tRNA carrying one amino acid, e.g., valine (val), binds to the corresponding codon of mRNA and transfers its amino acid to the growing polypeptide chain. As the ribosome moves along the strand of mRNA, successive amino acids are added to the polypeptide chain.

ribosome along the mRNA molecule. Energy is also consumed in the formation of mRNA, as well as in its processing and passage to the cytoplasm.

The same strand of mRNA can be used more than once to synthesize several molecules of the protein because the message in mRNA is not destroyed during protein assembly. As a ribosome moves away from the end of the mRNA molecule during the process of synthesizing a protein, a second ribosome may become attached to the end of mRNA and begin the synthesis of a second protein molecule. Thus, a number of ribosomes may be attached to the same strand of mRNA, forming a **polyribosome**. Polyribosomes containing as many as 70 ribosomes attached to a single strand of mRNA have been observed. Each ribosome has a growing peptide chain attached to it. Ribosomes near the beginning of the chain have short peptide chains

representing the first few amino acids in the protein; ribosomes near the end of mRNA have protein chains which are almost completed (Fig. 4-18).

The mRNA molecules do not remain in the cytoplasm indefinitely. Eventually they are broken down into nucleotides by cytoplasmic enzymes. Therefore, if a gene corresponding to a particular protein ceases to be transcribed into mRNA, the synthesis of that protein will eventually slow down and cease as the existing molecules of mRNA are broken down.

There are additional posttranslational changes that can occur in the structure of some polypeptide chains following their synthesis on a ribosome. Some of these changes, such as the phosphorylation of certain amino acid side chains, are associated with the modulation of protein activity and have been discussed earlier in the chapter. Also, certain

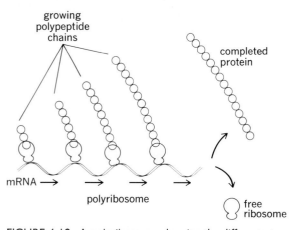

FIGURE 4-18. A polyribosome, showing the different stages of protein assembly as each ribosome moves along the strand of messenger RNA.

classes of proteins, the glycoproteins for example, are formed by the posttranslational addition of various carbohydrate groups to particular amino acid side chains. In a few cases, specific peptide bonds in a large polypeptide chain are broken to produce a number of shorter polypeptide chains, each of which may perform a different function. For example, as illustrated in Fig. 4-19, five different proteins present in a cell, are all derived from the same mRNA.

The existence of genes for rRNA and tRNA

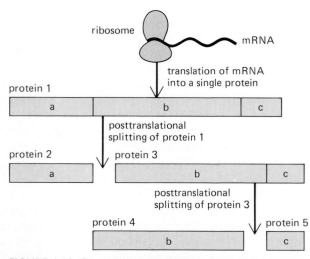

FIGURE 4-19. Posttranslational splitting of a protein can result in several proteins, each of which may perform a different function, but all derived from the same gene.

and the posttranslational processing of proteins have shown that the previously accepted generalization, "one gene—one polypeptide," is a little less precise than it was believed to be when it was initially formulated. It now appears that one gene can give rise to several related polypeptide chains.

The steps leading from DNA to a functional protein are summarized in Table 4-1.

TABLE 4-1. Sequence of events leading from DNA to protein synthesis

Transcription

1. The enzyme RNA polymerase binds to the promoter region of a gene and separates the two strands of the DNA double helix in the region of the gene to be transcribed.
2. Free ribonucleotides base-pair with the deoxynucleotides in DNA.
3. The nucleotides paired with one strand of DNA are linked by RNA polymerase to form mRNA containing a sequence of bases complementary to the DNA base sequence.
4. mRNA processing removes the intron regions of mRNA, which contain noncoding sequences, and splices together the exon regions that code for specific amino acids.

Translation

5. Processed mRNA passes from the nucleus to the cytoplasm, where one end of the mRNA binds to a ribosome.
6. Free amino acids are linked to their corresponding tRNAs by enzymes in the cytoplasm.
7. The three-base anticodon in an amino acid-tRNA complex pairs with the corresponding codon in the region of the mRNA bound to the ribosome.
8. The portion of the peptide chain that has already been synthesized (and is attached to an adjacent tRNA bound to the ribosome) is now linked by a peptide bond to the amino acid at the end of the tRNA next to it, thereby adding one more amino acid to the chain.
9. The tRNA that has been freed of the peptide chain is released from the ribosome.

TABLE 4-I. (continued)

10. The ribosome moves one codon step along mRNA.
11. Steps 7 to 10 are repeated over and over until the end of the mRNA message is reached.
12. The completed protein chain is released from the ribosome when the termination codon in mRNA is reached.
13. In some cases, the protein undergoes posttranslational processing in which various chemical groups are attached to specific side chains, or the protein is split into several smaller functional peptide chains.

Regulation of protein synthesis

Possible sites of regulation. As we mentioned earlier, some proteins are present in most cells of the body, but others are found only in specific types of differentiated cells. Furthermore, cells do not synthesize all of their proteins at the same rate. Cells thus have mechanisms which regulate both the types and rates of protein synthesis.

The complex interactions that lead to the regulation of protein synthesis in the cells of higher organisms are only beginning to be worked out. In those cases where the mechanism is at least partially understood, the basis for regulation is the binding of a molecule to a specific site, resulting in either the inhibition or activation of the activity occurring at that site. A brief discussion of enzyme induction and repression will serve to illustrate one important type of regulatory mechanism.

Induction and repression. One way of controlling protein synthesis, initally discovered in bacteria, is a mechanism by which the synthesis of particular proteins can be turned *on* and *off* by controlling the synthesis of mRNA in the nucleus, i.e., transcription. An excellent example concerns the enzyme **galactosidase**, which catalyzes the splitting of lactose into a molecule of glucose and a molecule of galactose:

$$\text{Lactose} \xrightarrow{\text{galactosidase}} \text{glucose} + \text{galactose}$$

When the bacteria *Escherichia coli* are grown in a nutrient medium that does not contain lactose, the cells synthesize little or no galactosidase. If lactose is then added to the medium, a rapid synthesis of galactosidase takes place. The presence of lactose has **induced** the synthesis of the enzyme required for its metabolism (Fig. 4-20). Such a system has many advantages since the cell need not expend energy synthesizing an enzyme unless there is a substrate present for the enzyme to act upon.

Inhibition, rather than induction, of enzyme synthesis in the presence of substrate is also observed in *E. coli*. The synthesis of the amino acid histidine, which is required for protein synthesis, involves a series of reactions catalyzed by 10 different enzymes. If *E. coli* are grown in a medium to which histidine has been added, the cells do not synthesize the 10 enzymes needed for histidine synthesis but rather utilize the histidine in the medium.

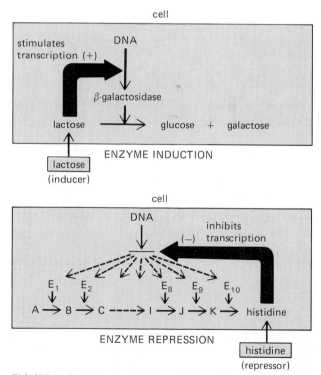

FIGURE 4-20. Induction and repression of protein (enzyme) synthesis by lactose and histidine supplied by the plasma. A plus sign indicates stimulation, and a minus sign indicates inhibition.

When histidine is removed from the medium, the 10 enzymes are rapidly synthesized to provide the histidine needed for growth. In this example, it appears that the presence of histidine has **repressed** the synthesis of the enzymes used in its synthesis (Fig. 4-20). This type of system is useful to a cell because the end product of a biochemical pathway may be able to regulate the synthesis of the enzymes in that pathway. As the concentration of the end product is increased, enzyme synthesis is reduced and the rate of product formation is decreased.

In general, both induction and repression involve a number of enzymes simultaneously. Mapping the location of the genes which code for each of the enzymes in the histidine synthetic pathway has shown that the genes are located next to each other and that one strand of mRNA is formed which contains the information for the synthesis of all 10 enzymes. Repression of enzyme synthesis involves blocking the synthesis of this mRNA molecule. A collection of genes which can be induced or repressed as a unit is known as an **operon**.

Enzyme induction and enzyme repression both involve repressor genes and repressor proteins. A **repressor gene** contains the coded information for the synthesis of a **repressor protein** (Fig. 4-21A). The repressor protein is synthesized in the standard manner, on the surface of a ribosome, from mRNA transcribed at the site of the repressor gene. Once formed, the repressor protein can bind to the initial gene of an operon, which is known as the **operator gene**. The specific binding of the repressor protein to the operator gene prevents synthesis of the operon's mRNA by blocking the binding of RNA polymerase, which is necessary for RNA synthesis.

The induction of enzyme synthesis involves the binding of an inducer molecule, such as lactose in the previous example, to a repressor protein, and thereby preventing its binding to the operator gene (Fig. 4-21B). Since the operator gene is no longer inhibited, it can initiate the synthesis of the mRNA. This mRNA then can initiate enzyme synthesis on the ribosomes. Thus, induction of enzyme synthesis really involves the inhibition of an inhibitor (the repressor molecule) usually present in the cell.

Enzyme repression by the end product of a biochemical pathway involves a set of interactions similar to enzyme induction except that, in this case, the repressor protein is unable to bind to the operator gene unless a molecule, such as an end product, is bound to the repressor protein. The end-product molecule, histidine in the example above, binds to the inactive repressor protein, forming an active repressor molecule which is able to block the synthesis of mRNA by binding to the operator gene.

Many of the genes in human cells are controlled by the mechanism of induction and repression in the presence or absence of certain molecules. For example, one of the mechanisms by which some hormones influence the activities of cells is by inducing the synthesis of specific proteins, as will be described in Chap. 9. However, not all genes are subject to induction and repression; many appear to lack operator genes and are thus continuously active, independent of the presence or absence of inducers or repressors. (However, the proteins synthesized by these genes may be subject to other types of control.)

Replication and Expression of Genetic Information

The development of the human body from a single fertilized egg cell involves cell growth, cell replication, and cell differentiation. The maintenance of structure and function in the adult body also makes use of these same processes. Each of these processes depends on the transcription and translation of genetic information into protein synthesis, whereas cell replication requires, in addition, the duplication of DNA and the transmission of identical copies of this genetic information to each of the two resulting cells.

Replication of DNA

DNA is the only type of molecule in a cell which is able to form a duplicate copy of itself without structure-specifying information from some other component in the cell. In contrast, mRNA molecules can be formed only in the presence of DNA, which provides the information for the ordering of the mRNA base sequence. Likewise, protein can

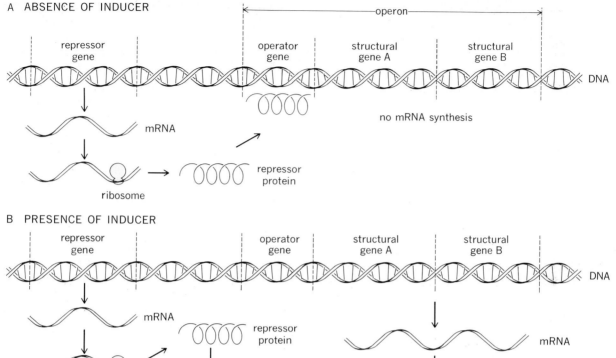

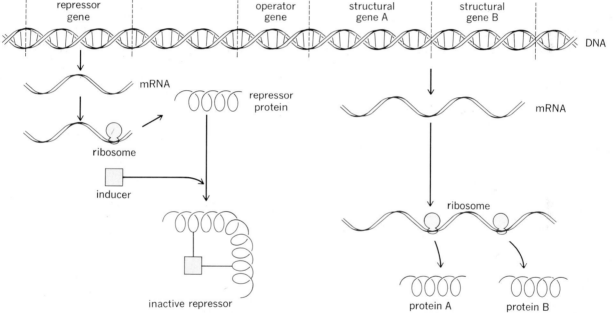

FIGURE 4-21. Enzyme induction. (A) No enzyme is formed in the absence of an inducer molecule. (B) Enzyme is formed in the presence of an inducer molecule due to the inactivation of the repressor protein.

be formed only if mRNA is present to provide the information necessary for ordering the sequence of amino acids in the protein. Finally, all other molecules synthesized by a cell are formed by metabolic pathways that use enzymes whose chemical specificity depends on the information obtained from DNA and used to build a specific enzyme.

The replication of DNA is, in principle, very similar to the process whereby mRNA is synthe-sized on the surface of DNA by base pairings. During DNA replication (Fig. 4-22), the two strands of DNA separate, and the exposed bases in each strand base-pair with free deoxynucleotide triphosphates, forming a complementary strand. The enzyme **DNA polymerase** joins the free nucleotides together to form a new strand of DNA in a reaction which is similar to the reaction which linked together the nucleotides in RNA. In contrast to the

NUCLEOTIDE

base

deoxyribose

phosphate

pyrophosphate

FIGURE 4-22. Replication of DNA involves the pairing of free nucleotides with the bases of each DNA strand, giving rise to two new DNA molecules each containing one old and one new nucleotide strand.

synthesis of mRNA, however, both strands of the original DNA act as templates for the synthesis of new strands. The end result is two identical molecules of DNA, each containing one strand of nucleotides present in the DNA molecule before duplication and one strand of newly synthesized DNA. Prior to cell division, the DNA molecules in the cell nucleus are replicated by the above process; one copy (i.e., one double-strand DNA molecule) is then passed on to each new cell. Thus, each replicated cell receives the same set of instructions as was originally present in the parent cell.

Cell division

Starting with a single fertilized egg cell, the first cell division produces 2 cells. When these daughter cells divide, they each produce 2 cells, giving a total of 4 cells. When these 4 cells divide, they produce a total of 8 cells, and these 8 produce 16 cells, and so forth. Thus, starting from a single cell, 4 division cycles will produce 16 cells (2^4), 10 division cycles will produce $2^{10} = 1024$ cells, and 20 division cycles will produce $20^{20} = 1,048,576$ cells. If the development of the human body involved only the repeated cycle of cell division and growth without any cell death, it would require only about 46 division cycles to produce all of the cells in the adult body, starting from a single cell.

The ability of a cell to divide and thereby reproduce itself is a characteristic of most (but not all) living cells. We have described above how DNA is able to reproduce itself. A cell, however, is much more than a single molecule; it is a complex organized collection of membrane structures and molecules. The division of such a complex structure into two roughly equivalent parts involves marked alterations in the structure and metabolism of the cell. Aside from the morphological description of the various stages of the division process as visualized with a light or electron microscope, surprisingly little is known about the underlying molecular events which occur during division or how the process is initiated and regulated. We will therefore limit our discussion of cell division to a brief description of the various stages of the process and focus our attention on the net result, which is the duplication, packaging, and distribution of DNA molecules to the daughter cells.

Although the time between cell divisions varies considerably in different types of cells, (some differentiated cells do not divide at all), the more rapidly growing cells divide about once every 24 h. During most of this period there is no visible evidence that the cell will undergo a division at some later time. For example, in a "24-h" cell, visible changes in cell structure begin to appear 23 h after the last division. The period between the end of one division and the structural changes that indicate the beginning of the next is known as **interphase**. Since the process of cell division lasts only about 1 h, the cell spends most of its time in interphase, and most of the cell properties described in this book are properties of interphase cells. However, one very important event related to the subsequent cell division does occur during the interphase, namely, the replication of DNA. A few hours after cell division, the process of DNA replication begins; it proceeds over a period which may last 10 to 12 h.

Several types of proteins are bound to the DNA molecules in the nucleus forming complexes of DNA and protein known as **chromatin**. The chromatin forms threadlike structures known as **chromatids**. After each replication of DNA, each DNA molecule forms two identical chromatids, which remain bound to each other until the time of nuclear division. A group of positively charged chromatin proteins, known as **histones**, appear to be responsible for the various conformational states of the chromatin threads.

When a cell is going to divide, each set of replicated chromatids condenses to form a highly coiled, rod-shaped body known as a **chromosome**. Within each chromosome, the two chromatids are held together at a single point known as the **centromere**. Human cells, with the exception of the male and female reproductive cells, each contain 46 chromosomes. Each chromosome consists of different DNA molecules and therefore carries a different set of genes. The amount of DNA found in a single human cell during interphase (when the chromatin threads are in the uncoiled state) could

form a thread about 180 cm long, a distance which is about 100,000 times the diameter of a typical cell; thus, the coiling and condensation of chromatin to form chromosomes provides a way of transferring these long threads to daughter cells.

Cell division consists of two parts: nuclear division (**mitosis**) and cytoplasmic division (**cytokinesis**). Mitosis is usually, but not always, followed by cytokinesis. During mitosis the duplicated chromatids separate and move to opposite poles of a cell, whereas during cytokinesis the entire cell divides into two, each half receiving an identical set of chromatids.

As mentioned above, the first sign that a cell is going to divide is the appearance of the chromosomes in the nucleus as the chromatin threads begin to coil. As the chromatids condense, the nuclear membrane breaks down, and a new structure appears in the cell; this is the **spindle apparatus**, which is composed of microtubules (spindle fibers). Some of these microtubules extend between two small cylindrical bodies known as **centrioles** (Fig. 4-23) located on opposite sides of the nucleus. Other spindle fibers pass from the centrioles to each chromosome where they are attached to the centromere region. As division proceeds, the two identical chromatids of each chromosome separate at the centromere and move toward the opposed centrioles (Fig. 4-23). The spindle fibers act as though they are pulling the chromatids toward the poles.

As the chromatids move toward opposite poles of the cell, cytokinesis begins: The cell begins to constrict along a plane perpendicular to the spindle apparatus, and constriction continues until the cell has been pinched in half, forming two daughter cells. Following cytokinesis the spindle fibers dissolve, the chromatids uncoil, and a new nuclear membrane is formed in each daughter cell (Fig. 4-23).

Most of the cell organelles are distributed randomly between the two daughter cells during cytokinesis. As the daughter cells grow, new membrane material is synthesized to form the membranes of the nuclear envelope, the Golgi apparatus, and the lysosomes. The mitochondria, on the other hand, appear to be able to duplicate them-

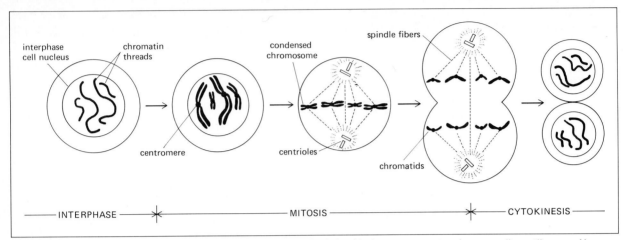

FIGURE 4-23. Sequence of events during cell division. (Only 4 of the 46 chromosomes in a human cell are illustrated.)

selves. Small amounts of DNA are found in these organelles, and new mitochondria are formed by the growth of a membraneous partition across an old mitochondrion, separating it into two new organelles.

Not only do mitochondria contain DNA but they also have ribosomes that are able to synthesize proteins within the mitochondria. However, mitochondria only synthesize some of the mitochondrial proteins, whereas others are synthesized in the cytoplasm from information carried in nuclear genes. The process of protein synthesis in the mitochondria resembles the specific steps involved in bacterial protein synthesis more closely than those of cytoplasmic protein synthesis, which has lead to the speculation that at some time in the distant past mitochondria have evolved from bacteria engulfed by cells.

Cell differentiation

Since identical sets of DNA molecules pass to each of the replicated cells during cell division, each cell in the body (with the exception of the reproductive cells) contains the same genetic information. How, then, is it possible for one cell to become a muscle cell (and synthesize muscle proteins) while another cell containing the same genetic information differentiates into a nerve cell (and synthesizes a different set of proteins)? We have already given a partial

answer—different combinations of genes are active in the different cells. The genes which contain the information responsible for the synthesis of muscle proteins are able to synthesize mRNA in muscle cells. These same genes are also present in nerve cells but do not form mRNA. Other genes are active in the nerve cell but are not active in muscle cells (Fig. 4-24). The problem of cell differentiation is thus related to the general problem of regulating protein synthesis. Certain genes appear to be ''turned on'' or ''turned off'' during cell differentiation. Mechanisms similar to those employed in enzyme induction and repression may be involved. To understand the complex process of cell differentiation a great deal more must be learned about the processes which regulate protein synthesis in the cells of higher organisms.

Mutation

In order to form the approximately 40 trillion cells of the adult human body a minimum of 40 trillion *individual* cell divisions must occur (actually 40 trillion minus one), and the DNA molecules in the original fertilized egg cell must be replicated at least 40 trillion times. If a secretary typed the same letter 40 trillion times, one would expect to find some typing errors. It is also not surprising to find that during the duplication of DNA errors occur which result in an altered sequence of bases and a change

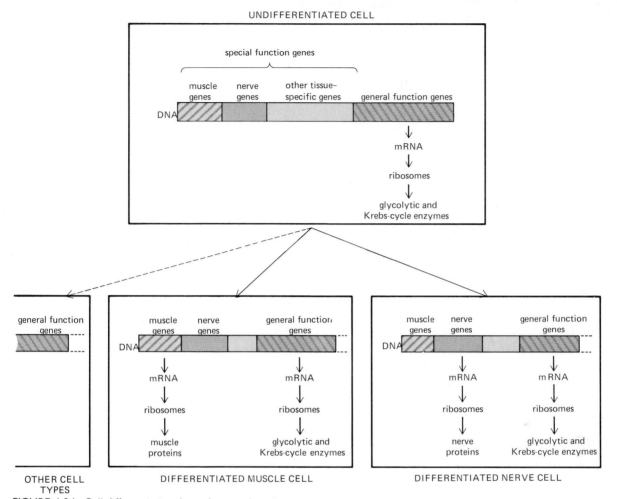

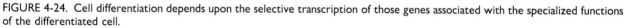

FIGURE 4-24. Cell differentiation depends upon the selective transcription of those genes associated with the specialized functions of the differentiated cell.

in the genetic message. What is surprising is that DNA can be duplicated so many times with relatively few errors. Any alteration in the genetic message carried by DNA is known as a **mutation**.

Mutations may result from a multitude of factors, all of which ultimately lead to an alteration of the base sequence in DNA and thus a change in the genetic information. The simplest type of mutation occurs when a single base is inserted at the wrong position in DNA. For example, the base sequence C-G-T forms the DNA code word for the amino acid alanine; if guanine (G) is replaced by adenine (A) in this sequence, it becomes C-A-T, which is

the code word for a different amino acid, valine. On the other hand, mutations that alter the base sequence in the intron regions of genes appear to have little effect on a cell, since this region does not contain information concerning the structure of specific proteins.

It is during replication of DNA, when free nucleotides are being incorporated into new strands of DNA, that incorrect base substitution is most likely to occur. Even in the absence of specific agents which increase the likelihood of mutations, mistakes in copying DNA do occur, so that the mutation rate is never zero. The major factors (mutagens) in the environment which increase the mutation rate are certain chemicals and various forms of ionizing radiation, such as x-rays, cosmic rays, and atomic radiation. Most of these factors cause the breakage of chemical bonds, so that incorrect pairings between bases occur or the wrong base is incorporated when the broken bonds are reformed.

Cells possess mechanisms for protecting themselves against certain types of mutation. For example, if an abnormal base pairing occurs in DNA, such as C with T (C normally pairs with G), a special set of enzymes will cut out the segment containing the abnormal base T, allowing the normal strand to resynthesize the deleted segment by normal base pairing. Certain diseases in which the cells of the skin have an increased tendency to become cancerous when exposed to the ultraviolet radiation in sunlight appear to result from a lack of the DNA repair enzymes in these cells.

When a single base is incorrectly substituted in a code word, the amino acid in the protein coded by the gene containing the mutation may be affected. However, because the genetic code has several different code words representing the same amino acid, a mutation in a single code word does not always alter the type of amino acid coded. We saw that substituting A for G in the alanine code word C-G-T produced the valine code word C-A-T. If, instead, the mutation caused cytosine (C) to be substituted for thymine (T), the new codon C-G-C is one of several that correspond to alanine, and the protein formed by the mutant gene will not be altered in amino acid sequence in spite of the change in base sequence.

Mutations can also occur in which large sections of DNA are deleted from the molecule or in which single bases are added or deleted. Such mutations may cause the loss of an entire gene or group of genes or may cause the misreading of a large sequence of bases. Figure 4-25 shows the effect of removing a single base on the reading of the genetic code. Since the code is read in sequences of three bases, the removal of one base not only alters the code word containing that base but also causes a misreading of all subsequent bases by shifting the reading sequence. Addition of an extra base would also cause an extensive misreading.

Assume that a mutation has altered the code word so that it now codes for a different amino acid, say the alanine C-G-T to valine C-A-T mutation. What effect does such a mutation have upon the cell? The effect depends upon both the type of gene and where in the gene the mutation has occurred. Although proteins are composed of many amino acids, the properties of a protein often depend upon only a very small region of the total molecule, such as the binding site of an enzyme. If the mutation altering an amino acid is not in the region of a binding site, there may be little or no change in the properties of the protein. On the other hand, if the mutation alters the binding site, a marked change in the properties of the protein may occur. If the protein is an enzyme, a mutation may render it totally inactive or change its specificity for substrate. The mutated enzyme may even catalyze an entirely different type of reaction. Thus, the alteration of a single amino acid may lead to a protein whose activity is unchanged, increased, or decreased or to a protein with entirely new substrate specificities.

Let us assume that the mutation leads to an enzyme that is totally inactive. If the enzyme is in a pathway supplying most of the cell's chemical energy, the loss of the enzyme may lead to the death of the cell. On the other hand, the enzyme may be involved in the synthesis of a particular amino acid,

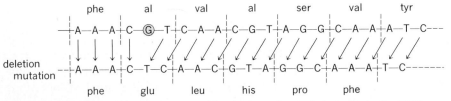

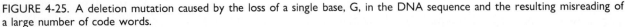

FIGURE 4-25. A deletion mutation caused by the loss of a single base, G, in the DNA sequence and the resulting misreading of a large number of code words.

and if the cell exists in a medium containing that amino acid, the cell's functioning will not be impaired by the loss of the ability to synthesize its own amino acid from raw materials. The effect of such a mutation, however, becomes apparent if the cell is deprived of preformed amino acids; in such an environment, the enzyme performs an essential function, and the loss of its activity by mutation may lead to the death of the cell.

To generalize, a mutation may have any one of three effects upon a cell: (1) It may cause no noticeable change in the cell's functioning, (2) it may modify cell function but still be compatible with cell growth and reproduction, or (3) it may lead to the death of the cell. Any one of these three types of mutation can occur in any of the cells in the human body, but the overall consequences of mutation depend upon the type of cell in which it occurs. If a liver cell in an adult undergoes a mutation that causes it to function abnormally or even die, the effect upon the total organism is usually negligible since there are thousands of similar liver cells performing similar functions. In contrast, since all the cells in the body are descended from one fertilized egg cell, any mutation occurring in the fertilized egg or its early descendents which does not lead to the immediate death of the cell is passed on to most of the cells in the developing organism. If the mutation is in an egg or sperm cell, all the cells descended from them inherit the mutation. Thus, mutations in the reproductive cells do not affect the individual in which they occur but do affect the children produced by these cells.

Many abnormal functions in the body are the result of such genetic mutations which are passed on from generation to generation. Inherited diseases of this type, which are often due to a single gene mutation that either fails to produce an active protein or produces an abnormally active protein, have been termed **inborn errors in metabolism**. An example is phenylketonuria, a disorder that can lead to one form of mental retardation in children. Because of a single abnormal enzyme, the victim is unable to convert the amino acid phenylalanine into the amino acid tyrosine at a normal rate. Phenylalanine is therefore diverted into other biochemical pathways in large amounts, giving rise to products

which interfere with the normal activity of the nervous system. These products, excreted in large amounts in the urine, account for the name of the disease. The symptoms of the disease are prevented if the content of phenylalanine in the diet is restricted during childhood, thus preventing the accumulation of the products formed from phenylalanine. As our knowledge of biochemistry, genetics, and disease increases, more and more diseases are being found that are the result of abnormally functioning enzymes derived from mutated genes.

Although mutations may alter the properties of the protein coded for by the mutant gene, not all such mutations are necessarily harmful to the cell in which they occur. In fact, mutation is the mechanism by which evolution occurs. Mutations may alter the activity of an enzyme in such a way that it is more, rather than less, active, or they may introduce an entirely new type of enzyme activity into a cell. If an organism carrying such a mutant gene is able to perform some function more effectively than an organism lacking the mutant gene, it has a better chance of surviving and passing the mutant gene on to its descendants. If the mutation produces an organism which functions less effectively than organisms lacking the mutation, the organism is less likely to survive and pass on the mutant gene. This is the principle underlying **natural selection**. Although any one mutation, if it is able to survive in the population, may cause only a very slight alteration in the properties of a cell, given enough time a large number of small changes can accumulate to produce very large changes in the structure and function of an organism. The evolution of life on earth from the first cell to human beings has proceeded over a period of some 3 billion years. During this long period of time, the environment has selected those mutations which best enable the cells carrying them to survive and propagate.

Cancer

Uncontrolled growth, usually accompanied by a loss of specialized cell functions, characterizes most cancers. As cancer cells grow and divide, they form an enlarging mass of tissue known as a **malig-**

nant tumor, which invades the surrounding normal tissues and disrupts the functioning of organs to the point where the disease is eventually fatal in most untreated cases. Cancer cells also fail to adhere firmly to neighboring cells, which can result in their breaking away from the parent tumor and spreading, by way of the circulatory system, to other parts of the body (**metastasis**), where they continue to grow, forming multiple tumor sites. If cancer is detected in the early stages of its growth, before it has metastasized, the tumor may be removed by surgery. Once it has metastasized to many organs, surgery is no longer possible. Drugs and radiation can be used to inhibit cell multiplication and destroy malignant cells, both before and after metastasis; unfortunately these treatments also damage the growth of normal cells.

The molecular mechanisms that produce the unrestrained multiplication of cancer cells are unknown, as are the regulatory systems that control normal tissue growth. Any normal cell in the body may at some point undergo a transformation to a malignant cell. A number of agents known as **carcinogens**, such as radiation, viruses, and certain chemicals, can induce the cancerous transformation of cells. It has long been suspected that these agents probably act by altering or activating various genes, particularly those associated with cell growth. Recently, several genes known as **oncogenes** have been discovered, which appear to be associated with the cancer process. These genes are present in normal cells, and malignant transformation appears to be associated either with the turning *on* of these genes, or with an increase in the rate at which these genes form their specific proteins. In some cases, the protein products of oncogenes have been identified as protein kinases which, as we have seen, can alter various cell functions through the process of covalent modulation by protein phosphorylation.

It is important to note that mutagenic agents also tend to be carcinogenic. These agents may be acting to alter the expression of oncogenes.

Recombinant DNA

In the early 1970s a number of bacterial enzymes were discovered that split molecules of DNA into a number of smaller fragments in a unique manner.

These enzymes, called **restriction nucleases**, identify specific nucleotide sequences, usually four to six nucleotides long, in DNA and in this region cut the two strands of DNA at a different site on each strand (Fig. 4-26). The two cut ends are therefore complementary to each other and can bind together by base pairing. Since the same nucleotide sequence is split by a given restriction nuclease at many points along a DNA molecule, the result is a number of DNA fragments, each having cohesive ends that can bind to the ends of other DNA segments which have been split by the same restriction nuclease. If the DNA molecules from two organisms are cut by the same restriction nuclease, then a DNA fragment from one organism can base-pair with the split end of DNA from a second organism. Another enzyme, known as a **ligase**, can covalently link the two molecules of DNA together. By this procedure, DNA fragments (genes) from one organism can be inserted into the DNA of a second organism to form **recombinant DNA**. If the recombinant DNA can be inserted into a living cell, the genetic message may be transcribed into mRNA along with the messages from the host's genes and translated into protein.

Recombinant DNA technology has greatly increased our knowledge of genes and how they work. It has led to the isolation of specific genes that can be transferred by recombinant DNA into bacteria, so that the multiplication of the bacteria will replicate the recombinant DNA along with the host DNA, producing large amounts of "cloned DNA." This cloned DNA can then be used to determine the nucleotide sequence of the gene, and this information can be used to determine the amino acid sequence of the structural protein coded for by the gene. The amino acid sequences of many proteins are now being determined by analyzing the nucleotide sequence of DNA genes cloned from recombinant DNA, a technical procedure that is much easier than determining the amino acid sequence from a protein directly.

The potential benefits and hazards of this recombinant technique are many. For example, splicing the gene that codes for human insulin onto a bacterial DNA and inserting it into a bacterium would allow the organism to synthesize insulin; this

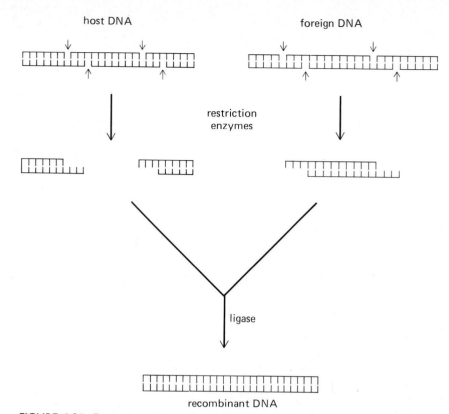

FIGURE 4-26. Formation of recombinant DNA by the use of bacterial enzymes.

hormone could then be extracted from the bacterium and used to treat diabetic patients who are unable to synthesize this hormone. On the other hand, the technology also bears potential hazards; for example, it would be theoretically possible to insert a gene that codes for a toxic protein into a harmless bacterium that commonly inhabits the human gastrointestinal tract and thereby produce a widespread epidemic. Only the future will tell whether the benefits of genetic engineering will outweigh the hazards of being able to manipulate the genetic constitution of organisms.

5 ENERGY AND CELLULAR METABOLISM

Metabolism refers to the total collection of chemical reactions that occur within a living organism. Some of these reactions result in the breakdown of organic molecules (such as carbohydrates, fats, and proteins), and such breakdown is known as **catabolism**. Most catabolic reactions are accompanied by the release of chemical energy, some of which cells use for the performance of functions. In contrast, other reactions combine small molecules to form larger molecules which a cell uses as structural elements or to perform specific functions; this overall synthesis of organic molecules by cells is known as **anabolism**. Thus, metabolism is made up of both catabolic and anabolic reactions.

The organic molecules in a cell undergo continuous transformations as some molecules are broken down while others of the same type are being synthesized. Chemically, no person is the same at noon as at 8 o'clock in the morning, since during even this short period much of the body's structure has been torn apart and replaced with newly synthesized molecules. In adults the composition of the body is in a dynamic steady state in which the rates of catabolism and anabolism are in balance. In such a dynamic steady state the adult body appears to remain unchanged even though the individual molecules which comprise its structure and determine its functions are continuously broken down and replaced. For example, a cell that contains 300 molecules of a specific protein may be synthesizing it at the rate of 20 molecules per minute, but if it is also breaking down the protein at the same rate, then the cell will always contain 300 molecules of that specific protein. The turnover rate (the time it takes to replace all the molecules of a specific type with newly synthesized molecules) of various molecules in cells varies considerably. Some are totally replaced every few minutes, whereas other replacements may take years. The dynamic steady state exists in adults only when there is no net gain or loss of weight. In children, in order for growth to occur, anabolism must exceed catabolism.

In addition to the overall dynamic steady state within a cell, there is a continuous input to and loss from a cell of organic molecules. The new molecules provide the raw material used to replace those molecules that have been broken down and lost from the cell, and most important, they are the source of the chemical energy used to maintain the structure and function of the cell.

In this chapter we focus attention on three basic aspects of metabolism: (1) the mechanisms for achieving highly specific types of chemical transformation, (2) the mechanisms for releasing energy from molecules and making it available for the performance of cellular function, and (3) the mechanisms for regulating the chemical reactions within a cell, i.e., the mechanisms by which the rates of energy production are increased or decreased to meet the varying energy requirements of the cell during different states of activity.

Energy

The concept of energy is the key to understanding the properties of chemical reactions. **Energy** is defined in dynamic terms as the ability to produce change and is measured by the amount of work performed during a given change. All physical and chemical change involves a redistribution of energy. Indeed, the presence of energy is revealed only when change is occurring. Whenever a change occurs in a system, energy must be transferred from one part of the system to another, but the total energy content of the system remains constant. That energy is neither created nor destroyed during any physical or chemical process is one of the most important axioms of science, the **law of the conservation of energy**.

The total energy content of any physical object consists of two components: the energy associated with the object because of its motion, called **kinetic energy**, and the energy associated with the object because of its position or internal structure, called **potential energy**. Movement is therefore a form of energy, and energy must be transferred to an object in order to produce movement. The amount of kinetic energy a moving object has is determined by its mass and its velocity. The faster an object moves, the greater its kinetic energy; the larger the

mass of an object, the greater is its kinetic energy at any given velocity. Kinetic energy is also associated with the motion of individual molecules, not just large objects. This kinetic energy of molecular motion is what we perceive as heat; the hotter an object is, the faster its molecules move and the greater their kinetic energy. Thus, heat is a form of energy—the kinetic energy of molecular motion.

Chemical energy

Potential energy has the potential of becoming kinetic energy when it is released. **Chemical energy** is a form of potential energy that is locked within the structure of molecules. It can be released during a chemical reaction in which chemical bonds are broken or formed. Since energy can be neither created nor destroyed during a chemical reaction, the difference in energy content between the reactant molecules and the product molecules must equal the amount of energy added or released during the reaction. For example, the reaction between hydrogen and oxygen to form water proceeds with the release of a considerable amount of heat energy, measured in units of **calories**. [One calorie (1 cal) is the amount of heat energy required to raise the temperature of one gram of water one degree Celsius.] Energies associated with chemical reactions are generally of the order of several thousand calories per mole and are reported as kilocalories (1 kcal = 1000 cal). The formation of 1 mol of water from hydrogen and oxygen releases 68 kcal of heat energy, the full reaction being

$$H_2 + O \longrightarrow H_2O + 68 \quad kcal/mol$$

In other words, the chemical energy stored in 1 mol of water molecules is less by 68 kcal than that originally present in the hydrogen and oxygen molecules.

A chemical bond (such as that between hydrogen and oxygen in the above equation) is formed between two molecules when they come close enough to each other for their electrons to interact. This happens when the random motion of the molecules brings them together in a collision.

For bond formation to occur, an adequate amount of energy must be transferred during the collision. For example, if hydrogen and oxygen are mixed together at room temperature, water is formed at an extremely slow rate, but if the mixture is heated, water is formed rapidly; the explanation for this phenomenon is that heating the mixture increases the rate of molecular motion (kinetic energy) so that not only is the frequency of collisions increased, but so is the magnitude of the kinetic energy transferred during each collision. The importance of this latter effect leads us to the concept of activation energy.

Activation energy

In the formation of new chemical bonds, the electric forces holding the atoms of a molecule together must be disrupted before the altered arrangement of electrons in the new chemical bond can occur. To upset the balance of forces in the molecule, a quantity of energy known as **activation energy** must be added. The activation energy can be acquired by the transfer of kinetic energy during collisions with other molecules. Heating the mixture of hydrogen and oxygen in our example above increased the magnitude of the kinetic energy transferred during each collision and provided the activation energy necessary for the reaction to occur.

As a mechanical analogy to a chemical reaction, consider the potential energy of a ball resting in a depression on top of a hill as representing the chemical potential energy stored in a molecule (Fig. 5-1). If the ball rolls down the hill, potential energy is converted into kinetic energy, just as chemical energy can be released as kinetic energy (heat) during a reaction. However, the ball will not roll down the hill spontaneously; it must be given a slight push to overcome the small ridge at the top of the hill. This push represents the activation energy necessary to initiate a chemical reaction. The magnitude of the activation energy can be represented by the size of the hump which the ball must roll over before it can roll down the hill. It is an important determinant of the rate of the reaction, since only molecules which have acquired this amount of en-

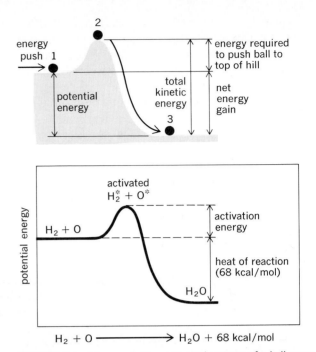

FIGURE 5-1. Changes in the potential energy of a ball moved from the top to the bottom of a hill compared with changes in chemical potential energy during a chemical reaction.

ergy are able to react. At any given temperature the larger the activation energy, the fewer the number of molecules in the population that will have this amount of energy, and thus the slower the reaction.

Determinants of Biochemical Reaction Rates

The rate (velocity) of a chemical reaction is measured by determining the amount of product that is formed per unit of time. Reaction rates have units of grams per second, molecules per second, or mols per second. The factors which determine the rate of a chemical reaction are understandable in terms of the basic requirements for molecular interaction described above: the frequency of collisions and the achievement of the required activation energy. On several occasions in the previous sections we have referred to heating a solution as a method for increasing collision frequency and supplying activation energy. The body temperature of 37°C

(98.6°F) provides a warm environment in which molecules have a fairly high kinetic energy, but since the temperature remains nearly constant, *changes* in chemical reaction rates are not due to *changes* in temperature. Rather, the major factors influencing reaction rates are (1) molecular concentrations, which alter collision frequency, and (2) the class of protein molecules known as enzymes, the function of which is to facilitate the acquisition of the required activation energy.

Chemical concentration: Law of mass action

The **law of mass action** states that an increase (decrease) in the concentration of one of the molecules in a chemical reaction will increase (decrease) the rate of the reaction in which that molecule is involved. Consider the following general reaction:

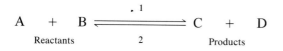

Any chemical reaction, if given sufficient time, will reach a state of **chemical equilibrium** in which the rates of the forward (1) and reverse (2) reactions become equal. Once equilibrium has been reached, the chemical composition of the reaction mixture will not change. Under these conditions, molecules of A and B still react to form C and D, but since the latter molecules also react to form A and B at the same rate, no net change in the concentrations of any of the molecules occurs.

If, after chemical equilibrium has been reached, the concentration of A is increased by adding more A, the rate of the forward reaction (1) increases, since the larger number of A molecules increases the probability of a collision of an A molecule with a B molecule. The higher rate of the forward reaction leads to greater formation of C and D. As their concentrations rise, the rate of the reverse reaction (2) also rises until a new chemical equilibrium is achieved. At equilibrium the concentration of B will be decreased. The conversion of A molecules into product molecules can also be

achieved by increasing the concentration of B, rather than A. However, note that in this case, at equilibrium, the concentration of A will be decreased. If the concentration of B molecules is made very large, almost all of the A molecules present can be converted into product molecules.

The conversion of A molecules into product molecules can also be influenced by initially changing the concentration of product, rather than reactant, molecules. If the concentration of C is lowered by removing it from the reaction mixture, the rate of the reverse reaction (2) is lowered. Therefore, A and B molecules are not replaced as rapidly as they react to form C and D. The net effect, when a new equilibrium is reached, is a decrease in the concentration of A and B.

Enzymes

During the nineteenth century chemists began to accelerate the rates of chemical reactions in ways which did not involve high temperatures or concentrations. Adding small amounts of certain substances, such as powdered platinum, was found to accelerate some reactions even though the added material did not undergo any net chemical change during the reaction. Such substances are called **catalysts**. The following are three basic characteristics of catalysts:

1. A catalyst accelerates the rate of a chemical reaction, but it will not cause a reaction to occur that would not occur in the absence of the catalyst.
2. A catalyst accelerates both the forward and reverse rates of a chemical reaction, and thus does not change the chemical equilibrium that is finally reached; it only increases the rate at which equilibrium is achieved.
3. In performing its function a catalyst undergoes no net chemical change; i.e., it is the same molecule after it accelerates a reaction as it was before.

Thus, a molecule of catalyst is not consumed and can be used over and over again. A few molecules of catalyst are sufficient to transform large quantities of reactants into products.

Enzymes are biological catalysts, and since all enzymes are proteins, an enzyme can be defined as a protein catalyst. In order to accelerate a reaction an enzyme must come into contact with the reactant molecules, known as the **substrates**. The reaction between enzyme and substrate can be written

$$\underset{\text{Substrate}}{S} + \underset{\text{Enzyme}}{E} \longrightarrow \underset{\substack{\text{Enzyme-}\\\text{substrate}\\\text{complex}}}{ES} \longrightarrow \underset{\text{Product}}{P} + \underset{\text{Enzyme}}{E}$$

The enzyme combines with substrate to form an enzyme-substrate complex, which breaks down to release product molecules and enzyme. At the end of the reaction, the enzyme molecule is free to undergo the same reaction with additional substrate molecules. The overall effect is to accelerate the conversion of substrate molecules into product molecules with the enzyme acting as a catalyst:

$$\text{Substrates} \xrightarrow{\text{enzyme}} \text{products}$$

Enzymes are generally named by adding the suffix *-ase* to the name of the substrate or to the name of the type of reaction catalyzed by the enzyme. For example, the enzyme lactic dehydrogenase catalyzes the removal of hydrogen atoms from lactic acid.

The interaction between substrate and enzyme has all the characteristics described in Chap. 4 for the binding of a ligand to a protein binding site—specificity, affinity, and saturation. The region of the enzyme molecule to which the substrate (ligand) binds is known as the enzyme's **active site** (a term equivalent to "binding site"). The shape of the enzyme molecule in the region of the active site provides the basis for its chemical specificity, since the shape of the active site is complementary to the shape of the substrate. (How strongly the substrate is bound to the active site determines not the specificity, but the affinity of the enzyme for its substrate.) Enzymes are highly specific for the type of substrate molecule they act upon. Thousands of reactions occur in a cell, and almost all of them are catalyzed by specific enzymes. Some enzymes are so specific that they interact with only one particular type of substrate and no other. At the other extreme, some enzymes interact with a wide range of different substrates, all of which contain a particular type of chemical bond or grouping.

The rate at which substrate is converted into product molecules depends upon both substrate concentration and enzyme concentration. An enzyme can catalyze a reaction only when substrate is bound to its active site. Whether substrate is bound depends on the substrate concentration and the affinity of the enzyme binding site for substrate. As with any protein binding site, at high substrate concentrations the binding site becomes saturated. The maximum catalytic activity, i.e., the maximal rate at which an enzyme can convert substrate to products, will occur at a substrate concentration equal to the concentration necessary to saturate the enzyme (Fig. 5-2). Increasing the substrate concentration beyond saturation levels will not produce any further increase in the rate of an enzymatic reaction. The maximal catalytic activity of a single enzyme molecule in the presence of a saturating substrate concentration varies, depending on the enzyme, from about one to several hundred thou-

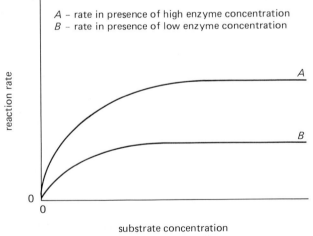

FIGURE 5-2. Rate of an enzyme-catalyzed reaction as a function of substrate concentration at two different enzyme concentrations, A and B. In the example, the higher enzyme concentration is twice that of the lower enzyme concentration and produces twice the maximal reaction rate.

sand molecules of product formed per second per molecule of enzyme.

At any substrate concentration, however, including saturating concentrations, the rate of the reaction can be increased by increasing enzyme concentration. If the number of enzyme molecules is doubled, twice as many active sites will be available to convert twice as much substrate to product.

Cofactors. Many, but not all, enzymes are inactive in the absence of certain nonprotein substances known as **cofactors**. Some cofactors are metal ions (such as magnesium, iron, zinc, and copper) which bind to the enzyme. Many of the essential trace elements (Chap. 3) appear to function as cofactors for specific enzymes. Since only a few molecules of each specific enzyme need be present to catalyze the conversion of a large amount of substrate to product, very small quantities of these essential metallic cofactors are sufficient to maintain enzymatic activity.

In addition to metal ions, a number of nonprotein organic molecules, known as **coenzymes**, also function as cofactors. Coenzymes may be covalently linked to enzymes, or they may be loosely bound and able to dissociate from the enzyme binding site. Enzymes which require coenzymes catalyze reactions in which a few atoms, such as two hydrogen atoms, a methyl group (—CH$_3$), or an

acetyl group (—$\overset{\overset{\displaystyle O}{\displaystyle \|}}{C}$—CH$_3$) are either removed from or added to a substrate. During the reaction these molecular fragments are either transferred from the substrate to the coenzyme molecule or from the coenzyme molecule to the substrate. The coenzyme molecule actually becomes one of the substrates of the reaction, as illustrated below for the removal of two hydrogen atoms from a substrate:

$$R\text{–}2H + \text{coenzyme} \xrightarrow{\text{enzyme}} R + \text{coenzyme–}2H$$

What makes a coenzyme different from another substrate is the fate of the coenzyme product. In the example above, the two hydrogens that were transferred to the coenzyme can now be transferred

from the coenzyme to another substrate with the aid of a second enzyme. This second reaction converts the coenzyme back to its original form, and it again becomes available to accept two more hydrogens (Fig. 5-3). A single molecule of coenzyme can be used over and over again to transport molecular fragments from one reaction to another. Thus, as with the metallic cofactors, only small quantities of the coenzymes are necessary to maintain those enzymatic reactions in which they participate.

The organic portions of coenzyme molecules are derived from the special class of molecules known as **vitamins**. Table 5-1 lists a few of the coenzymes, the vitamins from which they are derived, and the type of chemical group they transfer. Vitamins will be discussed further later in this chapter.

Regulation of enzyme-mediated reactions

The rate of an enzyme-mediated reaction depends on three factors: (1) substrate (and product) concentrations, (2) enzyme concentration, and (3) enzyme activity (Fig. 5-4). Altering any of these factors can be used to control the rates of metabolic reactions.

First, reaction rates can be altered by altering concentration. The concentration of a given substrate in a cell may change because of changes in the delivery of the substance to a cell. This may vary because of changes in the diet, changes in the rate of absorption of the substance from the intestine, or changes in the blood flow to a given tissue. Once a substance has reached the region of a cell, its entry into the cell through the plasma membrane

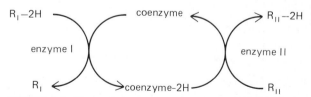

FIGURE 5-3. The transfer of two hydrogen atoms from reaction I to II by way of a coenzyme molecule. Reaction II regenerates the original coenzyme molecule that can now repeat the cycle.

TABLE 5-1. Some coenzymes

Coenzyme	Vitamin from which coenzyme derived	Transferred chemical group
Nicotinamide adenine dinucleotide (NAD$^+$)	Niacin	Hydrogen (—2H)
Flavine adenine dinucleotide (FAD)	Riboflavin	Hydrogen (—2H)
Coenzyme A (CoA)	Pantothenic acid	Acetyl groups ($-\overset{\overset{\textstyle O}{\|}}{C}-CH_3$)
Tetrahydrofolate	Folic acid	Methyl groups (—CH$_3$)

can be controlled by mechanisms that will be discussed in the next chapter. In addition to factors that alter the supply of a substrate from outside a cell, intracellular substrate concentration can be altered by other reactions that either utilize the same substance (and thus tend to lower its concentration) or synthesize the substance (and thereby increase its intracellular concentration.)

Second, a reaction rate can be altered by changing the concentration of the enzyme which catalyzes the reaction; this has the effect of providing more (or fewer) sites with which substrate can react. Certain reactions proceed faster in some cells than in others because more enzyme is present. The amount of enzyme depends on the rate at which enzyme is synthesized and its rate of degradation. For most enzymes these rates are fairly constant, and thus their concentrations do not change appreciably during different states of cell activity. The concentrations of certain enzymes, however, can be altered by controlling the transcription of these genes into messenger RNA molecules (Chap. 4). Control of the rate of protein breakdown is also used to control enzyme concentration.

Third, reaction rate can be altered by changes in the catalytic activity of the individual enzyme

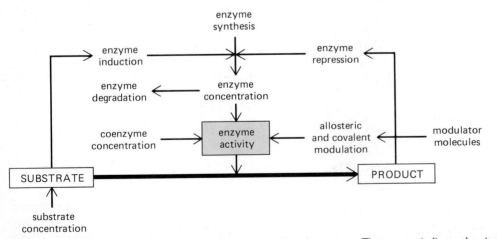

FIGURE 5-4. Factors that affect the rate of enzyme-mediated reactions. The arrows indicate the sites at which the various factors affect the reaction rate.

molecules, i.e, by changes in the rate at which an enzyme converts substrate to product. For certain enzymes, the concentration of cofactors directly affects the activity of the enzyme, and these enzymes are completely inactive in the absence of a cofactor. As we shall see, the concentrations of cofactors change during different states of metabolic activity.

Catalytic activity can also be altered by the processes of allosteric and covalent modulation described in Chap. 4, whereby the binding of a modulator molecule to one site on an enzyme or the phosphorylation of an enzyme alters the shape and thereby the activity of a second binding site (the active site) on the same enzyme. The results of such modulation may be either to increase or to decrease the catalytic activity of the enzyme. The products of certain reactions function as modulator molecules, binding to allosteric enzymes and thereby altering enzyme activity. In other cases, chemical messengers, hormones for example (see Chaps. 7 and 9), lead to the activation of certain protein kinase enzymes which, in turn, modulate specific enzyme molecules by covalent modulation.

Multienzyme metabolic pathways and their regulation

The sequence of enzyme-mediated reactions leading to the formation of a particular product is known as a **metabolic pathway**. For example, the 19 reactions which convert glucose to carbon dioxide and water constitute the metabolic pathway for glucose catabolism. Each reaction produces only a small change in the structure of the substrate, usually involving the breaking of a single chemical bond followed by the formation of a new bond with a different group (see, for example, Figs. 5-10 and 5-11). By such a sequence of small steps, a complex chemical structure, glucose, is transformed into relatively simple molecular structures, carbon dioxide and water.

Consider a general metabolic pathway consisting of four enzymes (e), leading from an initial substrate A to the end product E through a series of intermediates B, C, and D:

$$A \xrightarrow{e_1} B \xrightarrow{e_2} C \xrightarrow{e_3} D \xrightarrow{e_4} E$$

By mass action, increasing the concentration of A will lead to an increase in the concentration of B, (provided e_1 is not already saturated with substrate) and so on until eventually there is an increase in the concentration of the end product E. As was discussed earlier, the rate at which substrate is converted to product for any one step in the reaction sequence depends on a number of factors, in particular, enzyme concentration and catalytic activity; it will therefore be extremely unlikely that reaction rates will be exactly the same at each step. Thus, one step is very likely to be inherently slower than all the others. This step is known as the **rate-limiting reaction** in a metabolic pathway; none of the reactions that occur later in the sequence, including the formation of end product, can proceed more rapidly since they depend upon the rate at which substrate becomes available, which is determined by the rate-limiting reaction. By regulating the activity of the rate-limiting enzyme, the rate of flow through the whole pathway can be increased or decreased. Thus, it is not necessary to alter the activities of all the enzymes in a metabolic pathway in order to control the rate at which the final end product is produced.

Rate-limiting enzymes are usually the sites of allosteric and covalent regulation in metabolic pathways. Let us assume that enzyme e_2 is rate-limiting in this pathway. The end product E may interact with enzyme e_2 to produce an allosteric inhibition of its activity (Fig. 5-5); i.e., E may be a modulator molecule for this enzyme. This form of **end-product inhibition** is quite common in many synthetic pathways. It effectively prevents an excessive accumulation of end product when the end product is not being utilized.

Let us now add another level of complexity. Suppose the first reaction is reversible:

$$A \underset{}{\overset{e_1}{\rightleftharpoons}} B$$

Note that the same enzyme (e_1) catalyzes the reaction in both directions. This is a very important

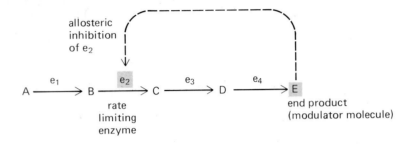

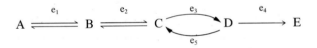

FIGURE 5-5. End-product inhibition of rate-limiting enzyme in a metabolic pathway. The end product E becomes the modulator molecule which produces an allosteric inhibition of enzyme e_2.

point: An enzyme speeds up a reaction but does not determine the direction in which the reaction proceeds. Therefore, in a reversible reaction, increasing the concentration or activity of the mediating enzyme will speed the reaction in both directions.

In theory, all chemical reactions are reversible. However, whether a particular reaction is spontaneously reversible to any extent, given a finite amount of time, depends upon the amount of chemical energy exchanged during the reaction. If there has been little loss of energy when substrate is converted to product, the reaction will be readily reversible. However, if large losses of energy have occurred, this energy must be made available if the reaction is to be reversed. Therefore, in practice, reactions can be classified into two categories, those that are reversible because they involve only small losses of energy and those that are irreversible because of the release of large quantities of energy. (It should be emphasized that it is the quantity of energy released and not a property of the enzymes which determines the reversibility of a given reaction.)

Reversible reaction: $A \xrightleftharpoons[]{e_1} B + $ small amount of energy

Irreversible reaction: $C \xrightarrow{e_3} D + $ large amount of energy

The product D in the irreversible reaction above can, however, be converted into C, provided a source of sufficient energy is added to the system. Often this energy is provided by coupling the reaction to the simultaneous breakdown of another substrate which releases large quantities of energy. Thus, an irreversible step can be ''reversed''

through an alternative route, using a second enzyme and an additional substrate to provide the required energy. Two such high-energy irreversible reactions catalyzed by separate enzymes are indicated by bowed arrows to emphasize that two distinct enzymes are involved in the two directions:

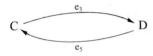

In summary, consider the following modification of our general metabolic pathway:

$$A \xrightleftharpoons[]{e_1} B \xrightleftharpoons[]{e_2} C \underset{e_5}{\overset{e_3}{\rightleftharpoons}} D \xrightarrow{e_4} E$$

The first two reactions are low-energy reversible reactions, each catalyzed by a single enzyme. The third reaction is a high-energy reaction having separate enzymes catalyzing the reaction in the two directions. The fourth reaction (D to E) is a high-energy irreversible reaction catalyzed by a single enzyme. In this metabolic pathway there is no route by which E can be converted back into A because of the irreversible last step. However, D and other intermediates can lead to the formation of A by reversal of the pathway. The third step in the pathway is mediated by two separate enzymes and provides an example of how the direction of flow in a metabolic pathway can be regulated. If the two separate enzymes e_3 and e_5 are subject to separate allosteric or covalent control, it would be possible to inhibit e_5, allowing the reactions to proceed in the direction of product E; alternatively, inhibiting e_3 would allow D to be converted into A.

The combination of reversible and irreversible reactions, together with a variety of mechanisms for controlling enzyme activities, provides the cell with a wide range of controls for regulating metabolic activity. When one considers the thousands of reactions that occur in the body and the permutations and combinations of possible control points, the overall result is indeed staggering. The details of metabolic regulation at the enzymatic level are beyond the scope of this book. In the remainder of the chapter we shall consider only the general pathways by which cells metabolize carbohydrates, fats, and proteins and indicate some of the key control points in these pathways.

ATP and Cellular Energy Transfer

The role of ATP

The functioning of a cell depends upon its ability to extract and use the chemical energy locked within the structure of organic molecules. When 1 mol of glucose is broken down in the presence of oxygen into carbon dioxide and water, 686 kcal of energy is released. In a test tube this energy appears as heat which could be used to perform work in a steam engine, for example. However, a cell cannot transform heat energy into work. In a cell, some of the released energy is captured by transferring it to another molecule; i.e., energy lost by one molecule is transferred to the chemical structure of another and thus does not appear as heat. The molecule to which the energy is initially transferred can, in turn, transfer this energy to yet another molecule or to energy-requiring processes in a cell.

In all cells, from bacteria to cells of human beings, the molecule which accepts energy from the breakdown of fuel molecules and donates this energy to cell functions is **adenosine triphosphate (ATP)** (Fig. 5-6). (Note that AMP, the monophosphate derivative of ATP, is one of the four nucleotides found in DNA and RNA.) ATP is synthesized from adenosine diphosphate (ADP) and inorganic phosphate (P_i), and 7 kcal/mol must be added to achieve this reaction:

$$ADP + P_i + 7 \text{ kcal/mol} \longrightarrow ATP + H_2O$$

This energy is obtained from the catabolism of carbohydrates, fats, and protein. It is then released during the subsequent hydrolysis of ATP,

$$ATP + H_2O \longrightarrow ADP + P_i + 7 \text{ kcal/mol}$$

and is used by cells to perform energy-requiring functions (muscle contraction, the active transport of molecules across cell membranes, and the synthesis of organic molecules).

Energy is continuously cycled through ATP molecules in a cell. A typical ATP molecule may exist for only a few seconds before it is broken down to ADP and P_i, with the released energy used to perform a cell function. Equally rapidly the products of ATP hydrolysis, ADP and P_i, are converted back into ATP through coupling to energy-releasing reactions, i.e., the breakdown of carbohydrates, lipids, or proteins (Fig. 5-7). Thus, although energy is present in the structure of the ATP molecule, its function is not to store energy but to transfer it from fuel molecules to cell functions. The total energy stored in all the ATP molecules of a cell can supply the energy requirements of most cells for only a fraction of a minute. The actual storage of energy in the body is performed by the molecules of carbohydrate, lipid, and protein whose energy is transferred to ATP.

There are two mechanisms by which energy is coupled to the formation of ATP: (1) **substrate phosphorylation** and (2) **oxidative phosphorylation**. Substrate phosphorylation occurs when one of the phosphate groups in a metabolic intermediate is transferred to ADP to form ATP. In contrast, oxidative phosphorylation, a chemically more complex process, couples inorganic phosphate to ADP directly rather than by using a phosphorylated intermediate to provide the phosphate.

Oxidative phosphorylation

In the breakdown of carbohydrates, fats, and proteins many of the chemical reactions involve the removal of hydrogen atoms from various intermediates. These reactions require a coenzyme, to which the hydrogen atoms are transferred. As a result of such hydrogen transfers, some of the

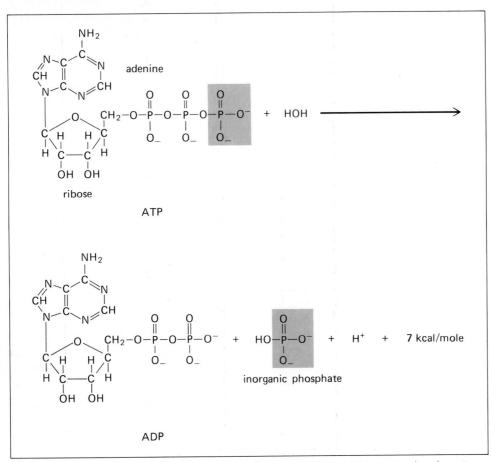

FIGURE 5-6. Chemical structure of ATP. Its breakdown to ADP and inorganic phosphate is accompanied by the release of 7 kcal/mol of energy.

chemical energy in molecules of carbohydrate, fat, and protein is transferred to the coenzyme molecules. The process of oxidative phosphorylation then uses this energy to synthesize ATP. Since the hydrogen atoms can be obtained from a wide variety of organic molecules, the process illustrates how energy obtained from many sources can be funneled into the common energy carrier, ATP.

How does the energy transferred with hydrogen to the coenzyme molecules actually get passed on to ATP? This involves a complex series of reactions which take place on the inner membranes of mitochondria. In most cells, oxidative phosphorylation is responsible for forming the vast majority of ATP; the mitochondrion, for this reason, is often referred to as the "powerhouse" of the cell. These membranes contain a group of iron-containing proteins known as **cytochromes** (they are brightly colored) which accept a pair of electrons from two hydrogens (Fig. 5-8). These electrons pass through a sequence of different cytochromes, which eventually donate them to molecular oxygen, which then reacts with the hydrogen ions (released at the beginning of the sequence of reactions) to form water. The overall reaction can be written

$$\text{Coenzyme–2H} + \frac{1}{2}O_2 \xrightarrow{\text{cytochromes}} \text{coenzyme} + H_2O + 52 \text{ kcal/mol}$$

The total result is energetically similar to the reac-

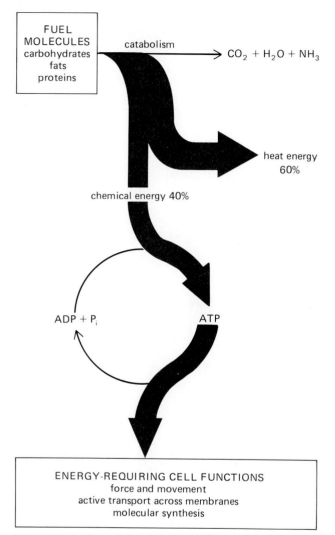

FIGURE 5-7. Flow of chemical energy from fuel molecules to ATP and heat, and then from ATP to energy-requiring cell functions.

tion between free hydrogen and oxygen to form water, with the release of heat energy.

Why must the cell proceed through a complex sequence of reactions to achieve what is essentially the formation of water from hydrogen and oxygen? If a mole of coenzyme–2H reacted directly with oxygen, all 52 kcal of energy would be released, but since only 7 kcal of energy is necessary to form 1 mol of ATP from ADP and inorganic phosphate,

the remaining 45 kcal of energy would appear as wasted heat. In contrast, by passing through the cytochrome chain, the same total energy is released in small steps, along with each electron transfer between successive cytochromes. The energy released at a number of these steps can be coupled to ATP synthesis. Either two or three molecules of ATP can be formed from each pair of hydrogen atoms delivered to the cytochromes (Fig. 5-8), depending on which coenzyme delivers the hydrogen atoms and thus on the point at which the electrons enter the cytochrome chain. Of the total energy released during electron transport along the cytochrome chain, 40 percent is transferred to ATP, while the remaining 60 percent appears as heat.

The mechanism by which the released energy is actually coupled to the synthesis of ATP during electron transport is complex. It involves the movement of ions across the inner mitochondrial membrane according to principles that will be discussed in Chap. 6.

It should be emphasized that, aside from the supply of energy, an important result of oxidative phosphorylation is the removal of hydrogen atoms from the coenzyme, which then becomes available to pick up more hydrogen atoms and transfer them to the cytochrome chain. Without the return of the coenzymes to their original state, the enzymes which require these coenzymes would become inactive, and the metabolic pathways containing them would cease to function.

Note that if oxygen is not available to the cytochrome system, ATP cannot be formed, since oxygen atoms are the final acceptors of the hydrogen atoms. Cyanide reacts with the final cytochrome in the chain, preventing electron transfer to oxygen and thus blocking the production of ATP by the mitochondria. The cause of death from cyanide poisoning is the same as that from a lack of oxygen—failure to form adequate amounts of ATP to maintain cell functions.

The consumption of oxygen by mitochondrial oxidative phosphorylation accounts for almost all of the oxygen used by the body. However, there are a small number of enzymes in most cells that will catalyze the incorporation of molecular oxygen

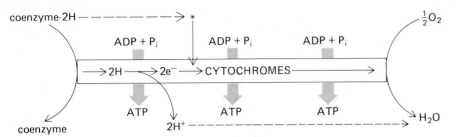

FIGURE 5-8. Energy is coupled to the formation of ATP at three points in the cytochrome chain during oxidative phosphorylation. Some coenzyme–2H (marked *) enters this chain beyond the point at which the first ATP is formed and gives rise to two rather than three ATP molecules for each pair of hydrogen atoms that react with oxygen to form water.

into organic molecules or catalyze the transfer of hydrogen atoms to molecular oxygen to form water or hydrogen peroxide (H_2O_2). These enzymes are not associated with ATP production, and most of them are not located in the mitochondria. One group is located in the cell organelles known as peroxisomes.

The Krebs cycle

As we have just seen, the body's survival depends on ATP formed by the process of oxidative phosphorylation, in which hydrogen atoms derived from the catabolism of fuel molecules are transferred by way of coenzymes to the cytochromes and ultimately combined with molecular oxygen. The source of most of these hydrogen atoms is a pathway known as the **Krebs cycle** (also called the **citric acid** or **tricarboxylic acid cycle**).

The Krebs-cycle enzymes are soluble enzymes located in the inner compartment of mitochondria, whereas the enzymes of oxidative phosphorylation are embedded in the membrane which surrounds this compartment (the inner mitochondrial membrane). Intermediates derived from all three major fuel molecules—carbohydrates, fats, and proteins—can be metabolized by the Krebs-cycle enzymes. The Krebs-cycle reactions give rise to three end products: (1) ATP, (2) carbon dioxide, and (3) hydrogen bound to coenzyme molecules (Fig. 5-9).

The most important intermediate entering the Krebs cycle is **acetyl coenzyme A**. Just as hydrogen is transferred from one reaction to another by way of a coenzyme carrier molecule, 2-carbon acetyl

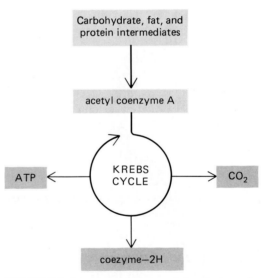

FIGURE 5-9. Inputs and outputs of the Krebs cycle reactions, which occur in mitochondria.

fragments are transferred from one reaction to another by the carrier molecule coenzyme A. Carbohydrates and fatty acids provide the major source of the acetyl groups that enter the Krebs cycle.

The products of protein (amino acid) catabolism feed into the Krebs cycle in several ways. In some cases they are metabolized to molecules which are identical to various intermediates in the Krebs cycle, whereas in other cases the metabolites enter the pathway for carbohydrate breakdown and ultimately find their way into molecules of acetyl coenzyme A (as will be discussed later in this chapter).

The seven reactions which constitute the Krebs cycle are illustrated in Fig. 5-10. This sequence of reactions converts each 2-carbon acetyl fragment that enters the cycle into two molecules of carbon dioxide and transfers four pairs of hydrogen atoms to coenzyme molecules. In reaction (1), acetyl coenzyme A donates its 2-carbon acetyl fragment to the 4-carbon molecule, oxaloacetic acid, to form the 6-carbon molecule, citric acid. The two molecules of carbon dioxide are formed when a carboxyl group (—COOH) is split off in reaction (3) and again in reaction (4), leaving a 4-carbon molecule, which after a few steps, becomes the 4-carbon acceptor (oxaloacetic acid) of another acetyl group to begin the cycle again.

The carbon dioxide formed in the Krebs cycle is the major waste product of metabolism. (Only small amounts of carbon dioxide are formed by other non-Krebs cycle reactions.) Thus, the two gases oxygen and carbon dioxide, which enter and leave the body by way of the lungs, take part in reactions that are located almost exclusively in mitochondria. Note, however, that the oxygen that appears in carbon dioxide is derived not from molecular oxygen but from the oxygen in the carboxyl groups of the Krebs-cycle intermediates. As we have seen, molecular oxygen ends up in molecules of water after combining with hydrogen during oxidative phosphorylation.

One molecule of ATP is formed in reaction (4) from each turn of the Krebs cycle. During this reaction a molecule of inorganic phosphate is linked by substrate phosphorylation to guanosine diphosphate (GDP) to form the high-energy molecule, guanosine triphosphate (GTP). GTP, like ATP, is an energy-transporting molecule, and the energy in GTP can be transferred to ATP by the following reaction:

$$GTP + ADP \longrightarrow GDP + ATP$$

In addition to the 1 molecule of ATP that is formed directly in the Krebs cycle, 11 additional ATP molecules can be formed by oxidative phosphorylation from the 4 pairs of hydrogen atoms that are derived from each 2-carbon fragment entering the cycle.

Pairs of hydrogen atoms are transferred from intermediates to coenzyme molecules at reactions (3), (4), (5), and (7), producing a total of 4 coenzyme–2H molecules. As we have seen, there are three sites along the electron transport chain at which ATP can be formed from ADP and inorganic phosphate. Three of the coenzymes from the Krebs cycle donate their hydrogen atoms to the cytochromes at the beginning of the chain and thus each forms 3 molecules of ATP. However, the type of coenzyme involved in reaction (5) of the Krebs cycle donates its hydrogen at a point beyond the site where the first ATP is formed (Fig. 5-8) and thereby gives rise to only 2 molecules of ATP.

Combining the reactions of the Krebs cycle and oxidative phosphorylation, we see that for every 2-carbon fragment (derived from the breakdown of carbohydrates, fats, or proteins) that enters the mitochrondria, 12 molecules of ATP are formed (1 in the Krebs cycle and 11 during oxidative phosphorylation). At the same time, the carbon and oxygen atoms in the 2-carbon fragment are converted into 2 molecules of carbon dioxide (via the Krebs cycle), and during oxidative phosphorylation 2 molecules of oxygen are consumed in reacting with the 4 pairs of hydrogen atoms that were obtained from the Krebs cycle.

Oxygen-independent ATP synthesis: Glycolysis

As was described above, the formation of ATP by oxidative phosphorylation is dependent on the transfer of hydrogen to oxygen and is thus dependent on the availability of oxygen. However, ATP can also be formed in the absence of oxygen by substrate phosphorylation in a series of reactions that is linked to the breakdown of carbohydrate.

When the first cells were formed some 3 to 4 billion years ago, there was no molecular oxygen in the earth's atmosphere (since atmospheric oxygen is a product of photosynthesis occurring in plant cells, which did not exist then). These first cells obtained their energy from the **anaerobic** (without oxygen) breakdown of organic molecules which had accumulated in the oceans over millions of years. An anaerobic pathway for energy release has

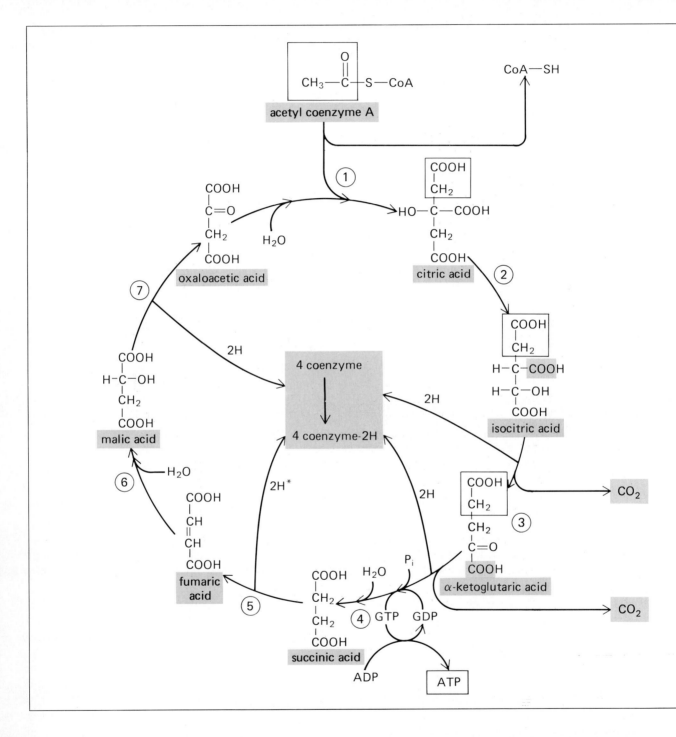

FIGURE 5-10. Krebs-cycle pathway by which fuel molecules are catabolized to CO_2 and hydrogen is transferred to coenzyme carrier molecules. The two hydrogen atoms (2H*) transferred to coenzyme in reaction 5 produce only two ATP molecules during oxidative phosphorylation, the other three pairs of hydrogen atoms each give rise to three ATP molecules.

been retained throughout evolution, and some unicellular organisms today are still able to grow and multiply in the absence of oxygen, using only this pathway to provide the energy for ATP synthesis. As cells evolved, some modifications did occur in the pathway with regard to the final end products formed, but the first 10 enzymatic steps in the pathway (from glucose to pyruvic acid) remain the same. In **alcoholic fermentation**, which occurs in some species of bacteria and yeast, the final end products formed from pyruvic acid are ethyl alcohol and carbon dixoide. In other microorganisms and in all higher organisms, including human beings, the final end product formed from pyruvic acid in the absence of oxygen is lactic acid, and the pathway is known as **glycolysis**. Unlike the Krebs cycle, which can be entered by any fuel type—carbohydrate, fat, or protein—capable of generating acetyl CoA (or another entry molecule), glycolysis is strictly a carbohydrate pathway.

The enzymes for the 11 reactions of glycolysis (Fig. 5-11) are dissolved in the cytosol. The net chemical change that occurs during glycolysis can be written

$$\text{Glucose} + 2\ \text{ADP} + 2P_i \longrightarrow$$
$$2\ \text{lactic acid} + 2\ \text{ATP} + 2H_2O$$

One molecule of glucose is converted to two molecules of lactic acid, and during this process a net total of two molecules of ATP are formed.

In the early stage of glycolysis, shown in reactions (1) and (3) in Fig. 5-11, two molecules of ATP are used to add phosphate groups to glucose, but this energy debt is repaid with interest when four molecules of ATP are formed later in the pathway, resulting in an overall gain of two molecules of ATP. The first synthesis of ATP in glycolysis occurs by substrate phosphorylation during reaction (7), in which one of the phosphate groups on a glycolytic intermediate is transferred to ADP to form ATP. (Recall that substrate phosphorylation is the transfer of phosphate from a substrate to ADP,

in contrast to oxidative phosphorylation, which uses the energy released from the combination of hydrogen and oxygen to couple free inorganic phosphate to ADP in the mitochondria.) A similar substrate phosphorylation occurs at reaction (10). Since reactions (4) and (5) convert one 6-carbon molecule into two 3-carbon molecules, and since each 3-carbon intermediate forms one molecule of ATP at reaction (7) and another at reaction (10), a total of four molecules of ATP are formed by these steps.

If glucose is to be broken down to lactic acid, not only must the glycolytic enzymes be present but ATP, ADP, and inorganic phosphate must be available to participate as substrates in their appropriate reactions. In addition, a coenzyme, to which hydrogen atoms are transferred in reaction (6), is required. Without this coenzyme, reaction (6) would not proceed and the subsequent formation of ATP at reactions (7) and (10) would not occur. Therefore, once the coenzyme has accepted a pair of hydrogen atoms, it must transfer them to another reaction in order to regenerate the coenzyme if additional glucose molecules are to proceed through reaction (6). We have already seen that one way to regenerate the coenzyme is to donate the hydrogens to oxygen in the process of oxidative phosphorylation. However, glycolysis can occur in the absence of oxygen. Thus, there must be a molecule other than oxygen to which hydrogen can be transferred. This acceptor molecule is pyruvic acid, which is converted to the end product, lactic acid, by the addition of hydrogen from the coenzyme in reaction (11). Thus, during glycolysis in the absence of oxygen, coenzymes transfer hydrogens from reaction (6) to (11) (Fig. 5-12).

We have emphasized that glycolysis is a biochemical pathway that can form ATP in the absence of oxygen, but in fact, the first 10 reactions from glucose to pyruvic acid also occur when oxygen is present (in **aerobic** conditions). Under such conditions the coenzyme–2H that is formed by reaction (6) is regenerated by oxidative phosphorylation in

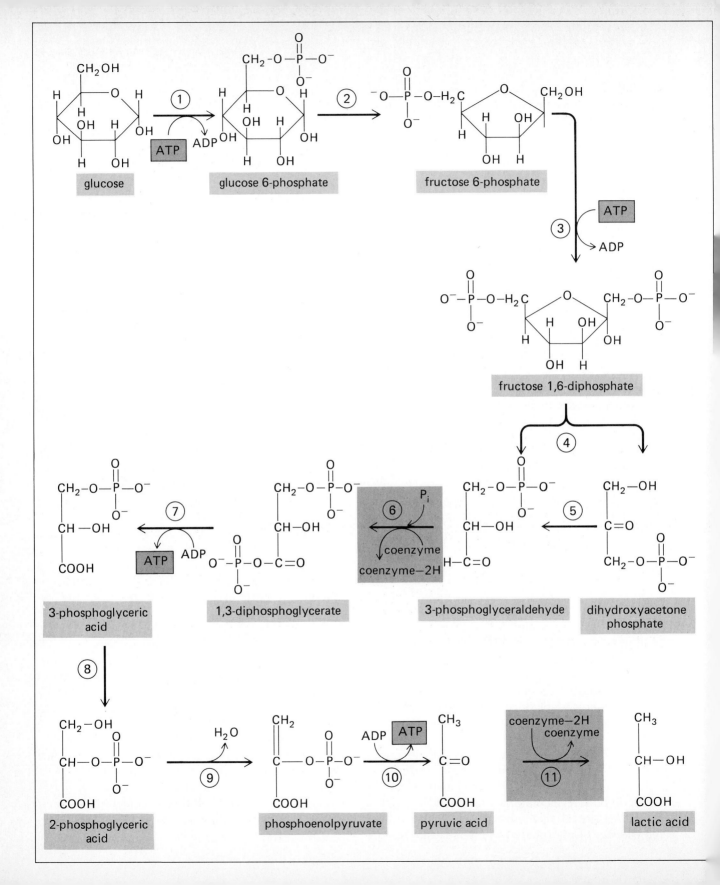

FIGURE 5-11. Reactions of the glycolytic pathway by which glucose is broken down into two molecules of lactic acid with the net formation of two molecules of ATP.

the mitochondria. Instead of being converted to lactic acid, pyruvic acid is broken down to acetyl CoA, which is then metabolized by the Krebs cycle. Whether oxygen is present or not, a net synthesis of two molecules of ATP still results from the breakdown of glucose to pyruvic acid. To reiterate, in the absence of oxygen the end product of glucose metabolism will be lactic acid, while in the presence of oxygen glucose will be completely metabolized to carbon dioxide and water.

In higher organisms, the percent of the total ATP production that occurs by glycolysis is very small compared to the amount formed by oxidative phosphorylation. However, there are special cases in which glycolysis supplies most, and in some cases all, of a cell's ATP. For example, red blood cells contain the enzymes for glycolysis but have no mitochondria; all their ATP production occurs by glycolysis. Also, certain types of skeletal muscles contain considerable amounts of glycolytic en-

zymes but have few mitochondria; during intense muscle activity, glycolysis can become the major source of ATP. Moreover, most, if not all, cells are able to withstand very short periods of low oxygen by using anaerobic glycolysis. However, most cells do not have sufficient numbers of glycolytic enzymes or sufficient amounts of glucose to provide by anaerobic glycolysis the high rates of ATP production required to maintain their activities.

Carbohydrate, Fat, and Protein Metabolism

In this section, we shall focus upon the major pathways by which each of the classes of fuel molecules—carbohydrates, fats, and proteins—are catabolized to provide the energy for ATP formation. We shall also consider the storage of these fuel molecules and the pathways and restrictions governing the conversion of molecules from one class of molecule to another, as in the conversion of carbohydrate into fat. Although these catabolic and anabolic pathways are present in almost all cells, a particular type of cell may contain higher concentrations of the enzymes in one pathway than do other cells or may be subject to different intracellular and extracellular control mechanisms.

Carbohydrate metabolism

Carbohydrate catabolism. Carbohydrates function primarily as fuel molecules which can be catabolized to release chemical energy that is coupled to the formation of ATP. We have already described, in the last sections, the major pathways of carbohydrate catabolism, namely, the breakdown of glucose to lactic acid by way of the glycolytic pathway and the breakdown of carbohydrate intermediates to carbon dioxide and hydrogen by way of the Krebs cycle. We have only to identify the link between these two pathways to provide the complete pathway for the catabolism of carbohydrate. Pyruvic acid, the carbohydrate intermediate preceding lactic acid in the glycolytic pathway, provides this link (Fig. 5-13).

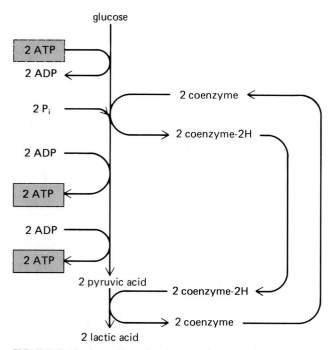

FIGURE 5-12. Anaerobic glycolytic pathway leading to lactic acid. The coenzymes are regenerated when they transfer hydrogen to pyruvic acid during the formation of lactic acid.

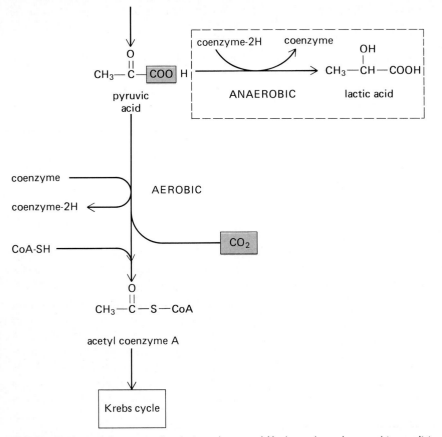

FIGURE 5-13. Link between glycolytic pathway and Krebs cycle under aerobic conditions.

We have emphasized that the full glycolytic pathway from glucose to lactic acid occurs under anaerobic conditions (since, in the absence of oxygen, the conversion of pyruvic acid to lactic acid provides the only means of regenerating hydrogen-carrying coenzyme molecules). However, when the oxygen supply is adequate (the far-more-common condition), pyruvic acid does not go to lactic acid to any great extent, since the coenzyme molecules formed during glycolysis can donate their hydrogens to oxygen. Instead, pyruvic acid enters the mitochondria, where enzymes transfer the 2-carbon acetyl fragment of pyruvic acid to coenzyme A and release a molecule of carbon dioxide from the carboxyl group. (These reactions are also associated with the transfer of two hydrogen atoms to a coenzyme molecule, which will pass them on to the oxidative phosphorylation pathway, thereby regenerating the original coenzyme.) The acetyl group can now enter the Krebs cycle, in which its two carbon atoms are degraded to carbon dioxide and its hydrogen atoms passed on to the oxidative phosphorylation pathway.

We have illustrated the catabolism of carbohydrates in terms of the breakdown of glucose because glucose is the most important carbohydrate in the body. Other sugars, such as fructose (found in table sugar) and galactose (found in milk), are catabolized by the same pathway as glucose after first being converted into glucose or one of the early intermediates in the glycolytic pathway.

The total amount of chemical energy that is

released during the catabolism of glucose to carbon dioxide and water equals 686 kcal per mole of glucose:

$$C_6H_{12}O_6 + 6O_2 \rightarrow 6H_2O + 6CO_2 + 686 \text{ kcal/mol}$$

How much of this energy can be transferred to ATP? Figure 5-14 illustrates the overall pathway of glucose catabolism and the sites at which ATP is formed. As we have seen, a net formation of 2 molecules of ATP occurs by substrate phosphorylation during glycolysis, and 2 more are formed by substrate phosphorylation in the Krebs cycle. Hydrogen atoms derived from glycolysis, the conversion of pyruvic acid to acetyl coenzyme A, and the Krebs-cycle reactions provide the means of forming an additional 34 molecules of ATP by oxidative phosphorylation. Since 7 kcal is required to form 1

mol of ATP, and a total of 38 mol of ATP is formed per mole of glucose catabolized, a total of 266 kcal (7 × 38) is transferred from glucose to ATP. The remaining 420 kcal (686 − 266) is released as heat. Thus, 39 percent (266/686 × 100) of the energy in glucose is transferred to ATP under aerobic conditions.

In the absence of oxygen, only 2 molecules of ATP can be formed by glycolysis; this is only 2 percent of the energy that is stored in glucose. Thus, the evolution of aerobic metabolic pathways greatly increased the amount of energy that could be made available to a cell from glucose catabolism. For example, if a muscle consumed 38 molecules of ATP during the process of contraction, this amount of ATP could be supplied by the breakdown of 1 molecule of glucose in the presence of oxygen. However, the catabolism of 19 molecules of glucose

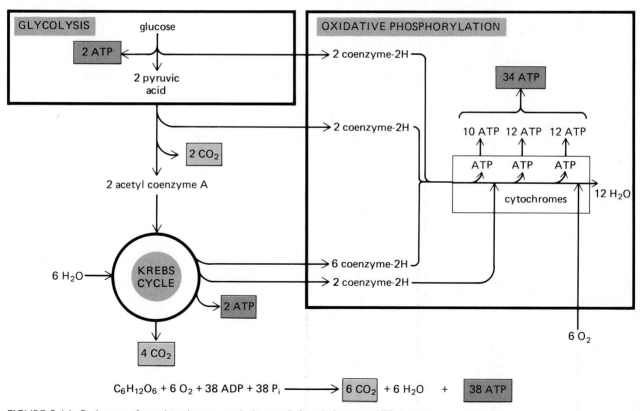

FIGURE 5-14. Pathways of aerobic glucose catabolism and their linkage to ATP formation.

to lactic acid under anaerobic conditions would be required to supply the same 38 molecules of ATP. Nevertheless, although only 2 molecules of ATP are formed per molecule of glucose under anaerobic conditions, large amounts of ATP can still be supplied by the glycolytic pathway if large amounts of glucose are broken down to lactic acid.

Carbohydrate storage. Although the major role of carbohydrate is to provide energy for ATP formation, very little carbohydrate is stored in the body to provide a reserve supply of fuel (most fuel is stored in the form of fat, as we shall see). The small amount of carbohydrate that is stored is in the form of the branched-chain glucose-polysaccharide known as **glycogen**. Most of the glycogen is stored in skeletal muscles and in the liver, with smaller amounts found in most other tissues.

Glycogen is synthesized from glucose by the pathway illustrated in Fig. 5-15, the enzymes of which are in the cytosol. The first step, the transfer of phosphate from a molecule of ATP to glucose, forming glucose 6-phosphate, is the same as the first step in glycolysis (thus, glucose 6-phosphate can either be broken down to pyruvic acid or be incorporated into glycogen).

Note that (as indicated by the bowed arrows in Fig. 5-15) different enzymes are used to synthesize and break down glycogen. The existence of

these different enzymes, both of which are subject to covalent modulation, provides a mechanism for regulating the flow of glucose to and from glycogen. When an excess of glucose is available to a liver or muscle cell, the enzymes in the glycogen-synthetic pathway are activated (by chemical signals that will be described in Chap. 15), and the enzyme that breaks down glycogen is simultaneously inhibited, leading to the net storage of glucose in the form of glycogen. When less glucose is available, the reverse combination of enzyme stimulation and inhibition occurs, and there is a net breakdown of glycogen into glucose 6-phosphate.

At this point, two paths are available to glucose 6-phosphate: (1) in most cells, including skeletal muscle cells, it enters the glycolytic pathway and, in the presence of oxygen, passes on through the Krebs cycle to provide the energy for ATP formation; (2) in liver cells, however, glucose 6-phosphate is mainly converted back into free glucose, which can be released into the blood and become available to be used as fuel by other cells in the body (see Chap. 15).

Glucose synthesis. In addition to being able to form glucose from glycogen, liver (and kidney) cells can also synthesize glucose from intermediates derived from other metabolic pathways. Figure 5-16 illustrates the pathways used in the synthesis of glucose. The first important point to note is that the pathway leading from pyruvic acid to glucose 6-phosphate uses most, but not all, of the same cytoplasmic enzymes used in the glycolytic breakdown of glucose to pyruvic acid, since these enzymes catalyze reversible reactions. However, two reactions in the glycolytic pathway, (3) and (10), are irreversible. Reaction (3) is a major control point in the carbohydrate pathway because it involves two regulated enzymes, one mediating the flow of metabolites from glucose to pyruvic acid and the other mediating the flow in the opposite direction. By regulating the activities of these two enzymes, carbohydrate metabolism can be shifted between net glucose breakdown and net synthesis, depending on the metabolic state of the cell and its energy requirements.

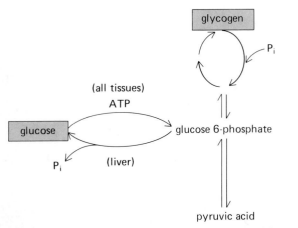

FIGURE 5-15. Pathways for the formation and breakdown of glycogen.

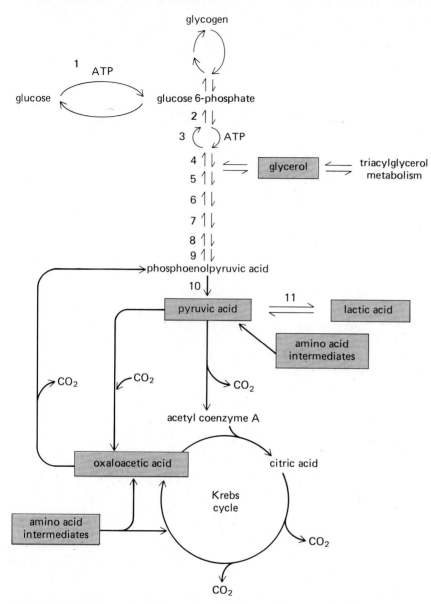

FIGURE 5-16. Pathways and intermediates leading to glucose synthesis.

The second irreversible reaction in the glycolytic pathway, reaction (10), does not have a corresponding enzyme that can mediate the direct conversion of pyruvic acid back to phosphoenolpyruvic acid; however, two additional enzymes accomplish this conversion by an important indirect route. This route has two steps: First, a molecule of carbon dioxide is combined with pyruvic acid to form a 4-carbon intermediate in the Krebs cycle, oxaloacetic acid; second, another enzyme catalyzes the removal of carbon dioxide from oxaloacetic acid (in the presence of ATP or a related phosphate-donating molecule) to form phosphoenolpyruvic acid. Note that as a result of these two enzymes, any organic molecule whose carbon atoms can be converted to pyruvic or oxaloacetic acid can give rise to glucose 6-phosphate. Regulation of these enzymes, therefore, provides an important mechanism for controlling glucose synthesis.

The conversion of pyruvic acid to oxaloacetic acid not only provides a pathway that can eventually lead to glucose synthesis but also provides a means of forming this very important Krebs-cycle intermediate, which is the acceptor of acetyl fragments as they enter the Krebs cycle.

Having identified the enzymes and the pathways leading to glucose synthesis, we can now ask what types of molecules can be fed into this synthetic pathway. First, one of the most important sources of carbon atoms is amino acids, some of which (as we shall see) can be converted into either pyruvic acid or one of the intermediates in the Krebs cycle which leads to oxaloacetic acid. Thus, proteins can, by way of their amino acids, be converted into carbohydrates. Second, lactic acid can be converted into pyruvic acid by the removal of hydrogen and thus can be used to synthesize glucose 6-phosphate. The large quantities of lactic acid released into the blood by muscles during intense exercise can be taken up by liver cells and used to synthesize glucose, which is then released into the blood where it becomes available to other cells, including muscle cells. Third, one other important source for glucose synthesis is glycerol, the 3-carbon backbone to which fatty acids are attached to form triacylglycerols. Glycerol can be converted into one of the 3-carbon intermediates in glycolysis formed by reaction (4), from which point it can be converted into glucose 6-phosphate. Glycerol becomes available for glucose synthesis during periods when stored fat is catabolized to glycerol and fatty acids.

Fat metabolism

Triacylglycerol (neutral fat) consists of three fatty acids linked to glycerol (Chap. 3) and constitutes the major stored fuel in the body (Table 5-2). Under resting conditions, approximately half of the energy used by tissues such as muscle, liver, and kidney

TABLE 5-2. Fuel content of a 70-kg man

	Total body content, kg	Energy content, kcal/g	Total body energy content	
			kcal	%
Triacylglycerols	15.6	9	140,000	78
Proteins	9.5	4	38,000	21
Carbohydrates	0.5	4	2,000	1

is derived from the catabolism of fatty acids. Although most cells store small amounts of fat in the form of fat droplets in their cytosol, most of the body's fat is stored in specialized cells known as **adipocytes**. Clusters of adipocytes form **adipose tissue**, most of which is located in deposits underlying the skin. The function of adipocytes is to synthesize and store triacylglycerols during periods of food uptake. Between meals, these fat stores are broken down into glycerol and fatty acids, which are released into the blood and delivered to other cells which catabolize them to provide the energy for ATP formation.

The glycerol portion of triacylglycerol is a carbohydrate and, as we have seen, can be converted into one of the 3-carbon intermediates in the glycolytic pathway by which it can generate acetyl coenzyme A or be used to synthesize glucose. The main pathway for the catabolism of fatty acids to acetyl coenzyme A (Fig. 5-17) is located in the inner compartment of the mitochondria along with the Krebs-cycle enzymes. An additional pathway for fatty acid catabolism is located in the peroxisomes. This process involves the transfer of the last two carbon atoms in a fatty acid to coenzyme A. One molecule of ATP is consumed in the initial reaction, and two pairs of hydrogen atoms are transferred to coenzymes for each pair of carbon atoms split from the end of the fatty acid chain. Each passage through this sequence shortens the fatty acid chain by two carbon atoms until all of the carbon atoms have been transferred to acetyl coenzyme A. These can proceed through the Krebs cycle and through oxidative phosphorylation, each 2-carbon fragment

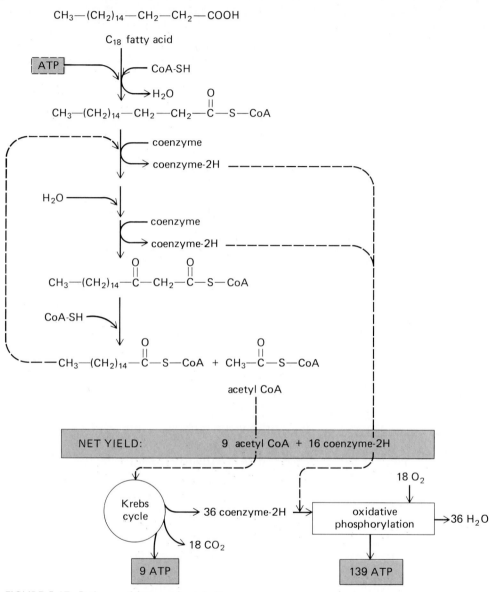

FIGURE 5-17. Pathway of fatty acid catabolism.

producing 2 molecules of CO_2 and 12 molecules of ATP. The 2 coenzyme–2H molecules produced directly during the formation of each acetyl coenzyme A also enter the oxidative phosphorylation pathway, one forming 3 molecules of ATP, the other 2.

Most fatty acids in the body contain 14 to 22 carbons, 16 and 18 being most common. The catabolism of an 18-carbon saturated fatty acid (molecular weight = 284) yields 147 molecules of ATP (Fig. 5-17), while the catabolism of 1 glucose molecule (molecular weight = 180) yields 38 molecules of ATP. Thus, the catabolism of 1 g of fat (approximately 90% fatty acids and 10% glycerol by weight) yields about two and one-half times as much ATP as does 1 g of carbohydrate. If an average man stored most of his fuel as carbohydrate rather than as fat, his body weight would be approximately 30 percent greater (89 kg versus 70 kg), and he would consume more energy moving this extra weight around. Thus, a major step in fuel economy occurred when animals evolved the ability to store fuel as fat; in contrast, plants store almost all of their fuel as carbohydrate (starch).

The synthesis of fatty acids occurs by reactions which are almost the reverse of those that degrade them. The enzymes in the synthetic pathway are located in the cytosol, whereas (as we have just seen) the enzymes catalyzing the breakdown of fatty acids are located in the mitochondria and peroxisomes. Fatty acid synthesis begins with acetyl coenzyme A which, through a series of reactions requiring ATP and coenzyme–2H, is combined with another molecule of acetyl coenzyme A to form a 4-carbon chain. By repetition of this process, long-chain fatty acids are built up two carbons at a time, which accounts for the fact that all of the fatty acids synthesized in the body contain an even number of carbon atoms. The overall reaction for the synthesis of an 18-carbon fatty acid can be written

$$9CH_3\overset{\overset{\displaystyle O}{\|}}{C}-S-CoA + 8ATP + 16coenzyme-2H \longrightarrow$$

$$CH_3(CH_2)_{16}COOH + 8ADP + 8P_i + 16coenzyme$$
$$+ 7H_2O + 9CoA-SH$$

Finally, triacylglycerols are formed by linking fatty acids to each of the three hydroxyl groups in glycerol.

A comparison of the molecules required for fatty acid and glycerol synthesis with those produced by glucose catabolism should reveal why glucose is so easily converted to triacylglycerol. First, note that acetyl coenzyme A, the starting material for fatty acid synthesis, can be formed from pyruvic acid, an intermediate in glycolysis. Second, the other ingredients required for fatty acid synthesis—coenzyme–2H and ATP—are produced during carbohydrate catabolism. Third, glycerol can be

formed from a glucose intermediate, 3-phosphoglyceraldehyde (Fig. 5-11). It should not be surprising, therefore, that much of the carbohydrate in food is converted into fat and stored in adipose tissue shortly after its absorption from the gastrointestinal tract. Mass action resulting from the increased concentration of glucose intermediates, as well as the specific hormonal regulation of key enzymes, promotes the conversion of glucose to fat following a meal, as will be described in Chap. 15.

It is very important to note that fatty acids, or more specifically, the acetyl coenzyme A derived from fatty acid breakdown, cannot be used to synthesize new molecules of glucose. The reasons for this can be seen by examining the pathways for glucose synthesis (Fig. 5-16). First, because the reaction in which pyruvic acid is broken down to acetyl coenzyme A and carbon dioxide is irreversible, acetyl coenzyme A cannot be converted into pyruvic acid, which by "reverse glycolysis" could yield glucose. Second, the equivalent of the two carbon atoms in acetyl coenzyme A are converted into two molecules of carbon dioxide during their passage through the Krebs cycle before reaching oxaloacetic acid, another takeoff point for glucose synthesis. Thus, glucose can readily be converted into fat, but the fatty acid portion of fat cannot be converted to glucose. The glycerol backbone of fat, as we mentioned earlier, can be converted to glucose.

Protein metabolism

In the last chapter we described how amino acids are linked together to form proteins and how DNA and RNA are involved in directing the process. In contrast to the complexities of this synthetic pathway, protein catabolism requires only a few enzymes (**proteases**) to break the peptide bonds between amino acids. Some of these enzymes split off single amino acids, one at a time, from the ends of the protein chain, while others break peptide bonds between specific amino acids within the chain, thereby forming shorter peptides. The overall pathways for protein synthesis and degradation can thus be summarized as

$$\text{Amino acids} \underset{\text{proteases}}{\overset{\text{nucleic acids + ribosomes}}{\rightleftharpoons \rightarrow \rightarrow \rightarrow}} \text{protein}$$

In addition to their incorporation into proteins, amino acids can be catabolized to provide energy for ATP synthesis and can provide the intermediates for the synthesis of a number of chemical messengers used in the nervous and hormonal systems. A few basic types of reactions common to most of these pathways can provide an overview of the general metabolism. Unlike carbohydrates and fats, amino acids contain nitrogen atoms (in their amino groups) in addition to carbon, hydrogen, and oxygen atoms. Once this amino group is removed, the remainder of the molecule can be metabolized to intermediates capable of entering the glycolytic or Krebs-cycle pathways. Since there are 20 different amino acids, there are a large number of intermediates that can be formed and many pathways for processing these intermediates.

The two types of reactions by which the amino group is removed are illustrated in Fig. 5-18. In the first type, known as **oxidative deamination**, the amino group is released as a molecule of ammonia (NH_3) and replaced by an oxygen atom (derived from water) to form a **keto acid**. (During this reaction the two hydrogen atoms in water are transferred to a coenzyme.) The second type of reaction, known as **transamination**, involves the transfer of the amino group from an amino acid to a keto acid. Note that the keto acid to which the amino group is transferred becomes an amino acid. Thus, by means of transamination, some new amino acids can be formed from certain carbohydrate intermediates (keto acids).

Figure 5-19 shows the oxidative deamination and transamination of two amino acids, glutamic acid and alanine. Note that the keto acids formed are intermediates either in the Krebs cycle (as is α-ketoglutaric acid) or glycolytic pathway (as is pyruvic acid). Once formed, these keto acids can be metabolized to produce carbon dioxide and form ATP, or they can be used as intermediates in the synthetic pathway leading to the formation of glucose. As a third alternative, they can be used to synthesize fatty acids after their conversion to acetyl coenzyme A, by way of pyruvic acid. Thus, amino acids can be used as a source of energy, and some can be converted into carbohydrate and fat.

It must be recognized that, unlike fat (and to a lesser extent carbohydrate), excess protein is not really "stored" in cells; i.e., the proteins of the

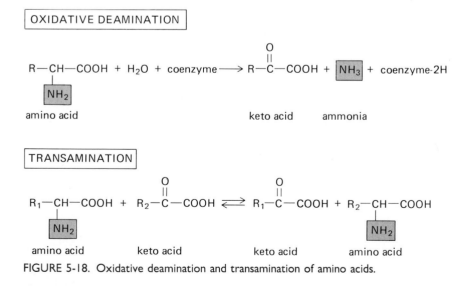

FIGURE 5-18. Oxidative deamination and transamination of amino acids.

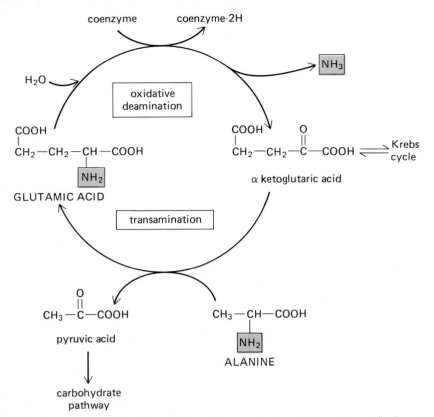

FIGURE 5-19. Oxidative deamination and transamination of the amino acids glutamic acid and alanine lead to keto acids that can enter the carbohydrate pathways.

body participate directly in the ongoing structure and function of cells. Therefore, when endogenous protein is catabolized to provide energy or to serve as a precursor for carbohydrate and fat, functioning proteins are being consumed. During starvation, for example, muscle cells become progressively smaller as the contractile proteins are degraded to provide amino acids used by the liver to form glucose.

When carbohydrate or fat molecules are catabolized to provide energy for ATP formation, the carbon, hydrogen, and oxygen atoms they contain end up in carbon dioxide and water molecules, which are the waste products of metabolism. The carbon dioxide is eliminated from the body by the lungs, and water is excreted primarily in the urine. As we have seen, the catabolism of amino acids can lead, in addition, to the nitrogen-containing product ammonia. (The nitrogen derived from

amino groups can also be used by cells to synthesize other important nitrogen-containing molecules, such as the purine and pyrimidine bases found in nucleic acids.) Ammonia, which can be highly toxic to cells if allowed to accumulate, readily leaks into the blood and is carried to the liver (Fig. 5-20). Here, in a series of reactions, two molecules of ammonia are linked with carbon dioxide to form **urea**. Thus urea, which is relatively nontoxic to cells, is the major nitrogenous waste product of protein catabolism; it leaves the liver and is excreted by the kidneys into the urine. Two of the 20 amino acids also contain atoms of sulfur, which can be converted to sulfate, SO_4^{2-}, and be excreted in the urine.

Thus far, we have discussed mainly amino acid catabolism. Figure 5-19 also demonstrates how amino acids can be formed. Keto acids, such as α-ketoglutaric and pyruvic acids, can be formed from

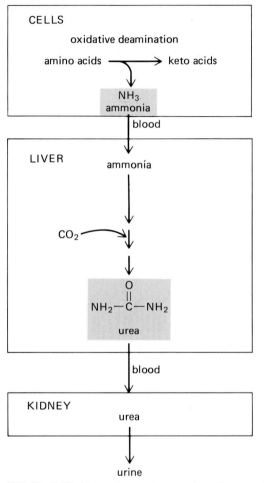

FIGURE 5-20. Formation and excretion of urea, the major nitrogenous waste product of protein catabolism.

the breakdown of glucose and then be transaminated, as described above, to form glutamic acid and alanine; thus glucose can be used to form certain amino acids, provided other amino acids are available in the diet to supply amino groups for transamination. However, not all of the amino acids can be formed by the body. Those that must be obtained from the food we eat (9 of the 20 amino acids) are known as **essential amino acids**.

Summary of fuel metabolism

Having discussed the metabolism of the three major classes of organic molecules, we can now briefly review how each class is related to the others and to the process of synthesizing ATP. Figure 5-21 shows the major pathways we have discussed and the relations of the common intermediates. All three classes of molecules can enter the Krebs cycle through some intermediate, and thus, all three can be used as a source of chemical energy for the synthesis of ATP by oxidative phosphorylation. Hydrogen and oxygen provide the substrates for this oxidative phosphorylation. Some of the hydrogens are derived from the Krebs-cycle reactions directly; others may come from glycolysis or from the breakdown of fatty acids and amino acids. Glucose can be converted into fat or into some amino acids by way of the common intermediates such as pyruvic acid, α-ketoglutaric acid, and acetyl CoA. Similarly, some amino acids can be converted into glucose and fat. Fatty acids cannot be converted into glucose because of the irreversibility of the reaction converting pyruvic acid to acetyl CoA, but the glycerol portion of triacylglycerols can be converted into glucose. Metabolism is, thus, a highly integrated process in which all classes of molecules can be used, if necessary, to provide energy for the cell through ATP synthesis and in which each class of molecule can, to a large extent, provide the raw materials required to synthesize members of other classes.

We have specifically noted in the glycolytic pathway a number of enzymes whose activity can be altered by allosteric and covalent modulation, thereby regulating the net anabolism or catabolism of glucose. Other regulated enzymes occur in the Krebs cycle and in the pathways for fatty acid and amino acid metabolism. For example, ATP and its products, ADP and P_i, are modulator molecules able to activate or inhibit various allosteric enzymes. If a cell is utilizing large amounts of energy, the rate of ATP breakdown to ADP and P_i will be rapid, leading to a fall in ATP concentration and a rise in the concentrations of ADP and P_i. The changes in the concentrations of these molecules allosterically activate or inhibit various enzymes in such a pattern that the overall rates of glucose and fat catabolism are increased, thereby providing the energy necessary to form more ATP. Likewise, when the cell is using less energy, the rise in ATP and the fall in ADP and P_i inhibit the appropriate

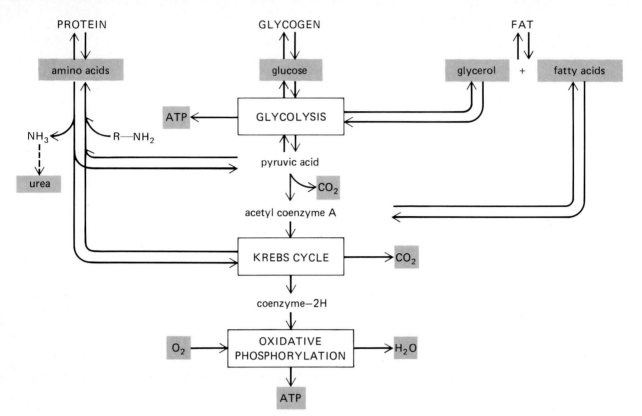

FIGURE 5-21. Interrelations between the pathways for the metabolism of carbohydrates, fats, and proteins.

catabolic enzymes and activate the anabolic enzymes, leading to a net storage of fuel as glycogen or fat. Similar changes in the concentrations of the hydrogen-carrying coenzymes (high concentration of coenzyme–2H when energy consumption is low and low concentration of coenzyme–2H when energy consumption is high) can also alter the rates of reactions using coenzyme-dependent enzymes. Other intermediates, including glucose 6-phosphate, citric acid, and acetyl coenzyme A, can function as modulator molecules for allosteric enzymes, thereby providing the cell with a variety of interacting control systems.

Essential Nutrients

There are many substances which are required for normal or optimal body function but which are synthesized by the body either not at all or in amounts inadequate to keep pace with the rates at which they are broken down or excreted. They are known as **essential nutrients**. Because they are all removed at some finite rate, a continual new supply must be provided in the diet.

It should be emphasized that the term "essential nutrient" is reserved for substances that fulfill two criteria: (1) They must be essential for good health, and (2) they must not be synthesized by the body in adequate amounts. Thus, glucose, although essential for normal metabolism, is not classified as an essential nutrient because the body normally can synthesize all it needs (from amino acids, for example).

Approximately 50 substances are classified as essential nutrients (Table 5-3 and Appendix 2). Water is an essential nutrient because far more water is lost in the urine and from the respiratory tract than is synthesized by the body. (Recall that water is formed as an end product of oxidative phosphorylation as well as from other metabolic

TABLE 5-3. Essential nutrients

Water
Mineral elements
   Major minerals (7)
   Trace elements (13)

Growth factors
   Amino acids (9)
   Unsaturated fatty acids (2)
   Inositol
   Choline
   Carnitine

Vitamins
   Water-soluble (10)
   Fat-soluble (4)

reactions.) Therefore, to maintain water balance, an intake of water is essential.

The mineral elements provide an example of substances which are not metabolized but are continually lost from the body in the urine, feces and various secretions. The major minerals (see Table 3-3), such as Na, K, and Ca, must be supplied in fairly large amounts, whereas only small quantities of the trace elements, such as Zn, I, and Cu are required.

The amount of an essential nutrient that must be ingested forms the major distinction between **growth factors** and **vitamins**. Growth factors are organic molecules which are required in relatively large amounts in the diet, whereas vitamins are only required in trace quantities. For example, a daily intake of approximately 2 g of the essential amino acid methionine is required, compared to only about 1 mg of the vitamin thiamine.

Vitamins

Fourteen different vitamins have been shown to be essential for normal growth and health in humans. (The same set of vitamins is not essential for all species.) The exact chemical structures of the first vitamins to be observed were unknown. Therefore, they were simply identified by letters of the alphabet (vitamin A, B, C, etc.) and only later were their specific chemical structures identified (e.g., vitamin A = retinol, vitamin C = ascorbic acid). Vitamin

B turned out to be composed of eight different substances now known as the vitamin B complex. These are thiamine or B_1, riboflavin or B_2, niacin, pyridoxine or B_6, pantothenic acid, folic acid, biotin, lipoic acid, and cobalamine or B_{12}.

The vitamins as a class have no particular chemical structure in common, but they can be divided into the **water-soluble vitamins** (the B complex and vitamin C) and the **fat-soluble vitamins** (A, D, E, and K). The water-soluble vitamins function as coenzymes, as described earlier. The fat-soluble vitamins in general do not function as coenzymes, and their specific roles in the body will be described in later chapters.

We have mentioned that vitamins are required in the diet because the body is either unable to synthesize them or unable to produce adequate amounts of them. However, plants and bacteria have the enzymes necessary for vitamin synthesis, and it is from these sources that we ultimately obtain vitamins. By themselves vitamins do not provide chemical energy, although they may participate as coenzymes in chemical reactions which release energy from other molecules. Increasing the amount of vitamins in the diet does not necessarily increase the activity of those enzymes for which the vitamin functions as a coenzyme. Only very small quantities of coenzymes are necessary to saturate their binding sites on the enzyme molecules, and increasing the concentration above this level does not increase the enzyme's activity. Obviously, a lack of vitamins in the diet causes ill effects, since they are essential for the activity of many enzymes.

Quantities of water-soluble vitamins in the diet that exceed the small amounts that are continually lost have not yet been proven to have beneficial effects. As the amount of these water-soluble vitamins in the diet is increased so is the amount excreted in the urine, with the result that accumulation of these vitamins in the body is limited. On the other hand, in the case of the fat-soluble vitamins, the intake of very large quantities is known to produce toxic effects that damage various tissues. These effects are related to the fact that very large quantities of these fat-soluble vitamins can accumulate in the body because they dissolve readily in the fat stores in adipose tissue.

6

MOVEMENT OF MOLECULES ACROSS CELL MEMBRANES

The contents of a cell are separated from the surrounding extracellular fluid by a thin layer of lipids and protein—the **plasma membrane**. In addition, membranes associated with mitochondria, endoplasmic reticulum, lysosomes, the Golgi apparatus, and the nucleus divide the intracellular fluid into separate compartments. The term ''membrane'' refers to any of these cell membranes, whereas the term ''plasma membrane'' denotes specifically the surface membrane.

The contributions of this array of membranes to cell function fall into two general categories: (1) They provide barriers to the movements of molecules and ions both between the various cell organelles within the cell and between the cell and the extracellular fluid and (2) they provide a scaffolding to which various cell components are anchored.

The chemical compositions of the intracellular and extracellular fluids (Table 6-1) are quite different. This fact demonstrates that the plasma membrane prevents free mixing of the molecules on the two sides of the membrane. However, since cells are able to take in some substances from the extracellular fluid and give off other substances to it, some molecules are able to cross the plasma membrane. Thus, the plasma membrane is selectively permeable. Similarly, the selective permeability of organelle membranes provides the means for confining various chemical reactions to specific cell organelles while allowing key substrates and products to move between the organelles and the cytosol. Moreover, the permeability of cell membranes to specific substances can be increased or decreased. As we shall see, these alterations in membrane permeability provide an important mechanism for regulating various cell functions.

Many membrane functions are associated with the binding of substances to membranes. A variety of specific binding sites are located on the outer surface of the plasma membrane. These protein molecules function as ''recognition sites'' (receptors) for hormones and other chemical messengers;

TABLE 6-1. Composition of interstitial and intracellular fluids

	Concentration in interstitial fluid, mM	Concentration in intracellular fluid,* mM
Na^+	140	15
K^+	4	150
Ca^{2+}	1	1.5
Mg^{2+}	1.5	12
Cl^-	110	10
HCO_3^-	30	10
P_i	2	40
Amino acids	2	8
Glucose	5.6	1
ATP	0	4
Protein	0.2	4.0

*The intracellular concentrations differ in various tissues, but the pattern shown above holds for essentially all cells. Furthermore, the intracellular concentrations listed above may not reflect the free concentration of the substance in the cytosol, since some may be bound to proteins or confined within cell organelles. For example, the free cytosolic concentration of calcium is only about 0.0001 mM. (Plasma concentrations differ only slightly from interstitial concentrations. See Table 11-1 for a listing of plasma concentrations.)

thus the plasma membrane serves as a signal-receiving device for the detection of chemical signals which regulate the cell's activity. Surface binding sites also participate in the arrangement of cells into tissues during growth and development by providing sites for specific cell-to-cell adhesions. Fibrous proteins are attached to the inner surface of the plasma membrane and function both in the maintenance of cell shape and in the changes in the surface configuration of cells during certain states of activity. In muscle cells, the contractile protein filaments are anchored to the plasma membrane, enabling them to transmit force to the outside of the cell. A number of enzymes are anchored to cell membranes; hence, certain chemical reactions occur at the outer or inner surface of the plasma membrane or in association with the membranes of a particular cell organelle. For example, the reactions of oxidative phosphorylation (as we saw in the last chapter) occur on the inner membrane of the mitochondria, in which are embedded the enzymes of the electron-transport chain.

Membrane Structure

All cell membranes have a similar general structure—they consist of a bimolecular layer of lipid molecules in which proteins are embedded. The lipid layer forms the barrier preventing the diffusion of most molecules through the membrane, whereas the proteins provide the pathways for the selective transfer of substances through the lipid barrier. The proteins also constitute the binding sites and enzymes associated with membranes. An electron micrograph of the plasma membrane of a red blood cell is shown in Fig. 6-1. The membrane appears as two dark lines separated by a light interspace. The dark lines correspond to the staining of the polar regions of the proteins and lipids, whereas the light interspace corresponds to the nonpolar regions of these molecules in the membrane.

The major membrane lipids are phospholipids. As described in Chap. 3, these are amphipathic molecules, one end having a charged region, whereas the remainder of the molecule, consisting of two long fatty acid chains, is nonpolar. When phospholipids are mixed with water in the laboratory, these amphipathic molecules aggregate into patterns in which the polar ends of the molecules are electrically attracted to the polar water molecules while the nonpolar fatty acid chains are clustered together out of contact with water. Two stable geometries are possible—a spherical cluster (a micelle, see Fig. 3-7) or a bimolecular layer (Fig. 6-2). In both cases the polar ends of the phospholipids are at the surface, while the nonpolar fatty acid chains are in the interior. The phospholipids in cell membranes are organized into bimolecular layers similar to those which can form spontaneously in phospholipid-water mixtures.

There are no chemical bonds linking phospholipids to each other; therefore, each molecule is free to move independently of the others. This results

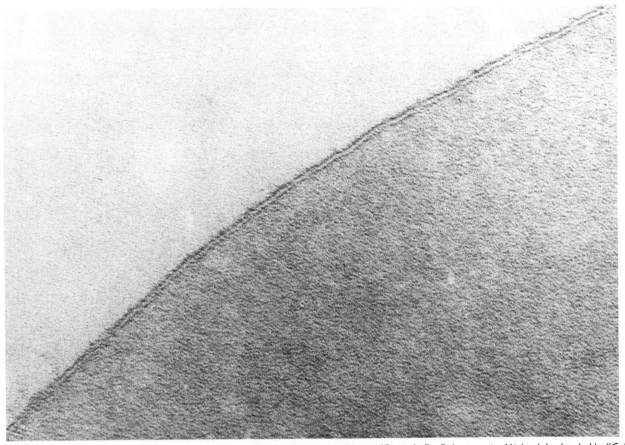

FIGURE 6-1. Electron micrograph of a human red-cell plasma membrane. [*From J. D. Robertson in Michael Locke (ed.), "Cell Membranes in Development," Academic Press, Inc., New York, 1964.*]

in considerable random lateral movement parallel to the surface of the bilayer. In addition, the long fatty acid chains can bend and wiggle back and forth (like a piece of rope that is shaken). Thus, the lipid bilayer in the membrane has the characteristics of a fluid, e.g., an oil. This fluidity makes the entire membrane quite flexible; it can easily be bent and folded, although horizontal stretching may cause it to rupture.

In addition to phospholipids, the plasma membrane contains a second type of lipid—that is, cholesterol, a steroid. The plasma membrane contains about one molecule of cholesterol for each molecule of the phospholipid, but intracellular membranes contain very little cholesterol. Cholesterol is slightly amphipathic due to a single polar hydroxyl group (see Fig. 3-13) on its nonpolar ring structure. Therefore, cholesterol, like the amphipathic phospholipids, is inserted into the lipid bilayer with its polar region at a bilayer surface and its nonpolar rings in the interior of the bilayer in association with the fatty acid chains. The cholesterol in the plasma membrane appears to play an important role in maintaining the fluidity and stability of the lipid bilayer.

Membrane proteins of one general class, known as **integral proteins**, are closely associated with the membrane lipids and cannot be extracted from the membrane without disrupting the lipid bilayer. Like the phospholipids, these integral proteins are amphipathic, having a molecular conformation in which the polar amino acid side chains

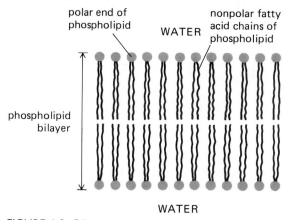

FIGURE 6-2. Bilayer organization of phospholipid molecules in water.

are located in one region of the molecule while the nonpolar side chains are clustered together in a separate region. Such amphipathic integral proteins are arranged in the membrane with the same orientation as the amphipathic lipids—the polar regions are at the surface in association with polar water molecules, and the nonpolar regions are in the interior of the membrane in association with the nonpolar fatty acid chains (Fig. 6-3).

Most integral proteins span the entire membrane. These proteins have two polar regions, one on each side of the membrane; these regions are connected by a nonpolar segment in the nonpolar interior region of the membrane. Some of these integral transmembrane proteins form channels through which ions can move across the membrane, whereas others are involved in more specialized mechanisms of membrane transport. Still others appear to be associated with the transmission of chemical signals from a surface binding site to the opposite side of the membrane. Some of these proteins function as enzymes, with their substrate-binding sites located at the surface of the membrane. Some integral proteins may be confined to one side or the other of the membrane (Fig. 6-3). Like the membrane lipids, many of the integral membrane proteins can move laterally in the plane of the membrane, while others are immobilized as a result of being linked to a fibrous network of peripheral proteins.

Membrane proteins of a second general class, **peripheral proteins**, can be extracted from the membrane without disrupting the lipid bilayer. These proteins are not amphipathic and do not associate with the nonpolar membrane lipids. Peripheral proteins appear to be located at the membrane surface where they are bound to the polar regions of the integral membrane proteins (Fig. 6-3). Most of the peripheral proteins appear to be located on the cytoplasmic surface of the plasma membrane.

In addition to lipids and proteins, the plasma

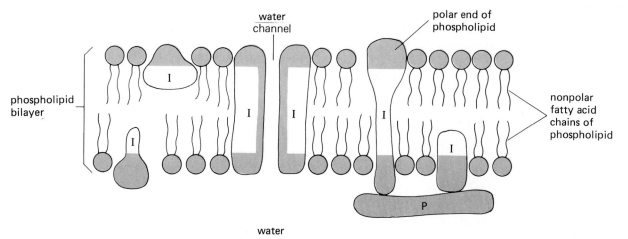

FIGURE 6-3. Arrangement of integral (I) and peripheral (P) membrane proteins in association with bimolecular layer of phospholipids. Shaded areas at the membrane surface indicate the polar regions of the proteins and phospholipids.

membrane contains small amounts of carbohydrate covalently linked to lipids and proteins to form glycolipids and glycoproteins. The carbohydrate portions of these molecules are all located at the extracellular surface. This is another example of the asymmetrical distribution of membrane components (the first being the distribution of peripheral proteins). In addition, the lipids in the outer half of the bilayer differ somewhat in kind and amount from those on the inner half of the bilayer, and as we have seen, the proteins or portions of proteins that are on the outer surface differ from those on the inner surface. Thus, plasma membranes are structurally asymmetrical and many membrane functions are related to this asymmetry.

Although all membranes appear to have the general structure described above, which has come to be known as the **fluid-mosaic model** (Fig. 6-4), plasma membranes and organelle membranes contain different proteins and lipids; moreover, the plasma membranes from different types of cells also differ in terms of their specific lipids and proteins. Cell membranes are 6 to 10 nm thick, most intracellular membranes being thinner than plasma membranes. This variation in thickness is primarily the result of differences in the protein composition and the amount of polar proteins at the membrane surfaces.

The description of the movements of molecules and ions across the fluid-mosaic structure of the membrane is the subject of the remainder of this chapter, which begins with a description of the random movements of molecules which give rise to diffusion.

Diffusion

Diffusion is the movement of molecules from one location to another by random thermal motion. The average velocity of molecular movement depends upon the temperature and the mass of the molecule. All molecules undergo continuous thermal motion, and the warmer an object is, the faster its molecules move. The energy for the movement comes from heat energy, hence the term "thermal motion." At body temperature, a molecule of water moves at about 2500 km/h (1500 mi/h), on an average, whereas a molecule of glucose, which is 10 times heavier, moves at about 850 km/h. In water, where individual molecules are separated from each other by about 0.3 nm, such rapidly moving molecules cannot travel very far before colliding with other molecules. They bounce off each other like rubber balls, undergoing millions of collisions every second. Each collision alters the direction of the molecule's movement, so that the path of any one molecule through a solution may look like that shown in Fig. 6-5. Since any molecule may at any instant be moving in any direction, such movement is said to be "random," meaning that it has no preferred direction of movement.

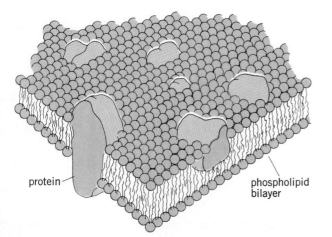

FIGURE 6-4. Fluid-mosaic model of cell membrane structure. (Redrawn from S. J. Singer and G. L. Nicholson, Science, **175**:723. Copyright 1972 by the American Association for the Advancement of Science.)

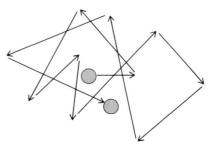

FIGURE 6-5. Random path traveled by a single molecule undergoing thermal motion in a solution. Changes in direction result from collisions with other molecules.

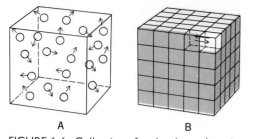

FIGURE 6-6. Collection of molecules undergoing random thermal motion in a small volume element (A) results in approximately equal numbers of molecules moving up and down, right and left, etc. A large volume of solution is equivalent to a number of small volume elements (B) in which the fluxes of the molecules between the volume elements depend upon the concentrations of molecules in each volume element.

Magnitude and direction of diffusion

Any volume of solution can be looked upon as being composed of a number of small volume elements joined together at their surfaces (Fig. 6-6). The movement of molecules across the imaginary surfaces of these volume elements represents the redistribution of molecules between different portions of the total solution. The amount of material crossing a surface in a unit of time is known as a **flux**. If the number of molecules in a unit of volume is doubled, the flux of molecules across each surface would also be doubled, since twice as many molecules would be moving in any one direction at a given time. To generalize, the concentration of molecules (number of molecules in a unit of volume) in any region of a solution determines the magnitude of the flux across the surfaces of this region.

The diffusion of glucose between two compartments of equal volume is illustrated in Fig. 6-7. Initially (Fig. 6-7A), glucose is present in compartment 1 and there is no glucose in compartment 2. The random movements of the glucose molecules in compartment 1 carry some of them into compartment 2. The flux of glucose from compartment

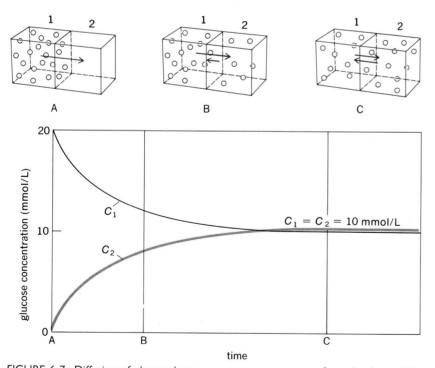

FIGURE 6-7. Diffusion of glucose between two compartments of equal volume. (A) At initial conditions, no glucose is present in compartment 2. (B) Some glucose molecules have moved into compartment 2, and some of them are moving at random back into compartment 1. (C) Diffusion equilibrium has been reached, the fluxes of glucose between the two compartments being equal in the two directions.

1 to 2 depends on the concentration of glucose in compartment 1, which is initially 20 mmol/L. After a short time some of the glucose molecules that have entered compartment 2 will randomly move back into compartment 1 (Fig. 6-7B). The magnitude of the glucose flux from compartment 2 to 1 depends upon the concentration of glucose in compartment 2 at the time. Thus, the **net flux** of glucose between the two compartments is the difference between the two one-way fluxes—from 1 to 2 and from 2 to 1. It is the net flux that determines the net gain of molecules by compartment 2 and the net loss from compartment 1.

As the concentration of glucose in compartment 2 increases, the flux of glucose from 2 to 1 increases; simultaneously, as the concentration in compartment 1 decreases, the flux from 1 to 2 decreases. Eventually the concentration of glucose in the two compartments becomes equal at 10 mmol/L, the two one-way fluxes become equal in magnitude but opposite in direction, and the net flux of glucose becomes zero (Fig. 6-7C). The system has now reached **diffusion equilibrium**, and no further change in the glucose concentration of the two compartments will occur.

Several important properties of the diffusion process should be reemphasized using this example. Three fluxes have been identified—the two one-way fluxes occurring in opposite directions from one compartment to the other and the net flux, which is the difference between them. It is the net flux which is the most important component in the diffusion process, since it is the net amount of material that is transferred from one location to another. Although the movement of individual molecules is random, the net flux will always proceed in a specific direction—from regions of higher to regions of lower concentration.

Both the direction and the magnitude of the net flux is determined by the **concentration gradient**—the difference in concentration between two adjacent regions. The larger the concentration gradient, the greater the magnitude of the net flux, which will always proceed in the direction from higher to lower concentration. For net diffusion to occur, there must be a concentration gradient. If there is

no concentration gradient (a condition that will occur when the two concentrations are equal) then the net flux between the two compartments will be zero and the system will be at diffusion equilibrium.

At any particular concentration gradient, the absolute magnitude of the net flux depends on several additional factors: (1) temperature—the higher the temperature, the greater the velocity of molecular movement and the greater the flux; (2) mass of the molecule—larger molecules (e.g., proteins) have a greater mass and slower velocity than smaller molecules (e.g., glucose) and thus have a smaller flux; (3) medium in which diffusion is occurring—collisions are more frequent in water than in a gas (this explains, for example, the slower rate of diffusion of oxygen dissolved in water than in air); (4) surface area—the greater the surface area between two regions the greater the space available for diffusion and thus the greater the total net flux. When the diffusion of two different molecules or of the same molecule in two different mediums, or at different temperatures, or through different geometries are compared, then the above factors contribute, along with the concentration gradient, to the differences in the magnitudes of the net fluxes.

Speed of diffusion

Many properties of living organisms are closely associated with the process of diffusion: Oxygen and nutrients enter the blood by diffusion and molecules leave the blood and enter the extracellular fluid of the tissues by diffusion; the movement of some molecules across the plasma membrane and across organelle membranes occurs by diffusion. The speed of diffusion is thus an important factor in determining the rate at which molecules can reach a cell and the rate at which metabolic products can leave it and gain entry to the blood.

Although individual molecules travel at high velocities, the number of collisions they undergo prevents them from traveling very far in a straight line. Figure 6-8 illustrates the diffusion of glucose from the blood, where it is maintained at a relatively constant concentration, into the extracellular fluid surrounding the cells of a tissue. The question we

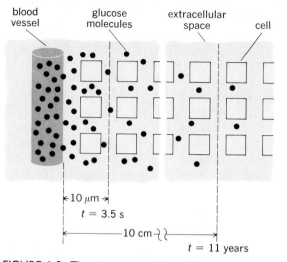

blood vessel   glucose molecules   extracellular space   cell

$\leftarrow 10\ \mu m \rightarrow$

$t = 3.5\ s$

$\longleftarrow 10\ cm \longrightarrow$

$t = 11\ years$

FIGURE 6-8. The time required for diffusion to raise the concentration of glucose at a point 10 μm (about one cell diameter) away from a blood vessel to 90 percent of the blood glucose concentration is about 3.5 s, whereas it would take over 11 years for glucose to reach the same concentration at a point 10 cm (3.9 in.) away.

wish to answer is: How long will it take the concentration of glucose to reach 90 percent of the glucose concentration in the blood at points 10 μm and 10 cm away from the blood vessel (assuming no metabolism of the glucose by cells)? The answers are 3.5 s and 11 years, respectively. Thus, although diffusion can distribute molecules rapidly over short distances (for example, within cells or the extracellular regions surrounding a few layers of cells), diffusion is an extremely slow process when distances of a few centimeters or more are involved. For an organism as large as a person, the diffusion of oxygen and nutrients from the body surface to tissues located several centimeters below the surface would be far too slow to provide adequate nourishment. Accordingly, the circulatory system provides the mechanism for rapidly moving materials over large distances (by blood flow using a mechanical pump, the heart), with diffusion providing movement into and out of the blood and through the extracellular fluid.

The rate at which diffusion is able to move molecules within a cell is one of the factors limiting the size to which a cell can grow. The larger the

cell, the greater is the volume of cytoplasm that must be supplied by diffusion from a given unit of cell surface area. A cell does not have to be very large before diffusion becomes too slow to provide sufficient nutrients to the large volume of intracellular fluid. For example, the center of a 20-μm diameter cell reaches diffusion equilibrium with oxygen in about 15 ms, whereas it would require 265 days for oxygen to come to equilibrium with the central portion of cytoplasm in a cell the size of a basketball. In some cases, a cell which has a fairly large cell volume has a shape that still provides a small volume/surface area ratio. For example, muscle cells are often more than 10 cm long but only 10 to 100 μm thick. Thus, a molecule has to diffuse only a short distance from the surface to reach the center of the muscle cell.

Diffusion across membranes

Some, but not all, molecules are able to diffuse across cell membranes. The pathways by which this diffusion takes place will be discussed in the next section of this chapter. The rate at which a substance diffuses across the plasma membrane can be measured by monitoring the rate at which the intracellular concentration increases when a cell is placed in a solution containing the substance. The intracellular concentration refers to the concentration of the substance in the cytosol, the intracellular fluid that is in contact with the inner surface of the plasma membrane. If we make the volume of solution outside the cell very large, then the concentration of the substance in this extracellular solution undergoes very little change as the substance diffuses into the small intracellular volume; it can be assumed to stay constant (Fig. 6-9). As with all diffusion processes, the net flux of material across the membrane, F, is from the region of high concentration (the extracellular solution in this case) to the region of low concentration. This net diffusion of material into the cell will continue until intracellular and extracellular concentrations become equal, $C_i = C_o$. The magnitude of the net flux depends on the concentration gradient across the mem-

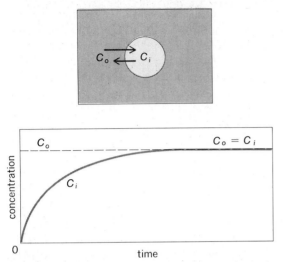

FIGURE 6-9. Change in intracellular concentration C_i as material diffuses from a constant extracellular concentration C_o across the plasma membrane of a cell.

brane $(C_o - C_i)$, the surface area of the membrane A, and the **membrane permeability constant k_p.**

$$F = k_p A (C_o - C_i) \qquad [6\text{-}1]$$

The numerical value of the permeability constant depends on the type of molecule, its molecular weight, the temperature, and the characteristics of the membrane through which the molecule is diffusing. It is an experimentally determined number, and for a given type of molecule at a given temperature, it reflects the ease with which the molecule is able to move through a given membrane. In other words, the greater the permeability constant, the greater the permeability of the membrane to a given substance.

The above description has been in terms of molecules diffusing into a cell from a region of high extracellular concentration. Obviously, molecules can also undergo a net diffusion out of a cell if the intracellular concentration is greater than the extracellular concentration, as may occur for substances that are being synthesized by cells. The same equation (Eq. 6-1) applies, but now the net flux will be negative, indicating a net movement out of the cell.

The rates at which a number of different mol-

ecules diffuse into a cell are shown in Fig. 6-10. Such measurements reveal several properties of the membrane. First, the rates of diffusion across membranes are found to be a thousand to a million times smaller than the diffusion of the same molecules in water; i.e., the molecules are entering the cell at a much slower rate than if there were no barrier at the surface. Second, not only is the membrane a barrier which slows down the entry of molecules, it is also a highly selective barrier which allows some molecules to enter rapidly while others, which may diffuse at a similar rate in solution, enter more slowly or not at all. In other words there is a different membrane permeability constant k_p for each substance.

Diffusion through the lipid bilayer. When the permeability constants of a number of different organic molecules are examined in relation to their molecular structure, a correlation emerges. Whereas most polar and ionized molecules enter cells very slowly or not at all, nonpolar molecules enter cells rapidly (i.e., they have large permeability constants). This observation is explained by the fact that nonpolar molecules can dissolve in the nonpolar regions of the membrane (regions occupied by the fatty acid chains of the membrane phospholipids); in contrast

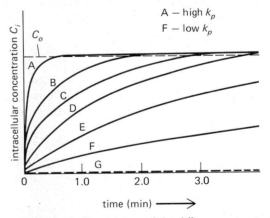

FIGURE 6-10. Comparison of the different rates at which various molecules, A through G, are able to diffuse across a plasma membrane from an extracellular medium of constant concentration C_o. Molecule A enters the cell rapidly and has a high permeability constant k_p. Molecule F enters the cell very slowly, and molecule G is unable to cross the membrane at all.

polar and ionized molecules are strongly attracted to the polar water molecules on either side of the membrane and are restricted from entering the nonpolar regions of the membrane. In other words, nonpolar molecules rapidly cross the membrane by directly diffusing through the lipid bilayer. The more lipid-soluble the substance (the fewer its polar or ionized groups), the greater will be the number of its molecules dissolved in the membrane lipids and the higher its concentration within the membrane, and thus, the greater will be its flux across the membrane, since the flux depends on concentration. Oxygen, carbon dioxide, fatty acids, and the steroid hormones are examples of nonpolar molecules which diffuse rapidly through the lipid portions of membranes. Also, the nonionized forms of some metabolic intermediates, such as lactic acid, can diffuse through the lipid bilayer, whereas the ionized form cannot.

That it is the phospholipid-bilayer portion of the membrane that provides the selective barrier, allowing nonpolar molecules to pass but excluding polar or ionized substances, is substantiated by experiments using artificial membranes consisting of a bimolecular layer of phospholipids but containing no protein; the rates of diffusion across these artificial lipid bilayers are very similar to the rates of diffusion through cell membranes.

Diffusion through protein channels. Ions such as Na^+, K^+, Cl^-, Ca^{2+}, etc., diffuse across plasma membranes at rates that are much faster than would be predicted from their very low solubility in lipid. Moreover, different cells have quite different permeabilities to these ions, whereas nonpolar substances which diffuse through the lipid bilayer have similar permeabilities when different cells are compared. The fact that artificial phospholipid bilayers are practically impermeable to these ions, whereas cell membranes have a considerable ion permeability, suggests that it is the protein component of the membrane that is responsible for the permeability of these ions.

As we have seen, many of the integral membrane proteins span the lipid bilayer. Such proteins can form **channels** (pores) through which ions can diffuse across the membrane. A single protein may have a conformation similar to a doughnut, with the hole in the middle providing the channel for ion movement. In other cases, several proteins aggregate to form the walls of a channel. The diameters of such protein channels are very small, only slightly larger than that of the ions that pass through them. The small size of the channels prevents larger, polar organic molecules, e.g., glucose, amino acids, and most ionized metabolic intermediates, from entering the channel. Only the mineral element ions (and possibly water) are small enough to pass through these protein channels.

The variation in ion permeability found in different cell membranes reflects differences in the number of ion channels in a given membrane; the greater the number of channels, the greater the ion flux across the membrane for any given ion concentration gradient. Furthermore, the ion channels show a selectivity for the type of ions that can pass through them. This selectivity is based partially on the diameter of the channel and partially on the charged and polar surfaces of the proteins that form the walls of the channel and that can electrically attract or repel the charged ions. Thus, for example, some channels (K-channels), will allow only potassium ions to pass, others are specific for sodium (Na-channels), and still others are slightly less specific and allow both sodium and potassium ions to pass (Na,K-channels). Since different proteins form the different types of ion channels, membranes which have the same permeability to potassium, because they have the same number of potassium channels, may have quite different permeabilities to sodium because of differences in the number of sodium channels.

Thus far we have described the diffusion of molecules in terms of their random thermal motion and with the concentration gradient determining the direction and magnitude of the net flux. When describing the diffusion of ions across membranes, one additional force must be considered, an electric force that acts upon (charged) ions. There is a separation of electric charge across most plasma membranes, the origin of which will be described in Chap. 8. Such a charge separation, known as a

membrane potential, affects the diffusion of ions across the membrane. For example, if the inside of a cell has a net negative charge (as it does in most cells) there will be an electric force attracting positive ions into the cell and repelling negative ions. Even if the ion concentration was initially the same on both sides of the membrane, so that there was no concentration gradient, there would still be a net flux of ions through the appropriate channels due to the membrane potential. Thus, the direction and magnitude of the ion fluxes across membranes depends on both the concentration gradient and the electrical gradient (the electric potential across the membrane). These two driving forces are collectively known as the **electrochemical gradient**.

Regulating diffusion through membranes. The net diffusional flux of a substance across a membrane may be altered by changes in one or more of the variables in Eq. 6-1: (1) the surface area (*A*), (2) the concentration gradient for the substance ($C_o - C_i$) (and the electrical gradient in the case of ions), and (3) the membrane permeability constant k_p for the substance. Of these three, the membrane permeability constant is the one over which cells have some direct control and the one which can be altered to effect rapid changes in fluxes (particularly of ions) across the membrane.

Changes in ion permeability can occur rapidly as a result of the opening or closing of ion channels. The proteins that form some (but not all) ion channels can undergo conformational changes that result in opening or closing the channel. Two general types of events are known to change the conformation of ion-channel proteins: (1) binding of chemicals to the channel protein producing either an allosteric or a covalent change in channel shape and (2) changes in membrane potential which act upon the charged regions of the channel proteins to produce a change in their shape. Thus, chemical and electric signals act upon the channel proteins to produce changes in membrane permeability.

Although there are many channels which are subject to regulation in the above manner, there are also many channels that remain open at all times. Moreover, there are also several classes of channels which are specific for the same ion. For example, in the same membrane there can be sodium channels that are permanently open, sodium channels that open in response to chemical messengers, and sodium channels that are sensitive to membrane potential. The roles of these various ion channels in determining the electrical activities of cells will be discussed in Chap. 8.

Mediated-Transport Systems

Thus far we have considered the diffusion of ions and nonpolar molecules through membranes. Although diffusion may account for some of the transmembrane movements of ions, it does not account for all; moreover, there are a number of other molecules, such as amino acids and glucose, which are able to cross membranes yet are too polar and too large to diffuse through the lipid bilayer or through protein channels. The passage of these molecules and the nondiffusional movements of ions through membranes are mediated by specific integral membrane proteins known as **carriers**. The characteristics of such **mediated-transport systems** are quite different from the passage of substances through membranes by diffusion.

A substance that crosses by mediated transport must first bind to a specific site on a membrane-protein carrier molecule (Fig. 6-11), which is exposed to the solution at one surface of the membrane. These carrier proteins, which span the membrane, then undergo a change in shape, which results in the exposure of the binding site to the solution on the opposite side of the membrane. The dissociation of the substance from the binding site and its entry into solution complete the process of moving the material through the membrane. Note that the actual translocation of the substance through the membrane is not by diffusion, although diffusion brings the substance to the carrier binding site on one side of the membrane and moves it away on the opposite side. The membrane translocation step involves (1) binding, (2) a change in carrier shape, and (3) dissociation. Using this carrier mechanism, molecules can move in either direction

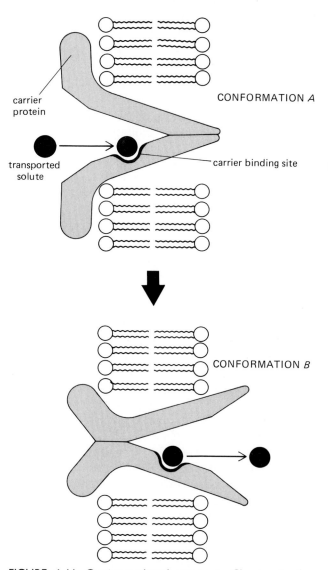

carrier protein

transported solute

CONFORMATION A

carrier binding site

CONFORMATION B

FIGURE 6-11. Carrier-mediated transport. Changes in the conformation of a membrane carrier protein from (A) to (B) alters the accessibility of the carrier binding site from one side of a membrane to the other, allowing a solute molecule that is bound to the binding site to move through the membrane. The carrier then returns to conformation (A) and can pick up another molecule of solute to repeat the process.

across the membrane, getting on the carrier at one side and off at the other.

At present, little is known about the specific shape changes that carrier proteins undergo during this process. It appears that many carriers are composed of more than one polypeptide chain, so that changes in the aggregation of these chains may lead to changes in the accessibility of the binding site from one side of the membrane to the other. The binding of a substance to a carrier may trigger the shape change, but in addition there is probably a thermal oscillation between the two conformations of a carrier that goes on continuously whether or not the binding site is occupied.

The fact that the transported substance binds to a specific site on a carrier protein accounts for a number of the characteristics of mediated-transport systems which are directly related to the characteristics of binding sites in general (described in Chap. 4). For example, mediated-transport systems have the characteristics of chemical specificity and saturation.

The binding site on the carrier will bind only specific ligands. For example, although both amino acids and sugars undergo mediated transport, the system which transports amino acids does not transport sugars, and vice versa. Thus, there are many different types of carrier proteins in membranes, each type specific for a particular substance or a specific class of related substances (a single carrier may bind more than one type of amino acid, for example). Different cells may transport different molecules because their membranes contain different carrier proteins.

The saturation of carrier binding sites sets a limit to the magnitude of the carrier-mediated flux across a membrane. There are a finite number of specific carrier proteins in a given membrane. As with any binding site, as the concentration of the ligand (the solute to be transported in this case) is increased, the number of occupied binding sites on the carrier molecules increases until the system becomes saturated (which occurs when all the sites are occupied.) The number of molecules of transported substance that move across the membrane depends on the number of occupied binding sites on the carrier molecules. At saturation, the maximal number of sites is occupied and the carrier-mediated flux of solute across the membrane is also maximal. Further increases in solute concentration will not increase the transport flux across the membrane. Moreover, the magnitude of the maximal flux depends upon the number of carrier molecules in a membrane—the greater the number of carrier proteins, the greater the maximal flux at saturation. Contrast this behavior of mediated-transport systems (Fig. 6-12) with the expected fluxes produced by diffusion. The flux due to diffusion will increase directly with increases in concentration and will not reach a maximal rate at high concentration, since diffusion does not involve the saturation of a fixed number of binding sites.

As we shall see in later chapters, the magnitude of a mediated-transport flux is subject to regulation by various control mechanisms, such as hormones. Changes in the mediated-transport flux are brought about by changing either the number of carriers in a membrane or the affinity of the binding sites for the transported solute. Changing the affinity will change the number of binding sites occupied at any given concentration less than a saturating concen-

tration and thus will change the flux through the membrane.

When a number of mediated-transport systems are examined, it becomes apparent that there are two general categories of carrier-mediated processes; these have come to be known as **facilitated diffusion** and **active transport**. Active transport requires energy, whereas facilitated diffusion does not.

Facilitated diffusion

Facilitated diffusion is an unfortunate term for this process of mediated transport since the mechanism has nothing to do with diffusion. The term arose because facilitated diffusion and diffusion have some characteristics in common: In both systems there is no direct use of metabolic energy other than thermal energy, the net flux across a membrane always proceeds from higher to lower concentration, and when equilibrium is reached the concentrations on the two sides of a membrane are equal. However, when the transported-solute concentration is increased, facilitated-diffusion systems reach a maximal flux because their carriers become saturated, whereas fluxes due to diffusion do not saturate (Fig. 6-13).

When a facilitated-diffusion carrier changes its shape, so that the accessibility of its binding site changes from one side of the membrane to the other, the binding properties of the site do not change. The flux from side 1 to side 2 of a membrane depends on the number of carrier binding sites accessible on side 1 that are occupied by solute, which in turn depends on the concentration of the transported solute in the medium on side 1. Movement in the opposite direction, from side 2 to 1, depends on the concentration on side 2. Since the binding characteristics are the same in the two carrier conformations, when the concentration of transported solute is the same on both sides of the membrane, equal numbers of molecules will be bound at both surfaces and the net flux will be zero. When the concentrations are not equal, the net flux will always proceed from higher to lower concentration, since there will be more bound solute on

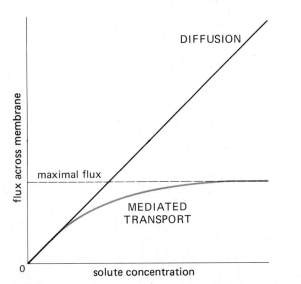

FIGURE 6-12. The net flux of molecules diffusing across a plasma membrane increases in proportion to the extracellular concentration, whereas the net flux of molecules entering by mediated-transport systems reaches a maximal value that does not increase with further increases in concentration. This maximal mediated-transport flux corresponds to the saturation of all the available carrier molecules by the transported solute.

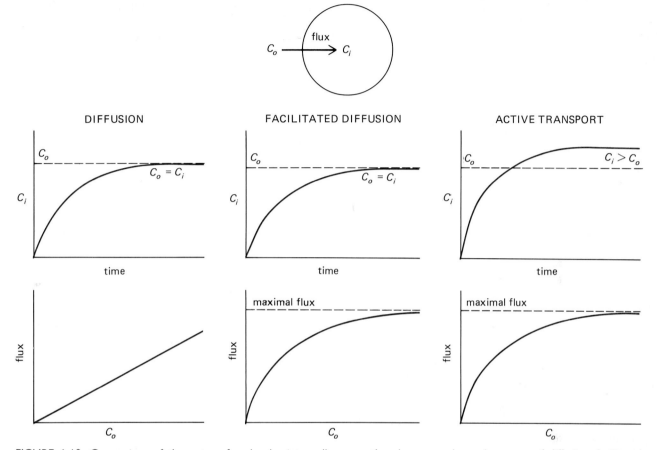

FIGURE 6-13. Comparison of the entry of molecules into cells across the plasma membrane by means of diffusion, facilitated diffusion, and active transport.

the high-concentration side than on the low-concentration side (provided the system is not saturated).

One of the most important facilitated-diffusion systems in the body is the one that moves glucose across most plasma membranes. Without glucose carriers in the plasma membrane, cells would be impermeable to glucose (a relatively large, polar molecule), and one of the most important nutrients would be unavailable to cells. One might expect that as a result of facilitated diffusion, the glucose concentration inside of cells would become equal to the extracellular concentration of glucose. This does not occur in most cells because glucose is metabolized almost as quickly as it enters the cell.

Thus, the intracellular glucose concentration is lower than extracellular concentration (Table 6-1), and there is a continuous net flux of glucose into most cells.

Active transport

Active transport differs from facilitated diffusion in its use of energy to make possible the movement of substances "uphill," i.e., from a region of lower concentration to one of higher concentration across a membrane (Fig. 6-14). Recall that in the case of ions, the electrical gradient as well as the concentration gradient influences fluxes across membranes. Thus, the active transport of ions involves

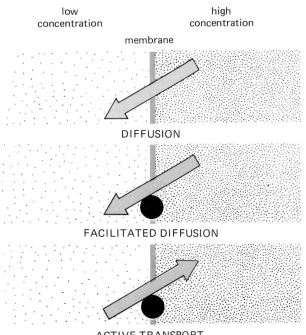

FIGURE 6-14. Direction of net flux for solute crossing membranes by (1) diffusion—high to low concentration, (2) facilitated diffusion—high to low concentration, and (3) active transport—low to high concentration.

the movement of ions against an electrochemical gradient. However, to simplify our discussion we shall only refer to the concentration gradient, by implication including the electrical gradient when ions are involved. Like facilitated diffusion, active-transport systems make use of protein carrier molecules in the membrane, exhibit chemical specificity, and reach a maximal flux when their binding sites are saturated. Because active-transport processes are able to move substances from lower to higher concentrations, up a concentration gradient, these carrier processes are often referred to as "pumps" in analogy with energy-requiring mechanical pumps.

The basic model for carrier-mediated active transport is the same as that for facilitated diffusion—a protein carrier molecule, with a binding site for a specific substance, that is able to change its shape so that the binding site is accessible first to one side of a membrane and then to the other.

However, we noted that in the case of facilitated diffusion the characteristics of the binding sites were the same when the binding site was exposed on either side of the membrane. In contrast, active-transport carriers have binding sites that differ in affinity when exposed on opposite sides of the membrane, and it is this difference in affinity which is responsible for the ability of the carrier to move substances from low to high concentration.

Figure 6-15 shows a membrane containing active-transport carriers for a particular solute, separating two solutions of equal solute concentration. If the binding sites exposed to side 1 have a greater affinity for the transported solute than do the sites on side 2, then the amount of solute bound on side 1 will be greater, and the flux from 1 to 2 will be greater than the flux from 2 to 1. The result is a net

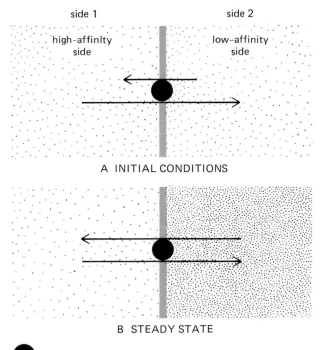

active transport carrier protein

FIGURE 6-15. Under initial conditions (A) when the concentrations on the two side of a membrane are equal, there is a net flux from side 1 to side 2 through an active-transport carrier from the high-affinity side of the membrane to the low-affinity side of the membrane. In the steady state (B) there is no net flux through the active-transport carrier.

flux of solute from 1 to 2 even though there is no concentration gradient across the membrane (Fig. 6-15A). In time, this net flux will increase the concentration of the transported substance on side 2 while decreasing the concentration on side 1. Note that molecules are moving in both directions through the active transport carrier, but the net flux is from lower to higher concentration. Eventually a concentration will be reached on side 2 which will produce an occupancy of the low-affinity carrier sites on side 2 equal to the occupancy of the high-affinity sites on the opposite side of the membrane. At this point a steady state is reached (Fig. 6-15B) in which the net flux through the active-transport carrier is zero.

The direction in which "uphill" transport takes place depends on which side of the membrane has the low-affinity binding sites and which side has the high-affinity binding sites, net "uphill" movement being directed towards the low-affinity side of the membrane. Depending on the arrangement of the carriers, active transport can be directed toward either side of a membrane. Some transport systems pump molecules into cells, whereas others pump them out; moreover, the same principle is true for the active-transport processes that exist across the membranes of cell organelles.

The role of energy in this process is to maintain a difference in the affinity of the carrier binding sites on the two sides of the membrane. If the carrier is deprived of its source of energy, no difference in carrier affinities can be produced and the carrier will take on the characteristics of a facilitated-diffusion system. Two means of linking energy to the active transport process are known: (1) the direct use of ATP in **primary active transport** and (2) the use of a concentration gradient for sodium (or other ions) across a membrane to supply the energy in a process known as **secondary active transport**.

Primary active transport. Chemical energy is transferred directly from ATP to carrier proteins during primary active transport. The carrier protein acts as an enzyme, an ATPase, which catalyzes the phosphorylation of a site on the carrier molecule itself (Fig. 6-16). Phosphorylation of the carrier al-

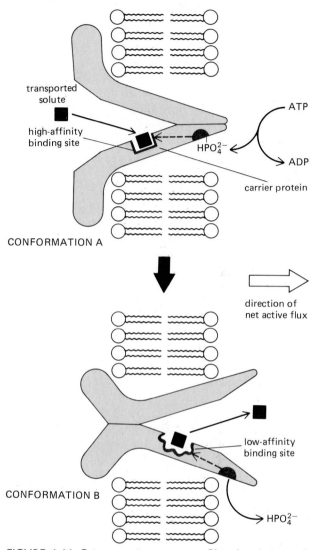

FIGURE 6-16. Primary active transport. Phosphorylation and dephosphorylation of the carrier alters the affinity of the transport binding site exposed on the two sides of the membrane.

ters the affinity of the binding site, either increasing or decreasing it, depending on the specific transport system. (Note that this is another example of a covalent modulation of a protein, as discussed in Chap. 4.) When the carrier protein changes its shape so that the binding site for the transported solute is now exposed to the opposite side of the membrane, the covalent bond between carrier and phosphate is broken, and the phosphate group is released into the intracellular fluid. The loss of the phosphate group changes the affinity of the transport-binding site. Thus, energy is used in the cycle of phosphorylation and dephosphorylation of the carrier, producing changes in the affinity of the binding site for the transported solute on opposite sides of the membrane. Metabolic inhibitors, which block the synthesis of ATP, also inhibit the primary active transport of solutes because they deprive the carrier of its source of energy.

The only primary active-transport carriers that have been clearly identified transport ions across membranes. (We have already established that ions can diffuse through protein channels.) The presence of both ion channels and active-transport carriers for ions in the same membrane results in "pump-leak" situations in which the pump moves ions up a concentration gradient while they leak (diffuse) down the concentration gradient through ion channels. (Recall that we are using the term "concentration gradient" broadly, to include the electric forces acting on ions.) The consequences of this pump-leak relation for cell electrical activity will be described in Chap. 8.

Three primary active transport carriers have been identified, one that pumps sodium and potassium ions (Na,K-pump), one that pumps calcium ions (Ca-pump), and a third that pumps hydrogen ions (H-pump). These carrier proteins all hydrolyze ATP in the process of phosphorylating the carrier protein, and hence these carrier proteins have become known as the Na,K-ATPase, the Ca-ATPase and the H-ATPase.

The Na,K-ATPase carrier is present in all plasma membranes; it is the pumping activity of this transport system that leads to the characteristic distribution of high intracellular potassium and low intracellular sodium relative to their respective extracellular concentrations (Table 6-1). This transport system pumps potassium ions into cells and sodium ions out of cells, resulting in an intracellular concentration of potassium that is about 40 times greater than that of extracellular potassium, and an extracellular sodium concentration that is about 10

times greater than intracellular sodium in most cells. This carrier moves three sodium ions out of a cell and two potassium ions into a cell for each molecule of ATP that is hydrolyzed. Thus, this carrier has multiple sodium and potassium ion-binding sites.

A Ca-ATPase carrier is located in the plasma membrane and in several organelle membranes, including the membranes of the endoplasmic reticulum and the mitochondria. In the plasma membrane, the direction of calcium transport is from intracellular to extracellular fluid, with the result that the cytosol of most cells has a very low calcium concentration (about 10^{-7} mol/L compared with an extracellular calcium of 10^{-3} mol/L, which is about 10,000 times greater). As will be discussed in Chap. 7, an increase in the cytosolic concentration of calcium ions is a major chemical signal that regulates a number of cell functions. There are two sources of this increased cytosolic calcium—extracellular fluid and cell organelles. In both cases, events which lead to an opening of specific calcium channels allow calcium ions to diffuse from high to low concentration into the cytosol (Fig. 6-17). The response to the increased cytosolic calcium is terminated as the Ca-ATPase lowers the cytosolic calcium by pumping it out of the cell or into the lumen of a cell organelle.

The H-ATPase carrier is located in the plasma membrane and in several organelle membranes, including the inner mitochondrial membrane. In the plasma membrane the H-pump transports hydrogen ions out of cells. It is this pump that is associated with the secretion of hydrochloric acid by the stomach, as will be discussed in Chap. 14.

If the splitting of ATP provides the energy for creating an ion-concentration gradient, could an ion-concentration gradient provide the energy for the synthesis of ATP? The answer is *yes,* as has been shown experimentally for each of the carrier ATPases. When ions flow through the pump from high to low concentration, ATP is synthesized. In the case of the mitochondrial H-ATPase, this is the normal mode of its operation. Hydrogen is not pumped across the mitochondrial membranes; in-

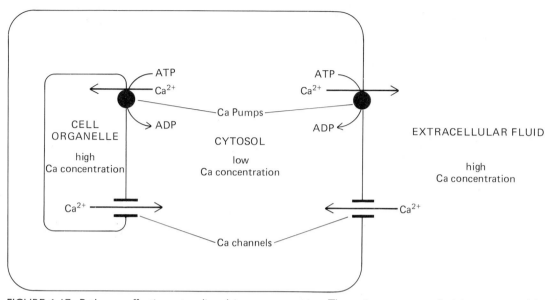

FIGURE 6-17. Pathways affecting cytosolic calcium concentration. The active transport of calcium moves calcium ions out of the cytosol, whereas calcium channels allow calcium to diffuse into the cytosol from the extracellular fluid or from cell organelles.

stead, a concentration gradient of hydrogen ions produced by the electron-transport chain in the inner mitochondrial membrane is used to provide the energy for ATP synthesis as the hydrogen ions move from high to low concentration through the H-ATPase. This is, very briefly, the mechanism by which oxidative phosphorylation forms ATP from ADP and inorganic phosphate.

Secondary active transport. The energy for secondary active transport is obtained from an ion-concentration gradient that is established across membranes by one of the primary pumps just described. The sodium gradient produced by the Na,K-ATPase is the one most commonly used, and our discussion will be restricted to it. A concentration gradient is a source of energy, and energy can be transferred to other systems provided the appropriate coupling device, such as a protein carrier molecule, is available. This system is analogous to a waterfall in which the fall of water from higher to lower levels can transfer energy to a waterwheel, which can then be used to perform work. In secondary active transport the energy released during the movement of sodium from higher to lower concentrations is transferred to the simultaneous movement of another solute from lower to higher concentration.

The carrier proteins for secondary active transport have at least two binding sites, one for sodium ions and one for the transported solute. The binding of sodium at the sodium site alters the shape of the transport site and thus alters the affinity of this site for the other transported solute (Fig. 6-18). (This is another example of allosteric modulation of proteins.) As a result of the pumping of sodium out of cells by the Na,K-ATPase, the extracellular concentration of sodium is much greater than the intracellular concentration. Therefore, the degrees to which the sodium sites are occupied on the two sides of the membrane are different: More sodium is bound to the carriers oriented toward the outside of the cell (which has a higher sodium concentration) than is bound to the sodium sites exposed to the lower intracellular sodium concentration. As a result of this difference in sodium binding on the

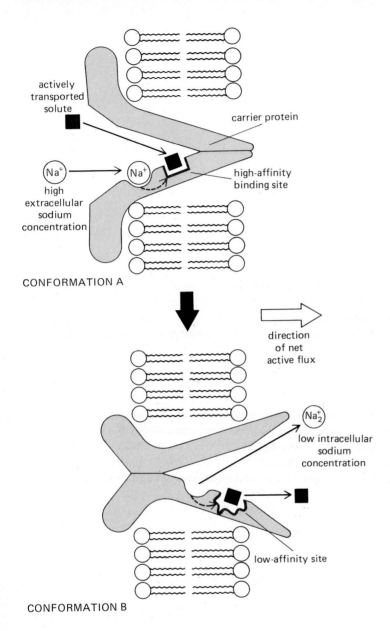

actively transported solute

carrier protein

Na⁺

high-affinity binding site

high extracellular sodium concentration

Na⁺

CONFORMATION A

direction of net active flux

Na_2^+

low intracellular sodium concentration

low-affinity site

CONFORMATION B

FIGURE 6-18. Secondary active transport. Differences in the binding of sodium to the carrier on the two sides of the membrane due to the differences in intracellular and extracellular sodium concentrations (the sodium gradient) alter the affinity of the binding site for the other transported solute.

two sides, the affinities of the binding sites for the other transported solute will also differ. It is this difference in affinity that produces active transport of the other solute from low to high concentration.

A net movement of sodium by a secondary-transport carrier is always from high extracellular concentration into the cell, where the concentration of sodium is low. Thus, this movement of sodium is "downhill" and there is not an active transport of sodium ions across the membrane. Net movements of the other transported solute from lower to higher concentration can be either in the same di-

EXTRACELLULAR FLUID   INTRACELLULAR FLUID

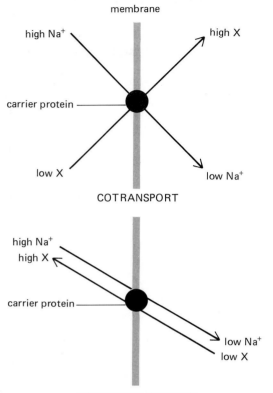

FIGURE 6-19. Direction of transported solute movement relative to sodium movement in a secondary active-transport system. Both move in the same direction through cotransport carriers but in opposite directions through countertransport carriers.

rection as the net sodium flux (cotransport) or in the opposite direction (countertransport), depending on the type of carrier (Fig. 6-19).

To reiterate, the energy for secondary active transport of solute is directly transferred to the carrier system by the sodium gradient. However, recall that the energy in the sodium gradient comes from metabolism by way of the ATP used to provide the energy for the pumping of sodium out of the cell by the Na,K-pump. Thus, the creation of the sodium concentration gradient across the plasma membrane is a means of ''storing'' energy which can then be used to drive secondary active-transport

pumps. Thus, it is not too surprising to find that between 10 and 40 percent of the ATP produced by a cell, under resting conditions, is used by the Na,K-pump to maintain the sodium gradient. Although secondary active transport of solutes is not *directly* linked to ATP, inhibition of ATP production will eventually lead to its inhibition, since the sodium concentration gradient cannot be maintained if sodium ions entering the cell (through protein channels and by the secondary active-transport carriers) are not being pumped out.

There are, thus, several routes by which sodium can cross membranes—by diffusion through protein channels from high to low concentration, by primary active transport from low to high concentration, and from high to low concentration on the carriers of secondary active-transport systems. Note that sodium enters cells during secondary active transport and must be pumped out by the Na,K-pump if the sodium gradient is to be maintained.

A variety of organic molecules and a few ions are moved across membranes by sodium-driven secondary active transport. In most cells, amino acids are actively transported from extracellular to intracellular fluid at the expense of the sodium gradient, attaining intracellular concentrations 2 to 20 times higher than in the extracellular fluid. As we shall see, the active absorption of organic molecules from the intestine and across the walls of the renal tubules is linked to the sodium gradient.

In addition to the previously described primary active transport of calcium from cytosol to extracellular fluid there also exists in many membranes an Na-Ca countertransport carrier which uses the movement of sodium ions into the cell to pump calcium ions out. In this system, an increase in the intracellular sodium concentration will decrease the sodium gradient across the plasma membrane and, hence, decrease the pumping of calcium out of the cell. This results in a rise in intracellular calcium which can lead to an alteration in many cell activities. What we see here are the cause-and-effect links by which changes in sodium pump activity, for example, may produce widespread effects on cell functions.

Table 6-2 provides a summary of the major characteristics of the different pathways by which substances move through cell membranes.

Osmosis

Water, which makes up the bulk of fluid inside and outside of cells, is a small, polar molecule about 0.3 nm in diameter that diffuses across most cell membranes very rapidly. One might expect that, because of its polar structure, water would not penetrate through the lipid bilayer of membranes. However, artificial phospholipid bilayers also have a high permeability to water, which suggests that this small molecule may be able to diffuse through the lipid regions of membranes. Water, because of its small size, can also pass through some protein channels. There is no evidence to suggest that water crosses membranes by way of mediated-transport systems.

Osmosis is the net movement of water from regions of high to low water concentration. Thus, osmosis is simply a special case of diffusion—the diffusion of water. As with any diffusion process, there must be a water concentration gradient across a membrane in order to produce a net water flux.

How can a difference in water concentration be established across a membrane? One usually thinks of the concentration of a substance in a so-

TABLE 6-2. Major characteristics of pathways by which substances cross membranes

	Diffusion		Mediated transport		
	Through lipid bilayer	Through protein channel*	Facilitated diffusion	Primary active transport	Secondary active transport
Direction of net flux	High to low concentration	High to low concentration	High to low concentration	Low to high concentration	Low to high concentration
Equilibrium or steady state	$C_o = C_i$	$C_o = C_i$	$C_o = C_i$	$C_o \neq C_i$	$C_o \neq C_i$
Use of integral membrane protein	No	Yes	Yes	Yes	Yes
Maximal flux at high concentration (saturation)	No	No	Yes	Yes	Yes
Chemical specificity	No	Yes	Yes	Yes	Yes
Use of energy and source	No	No	No	Yes; ATP	Yes; ion gradient (usually Na)
Typical molecules using pathway	Nonpolar: O_2, CO_2, fatty acids	Ions: Na^+, K^+, Ca^{2+}	Polar: glucose	Ions: Na^+, K^+, Ca^{2+}	Polar: amino acids, glucose, some ions

*Some of the characteristics, e.g. chemical specificity, of diffusion through protein channels resemble those of facilitated diffusion, and some classifications include protein channels under the heading of facilitated diffusion.

lution as referring to the amount of solute dissolved in a given volume of solvent; e.g., a one molar (1 *M*) solution of glucose contains one mole of glucose dissolved in enough water to form one liter of solution. However, the water in this same solution also has a concentration. A liter of pure water weighs about 1000 g. The molecular weight of water is 18; thus, the concentration of pure water is 1000/18 = 55.5 mol/L. However, if a solute molecule such as glucose is dissolved in water, the concentration of water in the resulting solution is less than that of pure water. If glucose molecules were similar in size and polarity to water, then the decrease in water concentration would be equal to the added glucose concentration, since each glucose molecule would now occupy a space formerly occupied by a water molecule (Fig. 6-20). Thus, the water concentration in a 1 *M* glucose solution would be 54.5 *M* rather than 55.5 *M,* the concentration in pure water. (Because molecules have different sizes and may contain charged or polar groups which interact with the polar water molecules, there is not an exact one-to-one displacement of one water molecule for each solute molecule in solution, but for most solutes this generalization is approximately correct.) The way to lower water concentration is, thus, to increase solute concentration; the greater the solute concentration, the lower the water concentration of a solution.

It is essential to recognize that the degree to which the concentration of water is decreased by the addition of solute depends upon the number of particles (molecules or ions) of solute in solution (the solute concentration) and not upon the chemical nature of the solute, because one particle of solute will approximately displace one molecule of water. For example, one mole of glucose added to one liter of water decreases the water concentration to approximately the same extent as does one mole of an amino acid, or one mole of urea, or one mole of any molecule that exists as a single particle in solution. A molecule which ionizes in solution, such as sodium chloride, decreases the water concentration in proportion to the number of ions formed. Hence, one mole of sodium chloride gives rise to one mole of sodium ions and one mole of chloride ions to produce a total of two moles of solute particles, which lowers the water concentration twice as much as does one mole of glucose. By the same reasoning, adding one mole of $MgCl_2$ will lower the water concentration by approximately 3 mol/L.

Since the concentration of water depends upon the number of solute particles in solution rather than their chemical properties, it is useful to have a concentration term which refers to the total concentration of all solute particles in a solution, regardless of their chemical composition. The total solute concentration of a solution is known as its **osmolarity**. One osmole is equal to one mole of solute particles. Thus, a 1 *M* solution of glucose has a concentration of 1 osmol/L, but a 1 *M* solution of sodium chloride contains 2 osmoles of particles per liter of solution. A liter of solution containing 1 mol of glucose and 1 mol of sodium chloride has an osmolarity of 3 osmol/L since it contains 3 mol of solute particles. The concentration of water in any two solutions having the same osmolarity is the same, since the total number of solute particles per unit volume is the same. A solution with an osmolarity of 3 osmol/L may contain 1 mol of glucose and 1 mol of sodium chloride, or 3 mol of glucose, or 1.5 mol of sodium chloride, or any other combination of solutes as long as the total solute concentration is equal to 3 osmol/L. It is important to realize that although osmolarity refers to the concentration of solute particles in solution, it is also a measure of the *water concentration* in the solution, since the *higher* the osmolarity of a solution, the *lower* its water concentration.

Figure 6-21 shows two compartments separated by a membrane that is permeable to both solute and water. The concentration of solute (4

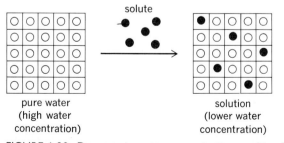

pure water
(high water
concentration)

solute

solution
(lower water
concentration)

FIGURE 6-20. Decrease in water concentration resulting from the addition of solute molecules to pure water.

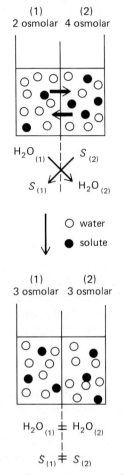

(1) (2)
2 osmolar 4 osmolar

$H_2O_{(1)}$ $S_{(2)}$

$S_{(1)}$ $H_2O_{(2)}$

○ water

● solute

(1) (2)
3 osmolar 3 osmolar

$H_2O_{(1)} \pm H_2O_{(2)}$

$S_{(1)} \mp S_{(2)}$

FIGURE 6-21. Changes in water concentration, solute concentration, and compartment volume when both water and solute are able to diffuse across a membrane separating two compartments that initially contained different concentrations of solute (osmolar = osmol/L).

osmol/L) in compartment 2 is higher than the concentration of solute (2 osmol/L) in compartment 1. This difference in solute concentration also means there is a difference in water concentration across the membrane, with the water in compartment 1 having a higher concentration than the water in compartment 2. Therefore, water undergoes a net diffusion from compartment 1 to 2, while solute is diffusing in the opposite direction, from 2 to 1. The net result is the redistribution of water and solute between the two compartments so that the final

concentration of solute in both is 3 osmol/L. Note that the final volumes of the two compartments remain unchanged, since some of the water and solute molecules have merely exchanged positions across the membrane, although transitory changes in the volume of the two compartments may occur because of difference in the rates at which solute and water diffuse.

If we make one alteration in this system, namely, replacing the membrane with one permeable to water molecules but impermeable to the solute, the final result is different (Fig. 6-22). Initially, the same concentration gradients are found across the membrane as in the previous example. Water diffuses from its high concentration in compartment 1 into compartment 2, but since there can be no movement of solute in the opposite direction, the volume of compartment 2 increases. Solute molecules cannot leave compartment 1 and the loss of water from the compartment decreases its volume, which has the effect of increasing the concentration of solute there, while the water entering compartment 2 dilutes the solute concentration on that side of the membrane. This system also comes to equilibrium when the concentrations of water and solute in the two compartments become equal, but in this case, the equilibrium is achieved by the transfer of water alone and leads to a change in the final volume of the two compartments. This example illustrates the following generalization: In order for a volume change to occur, the movement of solute molecules must be restricted by a semipermeable membrane.

We have treated the two compartments as if they were infinitely expandable, so that the net transfer of water does not create a difference in pressure across the membrane; this is essentially the situation that occurs across plasma membranes. In contrast, if the walls of compartment 2 could not expand, the movement of water into compartment 2 would raise the pressure in this compartment, which would oppose further water entry. The pressure that must be applied to one side of a membrane to prevent the osmotic flow of water across the membrane is often used as a measure of the solute concentration of a solution. Such a pressure is

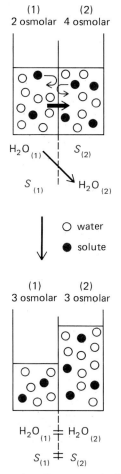

FIGURE 6-22. Changes in water concentration, solute concentration, and compartment volume when only water is able to diffuse across a membrane separating two compartments that initially contained different concentrations of solute. In this example the two compartments are treated as if they were infinitely expandable and thus there is no pressure gradient created across the membrane.

known as the **osmotic pressure** of a solution, and it is defined as the pressure that would have to be applied to a solution to *prevent* water movement into the solution from a compartment of pure water. The greater the osmotic pressure of a solution, the lower is the water concentration of that solution; the greater the osmotic flow between the solution and pure water, the greater is the pressure required to prevent the osmotic flow. Therefore, a solution

with a low water concentration (high solute concentration) is said to have a high osmotic pressure. It is very important to note that osmotic pressure does not actually exist in a solution; it is not a real pressure that causes water to flow across membranes during osmosis. Osmotic pressure is simply used as a measure of the water flow which would be produced by osmosis from a compartment of pure water across a semipermeable membrane into the solution in question.

The osmotic behavior of cells is similar to that described above. Most cells are surrounded by a plasma membrane that is permeable to water, and the intracellular and extracellular solutions contain many molecules which are unable to cross the membrane. The permeability constant of the plasma membrane to water is very large. Any change in the water concentration surrounding a cell will lead to a rapid flow of water across the membrane until a new state of osmotic equilibrium is reached at which the water concentration is again equal on both sides of the membrane. Since the resulting changes in cell volume are completed very rapidly, for all practical purposes the concentration of water in the intracellular fluid may be assumed to be equal to the concentration of water in the extracellular fluid.

When a cell is in osmotic equilibrium, the water concentration is the same on the two sides of the membrane. Thus, the total solute (osmolar) concentrations on the two sides of the membrane must also be the same. If a solute cannot cross the membrane, any change in its concentration produces a change in water concentration, which can lead to a net flux of water into or out of the cell and a change in cell volume. Thus, the volume of a cell ultimately depends upon the concentration of nonpenetrating solutes on the two sides of the plasma membrane.

About 85 percent of the solutes in the extracellular fluid are sodium and chloride ions, which can diffuse ("leak") through protein channels in the plasma membrane. However, as we have seen, the membrane contains an active-transport system for pumping sodium ions out of the cell, resulting in a pump-leak situation in which there is no net movement of sodium into or out of the cell. Changes in extracellular sodium concentration change the leak flux but also change the pump flux of sodium, so that there is very little change in the intracellular sodium concentration with changes in extracellular sodium. The restriction on net sodium movement also restricts the net movement of chloride ions because of the electrical attraction between opposite electric charges. (This concept will be developed more formally in Chap. 8.) Thus, sodium and chloride ions behave as if they were nonpenetrating solutes, and changes in their extracellular concentrations will alter the concentration of extracellular water and lead to changes in cell volume.

Inside the cell the major solute particles are potassium ions and a number of organic solutes. Most of the latter are unable to cross the membrane. Although potassium ions can leak out of the cell, they are actively transported back by the same pump that pumps out sodium, the Na,K-pump. The net effect, as for sodium, is that potassium behaves as if it cannot cross the membrane. Thus, sodium chloride outside the cell and potassium and organic solutes inside the cell represent the major effective nonpenetrating solutes determining the water concentrations on the two sides of the membrane. In contrast, solutes which are not actively transported across the membrane but which can cross by diffusion have no net effect upon cell volume since they eventually reach the same concentration on both sides of the membrane and, thus, do not produce any difference in water concentration across the membrane. It is only the nonpenetrating or effectively nonpenetrating solutes (Na^+, K^+, Cl^-) which can alter cell volume by producing a change in the water concentration across the membrane.

Let us take some examples. Red blood cells placed in solutions containing different concentrations of sodium chloride change their volume in each solution according to the water concentration inside relative to the concentration outside the cell (Fig. 6-23). If the outside water concentration is greater than the intracellular water concentration, water will move into the cell, causing it to swell. A cell will shrink if the concentration of water outside is less than it is inside. By comparing the volume of a cell as it exists in the blood with its volume in

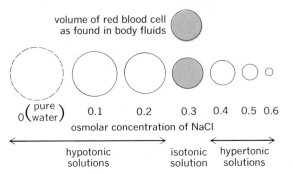

FIGURE 6-23. Comparison of the volumes of red blood cells in solutions of different sodium chloride concentrations with the volume of a red blood cell surrounded by body fluids.

each of several sodium chloride solutions, that sodium chloride solution in which the cell volume remains the same as it is when surrounded by extracellular body fluids can be determined. This solution must have the same number of nonpenetrating solute particles as the intracellular fluid in red cells. Such a solution is said to be **isotonic**. An isotonic solution is defined as any solution in which the cell has the same volume that it has when surrounded by normal extracellular body fluids. A 0.15 M solution of sodium chloride is found to be isotonic for the human red blood cell; its osmolar concentration is 0.3 osmol/L. Any other solution containing 0.3 osmol/L of *nonpenetrating* solutes will be isotonic for the same cell. Solutions in which cells swell, because the solutions contain less than 0.3 osmol/L of nonpenetrating solute particles, are called **hypotonic**. Solutions in which cells shrink, because the solutions contain more than 0.3 osmol/L of nonpenetrating solute particles, are called **hypertonic** solutions (Fig. 6-23).

Endocytosis and Exocytosis

Earlier in this chapter we described the various pathways by which molecules pass through cell membranes. There is an additional pathway which does not require the molecule to cross the structural matrix of the membrane in order to enter or leave a cell. When living cells are observed under a light microscope, small regions of the plasma membrane can be seen to invaginate into the cell and pinch off, forming small intracellular membrane-bound vesicles which enclose a small volume of extracellular fluid; this process is known as **endocytosis** (Fig. 6-24). A similar process in the reverse direction, known as **exocytosis**, occurs when membrane-bound vesicles in the cytoplasm fuse with the plasma membrane and release their contents to the outside of the cell.

Endocytosis

Several varieties of endocytosis can be identified. When the vesicle encloses a small volume of extracellular fluid, the process is known as **fluid endocytosis**. The composition of the vesicle contents is the same as that of the extracellular fluid. In other cases, specific molecules in the extracellular fluid bind to sites on the plasma membrane and are carried into the cell when the membrane invaginates; this is known as **adsorptive endocytosis**. In addition to taking in trapped extracellular fluid, adsorptive endocytosis leads to the selective concentration in the vesicle of the material bound to the membrane. Both fluid and adsorptive endocytosis are often referred to as **pinocytosis** (cell drinking). A third type of endocytosis occurs when large multimolecular particles, such as bacteria and debris from damaged tissues, are engulfed by the plasma membrane and enter certain types of cells. In this case, the membrane folds around the surface of the particle so

that little extracellular fluid is enclosed within the vesicle; this form of endocytosis is known as **phagocytosis** (cell eating). While most cells undergo pinocytosis, only a few special cells carry out phagocytosis. The properties of these phagocytic cells will be further discussed when we describe the defense mechanisms of the body in Chap. 17.

In some cells, fluid endocytosis occurs continuously, whereas adsorptive endocytosis and phagocytosis are stimulated by the binding of specific molecules or particles to sites on the cell surface; this binding appears to trigger the endocytosis locally, since adjacent regions of the membrane which lack bound molecules or particles do not increase their rate of endocytosis. The process of endocytosis requires metabolic energy in the form of ATP, and contractile proteins associated with the plasma membrane have been implicated in this process although the molecular mechanisms governing the membrane movements during endocytosis are unknown.

What is the fate of the endocytotic vesicles once they have entered the cell? In some cases, as in the capillary endothelium (which forms the walls of the smallest blood vessels), the vesicles pass through a thin layer of cytoplasm and fuse with the opposite plasma membrane, releasing their contents to the extracellular space on the opposite side of the cell by exocytosis (Fig. 6-25). This provides a pathway for transferring large molecules, such as proteins, from the blood to the interstitial fluid and vice versa. A similar process allows small amounts of protein to be absorbed across the intestinal epithelium. However, in most cells, endocytotic vesicles fuse with the membranes of lysosomes (Fig. 6-25), cell organelles containing digestive enzymes which break down large molecules such as proteins, polysaccharides, and nucleic acids. The fusion of the endocytotic vesicle with the lysosomal membrane exposes the contents of the vesicle to these digestive enzymes. The small molecular products of digestion, such as amino acids and sugars, then cross the membrane of the lysosome and enter the cytosol. The endocytosis of bacteria and their destruction in the lysosomes is one of the body's major defense mechanisms against germs.

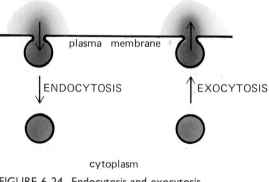

FIGURE 6-24. Endocytosis and exocytosis.

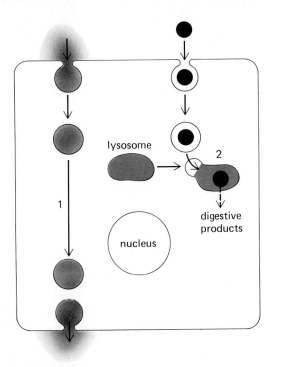

FIGURE 6-25. Fate of endocytotic vesicles. Pathway I transfers extracellular materials from one side of the cell to the other. Pathway 2 leads to fusion with lysosomes and the digestion of the vesicle contents.

The process of endocytosis removes a small portion of the membrane from the surface of the cell as this portion of the membrane is internalized as the membrane of the endocytotic vesicle. If this membrane were not replaced, the surface area of the cell would decrease. In cells which have a great deal of endocytotic activity, more than 100 percent of the plasma membrane may be internalized in an hour, yet the cell volume and membrane surface area remain constant. This is because the membrane that is removed is replaced at about the same rate by vesicle membrane that fuses with the plasma membrane during exocytotic processes. Thus, the plasma membrane is constantly recycled in these cells.

Exocytosis

Exocytosis performs two functions for cells: (1) It provides a way to replace portions of the plasma membrane that have been removed by endocytosis and in the process to replace or add new components, such as carrier proteins or specific binding sites for chemical messengers; and (2) it provides a route by which certain types of molecules that are synthesized by cells can be released into the extracellular fluid. Many of the specialized end products of metabolism secreted by cells into the extracellular fluid are unable to cross plasma membranes because they are large polar molecules which do not undergo mediated transport (proteins, for example). By being enclosed in vesicles, they can be released from the cell by exocytosis.

The secretion of substances by exocytosis is triggered in most cells by specific stimuli and does not occur continuously. For example, nerve cells release their chemical messengers by the process of exocytosis in response to a nerve impulse, and various gland cells secrete proteins in response to specific neural or hormonal stimuli. As we shall see in subsequent chapters, these various stimuli cause an increase in cytosolic calcium concentration by opening calcium channels either in the plasma membrane or in the membranes of certain intracellular organelles. It is the resulting increase in the concentration of cytosolic calcium which triggers the fusion of intracellular vesicles with the plasma membrane. In the absence of a rise in cytosolic calcium the secretory vesicles do not undergo fusion with the plasma membrane. How calcium brings about membrane fusion is unknown. One hypothesis is that calcium may activate contractile proteins in the membrane and surrounding cytoplasm which act upon the vesicles to produce their fusion with the plasma membrane.

How are substances that are secreted by exocytosis packaged into vesicles? In some cases, where relatively small organic end products are sequestered in vesicles, the enzymes for their synthetic pathways are located in the vesicles. In other cases, the vesicle membrane has special mediated-transport systems, which concentrate the product within the vesicle after it has been synthesized in the cytosol. Very high concentrations of certain products can be achieved within vesicles by a combination of mediated transport followed by their specific binding to protein binding sites within the vesicle. When large quantities of the material to be secreted are stored in secretory vesicles, the product is immediately available for rapid secretion in response to a stimulus, without delays that might occur if the material had to be synthesized after the arrival of the stimulus.

Protein secretion. The packaging of proteins into vesicles creates special problems for the cell because of the complex enzymatic machinery associated with protein synthesis and the inability of these large molecules to pass through membranes. As described in Chap. 4, ribosomes are the sites at which amino acids are assembled into proteins.

Upon arriving in the cytoplasm from the nucleus, a molecule of messenger RNA binds to a free ribosome and begins the process of translating its message into a polypeptide chain. The first portion of the protein to be synthesized appears to determine whether the protein will be released into the cytosol to function as an intracellular protein or be secreted from the cell by exocytosis. This initial portion of the protein chain is known as the **signal sequence**. If there is no signal sequence at the beginning of the protein, the protein synthesis is completed on the free ribosome and the protein is released into the cytosol. If there is a specific signal sequence, however, it acts as a ligand which binds to a specific binding site on the surface of the granular endoplasmic reticulum. It is by this mechanism that ribosomes become attached to the reticulum. Once the messenger RNA, with its attached ribosome and growing protein chain, is bound to the reticulum, the polypeptide chain is inserted into the reticulum membrane, passes through the membrane as the protein is being synthesized, and accumulates in the lumen of the endoplasmic reticulum. The mechanism by which the growing polypeptide chain passes through the reticulum membrane is unknown; this is the only stage along its route to the outside of the cell at which the protein that is to be secreted passes through a membrane. When the synthesis is complete, the ribosome is released from the surface of the reticulum.

Within the lumen of the endoplasmic reticulum several events take place which modify the structure of the newly formed protein. Enzymes are present which remove the signal sequence; therefore, this portion of the protein, which was necessary for directing it along the pathway leading to its secretion, is not present in the secreted protein. Almost all proteins that are secreted are found to be glycoproteins, and some of the carbohydrate groups are added to these proteins by enzymes in the endoplasmic reticulum. In addition, in some proteins, certain of the amino acid side chains are modified by the addition of various groups during the time the protein is in the lumen of the reticulum.

Following the modifications in the protein structure, portions of the reticulum membrane then bud off, forming small vesicles containing the newly formed protein. These vesicles migrate to the region of the Golgi apparatus (Fig. 6-26) where they fuse with the Golgi membranes. Within the Golgi apparatus the protein undergoes still further modification. Some of the carbohydrates that were added in the granular endoplasmic reticulum are now removed and new groups added. These carbohydrate groups appear to function as labels that can be recognized when the protein encounters various binding sites, both during the remainder of its trip through the cell and after its secretion into the extracellular fluid. Following this further modification in the Golgi apparatus, the proteins are again packaged into vesicles that bud off the surface of the Golgi membranes. These secretory vesicles, as they are now called, move to the region of the plasma membrane where, upon receiving the appropriate signal (calcium ions), they are secreted from the cell by exocytosis.

Membrane Formation

The endoplasmic reticulum is the site of membrane synthesis, and the Golgi apparatus determines the distribution of the membranes to their various destinations.

In addition to synthesizing the proteins that will be secreted from cells or passed on to various intracellular organelles, the ribosomes bound to the

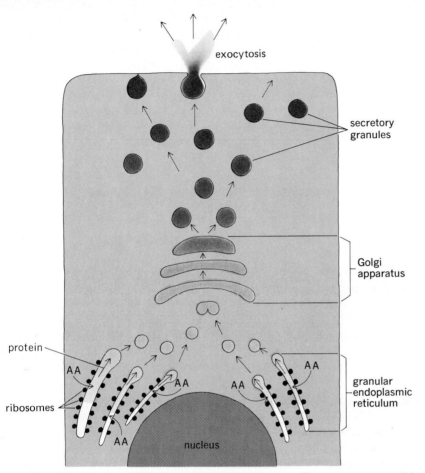

FIGURE 6-26. Pathway that leads to the release of protein from a secretory cell by exocytosis. (AA signifies free amino acids that are assembled into proteins on ribosomes.)

endoplasmic reticulum also synthesize the proteins that become integral membrane proteins. These proteins, rather than being released into the lumen of the reticulum, remain embedded in the membrane. These membrane proteins are then passed on to other membranes (either organelle membranes or the plasma membrane) as vesicle membranes migrate and fuse throughout the cell. The enzymes that synthesize phospholipids and cholesterol are also associated with the membranes of the smooth endoplasmic reticulum. Thus, both membrane lipids and membrane proteins are synthesized and incorporated into membranes in the endoplasmic reticulum.

Some of the vesicles that bud off the Golgi

apparatus fuse with other cell organelles rather than with the plasma membrane. It is by this means that some of the proteins reach the lumens of various cell organelles; an example is the digestive enzymes found in the lumen of the lysosomes. One of the important functions of the Golgi apparatus is to sort out the various proteins synthesized by the granular endoplasmic reticulum and package them in the appropriate vesicle that will direct them to their final destination. How this sorting and labeling of vesicles is performed by the Golgi apparatus is unknown.

Vesicle formation on the surface of one cell organelle, movement to another organelle, followed by fusion with the membranes of that organelle provides a major pathway for shuttling materials between organelles within a cell. The budding and fusing of vesicles are analogous to the processes of endocytosis and exocytosis.

Membrane Junctions

In addition to providing a barrier to the movements of molecules between the intracellular and extracellular fluids, plasma membranes are involved in interactions between cells to form organized tissues. Some cells, particularly those of the blood, do not associate with other cells but remain as independent cells suspended in a fluid (the blood plasma). Most cells, however, are packaged into tissues and organs and are not free to move around the body. But even the cells in an organized tissue are not packaged so tightly that the adjacent cell surfaces are in direct contact with each other. There usually exists a space of at least 20 nm between the opposing membranes of adjacent cells; this space is filled with extracellular fluid and provides the pathway for the diffusion of substances within the tissue.

The forces that bring about the organization of cells into tissues and organs are poorly understood but appear to depend, at least in part, on the ability of binding sites on the cell surface to recognize and associate with specific protein aggregation factors which bind cells together (Chap. 4).

Some tissues can be separated into individual cells by fairly gentle procedures such as shaking the tissues in a medium from which calcium ions have been removed. The absence of calcium ions decreases the ability of these cells to stick together, suggesting that the doubly charged Ca^{2+} may act as a link between cells by combining with negatively charged groups on adjacent cell surfaces. Many types of cells, however, require more drastic chemical procedures to separate them from their neighbors.

The electron microscope has revealed that a variety of cells are physically joined by specialized types of membrane junctions. One type, known as a **desmosome**, is illustrated in Fig. 6-27A. The desmosome consists of two opposed membranes that remain separated by about 20 nm but show a dense accumulation of matter at each membrane surface and between the two membranes. In addition, fibers extend from the inner surface of the desmosome into the cytoplasm and appear to be linked to other desmosomes of the cell. The function of the desmosome is to hold adjacent cells together in areas that are subject to considerable stretching, such as in the skin and heart muscle. Desmosomes are usually disk-shaped and thus could be likened to rivets or spot-welds as a means of linking cells together.

Another type of membrane junction, the **tight junction**, (Fig. 6-27B) is an actual fusing of the outer protein surfaces of two adjacent plasma membranes so that there is no space between adjacent cells in the region of the tight junction. This type of fusion extends around the circumference of the cell and greatly reduces (but does not completely eliminate) the extracellular route for the passage of molecules between the cells. Tight junctions are found between epithelial cells. Such epithelial layers usually separate two compartments having different chemical compositions; for example, the intestinal epithelium lies between the lumen of the intestinal tract, which contains the products of food digestion, and the blood vessels that pass beneath the epithelial layer. These epithelial layers generally mediate the passage of molecules between the two compartments. It was noted earlier that most cells are separated from each other by a space of at least 20 nm which is filled with extracellular fluid. However, such spaces would make a very leaky barrier and would allow even large protein molecules to diffuse between the two compartments by passing between adjacent cells. In fact, however, the intestinal epithelium in an adult is practically impermeable to protein, and the reason is the presence of tight junctions joining the epithelial cells near their luminal border. Small ions can pass through certain tight junctions but not others. Thus, some epithelial layers are leakier to ions than others. In order to cross an epithelial layer of cells, most molecules must first cross the plasma membrane of an epithelial cell, pass through the cytoplasm, and exit through the plasma membrane on the opposite side of the cell. Thus, the tight junction, in addition to helping to hold cells together, also seals the pas-

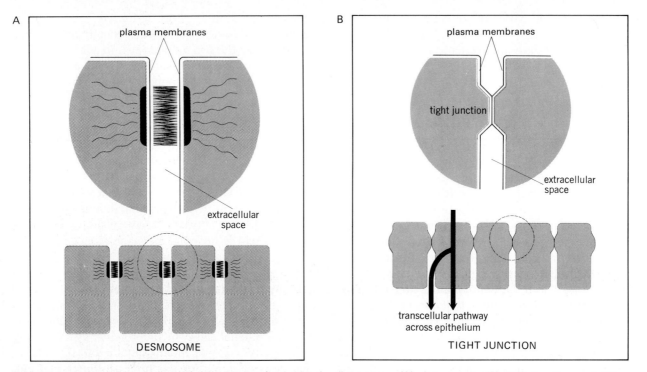

FIGURE 6-27. Schematic diagram of three types of specialized cell junctions: (A) desmosome, (B) tight junction, and (C) gap junction. (D) Electron micrograph of two intestinal epithelial cells joined by a tight junction near the luminal surface and a desmosome a short distance below the tight junction. [*Electron micrograph from M. Farquhar and G. E. Palade, J. Cell. Biol., **17**:375-412 (1963).*]

sageway between adjacent cells. Figure 6-27D shows both a tight junction and a desmosome located near the luminal border between the two epithelial cells.

There is yet another type of junction, known as a **gap junction** (Fig. 6-27C). When a molecule to which the plasma membrane is impermeable is injected into most cell types, through a micropipette inserted into the cell, it will remain within the cell and not pass into adjacent cells. When this experiment is performed on certain types of cells, however, the marker molecule is found in the cell adjacent to the injected cell but not in the extracellular medium, suggesting that there is a direct channel linking the cytoplasms of the two cells. In the region of the gap junction the two opposing plasma membranes come within 2 to 4 nm of each other. Small channels about 1.5 nm in diameter extend across this gap and directly link the cytoplasm of the two

cells. The small diameter of these channels limits the molecules that can pass between the connected cells to small molecules and ions, such as sodium and potassium, and excludes the exchange of large protein molecules. A variety of cell types possess gap junctions, including the muscle cells of the heart and smooth muscle cells. As we shall see, the gap junctions in these cells play a very important role in the transmission of electrical activity between adjacent muscle cells. In other cases, gap junctions are thought to coordinate the activities of adjacent cells by allowing chemical messengers to move from one cell to the next.

Epithelial Transport

Epithelial cells are located at the surfaces of hollow organs or tubes and regulate the movements of sub-

C

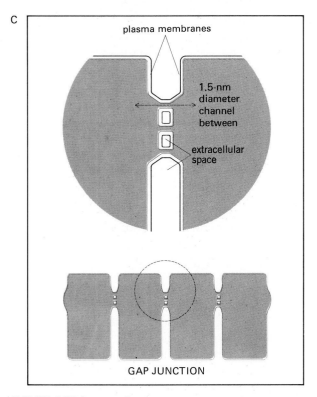

plasma membranes

1.5-nm diameter channel between

extracellular space

GAP JUNCTION

D

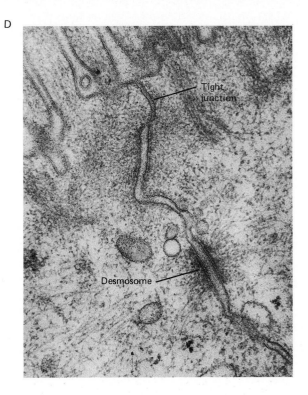

Tight junction

Desmosome

FIGURE 6-27 (continued)

stances across these surfaces. As we have just seen, epithelial cells are connected by tight junctions which restrict the diffusion of substances between the cells. Most substances that cross these epithelial cells must pass from an extracellular compartment on one side of the epithelial layer, through a plasma membrane into the cytosol, and then from the cytosol across a second plasma membrane on the other side of the epithelium into a second extracellular compartment. One surface of the epithelium generally faces a hollow fluid-filled chamber, and this side is referred to as the **luminal side** of the epithelium; the opposite surface, which is usually adjacent to an underlying network of blood vessels, is referred to as the **blood side**.

Movements of molecules through the plasma membranes at the luminal and blood surfaces of an epithelial cell proceed through the same types of pathways already described for movements across

membranes in general. However, the permeability characteristics of the luminal plasma membrane and the plasma membrane on the blood side of the cell are not the same; these membranes contain different channels for diffusion and different carrier proteins for mediated transport.

As a result of these differences in the two epithelial surfaces, it is possible for substances to undergo a net movement from a low concentration on one side of an epithelium to a higher concentration on the other side, or (in other words) to undergo active transport across the epithelial layer of cells. Examples of such epithelial active-transport processes are the absorption of materials from the gastrointestinal tract into the blood, the movements of substances between the kidney tubules and the blood during the formation of urine, and the secretion of salts and fluid by various glands.

Figure 6-28 illustrates two examples of epithe-

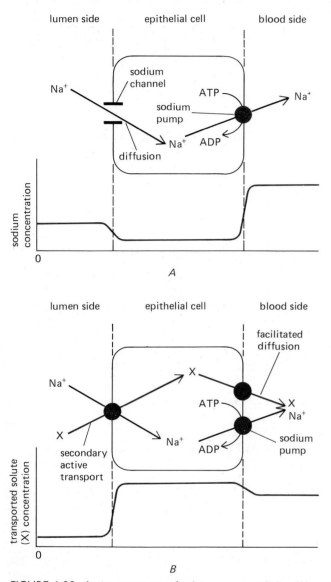

lumen side   epithelial cell   blood side

A

lumen side   epithelial cell   blood side

B

FIGURE 6-28. Active transport of solute across epithelia. (A) The transepithelial transport of sodium involves diffusion through channels at one surface followed by active transport across the opposite surface, out of the cell. (B) The transepithelial transport of most organic solutes involves their movement into a cell through a secondary active-transport pump that is coupled to sodium, followed by movement out of the cell, down a concentration gradient through a facilitated-diffusion carrier at the opposite surface. Shown below each part is the concentration profile of the transported solute across the epithelium.

lial active transport. Sodium is actively transported across most epithelia; net sodium movement from lower to higher concentration occurs from lumen to blood in absorptive processes and from blood to lumen during secretion. During absorption, the movement of sodium from a lumen into an epithelial cell occurs by diffusion through a sodium channel in the luminal membrane. Diffusion occurs into the cell because the intracellular concentration of sodium is kept low by the active transport of sodium out of the cell across the plasma membrane on the blood surface of the cell. All of the Na,K-pumps are located in the membrane on the blood side of the cell. The sodium concentration in the extracellular fluid on the blood side of the cells is much greater than in the cytosol and may even be greater than the sodium concentration on the luminal side. Note that even when the sodium concentrations are the same on the two sides of the epithelium, there will still be a net movement of sodium from lumen to blood (absorption of sodium).

A second example illustrates the absorption of organic molecules across an epithelial layer (Fig. 6-28B). In this case, entry of the organic molecule across the luminal plasma membrane occurs through a secondary active-transport carrier that is linked to the sodium gradient across the membrane. The absorbed substance moves from lower concentration in the luminal fluid to a higher concentration in the cell. Exit of the substance on the blood side occurs by a facilitated-diffusion carrier which moves the material from its high concentration in the cell to a lower concentration in the extracellular fluid on the blood side. The concentration of the substance on the blood side may be considerably higher than the concentration in the lumen since the blood-side concentration can approach equilibrium with the high intracellular concentration that was achieved by the secondary active-transport process in the luminal membrane.

Although water is not actively transported across cell membranes, a net movement of water across an epithelial layer can be achieved as a result of the active movements of solutes, especially sodium, across the epithelium. The active transport of sodium across an epithelial layer of cells results in a decrease in the sodium concentration on one side and an increase in sodium on the other. These changes in solute concentration will be accompanied by changes in the water concentration on the two sides, since a change in solute concentration, as we have seen, produces a change in water concentration. Provided that the epithelium is permeable to water, the water-concentration difference produced will cause water to move by osmosis from the low-sodium to the high-sodium side of the epithelium (Fig. 6-29). Thus, water moves in the same direction as net solute movement across the epithelium.

Many of the fluxes of material across epithelial cells can be regulated by various neural and hormonal signals which open or close channels or alter the numbers or affinities of the various carriers in the membranes of the epithelial cells.

Glands

Glands are formed during embryonic development by the invagination of epithelial surfaces. Many of these glands remain connected by ducts to the epithelial surfaces from which they were formed; others lose their connection with the epithelium of origin and become isolated clusters of cells. The first type of gland is known as an **exocrine gland**, and its secretions flow through ducts and are discharged into the lumen of an organ or, in the case of the skin glands, onto the surface of the skin. Sweat glands and salivary glands are examples of exocrine glands.

The second type of gland is known as an **endocrine gland** or a ductless gland, since its secretions do not flow through ducts but are released directly into the interstitial fluid surrounding the gland cells. From this point, the material eventually enters the blood, which carries it throughout the body. The endocrine glands secrete a major class of chemical messengers—the hormones.

In practical usage, the term "ductless gland" has come to be synonymous with endocrine gland. However, it should be noted that there are cells that secrete nonhormonal, organic substances into the blood; for example, fat cells secrete fatty acids and glycerol into the blood, and the liver secretes glucose, amino acids, fats, and proteins. These substances serve as nutrients for other cells or perform special functions in the blood, but they do not act as hormones.

The substances secreted by glands fall structurally into two categories: (1) the organic materials that are, for the most part, synthesized by the gland cells and released into the extracellular fluid and (2) salt and water, which are not synthesized by gland cells but are simply moved from one extracellular compartment to another across the glandular epithelium.

Organic molecules are secreted by gland cells

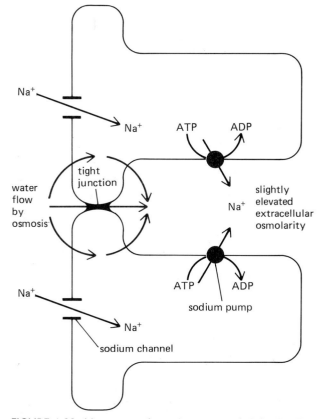

FIGURE 6-29. Movement of water across epithelial cells. The active transport of sodium out of epithelial cells, into the surrounding interstitial spaces on one side of the cell, produces an elevated osmolarity in this region, leading to the osmotic flow of water across the epithelial layer.

via all the pathways already described for movements across membranes: diffusion in the case of lipid-soluble materials, mediated transport for some polar materials, and exocytosis for the very large molecules such as proteins. Gland cells have a few additional means of secreting materials into the extracellular fluid. A type of secretion known as **apocrine secretion** involves the breaking-away of the entire top portion of the gland cell, which contains numerous secretory vesicles, followed by the resealing of the ruptured plasma membrane. An even more extreme form of secretion, known as **holocrine secretion**, involves the complete disintegration of the cell, thereby releasing the entire cell contents and at the same time destroying the cell.

The secretion of salts and water occurs by the mechanism described in the last section for the movements of salt and water across epithelia. Salts are actively transported across the epithelium, producing changes in the osmolarity of the surrounding extracellular fluid, which in turn causes the osmotic flow of water. In exocrine glands, as the secreted fluid passes along the duct connecting the gland to the luminal surface, the composition of the fluid may be altered as a result of absorption or secretion by the duct cells. Often the composition of the secreted material at the end of the duct will vary with the rate of secretion, reflecting the amount of time the fluid remains in the duct where it can be modified.

Most glands undergo a low, basal rate of secretion which can be greatly augmented in response to the appropriate signal, usually a nerve impulse or a hormone. The mechanism of the increased secretion is again one of altering some portion of the secretory pathway. This may involve increasing the rate at which a secreted substance is synthesized (by activating the appropriate enzymes), or providing the calcium signal for the exocytosis of already-synthesized material, or altering the pumping rates of a carrier-mediated process (by altering the affinity of the transport site or the number of transport carriers in the membrane). The volume of fluid secreted by an exocrine gland can be increased by directly increasing the sodium-pump activity or by controlling the opening of sodium channels at one surface of the epithelial cell, allowing more sodium to enter the cell; this increases the amount of sodium pumped out of the cell (via the Na,K-pump), which in turn increases the flow of water across the epithelium (since water osmotically follows the solute movement).

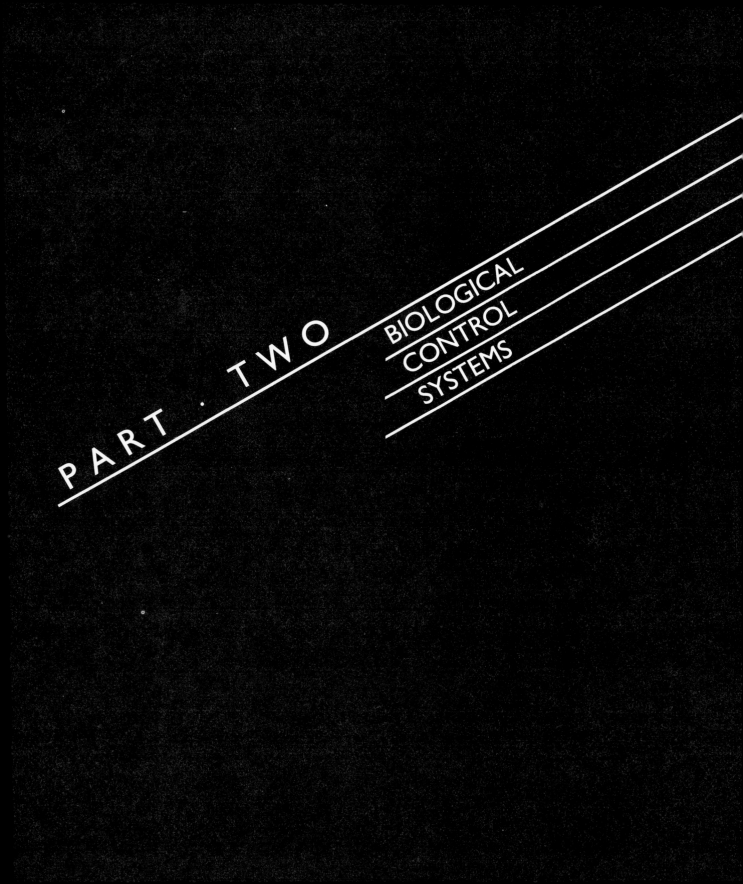

PART · TWO

BIOLOGICAL
CONTROL
SYSTEMS

HOMEOSTATIC MECHANISMS

As described in Chapter 1, Claude Bernard was the first to recognize the central importance of maintaining a stable internal environment. This concept was further elaborated and supported by the American physiologist W. B. Cannon, who emphasized that such stability could be achieved only through the operation of carefully coordinated physiological processes. The activities of cells, tissues, and organs must be regulated and integrated with each other in such a way that any change in the internal environment initiates a reaction to minimize the change. **Homeostasis** denotes the stable conditions of the internal environment which result from these compensating regulatory responses, which are known as **homeostatic mechanisms.** Changes in the composition of the internal environment do occur, but the magnitudes of these changes are small and are kept within narrow limits through multiple coordinated homeostatic processes, descriptions of which constitute the bulk of the remaining chapters.

Concepts of regulation have already been introduced in the context of a single cell. In Chap. 4, we described how the control of protein synthesis is regulated by the genetic apparatus. In Chap. 5,

we described how metabolic pathways in a cell are regulated by the principle of mass action and by changes in enzyme activity. Each individual cell exhibits some degree of self-regulation, but the existence of a multitude of cells organized into specialized tissues, which are further combined to form organs, imposes the need for overall regulatory mechanisms to coordinate the activities of all cells. To achieve this, communication between cells, often over relatively long distances, is essential.

General Characteristics of Homeostatic Control Systems

We shall define a **homeostatic system** as a control system which consists of a collection of interconnected cells and which functions to maintain a physical or chemical property of the body relatively constant. Consider the regulation of body temperature. Our subject is a resting, lightly clad person in a room having a temperature of 20°C and moderate humidity. Internal body temperature is a normal 37°C, and the person is losing heat to the external environment, which is at a lower temperature. The chemical reactions occurring within the cells of the body are producing heat at a rate equal to the rate of heat loss. Under these conditions the system is said to be in a **steady state,** the body undergoes no net gain or loss of heat, and the body temperature remains constant; this steady-state temperature is known as the **operating point** of the thermoregulatory system. This illustrates a crucial generalization about homeostasis (Table 7-1): Stability of an internal-environmental variable is achieved by balancing inputs and outputs. In this case, the variable—body temperature—remains constant because metabolic heat production (input) equals heat loss from the body (output).

Now we lower the temperature of the room rapidly, say to 5°C, and keep it there. This, of course, immediately increases the loss of heat from the person's warm skin, upsetting the dynamic balance between heat gain and loss; body temperature, therefore, starts to fall. But very rapidly, a variety of homeostatic responses occur which limit the fall (Fig. 7-1). First, the blood vessels to the skin narrow, reducing the amount of warm blood brought to the skin and thus reducing heat loss. However, at a room temperature of 5°C, the degree of blood vessel constriction undergone by our subject is not able to eliminate completely the extra heat loss from the skin. He or she curls up in order to reduce the surface area of skin available for heat loss; this helps a bit, but excessive heat loss still continues and body temperature keeps falling (although at a slower rate). The subject has a strong desire to put on more clothing (''voluntary'' behavioral responses are often crucial components in homeostasis), but none is available. Clearly, then, if excessive heat loss (output) cannot be prevented, the only way of restoring the balance between heat input and output is to increase input, and this is precisely what occurs. The subject begins to shiver, and the chemical reactions responsible for the rhythmic muscular contractions which constitute shivering produce large quantities of heat. The end result of all these responses is that heat loss from the body and heat production by the body are both high but are once again equal, so that body temperature stops falling. Indeed, heat production may transiently exceed heat loss so that body temperature may begin to go back toward the value originally existing before the room temperature was lowered. It will eventually stabilize at a temperature a bit below this original value. Here is another crucial generalization about homeostasis (Table 7-1): Stability of an internal-environmental variable depends not upon the absolute magnitudes of opposing inputs and outputs, but only upon the balance between them.

The thermoregulatory system just described is a **negative-feedback** system, defined as a system in which a change (increase or decrease) in the variable being regulated brings about responses which tend to push the variable in the direction opposite (''negative'') to the original change. Thus, in our example, the *decrease* in body temperature led to responses which tended to *increase* the temperature, i.e., return it to its original value. Negative-feedback control systems thus lead to stability and are the most common homeostatic mechanisms of the body.

TABLE 7-1. Some important generalizations about homeostatic control systems

---

1. Stability of an internal-environmental variable is achieved by balancing inputs and outputs; it is not the absolute magnitudes of the inputs and outputs that matter, but the balance between them.

2. In negative-feedback systems, a change in the variable being regulated brings about responses which tend to push the variable in the direction opposite to the original change.

3. Homeostatic control systems cannot maintain complete constancy of any given feature of the internal environment; therefore, any regulated variable will have a more or less narrow range of normal values depending on the external environmental conditions.

4. It is not possible for everything to be maintained relatively constant by homeostatic control systems; there is a hierarchy of importance, such that the constancy of certain variables is altered to maintain others relatively constant.

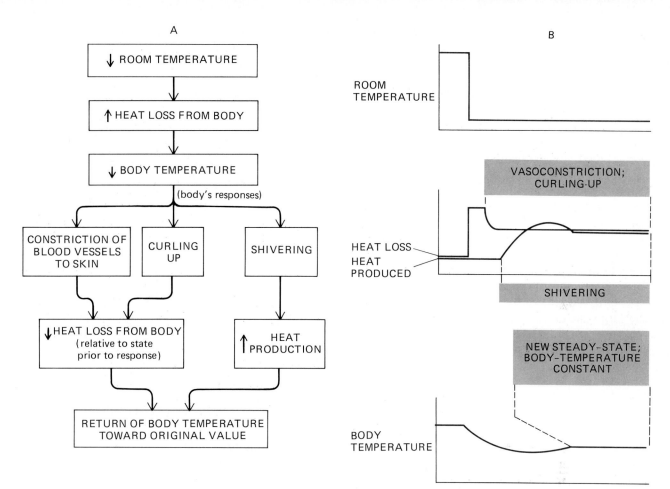

FIGURE 7-1. Example of a homeostatic control system. (A) Homeostatic system for maintaining body temperature relatively constant when room temperature decreases. This diagram is typical of those used throughout this book to illustrate homeostatic systems, and several conventions should be carefully noted. The arrows connecting any two terms (all in boxes in the figure) denote cause and effect; i.e., an arrow can be read as "causes" or "leads to" (e.g., shivering "leads to" increased heat production). The arrows modifying each term (e.g., the arrow next to heat production) denote directional changes (increases or decreases). In general, one should add the words "tends to" in thinking about cause-and-effect relationships; i.e., shivering "tends to" cause an increase in heat production, or curling up "tends to" cause a decrease in heat loss from the body. This is because variables like heat production and heat loss are under the influence of many factors, some of which oppose each other. (B) Time course of homeostatic adjustments to decreased environmental temperature. Note the opposing effects on heat loss: The decrease in room temperature tends to increase heat loss, whereas the body's responses—constriction of skin blood vessels (vasoconstriction) and curling up—tend to decrease heat loss. Note also that heat production (due to shivering) transiently exceeds heat loss, and body temperature actually starts back up toward its original value. At the new steady state, heat loss and heat production are equal (both are elevated), and body temperature remains constant, although lower than its original value.

There is, however, another type of feedback known as **positive feedback** in which an initial disturbance in a system sets off a train of events which increases the disturbance even further. Thus positive feedback does not favor stability but often abruptly displaces a system away from its steady-state operating point. As we shall see, several important positive-feedback relationships occur in the body, blood clotting being an example.

Much of the sequence of events shown in Fig. 7-1 is, of course, familiar, since the ultimate responses—blanching of the skin, curling up, and shivering—are visible, but what about the internal workings which underlie them? The blanching of the skin is the direct consequence of contraction of the smooth muscles which encircle the skin blood vessels, in response to signals sent to the muscles by nerves. Similarly, the nerves supplying many of the large skeletal muscles signal these muscles to produce the contractile activity known as shivering. But this begs the critical question. What initiated the signals in the nerves to the various muscles? The nerves are controlled by nerve centers in the brain which are, in turn, controlled by nerve signals coming from small groups of temperature-sensitive nerve cells in various parts of the body. The chain of events can be summarized: Decreased body temperature → increased activity of temperature-sensitive nerve cells in the body → increased activity of a series of nerve cells, the last of which stimulate the skin blood-vessel muscles and the skeletal muscles.

Another generalization about homeostatic control systems follows from this description of their components and the nature of negative feedback (Table 7-1): They cannot maintain *complete* constancy of the internal environment in the face of a persistent change in the external environment, but can only minimize changes. Imagine for the moment that the body's responses to a cold environment really did return body temperature completely to 37°C. What would happen to the output of the first component, the temperature-sensitive nerve cells (the "detectors")? Recall that it was the drop in body temperature that elicited the change in output from these detectors in the first place. If the body temperature were to return completely to its

original value, so would the output of the detectors, and the entire sequence of events would stop; the blood vessels would stop constricting, shivering would cease, and body temperature would once again fall since the compensations would no longer be present. In other words, as long as the exposure to cold continues, some decrease in body temperature must persist to serve as a signal to maintain the responses. (Of course, whenever the person is able to extricate herself or himself from the cold environment, then body temperature will return completely to the pre-cold value.) Just how big this so-called **error signal** is depends on the magnitude of the change in the external environment as well as on the sensitivity of the detectors, the efficiency of the connections between the system's components, and the responsiveness of the final apparatus in the control system. The temperature-regulating systems of the body are extremely sensitive, so that body temperature normally varies only by about 1°C even in the face of marked changes in the external environment.

Inherent in the last sentence is another generalization about homeostasis (Table 7-1): Even in reference to one individual (thus ignoring variation between persons), any regulated variable in the body (temperature, plasma glucose concentration, etc.) cannot be assigned a single "normal" value but will have a more or less narrow range of normal values, depending on the external environmental conditions. The more precise the homeostatic mechanisms for regulating this variable are, the narrower will be the range. Thus, the concept of homeostasis is an ideal, for total constancy of the internal environment can never be achieved.

One more generalization (Table 7-1): It is not possible for everything to be maintained relatively constant by homeostatic mechanisms. In our example, body temperature was maintained constant by means of large changes in skin blood flow and skeletal muscle contraction. Moreover, because so many properties of the internal environment are closely interrelated, it is often impossible to maintain one property relatively constant without moving others farther from their usual values. For example, we shall see in Chap. 12 that in order to maintain the acidity of the blood constant, the concentration of carbon dioxide in the blood may be greatly altered.

Acclimatization

Homeostatic control systems are inherited biological adaptations which favor survival in specific environments. However, an individual's ability to adapt to a particular environmental stress is not fixed but can be enhanced, with no change in genetic endowment, by prolonged exposure to that stress. This is known as **acclimatization**. Let us take sweating in response to heat exposure as an example and perform a simple experiment. On day 1 we expose a person for 30 min. to a high temperature and ask him or her to do a standardized exercise test. Sweating soon begins (as the result of another thermoregulatory system), and the volume of sweat produced is measured (simply by weighing the subject before and after the test). The sweating provides a mechanism for increasing heat loss from the body and thus tends to minimize the rise in body temperature in a hot environment. Then, for a week, the person enters the heat chamber for 1 or 2 h per day and exercises. On day 8, the sweating rate is again measured during the same exercise test performed on day 1; the striking finding is that the subject begins to sweat earlier and much more profusely than occurred on day 1. Accordingly, body temperature does not rise to nearly the same degree. The subject has acclimatized to the heat; that is, she or he has undergone an adaptive change induced by exposure to the heat and is now specifically better adapted to heat exposure without change in gene structure. In a sense, acclimatization may be viewed as an environmentally induced improvement in the functioning of an already existing, genetically based homeostatic system. Repeated or prolonged exposure to the environmental stress enhances (up to some maximal limit) the responsiveness of the system to that particular type of stress.

In the example above, the heat acclimatization of our subject is completely reversible since, if the daily exposures are discontinued, within a relatively short time the sweating rate during future tests will revert to the preacclimatized value. How-ever, not all acclimatization is reversible. If the acclimatization is induced very early in the individual's life, at the so-called critical period for development of that structure or response, it may be irreversible. For example, the barrel-shaped chests of natives of the Andes mountains do not represent a genetic difference between them and their lowland countrymen but rather an irreversible acclimatization induced during the first few years of their lives by their exposure to the low-oxygen environment of high altitude. It remains even though the individual moves to a lowland environment later in life and stays there. Lowland persons who have suffered oxygen deprivation from heart or lung disease during their early years show precisely the same chest shape.

These **developmental acclimatizations** take the form not only of anatomical changes, as in the example cited, but of physiological responses as well. For example, because of their developmental acclimatizations, the Andean natives have a greater maximal capacity for exercise at high altitude than do lowlanders who have moved to the mountains beyond childhood and remained there the rest of their lives. The latter never reach the level of acclimatization achieved by those born in the mountains.

It has proven very difficult to determine for any particular acclimatization just which components of the control system the acclimatization has sensitized, but the right questions are easy to phrase on the basis of our earlier description of homeostatic control systems. For our example of sweating we would ask the following questions: Is the sensitivity of the thermosensitive elements increased so that they respond to smaller deviations of temperature from normal? Are the nerve-cell wiring connections in the system more responsive to signals? Are the sweat glands enlarged and so more able to respond to a given signal? Are those sweat-gland enzymes responsible for the actual production of the sweat more numerous or more active?

Biological rhythms

A striking characteristic of many bodily functions is the rhythmical changes they manifest. Body temperature, for example, fluctuates considerably during a normal 24-h period. Some rhythms have periods that are much longer than 24 h. The menstrual cycle is the best-known longer cycle, but there may well be others with even greater time spans. Many body rhythms seem to represent changes in the operating points of the control systems regulating them, but we generally are uncertain as to their precise mechanism. Central to this field is the problem of "biological clocks" within the body which set the rhythms. We do know that modern life, with its stepped-up pace, rapid changes, and creation of artificial environments, may frequently disrupt them with potentially harmful results. We also know that these rhythms have profound effects on the body's resistance to various stresses such as that produced by bacteria or drugs. For example, when rats are given a large dose of a potentially lethal drug, their survival depends profoundly on the time the drug is given. Differences in sensitivity of this type unquestionably also exist in people, and the hope is that a fuller understanding of them will provide a more rational basis for the timing of drug therapies.

Aging

Old age is not an illness; it is a continuation of life with decreasing capacities for adaptation. This view of aging in terms of a progressive failure of the body's various homeostatic adaptive responses has gained wide acceptance only recently, for there had been a strong tendency to confuse what we now recognize as a distinct aging process with those diseases frequently associated with aging. However, these diseases—notably atherosclerosis and cancer—are not necessary accompaniments of old age, even though their incidence does increase with age; rather, they seem to interact with the aging process in a positive-feedback cycle, each accelerating the other. For example, one theory of aging focuses on the fact that the collagen molecules of connective tissue develop increased cross-linkages with age, and that the resulting rigidity decreases tissue functioning. It is not difficult to imagine how this aging change in collagen, were it to occur in the large arteries, would also enhance the rate of

development of atherosclerosis, which, in turn, might enhance the rate of cross-linkage formation by altering the chemical environment of the collagen.

What is the nature of the aging process itself (in contrast to the "diseases of aging")? The physiological manifestations of aging are a gradual deterioration in function and in the capacity to respond to environmental stresses. Thus, such quantitative parameters of function as the amount of blood pumped by the heart per minute decrease, as does the ability to maintain the internal environment constant in the face of changes in temperature, diet, oxygen supply, etc. These manifestations of aging are related both to a decrease in the actual number of cells in the body (for example, we lose an estimated 100,000 brain cells each day) and to the disordered functioning of many of the cells which remain.

In thinking of how the total number of cells in the body diminishes, it is crucial to recognize that this number reflects the balance between new cell generation (by cell division) and the death of old cells. It had been known for many years that, in the adult human, certain specialized cells—notably nerve and muscle cells—lose most of their ability to divide, but it was not until recently that a limitation on cell division of other cell types was firmly established. In the crucial experiment, cells, when grown outside of the body, divided only a certain number of times and then stopped; moreover, the number of divisions correlated with the age of the donor. The fact that the number of divisions also correlated with the normal life span of the particular species from which the cells were obtained is strong evidence for the idea that cessation of mitosis is a normal, genetically programmed event. However, other experiments have demonstrated that manipulation of the chemical environment of some cells (in this case, by the addition of large quantities of vitamin E) can result in the cells dividing more times than usually observed before mitosis ceases. Thus, as in any gene-environment interaction, it is likely that environmental factors determine just how many divisions actually occur, within the limits set by the genetic program.

In addition to external-environmental factors

("wear-and-tear," radiation, chemicals, etc.) and inherent degeneration of macromolecules, it is very likely that cell death may be the result of the influence of one tissue or organ upon another. One of the most interesting possibilities in this last regard is that the body's cells might undergo subtle changes such that they would no longer be recognized as "self," with the failure of recognition leading to their destruction (Chap. 17).

The ultimate failure of mitosis may be due to an accumulation of copying errors in a cell's DNA molecules. The emphasis on the importance of DNA in aging applies not only to cell division but to all aspects of cell function. As mentioned above, aging is expressed not only by a decrease in the total number of cells but also by the deterioration of the functional capacity of those cells which remain. There is fairly general agreement that the immediate cause of this deterioration is an interference in the function of the cells' macromolecules—not just DNA, but RNA, cell proteins, and the flow of information between these macromolecules as well. There are probably many factors responsible for these macromolecular disturbances; as one recent reviewer put it, this is a field in which there are as many theories as there are investigators.

The Balance Concept and Chemical Homeostasis

As we have emphasized, one of the most important concepts in the physiology of homeostatic control systems is that of balance. Our first thermoregulatory example was really a study of heat balance in the body; i.e., the control system functioned to maintain a precise balance between the rates at which heat was produced by and lost from the body. Many homeostatic systems can be studied in terms of balance; some regulate the balance of a physical parameter (heat, pressure, flow, etc.), but most are concerned with the balance of a chemical component.

Figure 7-2 is a generalized schema of the possible pathways involved in the balance of a chemical substance. The **pool** occupies a position of central importance in the balance sheet; it is the body's readily available quantity of the particular substance and is frequently identical to the amount present in the extracellular fluid. The pool functions as "middleman," receiving from and contributing to all the other pathways.

The pathways on the left of the figure are sources of net gain to the body. A substance may be ingested and then absorbed from the gastrointestinal (GI) tract. It is important to realize that not all that is ingested must be absorbed from the gastrointestinal tract. The lungs offer another site of entry to the body for gases (particularly oxygen) and airborne chemicals. Finally, the substance may be synthesized by cells within the body.

The pathways to the right of the figure are sources of net loss from the body. A substance may be lost in the urine, feces, expired air, and menstrual fluid, as well as from the surface of the body (skin, hair, nails, sweat, tears, etc.). The substance may also be chemically altered and thus removed by metabolism.

The central portion of the figure illustrates the distribution of the substance within the body. From the readily available pool, it may be accumulated in storage depots; conversely, material may leave the storage depots to reenter the pool. Finally, the chemical may be incorporated into some other molecular structure (fatty acids into membranes, iodine into thyroxine, etc.). This process is reversible in that the substance is liberated again whenever the more complex molecule is broken down. For this reason, this pathway differs from metabolism, by which the substance is irretrievably altered. This pathway is also distinguished from storage in that the latter has no function other than the passive one of storage, whereas the incorporation of the substance into other molecules fulfills a specific function (e.g., the function of iodine in the body is to provide an essential component of the thyroxine molecule).

It should be recognized that not every pathway of this generalized schema is applicable to every substance; for example, the mineral electrolytes such as sodium cannot be synthesized, do not normally enter through the lungs, and cannot be removed by metabolism.

NET GAIN TO BODY   DISTRIBUTION WITHIN BODY   NET LOSS FROM BODY

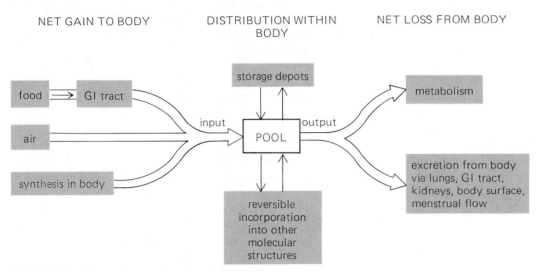

FIGURE 7-2. Balance diagram for a chemical substance.

The orientation of Fig. 7-2 illustrates two important generalizations concerning the balance concept: (1) The total-body balance depends upon the rates of total-body net gain and net loss, and (2) the pool concentration depends not only upon total-body losses and gains but upon exchanges of the substance within the body.

It should be apparent that, with regard to total-body balance of any chemical, three states are possible: (1) loss exceeds gain, the total amount of the substance in the body decreases, and the person is said to be in **negative balance**; (2) gain exceeds loss, the total amount of the substance in the body increases, and the person is said to be in **positive balance**; and (3) gain equals loss, and balance is stable.

Clearly a stable balance can be upset by alteration of the magnitude of any single pathway in the schema; e.g., severe negative water balance can be caused by increased sweating. Conversely, stable balance can be restored by homeostatic control of the pathways controlling water intake and output. Therefore, much research has been concerned with determining the key homeostatically controlled pathways for each substance, the specific mechanisms involved, and the limits of intake and output beyond which balance cannot be achieved.

Essential nutrients

As mentioned in Chap. 5, there are many substances which are required for normal or optimal body function but which are synthesized by the body either not at all or in amounts inadequate to achieve balance. They are known as essential nutrients. Because they are all excreted or catabolized at some finite rate, a continuous new supply must be provided by the diet. The physiology of each essential nutrient can be studied in terms of the schema of Fig. 7-2, i.e., by analyzing each pathway relevant to that nutrient. There is considerable variation in the relative importances of the pathways for the homeostatic regulation of the different nutrients. Thus, in Chap. 13 we shall see that ingestion and urinary excretion are the main controlled variables for water. In contrast, regulation of iron balance, as described in Chap. 11, is dependent largely upon the control of iron absorption by the gastrointestinal tract. Balance of essential amino acids is achieved in still another way; the rate of catabolism of these amino acids (specifically the loss of the NH_2 group from the amino acid) is reduced in the presence of amino acid deficiency and increased in the presence of excess.

The reason for placing so much emphasis on the pathways which are homeostatically controlled is that alteration of the nutrient flow via these pathways constitutes the mechanism by which balance is achieved whenever a primary change occurs in any of the other pathways. For example, when iron balance is upset either by a primary change in intake or by excretion, balance can be reestablished by a compensating change in the rate of absorption.

Let us look more closely at the spectrum of conditions possible as we alter primary intake of an essential nutrient, assuming all the other pathways are functioning normally. At zero intake, balance is impossible since, for essential nutrients, excretion or catabolism, or both, cannot be reduced to zero; accordingly negative balance persists as the body stores of the nutrient are progressively depleted. Thus the minimal combined rate of excretion and catabolism sets the lower limit for achievement of balance. At the other end of the spectrum, the maximal combined rate of excretion and catabolism sets the upper limit for achievement of balance, i.e., the maximal intake of nutrient that can be balanced by output. If intake is greater than this, body stores will continuously increase, with the potential for toxic effects. This occurs for the fat-soluble vitamins and is a problem for other nutrients as well.

Between the extremes of intakes there obviously exists a range of intakes at which balance can be maintained without overt manifestations of either deficiency or toxicity. How wide this range is depends upon how much the rates of excretion or catabolism can be altered. (For example, sodium balance can be achieved readily at intakes between 0.3 and 25 g/day.) It must be reemphasized that, despite the fact that balance is stable at all intakes in this range, there are some differences in body content and pool size over the range, since these differences are required to drive the homeostatic control systems.

Does it really matter for the health of the body at what point along its normal range the concentration of a particular nutrient exists? Let us take sodium as an example. The control systems for sodium balance have as their targets the kidneys, and they operate by inducing the kidneys to excrete into the urine an amount of sodium approximately equal to the amount ingested daily. Imagine now a person with a daily intake and output of 7 g of sodium and a stable amount of sodium in the body. Tomorrow the diet of the person changes, so that daily sodium consumption rises to 15 g and remains there indefinitely. On this same day, the kidneys excrete into the urine somewhat more than 7 g of sodium but certainly not all of the ingested 15 g. The result is that some excess sodium is retained in the body on that day. The kidneys do somewhat better on day 2, but it is probably not until day 3 or 4 that they are excreting 15 g. From this time on, output from the body once again equals input and sodium balance is stable. But, and this is the important point, the person has perhaps 2 to 3 percent more sodium in the body than was the case when the lesser amount of sodium was being ingested. It is this 2 to 3 percent of extra sodium which constitutes the signal for the control systems driving the kidneys to excrete 15 g/day rather than 7 g/day. An increase of 2 to 3 percent does not seem large, but it has been hypothesized that, over many years, this little extra might, in some persons, facilitate the development of high blood pressure.

The ultimate aim of nutrition is to determine the optimal intake for each nutrient. The critical question may be stated as follows: Does ingestion of more than the minimal amount of a nutrient produce greater health, growth, intelligence, and so forth? That is, is there an amount of nutrient which is optimal for health rather than merely adequate for avoiding disease? But this question raises the further question: Optimal for what aspect of health? For example, most Americans ingest very large quantities of protein, and this almost certainly has contributed to our increased body height. This might seem desirable, but in experimental animals it has been found that protein intakes which are optimal for maximal total body growth enhance the development of cancers. Obviously, "optimal for what?" cannot be answered without a clearly definable "what" and a careful consideration of many factors, including existing environmental influences and life style.

Finally, it must be emphasized that the widely used Recommended Daily Allowances (RDA) of the National Academy of Sciences are not meant to be recommendations for an optimal diet. To avoid this common confusion, some have proposed that the designation ''recommended allowance'' be changed to ''acceptable nutrient intakes,'' which emphasizes that these are the daily intakes of nutrients considered sufficient, within the limits of knowledge, to prevent nutritional inadequacy in most healthy persons. In other words, the RDAs are guides for preventing disease, not necessarily for optimizing health. (See Appendix 2 for a listing of RDAs.)

Components of Homeostatic Systems

Reflexes

The thermoregulatory systems we used as examples earlier in the chapter and many of the body's other homeostatic systems belong to the general category of stimulus-response sequences known as **reflexes**. We may be aware only of the stimulus and the final event in the sequence, the **reflex response**. Many reflexes regulating the internal environment occur without any conscious awareness. In the most narrow sense of the word, a reflex is an involuntary, unpremeditated, unlearned response to a stimulus; the pathway over which this chain of events occurs is ''built in'' to all members of a species. Examples of such basic reflexes would be pulling one's hand away from a hot object or shutting one's eyes as an object rapidly approaches the face. However, there are also many responses which appear to be automatic and stereotyped but which actually are the result of learning and practice. For example, an experienced driver performs many complicated acts in operating a car; to the driver these motions are, in large part, automatic, stereotyped, and unpremeditated, but they occur only because a great deal of conscious effort was spent to learn them. We shall refer to such reflexes as **learned** or **acquired**. In general, most reflexes, no matter how basic they may appear to be, are subject to alteration by learning; i.e., there is often no clear distinction between a basic reflex and one with a learned component.

The pathway mediating a reflex is known as the **reflex arc**, and its classical components are shown in Fig. 7-3.

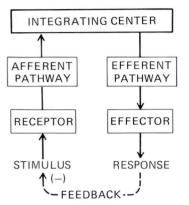

FIGURE 7-3. General components of a biological control system, the reflex arc. The response of the system has the effect of counteracting or eliminating the original stimulus. This phenomenon of negative feedback is emphasized by the minus sign in the feedback loop.

A **stimulus** is defined as a detectable change in the environment, such as a change in temperature, potassium concentration, pressure, etc. A **receptor** is the component which detects the environmental change (we had been referring to it as a ''detector'' in earlier sections). A stimulus acts upon a receptor to produce a signal which is relayed to an **integrating center**. The pathway traveled by the signal between the receptor and the integrating center is known as the **afferent pathway**.

An integrating center often receives signals from many receptors, some of which may be responding to quite different types of stimuli. Thus, the output of an integrating center reflects the net effect of the total afferent input; i.e., it represents an integration of numerous bits of information.

The output of an integrating center is then sent to the last component of the system, a device whose change in activity constitutes the overall response of the system. This component is known as an **effector**. The information going from an integrating center to an effector is like a command directing the effector to alter its activity. The pathway along which this information travels is known as the **efferent pathway**. In negative-feedback reflexes, the effector's response counteracts (at least in part) the original stimulus (environmental change) which triggered the entire sequence of events.

In summary, the components of a typical reflex arc are

1. Receptor
2. Afferent pathway
3. Integrating center
4. Efferent pathway
5. Effector

To illustrate the components of a reflex arc, let us apply these terms (Fig. 7-4) to the example of thermoregulation used previously. The temperature receptors are the endings of certain nerve cells in various parts of the body which generate electric signals in the nerve cells at a rate determined by the temperature. These electric signals are conducted by the nerve fibers (the afferent pathway) to a specific part of the brain (which acts as integrating center), which in turn determines the signals sent out along those nerve cells which cause skeletal muscles and the smooth muscles surrounding skin blood vessels to contract. The latter nerve fibers are the efferent pathway, and the muscles they innervate are the effectors.

Traditionally, the term "reflex" was restricted to situations in which the first four of the components are all parts of the nervous system, as in the thermoregulatory reflex. However, present usage is not so restrictive and recognizes that the principles are essentially the same when a blood-borne messenger known as a **hormone**, rather than a nerve fiber, serves as the afferent or (much more commonly) the efferent pathway, or when a hormone-secreting (**endocrine**) gland serves as integrating center. Thus, in the thermoregulation example above, the integrating centers in the brain not only send signals by way of the nerve fibers to muscle but also cause the release of several hormones which travel by the blood to many cells, producing an increase in their heat production by altering the rates of various heat-producing chemical reactions. These hormones therefore also serve as an efferent pathway in thermoregulatory reflexes.

Accordingly, in our use of the term "reflex," we include hormones as reflex components, so that afferent or efferent information can be carried by either nerve fibers or hormones. In any case, two different components must serve as afferent and

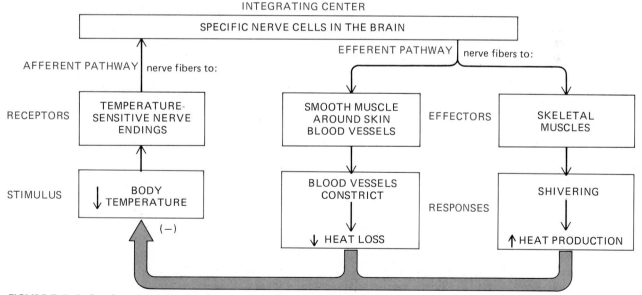

FIGURE 7-4. Reflex for minimizing the decrease in body temperature which occurs on exposure to a reduced external environmental temperature. This figure provides the internal components for the reflex originally shown in Fig. 7-1. The shaded arrow at the bottom indicates the negative-feedback nature of the reflex; i.e., the responses tend to eliminate the original stimulus.

efferent pathways; thus the input to and output from the integrating center may both be neural, but they must be two different nerve fibers. Of course, they may also both be hormonal, or one may be neural and the other hormonal. Moreover, depending on the specific nature of the reflex, the integrating center may reside in either the nervous system or an endocrine gland. Thus, intercellular communication over long distances is accomplished by means of nerve cells and hormones, and the specific mechanisms by which these two communication systems operate are the subject of the next two chapters.

Finally, we must identify the effectors, the cells whose outputs constitute the ultimate responses of the reflexes. Actually, most cells of the body act as effectors in that their activity is subject to control by nerves or hormones. There are, however, two specialized classes of tissues—muscle and gland tissues—which are the major effectors of biological control systems. Muscle cells are specialized for the generation of force and movement, and their physiology is described in Chap. 10. Glands consist of epithelial cells specialized for the function of secretion (Chap. 6).

There are problems which arise when one attempts to categorize the components of some complex reflex arcs according to the five terms listed above. For example, in the reflex arc shown in Fig. 7-5, the stimulus, receptor, and effector (muscle) are quite clear, as is the designation of the first nerve (A) in the chain and the last hormone (B) as afferent and efferent pathways, respectively. But what about the brain, nerve B, hormone A, and the two endocrine glands? It is fruitless to assign rigid terms to the interior components of such complex reflex chains. It is far more important for the reader to appreciate the sequence of events rather than to worry about the labels.

Another problem is that some hormonal reflex arcs seem to lack a receptor and afferent pathway. This may seem puzzling, but the following example, shown in Fig. 7-6, may help. Parathyroid hormone is a hormone which is secreted by the parathyroid glands. It acts upon bone, causing it to increase its release of calcium into the blood. When the blood-calcium concentration is decreased for any reason, an increased amount of parathyroid hormone is secreted into the blood; the blood-borne parathyroid hormone reaches bone throughout the body and induces it to release more calcium into the blood.

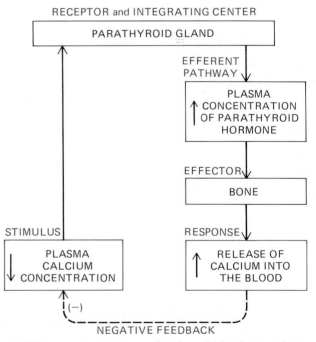

FIGURE 7-6. Homeostatic reflex by which plasma calcium concentration is controlled. Note the absence of an afferent pathway.

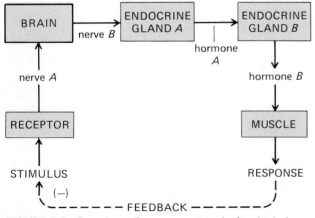

FIGURE 7-5. Complex reflex arc composed of multiple hormones and nerve fibers.

Clearly, in this reflex, plasma calcium concentration is the stimulus, blood-borne parathyroid hormone acts as the efferent pathway, and bone is the effector. But we have made no mention yet of the receptors, afferent pathway, or integrating center. Actually, the parathyroid gland cells are themselves sensitive to the calcium concentration of the blood supplying them; thus, there is no separate receptor or afferent pathway, and the parathyroid cells serve as the integrating center.

To summarize, many reflexes function to keep a physical or chemical variable of the body relatively constant. One may analyze any such system by answering a series of questions: (1) What is the variable (blood glucose, body temperature, blood pressure, etc.) which is being maintained relatively constant in the face of changing conditions? (2) Where are the receptors which detect changes in the state of this variable? (3) Where is the integrating center to which these receptors send information and from which information is sent out to the effectors, and what is the nature of these afferent and efferent pathways? (4) What are the effectors and how do they alter their activities so as to maintain the regulated variable at the operating point of the system?

Local homeostatic responses

In addition to reflexes, another group of biological responses is of great importance for homeostasis. We shall call them **local responses**. Local responses are initiated by a change in the external or internal environment, i.e., a stimulus, and induce an alteration of cell activity with the net effect of counteracting the stimulus. Thus, a local response is, like a reflex, a sequence of events proceeding from stimulus to response, but unlike a reflex, the entire sequence occurs only in the area of the stimulus, neither hormones nor nerves being involved.

Two examples should help clarify the nature and significance of local responses: (1) Damage to an area of skin causes the release of certain chemicals from cells in the damaged area which help the local defense against further damage (Chap. 17) and (2) an exercising muscle liberates into the extracel-

lular fluid chemicals which act locally to dilate the blood vessels in the area, thereby permitting the required inflow of additional blood to the muscle (Chap. 11). The significance of such local responses is that they provide individual areas of the body with mechanisms for local self-regulation.

Chemicals as intercellular messengers in homeostatic systems

It should be evident that essential to reflexes and local responses is the ability of cells to communicate with one another. When a hormone is involved in a reflex, it is obvious that the communication between cells, i.e., between hormone-secreting cell and effector, is accomplished by a chemical agent, the hormone (with the blood acting as the delivery service). What has not been said, however, is that the way most nerve cells communicate with each other or with effectors is also by means of chemical agents, which are known as **neurotransmitters**. Thus, one neuron alters the activity of the next neuron in a chain by releasing from its ending a neurotransmitter which diffuses through the extracellular fluid separating the two nerve cells and acts upon the second. Similarly, neurotransmitters released from the ends of the nerve cells going to effector cells constitute the immediate signal, or input, to the effector cells. As will be described more fully in Chap. 9, the chemical messengers released by certain neurons do not act on adjacent neurons or effector cells but, rather, enter the bloodstream to act on cells elsewhere in the body; these messengers therefore are properly termed hormones (or neurohormones), not neurotransmitters (Fig. 7-7).

The role of chemicals as intercellular messengers holds not only for reflexes but also for local responses (as the examples above illustrate). The specific chemical messengers involved in local responses are referred to as **paracrines**. Paracrines are synthesized by cells and released, given the appropriate stimulus, into the extracellular fluid, through which they diffuse to reach neighboring cells, upon which they exert their particular effects. (Note that, given this broad definition, neurotrans-

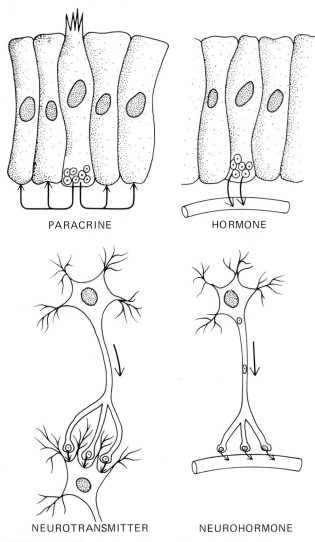

PARACRINE   HORMONE

NEUROTRANSMITTER   NEUROHORMONE

FIGURE 7-7. Categories of intercellular chemical messengers. A neurohormone is the product of a neuron but is secreted into the blood; i.e., it is a hormone. (Redrawn from Keuger and Martin.)

TABLE 7-2. A few important paracrines

Adenosine

Adenosine diphosphate (ADP)

Estrogen

Histamine

Nerve growth factor

Prostaglandins

Serotonin

Somatostatin

Testosterone

and secreted for the purpose of regulating local cells. During the course of metabolism, many substances, such as potassium and hydrogen ions, are released from cells and produce responses in neighboring cells, particularly vascular smooth muscle cells. Such substances serve as local chemical messengers but they are secondary by-products of cellular activity, not specific chemical messengers as is the case of paracrines, hormones, and neurotransmitters.

There is one more category of local chemical messenger which is not an intercellular messenger at all. Rather, the chemical is secreted by a cell into the extracellular fluid surrounding the cell and the chemical acts upon the very cell which secreted it. The general term for this is **autocrine**.

As we shall see, a given chemical messenger may be synthesized by a number of different cell types so that a single molecular type may serve as a neurotransmitter (released from neuron terminals), as a hormone (released from endocrine gland cells and reaching its target cells via the blood), and as a paracrine (or autocrine). For example, the chemical called somatostatin functions in each of the three categories in different parts of the body. A logical hypothesis to explain such diversification has been suggested by the finding in unicellular organisms of molecules similar to vertebrate hormones and neurotransmitters. In brief, it is likely that the chemical messengers which serve as hor-

mitters could be classified as a subgroup of paracrines, but by convention they are not.) Paracrines are generally inactivated rapidly by locally existing enzymes so that they do not gain entry to the bloodstream in large quantities. Table 7-2 presents a listing of some important paracrines.

Not all local responses are mediated by paracrines—*specific* chemical substances synthesized mones, neurotransmitters, and paracrines in vertebrates originated in unicellular organisms such as bacteria, in which they functioned as local factors, acting upon the organisms themselves and their neighbors. With evolution, the anatomic elements of the system have become more complex and specialized, but the biochemical elements have been preserved.

All of the types of intercellular communication described in this section have involved secretion of the chemical messenger into the extracellular fluid. However, there are two other important types of intercellular chemical communication which do not require such secretion. Via gap junctions (Chap. 6), chemicals can move from one cell to another adjacent cell without ever entering the extracellular fluid (electrical communication between cells, rather than chemical, can also occur via gap junctions, as will be described in Chaps. 8 and 10). Secondly, the chemical messenger may not actually be released from the signaling cell but may rather be located in the outer plasma-membrane of the cell; in these cases, the signaling cell actually links up with its target cell via the membrane-bound messenger. This type of signaling is of particular importance in the functioning of the cells which protect the body against microbes and other foreign cells (Chap. 17).

Prostaglandins. The most ubiquitous of chemical messengers are the **prostaglandins** and the related compounds, the **prostacyclins**, **thromboxanes**, and **leukotrienes** (for convenience, these substances are often all lumped together under the general name of prostaglandins). They are all unsaturated fatty acids which have a carbon ring at one end. They are divided into main groups, each designated by a letter—PGA, PGE, etc.—on the basis of structural differences in their rings. Within each group, further subdivision occurs according to the number of double bonds in the fatty acid side chains; for example, PGE_2 has two double bonds, and this fact is denoted by the subscript in its name.

The major chemical precursor of the prostaglandins is the fatty acid **arachidonic acid**, which is present in plasma-membrane phospholipid. The ap-propriate stimulus (hormone, neurotransmitter, paracrine, drug, or toxic agent) activates a membrane-bound enzyme which splits off the arachidonic acid from the phospholipid. The arachidonic acid is then metabolized by several enzyme-mediated steps to yield the specific prostaglandins (Fig. 7-8). One of the initial enzymes crucial to the entire system is blocked by aspirin.

Once synthesized in response to a stimulus, the prostaglandins are not stored in cells to any extent but are released immediately and act locally; accordingly, the prostaglandins fit into the category of paracrines. Interestingly, a prostaglandin often acts not only on neighboring cells (i.e., as a paracrine) but on the very cell which synthesized it (i.e., as an autocrine). The common denominator of much of the autocrine type of prostaglandin action is to modulate the response of the target cell to the original stimulus. For example, a neurotransmitter might act upon a muscle cell to trigger a particular response, say increased contraction; simultaneously, the neurotransmitter might also trigger production by that same muscle cell of a particular prostaglandin which has the effect of dampening the cell's response to the neurotransmitter, i.e., preventing excessive contraction.

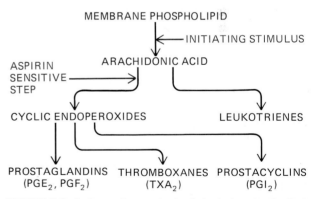

FIGURE 7-8. Pathways for synthesis of classical prostaglandins, thromboxanes, prostacyclins, and leukotrienes. Different enzymes mediate each of the steps; for example, phospholipase catalyzes the formation of archidonic acid, and cyclo-oxygenase the formation of the cyclic endoperoxides. Various cell types in the body possess different enzyme profiles and so produce different final substances. The abbreviations in parentheses denote major examples of each of the families.

After their action, the prostaglandins are quickly metabolized by local enzymes to inactive forms. Whether certain prostaglandins can also function as hormones, i.e., can be carried by the blood to exert effects on sites distant from their tissues of origin, remains unsettled.

The prostaglandins exert a bewildering array of effects; for example, they are important in blood clotting, regulation of smooth muscle contraction, modulation of neurotransmitter release and action, multiple processes in the reproductive system, control of hormone secretion, and the body's defenses against injury and infection. We shall describe these in future chapters.

Receptors

The first step in the action of any intercellular chemical messenger on its target cell is its binding to specific protein molecules of that cell; these molecules are known as **receptors** (in the general language of Chap. 4, the chemical messenger is a "ligand" and the receptor a "binding site"). This term can be the source of confusion because the same word is used to denote the "detectors" in a reflex arc described earlier in this chapter; the reader must keep in mind the fact that "receptor" has two totally distinct meanings, but the context in which it is used usually makes it quite clear which is meant.

What is the nature of these receptors with which chemical messengers combine? They are proteins located either in the cell's plasma membrane or inside the cell. It is the combination of chemical messenger and receptor which initiates the events leading to the cell's response. The existence of receptors explains a very important characteristic of intercellular communication—**specificity**. Although a chemical messenger—hormone, neurotransmitter, or paracrine—may come into contact with many different cells, it influences only certain cells and not others. The explanation is that cells differ in the types of receptors they contain. Accordingly, only certain cell types, frequently just one, possess the receptor required for combination with a given chemical messenger. (This is another example of how protein structure confers specificity upon a biological process.)

It is important to realize that in the many cases where different cell types possess the same receptor type for a particular messenger, the responses of the various cell types to that messenger may differ from each other. For example, the neurotransmitter norepinephrine causes blood-vessel smooth muscle to contract, but via the same receptor type, norepinephrine causes cells in the pancreas to secrete less insulin. In essence, then, the receptor functions as a molecular "switch" which when "switched on" by the messenger's binding to it, elicits the cell's response; just as identical types of switches can be used to turn on a light or a radio, a single type of receptor can be used to produce quite different responses in different cell types.

Similar reasoning explains perhaps a more surprising phenomenon: A single cell may contain several different receptor types for a single messenger; combination of the messenger with one of these receptor types may produce a cellular response quite different from (sometimes opposite to) that produced when the messenger combines with the other receptor type. (The degree to which the messenger molecules combine with the different receptor types on a single cell is determined by such factors as the affinity of the different receptor types for the messenger.)

Other characteristics of messenger-receptor interactions are **saturability** and **competition**. In most systems, a cell's response to a messenger increases as the extracellular concentration of messenger increases, because the number of receptors occupied by messenger increases. But there is an upper limit to this responsiveness, in part because only a finite number of receptors is available. Competition is the ability of molecules very similar in structure to the messenger to combine with the receptor and thereby prevent the normal messenger from doing so. Competition underlies much present-day use of drugs. If researchers or physicians wish to interfere with the action of a particular messenger, they can administer competing molecules (if available) which bind to the receptors for that messenger but do not trigger the cell's response. Such drugs are known as **antagonists** with regard to the usual chemical messenger. On the other hand, some drugs which bind to a particular receptor type do trigger the cell's response exactly as if the true chemical messenger had combined with receptor; such drugs are known as **agonists** and are used to mimic the messenger's action.

It should be emphasized that receptors are, themselves, subject to physiological regulation. The number of receptors a cell has and the affinity of the receptors for their specific messenger can be increased or decreased, at least in certain systems. An important example of such regulation is the phenomenon of **down-regulation**, which applies to many hormones and neurotransmitters. When a high extracellular concentration of one of these messengers is maintained chronically, the total number of target-cell receptors for that messenger may decrease (be down-regulated). Down-regulation has the effect of reducing target cells' responsiveness to high concentrations of messengers and thus represents a locally occurring negative-feedback mechanism. Change in the opposite direction (**up-regulation**) also occurs, although less commonly; cells exposed chronically to very low concentrations of a messenger may develop many more receptors for that messenger, thereby becoming **supersensitive** to it. Regulated changes in receptor number are achievable because there is a continuous degradation and synthesis of receptors. Membrane-bound receptors are internalized (i.e., taken into the cell by endocytosis) and either broken down or reinserted back into the membrane along with newly synthesized receptors; intracellular receptors are also continuously being degraded and resynthesized. Alteration of the rate of one or more of these processes will raise or lower the number of cell receptors of a particular type. For example, some hormones induce endocytosis of the membrane receptors to which they bind; this increases the rate of receptor degradation so that at high hormone concentrations the number of plasma-membrane receptors gradually decreases—this is an example of down-regulation.

Down-regulation and up-regulation are physiological responses, but there also seem to be many disease processes in which the number of receptors or their affinity for messenger becomes abnormal; the result is unusually large or small responses to any given level of messenger. For example, the disease called myasthenia gravis is due to a destruction of the skeletal muscle receptors for acetylcholine, the neurotransmitter which normally causes contraction of the muscle in response to nerve stimulation; the result is muscle weakness or paralysis.

We have pointed out that receptors may be located in the cell's plasma membrane or inside the cell. In fact, the plasma membrane is the much more common location, the major important exceptions being the intracellular receptors for lipid-soluble hormones (these hormones readily traverse the lipid-rich plasma membranes of cells). The plasma membrane of a single cell usually contains many different receptor types, each capable of binding only one type of chemical messenger.

Postreceptor Events

The combination of messenger with receptor is only the initial step leading to the cell's response, which ultimately takes the form of a change in (1) membrane permeability, transport, or electrical state; (2) the rate at which a particular substance is synthesized or secreted by the cell; or (3) the rate or strength of muscle contraction. Despite the seeming variety of these ultimate responses, there is a common denominator to all of them: They are all due to alterations of particular cell proteins. Let us take a few examples of messenger-induced responses, all of which are described fully in subsequent chapters. Changes in muscle contraction reflect altered conformation of the contractile proteins in the muscle. Changes in the rate of secretion of glucose by the liver reflect altered activity and concentration of enzymes in the metabolic pathways for glucose synthesis. (Recall from Chap. 4 how the activity of an enzyme can be altered by allosteric or covalent modulation, and how the concentration of the enzyme can be altered by changes in the rate at which the protein is synthesized.) Generation of electric signals in nerve cells reflects altered conformation of membrane proteins constituting ion channels.

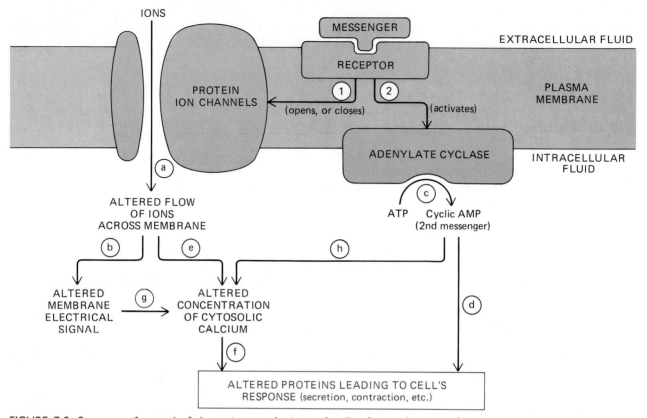

FIGURE 7-9. Summary of several of the various mechanisms whereby the combination of messenger with its specific plasma membrane receptor elicits the cell's response to that messenger. Note that, in the sequences shown, the receptor, once messenger has combined with it, does only one of two things—it either directly influences adjacent membrane channels or it activates adenylate cyclase. The mechanisms by which either electrical changes in the membrane or cAMP alter cytosolic calcium concentration are not shown—they include alteration of plasma-membrane calcium channels or active transport systems, as well as release or uptake of calcium by cell organelles. See text for description of numbered and lettered steps.

Thus, the question of postreceptor events can be stated as follows: What are the sequences of events leading from the binding of messenger by receptor to those changes in the conformation, activity, or concentration of cellular proteins which underlie the cell's response to the messenger? The answer to this question for messengers which bind to intracellular receptors (notably the steroid hormones) will be given in Chap. 9, and we will restrict the present discussion to the case of plasma-membrane receptors. (Figure 7-9 provides an overview, coded to the text discussion.)

The first step in the sequence is that the binding of a messenger to a plasma-membrane receptor produces a change in the conformation or structure of the receptor. What happens next is by no means identical for all plasma-membrane receptors (indeed it is unknown for many), and we will describe just two of the best understood sequences, those involving receptor-operated channels and those involving "second-messengers"—cyclic AMP and calcium. (As we shall see, these two categories— receptor-operated channels and second messengers—do not exist in isolation but interact.)

Receptor-operated channels

Adjacent to some plasma-membrane receptors are membrane proteins which constitute ion channels through the membrane. The alteration in membrane-receptor structure produced by the binding of messenger by receptor may permit the receptor to interact with the adjacent ion-channel protein (1 in Fig. 7-9). The effect of the interaction with receptor is to open or close the channel, which results in an increase or decrease in the diffusion of the ion (or ions) specific to the channel across the plasma membrane (*a* in Fig. 7-9). Such channels are known appropriately as **receptor-operated channels**. We shall see in Chap. 8 how such ion flows generate electric signals in nerve, muscle, and some gland cells (*b* in Fig. 7-9). But generation of electric signals is not the only function of receptor-operated channels; when the channel opened is a calcium channel, the resulting flow of calcium ions into the cell triggers intracellular events leading to the cell's response. We will pick up this story in a moment, after discussing adenylate cyclase and cyclic AMP.

Adenylate cyclase and cyclic AMP

The binding of messengers to certain plasma-membrane receptors causes the receptors to activate the enzyme **adenylate cyclase**, which is located on the inner surface of the membrane (2 in Fig. 7-9; Fig. 7-10). The activated adenylate cyclase then catalyzes the conversion of intracellular ATP to **cyclic 3',5'-adenosine monophosphate**, called simply **cyclic AMP (cAMP)** (*c* in Fig. 7-9 and Fig. 7-10). Cyclic AMP can then diffuse throughout the cell to trigger the intracellular sequences of events (*d* in Fig. 7-9) leading ultimately to the cell's overall response. Cyclic AMP, in fulfilling this function, has been termed a "second messenger." In this usage, cyclic AMP was thought at first to be a second messenger only for hormones. However, it is now recognized to serve this function in the responses to certain neurotransmitters and paracrine agents, as well. The action of cyclic AMP is terminated by its breakdown to noncyclic AMP, a reaction catalyzed by the enzyme **phosphodiesterase** (Fig. 7-11).

Thus far we have been concerned with getting cAMP generated. What does this intracellular messenger do? It activates one or more of the enzymes known collectively as **cAMP-dependent protein kinases**, which may be free in the cytoplasm or membrane-bound. In other words, the event common to

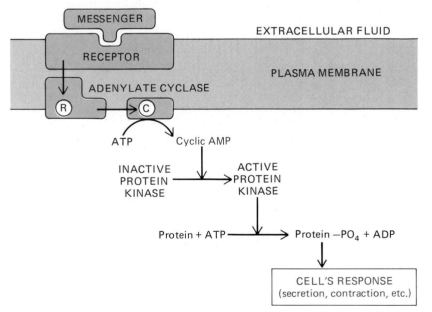

FIGURE 7-10. Cyclic AMP "second-messenger" system. Combination of the "first messenger" (hormone, neurotransmitter, or paracrine) with its specific receptor permits the receptor to bind to the regulatory unit (R) of adenylate cyclase. This, in turn, binds to the catalytic unit (C), activating it and causing it to catalyze formation of cyclic AMP (cAMP) inside the cell. This cAMP activates cAMP-dependent protein kinases, which in turn phosphorylate specific intracellular proteins, thereby triggering the biochemical reactions leading ultimately to the cell's response.

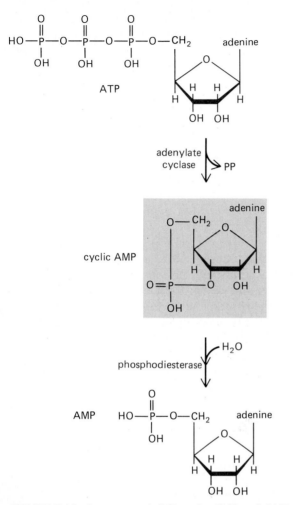

FIGURE 7-11. Structures of ATP, cyclic AMP, and AMP, the latter resulting from enzymatic alteration of cyclic AMP.

cascade. A single enzyme molecule, during the period when it is active, is capable of transforming into product not just one but many substrate molecules, let us say 100; therefore, one active molecule of adenylate cyclase may catalyze the generation of 100 cAMP molecules. At the two subsequent enzyme-activation steps in our example, another 100-fold amplification occurs. Therefore, the end result is that a single molecule of the original first messenger could theoretically trigger the generation of one million product molecules. This fact helps to explain how hormones and other messengers can be effective at extremely low extracellular concentrations.

How is it possible that activation of protein kinase by cAMP can be an event common to the great variety of biochemical sequences initiated by cAMP-generating messengers? As stated earlier, cAMP-dependent protein kinase is not just a single enzyme but a group of enzymes, each having distinct proteins for substrates and existing in different target cells and within different organelles of the same target cell. Thus, cAMP-dependent protein kinases can activate different proteins and so can trigger a variety of cascade activations simultaneously (Fig. 7-13). For example, this explains how a fat cell responds to epinephrine not only with glycogen breakdown (mediated by a particular enzyme) but also with triacylglycerol breakdown (mediated by a different enzyme). Moreover, whereas protein kinase activates certain enzymes, it inhibits others; this is true, for example, of the enzyme catalyzing the rate-limiting step in glycogen synthesis and explains how epinephrine inhibits glycogen synthesis at the same time that it stimulates glycogen breakdown (Fig. 7-14).

In summary, the biochemistry of the protein kinases and their substrates explains how a single molecule, cAMP, can produce so many different effects. It must be reemphasized that the ability of a cell to respond at all to a cAMP-generating first messenger depends upon the presence of specific receptors for that messenger in the plasma membrane. When messenger binds to these receptors, cAMP is generated and the ultimate response of that particular type of cell reflects the cell's types

all the sequences of biochemical events initiated by cAMP is activation of protein kinases. Protein kinases are enzymes which phosphorylate proteins by transferring to them a phosphate group (from ATP). Introduction of the phosphate typically modifies the activity of the protein (often itself an enzyme), and such reactions, therefore, are frequently critical regulatory points in metabolic pathways (Chap. 4).

Thus, the original activation of adenylate cyclase initiates a "cascade" in which proteins are converted in sequence from inactive to active forms. Figure 7-12 illustrates the benefit of such a

Number of
Molecules

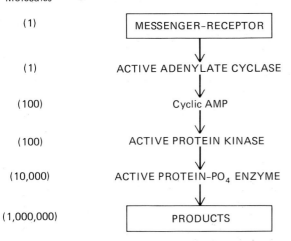

(1)	MESSENGER-RECEPTOR
(1)	ACTIVE ADENYLATE CYCLASE
(100)	Cyclic AMP
(100)	ACTIVE PROTEIN KINASE
(10,000)	ACTIVE PROTEIN-PO$_4$ ENZYME
(1,000,000)	PRODUCTS

FIGURE 7-12. Amplification in the cAMP system.

and locations of protein kinases and substrates. Not mentioned previously is the likelihood that, in some cases, the binding of messenger by receptor results in the inhibition of adenylate cyclase and less, rather than more, generation of cAMP. Finally, it should come as no surprise that a variety of drugs and diseases have alteration of the cAMP system as a common denominator.

Calcium as a "second messenger"

We previously described how the combination of messengers with certain plasma-membrane receptors causes the receptors to open adjacent plasma-membrane calcium channels. The cytosolic concentration of calcium is extremely low in virtually all cells because of the active transport of calcium out of the cell and into cellular organelles (Chap. 6). Because of the low cytosolic calcium, there always exists a large gradient for the diffusion of calcium into the cytosol. Therefore, when more plasma membrane calcium channels open, diffusion of calcium into the cytosol increases and an elevation of cytosolic calcium occurs (e in Fig. 7-9). How does an increased cytosolic calcium concentration lead to altered cell function (f in Fig. 7-9)? Often, a common initial event is combination of the addi-

tional calcium with a particular protein found in all cells, known as **calmodulin** (Fig. 7-15). On binding calcium, the calmodulin changes shape, becoming able to bind to an enzyme (or other specific protein) and activating or inhibiting it in turn. At this point, the story for calcium-activated calmodulin becomes analogous to that for cyclic AMP, since many of the enzymes influenced by calmodulin are protein kinases, and activation or inhibition of these calmodulin-dependent kinases leads to the cell's ultimate responses (contraction, secretion, etc.).

Cytosolic calcium's role as a second messenger has been introduced in this section in the context of receptor-operated calcium channels. However, this role has much greater scope, for there are very important physiological controls over cytosolic calcium concentration not mediated by the direct interaction of plasma-membrane receptors with calcium channels. First, there are many calcium channels (both in plasma membrane and cell organelles) which are not directly operated by receptors; rather, they open or close in response to membrane electric signals (g in Fig. 7-9) and other poorly understood chemical signals. Second, cAMP has important influences over cytosolic calcium concentration, as will be described in the next section (h in Fig. 7-9).

In conclusion, cytosolic calcium, controlled by a variety of inputs, acts as a second messenger in many different cells, including muscle, nerve, and gland cells (each of these is described in a subsequent chapter). Its role as a second messenger is frequently expressed by the term "coupler;" thus calcium is an **excitation-contraction coupler** in muscle (this means that it carries the "message" from the "excited" membrane to the contractile apparatus inside the muscle). Similarly, it is termed an **excitation-secretion coupler** in those cells in which it is the second messenger between the membrane events and the secretory apparatus.

Relationship between cyclic AMP and calcium

The preceding sections may well have given the impression that when a known second messenger is involved in a cell's response to a neurotransmit-

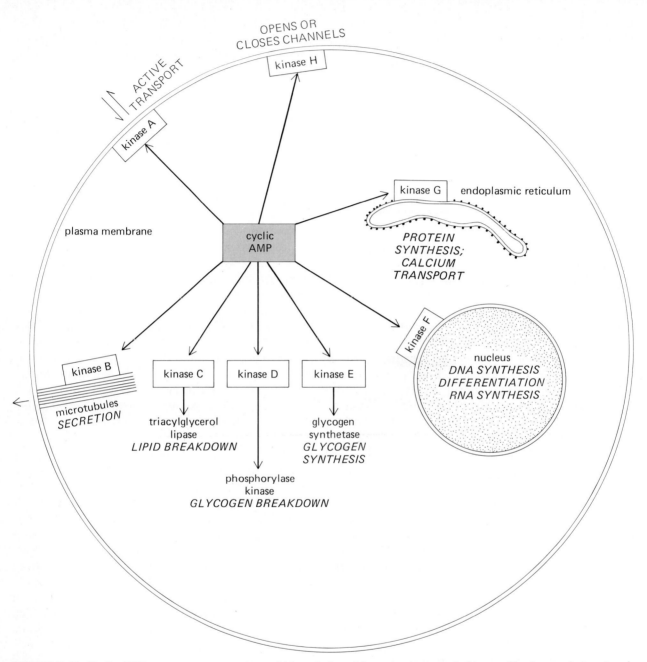

FIGURE 7-13. Cyclic AMP may exert many actions within a single cell by activating protein kinases. The letters designating the different kinases are arbitrary. It is extremely unlikely that any one cell would exhibit all these responses to cyclic AMP. The sequence leading from first messenger to activation of adenylate cyclase and, thereby, to formation of the cyclic AMP is not shown. (Adapted from Goldberg.)

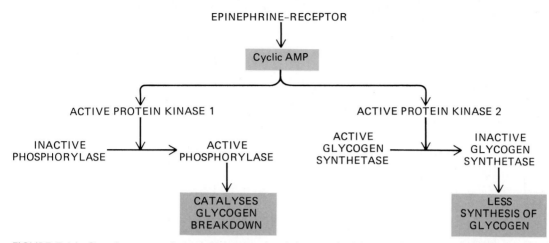

FIGURE 7-14. Coordinate stimulation of glycogen breakdown and inhibition of glycogen synthesis by epinephrine-induced generation of cAMP in liver cells. (The designation of the kinases as 1 and 2 is arbitrary.) The key point is that kinase-mediated addition of phosphate activates the enzyme phosphorylase, but inhibits the enzyme glycogen synthetase.

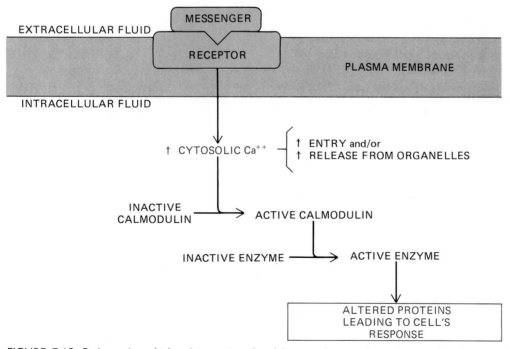

FIGURE 7-15. Pathway by which calcium, via calmodulin, couples membrane events (combination of "first messenger" with membrane receptor) to the cell's ultimate response. Many of the proteins influenced by calcium-activated calmodulin are kinases.

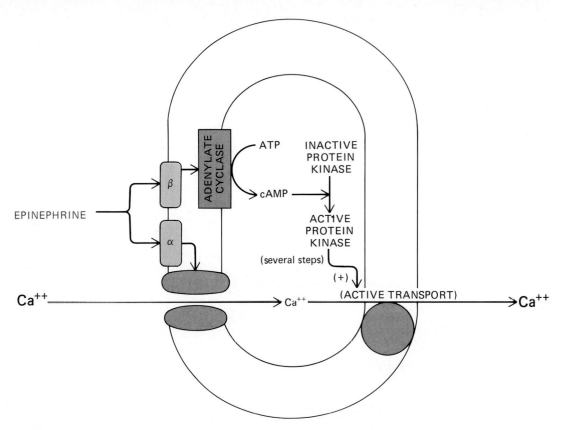

VASCULAR SMOOTH MUSCLE CELL

FIGURE 7-16. Diagrammatic summary of how epinephrine, via alpha (α) and beta (β) receptors, exerts opposing effects on cytosolic calcium concentration in the smooth muscle cells of blood vessels. Binding to the alpha receptor results in the opening of membrane calcium channels and diffusion of calcium into the cell. Binding to beta receptors results, via the cAMP system, in stimulation of the active transport system which moves calcium out of the cell. The roles played by these receptors in determining contraction of vascular smooth muscle cells are described in Chap. 11.

ter, hormone, or paracrine it is either calcium or cAMP. However, the fact is that both may be involved. In some cases, stimulation of a cell by first messenger causes cAMP and cytosolic calcium concentrations to change in the same direction; in other cases, these two second messengers change in opposite directions. Moreover, their effects are complexly intertwined. For example, active calmodulin may, itself, activate adenylate cyclase, and cAMP can, in certain systems, cause changes in cytosolic calcium concentration by influencing the transport of calcium by the plasma membrane or cell organelles (h in Fig. 7-9).

An example of such interactions as well as other concepts developed in the last sections of this chapter is provided by the effects of epinephrine (a hormone) on the muscle (smooth muscle cells) surrounding blood vessels (Fig. 7-16). There are two distinct types of receptors (designated alpha and beta) for epinephrine on these muscle cells, and combination of epinephrine with the two kinds produces diametrically opposed responses—contraction when epinephrine binds to alpha receptors and relaxation when it binds to beta receptors. The alpha-receptor effect is quite straightforward; the binding of epinephrine to this receptor causes, in sequence, the opening of receptor-operated calcium channels, an increase in cytosolic calcium concentration, activation of calmodulin, modulation of the contractile proteins, and increased contraction (as will be described in Chap. 10). In contrast, the binding of epinephrine to a beta receptor results, sequentially, in activation of adenylate cyclase, generation of cAMP, and activation of several protein kinases; the biochemical events triggered by these kinases (among other effects) induce increased active transport of calcium out of the smooth muscle cells, thereby causing cytosolic calcium concentration to decrease, which favors muscle relaxation. Whether the alpha effect (contraction) or the beta effect (relaxation) predominates depends upon the relative numbers of the two receptor types on the smooth muscle and their affinities for epinephrine.

Other second messengers

It would be incorrect to leave the impression that cAMP and calcium are the only possible second messengers, for it is very likely that others exist.

One candidate for such a role is a second cyclic nucleotide—**cyclic 3′,5′-guanosine monophosphate (cGMP)**—present in most, if not all, cells. Cyclic GMP exerts biochemical actions opposite to those of cAMP, and it has been hypothesized that the combination of some hormones, neurotransmitters, or paracrines with their plasma-membrane receptors may cause the generation of cGMP.

Receptors as protein kinases

In the sections above, we emphasized the central role of protein kinases and how both cAMP and calcium-calmodulin activate these important enzymes. Recently, it has become clear that the binding of messenger to some receptors causes these receptors, themselves, to become active protein kinases, phosphorylating specific cytosolic and plasma-membrane proteins (including the receptor, itself!). This provides yet another mechanism for transferring the first messenger's signal into cellular responses.

8 NEURAL CONTROL MECHANISMS

One of the greatest achievements of biological evolution has been the development of the human nervous system. It is made up of the brain and spinal cord as well as the many nerve processes that pass between these structures and the muscles, glands, or receptors they innervate. With the other, interrelated communication system, the endocrine system, the nervous system regulates many internal functions of the body and also coordinates the activities we know collectively as human behavior. These activities include not only easily observed acts such as smiling or reading, for example, but also less apparent phenomena such as feeling angry, being motivated, having an idea, or remembering a long-past event. These phenomena are functions we attribute to the ''mind,'' and they are believed to be related to the integrated activities of nerve cells.

In Chaps. 18 to 20 we shall consider specific aspects of neural activity concerned with sensation, the coordination of muscle activity in posture and movement, and the phenomena of consciousness, memory, and emotions. In this chapter we are concerned with the basic components common to all neural mechanisms. We begin by summarizing the structure of individual nerve cells and the basic organization and major divisions of the nervous system. Next, we describe the mechanisms underlying neural function, including the electric signals generated in nerve cells, the passage of signals from one nerve cell to another, and the processes by which external and internal stimuli initiate electric signals. Finally, we describe some of the basic patterns of neural interaction.

SECTION A.
STRUCTURE OF THE NERVOUS SYSTEM

Functional Anatomy of Neurons

The basic unit of the nervous system is the individual nerve cell or **neuron**.

Nerve cells occur in a wide variety of sizes and shapes; nevertheless, as shown in Fig. 8-1, most of them consist of four basic parts: (1) the cell body, (2) the dendrites, (3) the axon, and (4) the axon terminals. The **dendrites** form a series of highly branched outgrowths from the **cell body**. The dendrites and the cell body are the sites of most of the specialized junctions where signals are received from other neurons. The dendrites, in effect, increase the membrane surface area of the cell body, increasing the room available for incoming signals from other neurons.

The **axon**, or **nerve fiber**, is a single process extending from the cell body. The first portion of the axon plus the part of the cell body where the axon is joined is known as the **initial segment**. As we shall see, it is at the initial segment that the electric signals are initiated in many neurons, and they then propagate away from the cell body along the axon. The axon may give off branches called **collaterals** along its course, and near their ends, both the main axon and the collaterals undergo considerable branching, each branch ending in an **axon terminal**. These terminals are responsible for transmitting signals (chemical signals in most cases) from the neuron to the cells contacted by the axon terminals. However, the axons of some neurons release their chemical signals not from the axon terminal but from areas along the axon where the axon bulges. These regions are known as **varicosities**. Axons may be very long (more than a meter in length) in neurons that connect with distant parts of the nervous system or with peripheral organs; such neurons are called **projection neurons**. Axons may be very short in **local circuit neurons**, which connect only with cells in their immediate vicinity.

The axons of some (but not all) neurons are covered by **myelin** (Fig. 8-2), a fatty material formed from the plasma membranes of specialized cells which are closely associated with the axons. The spaces between the myelin-forming cells are **nodes of Ranvier**. Myelin insulates the membrane, making

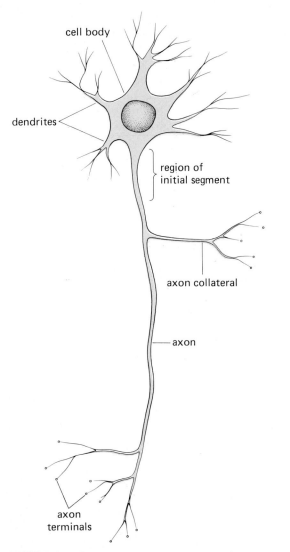

cell body

dendrites

region of
initial segment

axon collateral

axon

axon
terminals

FIGURE 8-1. Diagrammatic representation of a neuron.

it more difficult for electric currents to flow between the intracellular and extracellular fluid compartments.

As in other types of cells, a neuron's cell body contains the nucleus and thus has the genetic information necessary for protein synthesis. In order to maintain the structure and function of the processes extending from the cell body, materials, especially proteins, must be moved from the cell body where they are made to the peripheral processes. This occurs by a process known as **axon transport**, which appears to involve contractile proteins, microtubules, and filamentous structures (neurofilaments). However, axon transport of certain materials also occurs in the opposite direction, i.e., from axon terminals to the cell body, and this permits chemicals picked up at the terminals to influence metabolism in the cell body.

The many different shapes of neurons can be correlated with their location and function (Fig. 8-3). Yet, regardless of their shape, neurons can be divided into three functional classes: **afferent neurons, efferent neurons**, and **interneurons** (Fig. 8-4). Portions of the afferent and efferent neurons lie outside the central nervous system (the brain and spinal cord), whereas interneurons lie entirely within the central nervous system. At their peripheral endings afferent neurons have **receptors** which, in response to various physical or chemical changes in their environment, cause electric signals to be generated in the afferent neuron. (Recall from Chap. 7 that the term "receptor" has two totally distinct meanings, the one as defined here and the other referring to the specific proteins with which a chemical messenger combines to exert its effects on a target cell; both types of receptors will be referred to frequently in this chapter.) The afferent neurons propagate these electric signals from the receptors *into* the brain or spinal cord.

Efferent neurons transmit electric signals *from* the central nervous system out to the effector cells (muscle or gland cells).

The interneurons account for about 99 percent of all nerve cells. The number of interneurons in a pathway between afferent and efferent neurons varies according to the complexity of the action. One type of basic reflex, the stretch reflex, has no interneurons, the afferent neurons ending directly upon the efferent neurons; however, stimuli invoking memory or language may invoke millions of interneurons. As a rough estimate, for each afferent neuron entering the central nervous system, there are about 10 efferent neurons and about 200,000 interneurons.

One neuron can alter the activity of another neuron at anatomically specialized junctions known

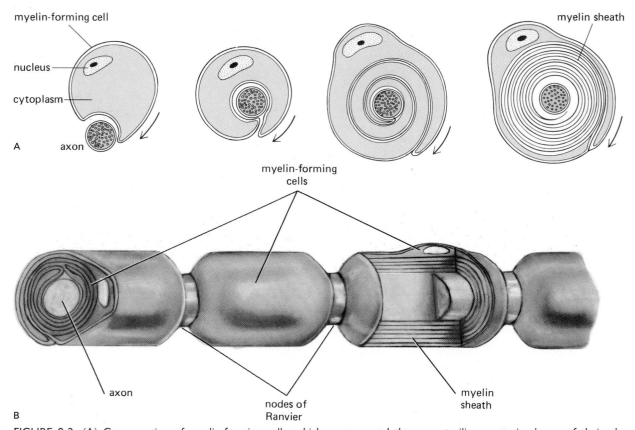

myelin-forming cell

nucleus

cytoplasm

myelin sheath

A

axon

myelin-forming
cells

axon

nodes of
Ranvier

myelin
sheath

B

FIGURE 8-2. (A) Cross section of myelin-forming cells, which wrap around the axon, trailing successive layers of their plasma membranes. (B) The myelin-forming cells are separated by a small space, the node of Ranvier.

as **synapses**. At most synapses, the activity is transmitted from one neuron to another by chemical messengers known as **neurotransmitters**. (This term also includes the chemicals by which efferent neurons communicate with effector cells.) Although there are many different transmitter substances in the nervous system, they are highly specific; in fact, neurons are often identified by the type of transmitter they release at their synapses with other neurons. The neurotransmitter released from one neuron alters the second cell by binding with a specific membrane protein known as a **receptor**. (The reader is cautioned one last time not to confuse this use of the term ''receptor'' with the receptors on afferent neurons mentioned above.)

Synapses occur between the axon terminal of one neuron and the cell body or dendrites of a second; however, in certain areas, synapses also occur between dendrites, between a dendrite and a cell body, or between an axon terminal and a second axon terminal. A neuron conducting electric signals toward a synapse is called a **presynaptic neuron**, whereas neurons conducting signals away from a synapse are **postsynaptic neurons**. Figure 8-5 shows how, in a multineuronal pathway, a single neuron can be postsynaptic to one group of cells and, at the same time, presynaptic to another.

A postsynaptic neuron may have thousands of synaptic junctions on the surface of its dendrites and cell body, so that signals from many presynaptic neurons converge upon it (Fig. 8-6). A single motor neuron in the spinal cord probably receives some 15,000 synaptic endings, and it has been calculated that certain neurons in the brain receive

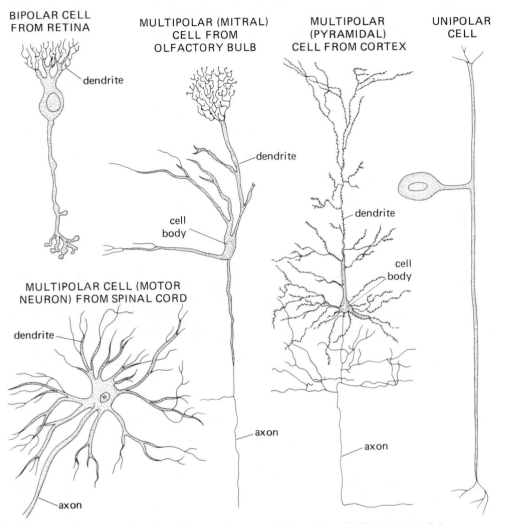

BIPOLAR CELL
FROM RETINA

dendrite

MULTIPOLAR (MITRAL)
CELL FROM
OLFACTORY BULB

dendrite

cell
body

axon

MULTIPOLAR
(PYRAMIDAL)
CELL FROM CORTEX

dendrite

cell
body

axon

UNIPOLAR
CELL

MULTIPOLAR CELL (MOTOR
NEURON) FROM SPINAL CORD

dendrite

axon

FIGURE 8-3. Examples of different shapes of neurons. (*Adapted from Kuffler and Nicholls.*)

more than 100,000. At a synapse a brief, small electric signal is produced in the postsynaptic cell.

Only about 10 percent of the cells in the nervous system are neurons; the remainder are **glial cells**, or **neuroglia**, which sustain the neurons metabolically, support them physically, and help regulate the ionic concentrations in the extracellular space. (Since the glia do not, however, branch as extensively as the neurons do, the neurons actually occupy about 50 percent of the volume of the central nervous system.)

Divisions of the Nervous System

The various parts of the nervous system are interconnected, but for convenience they can be divided into the **central nervous system**, composed of the brain and spinal cord, and the **peripheral nervous system**, consisting of the nerves extending from the brain and spinal cord (Fig. 8-7). We shall survey the anatomy and the broad functions of the various major structures of the nervous system; future chapters will describe these functions in more detail.

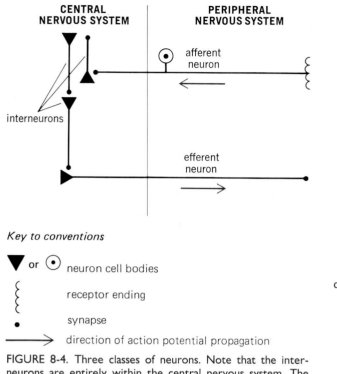

CENTRAL NERVOUS SYSTEM

PERIPHERAL NERVOUS SYSTEM

afferent neuron

interneurons

efferent neuron

Key to conventions

▼ or ⊙ neuron cell bodies

〰 receptor ending

• synapse

⟶ direction of action potential propagation

FIGURE 8-4. Three classes of neurons. Note that the interneurons are entirely within the central nervous system. The conventions that we will use for the different parts of nerve cells are listed in the key.

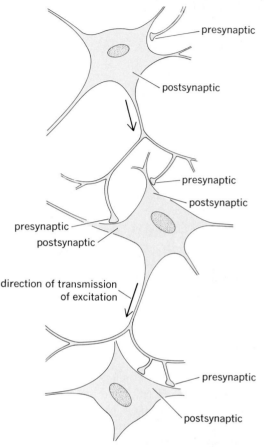

presynaptic

postsynaptic

presynaptic

postsynaptic

presynaptic

postsynaptic

direction of transmission of excitation

presynaptic

postsynaptic

FIGURE 8-5. A single neuron postsynaptic to one group of cells can be presynaptic to another.

Central nervous system: Spinal cord

The spinal cord is a slender cylinder about as big around as the little finger. Figure 8-8 shows the basic divisions of a cross section of the spinal cord. The central butterfly-shaped area is called the **gray matter**; it is filled with interneurons, the cell bodies and dendrites of efferent neurons, the entering fibers of afferent neurons, and glial cells. Cell bodies of neurons having similar functions are often clustered together into groups called **nuclei**. The gray region is surrounded by **white matter**, which consists largely of myelinated axons, the fatty myelin giving the region a white appearance. These axons are organized into groups called **tracts** and **pathways**, which run longitudinally through the cord, some descending to convey information from the brain to the spinal cord (or from upper to lower levels of the cord), others ascending to transmit in

the opposite direction. The fiber pathways are so organized that a given tract or pathway generally contains fibers subserving only one function. For example, the fibers that transmit information from light-touch receptors in the skin travel together in one pathway. Some of these spinal-cord pathways are illustrated in Fig. 8-9. The tracts and even the nuclei can be visualized as columns extending the length of the spinal cord.

Groups of afferent fibers (from peripheral receptors) enter the spinal cord on the dorsal side (the side toward the back of the body); these groups form the **dorsal roots** and contain the **dorsal root ganglia** (the cell bodies of the afferent neurons).

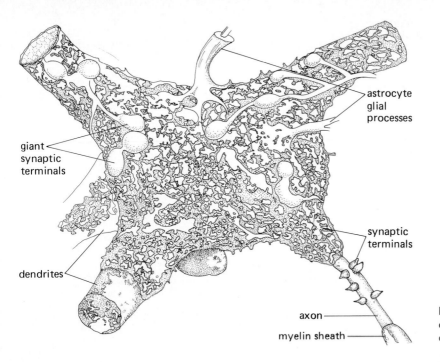

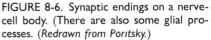

FIGURE 8-6. Synaptic endings on a nerve-cell body. (There are also some glial processes. (*Redrawn from Poritsky.*)

Efferent fibers (to effector cells) leave the spinal cord on the ventral side (the side toward the front of the body) via the **ventral roots**. Shortly after leaving the cord, the dorsal and ventral roots from the same level of the spinal cord combine to form a pair of **spinal nerves**, one on each side of the spinal cord. (The spinal nerves are discussed with the peripheral nervous system.)

Central nervous system: Brain

The brain is composed of six subdivisions: **cerebrum, diencephalon, midbrain, pons, medulla**, and **cerebellum** (Fig. 8-10 and Table 8-1). The midbrain, pons, and medulla together form the **brainstem**, whereas the cerebrum and diencephalon together constitute the **forebrain**.

Brainstem. The brainstem is literally the stalk of the brain; through it pass all the nerve fibers relaying signals of afferent input and efferent output between the spinal cord and higher brain centers. In addition, the brainstem gives rise to 10 of the 12

pairs of **cranial nerves**, whose axons innervate the muscles and glands of the head and many organs in the thoracic and abdominal cavities. The cranial nerves also innervate sensory receptors in these same areas.

Running through the entire brainstem is a core of tissue called the **reticular formation**, which is composed of small, highly branched neurons. These neurons receive and integrate information from many afferent pathways as well as from many other regions of the brain. Some reticular-formation neurons are clustered together, forming certain of the brainstem nuclei and integrating centers. These include the cardiovascular, respiratory, swallowing, and vomiting centers, all of which are discussed in subsequent chapters. The reticular formation can be divided functionally into descending and ascending systems. The descending pathways influence efferent neurons and frequently afferent neurons as well. The ascending components affect such things as wakefulness and the direction of attention to specific events. Both components function in controlling muscle activity.

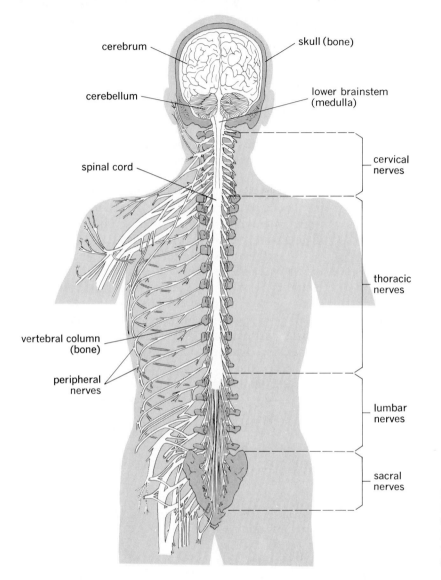

cerebrum

skull (bone)

cerebellum

lower brainstem
(medulla)

spinal cord

cervical
nerves

thoracic
nerves

vertebral column
(bone)

peripheral
nerves

lumbar
nerves

sacral
nerves

FIGURE 8-7. Nervous system viewed from
behind. The cervical, thoracic, lumbar, and
sacral nerves, which constitute the spinal
nerves, are shown only on the left side of
the figure, although they occur on both
sides. (*Redrawn from Woodburne.*)

Cerebellum. The cerebellum is chiefly involved with skeletal muscle functions; it helps to maintain balance and to provide smooth, directed movements. The location of the cerebellum relative to the brainstem and forebrain can be seen in Fig. 8-10.

Forebrain. The large part of the brain remaining when the brainstem and cerebellum have been excluded is the forebrain. It consists of a central core

(the diencephalon) and the right and left **cerebral hemispheres** (the cerebrum). The hemispheres, although largely separated by a longitudinal division, are connected to each other by axon bundles known as **commissures**, the **corpus callosum** being the largest (Fig. 8-11). Areas within a hemisphere are connected to each other by **association fibers**.

The outer part of the cerebral hemispheres, the **cerebral cortex**, is a cellular shell about 3 mm thick,

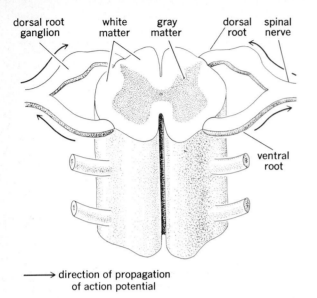

dorsal root ganglion white matter gray matter dorsal root spinal nerve

ventral root

⟶ direction of propagation of action potential

FIGURE 8-8. Section of the spinal cord (ventral view, i.e., from the front).

complex integrating area of the nervous system and is necessary for the bringing together of basic afferent information into meaningful perceptual images and for the ultimate refinement of control over efferent systems. The cortex is divided into several parts, or **lobes** (the frontal, parietal, occipital, and temporal lobes, Fig. 8-10). The cortex is an area of gray matter, so called because of the predominance of cell bodies. The **subcortical nuclei** form other areas of gray matter which lie, as the name suggests, under the surface of the cortex (Fig. 8-11) and contribute to the coordination of muscle movements. In other parts of the forebrain, nerve-fiber tracts predominate, their whitish myelin coating distinguishing them as white matter.

The **thalamus,** part of the diencephalon (Fig. 8-11), is a relay station and important integrating center for sensory input on its way to the cortex and is also important in motor control. Like the brainstem, it contains a central core that is part of the reticular formation. The **hypothalamus,** which lies below the thalamus (Fig. 8-11), is a tiny region whose volume is only 5 to 6 cm³. Yet it is responsible for the integration of many basic homeostatic mechanisms and behavioral patterns that involve correlation of neural and endocrine functions. In-

which forms the outer rim of the cross section in Fig. 8-11. The cortical surface of the cerebrum is highly folded, which increases the area available for cortical neurons without increasing appreciably the volume of the skull. The cortex is the most

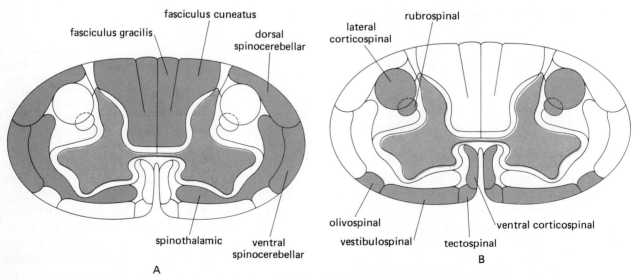

fasciculus gracilis fasciculus cuneatus dorsal spinocerebellar

spinothalamic ventral spinocerebellar

A

lateral corticospinal rubrospinal

olivospinal vestibulospinal tectospinal ventral corticospinal

B

FIGURE 8-9. Major tracts of the spinal cord; tracts are shaded and labeled. (A) Ascending tracts; (B) descending tracts.

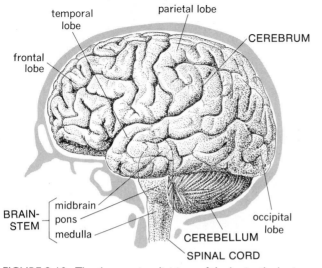

temporal lobe

parietal lobe

frontal lobe

CEREBRUM

BRAIN-STEM
- midbrain
- pons
- medulla

occipital lobe

CEREBELLUM

SPINAL CORD

FIGURE 8-10. The three major divisions of the brain: the brainstem, cerebellum, and forebrain. The outer layer of the forebrain (the cerebrum) consists of four lobes, as shown. The diencephalon, which is an extension of the brainstem and lies deep within the brain, is not shown.

TABLE 8-1. Divisions of the central nervous system		
Brain	Cerebrum; cerebral hemispheres, including cerebral cortex and basal ganglia	Forebrain
	Diencephalon, including thalamus and hypothalamus	
	Midbrain	Brainstem
	Pons	
	Medulla	
	Cerebellum	
Spinal cord		

deed, the hypothalamus appears to be the single most important control area for the regulation of the internal environment.

The **limbic system** is not a single brain region but is an interconnected group of brain structures within the forebrain, including portions of frontal-lobe cortex, temporal lobe, thalamus, and hypothalamus, as well as the circuitous neuron pathways connecting these parts (Fig. 8-12). The limbic system is associated with learning and emotional behavior; stimulation of some of the limbic areas leads to behavior interpreted as rewarding or pleasurable, whereas stimulation of other areas of the limbic system is associated with unpleasant feelings. Besides being connected with each other, the parts of the limbic system have connections with many other parts of the central nervous system. For example, it is likely that information from all the different types of receptors can influence activity in the limbic system, whereas activity of the limbic system can result in a wide variety of responses, such as sweating, blushing, heart-rate changes, laughing, and sobbing.

Peripheral nervous system

Nerve processes (axons) in the peripheral nervous system, which transmit signals to and from the central nervous system, are grouped into bundles called **nerves**, whereas the individual axons are called nerve fibers. (There are no ''nerves'' in the central nervous system; recall that axon bundles there are called tracts or pathways.) The peripheral nervous system consists of 43 pairs of nerves; 12 pairs connect with the brain and are called the **cranial nerves**, and 31 pairs connect with the spinal cord as the **spinal nerves**. The spinal nerves have been shown in Fig. 8-7.

Each nerve fiber is surrounded by a type of satellite cell called a Schwann cell. Some of the axons are wrapped in layers of Schwann-cell membrane, and these tightly wrapped membranes form the myelin sheath (see Fig. 8-2). Other fibers are unmyelinated and are simply tucked into invaginations of the Schwann cells (Fig. 8-13).

The individual nerve fibers in a nerve may be processes of either afferent or efferent neurons; accordingly they may be classified as belonging to the **afferent** or the **efferent division** of the peripheral nervous system (Table 8-2). All of the spinal nerves contain processes of both afferent and efferent neurons, whereas some of the cranial nerves (the optic nerves from the eyes, for example) contain only afferent fibers.

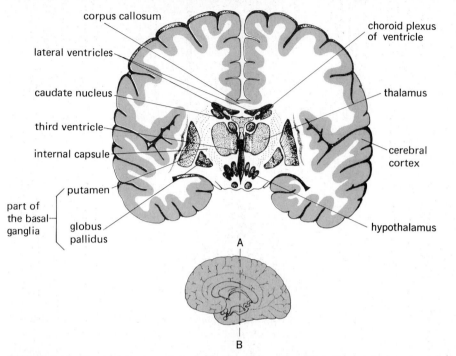

FIGURE 8-11. Coronal section of the brain. The dashed line AB indicates the location of the section. The caudate nucleus is also part of the basal ganglia.

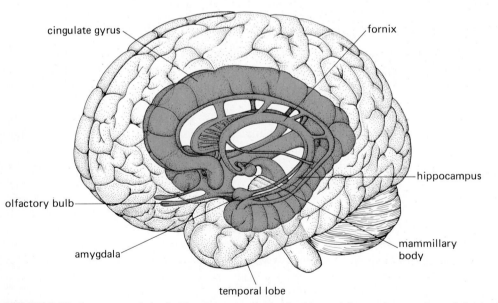

FIGURE 8-12. Structures of the limbic system are shaded in this partially transparent view of the brain.

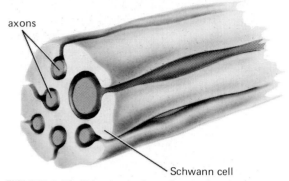

FIGURE 8-13. Schwann cell enclosing several unmyelinated nerve fibers. (*Redrawn from Warwick and Williams.*)

TABLE 8-2. Divisions of the peripheral nervous system

I   Afferent division

II Efferent division

   a. Somatic nervous system

   b. Autonomic nervous system

      1. Sympathetic division

      2. Parasympathetic division

Peripheral nervous system: Afferent division. Afferent neurons convey information from receptors in the periphery to the central nervous system. The cell bodies of afferent neurons are outside of the brain or spinal cord in structures called **ganglia**. From the region of the cell body, one long process (the peripheral process) extends away from the ganglion out to the receptors; commonly, the process branches several times as it nears its destination, each branch containing or innervating one receptor. A second process (the central process) passes from the cell body into the central nervous system, where it branches; these branches terminate in synapses on other neurons.

Afferent neurons are sometimes called **primary afferents** or **first-order neurons** because they are the first cells entering the central nervous system in the synaptically linked chains of neurons that handle incoming information. They are also frequently called "sensory neurons," but we prefer not to use this term because it implies that the signals transmitted by these neurons are destined to reach consciousness, whereas only a small fraction of the total afferent signals entering the central nervous system give rise to conscious sensations. For example, we have no conscious awareness of our blood pressure even though we have receptors sensitive to this variable.

Peripheral nervous system: Efferent division. The efferent division is more complicated than the afferent, being subdivided into a **somatic nervous system** and an **autonomic nervous system**. These terms are somewhat unfortunate because they conjure up additional "nervous systems" distinct from the central and peripheral nervous systems. Keep in mind that the terms refer simply to the efferent divisions of the peripheral nervous system.

Although separating these two divisions is justified by many anatomical and physiological differences, the simplest distinction between the two is that the somatic nerve fibers innervate skeletal muscle, whereas the autonomic nerve fibers innervate smooth and cardiac muscle and glands. Other differences are listed in Table 8-3.

Somatic nervous system. The somatic division of the peripheral nervous system is made up of all the fibers going from the central nervous system to skeletal muscle cells. The cell bodies of these neurons are located in groups in the brain or spinal cord; their large-diameter, myelinated axons leave the central nervous system and pass directly (i.e., without any synapses) to skeletal muscle cells. The neurotransmitter substance released by these neurons is **acetylcholine**. Because activity in the somatic efferent neurons leads to contraction of the innervated skeletal muscle cells, these neurons are often called **motor neurons**. Excitation of motor neurons only leads to the *contraction* of skeletal muscle cells; there are no *inhibitory* somatic neurons.

Autonomic nervous system. Fibers of the autonomic division of the peripheral nervous system innervate

---

TABLE 8-3. Differences between somatic and autonomic nervous systems

---

Somatic nervous system

1. Consists of a single neuron between the central nervous system and the effector organ

2. Innervates skeletal muscle

3. Always leads to excitation of the muscle

Autonomic nervous system

1. Has a two-neuron chain (connected by a synapse) between the central nervous system and the effector organ

2. Innervates smooth or cardiac muscle or gland cells

3. Can lead to excitation or to inhibition of the effector cells

---

cardiac muscle, smooth muscle, and glands. Anatomical and physiological differences within the autonomic nervous system are the basis for its further subdivision into **sympathetic** and **parasympathetic** components. The cell bodies of the first neurons in the two subdivisions are located in different areas of the central nervous system, and their fibers leave at different levels—the sympathetic fibers from the thoracic and lumbar regions of the spinal cord, and the parasympathetic from the brain and the sacral portion of the spinal cord (Fig. 8-14). For example, the vagus nerve, which is cranial nerve X, carries the parasympathetic fibers which innervate the viscera of the thoracic and abdominal cavities. Thus, the sympathetic division is also called the **thoracolumbar** division, and the parasympathetic the **craniosacral division**.

In the autonomic nervous system, the fibers between the central nervous system and the effector cells consist of two neurons and one synapse (Fig. 8-15). (This is in contrast to the single neuron of the somatic system.) The synapse between these two autonomic neurons is outside the central nervous system, in cell clusters called **autonomic ganglia**. The two divisions of the autonomic nervous system differ with respect to the locations of their ganglia. Most of the sympathetic ganglia lie close to the spinal cord, forming the paired chains of ganglia known as the **sympathetic trunks**, though others lie far away from the spinal cord, closer to the innervated organ. In contrast, the parasympathetic ganglia lie within the effector organ. The fibers passing between the central nervous system and the sympathetic or parasympathetic ganglia are called **preganglionic** autonomic fibers; those passing between the ganglia and the effector cells are the **postganglionic** fibers.

In both sympathetic and parasympathetic divisions, the major neurotransmitter released at the ganglionic synapse between pre- and postganglionic fibers is acetylcholine. In the parasympathetic division, the neurotransmitter between the postganglionic fiber and the effector cell is also acetylcholine. In the sympathetic division, the transmitter between the postganglionic fiber and the effector cell is usually **norepinephrine** (Fig. 8-15). (Other transmitters, such as epinephrine, acetylcholine, ATP, dopamine, and several peptides, are released by certain sympathetic neurons, but they play a smaller role in sympathetic nervous-system activity than norepinephrine does.)

Because historically it was believed that epinephrine rather than norepinephrine was the major sympathetic postganglionic neurotransmitter, and because epinephrine was called by its British name, "adrenaline," fibers that release norepinephrine came to be called **adrenergic** fibers. Fibers that release acetylcholine are called **cholinergic** fibers. Thus, the neurons in the efferent nervous system, both somatic and autonomic, are primarily cholinergic neurons with the major exception of the postganglionic sympathetic fibers, which are usually adrenergic.

Many drugs stimulate or inhibit various components of the autonomic nervous system. Important sites of action for many of them are the receptors for acetylcholine and norepinephrine. Indeed, experiments with these drugs have revealed that there are several different types of receptors for each transmitter. For example, acetylcholine receptors on the postsynaptic neurons of all autonomic ganglia respond to low doses of the drug nicotine and are therefore called **nicotinic receptors**. In contrast, the acetylcholine receptors on smooth mus-

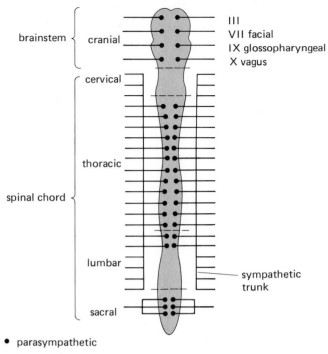

brainstem { cranial
- III
- VII facial
- IX glossopharyngeal
- X vagus

cervical

thoracic

spinal chord {

lumbar

sympathetic trunk

sacral

- parasympathetic
- sympathetic

FIGURE 8-14. Origins of the sympathetic and parasympathetic divisions of the autonomic nervous system.

cle, cardiac muscle, and gland cells are not stimulated by nicotine but are stimulated by the mushroom poison muscarine; they are called **muscarinic receptors**. (To complete the story of the peripheral cholinergic receptors, the receptors on the neuromuscular junctions of skeletal muscle fibers, innervated by the somatic motor neurons, are nicotinic.) Similarly, there are two classes of adrenergic receptors (i.e., receptors for epinephrine and for norepinephrine) also distinguished largely by the specific drugs that stimulate or block them. They are called **alpha-adrenergic** and **beta-adrenergic receptors**. Both alpha- and beta-adrenergic receptors can be subdivided into still further categories, again according to the drugs that influence them.

One set of postganglionic cells in the sympathetic division never develops axonal fibers; instead, upon activation by preganglionic axons, the cells of this "ganglion" release their transmitter directly into the blood stream. This "ganglion," called the **adrenal medulla**, is therefore an endocrine gland whose secretion is controlled by the sympathetic preganglionic fibers (Chap. 9). It releases a mixture of about 80 percent epinephrine and 20 percent norepinephrine into the blood. These substances, properly called hormones rather than neurotransmitters in these circumstances, are transported via the blood to receptor sites on effector cells sensitive to them. The receptor sites may be the same adrenergic receptors that are located near the varicosities of sympathetic postganglionic neurons and that are normally activated by the norepinephrine released from these varicosities, or the receptor sites may be located at other places that are far from the varicosities and are therefore activated only by the circulating epinephrine or norepinephrine.

The heart and many glands and smooth muscles are innervated by both sympathetic and parasympathetic postganglionic fibers, i.e., they receive **dual innervation**. Whatever effect one division has on the effector cells, the other division frequently (but not always) has just the opposite effect (Table 8-4). For example, norepinephrine released from

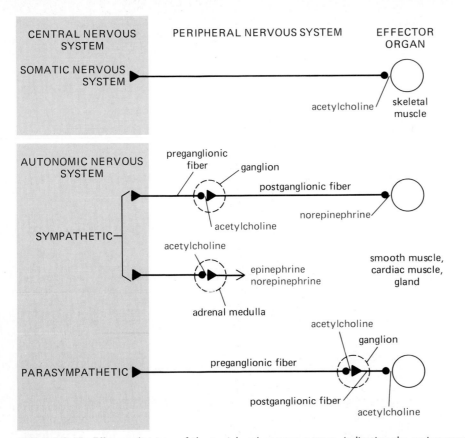

FIGURE 8-15. Efferent divisions of the peripheral nervous system, indicating the major neurotransmitter released at each site. In a few cases sympathetic neurons release a transmitter other than norepinephrine.

the sympathetic nerves to the heart increases the heart rate, whereas acetylcholine released from the parasympathetic fibers to the heart decreases it. In the intestine, activation of the sympathetic fibers reduces contraction of the smooth muscle in the intestinal wall, whereas the parasympathetics increase contraction. Normally, both systems are partially active. Dual innervation by fibers which cause opposite responses in effector cells provides a very fine degree of control over the effector organ, since it is like equipping a car with both an accelerator and a brake. With only an accelerator, one could slow the car simply by decreasing the pressure on the accelerator, but the combined effects of releasing the accelerator and applying the brake provide faster and more accurate control. In keeping with

this, the two divisions are usually activated reciprocally; i.e., as the activity of one division is increased, the activity of the other is decreased. However, as shown in Table 8-4, there are important cases in which effector cells are innervated by only one division; for example, the smooth muscle of most of the small blood vessels known as arterioles receives only sympathetic innervation. It should also be emphasized that skeletal muscle cells do not receive dual innervation; they are activated by somatic motor neurons, which are only excitatory.

Because glands, smooth muscle, and the heart participate as effectors in almost all bodily functions, it follows that the autonomic nervous system has an extremely important role in the homeostatic control of the internal environment. It exerts a wide array of effects (Table 8-4) which will be discussed in relevant places throughout this book.

Autonomic responses usually occur without conscious control or awareness, as though they were indeed autonomous (in fact, the autonomic nervous system has been called the "involuntary" nervous system). However, it is wrong to assume that this is always the case, for it has been shown that discrete visceral or glandular responses can be learned and thus, to this extent, voluntarily controlled. For example, to avoid an electric shock, a rat can learn to increase or decrease its heart rate selectively, and a rabbit can learn to constrict the blood vessels in one ear while dilating those in the other. The implications of such control of autonomic functions in human medicine are enormous. These experiments are also important in showing that small segments of the autonomic response can be regulated independently; thus, overall autonomic responses, made up of many small components, are quite variable. Rather than being gross, undiscerning discharges, they are finely tailored to the specific demands of any given situation.

Evolution of the Nervous System

During an early stage of evolution, the nervous system was probably a simple system with a limited number of interneurons interposed between the afferent and efferent nerve cells. The interneuronal component rapidly expanded until it came to be by far the largest part. The interneurons formed networks of increasing complexity, at first involved mainly with the stability of the internal environment and the position of the body in space. Those cells with increasing specialization of function came to be localized at one end of the primitive nervous system, and the brain began to evolve.

The brainstem, which is the oldest part of the brain in this evolutionary sense, retains today many of the anatomical and functional characteristics typical of those most primitive brains. With continued evolution, newer, increasingly complex structures were added on top of (or in front of) the older ones.

They developed as paired symmetrical tissues, the cerebral hemispheres, and reached their highest degree of sophistication with the formation of the cerebral cortex.

The newer structures served in part to elaborate, refine, modify, and control already existing functions. For example, it is possible to perceive the somatic stimuli of pain, touch, and pressure with only a brainstem, but the stimulus cannot be localized without a functioning cerebral cortex. Perhaps even more important is that these newer, more anterior parts of the brain came to be involved in the perception of goals, the ordering of goal priorities, and the patterning of behaviors to serve in the pursuit of these goals.

Blood Supply, Blood-Brain Barrier Phenomena, and Cerebrospinal Fluid

Usually, glucose is the only substrate metabolized by the brain to supply its energy requirements, and most of the energy from glucose is transferred to high-energy ATP molecules during its oxidative breakdown. The glycogen stores of brain are negligible; thus the brain is completely dependent upon a continuous blood supply of glucose and oxygen. Although the adult brain is only 2 percent of the body weight, it receives 15 percent of the total blood supply to support the high oxygen utilization. If the oxygen supply is cut off for 4 to 5 min, or if the glucose supply is cut off for 10 to 15 min, brain damage will occur. In fact, the most common cause of brain damage is a stoppage of the blood supply (a **stroke**). The cells in the region deprived of nutrients cease to function and die.

The exchange of substances between the blood and neurons in the central nervous system is different from the more or less unrestricted diffusion of nonprotein substances from capillaries in other organs. A complex group of **blood-brain barrier** mechanisms closely controls both the kinds of substances that enter the extracellular fluid of the brain and the rate at which they enter. This capillary barrier comprises both anatomical structures and physiological transport systems that handle differ-

TABLE 8-4. Some effects of autonomic nervous system activity*

Effectors	Effect of sympathetic nervous system	Effect of parasympathetic nervous system
Eyes		
Muscles of the iris	Contraction of radial muscle (widens pupil)	Contraction of sphincter muscle (makes pupil smaller)
Ciliary muscle	Relaxation (tightens suspensory ligaments, thus flattening lens for far vision)	Contraction (relaxes ligament, allowing lens to become more convex for near vision)
Heart		
S-A node	Increase in heart rate	Decrease in heart rate
Atria	Increase in contractility	Decrease in contractility
A-V node	Increase in conduction velocity	Decrease in conduction velocity
Ventricles	Increase in contractility	Decrease in contractility
Arterioles		
Coronary	Constriction	Dilation
Skin and mucous membrane	Constriction	———†
Skeletal muscle	Constriction or dilation	———
Abdominal viscera and kidneys	Constriction	———
Salivary glands	Constriction	
Penis or clitoris	Constriction	Dilation (causes erection)
Veins	Constriction	———
Lungs		
Bronchial muscle	Relaxation	Contraction
Bronchial glands	Inhibition of secretion	Stimulation of secretion
Salivary glands	Stimulation of secretion	Stimulation of secretion
Stomach		
Motility	Decrease	Increase
Sphincters	Contraction	Relaxation
Secretion	Inhibition	Stimulation
Gallbladder and ducts	Relaxation	Contraction
Liver	Glycogenolysis, gluconeogenesis	———
Pancreas		
Exocrine glands	Decrease in secretion	Stimulation of secretion
Endocrine glands (islets)	Inhibition of insulin secretion, stimulation of glucagon secretion	Stimulation of insulin secretion
Fat cells	Stimulation of lipid breakdown	———
Urinary bladder	Relaxation	Contraction
Uterus	Pregnant: contraction Nonpregnant: relaxation	Variable
Reproductive tract (male)	Ejaculation	———

TABLE 8-4. *(continued)*

Effectors	Effect of sympathetic nervous system	Effect of parasympathetic nervous system
Skin		
Muscles causing hair to stand erect	Contraction	———
Sweat glands	Stimulation of secretion	Stimulation of secretion
Lacrimal glands	———	Stimulation of secretion

* Specific effectors will be discussed in later chapters.
† Dash means that these cells are not innervated by this branch of the autonomic nervous system.
Table adapted from Louis S. Goodman and Alfred Gilman, "The Pharmacological Basis of Therapeutics," 6th ed., Macmillan, New York, 1980.

ent classes of substances in different ways. The blood-brain barrier mechanisms precisely regulate the chemical composition of the extracellular fluid of the brain and minimize the ability of many harmful substances to reach neural tissue. However, certain areas of the brain lack a blood-brain barrier and can be readily influenced by blood-borne substances.

In addition to its blood supply, the central nervous system is perfused by a second fluid, the **ce-** **rebrospinal fluid**. This clear fluid fills the four cavities (**ventricles**, Fig. 8-16) within the brain and surrounds the outer surfaces of the brain and spinal cord so that the central nervous system literally floats in a cushion of cerebrospinal fluid. Since the brain and spinal cord are very soft, delicate tissue with about the consistency of jelly, they are thus protected from sudden and jarring movements. Most of the cerebrospinal fluid is secreted into the ventricles by highly vascular tissues, the **choroid**

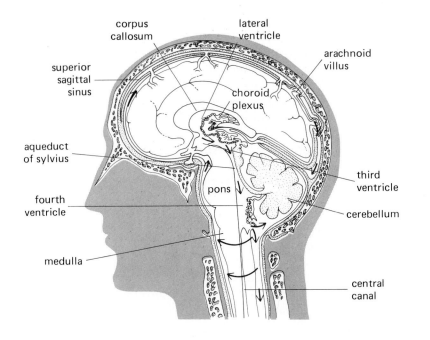

FIGURE 8-16. The ventricular system of the brain and distribution of the cerebrospinal fluid. (The second ventricle is not shown.) Cerebrospinal fluid is formed in the ventricles (much of it by the choroid plexuses), passes to the subarachnoid space outside the brain, and then passes through small valvelike structures into the large veins of the head. (*Adapted from Kuffler and Nicholls.*)

plexuses. A blood-brain barrier is present here, too, between the capillaries of the choroid plexuses and the cerebrospinal fluid that flows next to the neural tissue. Consistent with the barrier mechanisms, cerebrospinal fluid is a selective secretion, not a simple filtrate of plasma. For example, protein, potassium, and calcium concentrations are slightly lower in cerebrospinal fluid than in plasma, whereas the sodium and chloride concentrations are slightly higher.

The cerebrospinal fluid moves from its origin, through the interconnected ventricular system, to the brainstem, where it passes through small openings out to the surface of the brain and spinal cord. Aided by circulatory, respiratory, and postural pressure changes, the cerebrospinal fluid finally flows to the top of the outer surface of the brain, where it enters large veins through one-way valves between the cerebrospinal fluid and the blood. If the path of flow is obstructed at any point between its site of formation and its final reabsorption into the vascular system, cerebrospinal fluid builds up, causing hydrocephalus or "water on the brain." The resulting pressure in the ventricles may be high enough to damage the brain and cause mental retardation in severe, untreated cases.

As the cerebrospinal fluid flows from its origin in the ventricles, substances flow between it and the extracellular fluid of the neural tissue, since the walls of the ventricles are permeable to most substances in the cerebrospinal fluid. This exchange across the ventricular walls allows some nutrients to enter brain tissue from cerebrospinal fluid and allows some end products of brain metabolism to leave. However, the major sites of nutrient and end-product exchange are the cerebral capillaries.

This completes our survey of the organization of the nervous system. We now turn to the mechanisms by which neurons, synapses, and receptors on afferent neurons actually function, beginning with the electrical properties which underlie all these events.

SECTION B.
MEMBRANE POTENTIALS

Basic Principles of Electricity

All atoms contain negatively and positively charged particles—electrons and protons. Some molecules have no net electric charge since they contain equal numbers of electrons and protons. However, many molecules, as we have seen, have a net electric charge due to such components as the negative carboxyl group ($RCOO^-$) and the positive amino group (RNH_3^+). Moreover, inorganic substances such as sodium, potassium, and chloride (Na^+, K^+, and Cl^-) are present in solution as the charged particles known as ions.

With the exception of water, the major chemical substances in the extracellular fluid are, as we have seen (Chap. 6), sodium and chloride ions, whereas the intracellular fluid contains high concentrations of potassium and organic molecules, particularly proteins and phosphate compounds, containing ionized groups. Since many charged particles are found both inside and outside of cells, it is not surprising to discover that electrical phenomena resulting from the distribution of these charged particles play a significant role in cell function.

An electric force draws positively and negatively charged substances together. In contrast, like charges repel each other; i.e., positive charge repels positive charge and negative charge repels negative charge. The amount of force acting between electric charges increases when the charged particles are moved closer together and when the quantity of charge is increased (Fig. 8-17).

When oppositely charged particles come together, the electric force of attraction moving the particles can be used to perform work. Conversely, to separate oppositely charged particles, energy must be used to overcome this attractive force. Thus, separated electric charges of opposite sign

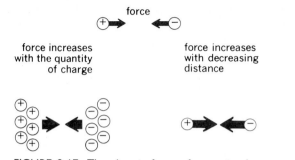

FIGURE 8-17. The electric force of attraction between positive and negative charges increases with the quantity of charge and with decreasing distance between charges.

have the "potential" of doing work if they are allowed to come together again. The potential of separated electric charges to do work when moving from one point in a system to another is called an **electric potential**, or the **potential difference**, which we shall often shorten to simply **potential**. The units of electric potential are **volts (V)**, but since the total amount of charge that can be separated in most biological systems is very small, the potential differences are small and are measured in **millivolts**. (A millivolt is 0.001 V.)

The movement of electric charge is called **current**. The electric force of attraction between charges of opposite sign tends to make them flow, producing a current. The amount of charge that moves, the current, depends upon the potential difference and the nature of the material lying between the separated charges. The hindrance to the movement of electric charge through a particular material is known as **resistance**. The relationship between current (I), voltage (V), and resistance (R) is given by Ohm's Law, $I = V/R$. The higher the resistance of the material, the lower the amount of current flow for any given voltage. Materials that have a high electrical resistance are known as **insulators**, whereas materials having a low resistance to current flow are **conductors**.

Pure water is a poor conductor of electricity, because it contains very few charged particles. But when sodium chloride is added, the solution becomes a relatively good conductor with a low resistance, because the sodium and chloride ions provide charges that can carry the current. The intracellular and extracellular fluids contain numerous ions that are able to move between areas of charge separation; i.e., they can carry current. Lipids, which contain very few charged groups, have a high electrical resistance. Therefore, the lipid components of the plasma membrane are regions of high electrical resistance separating two water compartments of low resistance.

Electrical and Chemical Properties of a Resting Cell

All cells under resting conditions have an electric-potential difference across the plasma membrane oriented so that the inside of the cell is negatively charged with respect to the outside. This potential is the **resting membrane potential**; its magnitude varies from about 5 to 100 mV, depending upon the type of cell. By convention, the polarity (positive or negative) of the membrane potential is stated in terms of the sign of the excess charge on the inside of the cell. As we shall see, the membrane potential of some cells, for example nerve and muscle cells, can change rapidly in response to stimulation. Membrane potentials are measured by inserting a very thin electrode (an electrical conductor) into the cell, another into the extracellular fluid, and connecting the two to a voltmeter (Fig. 8-18).

This membrane potential can be accounted for by the fact that there is a slightly greater number of negative charges than positive charges inside the cell and a slightly greater number of positive charges than negative charges outside. The excess negative charges inside the cell are electrically attracted to the excess positive charges outside the cell, and vice versa. Thus, these excess ions collect in a thin shell on the inner and outer surfaces of the plasma membrane (Fig. 8-19), whereas the bulk of the intracellular and extracellular fluid is electrically neutral. Thus, the total number of positive and negative charges that have to be separated across the membrane to account for the potential is an insignificant fraction of the total number of charges actually within the cell.

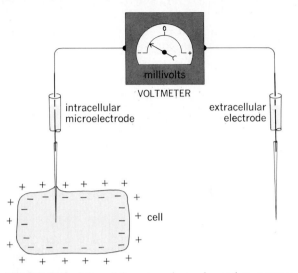

FIGURE 8-18. Intracellular microelectrode used to measure a membrane potential.

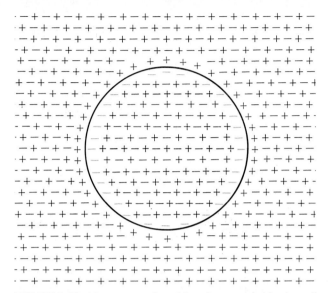

FIGURE 8-19. Net excess of positive charges outside the cell and the excess negative charges inside collect close to the plasma membrane. Note that these excess charges are but a very small fraction of the total number of ions inside and outside the cell.

The resting membrane potential is determined mainly by two factors: (1) the differences in ion concentration of the intracellular and extracellular fluids and (2) the permeabilities of the plasma membrane to the different ion species. The concentrations of sodium, potassium, and chloride ions in the extracellular and intracellular fluids of a nerve cell are listed in Table 8-5. There are many other charged substances in both fluid compartments, such as Mg^{2+}, Ca^{2+}, HCO_3^-, HPO_4^{2-}, SO_4^{2-}, amino acids, and proteins, but sodium, potassium, and chloride ions are present in highest concentrations and therefore generally play the most important roles in the generation of the resting membrane potential.

Note that the sodium and chloride ion concentrations are lower inside the cell than outside, and potassium concentration is greater inside the cell. As we mentioned in Chap. 6, the concentration differences for sodium and potassium are due to the action of a membrane active-transport system which pumps sodium out of the cell and potassium into it.

To understand how such concentration differences for sodium and potassium (maintained by the membrane pumps) create membrane potentials, let us consider the situation in Fig. 8-20. The assumption in this model is that the membrane contains open potassium channels but no sodium channels; therefore, potassium can diffuse through the membrane, but sodium cannot. Initially, there is no potential difference across the membrane because the two solutions are electrically neutral; i.e., they contain equal numbers of positive and negative ions (the positive ions are different—sodium versus potassium—on the two sides, but the numbers are the same, and each is balanced by an equal number of chloride ions). Because of the open potassium channels, potassium ions will diffuse down their concentration gradient from compartment 2 into compartment 1. There is also a concentration gradient favoring sodium diffusion in the opposite direction, but this membrane has no sodium channels. Accordingly, after a few potassium (positive) ions have moved into compartment 1, that compartment will have an excess of positive charge, whereas compartment 2 will have an excess of negative charge; i.e., a potential difference will exist across the membrane. Because it is due entirely to the diffusion of ions, it is called a **diffusion potential**.

TABLE 8-5. Distribution of major ions across the plasma membrane of a typical nerve cell

Ion	Extracellular concentration, mmol/L of water	Intracellular concentration, mmol/L of water
Na^+	150	15
Cl^-	110	10
K^+	5	150

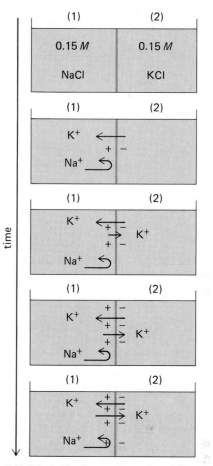

FIGURE 8-20. Generation of a diffusion potential across a membrane permeable only to potassium. Arrows represent ion movements.

In this example so far, the net diffusion of potassium was caused entirely by its concentration gradient across the membrane. Now we introduce a second factor which can cause net movement across a membrane, namely any electric potential across the membrane. Let us consider the example again. As the concentration gradient moves potassium from compartment 2 to compartment 1, and side 1 of the membrane becomes increasingly positive (Fig. 8-20), the membrane potential difference itself begins to influence the movement of the potassium ions; they are attracted by the negative charge of side 2 and repulsed by the positive charge of side 1. As long as the force due to the concentration gradient driving potassium from side 2 to side 1 is greater than the electric force driving it in the opposite direction, there will be net movement of potassium from side 2 to side 1, and the membrane potential will increase; side 1 will become more and more positive until the electric force opposing the entry of potassium into that compartment equals the force due to the concentration gradient favoring entry. The membrane potential at which the electric force is equal in magnitude but opposite in direction to the concentration force is called the **equilibrium potential** for that ion. At the equilibrium potential there is no net movement of the ion because the opposing forces acting upon it are exactly balanced.

The value of the equilibrium potential for any ion depends upon the concentration gradient for that ion across the membrane. If the concentrations on the two sides were equal, the force of the concentration gradient would be zero, and the equilib-

rium potential would also be zero. The larger the concentration gradient, the larger is the equilibrium potential since a larger electrically driven flux will be required to balance the larger flux due to the concentration difference. Using potassium concentration gradients typical for neurons, the equilibrium potential for potassium is close to -90 mV, the inside of the cell being negative with respect to the outside (Fig. 8-21).

If the membrane separating compartment 1 and compartment 2 is replaced with one permeable only to sodium, the initial net flow of the positively charged sodium ion will be from compartment 1 to compartment 2, and side 2 of the membrane will

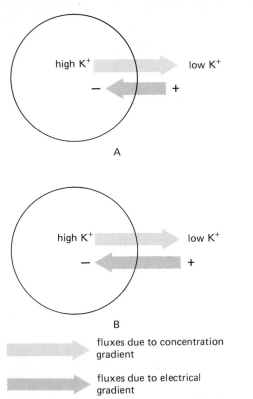

FIGURE 8-21. Forces acting on potassium when the membrane of a neuron is (A) at the resting potential (70 mV, inside negative) and (B) at the potassium equilibrium potential (90 mV, inside negative).

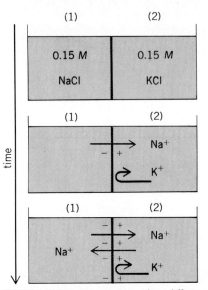

FIGURE 8-22. Generation of a diffusion potential across a membrane permeable only to sodium. Arrows represent ion movements.

become positive (Fig. 8-22). This movement of sodium down its concentration gradient is opposed by the electric force generated by that movement. A sodium equilibrium potential will be established with side 2 positive with respect to side 1, at which point net movement will cease. For most neurons, the sodium equilibrium potential is about $+60$ mV, inside positive (Fig. 8-23). Thus, the equilibrium potential for each ion species is different in magnitude and, possibly, even in direction from those for other ion species, since the concentration gradients are different.

The resting membrane potential

It is not difficult to move from these hypothetical experiments to a nerve cell at rest where (1) the potassium concentration is much greater inside the cell than outside it, and (2) the cell membrane is some 50 to 75 times more permeable to potassium than to sodium. A diffusion potential is generated across the plasma membrane largely because of the movement of potassium down its concentration gradient through open potassium channels, so that the inside of the cell becomes negative with respect to the outside. The experimentally measured membrane potential is not, however, equal to the potassium equilibrium potential because a small number of sodium channels are open in the resting membrane, and some sodium ions (which are positively charged) continually move into the cell, canceling the effect of an equivalent number of the potassium ions that have moved out of the cell.

Thus, when more than one ion species can move across the membrane, the membrane permeability for each ion as well as its concentration gradient must be considered when accounting for the membrane potential. The permeability of the membrane to a given ion species depends upon the number of open channels through which that ion can cross the membrane. If the membrane is impermeable to a given ion species, no ion of that

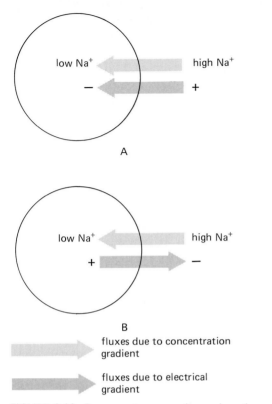

A

B

fluxes due to concentration gradient

fluxes due to electrical gradient

FIGURE 8-23. Forces acting on sodium when the membrane of a neuron is (A) at the resting potential (70 mV, inside negative) and (B) at the sodium equilibrium potential (60 mV, inside positive).

cellular sodium not progressively increase and of intracellular potassium decrease? The reason is that active-transport mechanisms in the membrane utilize energy derived from cellular metabolism to pump the sodium back out of the cell and the potassium back in. Actually, the pumping of these ions is linked because they are both transported by the Na,K-ATPase in the membrane (Chap. 6). In a resting cell, the number of ions moved by the pump equals the number of ions that diffuse across the membrane down their concentration and electrical gradients (Fig. 8-24), and the concentrations of sodium and potassium in the cell do not change. As long as the concentration gradients remain fixed and the ion permeabilities of the plasma membrane do not change, the electric potential across the resting membrane will remain constant. This constancy is not, however, an equilibrium state, because metabolic energy (in the form of the membrane pump) is required to maintain the status quo; rather, the constancy reflects a steady-state condition.

Thus far, we have described the membrane potential as purely a diffusion potential. There is, however, another component to it, reflecting the direct separation of charge achieved by the membrane Na,K-pump. If the membrane pump were to move equal numbers of sodium and potassium ions across the membrane in opposite directions, it would not contribute *directly* to the separation of charge and the pump would be electrically neutral.

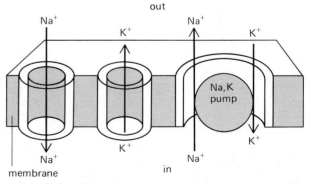

FIGURE 8-24. Fluxes of sodium and potassium ions across the plasma membrane of a resting neuron. The passive movements are exactly balanced by the active transport of ions in the opposite direction. (*Redrawn from Kandel and Schwartz.*)

species can cross the membrane and contribute to a diffusion potential, regardless of the electrical and concentration gradients that may exist. For a given concentration gradient, the greater the membrane permeability to an ion species, the greater the contribution that ion species will make to the membrane potential. Since the resting membrane is much more permeable to potassium than to sodium, the resting membrane potential is much closer to the potassium equilibrium potential than to that of sodium.

To reiterate, at a resting potential of -70 mV, neither sodium nor potassium is at its equilibrium potential, and there is a net movement of sodium into the cell and potassium out. If such net ion movements occur, why does the concentration of intra-

If the movement is unequal, however, and the pump does separate charge, it is known as an **electrogenic pump**. The Na,K-pump is electrogenic since it moves three sodium ions out of the cell for every two potassium ions that it brings in. This unequal transport directly increases the separation of charge across the plasma membrane; i.e., it makes the membrane potential larger (by transferring more positive charges to the outside) than it would be from the diffusion potential alone. In most cells (but by no means all), the electrogenic contribution of the pump to the membrane potential is quite small. It must be reemphasized that the pump makes an essential *indirect* contribution, because it maintains the concentration gradients down which the ions diffuse to produce most of the charge separation that makes up the membrane potential.

We have not yet dealt with chloride ions. The membrane permeability of most cells to chloride ions is relatively high, and many membranes do not contain chloride-ion pumps. Therefore, in these cells, the membrane potential (set up by the interactions just described) acts on chloride ions, moving them towards the positive side (the outside) of the cell until a concentration gradient develops (inside chloride-ion concentration lower than that outside), producing a diffusion of chloride ions that exactly opposes the electric force. The result is that the equilibrium potential for chloride ions is equal to the resting membrane potential, and chloride ions make no contribution to the magnitude of the membrane potential. Note that in this case for chloride ion, the concentration difference across the plasma membrane is generated passively by the electric force rather than by active pumps, as is the case for sodium and potassium.

The example just described was for a cell with no active chloride-ion transport, but some cells do, in fact, actively transport chloride ions. In these cells, the membrane potential is not at the chloride-ion equilibrium potential, and net chloride-ion diffusion contributes to the magnitude of the membrane potential. For example, certain neurons actively pump chloride ions out of the cell, and the net diffusion of chloride ions back in contributes to the excess negative charge inside the cell; i.e., it increases the magnitude of the membrane potential.

Electric Signals of the Nervous System: Graded Potentials and Action Potentials

Transient changes in the membrane potential from its resting level constitute electric signals which can alter cell activities. These signals occur in two forms: graded potentials and action potentials. Graded potentials are important in signaling over short distances; action potentials, on the other hand, are the long-distance signals of nerve (and muscle) membrane.

Adjectives such as "membrane" potential, "resting" potential, "diffusion" potential, "graded" potential, etc., define the conditions under which the potential is measured or the way it developed (Table 8-6). The terms "depolarize," "hyperpolarize," and "repolarize" will be used to describe the direction of changes in the membrane potential relative to the resting potential. The membrane is said to be **depolarized** when the magnitude of the membrane potential is smaller than the resting level, i.e, when the membrane potential is less negative or closer to zero than the resting level. (By convention, this includes states in which the polarity of the membrane potential actually reverses, i.e., in which the inside of a cell becomes positive (Fig. 8-25). The membrane is **hyperpolarized** when the membrane potential is greater than the resting potential; i.e., the inside of the cell is more negative or farther from zero than the resting level. When a membrane potential that has been depolarized returns toward the original resting potential, it is said to be **repolarizing**.

Graded potentials

Graded potentials are local changes in membrane potential in either a depolarizing or hyperpolarizing direction (Fig. 8-26, Part 1). They are usually produced by some specific change in the cell's environment acting on a specialized region of the membrane. They are called "graded potentials" because the amplitude of the potential change is variable (graded) and is related to the magnitude of the event which produces the change in membrane potential (Fig. 8-26, Part 2). We will encounter a number of graded potentials which are given various names

TABLE 8.6. A miniglossary

Potential = potential difference. The voltage difference between two points.

Membrane potential = transmembrane potential. The voltage difference between the inside and outside of a cell.

Diffusion potential. A voltage difference created by net diffusion of ions.

Equilibrium potential. The voltage difference across a membrane which is equal in force but opposite in direction to the concentration force affecting a given ion species.

Resting membrane potential = resting potential. The steady transmembrane potential of a cell that is not producing an electric signal.

Graded potential. A potential change of variable amplitude and duration that is conducted decrementally; it has no threshold or refractory period.

Action potential. A brief, all-or-none reversal of the membrane potential polarity; it has a threshold and refractory period and is conducted without decrement.

Postsynaptic potential. A graded potential change produced in the postsynaptic neuron in response to release of transmitter substance by a presynaptic terminal; it may be depolarizing (an *excitatory postsynaptic potential* or EPSP) or hyperpolarizing (an *inhibitory postsynaptic potential* or IPSP).

Receptor potential. A graded potential produced at the peripheral endings of afferent neurons (or in separate receptor cells) in response to an external stimulus.

Pacemaker potential. Spontaneously occurring graded potential change; it occurs only in certain specialized cells.

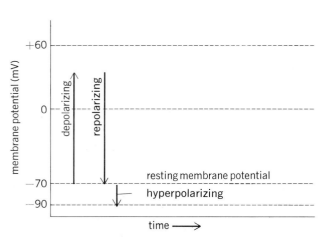

FIGURE 8-25. Depolarizing, repolarizing, and hyperpolarizing changes in membrane potential. As potentials become more positive inside than the resting potential, they are said to be "depolarizing;" those returning toward the resting membrane potential are said to be "repolarizing," and those becoming more negative inside than the resting membrane potential are "hyperpolarizing."

related to the location of the potential or the function it performs. These potentials include receptor potentials, synaptic potentials, end-plate potentials, and pacemaker potentials. The characteristics of each of these graded potentials will be described later in this chapter or in the chapter on muscle, Chap. 10.

Whenever a local change in membrane potential occurs, a current will flow between this region and the adjacent regions of the plasma membrane that are at the resting potential. Current always flows between two points if there is a difference in potential between the points and a conducting material lies between them. Thus, graded potentials cause current to flow in the intracellular and extracellular fluids in the region around them; the greater the potential change, the greater the current flow. (By convention, the direction in which positive ions move is designated the direction of current flow; negatively charged particles simultaneously move in the opposite direction.) In Fig. 8-27, a small region of a membrane has been depolarized by a stim-

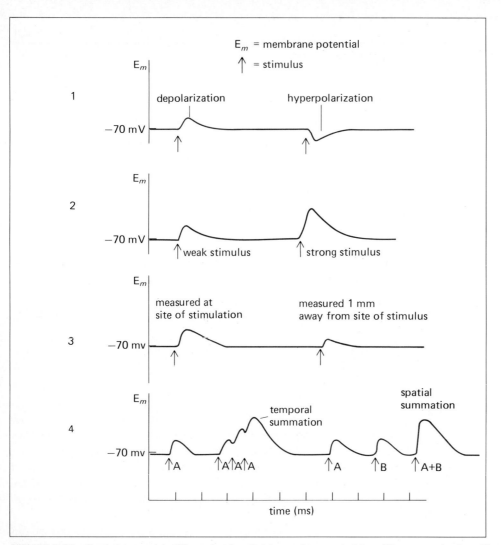

FIGURE 8-26. Graded potentials: (1) can be depolarizing or hyperpolarizing, (2) can vary in size, (3) are conducted decrementally, and (4) can be summed. Temporal and spatial summation will be discussed later in the chapter.

ulus. Accordingly, this area of the membrane has a potential less than that of adjacent areas. Inside the cell, current will flow through the intracellular fluid away from the depolarized region and toward the more negative, resting regions of the membrane; simultaneously, outside the cell, current will flow from the more positive region of the resting membrane toward the less positive region of depolarization. This local current flow (movement of positive charge) removes positive charges from nearby regions along the outside of the membrane, and it adds positive charge to adjacent sites along the inside of the membrane; thus it produces a decrease in the amount of charge separation (depolarization) of the adjacent membrane sites. Note that the local current is carried by ions such as K^+, Na^+, Cl^-, and HCO_3^-.

Local current flow is much like water flow through a leaky hose; charge is lost across the membrane because the membrane is permeable to ions, just as water is lost from a leaky hose, with the result that the magnitude of the current decreases

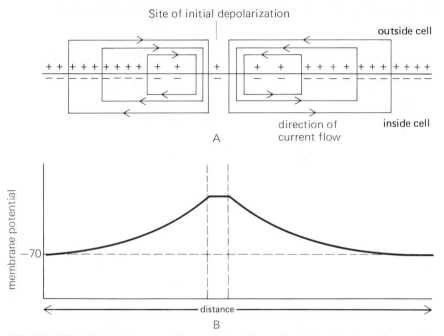

FIGURE 8-27. (A) Local current flows surrounding a depolarized region of a membrane produce a depolarization of adjacent regions of the membrane. (B) Potential changes result from local current flows.

with distance away from the initial site of potential change. In fact, plasma membranes are so leaky to ions that local currents almost completely die out within a few millimeters of their point of origin. There is another way of saying the same thing: Local current flow is **decremental**; i.e., its amplitude decreases with increasing distance. The resulting change in membrane potential, therefore, also decreases with distance from the site of origin of the potential (Fig. 8-26, Part 3).

Because the signal decreases with distance, graded potentials (and the current flows they generate) can function as signals only over very short distances (a few millimeters). Nevertheless, as we shall see, graded potentials play very important roles in the integration of signals by nerve cells (and other cells).

Action potentials

Action potentials are very different from graded potentials (Table 8-7). They are rapid alterations in the membrane potential that may last only 1 ms, during which the membrane may change from -70 to $+30$ mV and then repolarize to its original value (Fig. 8-28). Only a few types of cells, e.g., nerve, muscle, and some gland cells, have plasma membranes that are capable of producing action potentials. These membranes are called **excitable membranes**, and their ability to generate action potentials is known as **excitability**. The propagation of action potentials is the mechanism used by the nervous system to communicate over long distances.

How can an excitable membrane make rapid changes in its membrane potential? How does a change in the environment (a stimulus) interact with an excitable membrane to cause an action-potential response? How is an action potential propagated along an excitable membrane? These questions will be discussed in the following sections.

Ionic basis of the action potential. Action potentials can be explained by the concept already developed for describing the origins of the resting membrane potentials. This mechanism, known as the **ionic hy-**

TABLE 8-7. Differences between graded potentials and action potentials

Graded potentials	Action potentials
1. Graded response; amplitude varies with conditions of the initiating event	All-or-none response; once membrane is depolarized to threshold, amplitude independent of initiating event
2. Graded response; can be summed	All-or-none response; cannot be summed
3. Has no threshold	Has a threshold that is usually 10–15 mV depolarized relative to the resting potential
4. Has no refractory period	Has a refractory period
5. Is conducted decrementally; i.e., amplitude decreases with distance	Is conducted without decrement; the amplitude is constant
6. Duration varies with initiating conditions	Duration constant for a given cell type
7. Can be a depolarization or hyperpolarization	Is a depolarization (with overshoot)
8. Initiated by stimulus (receptor), by neurotransmitter (synapse), or spontaneously	Initiated by membrane depolarization

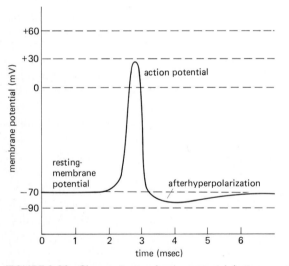

FIGURE 8-28. Changes in membrane potential during an action potential. (The sodium equilibrium potential is +60 mV; the equilibrium potential for potassium is −90 mV.)

pothesis, was developed by the English scientists A. L. Hodgkin and A.F. Huxley, who received the Nobel Prize in 1963 for this work.

We have seen that the magnitude of the resting membrane potential depends upon the concentration gradients of and membrane permeabilities to different ions, particularly sodium and potassium. This is true for the period of the action potential as well; the action potential results from a transient change in membrane ion permeability. In the resting state the open channels in the plasma membrane are predominantly those that are permeable to potassium ions (and chloride ions). Almost all the sodium-ion channels are closed, and the resting potential is, therefore, much closer to the potassium equilibrium potential than to the sodium equilibrium potential. During an action potential, however, the membrane permeabilities to sodium and potassium ions are markedly altered. In the depolarizing phase of the action potential, sodium channels open, increasing the membrane permeability to sodium ions several hundredfold. This allows sodium ions to rush into the cell. During this period more positive charge enters the cell in the form of sodium ions than leaves in the form of potassium ions, and thus the membrane potential decreases and eventually reverses polarity, becoming positive on the inside and negative on the outside of the membrane. In this phase the membrane potential approaches but does not quite reach the sodium equilibrium potential.

Action potentials in nerve cells last about 1 ms

(0.001 s). (They may be much longer in certain types of muscle cells.) What causes the membrane to return so rapidly to its resting level? The answer to this question is twofold: (1) The sodium channels, which were open during the depolarization phase, close and (2) a special set of potassium channels begin to open. The timing of these two events can be seen in Fig. 8-29. Closure of the sodium channels alone would restore the membrane potential to its resting level. However, the entire process is speeded up by the simultaneous increase in potassium permeability. These two events, closure of the sodium channels and opening of potassium channels, lead to a potassium diffusion out of the cell that is greater than the sodium diffusion in, rapidly returning the membrane potential to its resting level. In fact, after the sodium channels have closed, at least some of the voltage-sensitive potassium channels are still open, and there is generally a small hyperpolarizing overshoot of the membrane potential (**afterhyperpolarization**, Fig. 8-28).

One might think that large fluxes of ions across the membrane would be required to produce such large changes in membrane potential. Actually, only about one of every 100,000 potassium ions in a cell diffuses out to charge the membrane potential to its resting level, and approximately the same number

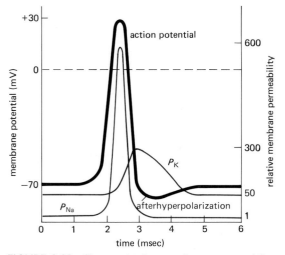

FIGURE 8-29. Changes in the membrane permeability to sodium ions (P_{Na}) and to potassium ions (P_K) during an action potential.

of sodium ions enters the cell during an action potential. These ion fluxes during an action potential are so small that they produce only infinitesimal changes in the intracellular ion concentration. Yet if this tiny number of ions crossing the membrane with each action potential were not eventually moved back across the membrane, the concentration gradients of sodium and potassium would gradually disappear and action potentials could no longer be generated. As described earlier, cellular accumulation of sodium and loss of potassium are prevented by the continuous action of the membrane active-transport system for sodium and potassium. The number of ions that cross the membrane during an action potential is so small that the pump need not keep up with the action-potential fluxes, and hundreds of action potentials can occur even if the pump is stopped experimentally.

Mechanism of permeability changes. Measurement of permeability changes reveals that the permeability to sodium is altered whenever the membrane potential changes; specifically, hyperpolarization of the membrane causes a decrease in sodium permeability, whereas depolarization causes an increase in sodium permeability.

In light of our previous discussion of the ionic basis of membrane potentials, it is very easy to confuse the cause-and-effect relationships of the statements just made. Earlier we pointed out that an increase in sodium permeability *causes* membrane depolarization; now we are saying that depolarization *causes* an increase in sodium permeability. Combining these two distinct causal relationships yields the positive-feedback cycle (Fig. 8-30) responsible for the rising phase of the action potential: Depolarization opens sodium channels so that the membrane permeability to sodium increases. Because of increased sodium permeability, sodium diffuses into the cell; this addition of net positive charge to the cell further depolarizes the membrane, which, in turn, produces a still greater increase in sodium permeability, which, in turn, causes. . . .

What is responsible for the behavior of the ion channels? They behave as though they are gov-

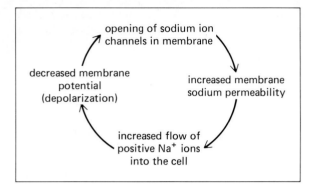

FIGURE 8-30. Positive-feedback relation between membrane depolarization and increased sodium permeability, which leads to the rapid depolarizing phase of the action potential.

erned by "gates," such that the channels are open under some conditions but closed at other times. (For example, at the resting membrane potential, most of the sodium channels are closed, which accounts for the low sodium permeability of the resting membrane.) The channels are thought to consist of assemblies of charged proteins that stretch through the membrane; their gating functions are probably achieved by changes in the shape or position of these proteins as a result of the electric forces acting upon them. For example, the "gates" to a channel would be closed when parts of the protein molecules effectively block the channel; under the influence of a reduced voltage across the membrane (depolarization), a rearrangement of the proteins would occur and open the channel. A more prolonged closing of the sodium channels may be due to another change in protein structure that blocks the channel again soon after it has opened, thus keeping it closed even if the membrane remains depolarized. Ion channels which can be opened or closed by changes in membrane potential are called **voltage-sensitive ion channels**. The potassium channels that open during an action potential are also voltage-sensitive.

Although we have mentioned only sodium and potassium channels, there are other types of voltage-sensitive channels that have varying degrees of importance in different cell types and explain at least some of the special properties of these cells. For example, as we shall discuss later in the chapter, some regions of nerve-cell membrane have calcium channels that open in response to membrane depolarization, a property important in communication between nerve cells. In some nonneural cells, voltage-sensitive calcium channels are so important that the action potentials of these cells are due entirely to calcium. In general, then, excitable membranes contain at least some voltage-sensitive channels, whereas other types of membranes, which do not generate action potentials, have only ion channels that do not vary with membrane polarization.

The generation of sodium action potentials is prevented by local anesthetics such as procaine hydrochloride (Novocaine) and lidocaine (Xylocaine) because these prevent opening of the sodium channels in response to depolarization. Without action potentials, afferent signals generated in the periphery in response to injury, for example, cannot reach the brain and give rise to the sensation of pain.

Threshold. Not all depolarizations trigger the positive-feedback relationship that leads to an action potential. Action potentials occur only when the membrane is depolarized enough (and, therefore, enough sodium channels are open) so that sodium entry exceeds the flux of potassium out of the cell. In other words, action potentials occur only when the *net* movement of positive charge is inward. The membrane potential at which the net movement of ions across the membrane first changes from outward to inward is the **threshold**, and stimuli which are just strong enough to depolarize the membrane to this level are **threshold stimuli** (Fig. 8-31). The threshold potential of most excitable membranes is a level about 15 mV more depolarized than the resting membrane potential. Thus, if the resting potential of a neuron is −70 mV, the threshold potential may be −55 mV; in order to initiate an action potential in such a membrane, the membrane potential must be depolarized by at least 15 mV. At depolarizations less than threshold, outward potassium movement still exceeds the sodium entry, and the positive-feedback cycle cannot get started despite the increase in sodium entry. In such cases, the membrane will return to its resting level as soon

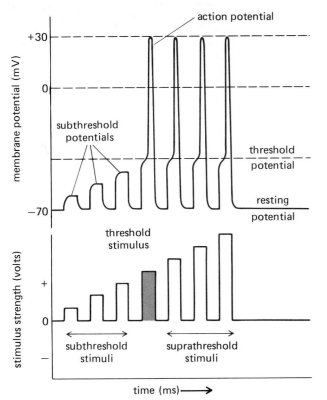

FIGURE 8-31. Decreasing membrane potential with increasing strength of depolarizing stimulus. When the membrane potential reaches the threshold potential, action potentials are generated. Increasing the stimulus strength above threshold level does not cause larger action potentials to be generated. (The after-hyperpolarization has been omitted from this figure.)

as the stimulus is removed, and no action potentials are generated. These weak depolarizations are **subthreshold potentials**, and the stimuli that cause them are **subthreshold stimuli**.

Stimuli of more than threshold magnitude (**suprathreshold stimuli**) also elicit action potentials, but as can be seen in Fig. 8-31, the action potentials resulting from such stimuli have exactly the same amplitude as those caused by threshold stimuli. This is explained by the fact that once threshold is reached, membrane events are no longer dependent upon stimulus strength. The depolarization continues on as an action potential because the positive-feedback cycle is operating. Action potentials either occur maximally, as determined by the electrochemical conditions across the membrane, or they

do not occur at all. Another way of saying this is that action potentials are **all-or-none**.

The firing of a gun is a mechanical analogy that shows the principle of all-or-none behavior. The magnitude of the explosion and the velocity at which the bullet leaves the gun do not depend on how hard the trigger is squeezed. The trigger is pulled hard enough to fire the gun, or it is not; one cannot fire a gun halfway.

Because of this all-or-none nature, a single action potential cannot convey information about the magnitude of the stimulus that initiated it, since a threshold-strength stimulus and one of twice threshold strength give the same response. Since one function of nerve cells is to transmit information, one may ask how a system utilizing all-or-none signals can convey information about the strength of the stimulus. How can one distinguish between a loud noise and a whisper, a light touch and a pinch? This information, as we shall see later, depends in part upon the number of action potentials transmitted per unit of time, i.e., the frequency of action potentials, and not upon their size.

Refractory periods. How soon after firing an action potential can an excitable membrane fire a second one? If we apply a threshold-strength stimulus to a membrane and then stimulate the membrane immediately with a second threshold-strength stimulus, the membrane does not respond to the second stimulus (Fig. 8-32). Even though identical stimuli are applied, the membrane is unresponsive for a short time, and is said to be **refractory** to a second stimulus.

Instead of applying the second stimulus at threshold strength, if we increase it to suprathreshold levels, we can distinguish two separate refractory periods associated with an action potential (Fig. 8-33). During the action potential, a second stimulus—no matter how strong—will not produce a second action potential, and the membrane is said to be in its **absolute refractory period**. Following the absolute refractory period there is an interval during which a second action potential can be produced, but only if the stimulus strength is considerably greater than the usual threshold level. This

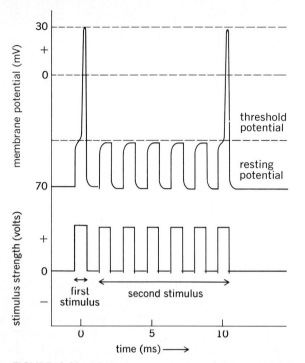

FIGURE 8-32. After generation of an action potential, the membrane remains refractory to a second threshold-level stimulus for several milliseconds.

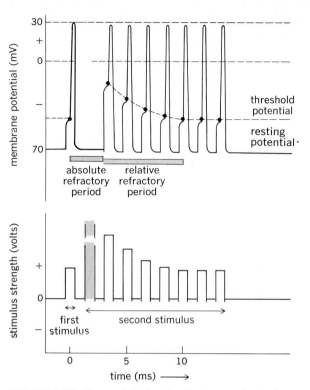

FIGURE 8-33. The magnitude of a single second stimulus necessary to generate a second action potential during the refractory period is greater than the initial stimulus and decreases in magnitude as the time between the first and second stimuli increases. Immediately following an action potential, the membrane is absolutely refractory to all stimulus strengths.

is the **relative refractory period**, and it can last some 10 to 15 ms or longer.

The mechanisms responsible for the refractory periods are related to the membrane mechanisms that alter the sodium and potassium permeabilities. The absolute refractory period corresponds roughly with the period of sodium permeability changes, and the relative refractory period corresponds with the period of increased potassium permeability. After an action potential, time is required to restore the proteins of the voltage-sensitive ion channels to their original resting configuration or position.

The absolute refractory period limits the number of action potentials that can be produced by an excitable membrane in a given period of time. Recordings made from intact nerve cells that are responding to physiological stimuli indicate that most nerve cells respond at frequencies up to 100 action potentials per second, and some may produce much higher frequencies for brief periods of time.

Action potential propagation. Once generated, one particular action potential does not itself travel along the membrane; rather, each action potential triggers, by local current flow, a new one at an adjacent area of the membrane. The local current flow is great enough to depolarize the adjacent membrane site to its threshold potential so that the sodium positive-feedback cycle takes over, and a new action potential occurs there (Fig. 8-34). Once the new site is depolarized to threshold, the action potential generated there is dependent solely upon the electrochemical gradients and membrane permeability properties at the new site. Since these factors are usually identical to those at the site of the initial action potential, the new action potential is virtually identical to the old. The new action

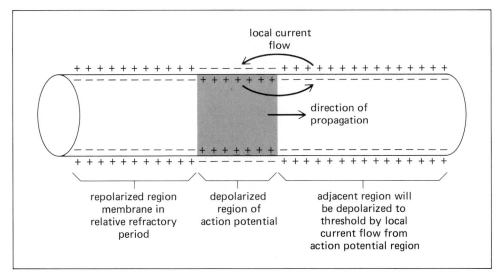

local current
flow

direction of
propagation

repolarized region
membrane in
relative refractory
period

depolarized
region of
action potential

adjacent region will
be depolarized to
threshold by local
current flow from
action potential region

FIGURE 8-34. Propagation of an action potential along a plasma membrane.

potential produces local currents of its own which depolarize the next region adjacent to it, producing yet another action potential at the next site, and so on along the length of the membrane. Thus, no distortion occurs as this process repeats itself along the membrane; the action potential arriving at the end of the membrane is identical to the initial one. In other words, action potentials are not conducted decrementally.

Local current flows between regions of different electric potential; therefore current will also flow toward the original site of stimulation from adjacent sites. However, the membrane areas that have just undergone an action potential are refractory and cannot undergo another; thus, the only direction of action-potential propagation is away from the stimulation site.

The action potentials in skeletal muscle cells are initiated near the middle of these cylindrical cells and propagate toward the two ends, but in most nerve cells, action potentials are initiated at one end of the cell and propagate toward the other end. This unidirectional propagation of action potentials along excitable membranes is determined by the stimulus location rather than an intrinsic inability of the membrane to conduct in the opposite direction.

The velocity with which an action potential propagates along a membrane depends upon fiber diameter and whether or not the fiber is myelinated. The larger the fiber diameter, the faster is the action-potential propagation, because a large fiber offers less resistance to local current flow, and adjacent regions of the membrane are brought to threshold faster.

Myelin, the fatty material mentioned earlier, is the second factor influencing propagation velocity. Since myelin makes it more difficult for currents to flow between intra- and extracellular fluid compartments, action potentials do not occur along the sections of membrane protected by myelin; they occur only where the myelin coating is interrupted at the nodes of Ranvier, which occur at regular intervals along the axon. Thus, the action potential jumps from one node to the next as it propagates along a myelinated fiber, and for this reason this method of propagation is called **saltatory conduction** (Latin *saltare,* to leap). Saltatory conduction causes a more rapid propagation of the action potential than occurs in nonmyelinated fibers of the same axon diameter. This occurs because less current passes out through the myelin-covered section of the membrane during local current flow and there is a smaller drop in the voltage gradient along the fiber (Fig. 8-35). Therefore, the membrane of the node next to the active node is brought to threshold faster and undergoes an action potential sooner than if the myelin were not present. (The larger the axon diameter, the faster the action potential propagates in both myelinated and nonmyelinated fibers.)

The range of conduction velocities is from about 1 mi/h for small-diameter unmyelinated fibers to about 400 mi/h for large-diameter myelinated fibers. At a velocity of 1 mi/h, an action potential will travel from the head to the toe of an average-size person in about 4 s; at a velocity of 400 mi/h, it would take about 10 ms.

Initiation of action potentials. We have emphasized that depolarization of a neuron to threshold will trigger an action potential which is relayed along the length of the nerve fiber. However, we have thus far avoided the question of how the initial depolarization to threshold occurs physiologically. It is achieved in afferent neurons by a graded potential generated in the receptors at the peripheral ends of the neurons. In all other neurons, the depolarization to threshold is due either to a graded potential generated by synaptic input to the neuron or to a spontaneously occurring change in the neuron's membrane potential, known as a **pacemaker potential**. How the synaptic and receptor potentials are produced is the subject of the next two sections. The spontaneous generation of action potentials by pacemaker potentials occurs in the absence of any identifiable external stimulus and thus appears to be an inherent property of certain neurons (and

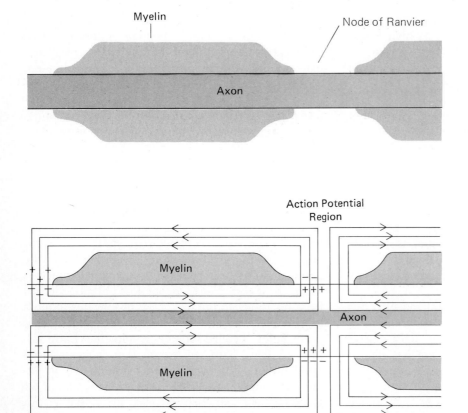

FIGURE 8-35. Ion current flow during an action potential in a myelinated axon. Compare this figure with Fig. 8-27.

other excitable cells including certain smooth muscle and cardiac cells). In these cells the ion channels in the plasma membrane appear to undergo spontaneous opening and closing, the net effect of which is to produce a graded depolarization of the membrane—the pacemaker potential. If threshold is reached, an action potential occurs; the membrane then repolarizes and again begins to depolarize (Fig. 8-36). There is no stable, resting membrane potential in such cells. The rate at which the membrane depolarizes determines the frequency of spontaneous firing of action potentials by these neurons.

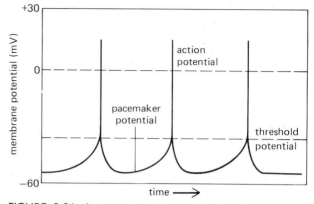

FIGURE 8-36. Action potentials resulting from spontaneous depolarizations of the membrane potential (pacemaker potentials).

SECTION C.
SYNAPSES

As defined earlier, a **synapse** is an anatomically specialized junction between two neurons, at which the electrical activity in one neuron influences the activity in the second. At a synapse a brief, graded change in membrane potential is produced in the postsynaptic cell. Its membrane potential is brought closer to threshold at an **excitatory synapse**, and it is brought farther from threshold at an **inhibitory synapse**. The level of excitability of the postsynaptic cell at any moment, i.e., how close its membrane potential is to threshold, depends on the number of synapses active at any one time and the numbers that are excitatory or inhibitory. If the membrane of the postsynaptic neuron reaches threshold it will generate action potentials. These are propagated along its axon to the terminal branches, which diverge to influence the excitability of other neurons or effector cells. The picture we are left with is one showing thousands of synapses from many different presynaptic cells converging upon a single postsynaptic cell, and the postsynaptic cell in turn branching to affect many other cells. Figure 8-37 demonstrates the neuronal relationships of **convergence** and **divergence**.

Thus, postsynaptic neurons function as neural integrators; i.e., their output reflects the sum of all the incoming signals arriving in the form of excitatory and inhibitory synaptic inputs.

Functional Anatomy of Synapses

There are two types of synapses: **electrical** and **chemical**. At electrical synapses, the plasma membranes of the pre- and postsynaptic cells are joined by gap junctions (Chap. 6), which allow the local currents resulting from action potentials in the presynaptic neuron to flow directly into the postsynaptic cell, depolarizing the membrane to threshold and thus initiating an action potential in the postsynaptic cell. Only a few electrical synapses have been identified in the mammalian nervous system, and the much more common chemical synapse is discussed in the following sections.

Figure 8-38 shows the structure of a single chemical synapse. The presynaptic neuron ends in a slight swelling, the **axon terminal**. A narrow extracellular space, the **synaptic cleft**, separates the pre- and postsynaptic neurons and prevents direct propagation of the current from the presynaptic

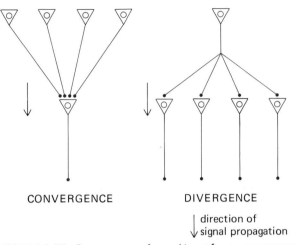

CONVERGENCE   DIVERGENCE

direction of
signal propagation

FIGURE 8-37. Convergence of neural input from many neurons onto a single neuron, and divergence of output from a single neuron onto many others.

neuron to the postsynaptic cell. Signals are transmitted across the synaptic cleft by means of a chemical **neurotransmitter** substance that is released from the presynaptic axon terminal (or from axon varicosities).

When an action potential in the presynaptic neuron reaches the end of the axon and depolarizes the terminal, small quantities of the neurotransmitter are released from the synaptic terminal into the synaptic cleft. Calcium is the link between depolarization of the presynaptic membrane by an arriving action potential and neurotransmitter release. Depolarization of the synaptic terminal causes voltage-sensitive calcium channels in the terminal membrane to open. Calcium diffuses into the presynaptic terminal from the extracellular space. The neurotransmitter in the axon terminals is stored in membrane-enclosed vesicles. The increase in calcium ions in the axon terminal causes some of these

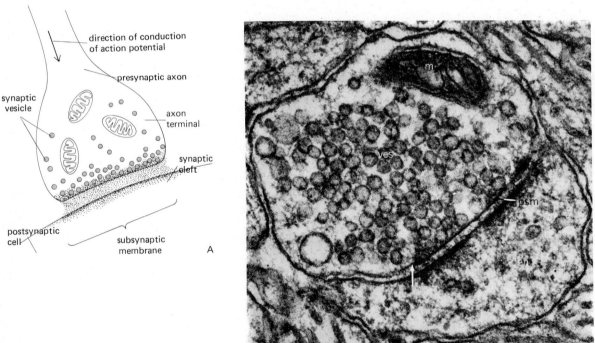

FIGURE 8-38. (A) Diagram of synapse. (B) Electron micrograph of synapse. Axon terminal contains vesicles (ves) and a mitochondrion (m), and dendrites show thickening of postsynaptic membrane (psm). Arrow marks beginning of synaptic cleft. (*From W. F. Windle, "Textbook of Histology," 5th ed., McGraw-Hill, New York, 1976.*)

storage vesicles to fuse with specific release sites on the inner surface of the plasma membrane and liberate their contents into the synaptic cleft by the process of exocytosis.

Once released from the axon terminal, the transmitter molecules diffuse across the cleft and bind to receptor sites on the plasma membrane of the postsynaptic cell. The combination of the transmitter with the receptor site opens specific ion channels in the postsynaptic membrane. Thus, they belong to the class of receptor-operated channels discussed in Chap. 7. These channels are distinct from the voltage-sensitive channels mentioned earlier during the discussion of the action potentials; the postsynaptic channels are closed unless they are opened as a result of a specific neurotransmitter combining with the receptor which operates them.

Because the transmitter is stored on the presynaptic side of the synaptic cleft and receptor sites are on the postsynaptic side, chemical synapses operate in only one direction. Because of this one-way transmission across synapses, action potentials move along a given multineuronal pathway in one direction.

There is a delay between the arrival of an action potential at a synaptic terminal and membrane-potential changes in the postsynaptic cell; it lasts less than a thousandth of a second and is called the **synaptic delay**. It reflects the time required for release of transmitter substance from the axon terminal. (The time required for the transmitter to diffuse across the synaptic cleft is negligible.)

The ion channels close, terminating the postsynaptic potential, when the transmitter leaves the postsynaptic receptors. The removal of neurotransmitter from the synaptic cleft may be accomplished by (1) the chemical transformation of the neurotransmitter into an ineffective substance, (2) diffusion of the transmitter away from the receptor site, or (3) active transport of the neurotransmitter back into the axon terminal (or, in some cases, into nearby glial cells).

The two kinds of chemical synapses, excitatory and inhibitory, are differentiated by the effects that the neurotransmitter has on the postsynaptic cell.

Whether the effect of the transmitter is inhibitory or excitatory depends on the type of receptor and the type of channel the receptor controls.

Excitatory synapse

At an excitatory synapse, the postsynaptic response to the neurotransmitter is a depolarization, bringing the membrane potential closer to threshold. Here the effect of the neurotransmitter-receptor combination is usually to open in the postsynaptic membrane ion channels that are permeable to sodium, potassium, and other small, positively charged ions. These ions then are free to move according to the electric and chemical gradients across the membrane. Recall that the cytosol of all nerve cells has a higher concentration of potassium and a lower concentration of sodium than does the extracellular fluid, and that the inside of a resting neuron is about 70 mV negative with respect to the outside. Thus, there is both an electric and a concentration gradient driving sodium into the cell, whereas in the case of potassium, the electric gradient is opposed by the concentration gradient. Opening channels to all small ions will therefore result in the simultaneous movement of a relatively small number of potassium ions out of the cell and a larger number of sodium ions into the cell. The *net* movement of positive ions is into the neuron; this slightly depolarizes the postsynaptic cell. This potential change is called an **excitatory postsynaptic potential** (**EPSP**, Fig. 8-39). The EPSP is a graded potential that spreads decrementally, by local current flow; its only function is to bring the membrane potential of the postsynaptic neuron toward or to threshold.

In some synapses it is possible that the combination of neurotransmitter with receptor causes an EPSP not by opening membrane channels, as described above, but by closing membrane channels to potassium.

Inhibitory synapse

Activation of an inhibitory synapse produces changes in the postsynaptic cell that lessen the likelihood that the cell will generate an action potential.

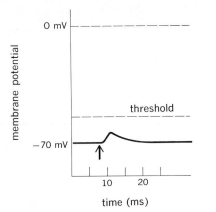

FIGURE 8-39. Excitatory postsynaptic potential (EPSP). Stimulation of the presynaptic neuron is marked by the arrow. Note the short synaptic delay before the postsynaptic cell responds.

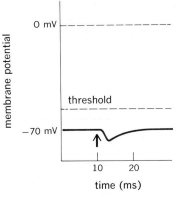

FIGURE 8-40. Inhibitory postsynaptic potential (IPSP). Stimulation of the presynaptic neuron is marked by the arrow. Note the short synaptic delay.

At an inhibitory synapse the combination of the neurotransmitter molecules with receptor sites on the postsynaptic membrane usually opens potassium or chloride channels, or both; sodium channels are not affected.

Earlier it was noted that if a cell membrane were permeable only to potassium ions, the resting membrane potential would equal the potassium equilibrium potential; i.e., the resting membrane potential would be −90 mV instead of −70 mV. Thus, the increased potassium permeability at an inhibitory synapse moves the membrane potential closer to the potassium equilibrium potential. This hyperpolarization is called an **inhibitory postsynaptic potential (IPSP,** Fig. 8-40).

In cells which actively transport chloride ions out of the cell, the chloride equilibrium potential is more negative than the resting potential (e.g., −80 mV) so that an increase in chloride permeability would have a hyperpolarizing effect similar to the effect of increasing potassium permeability; i.e., it would generate an IPSP. In cells in which the equilibrium potential for chloride ion is equal to the resting membrane potential, a rise in chloride ion permeability will not generate an IPSP but will still lessen the likelihood that the cell will reach threshold during excitatory input, since it increases the amount of sodium current necessary to displace the

membrane potential away from the resting potential.

In some synapses the combination of neurotransmitter with receptor elicits an IPSP not by opening potassium or chloride channels but by closing sodium channels.

Activation of the Postsynaptic Cell: Neural Integration

A feature that makes postsynaptic integration possible is that, in most neurons, one excitatory synaptic event by itself is not enough to bring the membrane potential of the postsynaptic neuron from its resting level to threshold; e.g., a single EPSP in a motor neuron is estimated to be only 0.5 mV, whereas changes of 15 to 25 mV are necessary to depolarize the membrane from its resting level to threshold. Since a single excitatory synaptic event does not bring the postsynaptic membrane to its threshold level, an action potential can be initiated only by the combined effects of many excitatory synapses. Of the thousands of synapses on any one neuron, probably hundreds are active simultaneously (or at least close enough in time so that the effects can summate), and the membrane potential of the neuron at any moment is the resultant of all the synaptic activity affecting it at that time. There is a depolarization of the membrane toward threshold when excitatory synaptic input predominates (this is known as **facilitation**), and a hyperpolarization when inhibitory input predominates (Fig. 8-41).

Let us perform a simplified experiment to see how two EPSPs, two IPSPs, or an EPSP plus an IPSP interact (Fig. 8-42). Let us assume that there are three synaptic inputs to the postsynaptic cell; A and B are excitatory and C is an inhibitory synapse. There are stimulators on the axons to A, B, and C so that each of the three inputs can be activated individually. A very fine electrode is placed in the cell body of the postsynaptic neuron and connected to record the membrane potential. In Part I of the experiment we shall test the interaction of the EPSPs by stimulating A and then, after a short while, stimulating A again. Part I of Fig. 8-42 shows that no interaction occurs between the two EPSPs. The reason is that the change in membrane potential associated with an EPSP is fairly short-lived. Within a few milliseconds (by the time axon A is restimulated), the postsynaptic cell has returned to its resting condition.

In Part II of the experiment, axon A is restimulated before the first EPSP has died away, and there is further depolarization. This is called **tem-**

poral summation since the input signals arrive at different times. (The same effect could have been achieved by stimulating first A, then B a short time later.) In Part III, axons A and B are stimulated simultaneously and the two EPSPs that result also summate in the postsynaptic neuron; this is called **spatial summation** since the two inputs occurred simultaneously at different locations. The summation of EPSPs can bring the membrane to its threshold so that action potentials are initiated.

So far we have tested only the patterns of interaction of excitatory synapses. What happens if an excitatory and inhibitory synapse are activated so that their effects occur in the postsynaptic cell simultaneously? Since EPSPs and IPSPs are due to local currents flowing in opposite directions, they tend to cancel each other, and there is little or no change in membrane potential (Fig. 8-42, Part IV). Inhibitory potentials can also show spatial and temporal summation.

As described above, the postsynaptic membrane is depolarized at an activated excitatory synapse and hyperpolarized at an activated inhibitory synapse. By the mechanisms of local current flow described earlier, the plasma membrane of the entire postsynaptic cell body, including the initial segment, becomes slightly depolarized during activation of an excitatory synapse and slightly hyperpolarized during activation of an inhibitory synapse. (Fig. 8-43).

In the above examples we referred to the threshold of the postsynaptic neuron. However, different parts of the neuron have different thresholds. The cell body and larger dendritic branches reach threshold when their membrane is depolarized about 25 mV from the resting level, but in many cells the initial segment has a threshold that is much closer to the resting potential. In cells whose initial-segment threshold is lower than that of the cell body and dendrites, the initial segment reaches threshold first whenever enough EPSPs summate, and the resulting action potential is propagated down the axon. The fact that the initial segment of the neuron has the lowest threshold explains why the location of individual synapses on the postsynaptic cell is

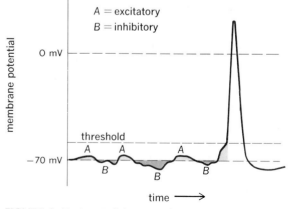

FIGURE 8-41. Intracellular recording from a postsynaptic cell during episodes when (A) excitatory synaptic activity predominates and the cell is facilitated and (B) inhibitory synaptic activity dominates.

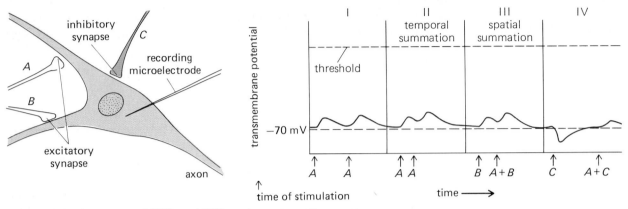

FIGURE 8-42. Interaction of EPSPs and IPSPs at the postsynaptic neuron.

important. A synapse located near the initial segment will have much more impact than will a synapse on the outermost branch of a dendrite.

Postsynaptic potentials last much longer than action potentials do. In the event that the initial segment is still depolarized above threshold after an action potential has been fired and the refractory period is over, a second action potential will occur. The greater the summed postsynaptic depolarization, the greater is the frequency of action potentials discharged. Neuronal responses almost always occur in bursts or trains of action potentials rather than as single isolated events.

This discussion has no doubt left the impression that one neuron can influence another only at a synapse. However, this view requires qualification concerning two points: (1) We have emphasized how local currents influence the membrane of the cell in which they are generated (as, for example, in action potential propagation and synaptic excitation). In addition, under certain circumstances, local currents from one neuron may directly affect the membrane potential of other neurons extremely close to it. This phenomenon is particularly important in areas of the central nervous system that contain a high density of unmyelinated neuronal processes. (2) Receptor sites which bind chemical messengers other than the neurotransmitters released by the presynaptic neuron exist on the postsynaptic cell at sites not associated with a synapse, and messengers binding to these receptors can influence the general responsiveness of the postsynaptic cell. Rather than controlling ion channels, the activation of these receptors usually involves a second messenger, such as cAMP or cGMP. These extrasynaptic receptors will be discussed more fully later.

Synaptic Effectiveness

The effectiveness of a given synapse can be influenced by any event that alters the amount of transmitter released in response to an action potential in the presynaptic neuron (the greater the amount of transmitter released, the larger the amplitude of the EPSP or IPSP in the postsynaptic cell). Factors that can affect the amount of transmitter released are (1) the membrane-potential of the axon terminal, (2) residual calcium within the axon terminal, and (3) the action of other neurons that form synapses on the axon terminal.

As we have seen, calcium channels in the plasma membrane of presynaptic axon terminals are voltage dependent; therefore the membrane potential of the terminal determines the amount of calcium that enters the terminal (calcium entry increases with depolarization of the terminal and decreases with hyperpolarization), and this, in turn, determines the amount of transmitter released. Accordingly, anything that alters the amplitude of an action potential will change the amount of calcium entering the terminal and, therefore, the amount of

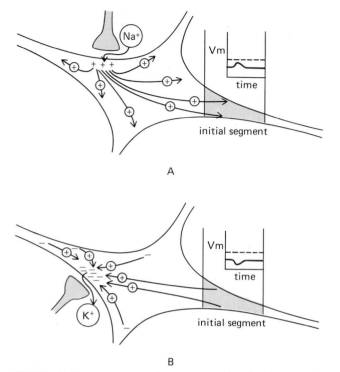

FIGURE 8-43. Comparison of excitatory (A) and inhibitory (B) synapses showing direction of current flow through the post-synaptic cell following synaptic activation. (A) Current flows through the cytoplasm of the postsynaptic cell away from the excitatory synapse, depolarizing the cell. (B) Current flows through the cytoplasm of the postsynaptic cell toward an inhibitory synapse, hyperpolarizing the cell. Arrows show direction of positive ion flow.

transmitter released. It is necessary to point out at this time a qualification in our definition of the action potential as an all-or-none event: The action potential fires maximally if it fires at all, but its "maximal" amplitude depends on the ion concentrations and state of the membrane at the time of activation. For example, if the extracellular sodium concentration is reduced by half, the action-potential amplitude will be smaller than usual because the concentration force driving sodium into the cell has been reduced. It will still be an all-or-none response, but it will be smaller in magnitude. Thus, any change in the immediate environment of the axon terminal may alter the amplitude of the action potential at that site.

Residual calcium within the synaptic terminal can also alter the amount of transmitter released during an action potential. Calcium that has entered the terminal during previous action potentials is normally removed by being pumped out of the cell. If the preceding action potentials have arrived in a rapid burst, the calcium removal systems may not be able to keep up with the entry, and calcium concentration in the terminal will increase. Consequently, the amount of calcium in the terminal following the next action potential (and, hence, the amount of transmitter released into the synaptic cleft) will be greater than normal.

A third way of altering the amount of transmitter released is by the action of axon-axon synapses, i.e., axon terminals which end on other axon terminals. In Fig. 8-44, there is a typical synapse between a presynaptic neuron's axon terminal (labeled B) and a postsynaptic neuron's cell body (labeled C). What is new in this figure is that the axon terminal B itself receives synaptic input from yet another axon (A). The neurotransmitter released by axon terminal A combines with receptor sites on axon terminal B, resulting in a change in the amount of transmitter released from this terminal in response to action potentials. (The mechanism of these effects is not yet clear but probably involves calcium.) Thus, neuron A has no *direct* effect on neuron C but has an important effect on the ability of B to influence C. Neuron A is said to be exerting a presynaptic effect on the synapse between B and C. Depending upon the nature of the transmitter released from neuron A and the type of receptor sites on neuron B, the presynaptic effect may decrease the amount of transmitter released from neuron B or increase it. If the presynaptic effect re-

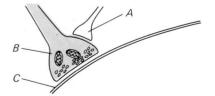

FIGURE 8-44. Axon-axon synapse (also called a presynaptic synapse) between axon terminal A and axon terminal B. C is the final postsynaptic cell.

duces the influence of neuron B on neuron C, the effect is called **presynaptic inhibition**; if it enhances the effect of neuron B on C, it is known as **presynaptic facilitation**. Generally, presynaptic synapses occur in nerve-fiber systems that transmit incoming afferent information.

Neurotransmitters

As described above, except for electrical synapses, presynaptic neurons influence postsynaptic neurons by means of chemical neurotransmitters; efferent neurons influence effector cells also by neurotransmitters. There are many different chemical substances known or suspected to be neurotransmitters; as shown in Table 8-8, they can be organized into several classes: (1) acetylcholine, which constitutes a class by itself, (2) monoamines, (3) amino acids, and (4) peptides. We have already pointed out (Chap. 7) that many of these chemicals serve not only as neurotransmitters but as hormones or paracrines. This is particularly true for certain of the peptides.

In a few cases more than one neurotransmitter may be released from a given axon terminal, but in any case, all axon terminals from a given neuron probably liberate the same transmitter (or transmitters). When more than one transmitter is released, one of them probably serves to enhance the effectiveness of the other.

We also emphasize that it is possible for the same neurotransmitter to produce an EPSP at one synapse but an IPSP at another. The response of a postsynaptic cell to a neurotransmitter depends on the type of ion channel that is opened or closed when the transmitter binds to a receptor. Thus, a neurotransmitter is merely a chemical messenger that triggers a response that is built into its target cell.

Acetylcholine

Acetylcholine (ACh) is synthesized from choline and acetyl coenzyme A in the cytoplasm of synaptic terminals and is stored in synaptic vesicles. The

TABLE 8-8. Classes of some chemicals known or presumed to be neurotransmitters

1. Acetylcholine

2. Monoamines
   Catecholamines
      Dopamine
      Norepinephrine
      Epinephrine
   Serotonin (5-hydroxytryptamine, 5-HT)
   Histamine

3. Amino acids
   Glycine
   Glutamate
   Aspartate
   Gamma-aminobutyric acid (GABA)

4. Peptides
   Enkephalins
   Endorphin
   Substance P
   Somatostatin
   Vasoactive intestinal peptide (VIP)
   Cholecystokinin-like peptide
   Neurotensin
   Insulin
   Gastrin
   Glucagon
   Thyrotropin-releasing hormone (TRH)
   Gonadotropin-releasing hormone (GnRH)
   Adrenocorticotropin (ACTH)
   Angiotensin II
   Bradykinin
   Vasopressin (antidiuretic hormone, ADH)
   Oxytocin
   Epidermal growth factor
   Prolactin
   Bombesin

concentration of acetylcholine in the synaptic cleft is reduced (thereby stopping receptor activation) by an enzyme, **acetylcholinesterase**, which is located on the postsynaptic membrane. This enzyme rapidly destroys acetylcholine, releasing choline, which is actively transported into the axon terminals where it is reused in the synthesis of ACh.

Despite evidence that acetylcholine is widely

distributed throughout the brain and spinal cord, relatively little is known about the ACh-secreting neurons in the central nervous system. In contrast, as we have seen, acetylcholine is one of the major neurotransmitters in the efferent divisions of the peripheral nervous system.

Monoamines

The monoamines are of the basic type R—NH_2. (Amino acids also fit this chemical type, but neurophysiologists generally put them into a category of their own; in any case, the monoamines are synthesized from amino acids.) The most common neurotransmitter molecules of this type are dopamine, norepinephrine, epinephrine, serotonin, and histamine.

Dopamine, norepinephrine, and **epinephrine** all contain a catechol ring (a six-sided carbon ring with two adjacent hydroxyl groups) and an NH_2 (amine) group; thus they are called **catecholamines**. They share the synthetic pathway shown in Fig. 8-45, which begins with the uptake of the amino acid tyrosine by axon terminals. Depending on the enzymes present in the terminals, the neurotransmitter that is finally formed may be any one of the three catecholamines. Transmitter activity ceases when the catecholamine concentration in the synaptic cleft declines, mainly because the catecholamine is actively transported back into the axon terminal.

There are relatively few catecholamine-releasing neurons in the central nervous system, and their cell bodies all lie in the brainstem. However, their axons go to virtually all parts of the brain and spinal cord, and they exert a much greater influence than their numbers alone would suggest. Their importance may be due to the divergence of their axon branches (it has been estimated that a single dopamine-releasing neuron in the human nervous system may have as many as 5,000,000 axon terminals), or it may be due to their unusual effects on postsynaptic neurons. The time-course of the action of the monoamines is often much slower than that of other neurotransmitters (their receptor action in some cases seems to be linked to cyclic nucleotides—

cAMP or cGMP—as second messengers). Thus, they may alter the actions of the postsynaptic neurons as these neurons respond to other transmitters which act more rapidly. Like acetylcholine, the catecholamines (with norepinephrine being by far the most important) are also transmitters in the autonomic nervous system.

Serotonin (5-hydroxytryptamine, 5-HT, Table 8-8) is produced in axon terminals from tryptophan, an essential amino acid. Changes in the blood levels of tryptophan or in the rate of tryptophan transport from the blood into the extracellular space of the nervous system and into the synaptic terminals can affect the rate at which serotonin is produced. Serotonin is important in the neural pathways controlling states of consciousness and mood, and several substances chemically related to it, e.g., psilocybin, a hallucinogenic agent found in some mushrooms, have potent psychic effects. Moreover, the drug LSD (lysergic acid diethylamide), a hallucinogen that induces a state resembling schizophrenia, is a serotonin antagonist. Although functioning in the central nervous system as a transmitter, serotonin is found in greatest concentration in blood platelets and specialized cells lining the digestive tract.

Histamine is synthesized from the amino acid histidine and appears to be concentrated in the hypothalamus. Histamine, like serotonin, is found in much greater amounts outside the nervous system (Chap. 17).

Amino acids

As we have just seen, a number of neurotransmitters are synthesized from amino acids; in addition, several amino acids function directly as transmitters (Table 8-8). Three such amino acids are **glycine, glutamate**, and **aspartate**. **GABA** (gamma-aminobutyric acid), which is not an amino acid but is synthesized from glutamic acid and is classified with the amino acids, also serves as a neurotransmitter and is the major inhibitory transmitter in the brain. Glycine, the simplest of the amino acids, is also an inhibitory transmitter, whereas aspartate and glutamate are among the most potent excitatory chemicals in the central nervous system.

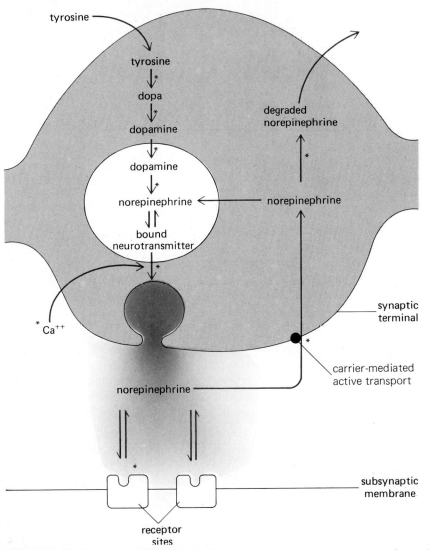

FIGURE 8-45. Sites of catecholamine synthesis, storage, release, action, re-uptake, and degradation. The asterisks indicate steps that can be modified by currently available drugs. This diagram shows the release of transmitter from a varicosity, but the same steps apply to the release from axon terminals.

Peptides

More than 25 small peptides that have potent excitatory or inhibitory effects on neurons have been found in the nervous system. Many of these molecules had been previously identified in nonneural tissue where they are known to function as hormones and paracrine agents. In other words, evolution has selected the same chemical messengers for use in widely differing circumstances. Synthesis of peptide transmitters in the brain is like the synthesis of peptides in other parts of the body: The peptides are derived from the enzymatic cleavage of larger precursor molecules which in themselves have little, if any, inherent biological activity. These precursors are formed from ribosomes in the cell body, often at a considerable distance from the axon terminals where the peptides are released. This is in contrast to acetylcholine and the monoamines, which are synthesized by enzymes in the axon terminals, close to the site of their storage, release, and re-uptake. Some of the neural peptides are listed in Table 8-8.

Certain of these peptides, the **endorphins** (endogenous morphines) and the **enkephalins**, are known as the **opioid peptides**; they have gained much interest since their discovery because receptors for these substances are the sites of action of the opiate drugs (e.g., morphine and codeine). Moreover, the endorphins and enkephalins have analgesic properties (i.e., they diminish the awareness of pain) when injected into the central nervous system. The receptors for these transmitters occur in discrete locations in the central nervous system, for example, in areas containing pathways that convey pain information and in parts of the brain that are involved in mood and emotions.

The endorphins and enkephalins are synthesized as part of larger precursor molecules. For example, **proenkephalin** gives rise to the enkephalins, whereas **pro-opiomelanocortin** gives rise to beta-endorphin. (Pro-opiomelanocortin is also the precursor of several hormones, as will be discussed in Chap. 9.)

Substance P, another of the neuroactive peptides, is the transmitter for neurons that transmit sensory information into the central nervous system and that are involved in mediating pain.

Modification of Synaptic Transmission by Drugs and Disease

The great majority of drugs that act on the nervous system do so by altering synaptic mechanisms, and all the synaptic mechanisms labeled in Fig. 8-46 are vulnerable. Drugs that, upon binding to a receptor, produce a response similar to that caused by the transmitter released at the synapse are called **agonists**. On the other hand, agents that bind to the receptor but are unable to activate it are **antagonists**. By occupying the receptors, antagonists prevent binding of the normal neurotransmitter when it is released at the synapse.

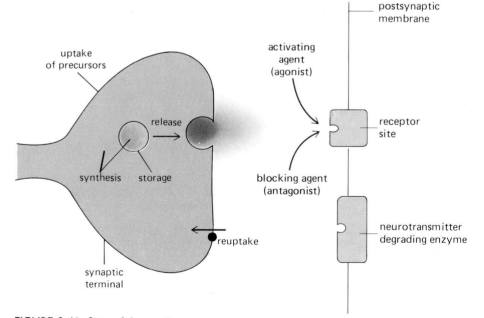

FIGURE 8-46. Sites of drug action at synapses.

Diseases can also affect synaptic mechanisms. For example, the toxin produced by the tetanus bacillus blocks at inhibitory synapses on the neurons supplying skeletal muscles, presumably by blocking the receptors. This eliminates inhibitory input to the neurons and leaves unchecked the excitatory inputs, resulting in muscle spasticity and seizures. The spasms of the jaw muscles appearing early in the disease are responsible for the common name of the condition, lockjaw.

Neuromodulators

Some chemical messengers elicit complex responses on neurons which cannot be simply described as EPSPs or IPSPs. The word "modulation" is used to describe these complex responses, and the messengers which cause them are called **neuromodulators**. The distinctions between neuromodulators (specifically, those released by neurons) and neurotransmitters, however, are not always clear. In general, neuromodulators bring about biochemical changes in a neuron, whereas neurotransmitters influence ion channels. The actions of both neuromodulators and neurotransmitters may result in changes in the membrane potential, but the changes induced by neurotransmitters are faster and more direct. Thus, neuromodulators tend to be associated with long-term effects (measured in minutes, hours or even days), whereas neurotransmitter actions occur within milliseconds.

Neuromodulators do not, themselves, function as specific transmitters but rather amplify or dampen the effectiveness of ongoing synaptic activity by altering the neurons at nonsynaptic sites. Extrasynaptic receptors, many of which activate a second-messenger system, were mentioned earlier in this chapter and are often the sites to which neuromodulators bind; this helps explain two of the characteristics of neuromodulation, i.e., the long time course of their activation and their biochemical as opposed to electrochemical effects.

So far in this section, we have been speaking of neuromodulators and the postsynaptic cell, but neuromodulators can also activate receptors on the axon terminals of the presynaptic neurons. By these presynaptic effects, neuromodulators may alter the neurotransmitter's synthesis, release, re-uptake, or metabolism. For example, the hormone cortisol influences the levels of the rate-limiting enzyme in the synthesis of norepinephrine and can be considered a neuromodulator.

From this example, it is clear that neuromodulators may be hormones. However, not all hormones act as neuromodulators, nor are all neuromodulators hormones. Some neuromodulators are synthesized and released locally; some are even synthesized by the presynaptic cell and released with the neurotransmitter from that cell. Moreover, a given substance may act as a neurotransmitter in one area of the brain and as a neuromodulator in a different (or even the same) area.

Table 8-9 lists some of the substances thought

TABLE 8-9. Some chemicals known or presumed to be neuromodulators

Histamine

Prostaglandins

Enkephalins

Endorphins

Substance P

Somatostatin

Aldosterone

Cortisol

Estrogens

Testosterone

Thyroid hormone

Gastrin

Vasoactive intestinal peptide (VIP)

Adrenocorticotropic hormone (ACTH)

Thyrotropin-releasing hormone (TRH)

Gonadotropin-releasing hormone (GnRH)

Vasopressin (antidiuretic hormone, ADH)

Angiotensin

Oxytocin

presently to function as neuromodulators. Trying to memorize such a table is of no value. We have presented it only to emphasize a particular point: Comparison of this table with Table 8-8 (neurotransmitters, particularly the peptides) and Table 9-1 (hormones) emphasizes the overlapping functions of these molecules. Indeed, all known neuromodulators also exert nonneural effects.

Neuroeffector Communication

Thus far we have described the effects on postsynaptic neurons of neurotransmitters released from the axon terminals, i.e., synaptic transmission. However, the axon terminals of efferent neurons do not synapse upon other neurons but rather terminate in the region of effector cells—muscle and gland cells. These junctions are called **neuroeffector junctions**, and the events that occur at these junctions are similar to those at a synapse. Neurotrans-

mitter is released upon the arrival of an action potential at the axon terminals (or the varicosities along the axons), and the transmitter then diffuses to the surface of the effector cell where it binds to a receptor on the plasma membrane of the effector cell. This receptor may be associated with ion channels that alter the membrane potential of the effector cell, or it may be associated with an enzyme that results in the formation of a second messenger in the effector cell. The response (muscle contraction or glandular secretion) of the effector cell to these changes will be described in later chapters. Suffice it here to reemphasize that it is the chemical messenger (the neurotransmitter) released by the efferent neuron's terminals that provides the link by which electrical activity of the nervous system is able to regulate effector cell activity. As described earlier, the major known neurotransmitters released by efferent neurons are acetylcholine and norepinephrine.

SECTION D.
RECEPTORS AND PROCESSING OF AFFERENT INFORMATION

Receptors

Information about the external world and the internal environment exists in different energy forms—pressure, temperature, light, sound waves, etc. It is the function of **receptors** to translate these energy forms into action potentials. The rest of the nervous system can extract meaning only from action potentials or, over very short distances, from graded potentials. Regardless of its original energy form, information from receptors must be translated into the language of action potentials. The process by which stimulus energy is transformed into an electrical response in a receptor is known as **transduction**.

Receptors are either specialized peripheral endings of afferent neurons or separate cells that affect the peripheral ends of afferent neurons. In the former case, the afferent neuron is activated directly when the stimulus energy impinges on the specialized plasma membrane of the receptor (Fig.

8-47A); in the latter case, the separate receptor cell contains the specialized membrane that is activated by the stimulus (Fig. 8-47B), and upon stimulation, the receptor cell releases a chemical messenger. This substance diffuses across the extracellular cleft separating the receptor cell from the afferent neuron and binds to specific sites on the afferent neuron. The relation between a separate receptor cell and the afferent nerve ending is similar in mechanism to the transfer of signals across a synapse. In the examples of Fig. 8-47A and B, the result is an electrical change (a graded potential) in the afferent neuron, which leads to the generation of action potentials.

There are many types of receptors, each of which is specific; i.e., each responds much more readily to one form of energy than to others, although virtually all receptors can be activated by several different forms of energy if the intensity is sufficient. For example, the receptors of the eye normally respond to light, but they can be activated

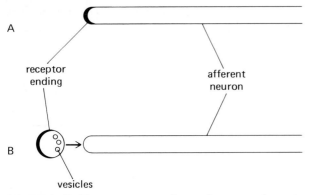

FIGURE 8-47. The sensitive membrane that responds to the stimulus is either (A) an ending of the afferent neuron itself or (B) on a separate cell adjacent to the afferent neuron (highly schematized).

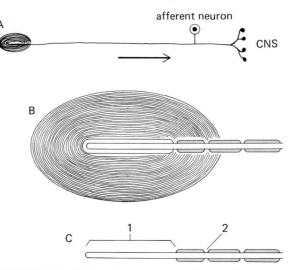

FIGURE 8-48. (A) An afferent neuron with a mechanoreceptor (pacinian corpuscle) ending. (B) A pacinian corpuscle showing the nerve ending modified by cellular structures. (C) The naked nerve ending of the same mechanoreceptor. The receptor potential arises at the nerve ending (1), and the action potential arises at the first node of the myelin sheath (2).

by an intense mechanical stimulus, like a poke in the eye. Note, however, that one still experiences the sensation of light in response to the poke in the eye; regardless of how the receptor is stimulated, it always gives rise to the same sensation. Thus, the sensation experienced depends on the type of receptor stimulated and not on the nature of the stimulus. This concept, that for every kind of sensation there is a special type of receptor whose activation always gives rise to that sensation, is stated as the **doctrine of specific nerve energies**.

Most receptors are exquisitely sensitive to their specific energy forms. For example, olfactory receptors can respond to as few as three or four odorous molecules in the inspired air, and visual receptors can respond to a single photon, the smallest quantity of light.

Receptor activation: The receptor potential

The general mechanisms of receptor activation are believed to be the same for all types of receptors. The stimulus acts on a specialized receptor membrane to alter its permeability and thereby produce a graded change in potential, called a **receptor potential**. In describing the general mechanisms for receptor activation we will use as an example the pacinian corpuscle, a receptor which is the peripheral ending of an afferent neuron and responds to mechanical deformation, such as touch (Fig. 8-48).

In a mechanoreceptor, a stimulus may bend, stretch, or press upon the receptor membrane and somehow, presumably by causing ion channels in the membrane to open, increase its permeability to certain ions. Because of the increased permeability, ions move across the membrane down their electrical and concentration gradients. Although these ion movements have not been worked out as clearly as those associated with action potentials, the permeability increases appear to be nonselective and to apply to all small ions.

Just as at an activated excitatory synapse, the result of a nonselective increase in membrane permeability is a net outward movement of a small number of potassium ions (small net driving force) and the simultaneous movement in of a larger number of sodium ions (large driving force). The result is net movement of positive charge into the cell, producing a decrease in membrane potential (depolarization). The movement of sodium into the receptor thus plays the major role in generating the receptor potential in the mechanoreceptor and very likely in most other receptor types as well. (A major exception is in the eye, as we shall see in Chap. 18.)

In the receptor region the plasma membrane has a very high threshold, so that the receptor potential does not depolarize this region enough to cause an action potential. Instead, this depolarization (i.e., the receptor potential) is conducted by local current flow a short distance from the receptor membrane to a nearby region where the membrane threshold is lower. In myelinated afferent neurons, this region is usually at the first node of the myelin sheath. As is true of all graded potentials, the magnitude of the receptor potential decreases with distance from its site of origin, but if the amount of depolarization that reaches the first node is large enough to bring the membrane there to threshold, action potentials are initiated. The action potentials then propagate along the nerve fiber; the only function of the receptor potential is to trigger action potentials.

Receptor potentials, being graded potentials, are not all-or-none; their magnitude and duration vary with stimulus strength and other variables to be discussed shortly. A change in the magnitude of the receptor potential causes, via local current flow, a similar (but smaller) change in the degree of depolarization at the first node. As long as the first node remains depolarized to threshold, action potentials continue to fire and propagate along the afferent neuron; the greater the depolarization (up to a point), the greater the frequency of action potentials (Fig. 8-49). Therefore, the magnitude and duration of the receptor potential determine the action-potential frequency in the afferent neuron. Although the receptor-potential amplitude is a determinant of action-potential frequency (i.e., the number of action potentials fired per unit time) it does not determine action-potential magnitude. Since the action potential is all-or-none, its magnitude is independent of the strength of the initiating stimulus.

Magnitude of the receptor potential

Factors that control the magnitude of the receptor potential include the stimulus intensity, rate of

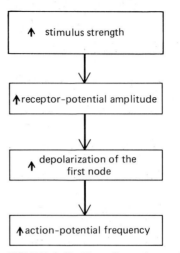

FIGURE 8-49. The effect of stimulus strength on action-potential frequency.

change of stimulus application, summation of successive receptor potentials, and adaptation.

Intensity and rate of change of stimulus application. Increasing the intensity of the stimulus increases the amplitude of the receptor potential (Fig. 8-50) because the ion permeability changes are greater. The amplitude of the receptor potential also increases with a greater rate of change of stimulus application; i.e., the faster the stimulus is applied, the larger the receptor potential.

Summation of receptor potentials. Another way of varying the amplitude of the receptor potential is by adding two or more receptor potentials together. If the receptor membrane is stimulated again before the receptor potential from a preceding stimulus has died away, the two potentials sum and make a larger depolarization (see Fig. 8-26, Part 4). This is because applying the second stimulus opens additional ion channels, producing more local current.

Adaptation. **Adaptation** is a decrease in the frequency of action potentials in an afferent neuron despite maintenance of the stimulus at constant strength. The mechanisms underlying adaptation vary in different receptor types. Adaptation of an afferent neuron can be seen in Fig. 8-51. Some

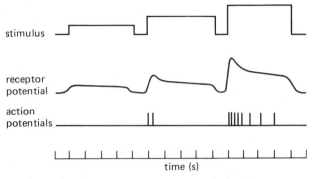

FIGURE 8-50. Relationship between stimulus intensity, receptor-potential magnitude, and action-potential frequency.

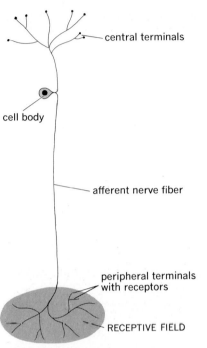

FIGURE 8-52. Sensory unit and receptive field.

receptors adapt completely so that in spite of a constantly maintained stimulus, the generation of action potentials stops. In extreme cases, the receptors fire only once at the onset of the stimulus. In contrast to these rapidly adapting receptors, slowly adapting types merely drop from an initial high action-potential frequency to a lower level, which is then maintained for the duration of the stimulus.

Sensory Units

The information from receptors travels over discrete neural pathways. These pathways, begin at the receptor, which is part of a sensory unit.

A single afferent neuron with all of its receptor endings makes up a **sensory unit**. In a few cases the afferent neuron terminates at a single receptor, but generally the peripheral end of an afferent neuron divides into many fine branches, each terminating at a receptor. All the receptors of a sensory unit are preferentially sensitive to the same stimulus modality (quality); for example, they are all sensitive to cold or all to vibration. The **receptive field**

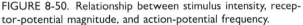

↑ stimulus on   stimulus off ↑

FIGURE 8-51. Action potentials in a single afferent nerve fiber, showing adaptation.

of a neuron is that area of the body which, if stimulated, leads to activity in the neuron (Fig. 8-52).

The central processes of the afferent neurons terminate in the central nervous system, diverging to terminate on several (or many) interneurons (Fig. 8-53A) and converging so that the processes of many afferent neurons terminate upon a single interneuron (Fig. 8-53B).

Basic Characteristics of Sensory Coding

The pathways in the central nervous system that convey signals which have originated in afferent neurons serve as the route for several different kinds of information: stimulus modality, intensity, and localization.

Stimulus modality

As described earlier, different receptors have different sensitivities; i.e., each receptor type responds more readily to one form of energy than to

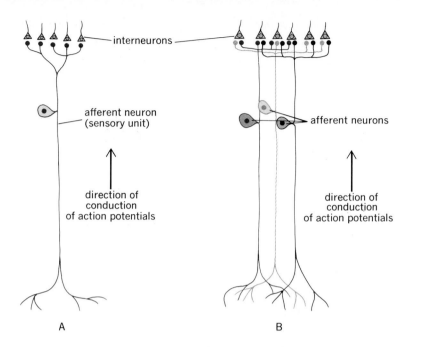

FIGURE 8-53. (A) Divergence of afferent neuron (sensory unit) terminals. (B) Convergence of input from several afferent neurons onto single interneurons.

others. These different forms of stimulus energy are called **modalities**. Therefore, the type of receptor activated by a stimulus constitutes the first step in the coding of different types (modalities) of stimuli. The identity of the stimulus modality is retained during transmission of the information to the cerebral cortex, as will be discussed in Chap. 18. Thus, the stimulus modality is indicated not only by the specific sensitivities of individual receptors, but also by the specificity of the pathways that convey the information to the brain and to the region of brain activated. This means that stimulation of a specific nerve pathway at any point along its course will give rise to the sensation that would have occurred if the receptor at the beginning of the pathway had been activated by its natural stimulus. For example, regardless of how or where the neural pathways from the eye are stimulated, one always perceives some form of light, and activation of the pain pathway by the pressure of a growth in the spinal column results in the perception of pain.

Stimulus intensity

We are certainly aware of different stimulus intensities. How is information about stimulus strength relayed by action potentials of constant amplitude?

One way is related to the frequency of action potentials; increased stimulus strength means a larger receptor potential and higher frequency of firing of action potentials. A record of an experiment in which increased stimulus intensity is reflected in increased action-potential frequency in a single afferent nerve is shown in Fig. 8-54.

Moreover, as stimulus strength increases, more and more receptors on other branches of the afferent neuron begin to respond (the receptors on different branches do not respond with equal ease to a given stimulus). The action potentials generated by these several receptors propagate along the branches to the main afferent nerve fiber and increase the frequency of action potentials there (Fig. 8-55).

In addition to the increased frequency of firing in a single neuron, similar receptors on the endings of other afferent neurons are also activated as stimulus strength increases, because stronger stimuli usually affect a larger area. For example, when one touches a surface lightly with a finger, the area of skin in contact with the surface is small, and only receptors in that area of skin are stimulated; pressing down firmly increases the area of skin stimulated. This "calling in" of receptors on additional

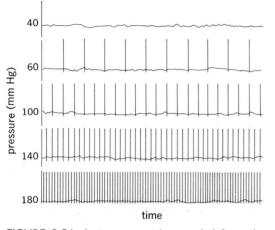

FIGURE 8-54. Action potentials recorded from the afferent fiber leading from a pressure receptor as the receptor was subjected to pressures of different magnitudes. Little or no adaptation is seen in these responses.

nerve cells is known as **recruitment**. These generalizations are true of virtually all afferent systems: Increased stimulus intensity is signaled both by an increased firing rate of action potentials in a single nerve fiber and by recruitment of receptors on additional afferent neurons.

Current techniques make it possible to stimulate a single receptor or afferent fiber in an awake person and get accurate reports of the perceived sensations as the strength of the stimulus is varied (Fig. 8-56). The results of such experiments show that the perception of the magnitude of the stimulus varies with stimulus intensity in a quantitative way; i.e., the magnitude of the stimulation, the afferent nerve response, and the perceptual estimation are all closely related. However, the relationship varies with the sensory system being studied. For example, in the visual system one photon (the smallest known quantity of light) can activate a single rod receptor cell, but no perception occurs; the simultaneous activation of about seven receptors is necessary for perception of the stimulus to occur. In contrast, in the case of certain mechanoreceptors, a single action potential in the afferent neuron can be perceived. However, in general, the threshold for perception is higher than the threshold of an individual receptor, and several receptor responses must occur simultaneously for perception to occur.

Stimulus localization

A third type of information to be signaled is the location of the stimulus. Since sensory units from only a restricted area converge upon one interneuron of the pathways specific for that modality, the

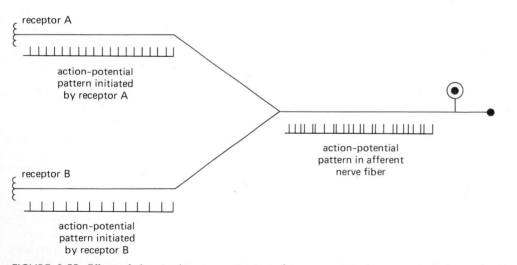

FIGURE 8-55. Effect of the simultaneous activation of two receptors in a sensory unit on the number of action potentials transmitted along the afferent neuron.

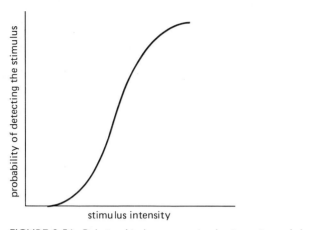

FIGURE 8-56. Relationship between stimulus intensity and the probability of detecting the stimulus. The actual shape of the curve varies with different sensory systems and under different conditions.

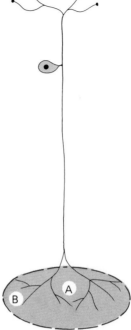

FIGURE 8-57. Two stimulus points (A and B) in the receptive field of a single afferent neuron.

pathway which begins with a particular interneuron transmits information about only that restricted area. Thus, the specific pathway indicates stimulus location as well as stimulus modality.

Despite the fact that the receptive fields of individual sensory units may vary considerably, e.g., from 2 to 200 mm^2 in skin, and the fact that receptive fields of adjacent neurons overlap considerably so that stimulation of only a single sensory unit almost never occurs, it is nevertheless possible for a person to pinpoint the location of a stimulus because of the interactions between the activated sensory units. Let us examine more closely how such interaction occurs.

The density of the receptors of a single sensory unit (i.e., the number of receptors in a given area) varies within the receptive field of that unit, usually being greatest at its geometric center. Thus, a stimulus of a given intensity activates more receptors and generates more action potentials in the afferent neuron of the sensory unit if it occurs at the center of the receptive field (point A, Fig. 8-57) than at the periphery (point B). But since the peripheral terminations of different afferent neurons overlap to a great extent, the placement of a stimulus determines not only the rate at which one nerve fiber fires but also the activity in sensory units with overlapping receptive fields. In the example in Fig.

8-58, neurons A and C, stimulated near the edge of their receptive fields where the receptor density is lower, fire at a lower frequency than neuron B, stimulated at the center of its receptive field. Thus, the information content of the pattern of activity in a population of afferent neurons is great.

As we have seen, stimulus strength is related to the firing frequency of the afferent neuron, but a high frequency of impulses in the single afferent nerve fiber of Fig. 8-57 could mean either that a stimulus of moderate intensity was applied at the center of the receptive field (point A) or that a strong stimulus was applied at the periphery (point B). Neither the intensity nor the localization of the stimulus can be detected precisely with a single neuron. But in a group of sensory units (Fig. 8-58), a high frequency of action potentials in neuron B arriving simultaneously with a lower frequency of action potentials in neurons A and C permits more accurate localization of the stimulus near the center of neuron B's receptive field. Once the location of the stimulus within the receptive field of neuron B

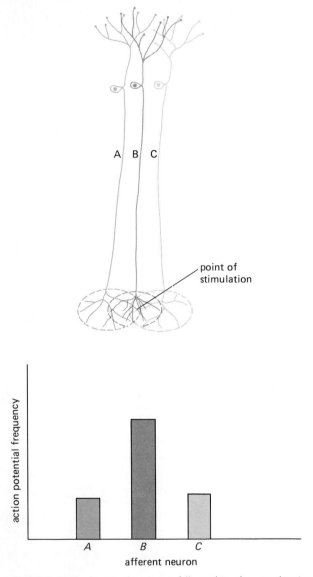

A B C

point of
stimulation

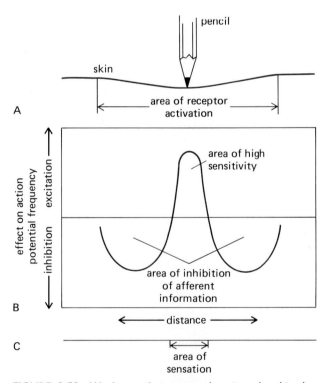

FIGURE 8-58. A stimulus point falls within the overlapping receptive fields of three afferent neurons.

example, the ability to discriminate between two adjacent mechanical stimuli to the skin is greatest on the thumb, fingers, and lips, where the sensory units are small and overlap considerably. The localization of visceral sensations is less precise than that of somatic stimuli because there are fewer afferent neurons and each has a larger receptive field.

Central control of afferent information

All incoming afferent information is subject to extensive control at synaptic junctions before it reaches higher levels of the central nervous system. Much of it is reduced or even abolished by inhibition from other neurons. Some of the neural elements mediating these controls are collaterals from

FIGURE 8-59. (A) A pencil tip pressed against the skin depresses surrounding tissues. Receptors are activated under the pencil tip and in the adjacent tissue. (B) Because of lateral inhibition, the central area of excitation is surrounded by an area where the afferent information is inhibited. (C) The sensation is localized to a more restricted region than that in which mechanoreceptors were actually stimulated.

is known, the firing frequency of neuron B can be used as a meaningful measure of stimulus intensity.

The precision with which a stimulus can be localized and differentiated from an adjacent stimulus depends both on the size of the receptive field covered by a single afferent neuron and on the amount of overlap of nearby receptive fields. For afferent neurons, interneurons in the local vicinity, and descending pathways from higher regions, particularly the reticular formation and cerebral cortex. The inhibitory controls are exerted via synapses mainly upon two sites: (1) the axon terminals of the afferent neurons and (2) the second-order neurons, i.e., the interneurons directly activated by the afferent neurons. The afferent terminals are influenced by presynaptic inhibition, whereas the second-order cells receive both pre- and post-synaptic inputs.

In some cases the afferent input is continually (i.e., tonically) inhibited to some degree. This provides the flexibility of either removing the inhibition (**disinhibition**) so as to allow a greater degree of signal transmission or of increasing the inhibition so as to block the signal more completely. The neurons of the pain pathways seem to be under such tonic inhibition.

Afferent information can be modified by facilitation as well as by inhibition. Both methods increase the contrast between ''wanted'' and ''unwanted'' information, thereby increasing the effectiveness of selected pathways and focusing sensory-processing mechanisms on ''important'' messages.

In some afferent systems, the control is organized in such a way that stronger inputs are enhanced and the weaker inputs of adjacent sensory units simultaneously inhibited. Such **lateral inhibition** can be demonstrated in the following way. While pressing the tip of a pencil against your finger with your eyes closed, you can localize the pencil point precisely, even though the region around the pencil tip is also indented and mechanoreceptors within this entire area are activated (Fig. 8-59); this is because the information from the peripheral region is removed by mechanisms of lateral inhibition. Lateral inhibition occurs in the pathways of most sensory modalities but is utilized to the greatest degree in the pathways providing the most accurate localization. For example, a stimulus can be localized very precisely in the pathways relaying information about hair deflection, whereas stimuli activating temperature receptors, whose pathways lack lateral inhibition, are localized only poorly.

SECTION E.
PATTERNS OF NEURAL ACTIVITY: THE FLEXION REFLEX

The purpose of this section is to illustrate some of the varying degrees of complexity exhibited by neural control systems and to review the neural mechanisms presented in the previous sections of this chapter. As an example, we will describe one of the simpler reflexes—the flexion reflex.

The sequence of events elicited by a painful stimulus to the toe leads to withdrawal of the foot from the source of injury; this is an example of the **flexion**, or **withdrawal, reflex** (Fig. 8-60). Receptors in the toe are stimulated and, via receptor potentials, transform the energy of the stimulus into action potentials in afferent neurons. Information about the intensity of the stimulus is coded by the frequency of action potentials in the afferent nerve fibers and by the number of afferent neurons activated. The afferent fibers branch after entering the spinal cord, diverging to form synaptic junctions with other neurons. In the flexion reflex, the second neurons in the pathway are interneurons. Because interneurons are interspersed between the afferent and efferent limbs of the reflex arc, the flexion reflex is one of the very large class of **polysynaptic** reflexes.

The afferent-neuron branches serve different functions. Some branches synapse with interneurons whose processes carry the information into the brain; it is only after the information transmitted in these pathways reaches the brain that the conscious correlate of the stimulus, i.e., the sensation of pain, is experienced. Other branches synapse with interneurons that, in turn, synapse upon the efferent neurons innervating flexor muscles. These muscles, when activated, cause flexion (bending) of the ankle and withdrawal of the foot from the stimulus.

Still other branches of the afferent neurons ac-

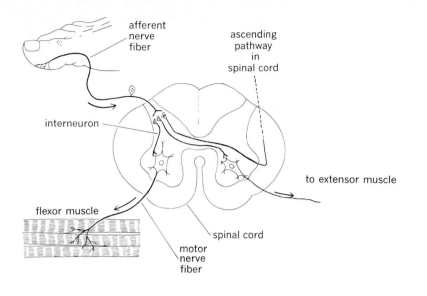

FIGURE 8-60. Flexion, or withdrawal, reflex pathway. Arrows indicate direction of action potential propagation.

tivate interneurons that inhibit the motor neurons of antagonistic muscles whose activity would oppose the reflex flexion. This inhibition of antagonistic motor neurons is called **reciprocal innervation**. In this case, the motor neurons innervating the foot and leg extensor muscles (which straighten the ankle and leg) are inhibited. Other interneurons excite extensor and inhibit flexor motor neurons controlling movement of the opposite leg, so that as the injured leg is flexed away from the stimulus, the opposite leg is extended more strongly to support the added share of the body's weight. This involvement of the opposite limb is called the **crossed extensor reflex**.

The neural networks activated in the flexion reflex are among the simplest in humans, yet the reflex response is coordinated and purposeful. The foot flexion away from the stimulus removes the receptors from the stimulus source, and further damage is prevented (an example of negative feedback). The purposefulness of the reflex response,

however, is independent of the sensation of pain, for the withdrawal occurs before pain is experienced. In fact, the reflex response can still be elicited after the spinal cord pathways that normally transmit the afferent information to the brain have been severed.

An important characteristic of this type of control system is that the observed response varies with the stimulus strength (between threshold and maximal stimulus strength). In general, low levels of stimulation lead to responses localized to the area of stimulus application, whereas more intense stimulation causes more widespread responses. Another characteristic of the flexion reflex is that the stimulus induces a very brief and nearly simultaneous response. However, in other types of reflexes (the swallow reflex, for example), a single brief stimulus may trigger off a chain of responses that last more than 10 s, and once initiated, the course of the reflex response is largely unaffected by stronger sensory input.

HORMONAL CONTROL MECHANISMS

A **hormone** is a specialized chemical messenger which is secreted by specific **endocrine** gland cells in response to certain stimuli and is carried by the blood to other sites in the body where it alters cell activity. Hormones are effective at very low blood concentrations.

The endocrine system constitutes the second major communication system of the body, the hormones serving as blood-borne messengers which control and integrate many functions: circulation (Chap. 11), water and electrolyte balance (Chap. 13), digestion and absorption of food (Chap. 14), organic metabolism and energy balance (Chap. 15), reproduction (Chap. 16), and responses to stress (Chap. 17). Figure 9-1 illustrates the location of some of the major endocrine glands. Clearly, the endocrine system differs from most of the other organ systems of the body in that the various en-docrine glands are not in anatomical continuity with each other; however, they do form a system in the functional sense.

In this last regard, note that the hypothalamus, a part of the brain, is an endocrine "gland." This is because the chemical messengers released by some of the terminals of neurons in the hypothalamus (and its extension, the posterior pituitary) do not function as neurotransmitters but rather are released into the bloodstream, which carries them to their sites of action elsewhere. Such hormones, released from neuron terminals, are sometimes called **neurohormones**.

This is only one of the ways in which the nervous and endocrine systems actually function as a single, interrelated system. The nervous system not only secretes hormones but controls the secretion of hormones by several endocrine glands; con-

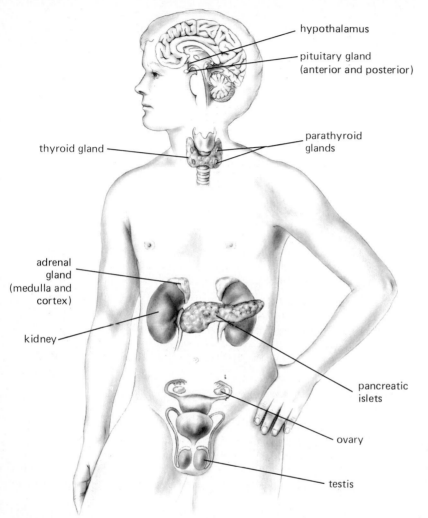

hypothalamus

pituitary gland
(anterior and posterior)

parathyroid
glands

thyroid gland

adrenal
gland
(medulla and
cortex)

kidney

pancreatic
islets

ovary

testis

FIGURE 9-1. Location of some endocrine glands. This hypothetical bisexual figure has both ovaries and testes; this does not occur in normal human beings. Not shown in the figure are the liver, gastrointestinal tract, pineal gland, thymus, and placenta, all of which secrete hormones.

versely, hormones markedly alter neural function and strongly influence many types of behavior. All these interrelationships between nervous and endocrine systems form the area of study known as **neuroendocrinology**.

The fact that the endocrine and nervous systems influence each other has been known for many years, but only recently has it become apparent that, as mentioned in Chaps. 7 and 8, some of the same chemicals serve as messengers in both systems, i.e., as both neurotransmitters (or neuromodulators) and hormones. For example, several of the hormones secreted by the gastrointestinal tract are also produced by neurons in the brain and function there as neurotransmitters.

Table 9-1 summarizes, for reference and orientation, the major endocrine glands, the hormones they secrete, and the hormones' functions. One phenomenon clearly evident from this table is that of multiple hormone secretion by a single gland. In certain glands, such as the pancreas, each of the multiple hormones is secreted by a distinct cell type; in other endocrine glands, it is likely that, although some separation of function exists, a single cell may secrete more than one hormone.

A frequent source of confusion concerning the

endocrine system involves not multiple hormone secretion by a single gland, but the mixed general functions exhibited by the gonads, pancreas, kidneys, and gastrointestinal tract. All these organs contain endocrine gland cells, but they also perform other completely distinct, nonendocrine functions. For example, most of the pancreas is concerned with the production of digestive enzymes; these are produced by exocrine glands; i.e., the secretory products are transported not into the blood but into ducts leading, in this case, into the intestinal tract. The endocrine function of the pancreas is performed by nests of endocrine cells scattered throughout the pancreas and quite distinct from the exocrine glands. This type of separation of function is true for all the organs having mixed roles; i.e., the endocrine function is usually subserved by gland cells distinct from the other cells which constitute the organ.

Another point of interest illustrated by Table 9-1 is that a particular hormone may be produced by more than one type of endocrine gland. For example, somatostatin is secreted by endocrine cells in both the gastrointestinal tract and the pancreas and is also one of the "releasing hormones" secreted by the hypothalamus. To take another example, estrogens (this term refers to a family of female sex hormones) are produced not only by the ovaries but by the placenta (which makes almost any hormone one chooses to look for, albeit often in very small quantities). As the ability to detect hormones continues to improve, we will almost certainly find more and more cases of a single type of hormone being secreted by more than one type of gland.

Despite the size of Table 9-1, it is by no means complete. It omits several entire classes of hormones (see footnote for Table 9-1), and it also omits substances suspected of being hormones but for which no specific function has been conclusively established. Among the most interesting of the latter substances are peptides which are cosecreted with other, well-established peptide hormones, as described below.

Table 9-1 is also incomplete in its listing of functions. Physiologists are finding that hormones with well-established functions are turning out to have additional, quite unrelated functions. For example, vasopressin, known for many years to control water excretion by the kidneys, may also exert effects on memory and learning (see Chap. 20). Such a strange juxtaposition of functions may be explained by an adage coined by a physiologist with regard to evolution: "Never make a new hormone if you can use an old one." In other words, receptors for an "old" hormone may evolve in a new target tissue, and the hormone may therefore come to exert an entirely new action.

Incomplete or not, Table 9-1 certainly emphasizes that the endocrine system includes a very large number of hormones. One way of describing the physiology of these individual hormones is to present all relevant material in a systematic, gland-by-gland approach. However, in keeping with our emphasis on hormones as components of homeostatic control mechanisms, we have chosen to describe the physiology of specific hormones and the glands which secrete them in subsequent chapters, in the context of the control systems in which they participate. For example, the pancreatic hormones are described in Chap. 15 (which deals with regulation of organic metabolism), the parathyroid hormones in Chap. 13 (in the context of calcium metabolism), and so on. Therefore, the aims of the present chapter are more limited: (1) to present the general principles of endocrinology, i.e., a structural and functional analysis of hormones in general, which transcends individual glands; and (2) an analysis of the hypothalamus-pituitary hormonal system. The control systems for these particular hormones are so interconnected that they are best described as a unit to lay the foundation for subsequent descriptions in other chapters.

Hormone–Target-Cell Specificity

Hormones travel in the blood and are therefore able to reach virtually all tissues. This is obviously very different from the efferent nervous system, which can send messages selectively to specific tissues and organs. Yet, the body's response to hormones

TABLE 9-1. Summary of the major hormones*

Gland	Hormone	Major function is control of:
Hypothalamus	Releasing hormones	Secretion by the anterior pituitary
	Oxytocin	(See posterior pituitary)
	Vasopressin	(See posterior pituitary)
Anterior pituitary	Growth hormone (somatotropin, GH, STH)†	Growth; organic metabolism; secretion of somatomedin by liver
	Thyroid-stimulating hormone (TSH, thyrotropin)	Thyroid gland
	Adrenocorticotropic hormone (ACTH, corticotropin)	Adrenal cortex
	Prolactin	Breasts (milk synthesis)
	Gonadotropic hormones: Follicle-stimulating hormone (FSH) Luteinizing hormone (LH)	Gonads (gamete production and sex hormone secretion)
Posterior pituitary‡	Oxytocin	Milk "let-down"; uterine motility
	Vasopressin (antidiuretic hormone, ADH)	Water excretion by kidneys
Adrenal cortex	Cortisol	Organic metabolism; response to stresses
	Androgens	Growth; in women, sex drive
	Aldosterone	Sodium and potassium excretion by kidneys
Adrenal medulla	Epinephrine Norepinephrine	Organic metabolism; cardiovascular function; response to stresses
Thyroid	Thyroxine (T_4) ⎫ Thyroid Triiodothyronine (T_3) ⎬ hormone	Energy metabolism; growth
	Calcitonin	Plasma calcium
Parathyroids	Parathyroid hormone (parathormone, PTH, PH)	Plasma calcium and phosphate
Gonads		
Female: ovaries	Estrogens Progesterone	Reproductive system; breasts; growth and development
Male: testes	Testosterone	Reproductive system; growth and development
Pancreas	Insulin Glucagon Somatostatin	Organic metabolism; plasma glucose
Kidneys	Renin ($\rightarrow$Angiotensin)** Erythropoietin 1,25-Dihydroxyvitamin D_3	Adrenal cortex; blood pressure Erythrocyte production Calcium balance
Gastrointestinal tract	Gastrin Secretin Cholecystokinin Gastric inhibitory peptide Somatostatin	Gastrointestinal tract; liver; pancreas; gallbladder

TABLE 9-1. (*continued*)

Gland	Hormone	Major function is control of:
Liver	Somatomedin	Bone growth
Thymus	Thymic hormone (thymosin)	Lymphocyte development
Pineal	Melatonin	Sexual maturity
Placenta	Chorionic gonadotropin Estrogens Progesterone	Maintenance of pregnancy

*This table does not list all hormones. Not given are the hormones secreted by white blood cells and macrophages (Chap. 17), a large number of so-called growth factors (Chap. 15), and a variety of substances suspected of being hormones but with inconclusively documented functions. The table also does not list all functions of the hormones.
†The names and abbreviations in parentheses are synonyms.
‡The posterior pituitary stores and secretes these hormones; they are synthesized in the hypothalamus.
**Renin is an enzyme which initiates reactions that generate angiotensin (Chap. 13).

is not all-inclusive but highly specific, in some cases involving only one organ or group of cells. In other words, despite the ubiquitous distribution of a hormone via the blood, only certain cells are capable of responding to the hormone; they are known as the **target cells** for that hormone. This ability to respond depends upon the presence in (or on) the target cells of specific receptors for those hormones. As described in Chap. 7, a receptor is a protein to which a chemical messenger specifically binds, thereby initiating a sequence of events culminating in the target cell's response.

Chemical Structures and Synthesis

Hormones fall mainly into three general chemical classes: (1) amines, (2) peptides and proteins, and (3) steroids.

Amine hormones

The amine hormones, all derived from the amino acid tyrosine, include thyroid hormone as well as epinephrine and norepinephrine—the two major hormones produced by the adrenal medulla.

Thyroid hormone. The thyroid gland, located in the neck, actually secretes three different hormones, but two of them—**thyroxine (T_4)** and **triiodothyronine (T_3)**—are collectively known as **thyroid hor-**mone **(TH)** (the subscripts in the abbreviations denote the number of iodine atoms in the molecule, Fig. 9-2). (The third hormone, called calcitonin, is a peptide). Iodine is essential for the synthesis of T_4 and T_3. Most ingested iodine is absorbed by the gastrointestinal tract (which converts it to iodide, the ionized form of iodine), and it is then removed from the blood by the thyroid cells, which have a powerful active-transport mechanism for iodide. Once in the gland, the iodide is converted back to iodine, which is then used, along with tyrosine, for hormone synthesis. The normal thyroid gland may store several weeks' supply of thyroid hormones bound to a large protein known as **thyroglobulin**. Hormone secretion occurs by enzymatic splitting of T_4 and T_3 from the thyroglobulin and the entry of this freed hormone into the blood. The hormone T_4 is secreted in much larger amounts than is T_3, but most of the circulating T_4 is converted to T_3 (by enzymatic removal of one iodine atom) when the hormone reaches the peripheral tissues. This newly formed T_3, along with that secreted by the thyroid gland, then binds to receptors in the target cells and elicits a response. (T_4, itself, can probably also bind to these receptors but to a much lesser degree than T_3 can.) This is only one example of the fact that some hormones must undergo further metabolism following their secretion in order to be fully effective.

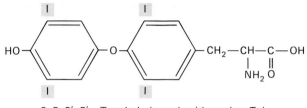

3, 5, 3′, 5′ – Tetraiodothyronine (thyroxine, T_4)

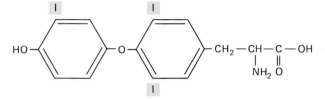

3, 5, 3′-Triiodothyronine (T_3)

FIGURE 9-2. Chemical structures of thyroxine and triiodothyronine, the hormones known collectively as thyroid hormone. The two molecules differ by only one atom of iodine.

Adrenal catecholamines. The other amine hormones are **epinephrine (E)** and **norepinephrine (NE)**, secreted by the adrenal medulla. Each of the paired **adrenal glands** really constitutes two distinct endocrine glands, an inner **adrenal medulla** and the surrounding **adrenal cortex**. As described in Chap. 8, the adrenal medulla is really like a sympathetic ganglion whose cell bodies do not send out nerve fibers but release their active substances directly into the blood, thereby fulfilling the criteria for an endocrine gland. The pathways for synthesis of epinephrine and norepinephrine (which belong to the family of catecholamines) were described in Chap. 8, when these substances were dealt with as neurotransmitters. In people, approximately 80 percent of the hormone produced by the adrenal medulla is epinephrine. Some norepinephrine (which is merely epinephrine without a methyl group) is also secreted, in amounts which are generally too small to exert significant actions on target cells; accordingly, future discussions of the functions of the adrenal medulla will focus on epinephrine only.

Peptide hormones

The great majority of all hormones are peptides or proteins. They range in size from very small peptides (having only three amino acids) to small proteins, but for the sake of convenience, we shall follow a common practice of endocrinologists and refer to them all simply as "peptides." Most of them are initially synthesized as parts of large peptides (**prohormones**) which are then cleaved (by proteolytic enzymes within the cell) to yield the active peptides. But these cleavages result in the generation of not only the classical hormone but also a variety of other peptides which may be stored along with the hormone in the cell's secretory vesicles. Upon stimulation of the cell, the other peptides are cosecreted with the hormone, and this phenomenon raises the possibility that they, too, may exert hormonal effects, ones which differ from those of the usual hormone. In other words, instead of just one peptide hormone, the cell may be secreting multiple hormones. Moreover, if the proteolytic enzymes performing the various cleavages are under physiological control, the relative proportions of the cosecreted substances could vary from situation to situation.

Steroid hormones

The third family of hormones is the **steroids**, whose ring structure was described in Chap. 3. Steroid hormones are produced by the **adrenal cortex**, the **testes**, the **ovaries**, and the **placenta**. (The active forms of the hormone vitamin D, while technically not steroids, can also be added to this category, because certain of their properties are so similar to those of steroids.) Cholesterol is the common precursor of all steroid hormones, whose individual biological activities reflect differences, often very subtle (Fig. 9-3), in the ring structure and the side chains attached to it.

Let us begin our discussion of steroid synthesis with the adrenal cortex, using as reference Fig. 9-4, which is highly simplified in that it leaves out many intermediate stages. The four hormones normally secreted in physiologically significant amounts by the adrenal cortex are **aldosterone, cortisol, corticosterone,** and **dehydroepiandrosterone.** Aldosterone is known as a **mineralocorticoid** because its effects are on mineral metabolism. Cortisol and corticosterone are called **glucocorticoids** be-

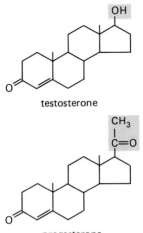

FIGURE 9-3. Structures of testosterone and progesterone. Testosterone is produced by metabolism of progesterone in several enzyme-mediated steps.

cause they have important effects on the metabolism of glucose (and other organic nutrients); cortisol is by far the more important of the two in people and so we shall deal only with it in future discussions. Dehydroepiandrosterone belongs to the general class of hormones known as **androgens**, all of which have effects similar to the dominant male sex hormone, testosterone.

A critical generalization about these biochemical pathways is that they (and the many intermediate steps left out of Fig. 9-4) involve small changes in the molecules (Fig. 9-3) and are mediated by specific enzymes; it is the relative concentrations of the key enzymes that determine flow along the different synthetic pathways. The adrenal cortex is not a homogenous gland but is composed of three distinct layers. One of these layers possesses very high concentrations of the enzymes re-

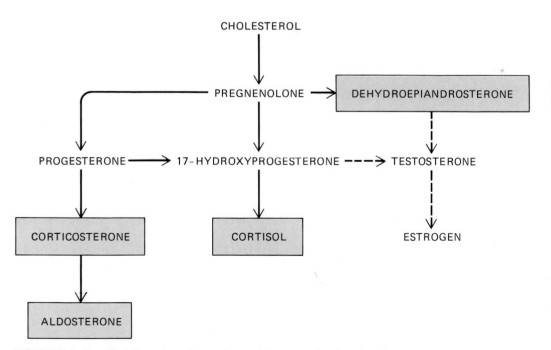

FIGURE 9-4. Simplified flow sheet for synthesis of hormones by the adrenal cortex; many intermediate steps have been left out. The four hormones in boxes are the major steroid hormones secreted. Dehydroepiandrosterone is an androgenic hormone; as denoted by the dashed lines, little testosterone or estrogen is normally produced by the adrenal cortex because the required enzymes are present in very low concentrations.

quired to convert corticosterone to aldosterone but lacks the enzymes required for the formation of cortisol and dehydroepiandrosterone. Accordingly, this layer synthesizes and secretes aldosterone but not the other major adrenal cortical hormones. In contrast, the other two layers of the adrenal cortex have just the opposite enzyme profile, and they secrete no aldosterone but much corticosterone, cortisol, and dehydroepiandrosterone. These two layers also have very low concentrations of the enzymes which carry the metabolism of dehydro-epiandrosterone further, which explains why only trivial amounts of testosterone and estrogen are secreted by the normal adrenal cortex.

In certain disease states, the adrenal cortex may secrete large amounts of a steroid that it does not normally secrete to any great extent. For example, an enzyme defect in the cortisol pathway may prevent the formation of cortisol by the adrenal cortex and result in the shunting of precursors into the dehydroepiandrosterone pathway instead; in a woman, the result of this would be masculinization (since dehydroepiandrosterone is an androgen). To take another example, the genes which code for the enzymes leading from dehydroepiandrosterone to testosterone are normally repressed in adrenal-cortical cells but may be derepressed in diseased cells, leading to secretion of large amounts of testosterone.

When we turn to the other steroid-producing endocrine glands, we find essentially the same potential biochemical pathways but very different concentrations of the key enzymes. Endocrine cells in the testes synthesize mainly testosterone, because they lack the enzymes in both the aldosterone and cortisol pathways but possess high concentrations of enzymes in the testosterone pathway (Fig. 9-5). As might be predicted, the ovarian cells, which synthesize estrogen, have high concentrations of the enzymes required to transform testosterone into estrogen; accordingly they rapidly convert testosterone into estrogen. The separation of enzyme activities is, however, not total, which explains why the normal testes do actually secrete small amounts of estrogen, whereas the ovaries secrete small amounts of testosterone. Thus, there are no uniquely male or female hormones although, of

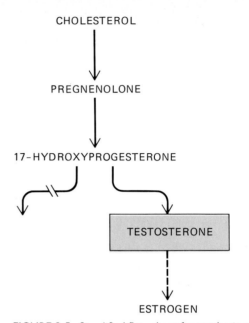

FIGURE 9-5. Simplified flow sheet for synthesis of testosterone by endocrine cells in the testes. Unlike the adrenal cortex, these cells do not possess the enzyme(s) needed to convert 17-hydroxyprogesterone to cortisol. Estrogen production is also very low for analogous reasons.

course, the relative concentrations of the "male" and "female" sex hormones are quite different in the two sexes.

Finally, following their secretion into the blood by the adrenal cortex or gonads, steroid hormones may undergo further interconversion in either the blood or other organs. For example, testosterone may be converted peripherally to estrogen, and this constitutes another source of estrogen in normal men. To take another example, dehydroepiandro-sterone, the androgen secreted by the adrenal cortex, can also be converted in peripheral tissues to estrogen; this accounts for the estrogen found in the blood of postmenopausal women, whose ovaries no longer secrete estrogen.

Transport of Hormones in the Blood

As we have pointed out in previous chapters, lipids are poorly soluble in water and largely circulate bound to plasma proteins. This is the case for most of the steroid hormones (and for the thyroid hormones even though they are not lipids). In such situations, the concentration of free hormone may be quite small and in equilibrium with the bound fraction:

$$\text{``Free'' hormone} + \text{plasma protein} \rightleftharpoons \text{hormone-protein complex}$$

It is important to realize that only the free hormone can cross most capillary walls to leave the plasma and exert effects on the target cells for that hormone; accordingly, for hormones which can bind to plasma proteins, the *effective* plasma concentration of the hormone is much less than the *total* plasma concentration. For example, less than 1 percent of plasma thyroid hormone is free, i.e., not bound to plasma proteins. As we shall see, the degree of protein binding also influences the rate of metabolism or excretion of the hormone.

Hormone Metabolism and Excretion

A hormone's concentration in the plasma depends not only upon its rate of secretion but also upon its rate of removal from the blood, either by excretion or metabolic transformation (Fig. 9-6). Sometimes the hormone is metabolized by the cells upon which it acts; for example, endocytosis of plasma-membrane hormone-receptor complexes (Chap. 7) enables cells to clear peptide hormone rapidly from their surfaces and degrade it intracellularly. But for most hormones, the major pathway for degradation or removal from the blood is by the liver or the kidneys.

Because the rate of urinary excretion of many hormones (and their metabolites) is directly proportional to the rate of secretion by their endocrine glands of origin, it is often used as an indicator of secretory rate.

In general, epinephrine, norepinephrine, and the peptide hormones, all of which circulate mainly free in the plasma, are comparatively easily excreted or attacked by blood and tissue enzymes, and so these hormones tend to remain in the bloodstream for only brief periods (minutes or hours)

after their secretion. In contrast, the circulating steroid hormones and thyroid hormones are, as mentioned earlier, mainly bound to plasma proteins; because bound hormone is relatively invulnerable to excretion or metabolism by enzymes, removal of these hormones from plasma generally takes much longer.

This section has emphasized enzyme-mediated metabolism of a hormone as one means of eliminating the original form of the hormone from plasma. Until recently, it was assumed that such metabolism always inactivated the hormone; i.e., it eliminated its ability to act on its target cells. Now it is clear that, in many cases, metabolism of the hormone may actually be critical for the hormone's ability to act (Fig. 9-6). We have already seen one example of this—the conversion of T_4 to T_3 by peripheral cells. Another example is provided by testosterone, which is converted to dihydrotestosterone in certain of its target cells; this latter steroid then elicits the response of the target cell.

An ``activating'' role of metabolism may be particularly important for certain of the peptide hormones, since circulating peptide fragments produced peripherally by metabolism of the originally secreted hormone exert important effects of their own on target cells. Indeed, it has been hypothesized that, in some cases, different fragments of the same original peptide hormone may be the mediators of different effects of that hormone.

An unusual aspect of several peptide hormone ``systems'' deserves mention here in the context of metabolism of hormones. In these cases, the peptides secreted by the endocrine cell do not, themselves, bind to receptors and exert actions. Rather, they function in the plasma as enzymes, acting upon other circulating proteins to catalyze formation of smaller peptides which function (sometimes after still further peripheral metabolism) as the active hormones in the ``systems'' (Fig. 9-6).

Finally, it should be noted that, whenever ``activation'' of a hormone through metabolism is required, deficiency of the enzymes which mediate the activation may constitute an important source of malfunction in disease. For example, a deficiency in the relevant target cells of the enzyme which converts testosterone to dihydrotestosterone would

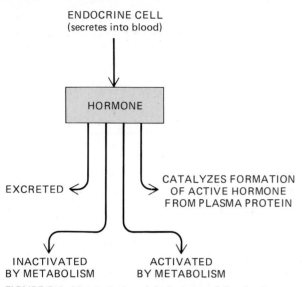

ENDOCRINE CELL
(secretes into blood)

HORMONE

EXCRETED

CATALYZES FORMATION
OF ACTIVE HORMONE
FROM PLASMA PROTEIN

INACTIVATED
BY METABOLISM

ACTIVATED
BY METABOLISM

FIGURE 9-6. Metabolic fates of a hormone following its secretion by an endocrine-gland cell. These fates are not mutually exclusive, and several often apply to any particular hormone.

result in the failure of testosterone to act on those target cells; the end result for those cells would be the same as if no testosterone were being secreted.

Mechanisms of Hormone Action

Hormone receptors

The end result of a hormone's stimulation of its target cells is an alteration in the rate at which the target cells perform a specific activity: Muscle cells increase or decrease contraction; epithelial cells (and other cell types as well) alter their rates of water or solute transport; gland cells secrete more or less of their secretory products. However, each of these responses is only the final event in a sequence that begins when the hormone interacts with the target cell. In all cases, this initial interaction occurs between hormone and target-cell receptors, i.e., proteins of the target cells which have a specific capacity to bind the hormone. As emphasized in earlier chapters, this "recognition" (binding) is made possible by the conformation of portions of the receptor molecules such that a "match" exists between them and the specific hormone. It is the

presence of the hormone-receptor combination which initiates the chain of intracellular biochemical events leading ultimately to the cell's response.

Where in the target cells are the hormone receptors located? The receptors for steroid hormones are soluble proteins within the cytosol of the target cells; steroids are lipid soluble and readily cross the plasma membrane to enter the cytosol and combine with their specific receptors. In contrast, the receptors for the lipid-insoluble peptide hormones are proteins located in the plasma membranes of the target cells. However, many peptide hormones do, in fact, gain entry to the interior of cells by internalization (endocytosis) of the plasma-membrane hormone-receptor complex. It is likely that most of the hormone molecules internalized in this way undergo catabolism, but it is also possible that they (or the peptide fragments resulting from their degradation) combine with intracellular receptors and exert additional effects upon the target cell.

The receptors for the lipid-insoluble amine hormones secreted by the adrenal medulla—epinephrine and norepinephrine—are on the plasma membrane, whereas the receptors for thyroid hormones are inside the cell (but in the nucleus rather than in the cytosol).

Changes in receptors may be important factors in altering hormone function. For example, certain men have a genetic defect manifested by the absence of receptors for dihydrotestosterone, the form of testosterone active in many target cells; these cells are, therefore, unable to bind dihydrotestosterone, and the result is lack of development of male characteristics, just as though the hormone were not being produced (note that this defect is quite distinct from the inability to synthesize dihydrotestosterone, described earlier).

Receptor modulation is not limited to disease states but is a normal component of hormonal control systems. Basic concepts of receptor modulation were described in Chap. 7, in the context of the receptors for all types of chemical messengers, because we wished to emphasize their general applicability. However, it is in the field of endocrinology that phenomena such as down-regulation have been most intensively studied. For example, a persistently elevated plasma insulin concentration induces a loss of insulin receptors, and this down-regulation prevents the target cells from overreacting to the high hormone concentration. This clearly constitutes an important negative-feedback control over insulin's actions. Similar controls are exerted by many other hormones over their specific receptors.

Hormones may alter not only their own receptors but the receptors for other hormones as well. If one hormone induces a loss of a second hormone's receptors, the result will be a reduction of the second hormone's effectiveness. For example, progesterone causes a loss of estrogen receptors on uterine smooth muscle. On the other hand, a hormone may induce an increase in the number or affinity of receptors for a second hormone; in this case the effectiveness of the second hormone will be increased. An example of this is the ability of thyroid hormone to increase the number of receptors for epinephrine (and norepinephrine) in the heart; this accounts for the fact that persons with excessive thyroid hormone (hyperthyroidism) are extremely sensitive to epinephrine.

Finally, the ability of one hormone to influence the receptors of a second hormone underlies certain situations in which physiological responses require actions of hormones in a sequence. For example, stimulation of release of an ovum (egg) from the ovary requires the action of two different hormones secreted by the anterior pituitary, and the first hormone paves the way for the second by increasing the number of the latter's receptors in the relevant cells of the ovary.

Events elicited by hormone-receptor binding

Steroid hormones. The common denominator of the effects of the steroid hormones (and the thyroid hormone) is an increased synthesis of specific proteins (enzymes, structural proteins, etc.) by their target cells. This increased protein synthesis is the result of stimulation by the hormone-receptor complex of mRNA synthesis. Recall that steroid hormones enter the cytosol of their target cells and combine with receptors (Fig. 9-7). This binding ac-

tivates the receptor so that it moves into the cell's nucleus, carrying with it the bound hormone. There the receptor combines with a DNA-associated protein specific for it. This interaction triggers transcription of the DNA segment, i.e., synthesis of mRNA, which can then enter the cytoplasm and serve as a template for synthesis of a specific protein. Note that this entire sequence depends upon several "recognitions"—hormone with a cytosolic receptor, then receptor with a specific nuclear protein. The latter step is essential for proper positioning of the receptor-hormone complex along the DNA so that the appropriate gene is transcribed.

Peptide hormones. As described above, the receptors for lipid-insoluble hormones (peptides and the hormones of the adrenal medulla) are located on the outer surface of the plasma membrane. For at least 12 of these hormones, the event triggered by hormone-receptor combination is activation of membrane-bound adenylate cyclase, with subsequent synthesis of cAMP inside the cell. Thus, these hormones utilize the cAMP-protein-kinase second-messenger system described in Chap. 7. However, it is clear that many actions of peptide hormones (and the catecholamines) are not mediated by activation of this system; in some cases, calcium is known to be the second messenger, whereas in others (insulin is a notable example), the relevant second messenger has not yet been identified. Finally, no second messenger at all is brought into play by certain hormone-receptor combinations; rather, as described in Chap. 7, the activated receptor may act directly on an adjacent protein involved in the opening or closing of membrane channels.

Hormone interactions on target cells

Cells are constantly exposed to the simultaneous effects of many hormones. This allows for complex hormone-hormone interactions on the target cells, including inhibition, synergism, and the important phenomenon known as **permissiveness**. In general terms, permissiveness means that hormone A must be present for the full exertion of hormone B's

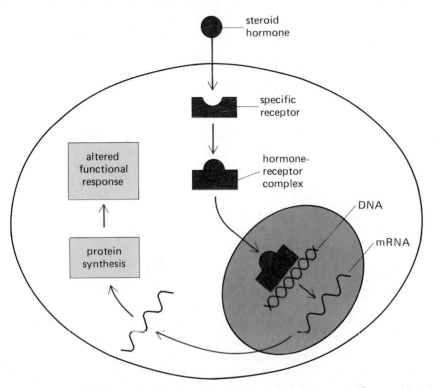

FIGURE 9-7. Steroid hormones enter the target cell and bind to a specific cytosolic receptor. The hormone-receptor complex is then transported into the nucleus, where the receptor binds to specific sites on the DNA molecule and activates transcription of new mRNA, which mediates the response characteristic of the cell by directing protein synthesis.

effect; a low, basal concentration of hormone A is all that is needed for this permissive effect. For example (Fig. 9-8), the hormone epinephrine causes marked release of fatty acids from adipose tissue, but only in the presence of permissive amounts of thyroid hormone. It is likely that many of these interactions may be the result of the influence of one hormone on the number or affinity of the other hormone's receptors. Many of the defects seen when an endocrine gland is removed or ceases to function because of disease actually result from loss of the permissive effects of the hormone secreted by that gland.

Pharmacological effects

Administration of very large quantities of a hormone for medical purposes may have results which are never seen in a normal person, although these same so-called **pharmacological effects** can occur in diseases when excessive amounts of hormones are secreted. Pharmacological effects are of great importance in medicine, since hormones in pharmacological doses are often used as therapeutic agents. Perhaps the most common example is that of the adrenal cortical hormone, cortisol, which is useful in suppressing allergic and inflammatory reactions.

Control of Hormone Secretion

Hormones are not secreted at constant rates. Indeed, as emphasized previously, it is essential that any regulatory system be capable of altering its output, and this is the case for the endocrine components of the body's regulatory systems, just as it is true for the neural components.

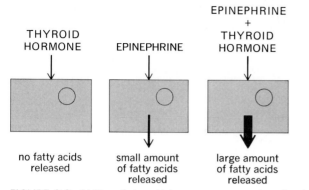

THYROID
HORMONE

EPINEPHRINE

EPINEPHRINE
+
THYROID
HORMONE

no fatty acids
released

small amount
of fatty acids
released

large amount
of fatty acids
released

FIGURE 9-8. Ability of thyroid hormone to permit epinephrine-induced liberation of fatty acids from adipose-tissue cells.

Most hormones are released in short bursts, with little or no release occurring between bursts; accordingly, the blood concentrations of hormones may fluctuate rapidly over brief time periods (Fig. 9-9). When an endocrine gland is stimulated by appropriate inputs, the bursts of hormone release occur more frequently so that the plasma concentration of the hormone increases; in contrast, in the absence of stimulation or in the presence of active inhibitory inputs, the bursts occur at a slower frequency or stop altogether, and the plasma concentration of the hormone decreases.

Many hormones manifest 24-h cyclical variations in their secretory rates, the circadian patterns being different for different hormones. Some are clearly linked to sleep, during which increased secretion may occur. For example, Fig. 9-9 illustrates how growth hormone's secretion is markedly increased during the early period of sleep. The mechanisms underlying these cycles are ultimately traceable to cyclical variations in the activity of neural pathways involved in the hormone's release.

The specific inputs which stimulate or inhibit the secretion of each particular hormone will be described in subsequent sections or chapters; suffice it here to emphasize that the direct inputs to endocrine cells fall into three general categories (Fig. 9-10): (1) changes in the concentrations of mineral ions or organic nutrients in the extracellular fluid surrounding the endocrine cells; (2) neural input, i.e., neurotransmitters released from neurons impinging on the endocrine cells; and (3) another hormone, i.e., one hormone often controls the secretion of another hormone.

Before looking more closely at each category, it must be stressed that, in many cases, the secretion of a hormone is influenced by more than one input; for example, insulin secretion is controlled by the extracellular concentrations of glucose and other nutrients, by both sympathetic and parasympathetic neurons to the secretory cells, and by several hormones acting on the secretory cells. Thus, endocrine cells, like neurons, may be subject to multiple, simultaneous (often opposing) inputs, the resulting output (rate of hormone secretion) reflecting the integration of all these inputs.

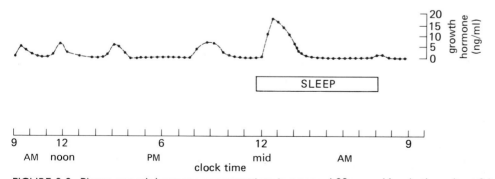

FIGURE 9-9. Plasma growth-hormone concentrations in a normal 23-year-old male throughout 24 h. Blood was sampled with an indwelling catheter so that sleep would not be disturbed. [Redrawn from Sassin et al. (1972).]

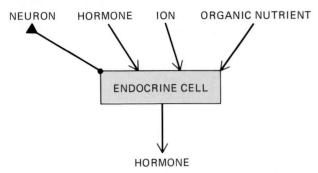

FIGURE 9-10. Types of inputs which act directly on endocrine-gland cells to stimulate (or inhibit) their secretion.

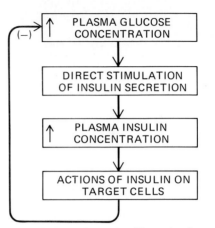

FIGURE 9-11. Example of how the direct control of hormone secretion by the plasma concentration of a substance (organic nutrient or ion) results in the negative-feedback control of that substance's plasma concentration.

Direct control by ions or organic nutrients

There are at least five hormones whose secretion is directly controlled, at least in part, by the plasma concentrations of specific ions or nutrients. In each case, a major function of the hormone is to regulate, in a negative-feedback manner, the plasma concentration of the ion or nutrient controlling its secretion (Fig. 9-11). For example, insulin secretion is stimulated by an elevated plasma glucose level, and the additional insulin then causes, by several actions, the plasma glucose concentration to decrease. Another group of hormones, those released by the gastrointestinal tract and involved in the control of digestion and absorption, is controlled by ions and nutrients not in the plasma but in the lumen of the stomach or small intestine; as we shall see, this permits the composition of a meal to influence the handling of that meal by the gastrointestinal tract.

Direct control by neurons

As described earlier, the adrenal medulla is really like a sympathetic ganglion whose cell bodies do not send out nerve fibers but release their active substances (mainly epinephrine) directly into the blood. The adrenal medulla behaves just like any sympathetic ganglion and is dependent upon stimulation by sympathetic preganglionic fibers. Destruction of these incoming nerves causes marked reduction of epinephrine release and failure of secretion to increase in response to the usual physi-

ological stimuli (Fig. 9-12).

Control of epinephrine secretion is not the only influence the autonomic nervous system has on the endocrine system. Some other endocrine glands receive a rich supply of sympathetic and (in some cases) parasympathetic neurons, whose activity influences rates of hormone secretion by the glands (Fig. 9-12). Examples are the secretion of glucagon and of the gastrointestinal hormones. In most cases studied, the autonomic input does not serve as the sole regulator of the hormone's secretion but serves rather as only one of multiple inputs.

Thus far, our discussion of direct neural control of hormone release has been limited to the role of the autonomic nervous system. However, one large group of hormones—those secreted by the hypothalamic neuroendocrine cells (including the posterior pituitary)—is under the direct control not of autonomic neurons but of brain neurons whose axon terminals synapse upon the secretory cells (Fig. 9-12). This category will be described in detail later in this chapter.

Direct control by other hormones

In many cases, the secretion of a particular hormone is directly controlled by the blood concentra-

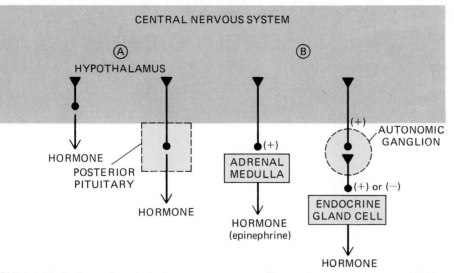

FIGURE 9-12. Pathways by which the nervous system influences hormone secretion. (A) Neurons in the hypothalamus (some of which terminate in the posterior pituitary) themselves secrete hormones. (B) The autonomic nervous system controls the secretion of hormones by the adrenal medulla and many other endocrine glands. Both parasympathetic and sympathetic inputs to these other glands may occur, some inhibitory and some stimulatory.

tion of yet another hormone. As we shall see, there are often complex sequences in which the only function of the first few hormones in the sequence is to stimulate the secretion of the next one. A hormone which stimulates the secretion of another hormone is often referred to as a **tropic** hormone.

Control Systems Involving the Hypothalamus and Pituitary

The pituitary gland lies in a pocket of the bone at the base of the brain, just below the hypothalamus (Fig. 9-13), to which it is connected by a stalk containing nerve fibers, and small blood vessels. In adult human beings it is composed of two adjacent lobes—**anterior** and **posterior**—each of which is a more or less distinct gland.

Posterior pituitary hormones

The posterior pituitary is actually an outgrowth of the hypothalamus and is neural tissue. The axons

of two well-defined clusters of hypothalamic neurons pass through the connecting stalk and end within the posterior pituitary in close proximity to capillaries (Fig. 9-14). The two peptide hormones **oxytocin** and **vasopressin** (the latter is also known as **antidiuretic hormone** or **ADH**), released from the posterior pituitary are actually synthesized in these hypothalamic neurons (only one of the hormone types is synthesized by any particular neuron); enclosed in small vesicles, they move down the cytoplasm of the neuronal axons to accumulate at the axon terminals. Release into the capillaries occurs in response to generation of an action potential in the axon. Therefore, the term "posterior pituitary hormones" is somewhat of a misnomer, since the hormones are actually synthesized in the hypothalamus, of which the posterior pituitary is merely an extension. Action potentials are generated in these hypothalamic hormone-secreting neurons as a result of synaptic input to them, just as in any other neuron. The reflexes controlling vasopressin release and the actions of this hormone are described in Chap. 13, those for oxytocin in Chap. 16. Finally,

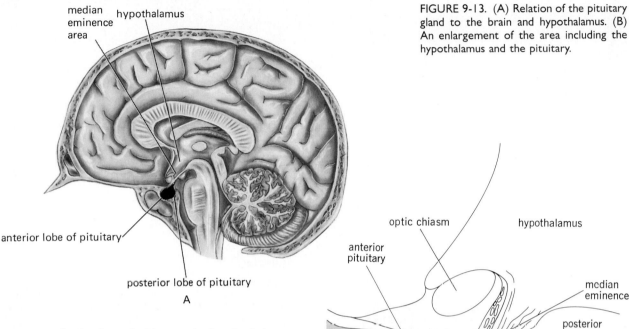

median
eminence
area

hypothalamus

anterior lobe of pituitary

posterior lobe of pituitary

A

FIGURE 9-13. (A) Relation of the pituitary gland to the brain and hypothalamus. (B) An enlargement of the area including the hypothalamus and the pituitary.

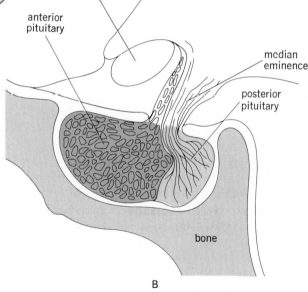

optic chiasm

hypothalamus

anterior
pituitary

median
eminence

posterior
pituitary

bone

B

vasopressin (and probably oxytocin, too) is produced by axon terminals other than those in the hypothalamus–posterior-pituitary complex; in these other locations the molecule serves as a neurotransmitter or neuromodulator.

The hypothalamus and anterior pituitary

The anterior pituitary has frequently been called the "master endocrine gland," because the multiple hormones it secretes control the secretion of four or five hormones (depending upon the person's gender) by other endocrine glands. However, if any "gland" deserves the title of "master," it is the hypothalamus, since this portion of the brain, in addition to producing oxytocin and vasopressin, secretes hormones which control the secretion of all the anterior pituitary hormones. The basic pattern for this overall system is (with one exception) sequences of three hormones; they begin with a hypothalamic hormone, which controls the secretion of an anterior pituitary hormone, which controls the secretion of a hormone from a peripheral endocrine gland (Fig. 9-15). This last hormone in the sequence then acts on its target cells. We begin our description of these sequences in the middle, i.e., with the anterior pituitary hormones.

Anterior pituitary hormones. As shown in Table 9-1 and Fig. 9-16, the anterior pituitary produces six different peptide hormones with well-established functions. They are **growth hormone (GH)**, **thyroid-stimulating hormone (TSH)**, **adrenocorticotropic hormone (ACTH)**, **prolactin**, **follicle-stimulating hormone (FSH)**, and **luteinizing hormone (LH)** (see Ta-

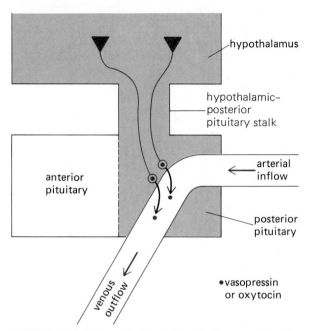

FIGURE 9-14. Schematized relationship between the hypothalamus and posterior pituitary.

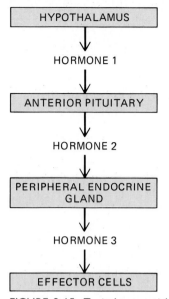

FIGURE 9-15. Typical sequential pattern by which a hypothalamic hormone controls the secretion of an anterior pituitary hormone, which in turn controls the secretion of a hormone by a peripheral endocrine gland.

ble 9-1 for other synonyms for all these hormones). Each of the first four is secreted by a distinct cell type in the anterior pituitary, whereas FSH and LH (collectively termed **gonadotropic hormones**) are probably both secreted by a single cell type.

The anterior pituitary secretes other peptides in addition to these six classical hormones. The most interesting of these are **β-lipotropin** and **β-endorphin**, the origin of which is via the pro-opiomelanocortin molecule previously described in Chap. 8. ACTH is originally synthesized as part of the large pro-opiomelanocortin molecule, which is then cleaved to form ACTH and another peptide known as β-lipotropin; the latter molecule can be cleaved, in turn, to form the powerful morphine-like substance β-endorphin. Then β-lipotropin and β-endorphin (and probably other cleavage products as well) are cosecreted from the anterior pituitary along with ACTH and may exert important hormonal effects quite different from those of ACTH (see Chaps. 17 and 18).

A function of five of the anterior pituitary hormones is to stimulate secretion of other hormones. **Thyroid-stimulating hormone** induces secretion of thyroid hormones (thyroxine and triiodothyronine) from the thyroid, and **adrenocorticotropic hormone**, meaning "hormone which stimulates the adrenal cortex," stimulates the secretion of cortisol by that gland. Whether TSH and ACTH have other hormonal functions besides these remains controversial. **Follicle-stimulating hormone** and **luteinizing hormone** also control the secretion of other hormones, in this case the sex hormones (estrogen, progesterone, and testosterone) by the gonads. However, FSH and LH differ from TSH and ACTH in that, besides controlling the secretion of other hormones, they have a second major role—regulating the growth and development of sperm and ova.

Growth hormone, like the gonadotropins, has dual roles: (1) it stimulates the release (mainly from the liver) of a group of growth-promoting peptide hormones known collectively as **somatomedin**; and

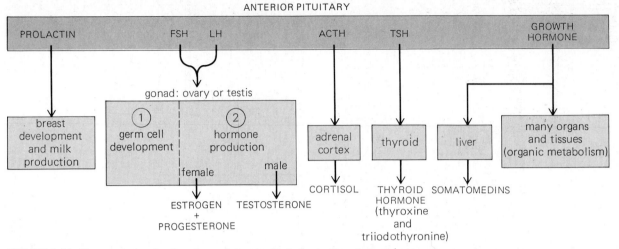

FIGURE 9-16. Targets and major functions of the six classical anterior pituitary hormones.

(2) it exerts direct effects of its own on the protein, lipid, and carbohydrate metabolism of various organs and tissues (see Chap. 15).

Prolactin is unique among the anterior pituitary hormones in that it does not exert major control over the secretion of a hormone by a peripheral endocrine gland; rather, its most important action is to stimulate milk production by a direct effect upon the breasts.

The target organs and functions of the six classical anterior pituitary hormones are summarized in Fig. 9-16. It should be emphasized that we are dealing in this chapter with these messengers only in terms of their functions as anterior pituitary hormones; recent studies strongly suggest that many, perhaps all, of them may be secreted by cells other than those in the anterior pituitary and may function in those other sites as neurotransmitters, neuromodulators, or paracrines.

Hypothalamic releasing hormones. We have seen that the secretion of various hormones by peripheral endocrine glands is controlled by hormones from the anterior pituitary. But secretion of the anterior pituitary hormones is, itself, largely regulated by still other hormones, produced by the hypothalamus. So important is this control that de-

struction of the hypothalamus results in marked reduction in the basal secretion of all anterior pituitary hormones (except one—prolactin—which manifests hypersecretion after hypothalamic destruction for reasons which will be subsequently apparent) and virtual elimination of the changes in secretion which normally occur in response to appropriate physiological stimuli.

An appreciation of the anatomical relationships between the hypothalamus and the anterior pituitary is essential for understanding how the hypothalamic hormones control the anterior pituitary. In contrast to the neural connections between the hypothalamus and posterior pituitary, there are no important neural connections between the hypothalamus and anterior pituitary. But there is an unusual capillary-to-capillary connection. Certain of the arteries supplying the hypothalamus end in its base (the **median eminence**) as intricate capillary tufts which recombine into the **hypothalamo-pituitary portal vessels** (the term "portal" denotes vessels which connect distinct capillary beds). These vessels pass down the stalk connecting the hypothalamus and pituitary and enter the anterior pituitary where they drain into a second capillary bed, the anterior pituitary capillaries. Thus, these portal vessels offer a local route for flow of capillary blood from hypothalamus to anterior pituitary.

The axons of neurons which originate in diverse areas of the hypothalamus (and possibly other adjacent brain areas) terminate in the median eminence around the capillary tufts, i.e., around the origins of the portal vessels. The generation of action potentials in these neurons causes them to release into the capillaries hormones which are carried by the portal vessels to the anterior pituitary, where they act upon the various pituitary cells to control their secretions of hormones (Fig. 9-17). (Thus, these hypothalamic neurons secrete hormones in a manner identical to that described previously for the neurons whose axons end in the posterior pituitary; the crucial difference between the two systems is that the posterior-pituitary capillaries drain into the systemic veins, whereas the blood in the median-eminence capillaries flows directly into the anterior pituitary.)

We are dealing with multiple discrete hypothalamic substances, each secreted by a unique group of hypothalamic neurons and influencing the release of one or, in some cases, several of the six classical anterior pituitary hormones (Table 9-2). Because most of these substances stimulate release of their relevant target hormones, they are all collectively termed **hypothalamic releasing hormones.** Each of the hypothalamic releasing hormones is named according to an anterior pituitary hormone whose secretion it controls (Table 9-2); for example, the secretion of ACTH (also known as corticotropin) is stimulated by **corticotropin releasing hormone (CRH).** However, as shown in Table 9-2, at least two of the hypothalamic hormones inhibit, rather than stimulate, release of anterior pituitary hormones. One of these inhibits secretion of growth hormone and is most commonly called **somatostatin.** The other inhibits secretion of prolactin and is termed **prolactin release inhibiting hormone (PIH).**

This system would be much easier to understand if there were a strict one-to-one correspondence of hypothalamic releasing hormone and anterior pituitary hormone, but such is not the case. Note in Table 9-2 that several of the hypothalamic hormones influence the secretion of more than one anterior pituitary hormone: **Gonadotropin releasing hormone (GnRH)** stimulates the secretion of both

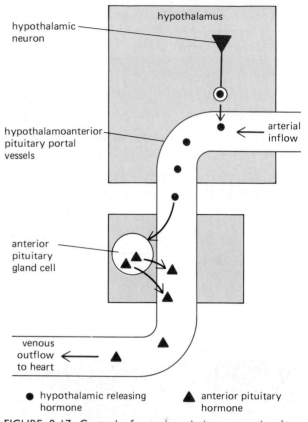

FIGURE 9-17. Control of anterior pituitary secretion by a hypothalamic releasing hormone.

FSH and LH; **thyrotropin releasing hormone (TRH)** stimulates the secretion not only of TSH (also known as thyrotropin), but of prolactin as well (it was named TRH because its effect on thyrotropin was discovered first); and somatostatin is capable of inhibiting the secretion not only of growth hormone but of three other anterior pituitary hormones as well, although to what degree the other inhibitory effects are physiological is still not clear.

These are cases of a single hypothalamic hormone controlling more than one anterior pituitary hormone, but Table 9-2 also shows that secretion of a single anterior pituitary hormone may be controlled by more than one hypothalamic releasing hormone. Moreover, the several releasing hormones involved may exert opposing effects. For example, the secretion of growth hormone is con-

TABLE 9-2. Hypothalamic releasing hormones and their major effects on the anterior pituitary

Hypothalamic releasing hormone	Effect on anterior pituitary
Corticotropin releasing hormone (CRH)	Stimulates secretion of ACTH (corticotropin)
Thyrotropin releasing hormone (TRH)*	Stimulates secretion of TSH (thyrotropin) and prolactin
Growth hormone releasing hormone (GHRH)	Stimulates secretion of GH
Somatostatin (also known as growth hormone release inhibiting hormone, GIH)	Inhibits secretion of GH and several other hormones
Gonadotropin releasing hormone (GnRH)	Stimulates secretion of LH and FSH
Prolactin releasing hormone (PRH)*	Stimulates secretion of prolactin
Prolactin release inhibiting hormone (PIH, dopamine)	Inhibits secretion of prolactin

*Note that TRH stimulates the secretion not only of TSH but of prolactin as well; therefore, TRH is also a "PRH." However, it is likely that there is a second PRH, which acts only on prolactin, and that is why PRH is listed separately here.

trolled by two releasing hormones—somatostatin (which inhibits its release) and **growth hormone releasing hormone** (which stimulates it). Clearly, in such multiple-control cases, the response of the anterior pituitary depends upon the relative amounts of the opposing hormones released by the hypothalamic neurons, as well as the relative sensitivities of the anterior pituitary to them.

Of the hypothalamic releasing hormones whose structures have been identified, one (PIH) is dopamine (the catecholamine which also serves as a neurotransmitter in various areas of the nervous system), and the others are peptides. These same peptides have been found in locations other than the hypothalamus, particularly in other areas of the nervous system and in the gastrointestinal tract. The finding of these peptides in areas of brain other than hypothalamus has raised the possibility that they may also serve as neurotransmitters or neuromodulators in the central nervous system.

Given that the hypothalamic releasing hormones control anterior pituitary function, we must now ask: What controls secretion of the releasing hormones? Some of the neurons which secrete releasing hormones may possess spontaneous autorhythmicity, but the firing of most of them requires neuronal and hormonal input to them. First, let us deal with the neuronal input.

The hypothalamus receives synaptic input, both stimulatory and inhibitory, from virtually all areas of the central nervous system, and specific neural pathways control, in large part, the secretion of the individual releasing hormones. In most cases, some degree of synaptic input (and/or spontaneous neuronal rhythmicity) is always occurring and determines the basal rate of releasing-hormone secretion. Secretion is then increased above the basal rate (or suppressed to levels below it) by additional synaptic input arising from reflexes or endogenous neural rhythms.

Figure 9-18 illustrates these points for the CRH-ACTH-cortisol system: Corticotropin releasing hormone, secreted by the hypothalamus, stimulates the anterior pituitary to secrete ACTH, which in turn stimulates the adrenal cortex to secrete the hormone cortisol. A wide variety of stresses, both physical (an injury, severe blood loss, etc.) and emotional (fright, for example) act, via neuronal pathways to the hypothalamus, to increase CRH secretion (and hence, that of ACTH and cortisol) markedly above its basal value; i.e., stress is the common denominator of reflexes leading to increased secretion of cortisol. But even in a completely unstressed person, the secretion of cortisol varies in a highly stereotyped manner during a 24-h period. This is because endogenous

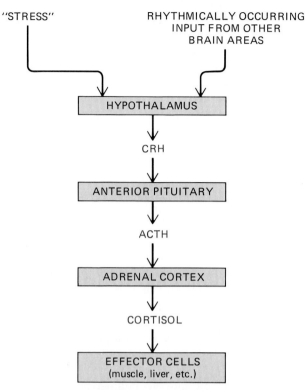

FIGURE 9-18. CRH-ACTH-cortisol hormonal sequence. The "stress" input to the hypothalamus is via neural pathways from other brain areas. CRH = corticotropin releasing hormone. ACTH = adrenocorticotropin. (See Chaps. 15, 17, and 20 for the many effects of cortisol on its effector cells.)

neural rhythms within the central nervous system impinge upon the hypothalamic neurons which secrete CRH.

Thus far, we have concentrated on neuronal input to the hypothalamus. Now we turn to how hormones control both releasing-hormone secretion and the sensitivity of the anterior pituitary to the releasing hormones.

Hormonal feedback. A prominent feature of each of the hormonal sequences initiated by the hypothalamic releasing hormones is negative feedback exerted upon the hypothalamic-pituitary system by one or more of the very hormones in the sequence. For example, in the CRH-ACTH-cortisol sequence (Fig. 9-19), the final hormone—cortisol—acts upon the hypothalamus to reduce secretion of CRH (by causing a decrease in the frequency of action-po-

tentials in the neurons secreting CRH). In addition, cortisol acts directly on the anterior pituitary to reduce the sensitivity of the ACTH-secreting cells to CRH. Thus, by a double-barreled action, cortisol exerts a negative-feedback control over its own secretion (Fig. 9-19).

It should be evident that such a system is highly effective in dampening hormonal responses, i.e., in limiting the extremes of hormone secretory rates. For example, when a painful stimulus elicits increased secretion, in turn, of CRH, ACTH, and cortisol, the resulting elevation in blood cortisol concentration feeds-back to partially inhibit the hypothalamus and anterior pituitary; therefore, cortisol secretion does not rise as much as it would without these negative feedbacks.

Negative feedback also serves to maintain the

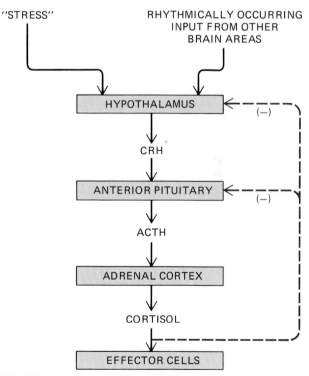

FIGURE 9-19. Negative feedback effects of cortisol. Cortisol not only acts on the hypothalamus to inhibit CRH secretion but acts on the anterior pituitary to reduce its responsiveness to CRH. The net effect is to reduce ACTH secretion and, hence, cortisol secretion. Thus, when the overall system is stimulated by stress, not as much additional cortisol is secreted as would be if the negative feedback did not exist.

plasma concentration of the final hormone in a sequence relatively constant whenever a primary change occurs in the excretion or metabolism of that hormone. An example of this is shown in Fig. 9-20.

The situation described above for cortisol, in which the hormone secreted by a peripheral endocrine gland exerts a negative-feedback effect over the anterior pituitary and/or hypothalamus is known as a **"long-loop" negative feedback** (Fig. 9-21). This type of feedback exists for each of the five three-hormone sequences initiated by a hypothalamic re-

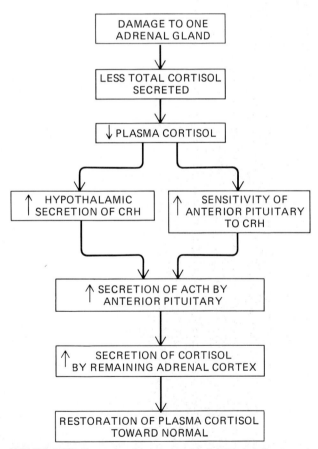

FIGURE 9-20. Example of how negative feedback in hormonal sequences helps maintain the plasma concentration of the final hormone when disease-induced changes in hormone secretion occur. The same analysis would apply if the original reduction in plasma cortisol were due to excessive metabolism of cortisol rather than deficient secretion.

leasing hormone (that for the gonadotropins also incorporates a long-loop *positive* feedback as we shall see in Chap. 16). Long-loop feedback does not exist for prolactin since this is the one anterior pituitary hormone which does not have major control over a peripheral endocrine gland; i.e., it does not participate in a three-hormone sequence. Nonetheless, there is negative feedback in this system, for prolactin itself acts upon the hypothalamus to influence the secretion of the hypothalamic releasing hormones which control its secretion. This is a **"short-loop" feedback** (Fig. 9-21), and it is quite likely that anterior pituitary hormones other than prolactin also exert such feedback on the hypothalamus. In this regard, recent evidence indicates that a significant portion of the blood flow from the pituitary actually goes back toward the central nervous system before entering the systemic circulation. Therefore, it is quite possible that pituitary hormones may reach the brain in high enough concentrations to influence the secretion of hypothalamic hormones.

Effects of other hormones. It must be emphasized that there are many hormonal influences (both stimulatory and inhibitory) on the hypothalamus and/or anterior pituitary other than those which fit the feedback patterns just described. In other words, a hormone which is not itself in a particular sequence may, nevertheless, exert important influence on the secretion of the hypothalamic or anterior pituitary hormones which are in that sequence. For example, estrogen markedly enhances the secretion of prolactin by the anterior pituitary, even though estrogen secretion is not, itself, controlled by prolactin. The significance of this is that one should not view these sequences as isolated units. This concept is true not just for the hypothalamus-pituitary system, but for the entire endocrine system; i.e., effects of one hormone upon the secretion of another are extremely common.

Specific systems

This concludes our general description of the hypothalamic–anterior-pituitary control systems. The specific subsets of this system involving the gonado-

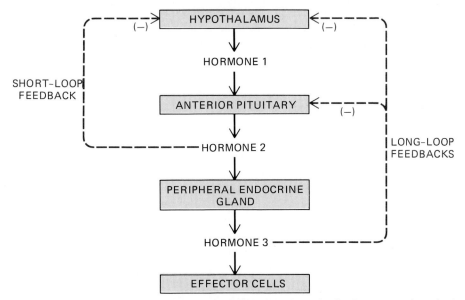

FIGURE 9-21. Short-loop and long-loop feedbacks. Long-loop feedback is exerted on the hypothalamus and/or anterior pituitary by the third hormone in the sequence. Short-loop feedback is exerted by the anterior pituitary hormone on the hypothalamus.

tropins and prolactin will be dealt with fully in Chap. 16. This section summarizes the control systems involving ACTH, TSH, and growth hormone.

CRH-ACTH-cortisol. The major hormonal interactions in this sequence have already been summarized in Fig. 9-19, but one additional point should be mentioned: ACTH stimulates not only the secretion of cortisol by the adrenal cortex but also the secretion of the other major adrenocortical hormones, aldosterone and dehydroepiandrosterone. This is because it acts on all cells of the adrenal cortex at a very early point in the steroid synthetic pathways; namely, it stimulates the conversion of cholesterol to pregnenolone, the precursor of all the adrenal steroids. However, as we shall see in Chap. 13, the control of aldosterone secretion by ACTH is relatively minor compared to other inputs.

Cortisol, the final hormone in the CRH-ACTH-cortisol sequence, can exert a wide variety of effects on different target cells. These effects will be described in greater detail in subsequent chapters. Here we simply state in very general terms cortisol's major functions: (1) Cortisol is required for the changes in organic metabolism which occur when a person has not eaten for more than a few hours; (2) during times of stress cortisol helps mobilize energy stores and maintains normal functioning of the cardiovascular system (this explains the adaptive value of stress being the major situation in which the CRH-ACTH-cortisol system is activated); and (3) in very large amounts, cortisol inhibits the body's responses to microbes and other tissue-injuring factors (achievement of this latter anti-inflammatory effect is the reason why physicians administer cortisol-like substances to persons suffering from diseases in which the body is being damaged by its own effective defense mechanisms).

In the discussion of the CRH-ACTH-cortisol system, we have thus far ignored the fact, mentioned earlier in this chapter, that several other peptides, including the morphine-like β-endorphin, are cosecreted with ACTH from the anterior pituitary. Thus, during stress, the neurally induced secretion of hypothalamic CRH causes the secretion of these peptides as well as of ACTH. They have been hypothesized to exert important effects on pain perception, learning, appetite, and many other bodily functions, but their precise contributions still remain unclear.

TRH-TSH-TH. TRH is the major hypothalamic hormone stimulating TSH secretion (Fig. 9-22). (Several other hypothalamic hormones—somatostatin and dopamine—probably exert some inhibitory effects on TSH release, but these effects are minor compared to the stimulation by TRH.) TSH, in turn, stimulates the release of thyroid hormone (TH, which includes T_4 and T_3), which, like cortisol, exerts multiple effects on many different target cells. The overall functions of all these effects (to be described in Chap. 15) are to facilitate bodily growth and normal maturation of the nervous system and to help set the rate at which the body metabolizes fuels.

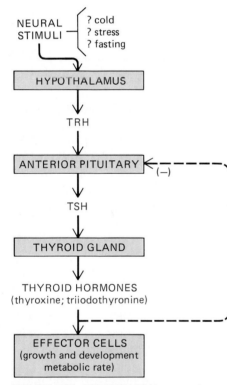

FIGURE 9-22. TRH-TSH-TH hormonal sequence. TRH = thyrotropin releasing hormone. TSH = thyrotropin. As denoted by the question marks, the physiological situations which influence TRH secretion (via neural inputs to the hypothalamus) are still unclear. Note that, in contrast to cortisol, thyroid hormones exert their negative feedback only upon the anterior pituitary. (See Chap. 15 for the effects of thyroid hormones on their effector cells.)

Thyroid hormone exerts a feedback inhibition of the hypothalamo-pituitary system by decreasing the sensitivity of TSH-secreting cells to the stimulatory effects of TRH. All the hormones in the system tend to be secreted in a relatively steady fashion; this is in keeping with the fact that the effects exerted by the thyroid hormones on growth and metabolism are quite long-lasting; i.e., there is no adaptive value to rapid increases or decreases in TH secretion.

It has been difficult to document, in people, physiological circumstances in which the secretion of TRH (and, therefore, the rest of the hormone sequence) is altered by neural input to the hypothalamus. Possible candidates include exposure to cold and fasting (or malnutrition), as described in Chap. 15.

The TRH-TSH-TH system offers one of the best examples of the role played by hormonal feedback in maintaining the plasma concentration of the final hormone in the sequence (TH in this case). Since iodine is essential for synthesis of TH, persons with iodine deficiency tend to have deficient production of TH. However, the resulting decrease in plasma TH concentration relieves some of the feedback inhibition TH exerts on the pituitary. Therefore, more TSH is secreted in response to the usual amount of TRH coming from the hypothalamus, and the increased plasma TSH stimulates the thyroid gland to enlarge and to utilize more efficiently whatever iodine is available. In this manner, plasma TH concentration may be kept quite close to normal. The enlarged thyroid is known as **iodine-deficiency goiter**.

Growth hormone–somatomedin. The secretion of growth hormone (GH) by the anterior pituitary is controlled mainly by two hormones from the hypothalamus (Fig. 9-23): (1) somatostatin, which inhibits GH secretion; and (2) growth-hormone releasing hormone (GHRH), which stimulates it. With such a dual control system, the rate of GH secretion at any moment reflects the relative amounts of simultaneous stimulation by GHRH and inhibition by somatostatin. For example, as shown in Fig. 9-9, very little growth hormone is secreted during the

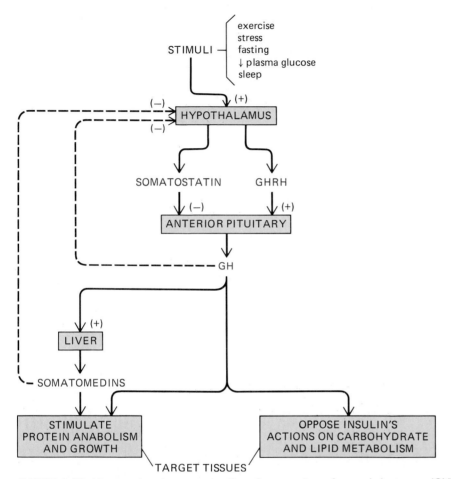

FIGURE 9-23. Hormonal pathways controlling the secretion of growth hormone (GH) and somatomedin. GHRH = growth hormone releasing hormone. See text for explanation of (+) and (−) effects on the hypothalamus.

day in nonstressed persons who are eating normally, and whether this reflects a very low secretion of GHRH or a very high secretion of somatostatin is not yet clear.

As mentioned earlier, growth hormone, in turn, stimulates the secretion of another hormone, somatomedin, from the liver. (Somatomedin should not be confused with somatostatin.) Somatomedin then stimulates bodily growth, particularly via actions on cartilage and bone (see Chap. 15).

So far, this sounds like the usual three-hormone sequence typical of the hypothalamus–anterior-pituitary system. The difference is that growth hormone, itself, exerts direct actions (i.e., actions

independent of somatomedin) on protein, carbohydrate, and lipid metabolism (as described in Chap. 15).

In contrast to the situation for the TRH-TSH-TH system, a large number of physiological states produce marked changes in GH secretion (and, in turn, somatomedin secretion). For example, GH secretion is stimulated by exercise, stress (both physical and emotional), fasting, and a rapidly decreasing plasma glucose concentration. But most striking of all are the very large surges of growth hormone secretion that occur within one to two hours of a person's falling asleep (Fig. 9-9). All these changes are mediated by the central nervous system acting via the hypothalamus (by some combination of decreased somatostatin secretion and increased GHRH).

There are two sources of hormonal negative feedback in this system: (1) A short-loop negative feedback exerted by GH, itself, on the hypothalamus; and (2) a classical long-loop negative feedback, exerted by somatomedin on the hypothalamus. It is not known which of these two feedbacks is the major one, nor is it known whether the dominant effect is stimulation of somatostatin release or inhibition of GHRH release.

Finally, the secretion of growth hormone is influenced by several other hormones. Some of these may do so indirectly by altering the release of somatostatin and/or GHRH from the hypothalamus, but others may act directly upon the anterior pituitary.

MUSCLE

The ability to produce force and movement is a characteristic of all living cells. During cell division chromosomes move to opposite sides of the cell and the cell constricts to form two daughter cells. Some cells can move through their environment; e.g., white blood cells migrate through tissues, and sperm cells migrate on their way to fertilize an egg. Still other cell movements involve hairlike processes known as **cilia**, which extend from the surfaces of specialized epithelial cells. By moving back and forth, these waving cilia propel materials through tubes lined with ciliated cells.

Muscle cells, with their specialized ability to generate force and motion, merely utilize an extension of a basic contractile process common to all cells. In multicellular organisms, control of the forces generated by muscle cells contributes not only to the regulation of the internal environment but also to the movement of the organism, itself, in its external environment and to the manipulation and control of objects in the external environment by which the human mind ultimately expresses itself.

On the basis of their structures, contractile properties, and control mechanisms, three types of muscle cells can be identified: (1) **skeletal muscle**, (2) **smooth muscle**, and (3) **cardiac muscle**. Most skeletal muscle, as the name implies, is attached to bones, and its contraction is responsible for the movements of the skeleton. The contraction of skeletal muscle is controlled by the somatic nervous system and is under voluntary control. The movements produced by skeletal muscle are primarily involved with interactions between the body and the external environment.

Smooth muscle surrounds hollow organs and tubes such as the stomach, intestinal tract, urinary bladder, uterus, blood vessels, and air passages to the lungs. It is also found as single cells distributed throughout organs (e.g., the spleen) and in small groups of cells attached to the hairs in the skin or the iris of the eye. Contraction of the smooth muscle surrounding hollow organs may propel the luminal contents through the hollow organ (e.g., the intestinal tract), or it may regulate the flow of the contents through tubes by changing the tube diameter and thus the frictional resistance to flow (e.g., in blood vessels). Smooth-muscle contraction is controlled by the autonomic nervous system, hormones, and factors intrinsic to the muscle itself; it is not usually under direct conscious control.

The third type of muscle, cardiac muscle, is the muscle of the heart, and its contraction propels blood through the circulatory system. Like smooth muscle, it is regulated by the autonomic nervous system, hormones, and intrinsic factors in the cardiac muscle.

Although there are significant differences in the structures, mechanical properties, and control mechanisms of the three types of muscle, the force-generation mechanism is similar in all of them. The muscle structure, molecular events associated with force generation, and control of force generation in skeletal muscle will be described first, followed by a description of smooth muscle. Cardiac muscle, which combines some of the properties of both skeletal and smooth muscle, will be described in the next chapter in connection with its role in the circulatory system.

Structure of Skeletal Muscle Fibers

A single muscle cell is known as a **muscle fiber**. Each skeletal muscle fiber is a cylinder, with a diameter of 10 to 100 μm and a length which may extend up to 300,000 μm (1 ft). The term ''muscle'' refers to a number of muscle fibers bound together by connective tissue (Fig. 10-1). The relation between a single muscle fiber (cell) and a muscle is analogous to that between a single nerve fiber (axon) and a nerve composed of many axons.

Figure 10-2 shows a section through a skeletal muscle as seen with a light microscope. The most striking feature is the series of transverse light-and-dark bands forming a regular pattern along each fiber. Both skeletal and cardiac muscle fibers have this characteristic banding and are known as **striated muscles**; smooth muscle cells show no banding pattern. Although the pattern appears to be continuous across the cytoplasm of a single fiber, the bands are actually confined to a number of approximately cylindrical elements (1 to 2 μm in diameter) known as **myofibrils** (Fig. 10-1). Myofibrils occupy about 80 percent of the fiber volume and vary in number from several hundred to several thousand per fiber, depending on the fiber's diameter (keep in mind that ''muscle cell'' and ''muscle fiber'' are synonymous).

Each myofibril consists of smaller **filaments** (Figs. 10-1 and 10-3), which are arranged in a repeating pattern along the length of the fibril. One unit of this repeating pattern is known as a **sarcomere** (little muscle). Each sarcomere contains two types of filaments: **thick filaments**, 12 to 18 nm in diameter, composed of the contractile protein **myosin**, and **thin filaments**, 5 to 8 nm in diameter, containing the contractile protein **actin**. The thick filaments are located in the middle of the sarcomere where their orderly parallel arrangement produces the **A band**. Two sets of thin filaments are present at opposite ends of each sarcomere. One end of a thin filament is anchored at the end of the sarcomere to a structure known as a **Z line**, whereas the other end overlaps one end of the thick filaments in the A band. Two successive Z lines define the limits of one sarcomere. The Z lines are a network of filaments which interconnect adjacent thin filaments within one sarcomere as well as thin filaments from two adjoining sarcomeres.

Between the ends of the A bands of two adjacent sarcomeres is the **I band** (Figs. 10-1 and 10-3) (note that, unlike the A band, the I band includes segments of two sarcomeres). The I band contains those portions of the thin filaments which do not overlap the thick filaments; it is bisected by the Z line. An additional band, the **H zone**, is a light band in the center of the A band; it corresponds to the space between the free ends of the two sets of thin filaments in each sarcomere; hence, only thick filaments are found in the H zone. Finally, a narrow,

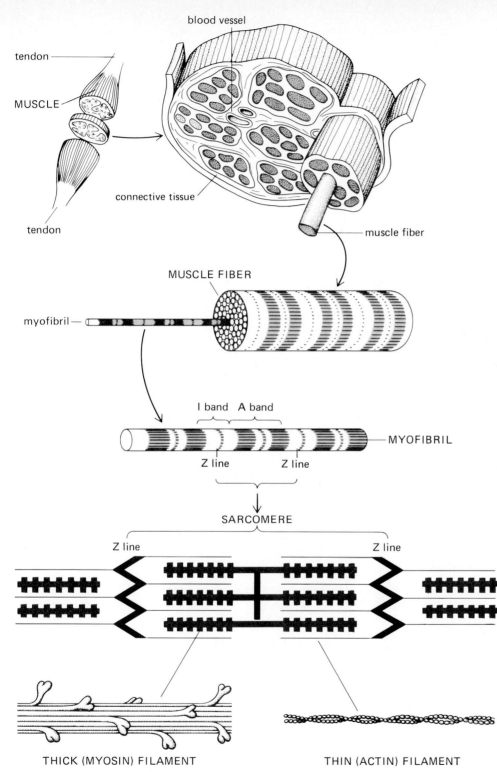

blood vessel

tendon

MUSCLE

connective tissue

tendon

muscle fiber

MUSCLE FIBER

myofibril

I band   A band

MYOFIBRIL

Z line   Z line

SARCOMERE

Z line   Z line

THICK (MYOSIN) FILAMENT   THIN (ACTIN) FILAMENT

FIGURE 10-1. Levels of structural organization in a skeletal muscle.

257

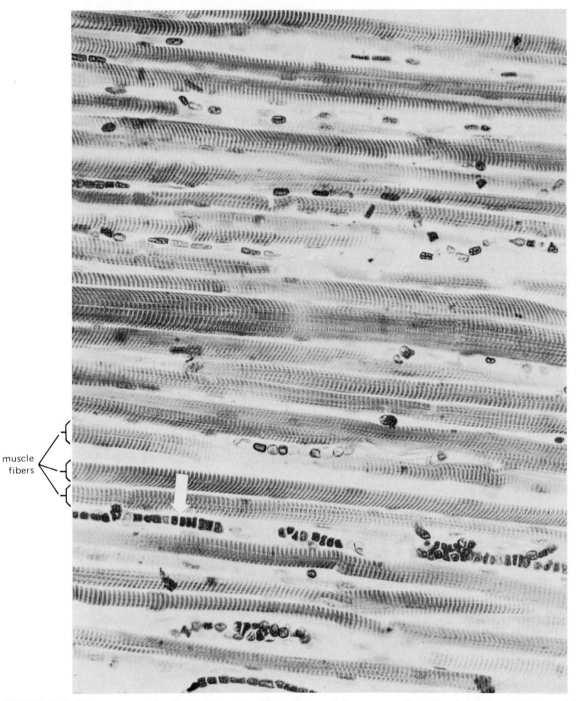

muscle
fibers

FIGURE 10-2. Photomicrograph of skeletal muscle fibers. Arrow indicates a capillary blood vessel containing red blood cells. *(From Edward K. Keith and Michael H. Ross, "Atlas of Descriptive Histology," Harper and Row, Publishers, Inc., New York, 1968.)*

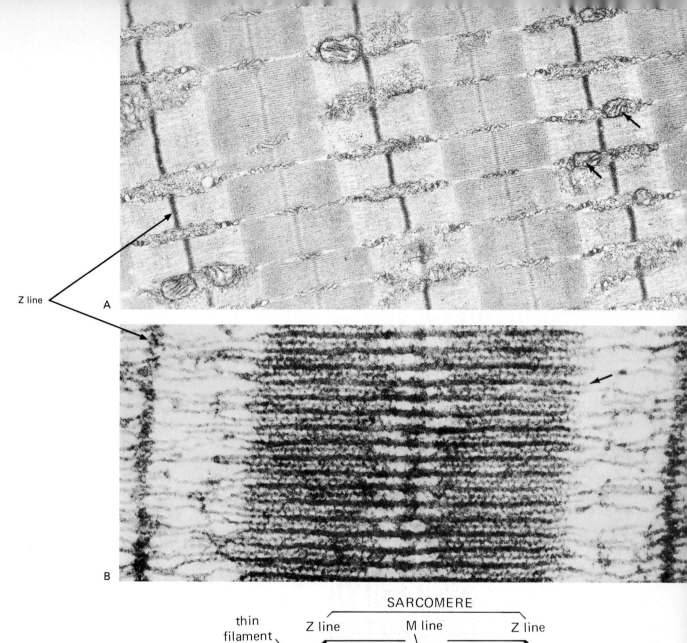

Z line

A

B

SARCOMERE

thin filament

Z line

M line

Z line

cross bridge

thick filament

H zone

A band

I band

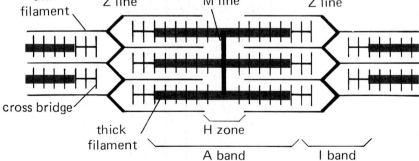

FIGURE 10-3. Filament organization of myofibrils. (A) Numerous myofibrils in a single skeletal muscle fiber (arrow indicates mitochondria located between the myofibrils). (B) High magnification of a single sarcomere within a single myofibril (arrow indicates end of thick filament). (C) Arrangement of thick and thin filaments that produces the striated banding patterns in myofibrils.

dark band can be seen in the center of the H zone; this is known as the **M line** and is produced by linkages between adjacent thick filaments. Thus, neither the thick nor the thin filaments are free-floating; the thin filaments are anchored to the Z line, whereas the thick filaments are linked together by the M line.

A cross section through several myofibrils in the region of their A bands, where both thick and thin filaments overlap (Fig. 10-4), shows their regular, almost crystalline, arrangement. Each thick filament is surrounded by a hexagonal array of six thin filaments, and each thin filament is surrounded by a triangular arrangement of three thick filaments. Altogether there are twice as many thin as thick filaments in the region of overlap.

In the region of overlapping filaments (Fig. 10-5), the space between adjacent thick and thin filaments is bridged by projections known as **cross bridges**. These cross bridges are portions of myosin molecules which extend from the surface of the thick filaments toward the thin filaments. During contraction, these cross bridges make contact with the thin filaments and exert force on the actin filaments. The cross bridges are the force-generating sites in muscle cells.

Molecular Mechanisms of Contraction

Sliding-filament mechanism

The lengths of neither the thick filaments nor the thin filaments change during shortening; rather, the two sets of filaments slide past each other. The following observations led to the **sliding-filament mechanism** of muscle contraction. When a skeletal muscle fiber shortens, the width of the A band in a sarcomere remains constant (Fig. 10-6). Since this width corresponds to the length of the thick filaments, these filaments do not change length during muscle shortening. In contrast, the widths of both the I band and the H zone decrease during shortening as the thin filaments move past the thick filaments. The H zone decreases as the ends of the thin filaments from opposing ends of a sarcomere approach each other, and it disappears when these

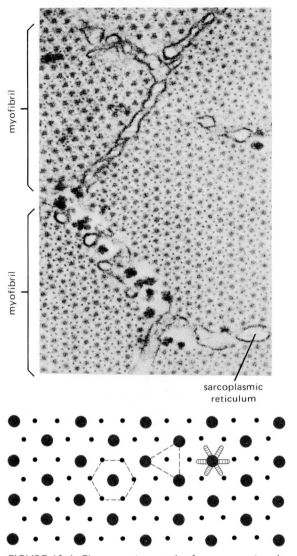

FIGURE 10-4. Electron micrograph of a cross section through six myofibrils in a single skeletal muscle fiber. Diagramed below are the hexagonal arrangements of the thick and thin filaments and the bridges extending from a thick filament to each of the surrounding thin filaments. [From H. E. Huxley, J. Mol. Biol., **37**:507–520 (1968).]

thin filaments meet in the center of the A band. The I band also decreases in width for the same reason, i.e., as more and more of the thin filament length overlaps the thick filaments. During shortening the length from the edge of the H zone to the nearest

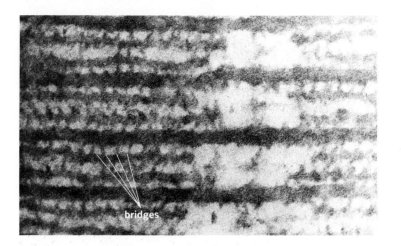

bridges

FIGURE 10-5. High-magnification electron micrograph in the region of the A band in a myofibril. The H zone can be seen at the right. Bridges between the thick and thin filaments can be seen at regular intervals along the filaments. [From H. E. Huxley and J. Hanson, in G. H. Bourne (ed.) "The Structure and Function of Muscle," vol. 1, Academic Press, Inc., New York, 1960.]

Z line (the length of a thin filament) remains constant. Thus, the lengths of the thin filaments also remain constant during shortening.

What produces the movement of these filaments? The answer is *the cross bridges*. When a cross bridge is activated, it moves in an arc parallel to the long axis of the thick filament, much like an oar on a boat. If, at this time, the cross bridge is attached to a thin filament, this swiveling motion slides the thin filament toward the center of the A band, thereby producing shortening of the sarcomere. One stroke of a cross bridge produces only a small displacement of a thin filament relative to a thick filament, but the cross bridges undergo many cycles of movement during a single contraction, each cycle requiring attachment of the bridge to the thin filament, angular movement of the bridge while attached, detachment from the thin filament, reattachment at a new location, and repetition of the cycle. Each cross bridge undergoes its own independent cycle of movement, so that at any one instant during contraction only about 50 percent of the bridges are attached to the thin filaments, while the others are at intermediate stages of a cycle.

Let us look more closely at the filaments and their cross bridges. Molecules of actin are arranged in two chains, helically intertwined to form the thin filaments (Fig. 10-7). Myosin is a much larger molecule with a globular end attached to a long tail. Approximately 200 myosin molecules comprise a single thick filament; the tails of the molecules lie along the axis of the filament and the globular heads extend out to the sides, forming the cross bridges. Each globular head contains a binding site able to bind to a complementary site on an actin molecule. The myosin molecules in the two halves of each thick filament are oriented in opposite directions, such that all their tail ends are directed toward the center of the fiber (Fig. 10-8). Because of this ar-

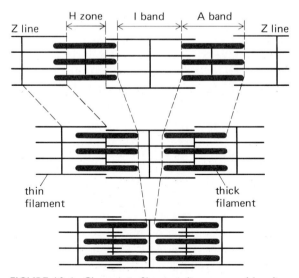

H zone · I band · A band

Z line · Z line

thin filament · thick filament

FIGURE 10-6. Changes in filament alignment and banding pattern in a myofibril during shortening.

THIN FILAMENT

FIGURE 10-7. Two helical chains of globular actin molecules form the primary structure of the thin filaments.

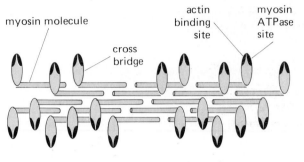

THICK FILAMENT

FIGURE 10-8. Orientation of myosin molecules in thick filaments, with the globular heads of the myosin molecules forming the cross bridges.

rangement, the power strokes of the cross bridges in the two ends of each thick filament are directed toward the center, thereby moving the thin filaments at both ends of the sarcomere toward the center of the sarcomere during shortening.

In addition to the binding site for actin, the globular end of myosin contains a separate enzymatic site that catalyzes the breakdown of ATP (adenosine triphosphate) to ADP (adenosine diphosphate) and inorganic phosphate, releasing the chemical energy stored in ATP. The splitting of ATP occurs on the myosin molecule before it attaches to actin, but the ADP and inorganic phosphate generated remain bound to myosin. The chemical energy released at the time of ATP splitting is transferred to myosin (M), producing an energized form of myosin (M*)

$$M + ATP \longrightarrow M^* \cdot ADP \cdot P_i$$
ATP breakdown

The subsequent binding of this energized myosin to actin (A) via a cross bridge triggers the discharge of the energy stored in myosin, with the resulting production of the force that causes movement of the cross bridge (the ADP and P_i are released from myosin at this time):

$$A + M^* \cdot ADP \cdot P_i \rightarrow A \cdot M^* \cdot ADP \cdot P_i \rightarrow A \cdot M + ADP + P_i$$
Actin binding   Energy release and
bridge movement

This sequence of energy storage and release by myosin is analgous to the operation of a mousetrap: Energy is stored in the trap by cocking the spring (by ATP splitting) and released by springing the trap (by binding to actin). This is, of course, only an analogy; the actual structural changes in myosin

which accompany energy storage and release are unknown.

During contraction, the myosin cross bridge binds very firmly to actin, and this linkage must be broken at the end of each bridge cycle; the binding of a new molecule of ATP to myosin is responsible for breaking this link:

$$A \cdot M + ATP \longrightarrow A + M \cdot ATP$$
A·M dissociation

Thus, upon binding (but not splitting) a molecule of ATP, myosin dissociates from actin. The free myosin bridge then splits its bound ATP, thereby re-forming the energized state of myosin, which can now reattach to a new site on the actin filament, and so on. Thus, ATP performs two distinct roles in the cross-bridge cycle: (1) The energy released from the splitting of ATP provides the energy for cross-bridge movement; and (2) the binding (not splitting) of ATP to myosin breaks the link between actin and myosin at the end of a cross-bridge cycle, allowing the cycle to be repeated. Figure 10-9 provides a summary of these chemical and mechanical changes that occur during one cross-bridge cycle. Many cross-bridge cycles occur during even the briefest contraction of a muscle fiber.

The importance of ATP in dissociating actin and myosin at the end of a bridge cycle is illustrated by the phenomenon of **rigor mortis** (death rigor), the stiffening of skeletal muscles after death. Rigor mortis begins 3 to 4 h after death and is complete

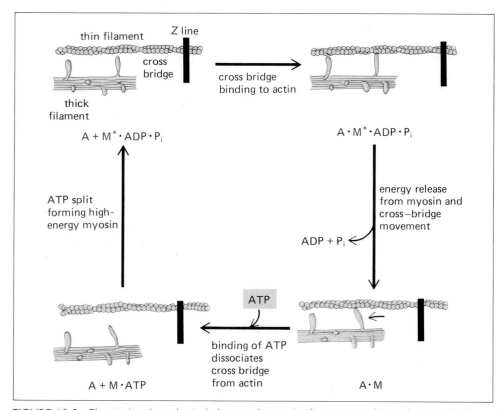

FIGURE 10-9. Chemical and mechanical changes during the four stages of a single cross-bridge cycle. Start reading the figure at the lower left.

after about 12 h, but then slowly disappears over the next 48 to 60 h. Rigor mortis occurs because of the lack of ATP. In the absence of ATP the myosin cross bridges are still able to bind to actin, but the subsequent breakage of the link between them and the energizing of myosin requires ATP, which is not produced in a dead muscle cell. The thick and thin filaments become linked to each other by the immobilized cross bridges, producing the rigid condition of a dead muscle. In contrast, in a living muscle at rest, the myosin bridges are not bound to actin, and the filaments readily slide past each other when the muscle is passively stretched.

Regulator proteins and calcium

Since a muscle fiber contains all the ingredients necessary for cross-bridge activity—actin, myosin, and ATP—the question arises: Why are muscles not in a continuous state of contraction? The answer is that in a resting muscle fiber, the cross bridges are unable to bind to actin and undergo the cross-bridge cycle that produces contraction. This inhibition of cross-bridge binding is due to two regulator proteins, **troponin** and **tropomyosin**, which are bound to the thin filaments (Fig. 10-10). Tropomyosin is a rod-shaped molecule with a length approximately equal to that of seven actin molecules. Tropomyosin molecules are arranged end-to-end along the chains of actin, so that they partially cover the myosin binding sites on actin, thereby preventing the cross bridges from making contact with these binding sites. Each tropomyosin molecule is held in this blocking position by a molecule of troponin, itself bound to both tropomyosin and actin.

Having accounted for the mechanism that prevents cross-bridge activity and thus keeps a muscle fiber turned *off*, we can now ask what turns the

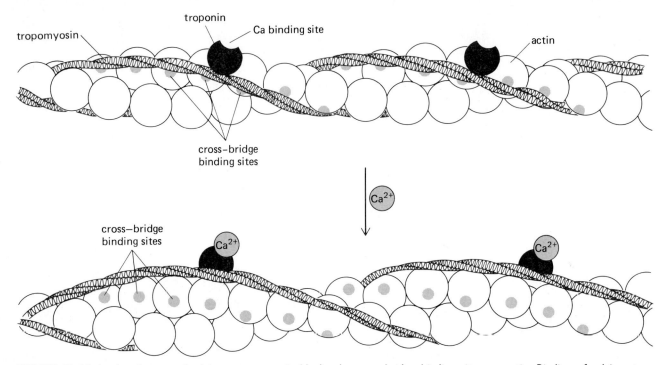

FIGURE 10-10. In the absence of calcium, tropomyosin blocks the cross-bridge binding sites on actin. Binding of calcium to troponin moves the tropomyosin to one side, exposing the binding sites and allowing myosin cross bridges to bind to actin.

muscle fiber *on*; i.e., what allows cross-bridge activity to proceed? In order for the cross bridges to bind to actin, the tropomyosin molecules must be moved away from their blocking positions on the thin filaments. This occurs when calcium binds to specific binding sites on the second of the two regulator proteins, troponin. The binding produces a change in the shape of troponin such that it pulls the tropomyosin bound to it to one side, uncovering the cross-bridge binding sites on actin (Fig. 10-10). Conversely, removal of calcium from troponin reverses the process, and tropomyosin moves back into its blocking position so that cross-bridge activity ceases.

In summary, cytosolic calcium-ion concentration determines the number of troponin sites occupied by calcium, which in turn determines the number of cross bridges that can bind to actin and exert force on the thin filaments to produce contraction of the muscle fiber. We now describe how

changes in cytosolic calcium concentration are coupled to the electrical events occurring in the muscle plasma membrane.

Excitation-contraction coupling

Excitation-contraction coupling refers to the sequence of events by which an action potential in the plasma membrane of a muscle fiber leads to cross-bridge activity via increased availability of calcium. The skeletal muscle plasma membrane is an excitable membrane capable of generating and propagating action potentials by mechanisms similar to those described for nerve cells (Chap. 8). An action potential in a skeletal muscle fiber lasts 1 to 2 ms and is completed before any signs of mechanical activity begin (Fig. 10-11). Once begun, the mechanical activity following a single action potential may last 100 ms or more. Note that the mechanical activity far outlasts the duration of the

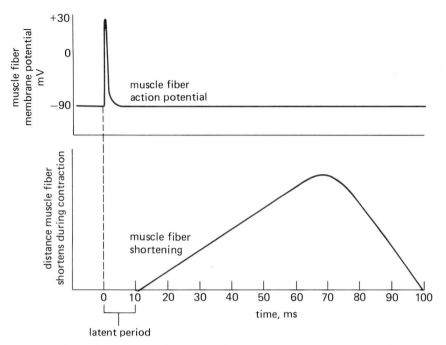

FIGURE 10-11. Time relations between a skeletal muscle-fiber action potential and mechanical contraction.

action potential; hence, the electrical activity in the plasma membrane does not *directly* act upon the contractile proteins but produces a state (increased cytosolic calcium concentration) which continues to activate the contractile apparatus long after the electrical activity in the membrane has ceased. In a resting muscle fiber the concentration of free calcium in the cytosol surrounding the thick and thin filaments is very low; calcium is not bound to troponin, and thus cross-bridge activity is blocked by tropomyosin. Following an action potential there is a rapid increase in cytosolic calcium concentration as calcium ions are released from the sarcoplasmic reticulum. The released calcium binds to troponin, removing the blocking effect of tropomyosin and thereby initiating contraction.

Sarcoplasmic reticulum. The sarcoplasmic reticulum in muscle is homologous to the endoplasmic reticulum found in most cells. The sarcoplasmic reticulum forms a sleevelike structure around each myofibril (Fig. 10-12). One segment of the sarcoplasmic reticulum surrounds the region of the A

band, and an identical but separate segment surrounds the I-band region. These segments of the reticulum are repeated, sarcomere by sarcomere, along the length of a myofibril. At each end of a sarcoplasmic-reticulum segment there are two enlarged saclike regions, the **lateral sacs**, which are connected to each other by a series of smaller tubular elements. The lateral sacs store the calcium that is released following membrane excitation. A separate, tubular structure, the **transverse tubule (t tubule)** crosses the fiber at the level of each A-I junction, passing between the adjacent lateral sacs and eventually joining the plasma membrane. The lumen of the t tubule is therefore continuous with the extracellular medium surrounding the muscle fiber.

The membrane of the t tubule, like the plasma membrane, is able to produce and propagate action potentials. An action potential, once initiated in the plasma membrane, is rapidly conducted over the surface of the cell and into the interior of the muscle fiber along the t tubules. As the action potential in a t tubule passes near the lateral sacs of the sar-

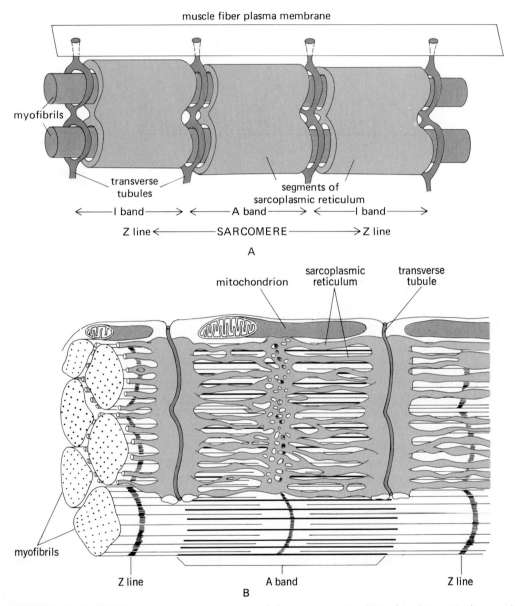

muscle fiber plasma membrane

myofibrils

transverse
tubules

segments of
sarcoplasmic
reticulum

←—I band—→ ←——A band——→ ←—I band—→

Z line ←————SARCOMERE————→ Z line

A

mitochondrion   sarcoplasmic   transverse
                 reticulum   tubule

myofibrils

Z line   A band   Z line

B

FIGURE 10-12. (A) Diagrammatic representation of the geometrical relationships between the membranes of the sarcoplasmic reticulum, the transverse tubules, and the myofibrils. (B) Three-dimensional view of transverse tubules and sarcoplasmic reticulum in a single human skeletal muscle fiber.

coplasmic reticulum, it triggers the release of calcium, probably as a result of opening calcium channels in the membrane of the lateral sacs. Since the reticulum surrounds the myofibrils, the released calcium has only a short distance to diffuse to reach troponin, where its binding initiates contraction.

One of the major unsolved problems in excitation-contraction coupling in skeletal muscle is the mechanism by which an action potential in the membrane of the t tubule triggers the release of calcium from the lateral sacs, since these sacs do not make physical contact with the t tubules. There is some evidence to suggest that calcium ions may also be involved in this linkage: An action potential in the t tubule increases the permeability of the t tubule membrane to calcium, which diffuses the short distance to the sarcoplasmic reticulum where it binds to a site that controls the opening of calcium channels in the reticulum membrane, thereby allowing large amounts of stored calcium to diffuse into the cytosol.

How is contraction, once initiated by the rise in cytosolic calcium concentration, turned *off*? As mentioned above, the contraction is turned *off* by removing the calcium from troponin, and this is achieved by lowering the calcium concentration in the region of the troponin-binding sites for calcium. The membrane of the sarcoplasmic reticulum contains a carrier-mediated primary active-transport system (Ca-ATPase) that pumps calcium ions from the cytosol into the lumen of the reticulum. Calcium is rapidly released from the reticulum upon arrival of an action potential, but the pumping of the released calcium back into the reticulum requires a much longer time. Therefore, the contractile activity continues for some time after the action potential.

To reiterate, just as contraction results from the release of calcium ions stored in the sarcoplasmic reticulum, so relaxation occurs as calcium is pumped back into the reticulum (Fig. 10-13). ATP is required to provide the energy for the calcium pump, and this is the third major role of ATP in the mechanics of muscle contraction; the others are the dissociation of actin and myosin at the end of each cross-bridge cycle and the provision of the energy for cross-bridge movement.

Membrane excitation: The neuromuscular junction. We have just seen that an action potential in a muscle-fiber plasma membrane is the signal that leads to mechanical activity by triggering the release of calcium from the sarcoplasmic reticulum. How are action potentials initiated in the plasma membranes of muscles fibers? There are three answers to this question, depending on the type of muscle that is being considered: (1) stimulation by a nerve fiber, (2) stimulation by hormones and local chemical agents, and (3) spontaneous electrical activity within the membrane itself. Stimulation by nerve fibers is the only mechanism by which skeletal muscles are normally excited, whereas, as we shall see, all three mechanisms are involved in initiating excitation in smooth and cardiac muscle.

The nerve cells whose axons innervate skeletal muscle fibers are known as **motor neurons** (somatic efferent), and their cell bodies are located in the brainstem or spinal cord. The axons of motor neurons are myelinated and are the largest-diameter axons in the body. They are therefore able to propagate action potentials at high velocities, allowing signals from the central nervous system to be transmitted with minimal delay to the site of force generation.

As the motor axon approaches a muscle, it divides into many branches, each of which forms a single junction with a muscle fiber (Fig. 10-14). Each motor neuron is connected through its branching axon to several muscle fibers. Although each motor neuron innervates many muscle fibers, each muscle fiber is innervated by only a single motor neuron. A motor neuron plus the muscle fibers it innervates is called a **motor unit**. The muscle fibers in a single motor unit are scattered throughout a muscle and do not all lie adjacent to each other. Since each muscle fiber in a motor unit is controlled by the same neuron, the electrical activity in a motor neuron controls the contractile activity of all the muscle fibers in that motor unit.

As an individual branch of the motor axon approaches a muscle fiber, it loses its myelin sheath and further divides into a fine terminal arborization which lies in grooves on the muscle-fiber surface. The region of the muscle membrane which lies directly under the terminal portion of the axon has special properties and is known as the **motor end plate**. The entire junction, including the axon terminal and motor end plate, is known as a **neuromuscular junction** (Fig. 10-15).

The axon terminals of a motor neuron contain

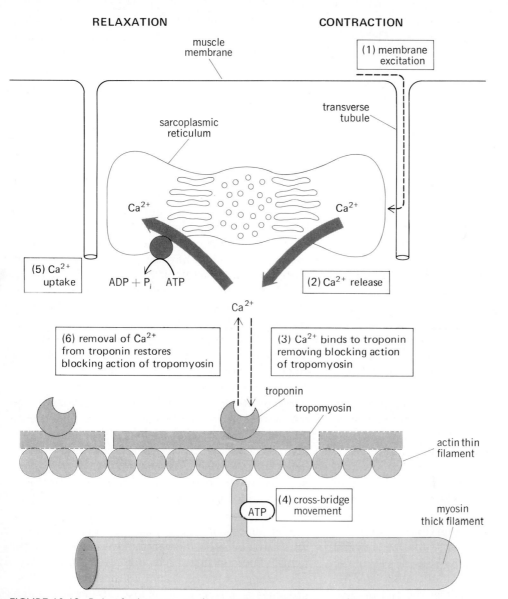

FIGURE 10-13. Role of calcium in muscle excitation-contraction coupling.

membrane-bound vesicles resembling the vesicles found at synaptic junctions. The vesicles contain the chemical transmitter acetylcholine (ACh). When an action potential in the motor neuron arrives at an axon terminal, located at the neuromuscular junction, it depolarizes the nerve membrane, opening voltage-sensitive calcium channels in the plasma membrane and allowing calcium ions to dif-fuse into the axon terminal. This calcium triggers a fusion of the transmitter vesicles with the nerve membrane (exocytosis), releasing acetylcholine into the extracellular cleft separating the nerve and muscle membranes.

The acetylcholine diffuses across this cleft and binds to receptor sites on the motor-end-plate membrane. The binding of ACh opens ion channels in the end-plate membrane that allow the passage of both sodium and potassium ions through the same channel opening. The net result is the movement of slightly more sodium in than potassium out, producing a depolarization of the motor end plate, which is known as an **end-plate potential (EPP)**. The mechanism responsible for producing an EPP in muscle is similar to that of an EPSP (excitatory postsynaptic potential) produced at a synaptic junction, but the magnitude of a muscle EPP is much larger than that of an EPSP, because much larger amounts of neurotransmitter are released over a larger surface area, opening many more ion channels. Local currents flow between the partially depolarized end-plate region and the muscle-fiber plasma membrane adjacent to this region, depolarizing the plasma membrane. The magnitude of a single EPP is normally sufficient to depolarize the adjacent plasma membrane to its threshold potential, initiating an action potential in the muscle plasma membrane. This action potential is then propagated over the surface of the muscle fiber by

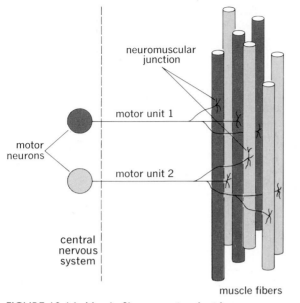

FIGURE 10-14. Muscle fibers associated with two motor neurons, forming two motor units within a muscle.

FIGURE 10-15. Neuromuscular junction.

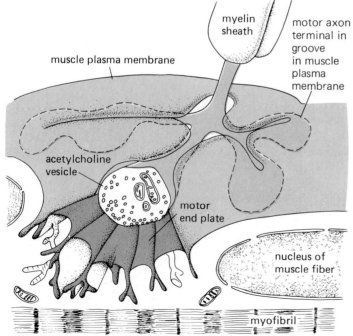

the same mechanism described for the propagation of action potentials along axon membranes (Fig. 10-16). However, in contrast to the location of synaptic junctions at one end of a neuron, most neuromuscular junctions are located in the middle of a muscle fiber, and it is from this region that a newly generated muscle action potential will be propagated in both directions toward the ends of the muscle fiber.

Because each EPP has sufficient magnitude to depolarize the muscle membrane to threshold, initiating an action potential, every action potential in a motor neuron produces an action potential in each of the muscle fibers in its motor unit. Thus, there is a one-to-one transmission of an action potential from the motor neuron to the muscle fiber. This is quite different from synaptic junctions, where multiple EPSPs resulting from multiple action potentials in presynaptic neurons must occur (undergoing temporal and spatial summation) in order for threshold to be reached and an action potential elicited in a postsynaptic membrane. A second differ-

ence between synaptic and neuromuscular junctions should be noted: At some synaptic junctions inhibitory postsynaptic potentials are produced; they hyperpolarize the postsynaptic membrane and decrease the probability of their firing an action potential, but such inhibitory potentials are not found in human skeletal muscle; all neuromuscular junctions are excitatory. Thus, the only way to reduce the electrical activity in the muscle membrane is to inhibit the initiation of action potentials in that muscle fiber's motor neuron.

In addition to receptor sites for acetylcholine, the motor end plate contains the enzyme acetylcholinesterase, which breaks down ACh (just as occurs at ACh-mediated synapses in the nervous system). ACh bound to the receptor sites is in equilibrium with free ACh in the cleft between the nerve and muscle membranes; therefore, as the concentration of free ACh falls (because of its breakdown by acetylcholinesterase), less ACh will bind to the receptor sites. When the receptor sites no longer contain bound ACh, the ion channels in the end plate

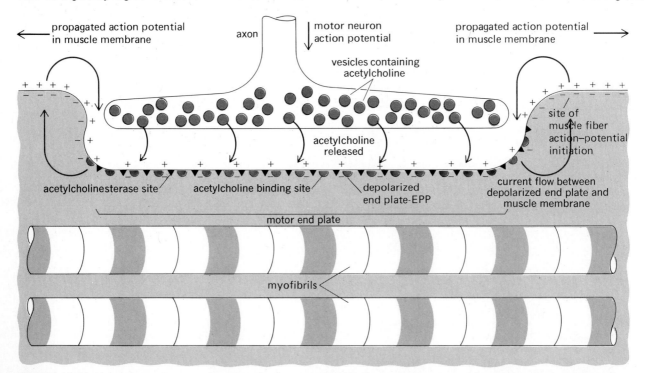

FIGURE 10-16. Events occurring at a neuromuscular junction which lead to an action potential in the muscle membrane.

close and the depolarized end plate returns to its resting potential.

There are many ways in which events at the neuromuscular junction can be modified by disease or drugs. For example, the deadly South American Indian arrowhead poison, **curare**, is strongly bound to the acetylcholine receptor site, but it does not open ion channels, nor is it destroyed by acetylcholinesterase. When a receptor site is occupied by curare, acetylcholine cannot bind to the receptor; therefore, although the motor nerves still conduct normal action potentials and release acetylcholine, there is no resulting EPP or contraction. Since the skeletal muscles responsible for breathing movements depend upon neuromuscular transmission to initiate their contraction, death comes from asphyxiation. Curare and similar drugs are used in small amounts to prevent muscular contractions during certain types of surgical procedures (such patients are artificially ventilated in order to maintain respiration).

Neuromuscular transmission can also be blocked by inhibition of acetylcholinesterase. Some organophosphates, which are the main ingredients in certain pesticides and "nerve gases" (the latter developed for biological warfare), inhibit this enzyme. In the presence of such agents acetylcholine is released normally upon the arrival of an action potential at the axon terminal but it is not destroyed because the acetylcholinesterase is inhibited; the ion channels in the end plate remain open, producing a maintained depolarization of the end plate. The failure of repolarization prevents the end plate from responding to the arrival of additional new action potentials; thus the muscle does not contract in response to further nerve stimulation, and the result is skeletal muscle paralysis and death from asphyxiation.

A third group of substances, such as botulinus toxin, produced by the bacterium *Clostridium botulinum*, blocks the release of acetylcholine from nerve terminals, preventing the transmission of the chemical signal from the nerve to the muscle fiber. Botulinus toxin is responsible for one type of food poisoning and is one of the most deadly poisons known. Less than 0.0001 mg is sufficient to kill a human being, and 500 g could kill the entire human population.

One form of neuromuscular disease, known as **myasthenia gravis**, which is associated with skeletal muscle weakness, is due to decreased numbers of ACh receptor sites at the motor end plate. The release of ACh from the nerve terminals is normal, but the magnitude of the muscle EPP is markedly reduced because of the decreased number of receptor sites. As described in Chap. 17, the destruction of the ACh receptors in this disease is brought about by the body's own defense mechanisms gone awry.

The sequence of events leading to contraction of a skeletal muscle fiber is summarized in Table 10-1.

Mechanics of Muscle Contraction

The force exerted by a contracting muscle on an object is known as muscle **tension**, and the force exerted on the muscle by the object is the **load**; muscle tension and load are opposing forces. The term **contraction** refers to the turning-on of the tension-generating process in muscle. Whether or not contraction produces movement depends on the magnitude of the load and the tension produced by the muscle. To move a load, muscle tension must be greater than the load.

When a muscle develops tension but does not change length, the contraction is said to be **isometric** (constant length). Such contractions occur when the muscle supports a load in a fixed position or attempts to move a load that is greater than the tension developed by the muscle. In contrast, a contraction occurring under conditions in which the load on a muscle remains constant but the muscle length is changing is said to be **isotonic** (constant tension). A muscle undergoes an isotonic contraction when moving a load.

The electrical and chemical events occurring in muscle fibers are the same in both isotonic and isometric contractions; i.e., the cross bridges are activated and exert force on the thin filaments. In an isotonic contraction the thin filaments move past

TABLE 10-1. Sequence of events between a motor-neuron action potential and contraction of a skeletal muscle fiber

1. An action potential is initiated and propagates in the axon of a motor neuron.

2. The action potential triggers the release of acetylcholine from the axon terminals at the neuromuscular junction.

3. Acetylcholine diffuses from axon terminals to the motor-end-plate membrane in the muscle fiber.

4. Acetylcholine binds to receptor sites on the motor-end-plate membrane, opening ion channels that are permeable to sodium and potassium ions.

5. Movement of ions across the motor-end-plate membrane depolarizes the membrane, producing the end-plate potential (EPP).

6. By local current flow the EPP depolarizes the adjacent plasma membrane to its threshold potential, generating an action potential that propagates over the surface of the muscle fiber and into the muscle fiber along the transverse tubules.

7. The action potential in the transverse tubules triggers the release of calcium ions from the lateral sacs of the sarcoplasmic reticulum.

8. The released calcium ions bind to troponin on the thin filaments, causing tropomyosin to move away from its blocking position uncovering the cross-bridge binding sites on actin.

9. The energized myosin cross bridges on the thick filaments bind to actin:

$$A + M^* \cdot ADP \cdot P_i \longrightarrow A \cdot M^* \cdot ADP \cdot P_i$$

10. This binding triggers the release of energy stored in myosin, producing an angular movement of each cross bridge:

$$A \cdot M^* \cdot ADP \cdot P_i \longrightarrow A \cdot M + ADP + P_i$$

11. ATP binds to myosin, breaking the linkage between actin and myosin and thereby allowing a cross bridge to dissociate from actin:

$$A \cdot M + ATP \longrightarrow A + M \cdot ATP$$

12. The ATP bound to myosin is split, transferring energy to the myosin cross bridge, producing an energized cross bridge:

$$M \cdot ATP \longrightarrow M^* \cdot ADP \cdot P_i$$

13. The cross bridges repeat the cycle (9 to 12), producing movement of the thin filaments past the thick filaments. These cycles of cross-bridge movement continue as long as calcium remains bound to troponin.

14. The concentration of calcium ion in the cytosol decreases as calcium is actively transported into the sarcoplasmic reticulum by a primary active-transport Ca-ATPase.

15. Removal of calcium ions from troponin restores the blocking action of tropomyosin, the cross-bridge cycle ceases, and the fiber relaxes.

the thick filaments, causing the muscle to shorten, whereas in isometric contractions, the cross bridges still exert force on the thin filaments, but the thick and thin filaments do not slide past each other, and the muscle length remains constant.

Figure 10-17 illustrates the general method of recording isotonic and isometric contractions. During an isotonic contraction the distance the muscle shortens, as a function of time, is recorded. The distance moved can be directly measured by a pen that is attached to a muscle (or to a muscle lever) and leaves a trace on a moving strip of paper as the muscle contracts and relaxes. During an isometric contraction, the force (tension) generated by the

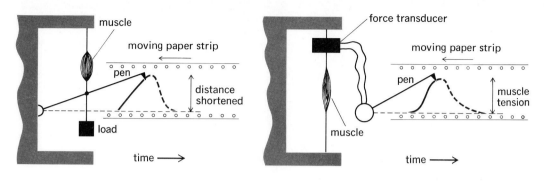

ISOTONIC CONTRACTION   ISOMETRIC CONTRACTION

FIGURE 10-17. Methods of recording isotonic and isometric muscle contractions.

muscle as a function of time is measured. To measure an isometric contraction the muscle is attached at one end to a rigid support and at the other to a force transducer, which controls the movement of a recording pen in proportion to the force exerted. The terminology and methods of recording contractions described above apply to both single fibers and whole muscles. We shall first describe the mechanics of single fiber contractions and later discuss the factors controlling the mechanics of whole muscle contraction.

Twitch contractions

The mechanical response of a muscle fiber to a *single* action potential is known as a **twitch**. Figure 10-18 shows the main features of an isometric and an isotonic twitch. In an isometric twitch there is an interval of a few milliseconds, the **latent period**, before the tension begins to increase. It is during this latent period that the processes associated with excitation-contraction coupling are occurring. The time interval from the beginning of tension development to the peak tension is the **contraction time**. Not all skeletal muscle fibers have the same contraction times. Some fast fibers have contraction times as short as 10 ms, whereas slower fibers may take 100 ms or longer. The time from peak tension until the tension has decreased to zero is known as the **relaxation time**.

Comparing an isometric twitch with an isotonic twitch in the same muscle fiber, one can see from

Fig. 10-18 that the latent period is longer, whereas the duration is shorter, in an isotonic twitch. Both the velocity of shortening and the duration of an isotonic twitch depend upon the magnitude of the load being lifted (Fig. 10-19). At heavier loads the

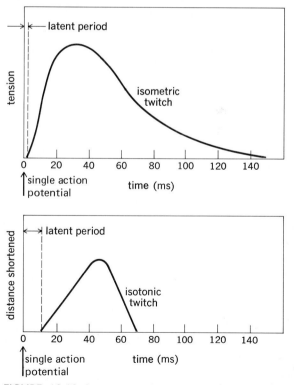

FIGURE 10-18. Isometric and isotonic skeletal muscle-fiber twitches following a single action potential.

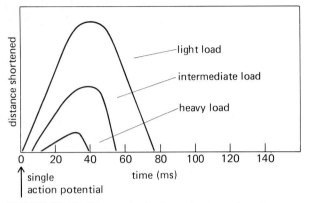

FIGURE 10-19. Changes in the isotonic shortening of a muscle fiber with different loads during a twitch.

latent period is longer but the velocity of shortening, the duration of the twitch, and the distance shortened all decrease. During the latent period of an isotonic contraction, the cross bridges begin to develop force, but actual shortening does not begin until the muscle tension slightly exceeds the load; therefore, the heavier the load, the longer the latent period. If the load on a fiber is increased, eventually a load will be reached that the muscle is unable to lift, the velocity of shortening will be zero, and the contraction becomes isometric.

Frequency-tension relation

Since a single action potential in a skeletal muscle fiber lasts 1 to 2 ms, whereas the mechanical response (a twitch) may last for 100 ms, it is possible for a second action potential to be initiated during the period of mechanical activity. Figure 10-20 illustrates the tension generated during the isometric contractions of a muscle fiber in response to three successive stimuli. In Fig. 10-20A the isometric twitch following the first stimulus S_1 lasts 150 ms. The second stimulus S_2, applied to the muscle fiber 200 ms after S_1, when the fiber has completely relaxed, causes a second identical twitch. In Fig. 10-20B the interval between S_1 and S_2 remains 200 ms, but a third stimulus is applied 60 ms after S_2, when the mechanical response resulting from S_2 is beginning to decrease. Stimulus S_3 induces a contractile response whose peak tension is greater than

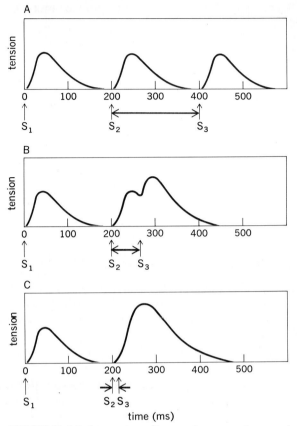

FIGURE 10-20. Summation of isometric contractions produced by shortening the time between stimulus S_2 and S_3.

that produced by S_2. In Fig. 10-20C the interval between S_2 and S_3 is further reduced to 10 ms, and the resulting peak tension is even greater; indeed the mechanical response to S_3 is a smooth continuation of the mechanical response already induced by S_2.

This increase in the mechanical response of a muscle fiber to a second action potential occurring during the mechanical response produced by a previous action potential is known as **summation**. The contraction of a muscle fiber can be sustained (in the absence of fatigue) if the fiber is repeatedly stimulated at a frequency sufficient to prevent complete relaxation between stimuli. A maintained contraction in response to repetitive stimulation is known as a **tetanus**. At low frequencies the tension may oscillate as the muscle fiber partially relaxes between stimuli, producing an **unfused tetanus**, whereas a **fused tetanus**, with no oscillations, is produced at higher frequencies of stimulation (Fig. 10-21).

As the frequency of action potentials increases, the level of sustained tension also increases until a maximal tetanic tension is reached, beyond which tension no longer increases with further increases in stimulation frequency. This maximal tetanic tension is about three to five times greater than the isometric twitch tension. A muscle fiber contracting isotonically can also undergo summation and tetanus, repetitive stimuli leading in this case to greater shortening. Since different muscle fibers have different contraction times, the stimulus frequency that will produce a maximal tetanic tension also differs from fiber to fiber. Frequencies of about 30 per second may produce a maximal tetanus in slowly contracting fibers, whereas frequencies of about 80 per second or more are necessary in very rapidly contracting fibers.

What is the mechanism responsible for the increase in the mechanical response of a muscle fiber to repetitive stimulation? The explanation involves the elastic properties of muscle fibers which cause a dissociation between the capacity to generate tension by cross-bridge activity, **internal tension**, and the **external tension** applied to the load on the muscle fiber. First let us define a new term: The maxi-

mal internal tension that can be generated by cross-bridge activity at any instant is known as the **active state** of a muscle fiber. In skeletal muscle fibers, the amount of calcium released from the sarcoplasmic reticulum by a single action potential is sufficient to produce nearly complete saturation of all troponin sites, so that all of the cross bridges are turned *on* and the filaments have the capacity for exerting their maximal internal tension during the early portion of a twitch (this is probably not the case in cardiac and smooth muscle). The active-state tension during a twitch rises rapidly to its maximal level when calcium is released from the sarcoplasmic reticulum and then declines as calcium is removed from troponin and pumped back into the sarcoplasmic reticulum.

As can be seen from Fig. 10-22A the active-state tension (the internal capacity to generate tension) is markedly different from that of the external tension that is exerted on a load during an isometric twitch. The cross-bridge tension is transmitted through the thick and thin filaments, across the Z lines, and eventually through extracellular connective tissue and tendons to the load. All these structures have a certain amount of elasticity and are collectively known as the **series elastic component**. The series elastic component has the properties of a spring placed between the force-generating cross bridges (the contractile component) and the exter-

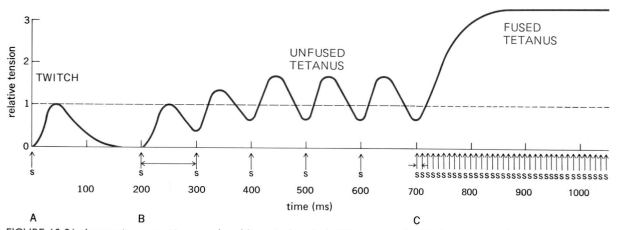

FIGURE 10-21. Isometric contractions produced by a single twitch (A), compared with those produced by 10 stimuli per second (B) and 100 stimuli per second (C).

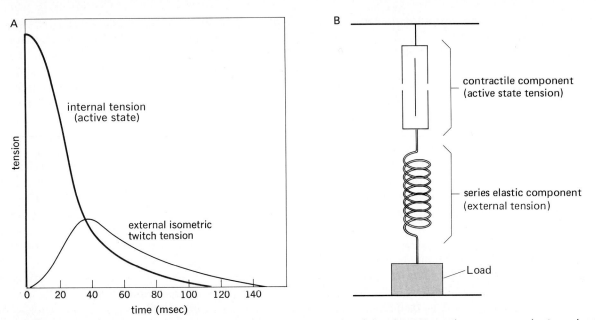

FIGURE 10-22. (A) Time course of the active-state tension produced by the contractile component (actin and myosin) in comparison with the development of external tension during an isometric twitch. (B) The tension in the contractile component (active-state tension) stretches the series elastic component. It is only the tension in the series elastic component that is transmitted to the exterior of the muscle as external tension.

nal load (Fig. 10-22B). The force generated by the cross bridges stretches this spring (the series elastic component), which in turn transmits its tension to the external load. The tension in the series elastic component is therefore the external tension, and its magnitude depends on the extent to which the series elastic component has been stretched by cross-bridge activity; the greater the stretch, the greater the tension in the series elastic component and the greater the tension transmitted to the load on the muscle fiber. Here is the crucial point: It takes time to stretch the series elastic component. In a single twitch, the tension in the series elastic component is still rising (Fig. 10-22A) at a time when the cross-bridge tension—the force stretching the series elastic component—is declining from its peak value as calcium is removed. Therefore, the peak tension developed by the series elastic component (the external tension) during a single twitch is never as great as the maximal tension that can be generated by the cross bridges.

Now we can explain summation and tetanus. When a muscle is stimulated repetitively, the cross bridges remain active for a longer period of time (i.e., the active state is prolonged) because the repetitive release of calcium from the sarcoplasmic reticulum maintains an elevated cytosolic calcium concentration. Because the cross bridges are active for a longer period of time, the series elastic component is stretched farther and the tension transmitted to the load is greater; i.e., summation occurs. In a maximal tetanus, the maintenance of the active state at its maximal level (all the cross bridges remain turned *on*) allows sufficient time to stretch the series elastic component to the point where its tension becomes equal to the force exerted by the cross bridges.

To summarize, the increase in external tension (summation and tetanus) which accompanies repeated stimulation is the result of an increase in the length of time the cross bridges remain active, thereby allowing more time to stretch the series elastic component and increase the amount of tension transmitted to the load on the muscle fiber.

Length-tension relation

One of the classic observations in muscle physiology is the relationship between a muscle's length and the tension it develops. A muscle fiber can be passively stretched to various lengths and stimulated at each of its new lengths; then the magnitude of the maximal isometric tetanus tension it can generate is measured at each length (Fig. 10-23). The length at which the fiber develops the greatest tension is termed the **optimal length**, l_o. When a muscle fiber is set at a length that is 60 percent of this optimal length, it develops no tension when stimulated. As the length of the fiber is increased by passive stretching, the tension generated during contraction increases to a maximum at l_o, and further lengthening of the fiber causes a drop in tension. When the muscle fiber is stretched to 175 percent of l_o or beyond, it develops no tension.

This relationship can be explained in terms of the sliding-filament mechanism. Stretching a muscle fiber changes the amount of overlap between the thick and thin filaments in the myofibrils. Stretching a fiber to 175 percent of l_o pulls the thick and thin filaments so far apart that there is no overlap between the two, and there can be no cross-bridge binding to actin and no tension developed. As the muscle length is decreased from 175 percent of l_o to the optimal length, a larger and larger region of each thick filament overlaps with thin filaments; the tension developed increases in proportion to the increased number of cross bridges in the overlap region. At l_o the maximal number of cross bridges bind to the thin filament, and tension is therefore maximal. At lengths less than l_o two mechanical factors lead to decreasing tension: (1) The thin filaments from the two halves of a sarcomere begin to overlap each other, interfering with cross-bridge binding and thereby decreasing the total number of active cross bridges; and (2) the thick filaments become compressed against the two Z lines, opposing tension development. In addition to these mechanical factors, at fiber lengths below about 80

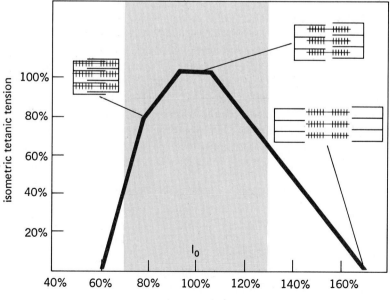

FIGURE 10-23. Variation in isometric tetanus tension with muscle fiber length. Maximal tension is produced at muscle length $l_o = 100$ percent. The shaded bands represent the range of length changes (from 70 to 130 percent) that can occur in the body while the muscles are attached to bones.

percent of l_o, calcium release from the sarcoplasmic reticulum in the central regions of the muscle fiber is decreased (for reasons that are not understood). This partial failure of excitation-contraction coupling also causes a decrease in the number of cross bridges that are producing tension.

In the body, where most skeletal muscles are attached to bones, the relaxed length of muscle fibers is near l_o and thus is near optimal for force generation. The total range of changes in length that a skeletal muscle can undergo is limited by the muscle's attachments to bones; it rarely exceeds a 30 percent change from the optimal length and is often much less. Over this range of lengths, the ability to develop tension never falls below about half of the tension that can be developed at the optimal length.

As we shall see, the length-tension relation plays a very important role in determining the magnitude of the force generated by cardiac muscle. The length of cardiac muscle fibers is not limited by an attachment to bones and thus they can undergo much larger changes in length with corresponding changes in their capacity to develop tension.

Load-velocity relation

The velocity at which a muscle fiber shortens while undergoing maximal tetanic stimulation decreases

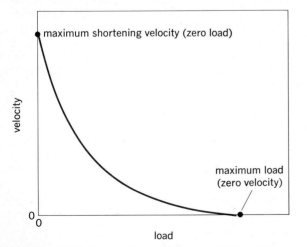

FIGURE 10-24. Shortening velocity as a function of load.

with increasing loads (Fig. 10-24). The maximal shortening velocity is achieved at zero load, and the velocity is zero (an isometric contraction) when the load is just equal to or greater than the maximal tetanic tension. The maximal velocity of shortening is not affected by the number of active cross bridges but appears to be determined by the rate at which individual cross bridges undergo their cycles of activity. The relation between load and velocity appears to be a fundamental property of the interactions between actin and myosin; specifically, the greater the load the slower the rate at which individual cross bridges repeat their cycles of activity.

Muscle Energy Metabolism

As we have seen, ATP performs three functions directly related to the contraction and relaxation of a muscle fiber: (1) The energy released from ATP-splitting by myosin is coupled to the movement of the cross bridges; (2) ATP-binding to myosin is necessary to break the link between the cross bridges and actin, allowing the cross bridges to operate cyclically; and (3) energy released from ATP-splitting is used by the calcium pump in the sarcoplasmic reticulum to reaccumulate calcium ions, producing relaxation.

In no other cell does the rate of ATP breakdown increase so much from one moment to the next as in a skeletal muscle fiber which goes from rest to a state of contractile activity. If a muscle fiber is to sustain contractile activity, molecules of ATP must be supplied by metabolism as rapidly as they are broken down by the contractile process. (If a muscle fiber had to rely on the supply of preformed ATP that exists at the start of contractile activity, this small amount of ATP would be completely consumed within a few twitches and the muscle would go into a state equivalent to rigor mortis.) There are three ways that a muscle fiber can form ATP (Fig. 10-25): (1) phosphorylation of ADP by **creatine phosphate**, (2) oxidative phosphorylation of ADP in the mitochondria, and (3) substrate phosphorylation of ADP, primarily by the glycolytic pathway in the cytosol.

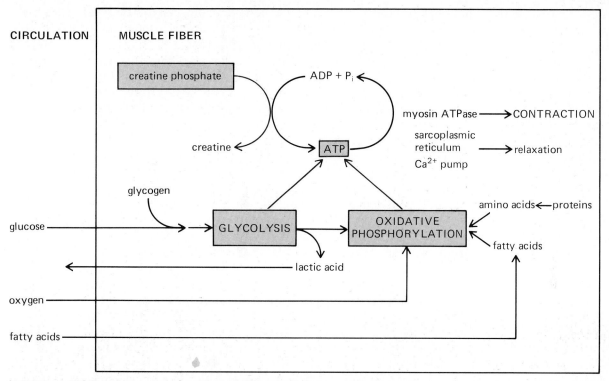

FIGURE 10-25. Metabolic pathways producing ATP utilized during muscle contraction.

Phosphorylation of ADP by creatine phosphate provides a very rapid means of forming ATP at the onset of contractile activity. When the chemical bond between creatine (C) and phosphate in a molecule of creatine phosphate (CP) is broken, an amount of energy is released that is similar in magnitude to the energy released when the terminal phosphate bond in ATP is broken. This energy, along with the phosphate group, can be transferred to ADP to form ATP in a reaction catalyzed by the enzyme **creatine kinase**:

$$CP + ADP \overset{\text{creatine kinase}}{\rightleftharpoons} C + ATP$$

This reaction is reversible, and ATP can phosphorylate creatine to form creatine phosphate and ADP. In a resting muscle fiber the increased concentration of ATP favors, by mass action, the formation of creatine phosphate. Therefore, during periods of

rest, muscle fibers build up a concentration of creatine phosphate approximately five times that of ATP. Then, at the beginning of contraction, when the concentration of ATP begins to fall and ADP concentration to rise owing to the increased rate of ATP breakdown, mass action favors the formation of ATP from creatine phosphate. This transfer of energy from creatine phosphate to ATP is so rapid that the actual concentration of ATP in a muscle fiber changes very little at the start of contraction, whereas the concentration of creatine phosphate falls rapidly.

Although the formation of ATP from creatine phosphate is very rapid, since it requires only a single enzymatic reaction, the amount of ATP that can be formed is limited by the initial concentration of creatine phosphate. If contractile activity is to be continued, the muscle must be able to form ATP from sources other than the limited creatine phosphate stores. The use of creatine phosphate at the start of contractile activity provides the time necessary for the slower, multienzyme pathways of oxidative phosphorylation and glycolysis to increase their rates of ATP formation to levels that match the rates of ATP breakdown.

At moderate levels of muscle activity (moderate rates of ATP breakdown) most of the ATP can be formed by oxidative phosphorylation, using fatty acids and, to a lesser extent, carbohydrates as the predominant source of fuel (see Chap. 5). A number of factors may limit the production of ATP by oxidative phosphorylation: (1) the quantity of oxygen delivered to a muscle by the blood; (2) the delivery of fuel molecules, such as glucose and fatty acids, to a muscle by the blood; and (3) the rates at which the enzymes in the metabolic pathways can process the fuel molecules. Factors (1) and (2) depend upon the capacity of the cardiovascular system to deliver the substances present in blood to a muscle (see Chaps. 11 and 12 for the role of the circulatory and respiratory systems in limiting maximal exercise). Factor (3) reflects the fact that the oxidative phosphorylation pathway is relatively slow and may not be able to produce ATP as rapidly as it is broken down, even when oxygen and fuel molecules are available.

Accordingly, when the level of exercise exceeds about 70 percent of maximum (70 percent of the maximal rate of ATP breakdown), glycolysis begins to produce an increasingly significant fraction of the total ATP generated by the muscle. The glycolytic pathway, although producing only small quantities of ATP from each molecule of glucose metabolized, can produce large quantities of ATP rapidly when enough enzymes and substrate are available. Not only can glycolysis produce ATP rapidly and in large amounts under such conditions but it also can occur in the absence of oxygen.

Although glycolysis can produce ATP very rapidly, it has the disadvantage of requiring very large quantities of glucose to produce relatively small amounts of ATP. This glucose can be obtained from two sources: from the blood or from the stores of glycogen within the muscle fibers. As the intensity of muscle activity increases, more and more of its ATP is formed by the anaerobic breakdown of muscle glycogen, with a corresponding increase in the production of lactic acid.

In summary (Fig. 10-25), creatine phosphate provides a very rapid mechanism for replacing ATP at the onset of contraction until the metabolic pathways can adjust to the increased demand for ATP production. During mild exercise, the oxidative metabolism of fatty acids (and glucose) delivered to the muscle by the blood provides most of the ATP (Chap. 15 describes the hormonal control of this process). As the intensity of exercise increases, more and more of the ATP is supplied by the anaerobic metabolism of glucose derived from the muscle's own glycogen stores.

At the completion of a prolonged period of exercise, creatine phosphate and glycogen levels are decreased. To return a muscle fiber to its original state, the glycogen stores and creatine phosphate must be replaced; both processes require energy. To provide the energy necessary for these synthetic processes, a muscle continues to consume increased amounts of oxygen for some time after it has ceased to contract (as evidenced by the fact that one continues to breathe deeply and rapidly for a period of time immediately following intense exercise). The longer and more intense the exercise, the longer it takes to restore a muscle fiber to its original metabolic state.

The overall efficiency with which a muscle fiber converts chemical energy (e.g., the energy content of glucose) into work (movement of a load) is only about 20 percent, the remaining 80 percent of the energy released during contraction appearing as heat. The greater the amount of exercise, the greater the amount of heat produced. This increased heat production may place a severe stress upon the ability to maintain a constant body temperature, especially on a hot day. On the other hand, the process of shivering (small, rapidly oscillating contractions of skeletal muscles) reflects the use of this same source of heat energy to maintain body temperature in a cold environment. The various mechanisms regulating body temperature will be discussed in Chap. 15.

Muscle fatigue

When a skeletal muscle fiber is continuously stimulated at a frequency that produces a maximal tetanic contraction, the tension developed by the fiber eventually declines. This failure of a muscle fiber to maintain tension as a result of previous contractile activity is known as **muscle fatigue**. Figure 10-26 illustrates several characteristics of the fatigue process. Fatigue is not limited to tetanic contractions but can occur with twitches if they are repeated frequently enough over an extended period of time. On the other hand, low-frequency twitches can be repeated indefinitely in certain types of skeletal muscles with no signs of fatigue. Thus, the onset of fatigue and its rate of development depend on the type of skeletal muscle and the intensity of contractile activity.

If, after the onset of fatigue, a muscle is allowed to rest for a period of time, it can recover its ability to contract upon restimulation. The extent of recovery after a period of rest varies depending upon the type of previous activity. A muscle will fatigue fairly rapidly if stimulated at a high frequency, but it will also recover rapidly from this fatigue. This is the type of fatigue that accompanies short-duration, high-intensity types of exercise, such as weight lifting. On the other hand, fatigue that develops more slowly with long-duration, low-intensity endurance exercise, such as long distance running, requires much longer periods of rest, often up to 24 h before complete recovery is achieved.

It might seem logical that depletion of available ATP would account for fatigue, but it does not. The ATP concentration in fatigued muscle is only slightly lower than in a resting muscle. Indeed it is neither the supply of energy nor the ability of the contractile proteins to generate tension that is responsible for fatigue. Rather, a failure of some step in excitation-contraction coupling appears to be the most likely cause of fatigue, but which step and what brings about its failure is currently unknown. Failure of the appropriate levels of the cerebral cortex to send excitatory signals to the motor neurons of a particular muscle can lead to its failure to contract. This type of **psychological fatigue** causes an individual to stop exercising even though the muscles themselves are not fatigued. An athlete's performance depends not only on the physical state of the appropriate muscles but also upon the "will to win," the ability to overcome psychological fatigue.

Muscle Fiber Differentiation and Growth

Each muscle fiber is formed during embryological development by the fusion of a number of small, mononucleated **myoblasts** to form a single cylindrical, multinucleated muscle fiber. It is only after the fusion of the myoblasts that the muscle fibers begin to form actin and myosin filaments and become capable of contracting. Once the myoblasts have fused, the resulting muscle fibers no longer have the capacity to divide but do have the ability to grow in length and diameter. This stage of muscle differentiation is completed around the time of birth. Therefore, after birth, if skeletal muscle fibers are destroyed, they cannot be replaced by the division of other existing, differentiated fibers. There can occur, however, the formation of new fibers from undifferentiated cells located adjacent to muscle fibers, which repeat the embryonic process of muscle-fiber formation. This capacity for forming new muscle fibers after birth is considerable but will not restore a severely damaged muscle to full strength. Additional compensation for a loss of muscle tissue occurs through the increased growth in the size of the remaining muscle fibers.

Following the fusion of myoblasts, motor neurons send axon processes into the embryonic muscle, forming neuromuscular junctions and bringing the muscle under the control of the central nervous system. From this point on there is a very critical dependence of the muscle fiber on its motor neuron, not only to provide the means of initiating contraction but also for the muscle's continued survival and development.

If the nerve fibers to a muscle are severed or otherwise destroyed, the denervated muscle fibers

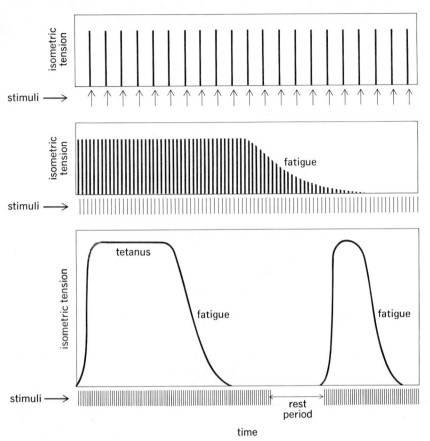

FIGURE 10-26. Muscle fatigue resulting from prolonged stimulation, and recovery of ability to contract after a period of rest.

become progressively smaller, and their content of actin and myosin decreases; this condition is known as **denervation atrophy**. A muscle can also atrophy with its nerve supply intact if the muscle is not used for a long period of time as when a broken arm or leg is immobilized in a cast; this condition is known as **disuse atrophy**. The extent of denervation atrophy can be diminished clinically by initiating action potentials in the muscle membrane using direct electric stimulation. This indicates that it is contractile activity that is the major factor preventing atrophy. In addition, growth-promoting factors released from nerve endings appear to contribute to the growth and maintenance of muscle tissue.

In contrast to the decrease in muscle mass that results from a lack of neural stimulation, increased amounts of contractile activity, as results from certain types of regular exercise, may produce a considerable increase in size (**hypertrophy**) of some muscle fibers as well as other changes in their chemical composition. Action potentials in nerve fibers appear to release chemical substances which influence the biochemical activities of the muscle fiber, but the identity of these tropic agents and their mechanisms of action are unknown.

Types of Skeletal Muscle Fibers

Not all skeletal muscle fibers have the same mechanical and metabolic capacities. Different types of fibers can be identified on the basis of both their maximal velocities of shortening and the types and amounts of the enzymes they contain, especially those enzymes associated with the production of ATP.

First let us deal with the maximal velocity of shortening. Recall that earlier in the chapter we referred to fast fibers and slow fibers. The myosin molecules in these fibers differ in the maximal rates at which they split ATP, which in turn determines the maximal rate at which cross bridges can cycle, and hence, determines the maximal velocity of shortening. Myosin with high ATPase activity is found in fast fibers, and myosin with a lower ATPase activity is associated with slow fibers.

The second major difference between skeletal muscle fibers involves the type of enzymatic machinery available for synthesizing ATP. Some fibers contain numerous mitochondria and thus have a high capacity for oxidative phosphorylation. Most of the ATP produced by such oxidative fibers is dependent upon blood flow to the muscle to deliver oxygen and fuel molecules, and these types of fibers are surrounded by numerous capillary blood vessels. These fibers also contain large amounts of an oxygen-binding protein known as **myoglobin**, which is similar to the hemoglobin which is present in red blood cells. Myoglobin increases the rate of oxygen diffusion into a muscle fiber, in addition to providing a small store of oxygen within the fiber.

Other types of fibers have few mitochondria but possess a higher concentration of glycolytic enzymes and a large store of glycogen. These fibers are specialized for the production of ATP by glycolysis in the absence of oxygen. Corresponding to their low requirement for oxygen, they are surrounded by relatively few capillaries and contain little myoglobin. Oxidative fibers containing large amounts of myoglobin have a dark red color and are often referred to as **red muscle** fibers; their color distinguishes them from the **white muscle** fibers which lack appreciable amounts of myoglobin.

Three types of skeletal muscle fibers can be distinguished, on the basis of differences in their myosin-ATPase activity and ATP-synthesizing capacity:

1. **Oxidative slow fibers**. These fibers combine a low myosin-ATPase activity with a high oxidative capacity.
2. **Oxidative fast fibers**. These fibers combine a high myosin-ATPase activity with a high oxidative capacity.
3. **Glycolytic fast fibers**. These fast fibers combine a high myosin-ATPase activity with a high glycolytic capacity.

These three types of fibers also show differences in their capacity to resist fatigue. Glycolytic fast fibers fatigue rapidly, while oxidative slow fibers are very resistant to fatigue, which allows them to maintain contractile activity for long periods with little loss of tension. The oxidative fast fibers have an intermediate capacity to resist fatigue.

Additional characteristics of the three types of fibers are summarized in Table 10-2.

All of the muscle fibers in a single motor unit are of the same type. Most muscles have a mixture of all three types of motor units interspersed with each other (Fig. 10-27). Depending on the proportions of the three fiber types in a muscle, the contractile properties of the muscles can have a wide range of contraction speeds and fatigability, as will be described later.

The muscles which support the weight of the body (the postural muscles of the back and legs) must be able to maintain their activity for long periods of time without fatigue. These muscles contain large numbers of oxidative slow fibers which are resistant to fatigue. In contrast the muscles in the arms may be called upon to produce large amounts of tension over a short time period, as in the lifting of heavy objects, and these muscles have a greater proportion of glycolytic fast fibers.

Contraction of Whole Muscles

The human body contains over 600 different skeletal muscles, which taken all together make up the largest tissue in the body, accounting for 40 to 45 percent of the body weight. Some muscles are very

TABLE 10-2. Characteristics of the three types of skeletal muscle fibers

	Oxidative slow fiber	Oxidative fast fiber	Glycolytic fast fiber
Speed of contraction	Slow	Fast	Fast
Myosin-ATPase activity	Low	High	High
Primary source of ATP production	Oxidative phosphorylation	Oxidative phosphorylation	Anaerobic glycolysis
Glycolytic enzyme activity	Low	Intermediate	High
Number of mitochondria	Many	Many	Few
Capillaries	Many	Many	Few
Myoglobin content	High	High	Low
Glycogen content	Low	Intermediate	High
Fiber diameter	Small	Intermediate	Large
Rate of fatigue	Slow	Intermediate	Fast

small, consisting of only a few hundred fibers; larger muscles may contain several hundred thousand fibers.

Surrounding the individual muscle fibers in a muscle is a network of collagen fibers and connective tissue through which pass blood vessels and nerves (Fig. 10-1). Collagen is a fibrous protein which has great strength but no active contractile properties. Each end of a muscle is usually attached to bone by bundles of collagen fibers known as **tendons**. The collagen fibers surrounding the fibers and in the tendons act as a structural framework which transmits the muscle tension to the bones.

In some muscles, the individual fibers extend the entire length of the muscle, but in most, the fibers are shorter. When fibers are shorter, their ends are anchored to the connective tissue network within the muscle. The transmission of force from muscle to bone is like a number of people pulling on a rope, each person corresponding to a single muscle fiber and the rope corresponding to the connective tissue and tendons.

Some tendons are very long, and the site of attachment of the tendon to bone is far removed from the muscle. For example, some of the muscles which move the fingers are in the forearm, as one can observe by wiggling one's fingers and feeling the movement of the muscles in the lower arm. These muscles are connected to the fingers by long tendons. If the muscles which move the fingers were located in the fingers themselves, we would have very fat fingers.

Control of muscle tension

The total tension an entire muscle can develop depends upon two factors (Table 10-3); (1) the number of muscle fibers in the muscle that are contracting at any given time and (2) the amount of tension developed by each contracting fiber.

The number of muscle fibers contracting at any time depends upon the number of motor neurons that are active and the number of muscle fibers associated with each neuron; in other words, it depends upon the number of motor units that are active and the size of each motor unit. The firing of action potentials in a motor neuron is determined by the activity of the synaptic inputs to the neuron. **Recruitment** is the increasing of the number of neurons that are firing action potentials by increasing the input of synaptic activity to more neurons. The greater the number of motor neurons (motor units) recruited, the greater the muscle tension.

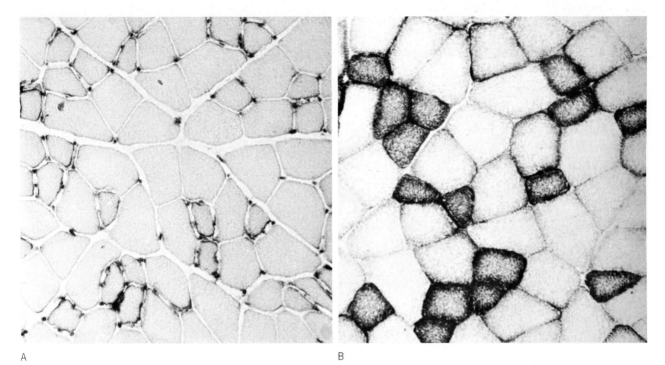

A   B

FIGURE 10-27. Cross sections of a skeletal muscle, showing individual muscle fibers which have been stained according to their chemical composition. (A) The capillaries surrounding the muscle fibers have been stained. Note the large number of capillaries surrounding the smaller-diameter oxidative fibers. (B) Darkly stained fibers reveal the presence of high concentrations of mitochondria in the oxidative, smaller-diameter fibers. *(Courtesy of John A. Faulkner.)*

The number of muscle fibers in a single motor unit varies considerably in different muscles. In muscles such as those in the hand and eye, which are able to produce very delicate movements, the size of the individual motor unit is small (e.g., in an eye muscle, one motor neuron innervates only about 13 muscle fibers). In contrast, in the more coarsely controlled muscles of the back and legs each motor unit contains hundreds and in some cases several thousand muscle fibers; e.g., a single motor unit in the large calf muscle of the leg contains about 1,700 muscle fibers. If a muscle is composed of small motor units, the total tension produced by the muscle can be increased in small steps by the recruitment of additional motor units, whereas if the motor units are large, large increases in tension occur as each additional motor unit is recruited. Thus, finer control of muscle tension is achieved in muscles with smaller motor units.

Not only can the number of recruited motor units be altered, but so can the type of motor unit. During weak contractions, only oxidative slow motor units are recruited. Additional motor units of the oxidative fast type are recruited as the force of contraction is increased, and the last motor units to be recruited during more forceful contractions are the glycolytic fast fibers. Thus, during moderate-strength contractions, such as are used in most endurance types of exercise, relatively few glycolytic fast fibers (which have a greater susceptibility to fatigue) are recruited, and most of the activity occurs in oxidative fibers (which are more resistant to fatigue).

The motor neurons to a given muscle fire asynchronously. Thus, some motor units may be active while others are momentarily inactive. In muscles which are active for long periods of time, such as the postural muscles which support the weight of the body, this asynchronous activity provides periods of rest between periods of contractile activity and allows a certain level of tension to be maintained while preventing fatigue.

The second factor determining total-muscle tension is the tension developed by each fiber in the active motor units. The factors that determine the tension produced in a single fiber have been discussed previously: (1) action potential frequency (the frequency-tension relation), producing summations of contractions and tetanus; (2) the extent to which a muscle fiber has been stretched by external forces (length-tension relation); (3) the previous extent of contractile activity, determining the state of fatigue; and (4) the fiber type (e.g., glycolytic fast fibers have a larger diameter and therefore produce more tension).

The neural control of skeletal muscle tension involves both the recruitment of motor units (to vary the number of active muscle fibers) and the frequency of action potentials in individual motor neurons (to vary the tension generated by each fiber). Of these two neural mechanisms, recruitment of motor units provides the primary means of varying tension. Most activity in motor neurons occurs in bursts of action potentials, which produce tetanic contractions rather than single twitches. Therefore, varying the frequency of action potentials in these neurons provides, at best, a way to make small adjustments in the tetanic tension of the recruited motor units.

Control of shortening velocity

As the recruitment of motor units increases, the amount of force applied to a load is increased, thereby moving the load with a greater velocity. Thus, velocity is controlled by the same factors that control total muscle tension.

Muscle adaptation to exercise

The frequency with which a muscle is used, as well as the duration and intensity of its activity, affects the properties of the muscle. We have already described how disuse leads to muscle atrophy,

TABLE 10-3. Factors determining total muscle tension

I. Number of muscle fibers contracting
   a. Recruitment of motor units
   b. Number of muscle fibers per motor unit

II. Tension produced by each contracting muscle fiber
   a. Action potential frequency (frequency-force relation)
   b. Changes in fiber length (length-tension relation)
   c. Duration of activity (fatigue)
   d. Fiber type

whereas at the other extreme, high-intensity, short-duration exercise, such as weight lifting, can produce considerable hypertrophy of muscle fibers. In both of these cases it is not the number of muscle fibers that changes but the size and metabolic capacity of each fiber.

Two types of change can occur as a result of exercise: (1) alterations in the ATP-forming capacity of muscle fibers as a result of increased enzyme synthesis and increased blood flow to the muscle and (2) changes in the diameter of the muscle fibers as a result of the formation of additional myofibrils.

Exercise that is of relatively low intensity but of long duration, such as long-distance running and swimming, produces increases in the number of mitochondria in the oxidative fast and slow fibers that are recruited in this type of activity. In addition, there is an increase in the number of capillaries around these fibers. All of these changes lead to an increase in the capacity for endurance activity with a minimum of fatigue. There is little change in fiber diameter and thus little change in the strength of muscles as a result of endurance exercise.

In contrast, high-intensity, short-duration exercise, such as weight lifting, affects primarily the glycolytic fast fibers that are recruited only during very powerful contractions. These fibers undergo a great increase in fiber diameter as a result of the increased synthesis of actin and myosin filaments forming an increased number of myofibrils. In addition, the number of glycolytic enzymes increases. The result of such high-intensity exercise is to increase the strength of the muscle, producing the bulging muscles of a professional weight lifter. Such muscles, although very powerful, have little capacity for endurance and fatigue rapidly.

Because different types of exercise produce quite different changes in the strength and endurance of a muscle, an individual performing regular exercises to improve muscle performance must be careful to choose a type of exercise that is compatible with the type of activity he or she ultimately wishes to perform. Thus, lifting weights will not improve the endurance of a long-distance runner, and jogging will not produce the increased strength desired by a weight lifter. A mixed program of exercises can improve both strength and endurance, but cannot achieve the maximal state of either strength or endurance that muscle is capable of with vigorous exercise training of one type. As we shall see in later chapters, endurance exercise produces changes not only in the skeletal muscles but also in the respiratory and circulatory systems, changes which improve the delivery of oxygen and fuel molecules to the muscle.

The molecular mechanisms responsible for the changes in muscle with different types of activity are unknown, but they appear to be related to the frequency and intensity of the contractile activity in the muscle fibers and thus to the pattern of action potentials produced in the muscle over an extended period of time. High-intensity exercise recruits most of the motor units in a muscle, whereas endurance exercise recruits primarily the oxidative motor units. If a regular pattern of exercise is stopped, the changes in the muscle that occurred as a result of the exercise will slowly return, over a period of months, to the condition before exercise began.

Immobilizing a muscle for a prolonged period of time not only produces disuse atrophy but also alters the type of myosin synthesized by the muscle fibers. The proportion of fast fibers in the muscle increases as some of the slow fibers begin to synthesize myosin with a high ATPase activity. Exercise produces little change in the types of myosin formed by the fibers and thus little change in the proportions of fast and slow fibers in a muscle. However, exercise can produce major changes in the proportion of oxidative fast fibers and glycolytic fast fibers, as a result of changes in the metabolic capacity of the muscle fibers.

Lever action of muscles and bones

A contracting muscle exerts a force on bones through its connecting tendons. When the force is great enough, the bone moves as the muscle shortens. A contracting muscle exerts only a pulling force, so that as the muscle shortens, the bones to which it is attached are pulled toward each other. **Flexion** refers to the bending of a limb at a joint, whereas **extension** is the straightening of a limb. These motions require at least two separate muscles, one to cause flexion and the other extension. Groups of muscles which produce oppositely directed movements at a joint are known as **antagonists**. From Fig. 10-28 it can be seen that contraction of the biceps causes flexion of the arm at the elbow, whereas contraction of the antagonistic mus-

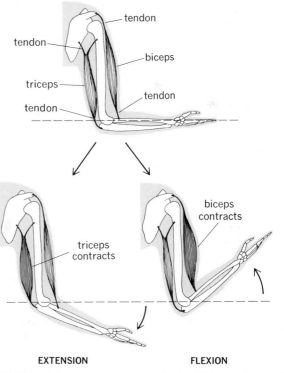

EXTENSION   FLEXION

FIGURE 10-28. Antagonistic muscles for flexion and extension of the forearm.

cle, the triceps, causes the arm to extend. Both muscles exert only a pulling force upon the forearm when they contract.

Sets of antagonistic muscles are required not only for flexion-extension, but for side-to-side movements or rotation of a limb. The contraction of some muscles leads to two types of limb movement; for example, contraction of the gastrocnemius muscle in the leg causes both extension of the foot and flexion of the leg at the knee, as in walking (Fig. 10-29). Contraction of the gastrocnemius mus-

cle at the same time as contraction of the quadriceps femoris (which causes extension of the lower leg) prevents the knee joint from bending, leaving only the ankle joint capable of moving; the foot is extended, and the body rises on tiptoe.

During smooth limb movements, either flexion or extension, both antagonistic muscles are undergoing simultaneous contractions. The muscle undergoing the strongest contraction shortens; this is known as a **concentric contraction**. The contraction of the antagonistic muscle opposes this shortening, but since it is generating less tension it actually undergoes a lengthening; that is, it is being stretched by the antagonistic muscle while undergoing an active contraction. Such a contraction is known as an **eccentric contraction**. Thus, during the flexion of the arm at the elbow, for example, the biceps undergoes a concentric contraction while the triceps is undergoing an eccentric contraction. During extension, the reverse would be true.

The arrangement of the muscles, bones, and joints in the body form lever systems. The basic principle of a lever is illustrated by the flexion of the arm by the biceps muscle (Fig. 10-30), which exerts an upward pulling force on the forearm about 5 cm away from the elbow. In this example, a 10-kg weight held in the hand exerts a downward force of 10 kg about 35 cm from the elbow. A law of physics tells us that the forearm is in mechanical equilibrium (no net forces are acting on the system) when the product of the downward force (10 kg) and its distance from the elbow (35 cm) is equal to

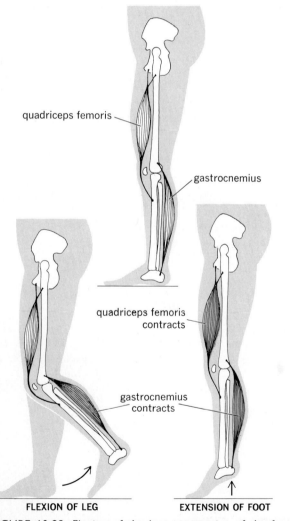

FIGURE 10-29. Flexion of the leg or extension of the foot follows contraction of the gastrocnemius muscle, depending on the activity of the quadriceps femoris muscle.

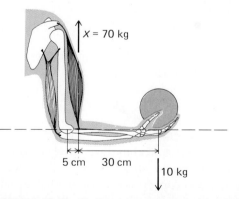

FIGURE 10-30. Mechanical equilibrium of forces acting on the forearm while it supports a 10-kg load.

the product of the upward force (X) exerted by the muscle and its distance from the elbow (5 cm); i.e., $10 \times 35 = 5X$; thus $X = 70$ kg. In other words, this sytem is working at a mechanical disadvantage since the force exerted by the muscle is considerably greater than the load it is supporting. In individuals with very powerful muscles, the large forces produced under conditions of maximum exertion sometimes tear the tendon away from the muscle or bone, or in rare cases, break the bone. However, the mechanical disadvantage under which most muscles operate is offset by increased maneuverability. In Fig. 10-31, when the biceps shortens 1 cm, the hand moves through a distance of 7 cm. Since the muscle shortens 1 cm in the same amount of time that the hand moves 7 cm, the *velocity* at which the hand moves is seven times greater than the rate of muscle shortening. The lever system amplifies the movements of the muscle so that short, relatively slow movements of the muscle produce faster movements of the hand. Thus, a pitcher can throw a baseball at 160 km/h even though his muscles shorten at only a fraction of this velocity.

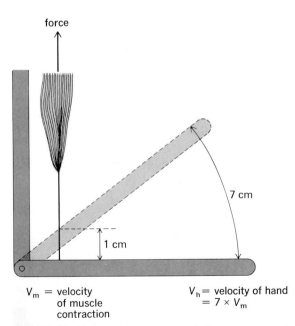

V_m = velocity of muscle contraction

V_h = velocity of hand = $7 \times V_m$

FIGURE 10-31. Small movements of the biceps muscle are amplified by the lever system of the arm, producing large movements of the hand.

Smooth Muscle

Smooth muscle, like skeletal muscle, uses cross-bridge movements between actin and myosin filaments to generate force and calcium ions to control cross-bridge activity. However, the structural organization of the contractile filaments, the process of excitation-contraction coupling, and the time-course of contraction are quite different in the two types of muscle. Furthermore, there is considerable diversity in the properties of smooth muscle, especially with respect to the mechanism of excitation-contraction coupling. Nevertheless, two characteristics are common to all smooth muscles: They lack the cross-striated banding pattern found in skeletal and cardiac fibers (hence the name "smooth muscle"), and the nerves to them are derived from the autonomic division of the nervous system rather than the somatic division; thus, smooth muscle is not normally under direct voluntary control.

Smooth muscle structure

Unlike skeletal muscle fibers, the precursor cells of smooth muscle fibers do not fuse during embryological development. Each differentiated smooth muscle fiber is a spindle-shaped cell containing a single nucleus. The diameter of these small fibers ranges from 2 to 10 μm, compared to a range of 10 to 100 μm for skeletal muscle fibers. In most hollow organs, the smooth muscle fibers are arranged in bundles organized in two layers—an outer longitudinal layer and an inner circular layer. In blood vessels, bundles of smooth muscle fibers are arranged in a circular or helical fashion around the vessel wall.

The cytoplasm of smooth muscle fibers is filled with filaments oriented approximately parallel to the long axis of the fiber (Fig. 10-32). Three types of filaments are present: thick myosin-containing filaments (which are longer than those of skeletal muscle); thin actin-containing filaments, which are anchored either to the plasma membrane or to cytoplasmic structures known as **dense bodies**, which have characteristics similar to the Z lines in skeletal muscle fibers; and intermediate-sized filaments that

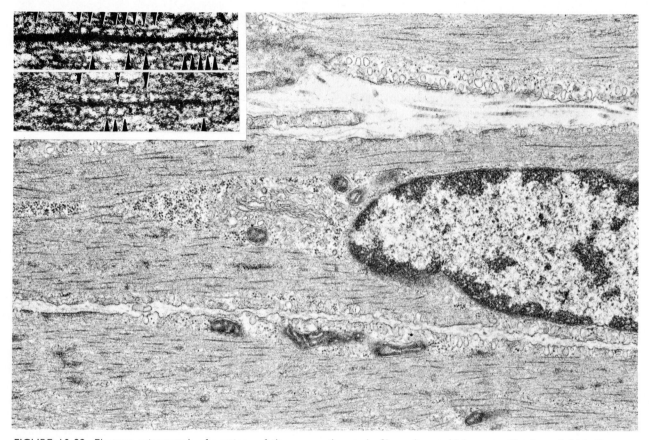

FIGURE 10-32. Electron micrograph of portions of three smooth muscle fibers. Insert, higher magnification of thick filaments, with projections (arrows) suggestive of cross bridges connecting adjacent thin filaments. *[From A. P. Somlyo, C. E. Devine, Avril V. Somlyo, and R. V. Rice, Phil. Trans. Roy. Soc. Lond. B, **365**:223–229, (1973).]*

do not seem to play a role in the active generation of force and probably function as an elastic framework or cytoskeleton to maintain the shape of the cell. There is no regular alignment of these filaments into sarcomeres and myofibrils, which accounts for the absence of a banding pattern in smooth muscle fibers.

The relative amounts of actin and myosin in smooth muscle differ from those found in skeletal muscle. The amount of myosin is only about one-third, whereas the actin content can be as much as twice that of skeletal muscle. These differences are reflected in the relative numbers of thick and thin filaments in the two types of muscle; in the region of thick- and thin-filament overlap, there are two thin filaments for every thick filament in striated muscle, whereas in smooth muscle there are 10 to 15 thin filaments for each thick filament. In spite of these differences, the maximal tension per unit of cross-sectional area developed by smooth muscles at their optimum length is comparable to that developed by skeletal muscle.

The tension developed by smooth muscle fibers varies with muscle length in a manner qualitatively similar to skeletal muscle. There is an optimal length at which tension development is maximal, and less tension is generated at lengths shorter or longer than this optimal length. However, the range of muscle lengths over which smooth muscle is able to develop tension is considerably greater than it is in skeletal muscle. The longer thick filaments, the lack of Z lines to limit shortening, and the lack of a regular alignment of overlapping filaments all contribute to this wider range of tension development. This property is highly adaptive, since smooth muscle surrounds hollow organs which undergo changes in volume and hence in the lengths of the smooth muscle fibers in their walls. Even with relatively large increases in volume (as during the accumulation of large amounts of urine in the bladder), the smooth muscle fibers in the wall retain some ability to develop tension, whereas such distortion would have stretched skeletal muscle fibers beyond the point of thick- and thin-filament overlap.

The presence of actin and myosin, overlapping thick and thin filaments with cross bridges, and a length-tension relationship, all suggest that smooth muscle contraction occurs by a sliding-filament mechanism similar to that in skeletal and cardiac muscle.

Excitation-contraction coupling

Changes in the cytosolic calcium concentration controls the contractile activity in smooth muscle fibers, as well as in skeletal muscle. However, there are significant differences between the two types of muscle in the way in which calcium exerts its effects on cross-bridge activity and the mechanisms by which its cytosolic concentration is controlled.

Smooth muscle cells do not have the calcium-binding protein troponin that is present in both skeletal and cardiac muscle and that mediates their response to changes in cytosolic calcium. In smooth muscle, calcium controls cross-bridge activity by regulating the activity of a protein kinase that phosphorylates myosin. The following sequence of events occurs after a rise in cytosolic calcium in a smooth muscle fiber: Calcium binds to calmodulin, a calcium-binding protein that is present in most cells (see Chap. 7) and whose structure is related to troponin; the calcium-calmodulin complex binds to a protein kinase thereby activating the enzyme; the active protein kinase then phosphorylates myosin. In smooth muscle, only the phosphorylated form of myosin is able to bind to actin and undergo a cross-bridge cycle that produces tension. Hence, cross-bridge activity in smooth muscle is turned *on*

as a result of myosin phosphorylation by an enzyme whose activity is controlled by calcium.

The calcium which triggers contraction in smooth muscle enters the cytosol from two sources: Some is released from the sarcoplasmic reticulum within the fiber, and some comes from the extracellular fluid by diffusion through calcium channels in the plasma membrane. Some types of smooth muscles rely more on one source of calcium than the other, but most make use of both sources to some extent. Relaxation is brought about by the removal of calcium from the cytosol through the action of calcium pumps either in the membrane of the sarcoplasmic reticulum or in the plasma membrane

In smooth muscle, the volume of the fiber occupied by sarcoplasmic reticulum is only a fraction of that in skeletal muscle. Furthermore, there are no transverse tubules connected to the plasma membrane (although regions of the sarcoplasmic reticulum are located near the plasma membrane, forming associations similar to those between a t tubule and the lateral sacs in skeletal muscle). Thus, skeletal muscle mechanisms for rapidly activating and relaxing the contractile apparatus are not present in smooth muscle. Furthermore, once the smooth muscle is activated by calcium, the rate at which myosin splits ATP is much slower than it is in skeletal muscle. This low myosin-ATPase activity accounts for the slow speed of smooth muscle contraction. A single twitch that lasts only a fraction of a second in skeletal muscle may last several seconds in smooth muscle.

Unlike skeletal muscle, in which a single input signal (a single action potential) releases sufficient calcium to turn on all the cross bridges in the fiber, only a portion of the cross bridges are activated in a smooth muscle fiber in response to most stimuli. Therefore, the tension generated by smooth muscle can be graded by varying the magnitude of the changes in cytosolic calcium concentration; i.e., the greater the increase in calcium concentration, the greater the number of cross bridges activated and the greater the tension.

In some smooth muscles, even in the absence of specific external stimuli, the cytosolic calcium concentration is sufficient to maintain a low level of cross-bridge activity and thus maintain a low level of tension that is known as **tone**. The intensity of this tonic tension can be varied by factors that alter the cytosolic calcium concentration. Sudden changes in cytosolic calcium can produce phasic contractions and relaxations on top of an ongoing tonic activity.

In a sense we have approached the question of excitation-contraction coupling in smooth muscle backward by first describing the "coupling" (the changes in cytosolic calcium), but now we must ask what constitutes the "excitation" which elicits these changes in calcium.

In contrast to skeletal muscle, in which contractile activity is dependent on a single input—the somatic neurons to the muscle—many inputs affect the contractile activity of smooth muscle fibers (Table 10-4). All of these inputs have the ultimate effect of increasing cytosolic calcium concentration, either by releasing calcium from the sarcoplasmic reticulum or by increasing the permeability of the plasma membrane to extracellular calcium. They do this either by changing the electric potential across the plasma membrane or by actions which are independent of changes in membrane potential (Fig. 10-33).

How is calcium influx or release from the sarcoplasmic reticulum altered by inputs which do not cause changes in membrane potential? In these cases an extracellular or intracellular chemical messenger acts on the sarcoplasmic reticulum or on

---

TABLE 10-4. Inputs influencing smooth muscle contractile activity

---

1. Spontaneous electrical activity in the smooth muscle plasma membrane

2. Neurotransmitters released by autonomic neurons

3. Hormones

4. Locally induced changes in the chemical composition (paracrines, acidity, oxygen, osmolarity, ion concentrations) of the extracellular fluid that surrounds a smooth muscle fiber

5. Stretching smooth muscle fibers

---

specific calcium channels in the plasma membrane to increase the calcium flux into the cytosol.

What is the relationship between the membrane potential and the cytosolic calcium concentration in a smooth muscle fiber? A decrease in membrane potential (depolarization) opens voltage-sensitive calcium channels in the plasma membrane, increasing the membrane's permeability to calcium. Since calcium is present at higher concentration outside the muscle fiber than inside, any increase in the number of open calcium channels results in an increased flux of calcium into the cell. The greater the depolarization, the greater the number of open calcium channels, the greater the rise in cytosolic calcium concentration, and the greater the contractile response. The amount of calcium released by the sarcoplasmic reticulum also increases with the degree of plasma-membrane depolarization, but the coupling mechanisms between the two membranes is not known.

Note that we have said nothing thus far of action potentials. In smooth muscle, unlike skeletal muscle, small changes in membrane potential, which do not lead to action potentials, produce changes in cytosolic calcium concentration and corresponding changes in contractile activity; indeed some smooth muscles do not normally produce action potentials. On the other hand, if action potentials *are* elicited by threshold-strength stimuli, there occurs a much greater increase in calcium and the force of contraction. Interestingly, in most smooth muscles in which action potentials occur, calcium ions, rather than sodium ions, carry positive charge into the cell during the rising phase of the action potential; i.e., depolarization of the membrane opens voltage-sensitive calcium channels, producing calcium action potentials rather than sodium action potentials.

Spontaneous electrical activity. Some types of smooth muscle fibers generate action potentials spontaneously in the absence of any neural or hormonal input. The plasma membrane in fibers exhibiting spontaneous activity does not maintain a constant resting potential; instead the membrane gradually depolarizes until it reaches the threshold

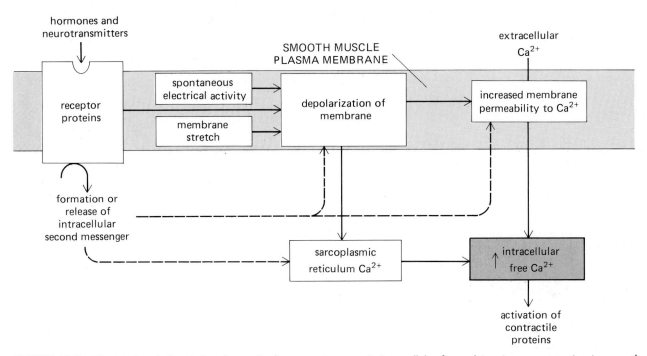

FIGURE 10-33. Electrical and chemical pathways leading to an increase in intracellular free calcium-ion concentration in smooth muscle. The dashed lines denote second-messenger pathways. Not shown are the pathways by which locally induced changes in extracellular composition influence the muscle.

potential and produces an action potential. Following repolarization, the membrane again begins to depolarize (Fig. 10-34), so that a sequence of action potentials occurs, the frequency being determined by the rate at which the membrane depolarizes and the nearness of the membrane potential to threshold. The potential change that occurs during the spontaneous depolarization to threshold is known as the **pacemaker potential**. (As will be described later, certain cardiac muscle fibers and some neurons in the central nervous system also have pacemaker potentials and can spontaneously generate action potentials.)

Some types of smooth muscle undergo slow oscillations in membrane potential, with cycle times of seconds or even minutes. These **slow wave potentials** appear to arise from cyclical changes in the rates at which sodium ions are actively transported out of the cell by electrogenic ion pumps. Increasing the pump rate hyperpolarizes the membrane, while decreasing the pump rate depolarizes the

membrane. Bursts of action potentials occur at the peaks of the slow wave potential if it passes the threshold potential (Fig. 10-35). It should be noted that pacemaker potentials and slow wave potentials do not occur in all types of smooth muscle cells.

Nerves and hormones. Some smooth muscles are sensitive to a variety of hormones in addition to being innervated by both sympathetic and parasympathetic fibers; others are innervated by fibers from only one autonomic division, and still others receive no innervation at all. Unlike skeletal muscle, smooth muscle fibers do not have a specialized motor-end-plate region. As the axon of a postganglionic autonomic neuron enters the region of smooth muscle fibers, it divides into numerous branches, each branch containing a series of swollen regions known as varicosities. Each varicosity contains numerous vesicles filled with neurotransmitter, which is released when an action potential conducted along the axon passes the varicosity.

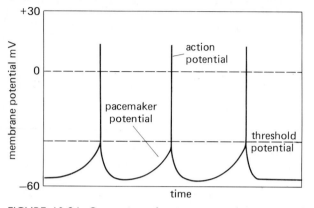

FIGURE 10-34. Generation of action potentials in a smooth muscle fiber resulting from spontaneous depolarizations of the membrane potential (pacemaker potentials).

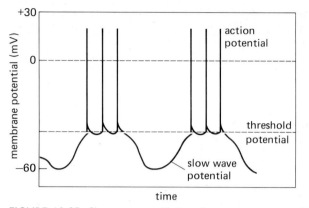

FIGURE 10-35. Slow wave oscillations in membrane potential trigger bursts of action potentials in smooth muscle plasma membrane.

Several varicosities from a single axon may be located along a muscle fiber, and a single muscle fiber may be located near varicosities belonging to postganglionic fibers of both sympathetic and parasympathetic neurons (Fig. 10-36). The concentration of released neurotransmitter that reaches the surface of a muscle fiber depends on the distance between the varicosity and the muscle surface, since the concentration of the neurotransmitter decreases as it diffuses away from its site of release. Since the response of the smooth muscle fiber depends on the number of receptor sites occupied by neurotrans-

mitter and thus upon the concentration of neurotransmitter, the magnitude of a smooth muscle response to nerve stimulation depends in part on how close the nerve varicosities are to the fiber surface.

Neurotransmitters, hormones, and local paracrine agents bind to specific receptor sites on the smooth muscle plasma membranes, and this binding leads to changes in cytosolic calcium by way of changes in the smooth muscle membrane potential and/or potential-independent alterations of membrane calcium channels.

When a particular hormone or neurotransmit-

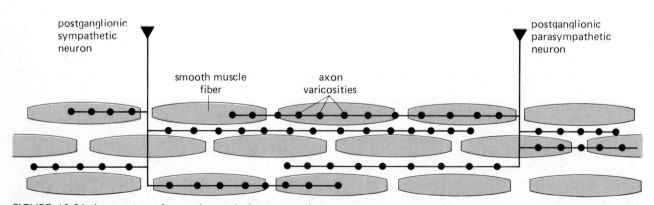

FIGURE 10-36. Innervation of smooth muscle by postganglionic autonomic neurons. Neurotransmitter is released from the varicosities along the branched axons adjacent to or overlapping smooth muscle fibers. (Pattern of innervation is typical of the multiunit type of smooth muscle.)

ter exerts its effect by depolarizing the plasma membrane, it may produce a range of responses, depending on its concentration at the membrane surface and the type of muscle fiber involved. Small depolarizations produced by a low concentration of neurotransmitter may increase the tonic tension in the fiber without producing action potentials or phasic contractions. If the fiber is spontaneously active, such a depolarization may increase the frequency of spontaneous action potentials, thereby increasing the intensity of the contractions. Finally, if the depolarization is great enough, it may initiate action potentials and phasic contractions in a fiber that does not have spontaneous electrical activity.

Whereas some hormones and neurotransmitters depolarize smooth muscle membranes, others produce a hyperpolarization which leads to a decreased cytosolic calcium concentration, producing a lessening of contractile activity. Thus, in contrast to skeletal muscle, which receives only excitatory input from its motor neurons, smooth muscle tension can be either increased or decreased by neural activity and circulating hormones. Moreover, a given neurotransmitter or hormone may produce opposite effects in different smooth muscle tissues. For example, norepinephrine, the neurotransmitter released from most postganglionic sympathetic neurons, depolarizes certain types of vascular smooth muscle, thereby initiating action potentials and contraction of the muscle; in contrast, it inhibits spontaneous activity in intestinal smooth muscle and produces a relaxation (Fig. 10-37). Thus, the nature of the response depends on events initiated by the binding of the chemical messenger to receptor sites in the membrane and is not an inherent property of the agent, itself.

Local factors. Local factors, including paracrines, acidity, oxygen concentration, osmolarity, and the ion composition of the extracellular fluid can also alter smooth muscle tension by affecting cytosolic calcium concentration. This responsiveness to local stimuli provides a means for altering smooth muscle tension in response to changes in the internal environment of a tissue, which in turn can lead to a

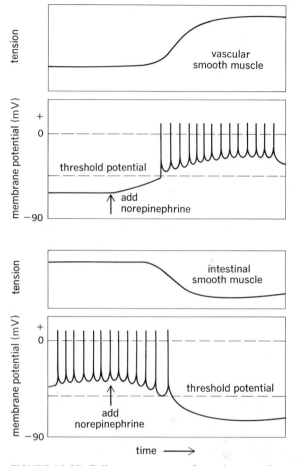

FIGURE 10-37. Different responses of vascular smooth muscle and intestinal smooth muscle to norepinephrine released from a sympathetic nerve ending.

local regulation of the internal environment that is independent of nerves and hormones. Examples involving the local regulation of blood flow in tissues and air flow in the lungs will be described in later chapters.

Classification of smooth muscle

The great diversity of smooth muscle characteristics has made it difficult to classify smooth muscles in a way comparable to the three types of skeletal muscle fibers. One of the broadest classifications divides smooth muscles into two groups based on the excitability characteristics of the muscle membrane and on the conduction of electrical activity from fiber to fiber within the muscle tissue: (1) *single-unit smooth muscles* whose membranes are capable of propagating action potentials from fiber to fiber and may manifest spontaneous action potentials and (2) *multiunit smooth muscles* which exhibit little, if any, propagation of electrical activity from fiber to fiber and whose contractile activity is closely coupled to the neural activity to them.

Single-unit smooth muscle. The smooth muscles of the intestinal tract, the uterus, and small-diameter blood vessels are examples of single-unit smooth muscles. All the muscle fibers in a single-unit smooth muscle undergo synchronous activity, both electrical and mechanical; i.e., the whole muscle responds to stimulation as a single unit. Each muscle fiber is linked to adjacent fibers by gap junctions (see Chap. 6), through which small ions, such as potassium and chloride ions, can move from cell to cell, carrying electric current. Thus, action potentials occurring in one cell are propagated by local current flows through these gap junctions into adjacent cells. Therefore, electrical activity occurring anywhere within a group of single-unit smooth muscle fibers can be conducted to all the other cells that are connected by gap junctions. Some of the fibers in a single-unit muscle are pacemaker cells that spontaneously generate action potentials, which are conducted by way of gap junctions into fibers which do not spontaneously generate action potentials (see Fig. 10-38).

The contractile activity of single-unit smooth muscles can be altered by nerves, hormones, and local factors using the variety of mechanisms described previously for smooth muscles in general. The extent to which these muscles are innervated varies considerably in different organs, and some are not innervated at all (Fig. 10-39). The nerve terminals are often restricted to the regions of the muscle containing pacemaker cells. By regulation of the activity of pacemaker cells, the activity of the entire muscle can be controlled.

One additional characteristic of single-unit smooth muscle is that a contractile response is often

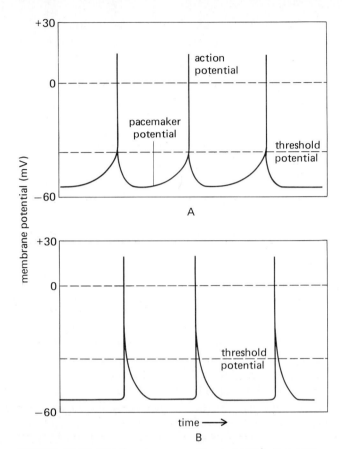

FIGURE 10-38. (A) Spontaneous action potentials in a pacemaker cell. (B) Action potentials in a nonpacemaker cell connected to above pacemaker cell by gap junctions.

induced by stretching the muscle. This may result from membrane depolarization caused by the stretch. In several hollow organs, stretching the smooth muscles in the walls of the organ as a result of increases in the volume of material in the lumen initiates a contractile response that opposes the increase in volume.

Multiunit smooth muscles. The smooth muscle in the large airways to the lungs, in large arteries, and attached to the hairs in the skin are examples of multiunit smooth muscles. These smooth muscles have few gap junctions, thus, each muscle fiber responds independently of its neighbors, and the muscle behaves as multiple units. Multiunit smooth muscles are richly innervated by branches of the

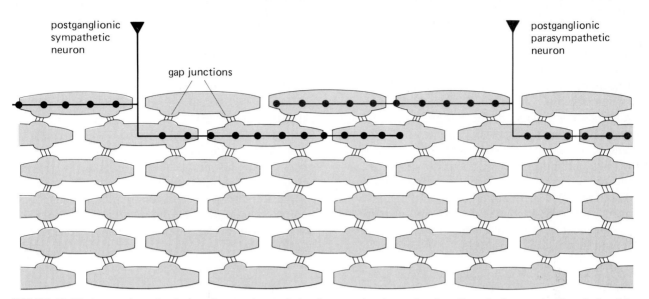

FIGURE 10-39. Innervation of a single-unit smooth muscle is often restricted to only a few fibers in the muscle. Electrical activity is conducted from fiber to fiber throughout the muscle by way of the gap junctions between the fibers.

TABLE 10-5. Characteristics of muscle fibers

Characteristic	Skeletal muscle	Smooth muscle		Cardiac muscle
		Single-unit	Multiunit	
Thick and thin filaments	Yes	Yes	Yes	Yes
Sarcomeres—banding pattern	Yes	No	No	Yes
Transverse tubules	Yes	No	No	Yes
Sarcoplasmic reticulum (SR)*	+ + + +	+	+	+ +
Gap junctions between fibers	No	Yes	Few	Yes
Source of activating calcium	SR	SR and extracellular	SR and extracellular	SR and extracellular
Site of calcium regulation	Troponin	Myosin	Myosin	Troponin
Speed of contraction	Fast-slow	Very slow	Very slow	Slow
Spontaneous production of action potentials by pacemakers	No	Yes	No	Yes
Presence of slow wave potentials	No	Yes	No	No
Tone (low levels of maintained tension in the absence of external stimuli)	No	Yes	No	No
Effect of nerve stimulation	Excitation	Excitation or inhibition	Excitation or inhibition	Excitation or inhibition
Physiological effects of hormones on excitability and contraction	No	Yes	Yes	Yes
Stretch of fiber produces contraction	No	Yes	No	Yes

* Number of plus signs (+) indicates the amount of endoplasmic reticulum present in a given muscle type.

autonomic nervous system. A number of smooth muscle fibers are influenced by the neurotransmitter released by a single nerve fiber, and a single smooth muscle fiber may be influenced by neurotransmitter from more than one neuron (Fig. 10-39). The contractile response of the whole muscle depends on the number of muscle fibers that are activated and on the frequency of nerve stimulation. Although stimulation of the nerve fibers to the muscle leads to a depolarization and a contractile response, action potentials do not occur in most multiunit smooth muscles. Circulating hormones can also increase or decrease contractile activity in multiunit smooth muscle, but stretch has no effect on contraction in this type of muscle.

Summary

Table 10-5 provides a comparison of some of the properties of the different types of muscle. Cardiac muscle has been included for completeness; its properties will be discussed in Chap. 11. In brief, cardiac muscle has a filament organization very similar to that of skeletal muscle, with myofibrils, sarcomeres, transverse tubules, and a well-developed sarcoplasmic reticulum. Like single-unit smooth muscle, some cardiac muscle fibers act as pacemakers to generate action potentials spontaneously which are then propagated throughout the heart by way of gap junctions between the cardiac fibers.

PART · THREE

COORDINATED
BODY
FUNCTIONS

CIRCULATION

11

Beyond the distance of a few cell diameters, diffusion is not sufficiently rapid to meet the metabolic requirements of cells. For multicellular organisms to evolve to a size larger than a microscopic cluster of cells, some mechanism other than simple diffusion was needed to transport molecules rapidly over the long distances separating the cells from the body's surface and between the various specialized tissues and organs. This problem was solved in the animal kingdom by the **circulation**, which comprises the **blood**, the set of tubes (**blood vessels**) through which the blood flows, and a pump (the **heart**) which produces this flow; the heart and blood vessels together are termed the **cardiovascular system**.

SECTION A.
BLOOD

Blood is composed of specialized cells and a liquid, **plasma**, in which they are suspended. The cells are the **erythrocytes** (or red blood cells), the **leukocytes** (or white blood cells), and the **platelets** (which are really cell fragments). Ordinarily, the constant motion of the blood keeps the cells well dispersed throughout the plasma, but if a sample of blood is allowed to stand (with clotting prevented), the cells slowly sink to the bottom. This process can be speeded-up by centrifuging (Fig. 11-1), which permits determination of the **hematocrit**, i.e., the percentage of total blood volume which is erythrocytes (the vast majority of all blood cells is erythrocytes). The normal hematocrit is approximately 45 percent in men and 42 percent in women.

From a knowledge of the hematocrit and an estimate of the total volume of blood in a person, one can calculate the total volumes of plasma and

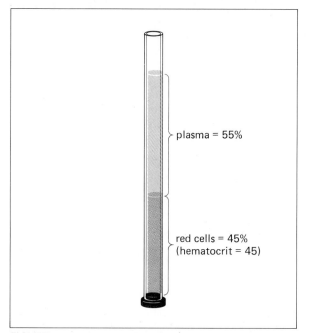

FIGURE 11-1. If a blood-filled capillary tube is centrifuged, the red cells become packed in the lower portion, and the percentage of red cells (hematocrit) can be determined.

erythrocytes. The total blood volume of an average person is approximately 8 percent of his or her total body weight. Accordingly, for a 70-kg person

$$\text{Total blood weight} = 0.08 \times 70 \text{ kg} = 5.6 \text{ kg}$$

One kilogram of blood occupies approximately one liter; therefore,

$$\text{Total blood volume} = 5.6 \text{ L}$$

If the hematocrit is 45 percent, then

$$\text{Total erythrocyte volume} = 0.45 \times 5.6 \text{ L} = 2.5 \text{ L}$$
$$\text{Plasma volume} = 5.6 \text{ L} - 2.5 \text{ L} = 3.1 \text{ L}$$

Plasma

Plasma consists of a large number of organic and inorganic substances dissolved in water (Table 11-1). The most abundant of these substances by weight are the proteins, which together compose approximately 7 percent of the total weight of plasma. These **plasma proteins** can be classified, according to certain physical and chemical reactions, into three broad groups: the **albumins**, the **globulins**, and **fibrinogen**. The albumins are by far the most abundant of the three groups. Most plasma proteins are synthesized by the liver, the major exception being the group known as **immunoglobulins**, which are formed in the lymph nodes and other lymphoid tissues (Chap. 17). The plasma proteins serve a host of important functions, which are summarized in Table 11-1 and which will be described in relevant chapters, but it must be emphasized that normally they are *not* taken up by cells and utilized as metabolic fuel. Accordingly, they must be viewed quite differently from most other organic constituents of plasma, such as glucose, which use the plasma as a vehicle for transport but function in cells. The plasma proteins function in the plasma itself or in the interstitial fluid. Finally, plasma is distinguished from **serum**, which is plasma from which fibrinogen and several other proteins involved in clotting have been removed as a result of clotting.

In addition to the organic solutes—proteins, nutrients, and metabolic end products—plasma contains a large variety of mineral electrolytes, the concentrations of which are also shown in Table 11-1. Comparison of the concentrations in millimoles per liter for these electrolytes and proteins may cause puzzlement in view of the previous statement that protein is the most abundant plasma solute by *weight*. Remember, however, that molarity is a measure not of the weight but the *number* of molecules or ions per unit volume. Protein molecules are so large in comparison with sodium ions that a very small number of them greatly outweighs a much larger number of sodium ions.

Cellular Elements of the Blood

Erythrocytes

Each milliliter of blood contains approximately 5 billion erythrocytes, and the total number of erythrocytes in the human body is about 25 trillion.

TABLE 11-1. Constituents of arterial plasma

Constituent	Amount/Concentration	Major Functions
Water	93% of plasma weight	Medium for carrying all other constituents
Electrolytes (inorganic):	Total < 1% of plasma weight	Keep H_2O in extracellular compartment; act
Na^+	142 mM	as buffers; function in membrane excitability
K^+	4 mM	
Ca^{2+}	2.5 mM	
Mg^{2+}	1.5 mM	
Cl^-	103 mM	
HCO_3^-	27 mM	
Phosphate (mostly HPO_4^{2-})	1 mM	
SO_4^{2-}	0.5 mM	
Proteins:	Total = 7.3 g/100 mL (2.5 mM)	Provide nonpenetrating solutes of plasma;
Albumins	4.5 g/100 mL	act as buffers; bind other plasma constitu-
Globulins	2.5 g/100 mL	ents (lipids, hormones, vitamins, metals,
Fibrinogen	0.3 g/100 mL	etc.); clotting factors; enzymes; enzyme pre-
		cursors; antibodies (immune globulins); hor-
		mones
Gases:		
CO_2	2 mL/100 mL plasma	
O_2	0.2 mL/100 mL	
N_2	0.9 mL/100 mL	
Nutrients:		
Glucose and other carbohydrates	100 mg/100 mL (5.6 mM)	
Total amino acids	40 mg/100 mL (2 mM)	
Total lipids	500 mg/100 mL (7.5 mM)	
Cholesterol	150–250 mg/100 mL (4–7 mM)	
Individual vitamins	0.0001–2.5 mg/100 mL	
Individual trace elements	0.001–0.3 mg/100 mL	
Waste products:		
Urea	34 mg/100 mL (5.7 mM)	
Creatinine	1 mg/100 mL (0.09 mM)	
Uric acid	5 mg/100 mL (0.3 mM)	
Bilirubin	0.2–1.2 mg/100 mL (0.003–0.018 mM)	
Individual hormones	0.000001–0.05 mg/100 mL	

These cells have the shape of a biconcave disk, i.e., a disk thicker at the edge than in the middle, like a doughnut with a center depression on each side instead of a hole (Fig. 11-2). Their shape and small size (7 μm in diameter) have adaptive value in that oxygen and carbon dioxide, carriage of which is the major function of the erythrocytes, can rapidly dif-

fuse throughout the entire cell interior. The plasma membranes of erythrocytes contain specific proteins which differ from person to person, and these confer upon the blood its so-called type, as will be described in Chap. 17.

The outstanding characteristic of erythrocytes is the presence of the iron-containing protein **hemo-**

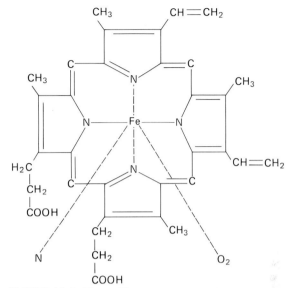

FIGURE 11-3. Heme. The nitrogen atom, N, at the lower left attaches to a polypeptide chain of globin to form a subunit of hemoglobin. Four of these make a single hemoglobin molecule.

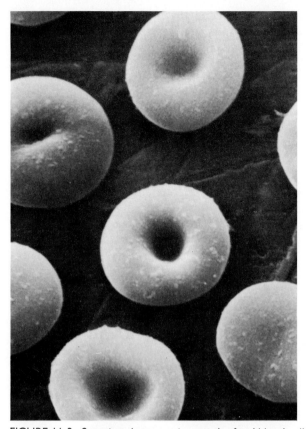

FIGURE 11-2. Scanning electron micrograph of red blood cells. *(From R. G. Kessel and C. Y. Shih, "Scanning Electron Microscopy in Biology," Springer-Verlag, New York, 1974, p. 265.)*

globin, which binds oxygen and constitutes approximately one-third of the total weight of the erythrocyte. Each hemoglobin molecule is made up of four subunits, each subunit consisting of an organic molecule known as **heme** attached to a polypeptide; the four polypeptides of each hemoglobin molecule are collectively called **globin**. It is the heme portions of the molecule (Fig. 11-3) which contain iron (Fe), and it is this iron that binds oxygen.

The amino acid sequences in the polypeptide chains of hemoglobin are, of course, genetically determined, and mutant genes cause the production of abnormal hemoglobins, some of which may cause serious malfunction. For example, **sickle-cell anemia** is caused by the presence of a particular form of abnormal hemoglobin molecules which, at the low oxygen concentrations existing in many capillaries, interact with each other to form fiberlike structures; these distort the erythrocyte membrane and cause the cell to form sickle shapes or other bizarre forms. This results both in the blockage of the capillaries, with consequent tissue damage, and in the destruction of the deformed erythrocytes, with consequent anemia. All this because of a change in one of 287 amino acids in the globin molecule!

There are other abnormal hemoglobins which cause changes in the structure of the erythrocyte, making it more fragile and likely to be damaged or destroyed prematurely. There are still others which alter the ability of the hemoglobin to bind oxygen; this leads not to erythrocyte destruction but to impairment of oxygen carriage and delivery to the tissues.

Another erythrocyte substance of great importance is the enzyme **carbonic anhydrase** which, as we shall see, facilitates the transportation of carbon dioxide.

Erythrocytes lack nuclei and organelles. Thus, they can neither reproduce themselves nor maintain their normal structure indefinitely. As essential enzymes in them deteriorate and are not replaced, the cells age and ultimately die. Fortunately, their oxygen-carrying ability is not significantly diminished during the aging period. The average life span of an erythrocyte is approximately 120 days, which means that almost 1 percent of all the erythrocytes in the body are destroyed every day. Destruction of erythrocytes is accomplished by a group of large phagocytic cells, the **macrophages**, found in liver, spleen, bone marrow, and lymph nodes, usually lining the blood vessels or lying close to them. These cells, the physiology of which is described in detail in Chap. 17, ingest and destroy the erythrocytes by breaking down their large, complex molecules.

In the process, hemoglobin molecules are catabolized to yield heme and the polypeptide chains of globin. Iron is then removed from the heme, and can be reused in the synthesis of new heme groups. The remainder of the heme is broken down in several steps to the yellow substance **bilirubin**, which is released into the blood, picked up by liver cells, and secreted into the bile; its further metabolism is described in Chap. 14.

The site of erythrocyte production is the soft, highly cellular interior of bones called **bone marrow**. The erythrocytes are descended from bone-marrow cells which contain no hemoglobin but do have nuclei and are therefore capable of cell division. After several cell divisions, cells emerge which are identifiable as immature erythrocytes, because they contain hemoglobin. As maturation continues, these cells accumulate increased amounts of hemoglobin and lose their nuclei (newly produced erythrocytes still contain a few ribosomes which produce a blue web when treated with special stains, an appearance which gives young erythrocytes the name **reticulocyte**). The mature erythrocytes leave the bone marrow and enter the general circulation.

This growth process requires the usual nutrients: amino acids, lipids, and carbohydrates. In addition, certain growth factors, including vitamin B_{12} and folic acid, are essential. Finally, the production of an erythrocyte requires the materials which go into the making of hemoglobin, particularly iron. A lack of any of these growth factors or raw materials results in the failure of normal erythrocyte formation. The substances which are most commonly lacking are iron, folic acid, and vitamin B_{12}.

Iron. Iron is obviously an essential component of the hemoglobin molecule since it is to this element that oxygen binds. The balance of iron and its distribution in the body are shown schematically in Fig. 11-4. About 70 percent of the total body iron is in hemoglobin, and the remainder is stored primarily in the liver, spleen, and bone marrow. As erythrocytes are destroyed, most of the iron released from hemoglobin is returned to these depots, from which it can again be released for hemoglobin synthesis. Small amounts of iron, however, are lost each day via the urine, feces, sweat, and cells sloughed from the skin. In addition, women lose a significant quantity of iron via menstrual blood. In order to remain in iron balance, the amount of this metal lost from the body must be replaced by ingestion of iron-containing foods. A significant upset of this balance may result either in iron deficiency leading to inadequate hemoglobin production or in an excess of iron in the body, with serious toxic effects.

The homeostatic control of iron balance resides

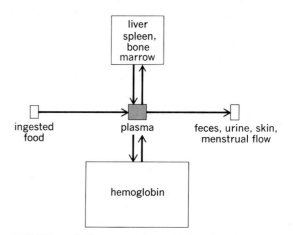

FIGURE 11-4. Summary of iron balance. The sizes of the boxes represent the quantities of iron involved. The magnitude of exchange in a particular direction varies according to conditions.

primarily in the intestinal epithelium, which actively absorbs the iron from ingested food. Only a small fraction of ingested iron is absorbed, but more important, this fraction is increased or decreased depending upon the state of body iron balance. These fluctuations appear to be mediated by changes in the iron content of the intestinal epithelium itself, as described in Chap. 14.

Iron has the tendency to form insoluble complexes with a variety of substances, but this problem has been circumvented in the body by the evolution of two iron-binding proteins which act as carrier molecules for transferring iron in the plasma (**transferrin**) and as the storage molecules for iron (**ferritin**).

Vitamin B_{12} and folic acid. Normal erythrocyte formation requires extremely small quantities (one-millionth of a gram per day) of a cobalt-containing molecule, **vitamin B_{12}**, which is required for final maturation of the erythrocyte. This substance is not synthesized in the body, which therefore depends for its supply upon dietary intake. (Absorption of vitamin B_{12} from the gastrointestinal tract into the blood is described in Chap. 14.) Dietary deficiency of vitamin B_{12} or impaired intestinal absorption of it leads to the failure of erythrocyte proliferation and maturation known as **pernicious anemia**. Since vitamin B_{12} is also needed for normal functioning of the nervous system, specifically for myelin formation, a variety of neurological symptoms may accompany the anemia.

Folic acid is a vitamin essential for the formation of DNA because it is involved in the synthesis of purines and thymine. Accordingly, its deficiency also results in failure of normal erythrocyte proliferation and maturation, and in poor growth and development of many other tissues, as well.

Regulation of erythrocyte production. In a normal person, the total volume of circulating erythrocytes remains remarkably constant. Such constancy is required for delivery of oxygen to the tissues. A constant number of circulating erythrocytes is maintained by balancing erythrocyte production with the sum of destruction and loss. During periods of severe erythrocyte destruction or hemorrhage, the rate of erythrocyte production can be increased more than sixfold.

It is easiest to discuss the regulation of erythrocyte production by first naming the mechanisms not involved. In the previous section, we listed a group of nutrient substances, such as iron and vitamin B_{12}, which must be present for normal erythrocyte production. However, none of these substances actually *regulates* the rate of erythrocyte production.

The direct control of erythrocyte production (erythropoiesis) is exerted by a hormone called **erythropoietin**, which is secreted by endocrine cells in the kidneys (and probably other organs as well). Erythropoietin acts on the bone marrow to stimulate the maturation and proliferation of erythrocytes. The common denominator of changes which increase the secretion of erythropoietin is a decreased oxygen delivery to the kidneys. As will be described in Chap. 12, decreased oxygen delivery can result from a decrease in the number of erythrocytes or their hemoglobin content, from decreased blood flow, or from decreased oxygen delivery by the lungs into the blood. As the result of the stimulus-induced increase in erythropoietin secretion, plasma erythropoietin increases, erythrocyte production increases, the oxygen-carrying capacity of the blood is increased, and oxygen delivery to the tissues is returned toward normal (Fig. 11-5). Testosterone, the major male sex hormone, also stimulates the basal release of erythropoietin, and this may account, at least in part, for the fact that the hemoglobin concentration in men is higher than the concentration in women.

Anemia. Anemia is a reduced total blood hemoglobin. It may be due to a decrease in the total number of erythrocytes (each having a normal quantity of hemoglobin), or to a diminished concentration of hemoglobin per erythrocyte, or to a combination of both. Anemia has a wide variety of causes, including dietary deficiencies (of iron, vitamin B_{12}, or folic acid), bone-marrow failure (due to toxic drugs and cancer, for example), excessive blood loss from the body, excessive destruction of erythrocytes (as in

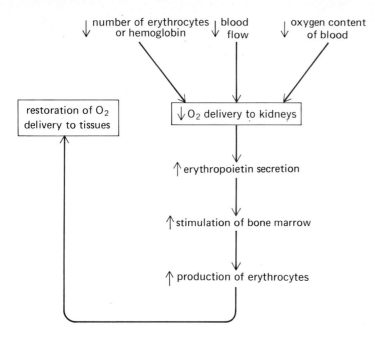

↓ number of erythrocytes   ↓ blood   ↓ oxygen content
   or hemoglobin   flow   of blood

↓ O₂ delivery to kidneys

restoration of O₂
delivery to tissues

↑ erythropoietin secretion

↑ stimulation of bone marrow

↑ production of erythrocytes

FIGURE 11-5. Role of erythropoietin in regulation of erythrocyte production and oxygen delivery.

sickle-cell anemia), and inadequate secretion of erythropoietin (as in kidney disease).

Leukocytes

If one takes a drop of blood, adds appropriate dyes, and examines it under a microscope, the various cell types can be seen (Fig. 11-6). Over 99 percent of all the cells are erythrocytes (Table 11-2). The remaining cells, the **leukocytes** (white blood cells), all have as their major function defense against foreign cells (such as bacteria) and other foreign matter. Unlike the erythrocytes, the major functions of leukocytes are exerted not within the blood vessels but in the interstitial fluid; i.e. leukocytes utilize the circulatory system only as the route for reaching a

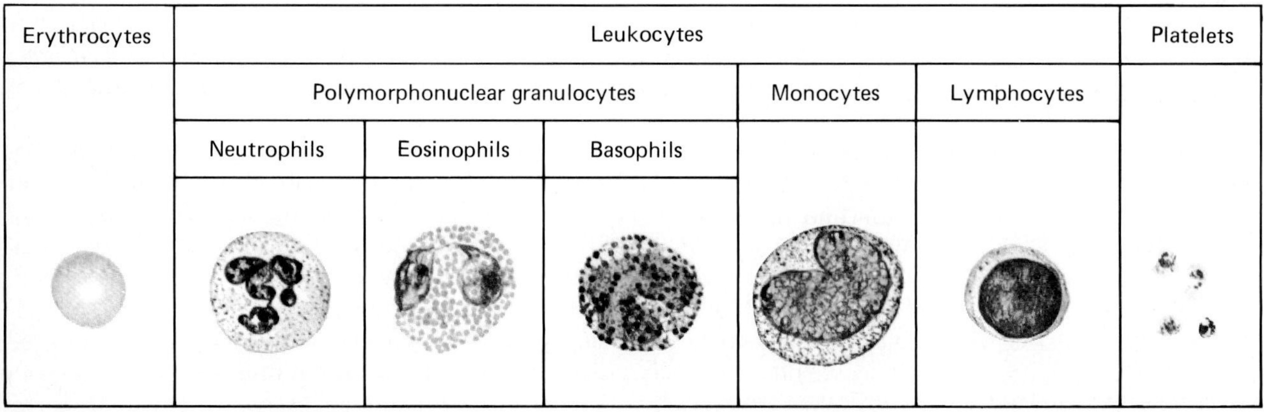

Erythrocytes	Leukocytes					Platelets
	Polymorphonuclear granulocytes			Monocytes	Lymphocytes	
	Neutrophils	Eosinophils	Basophils			

FIGURE 11-6. Normal blood cell types.

TABLE 11-2. Numbers and distributions of erythrocytes, leukocytes, and platelets in normal human blood

Total erythrocytes = 5,000,000,000 cells per milliliter of blood

Total leukocytes = 7,000,000 cells per milliliter of blood

Percent of total leukocytes:
    Polymorphonuclear granulocytes
        Neutrophils 50–70
        Eosinophils 1–4
        Basophils 0.1
    Mononuclear cells
        Monocytes 2–8
        Lymphocytes 20–40

Total platelets = 250,000,000 per milliliter of blood.

damaged or invaded area. Once there, they leave the blood vessels to enter the tissue and perform their functions (as we shall see in Chap. 17, the story for lymphocytes, one class of leukocyte, is more complicated than this). The specific roles of the various types of leukocytes are described in detail in Chap. 17, but we introduce them here to complete our survey of blood.

Leukocytes are classified according to their structure and affinity for various dyes. The name **polymorphonuclear granulocytes** refers to the three types of leukocytes with lobulated nuclei and abundant membrane-bound granules. The granules of one group show no dye preference, and the cells are therefore called **neutrophils**. The granules of the second group take up the red dye eosin, thus giving the cells their name **eosinophils**. Cells of the third group have an affinity for a basic dye and are called **basophils**. All three types of granulocytes are produced in the bone marrow, and their primary function is **phagocytosis**, the ingestion and destruction of particulate material. Basophils, in addition, can release powerful chemicals, such as histamine, which contribute to tissue damage and allergy. The basophil is virtually identical to the group of cells known as mast cells which are found in connective tissue throughout the body and do not circulate.

A fourth type of leukocyte quite different in appearance from the three granulocytic types is the **monocyte**. These cells, which are also produced by the bone marrow, are somewhat larger than the granulocytes and have a single oval or horseshoe-shaped nucleus and relatively few cytoplasmic granules. Upon entering an area which has been invaded or wounded, monocytes may function, themselves, as phagocytes, or they may be transformed into macrophages, yet another type of phagocytic cell.

The final class of leukocyte is the **lymphocyte**, of which there are two major types, B and T. Their outstanding structural features are a relatively large nucleus and scanty surrounding cytoplasm. Despite the fact that lymphocytes can originate in the bone marrow and thymus, most of them are housed in other lymphoid (i.e., lymphocyte-containing) tissues such as the lymph nodes, spleen, and tonsils. These tissues accumulate many of their lymphocytes by filtering them from the blood. In addition, the lymphoid tissues are able to form lymphocytes from cells that have migrated there from the bone marrow and thymus, and lymphoid tissues become very active lymphocyte producers as part of the body's response to infection. The circulating pool of lymphocytes continuously travels from the blood, across capillary walls to the lymph, thence to lymph nodes, and finally, back to the blood. Lymphocytes do not function as phagocytes but are responsible for antibody production and other specific defenses against foreign invaders.

Platelets

The circulating platelets are colorless corpuscles much smaller than erythrocytes and containing numerous granules (Fig. 11-7). There are approximately 250 million per milliliter of blood (compare this with the number of erythrocytes and leukocytes given in Table 11-2). The platelets, which are actually cell fragments and lack nuclei, originate from large cells (**megakaryocytes**) in bone marrow. Platelets are portions of the cytoplasm of these cells which are pinched off and which enter the circulation. They play crucial roles in blood clotting, as will be described in a subsequent section.

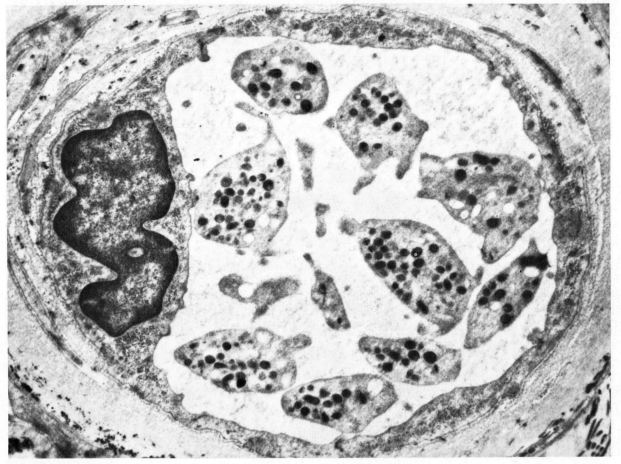

FIGURE II-7. Electron micrograph of a capillary containing several blood platelets in its lumen. Note the numerous granules and lack of nuclei.

Regulation of cell production

It should be clear by now that the bone marrow is the major site for formation of most blood cells. In children, the bone marrow of almost all bones produces blood cells, but by adulthood only the bones of the chest, the base of the skull, and the upper limbs remain active. Taken altogether, the bone marrow is one of the largest organs of the body, weighing almost as much as the liver, and it produces cells at an enormous rate.

What regulates the rate of production of each type of blood cell? We have already described the regulatory process for one cell type, the erythrocyte, production of which is controlled by the hormone erythropoietin. The production of each of the other cell types is also controlled by one or more specific circulating chemical messengers. However, at present these control systems for the different leukocytes and for platelets are still poorly understood.

SECTION B.
OVERALL DESIGN OF THE CARDIOVASCULAR SYSTEM

Rapid flow of blood through all parts of the body via the blood vessels is produced by pressures created by the pumping action of the heart. This flow is known as "bulk flow" since all constituents of the blood move together. The extraordinary degree of branching of these vessels assures that all cells of the body are within a few cell diameters of at least one of the smallest branches, the **capillaries**. These distances are small enough for diffusion across the walls of the capillaries (Fig. 11-8) and through the interstitial fluid to provide for the exchange of nutrients and metabolic end products between the capillary blood and the interstitial fluid immediately surrounding the cells near the capillary. (Exchanges between the interstitial fluid and the cells, themselves, are by both diffusion and mediated transport, as we shall see.)

Physiology as an experimental science began in 1628, when William Harvey demonstrated that the cardiovascular system forms a circle, so that blood is continuously being pumped out of the heart through one set of vessels and returned to the heart via a different set. There are actually two circuits (Fig. 11-9), both originating and terminating in the heart, which is divided longitudinally into two func-

tional halves. Blood is pumped via one circuit (the **pulmonary circulation**) from the right half of the heart through the lungs and back to the left half of the heart. It is pumped via the second circuit (the **systemic circulation**) from the left half of the heart through all the tissues of the body (except, of course, the lungs) and back to the right half of the heart. In both circuits, the vessels carrying blood away from the heart are called **arteries**, and the vessels carrying blood from the lungs or peripheral tissues back to the heart are called **veins**.

In the systemic circuit, blood leaves the left half of the heart via a single large artery, the **aorta**.

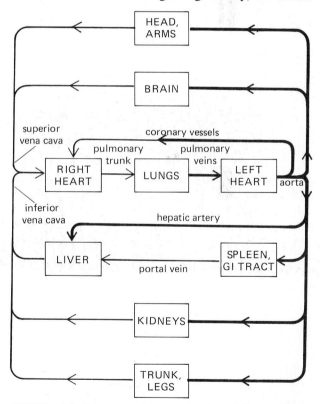

FIGURE 11-9. Diagrammatic representation of the cardiovascular system in the adult human being. Darker shading indicates blood with high oxygen content. The commonly used terms "right heart" and "left heart" refer to the right and left halves of the heart, which is a single organ.

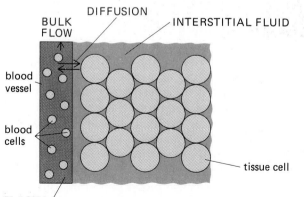

FIGURE 11-8. Bulk flow brings blood to the capillaries, whereas diffusion links plasma and the interstitial fluid.

The systemic arteries branch from the aorta, dividing in a highly characteristic manner into progressively smaller branches. The smallest arteries branch into **arterioles**. The arterioles branch into a huge number of very small, thin vessels, termed **capillaries**, which unite to form larger vessels—**venules**. (The arterioles, capillaries, and venules are collectively termed the **microcirculation**.) The venules unite to form fewer and still larger vessels, the **veins**. The veins from different organs and tissues unite to form two large veins, the **inferior vena cava** (from the lower portion of the body) and the **superior vena cava** (from the upper half of the body). By these two veins blood is returned to the right half of the heart. The entire systemic circuit can be visualized, therefore, as two trees, one arterial and the other venous, having the same origin (the heart) and being connected by fine twigs (capillaries) which unite the smallest branches of each tree (Fig. 11-10).

The pulmonary circulation is composed of a similar circuit. Blood leaves the right half of the heart via a single large artery, the **pulmonary trunk**, which divides into the two **pulmonary arteries**, one supplying each lung. In the lungs, the arteries continue to branch, ultimately dividing into capillaries. These capillaries unite to form small venules, which unite to form larger and larger veins. The blood leaves the lungs via the largest of these, the **pulmonary veins**, which empty into the left half of the heart. The blood flowing through the systemic veins, right half of the heart, and pulmonary arteries has a relatively low oxygen content. As this blood flows through the lung capillaries, it picks up large quantities of oxygen; therefore, the blood in the pulmonary veins, left heart, and systemic arteries has a high oxygen content. As this blood flows through the capillaries of tissues and organs throughout the body, some of this oxygen leaves the blood, resulting in the lower oxygen content of systemic venous blood.

In a normal person, blood can pass from the systemic veins to the systemic arteries only by first being pumped through the pulmonary circuit; thus all the blood returning from the body tissues is oxygenated before it is pumped back to them. Normally the total volumes of blood pumped through the pulmonary and systemic circuits during a given period of time are equal. In other words, the right heart pumps the same amount of blood as the left heart does. Only when blood flow is changing do these volumes differ from each other, and then only transiently.

Note that all the blood pumped by the right heart flows through the lungs; in contrast, only a fraction of the total left ventricular output flows through any single organ or tissue. In other words, the systemic circulation comprises numerous different pathways "in parallel." One significant deviation from this pattern is the blood supply to the liver, much of which is not arterial but venous blood from the **portal vein**, which drains the pancreas, the spleen, and the gastrointestinal tract. The typical distribution of the left ventricular output for a normal adult at rest is given in Fig. 11-11.

The last task in this survey of the overall design of the cardiovascular system is to introduce the concepts of pressure, flow, and resistance. Flow of blood throughout the entire cardiovascular system is always from a region of higher pressure to one of lower pressure. The units for this flow are volume per unit time, usually liters per minute (L/min). The units still in general use for the pressure difference (ΔP) driving the flow are millimeters of mercury (mmHg) because blood pressures were measured historically by determining how tall a column of mercury could be supported by the pressure. It must be emphasized that it is not the absolute pres-

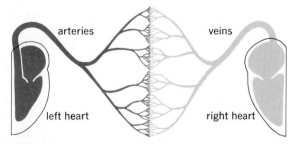

FIGURE 11-10. Systemic circulation shown as two trees connected by capillaries. As indicated by the color change, oxygen leaves the blood during passage through the systemic capillaries. *(Adapted from Rushmer.)*

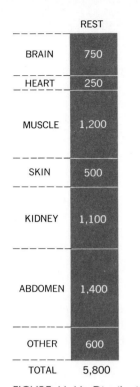

FIGURE 11-11. Distribution of blood flow to the various organs and tissues of the body at rest. The numbers show blood flow in milliliters per minute. *(Adapted from Chapman and Mitchell.)*

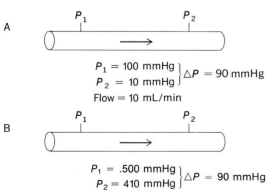

FIGURE 11-12. Flow between two points within a tube is proportional to the pressure difference between the points.

sure at any point in the cardiovascular system that determines flow but the *difference* in pressure between the relevant points (Fig. 11-12) (for example, between right atrium and right ventricle or between aorta and systemic capillaries).

But knowing the pressure difference between two points will not tell you the exact flow. For this, you need to know the **resistance** to flow *(R)*; i.e., how difficult it is for fluid to flow between two points at any given pressure difference. The basic equation relating these variables is, therefore:

$$F = \Delta P / R$$

In words, flow is directly proportional to the pres-

TABLE 11-3. Functions of components of the cardiovascular system

Component	Function
Heart	
Atria	Chambers through which blood flows from veins to ventricles; their contraction adds to ventricular filling but is not essential.
Ventricles	Chambers whose contractions produce the pressures which drive blood through the entire pulmonary and systemic vascular systems and back to the heart.
Vascular system	
Arteries	Low resistance conduits conducting blood to the various organs with little loss in pressure; they act as pressure reservoirs for maintaining blood flow between ventricular contractions.
Arterioles	Major sites of resistance to flow; responsible for the pattern of blood-flow distribution to the various organs.
Capillaries	Sites of exchange between blood and tissues.
Veins	Low-resistance conduits for blood flow back to the heart; their capacity for blood is adjusted so as to facilitate this flow.

sure difference and inversely proportional to the resistance. As we shall see, this equation is directly applicable to fluid dynamics in any portion of the cardiovascular system as well as to the overall system. We will describe the factors which determine resistance in a later section.

This completes our brief survey of the overall design of the cardiovascular system, and we now turn to a description of its components and their control. In so doing, we might very easily lose sight of the forest for the trees if we do not persistently ask of each section: "How does this component of the circulation contribute either to an adequate flow of blood to the various organs or to an adequate exchange of materials between blood and cells?" Table 11-3 provides a brief overview.

SECTION C.
THE HEART

Anatomy

The heart is a muscular organ located in the chest (**thoracic**) cavity and enclosed in a fibrous sac, the **pericardium**. The narrow space between the pericardium and the heart is filled with a watery fluid which serves as lubricant as the heart moves within the sac. The walls of the heart are composed primarily of cardiac muscle (**myocardium**). The inner surface of the myocardium, i.e., the surface in contact with the blood within the heart chambers, is lined by a thin layer of epithelial cells known as **endothelium**. Endothelium, as we shall see, lines the inner surface not only of the heart chambers but of the entire vascular tree as well.

The human heart is divided longitudinally into right and left halves (Fig. 11-13), each consisting of two chambers, an **atrium** and a **ventricle**. The cavities of the atrium and ventricle on each side of the heart communicate with each other, but the right chambers do not communicate directly with those on the left. Thus, right and left atria and right and left ventricles are distinct.

Perhaps the easiest way to picture the architecture of the heart is to begin with its fibrous skeleton, which comprises four interconnected rings of dense connective tissue (Fig. 11-14) which separate atria from ventricles. To the tops of these rings are anchored the muscle masses of the atria, pulmonary trunk (which divides into the pulmonary arteries), and aorta. To the bottom are attached the muscle masses of the ventricles. The connective tissue

rings form the openings between the atria and ventricles and between the ventricles and great arteries. To these rings are attached four sets of valves (Fig. 11-14).

Between the cavities of the atrium and ventricle in each half of the heart are the **atrioventricular valves (AV valves)**, which permit blood to flow from atrium to ventricle but not from ventricle to atrium (Fig. 11-15). The right AV valve is called the **tricuspid** and the left is the **mitral valve**. The opening

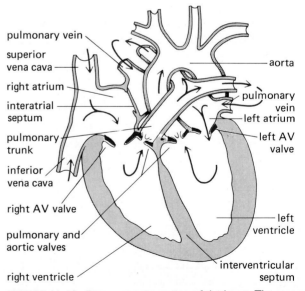

FIGURE 11-13. Diagrammatic section of the heart. The arrows indicate the direction of blood flow.

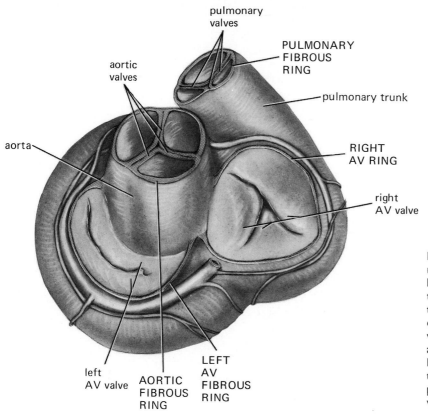

pulmonary
valves

PULMONARY
FIBROUS
RING

aortic
valves

pulmonary trunk

RIGHT
AV RING

aorta

right
AV valve

left
AV valve

AORTIC
FIBROUS
RING

LEFT
AV
FIBROUS
RING

FIGURE 11-14. Four connective-tissue rings compose the fibrous skeleton of the heart. To these are attached the valves, the trunks of the aorta and the pulmonary artery, and the muscle masses of the cardiac chambers. The right atrioventricular (AV) valve is formed of three cusps, or leaflets, and is also called the tricuspid valve. The left AV valve has two cusps and is also called the bicuspid or mitral valve. The aortic and pulmonary valves are also called semilunar valves because of their shape.

and closing of the valves is a passive process brought about by the difference in pressure across the valves. When the blood is moving from atrium to ventricle, the valves are pushed open, but when the ventricles contract, the valves are brought together by the increasing pressure of the blood in the ventricles, and the atrioventricular opening is closed. Blood is therefore forced into the pulmonary arteries (from the right ventricle) and into the aorta (from the left ventricle) instead of back into the atria. To prevent the valves themselves from being forced upward into the atrium, they are fastened by fibrous strands to muscular projections of the ventricular walls. These muscular projections do *not* open or close the valves; they act only to limit the valves' movements and prevent them from being everted.

The opening of the right ventricle into the pulmonary trunk, and of the left ventricle into the

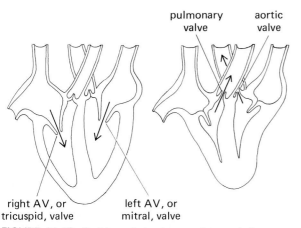

pulmonary
valve

aortic
valve

right AV, or
tricuspid, valve

left AV, or
mitral, valve

FIGURE 11-15. Position of the heart valves and direction of blood flow during ventricular relaxation (left) and ventricular contraction (right).

aorta, is also guarded by valves, the **pulmonary** and **aortic valves**, which permit blood to flow into these arteries during ventricular contraction but prevent reflux of blood in the opposite direction during ventricular relaxation. There are no valves at the entrances of the venae cavae into the right atrium and of the pulmonary veins into the left atrium; however, atrial contraction pumps very little blood back into the veins because atrial contraction compresses the veins at their sites of entry into the atria, increasing the resistance to backflow. (A little blood is ejected back into the veins and this accounts for the venous pulse which can often be seen in the neck veins when the atria are contracting.)

Fig. 11-16 summarizes the structures through which blood flows in passing from the systemic veins to the systemic arteries.

The myocardium is composed of layers of cardiac muscle which are tightly bound together and completely encircle the blood-filled chambers. Thus, when the walls of a chamber contract, they come together like a squeezing fist, thereby exerting pressure on the blood they enclose. Cardiac muscle cells combine certain of the properties of smooth and skeletal muscle. The individual cell is striated (Fig. 11-17) due to alignment of the repeating sarcomeres, which contain both thick myosin and thin actin filaments, as described for skeletal muscle. However, cardiac muscle cells are considerably shorter than the skeletal muscle fibers and have several branching processes. Adjacent cells are joined end to end at **intercalated disks**, within which are two types of membrane junctions: (1) desmosomes, which hold the cells together and to which the myofibrils are attached and (2) gap junctions, which allow the spread of action potentials from one cardiac muscle cell to another, in a manner similar to that in single-unit smooth muscle.

Besides the ordinary type of cardiac muscle shown in Fig. 11-17, certain areas of the heart contain specialized cardiac muscle fibers which are essential for normal excitation of the heart. They constitute a network known as the **conducting system** of the heart and are in contact with ordinary cardiac fibers via gap junctions, which permit passage of action potentials from the conducting system to the ordinary myocardial cells.

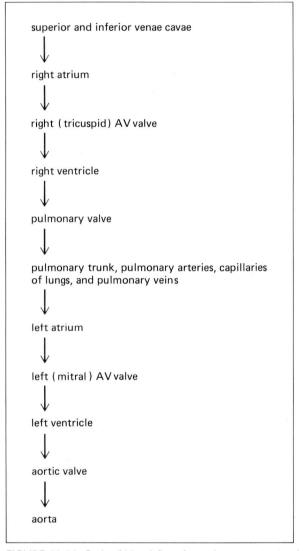

superior and inferior venae cavae

↓

right atrium

↓

right (tricuspid) AV valve

↓

right ventricle

↓

pulmonary valve

↓

pulmonary trunk, pulmonary arteries, capillaries of lungs, and pulmonary veins

↓

left atrium

↓

left (mitral) AV valve

↓

left ventricle

↓

aortic valve

↓

aorta

FIGURE 11-16. Path of blood flow from the systemic circulation through the pulmonary circulation and back to the systemic circulation.

The heart receives a rich supply of sympathetic and parasympathetic nerve fibers, the latter contained in the vagus nerves. The postganglionic sympathetic fibers terminate upon the cells of the specialized conducting system of the heart as well as on the ordinary myocardial cells of the atria and ventricles. The parasympathetic neurons also innervate the conducting system and the atrial myo-

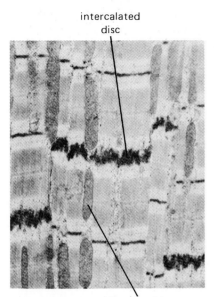

intercalated
disc

mitochondrion

FIGURE 11-17. Electron micrograph of cardiac muscle. Note the striations similar to those of skeletal muscle. The wide dark bands, called intercalated disks, occur where adjacent muscle fibers interdigitate. *(Courtesy of D. W. Fawcett.)*

cardial cells but not the ventricular myocardium to any great extent. The sympathetic postganglionic fibers release norepinephrine, and the parasympathetics release acetylcholine. The receptors for norepinephrine (and circulating epinephrine) on cardiac muscle are beta-adrenergic.

The blood within the heart chambers does not exchange nutrients and metabolic end products with the cells constituting the heart walls. The heart, like all other organs, receives its blood supply via arterial branches (**coronary arteries**) which arise from the aorta.

Heartbeat Coordination

Contraction of cardiac muscle, like that of other muscle types, is triggered by depolarization of the plasma membrane. What would happen if all the many muscle fibers in the heart were to depolarize and contract in a random manner? One result would be lack of coordination between pumping by each corresponding atrium and ventricle, but this defect is dwarfed by the more serious lack of muscle coordination in the ventricles. Random contraction of the many ventricular muscle cells would cause the blood to be sloshed back and forth within the ventricular cavities instead of being ejected into the aorta and pulmonary trunk. Thus, the complex muscle masses which form the ventricular pumps must contract more or less simultaneously for efficient pumping.

Such coordination is made possible by two factors already mentioned: (1) The gap junctions allow spread of an action potential from one muscle cell to the next, so that excitation in one cell spreads throughout the heart, and (2) the specialized conducting system in the heart facilitates the rapid and coordinated spread of excitation. Where and how does the action potential first arise, and what is the path and sequence of excitation?

Origin of the heartbeat

About 1 percent of cardiac muscle cells, like certain forms of smooth muscle, are autorhythmic; i.e., they are capable of autonomous rhythmical self-excitation. When individual cells of a heart are separated and placed in salt solution, some of the individual cells are seen to beat spontaneously. But they are beating at different rates. Figure 11-18 shows recordings of membrane potentials from two such cells, the most important feature of which is the gradual depolarization (**pacemaker potential**) causing the membrane potential to reach threshold, at which point an action potential occurs. Following the action potential, the membrane potential returns to the initial resting value, and the gradual depolarization begins again. It should be evident that the slope of this pacemaker potential, i.e., the rate of membrane potential change per unit time, determines how quickly threshold is reached and the next action potential elicited. Accordingly, cell *A* has a faster rate of firing than cell *B*.

When gap junctions are formed between two cells previously contracting autonomously at different rates, both the joined cells contract at the faster rate (Fig. 11-18). In other words, the faster cell sets the pace, causing the initially slower cell to contract at the faster rate. The action potential generated by

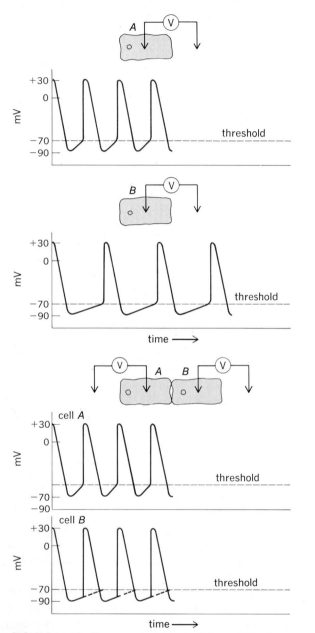

FIGURE 11-18. Transmembrane potential recordings from cardiac muscle cells grown in tissue culture. The dashed lines in the bottom recording of cell B indicate the course depolarization would have followed if the cells had not been joined.

the faster cell causes depolarization, via the gap junctions, of the second cell's membrane to threshold, at which point an action potential occurs in this second cell. Moreover, the second cell need not, itself, have been spontaneously active to be excited by the cell linked to it by gap junctions. The important generalization emerges that, because of the gap junctions between cardiac muscle cells, all the cells are excited at the rate set by the cell with the fastest autonomous rhythm.

Precisely the same explanation holds for the origination of the heartbeat in the intact heart. Several areas of the conducting system of the adult mammalian heart possess the capacity for autorhythmicity and pacemaking. The one with the fastest inherent rhythm (about 100 beats per minute) is a small mass of specialized myocardial cells embedded in the right atrial wall near the entrance of the superior vena cava (Fig. 11-19). Called the **sinoatrial (SA) node**, it is the normal pacemaker for the entire heart. Figure 11-20 is an intracellular recording from an SA-node cell; note the slow depolarization toward threshold (the **pacemaker potential**) which initiates the action potential. Compare this SA-nodal pattern to that of unspecialized, nonautorhythmical atrial cells, which fail to show a pacemaker potential.

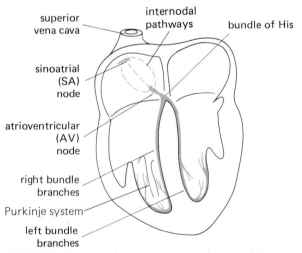

FIGURE 11-19. Conducting system of the heart. The interatrial conducting bundles are not shown in the figure. (Adapted from Rushmer.)

A

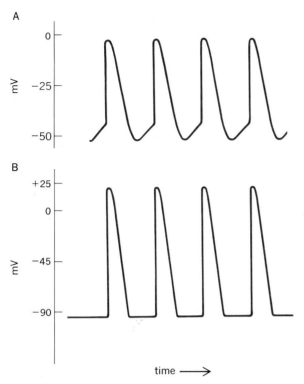

B

FIGURE 11-20. Transmembrane potential recordings from (A) SA-node cell and (B) atrial muscle fiber.

If the activity of the SA node is depressed or conduction from it blocked, another portion of the conducting system will take over as pacemaker. In contrast, ordinary atrial and ventricular muscle fibers, which constitute 99 percent of the total cardiac muscle, are not normally capable of generating pacemaker potentials.

What is the ionic basis of automaticity in the SA node; i.e., what is responsible for the rhythmically occurring pacemaker potential? The major factor for this gradual depolarization is a progressive "spontaneous" reduction in the membrane permeability of the SA-node cells to potassium.

Sequence of excitation

The cells of the SA node are connected by gap junctions with the surrounding atrial myocardial fibers. From the SA node, the wave of excitation spreads throughout both atria, passing from cell to cell by way of the gap junctions. Much of this spread is along ordinary atrial myocardial cells, but in addition there are several more-or-less specialized bundles of fibers which conduct the impulse directly from the SA node to another small mass of specialized cells, the **atrioventricular (AV) node**, located at the base of the right atrium. This node and the bundle of fibers (the **bundle of His**) leaving it constitute the only conducting link between the atria and ventricles, all other areas being separated by nonconducting connective tissue. This anatomical pattern ensures that excitation will travel from atria to ventricles only through the AV node and bundle of His, but it also means that malfunction of the AV node may completely dissociate atrial and ventricular excitation.

The AV node manifests one particularly important characteristic: The propagation of action potentials through the node is relatively slow (requiring approximately 0.1 s) mainly because of the small diameters of these cells. This delay allows the atria to contract and add additional blood to the ventricles before ventricular contraction.

After leaving the AV node (Fig. 11-19), the impulse enters the ventricles via the bundle of His (named for its discoverer); this divides into **right** and **left bundle branches**, which, in turn, branch into the **Purkinje fibers**, which then spread throughout much of the right and left ventricular myocardium. Finally, these fibers make contact with unspecialized myocardial cells through which the impulse spreads from cell to cell in the remaining myocardium. The rapid conduction along the specialized fibers and their highly diffuse distribution cause depolarization of all right and left ventricular cells more or less simultaneously and ensure a single coordinated contraction (Fig. 11-21).

There are many clinical examples of altered cardiac excitation which illuminate the normal physiological process. For example, disease may damage cardiac tissue, hampering conduction through the AV node, so that only a fraction of the atrial impulses are transmitted into the ventricles; thus, the atria may have a rate of 80 beats per minute and the ventricles only 60. If there is complete block at the AV node, none of the atrial im-

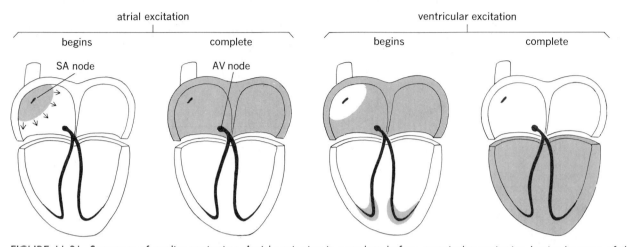

atrial excitation

begins | complete

SA node | AV node

ventricular excitation

begins | complete

FIGURE 11-21. Sequence of cardiac excitation. Atrial excitation is complete before ventricular excitation begins because of the delay at the AV node. For the sake of clarity, the internodal pathways (between SA and AV nodes) are not shown. *(Adapted from Rushmer.)*

pulses get through, and spontaneously excitable cells located just below the AV node begin to initiate excitation at their own spontaneous rate; i.e., a portion of the conducting system just below the AV node is capable of pacemaker action when released from capture by impulses from the SA node. This rate is quite slow, generally 25 to 40 beats per minute, and completely out of synchrony with the atrial contractions, which continue at the normal higher rate of the SA node. Under such conditions, the atria are totally ineffective as pumps since they are usually contracting against closed AV valves, but atrial pumping, as we shall see, is relatively unimportant for cardiac functioning (except during relatively strenuous exercise).

Partial AV block is not always caused by disease but may represent a normal life-saving adaptation. Imagine a person whose SA node has become hyperactive and is driving the atria at 300 beats per minute. Ventricular rates this high are very inefficient for pumping blood, because there is inadequate time to fill the ventricles between contractions. Fortunately, the long refractory period of the AV node may prevent passage of a significant fraction of the impulses, and the ventricles will beat at a slower rate.

Another group of abnormalities is characterized by a slow conduction velocity or lengthened

conduction route so that the impulse always arrives at an area which is no longer refractory, and the impulse keeps traveling around the heart in a so-called circus movement. This may lead to continuous, completely disorganized contractions (**fibrillation**), which are ineffective in producing flow and can cause death if they occur in the ventricles. For example, ventricular fibrillation is the immediate cause of death in many instances of heart attack (and in electrocution).

The electrocardiogram

The **electrocardiogram** (ECG or EKG) is primarily a tool for evaluating the electrical events within the heart. The action potentials of cardiac muscle can be viewed as batteries which cause current flow throughout the body fluids. These currents represent the integrated sum of many action potentials and can be detected at the surface of the skin. Figure 11-22 illustrates a typical normal ECG recorded as the potential difference between the right and left wrists. The first wave, P, represents atrial depolarization. The second complex, QRS, occurring approximately 0.1 to 0.2 s later, represents ventricular depolarization. The final wave, T, represents ventricular repolarization. No manifestation of atrial repolarization is evident because it occurs

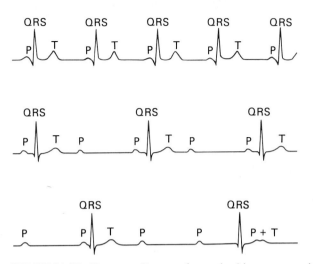

FIGURE 11-23. Electrocardiograms from a healthy person and from two persons suffering from atrioventricular block. Upper tracing: A normal ECG. Middle tracing: Partial block; one-half of the atrial impulses are transmitted to the ventricles. Lower tracing: Complete block; there is absolutely no synchrony between atrial and ventricular electrical activities.

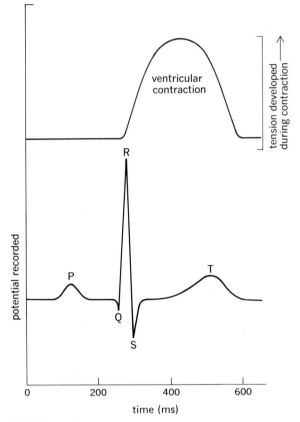

FIGURE 11-22. Typical electrocardiogram. P, atrial depolarization; QRS, ventricular depolarization; T, ventricular repolarization.

during ventricular depolarization and is masked by the QRS complex. Figure 11-23 gives one example of the clinical usefulness of the ECG. The patient having the middle tracing is suffering from partial AV-nodal block, so that only one-half the atrial impulses are being transmitted. Note that every second P wave is not followed by a QRS and T. Because many myocardial defects alter normal impulse propagation, and thereby the shapes and timing of the waves, the ECG is a powerful tool for diagnosing heart disease.

Excitation-contraction coupling

The depolarization (''excitation'') of cardiac muscle cells triggers their contraction. As described in Chap. 10, the mechanism coupling excitation and

contraction in all muscle is an increase in the cytosolic concentration of calcium; in cardiac muscle as in skeletal muscle, this calcium combines with the regulator protein, troponin, with the resulting removal of tropomyosin's inhibition of cross-bridge formation between actin and myosin. Again, as in skeletal muscle, the major source of the increased cytosolic calcium during the action potential is release of calcium from the sarcoplasmic reticulum. A second source is calcium which diffuses across the plasma membrane into the cardiac muscle cell during the action potential.

Figure 11-24 illustrates a typical action potential in a ventricular muscle cell. The distinctive feature of this action potential is the long, continued depolarization (plateau) following the initial depolarizing spike, a plateau not seen in the action potentials of neurons or skeletal muscle. The initial spike of the action potential is due to an explosive increase in membrane permeability to sodium. The sodium permeability then rapidly returns to very low values, and yet the membrane remains depolarized at the plateau; the major reason for this continued depolarization is a marked increase in the membrane permeability to calcium. The initial

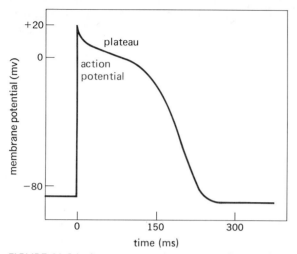

FIGURE 11-24. Action potential in a ventricular muscle cell.

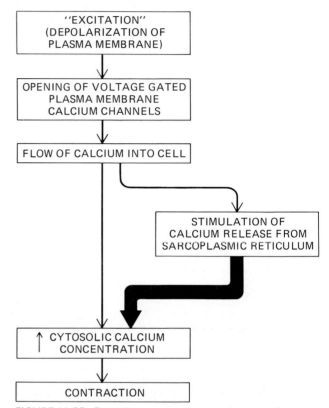

FIGURE 11-25. Excitation-contraction coupling in cardiac muscle. The thick arrow denotes that the sarcoplasmic reticulum is the major source of the increased cytosolic calcium. However, release of calcium from the sarcoplasmic reticulum is triggered by the calcium flow into the cell across the plasma membrane.

(sodium-dependent) membrane depolarization causes, in cardiac muscle cells, voltage-gated calcium channels in the membrane to open, which results in a flow of calcium ions into the cell.

The total amount of calcium entering the cell in this manner is actually very small compared with the amount of calcium simultaneously being released from the sarcoplasmic reticulum, but it is crucial for a simple reason: It is this new calcium entering across the plasma membrane via calcium channels which constitutes the signal for release of the much larger amount of calcium from the sarcoplasmic reticulum (Fig. 11-25). During repolarization, cytosolic calcium is restored to its original extremely low value by active transport of the calcium back into the sarcoplasmic reticulum and across the plasma membrane, and the muscle relaxes.

As we shall see, changes in cytosolic calcium account for many situations in which the strength of cardiac muscle contraction is altered. Thus, cardiac muscle differs greatly in this regard from skeletal muscle in which the change in cytosolic calcium occurring during membrane excitation is always adequate to produce maximal "turning-on" of cross bridges by binding to all the troponin sites. In cardiac muscle the calcium released is not sufficient to saturate all troponin sites, and thus the number of

active cross bridges can be graded if cytosolic calcium is increased.

Refractory period of the heart

Ventricular muscle, unlike skeletal muscle, is incapable of any significant degree of summation of contractions, and this is a very good thing. Imagine that cardiac muscle were able to undergo a prolonged tetanic contraction; during this period no ventricular filling could occur (since filling can only occur when the ventricular muscle is relaxed), and the heart would therefore cease to function as a pump. The inability of the heart to generate such tetanic contractions is the result of the long **refrac-**

tory period of cardiac muscle. Recall that in any excitable membrane an action potential is accompanied by a period during which the membrane cannot be reexcited. Following this absolute refractory period comes a second period during which the membrane can be depolarized again but only by a stimulus more intense then usual. In skeletal muscle, the absolute refractory periods are very short (1 or 2 ms) compared with the duration of contraction (20 to 100 ms), and a second contraction can be elicited before the first is over. In contrast, the absolute refractory period of cardiac muscle lasts almost as long as the contraction (250 ms) (because of the long plateau), and the muscle cannot be excited in time to produce summation (Fig. 11-26).

A common situation explainable in terms of the refractory period is shown in Fig. 11-27. In many people, drinking several cups of coffee causes increased excitability of areas other than the SA node owing to the action of caffeine. When such an area of the atria or ventricles actually initiates an action potential, the area is known as an **ectopic focus**. When one of these areas fires just after the completion of a normal contraction but before the next SA-nodal impulse, a premature wave of excitation occurs. As a result, the next normal SA-nodal impulse fires during the refractory period of the premature beat and is not propagated since the myocardial cells are refractory (the SA node still fires because it has a shorter refractory period). The second SA-nodal impulse after the premature contraction is propagated normally. The net result is an unusually long delay between beats. The contraction after the delay is unusually strong (for reasons to be described below) and the person is aware of the heart "skipping a beat."

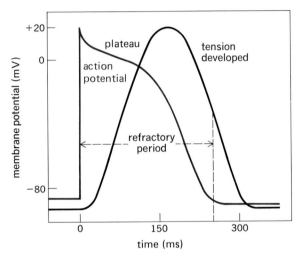

FIGURE 11-26. Relationship between membrane potential changes and contraction in a single ventricular muscle cell. The refractory period lasts almost as long as the contraction.

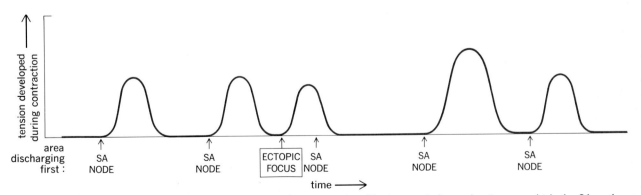

FIGURE 11-27. Effect of an ectopic discharge on ventricular contraction. The arrows indicate the times at which the SA node or ectopic focus fires. The premature beat induced by the ectopic discharge makes the ventricular muscle refractory at the time of the SA-node impulse. The failure of this impulse to induce a contraction results in a longer period than normal before the next beat.

Mechanical Events of the Cardiac Cycle

As pointed out earlier, fluid always flows from a region of higher pressure to one of lower pressure. The function of the heart is to generate the pressures which produce blood flow. The orderly process of depolarization described in the previous sections triggers contraction of the atria followed rapidly by ventricular contraction. The pressure and volume changes occurring in the systemic circulation during this cardiac cycle are summarized in Fig. 11-28. The cardiac cycle is divided into two major phases: the period of ventricular contraction (**systole**), followed by the period of ventricular relaxation (**diastole**). We start our analysis at the far left of Fig. 11-28 with the events of mid-to-late diastole, considering only the left heart and aorta; events on the right side have an identical sequence, but the pressure changes are smaller.

Mid-to-late diastole

The left atrium and ventricle are both relaxed, and left atrial pressure (not shown in the figure) is very slightly higher than left ventricular. Therefore, the AV valve is open, and blood entering the atrium from the pulmonary veins continues on into the ventricle. This is an important point: The ventricle receives blood throughout most of diastole, not just when the atrium contracts. Indeed, at rest, approximately 80 percent of ventricular filling occurs before atrial contraction.[1] Note that the aortic valve (between aorta and left ventricle) is closed, because the aortic pressure is higher than the ventricular pressure. The aortic pressure is slowly falling because blood is moving out of the arteries and through the vascular tree; in contrast, ventricular pressure is rising slightly because blood is entering it, thereby expanding the ventricular volume. At the very end of diastole, the SA node discharges, the atrium depolarizes (as signified by the P wave of the ECG), the atrium contracts, and a small volume of blood is added to the ventricle (note the

small rise in ventricular pressure and volume). The amount of blood in the ventricle just prior to systole is called the **end-diastolic volume**.

Systole

The wave of depolarization passes through the ventricle (as signified by the QRS complex of the ECG) and triggers ventricular contraction. As the ventricle contracts, ventricular pressure rises steeply. Almost immediately, this pressure exceeds the atrial pressure and closes the AV valve, thus preventing backflow into the atrium. Since for a brief period the aortic pressure still exceeds the ventricular, the aortic valve remains closed and the ventricle cannot empty despite its contraction. This early phase of systole is called **isovolumetric ventricular contraction** because ventricular volume is constant; i.e., the lengths of the muscle fibers remain approximately constant as in an isometric skeletal-muscle contraction. This brief phase ends when ventricular pressure exceeds aortic, the aortic valve opens, and **ventricular ejection** occurs. The ventricular-volume curve shows that ejection is rapid at first and then tapers off. Note that the ventricle does not empty completely; the amount remaining after ejection is called the **end-systolic volume**.

As blood flows into the aorta, the aortic pressure rises along with ventricular pressure. Note that peak aortic pressure is reached before the end of ventricular ejection; i.e., the pressure actually is beginning to fall during the last part of systole despite continued ventricular ejection. This is because the rate of blood ejection during the last part of systole is quite small (as shown by the ventricular volume curve) and is less than the rate at which blood is leaving the aorta (and other large arteries) via the arterioles; accordingly the volume and, therefore, the pressure in the aorta begin to decrease.

Early diastole

When ventricular contraction stops, the ventricular muscle relaxes rapidly. Ventricular pressure therefore almost immediately falls below aortic pressure, and the aortic valve closes. However, ventricular pressure still exceeds atrial so that the AV valve

---

[1] It is for this reason that the conduction defects discussed above, which eliminate the atria as efficient pumps, do not seriously impair cardiac function, at least at rest. Many persons lead relatively normal lives for many years despite atrial fibrillation. Thus, in many respects, the atrium may be conveniently viewed as merely a continuation of the large veins.

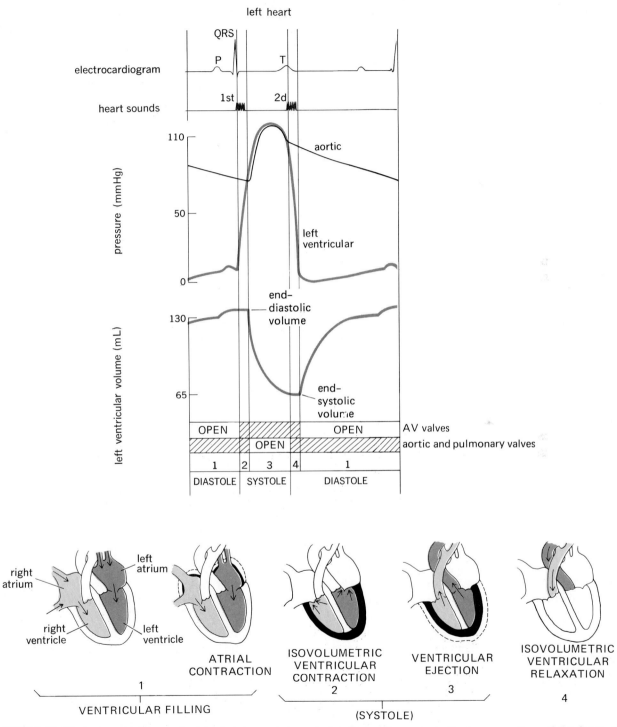

FIGURE 11-28. Summary of events in the left heart and aorta during the cardiac cycle. In the lower portions of the figure, the contracting portions of the heart are shown in black. *(Adapted from Ganong.)*

remains closed. This phase of early diastole, which is the mirror image of early systole, is called **isovolumetric ventricular relaxation**. It ends as ventricular pressure falls below atrial, the AV valve opens, and ventricular filling begins. Filling occurs rapidly at first and then slows down. The fact that ventricular filling is almost complete during early diastole is of the greatest importance; it ensures that filling is not seriously impaired during periods of increased heart rate, despite a reduction in the duration of diastole and, therefore, total filling time. However, when rates of approximately 200 beats per minute or more are reached, filling time does become inadequate so that the volume of blood pumped during each beat is decreased.

Pulmonary circulation pressures

The pressure changes in the right heart and pulmonary arteries are qualitatively similar to those just described for the left heart and aorta. However, there is a striking quantitative difference: The pressures in the right ventricle and pulmonary arteries are considerably lower than those in the left during systole. Thus, normal systolic pressures in the right ventricle and pulmonary arteries are approximately 30 mmHg. The pulmonary circulation is a low-pressure system (for reasons to be described in a later section). This difference is clearly reflected in the ventricular architecture, the right ventricular wall being much thinner than the left (Fig. 11-29). However, despite the lower pressure, the right ventricle ejects the same amount of blood as the left.

Heart sounds

Two heart sounds are normally heard through a stethoscope placed on the chest wall (see Fig. 11-28). The first sound, a soft low-pitched **lub**, is associated with closure of the AV valves at the onset of systole; the second, a louder **dup**, is associated with closure of the pulmonary and aortic valves at the onset of diastole. These sounds, which result from vibrations caused by closure of the valves, are perfectly normal, but other sounds known as **heart murmurs** are frequently (although not always) a sign of heart disease. Murmurs can be produced by blood flowing rapidly in the usual direction through an abnormally narrowed valve (stenosis), backward through a damaged leaky valve (regurgitation), or between the two atria or two ventricles via a small hole in the wall separating them. The exact timing and location of the murmur provide the physician with a powerful diagnostic clue. For example, a murmur heard throughout systole suggests a narrowed (stenotic) pulmonary or aortic valve, a leaky AV valve, or a hole in the interventricular septum, whereas a murmur heard throughout diastole suggests a stenotic tricuspid or mitral valve.

The Cardiac Output

The volume of blood pumped by each ventricle per minute is called the **cardiac output (CO)**, usually expressed as liters per minute. It must be remembered that the cardiac output is the amount of blood pumped by *each* ventricle, *not* the total amount pumped by both ventricles. The cardiac output is determined by multiplying the **heart rate (HR)** and **stroke volume (SV)**, the volume of blood ejected by each ventricle during each beat:

$$\underset{\text{L/min}}{CO} = \underset{\text{beats/min}}{HR} \times \underset{\text{L/beat}}{SV}$$

For example, if each ventricle has a rate of 72 beats per minute and ejects 70 mL with each beat, what is the cardiac output?

$$CO = 72 \text{ beats/min} \times 0.07 \text{ L/beat} = 5.0 \text{ L/min}$$

These values are approximately normal for a resting adult. (Since the total blood volume is also approximately 5 L, this means that essentially all the blood is pumped around the circuit once each minute.) During periods of exercise, the cardiac output may reach 30 to 35 L/min. Obviously, heart rate or stroke volume or both must have increased. Physical exercise is but one of many situations in which various tissues and organs require a greater flow of blood; e.g., flow through skin vessels increases

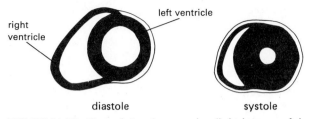

FIGURE 11-29. The relative shapes and wall thicknesses of the right and left ventricles during diastole and systole.

when heat loss is required, and flow through intestinal vessels increases during digestion. Some of the increased flow can be obtained merely by decreasing blood flow to some other organ, i.e., by redistributing the cardiac output, but most of the supply must come from a greater cardiac output. The following description of the factors which alter the two determinants of cardiac output—heart rate and stroke volume—applies in all respects to both the right and left heart, since stroke volume and heart rate are the same for both under steady-state conditions. Finally, it must be emphasized that heart rate and stroke volume do not always change in the same direction. For example, as we shall see, following blood loss stroke volume decreases, whereas heart rate increases. Clearly, these changes produce opposing effects on cardiac output.

Control of heart rate

Rhythmic beating of the heart at a rate of approximately 100 beats per minute will occur in the complete absence of any nervous or hormonal influences. This is the inherent autonomous discharge rate of the SA node. However, the heart rate may be much lower or higher than this, since the SA node is normally under the constant influence of nerves and hormones. As mentioned earlier, a large number of parasympathetic and sympathetic fibers end on the SA node. Stimulation of the parasympathetic (vagus) nerves causes slowing of the heart and, if strong enough, may stop the heart completely for a short time. The effects of the sympathetic nerves are just the reverse; their stimulation increases the heart rate. In the resting state, the parasympathetic influence is dominant and the rest-

ing heart rate in a normal person is well below 100 (the inherent rate).

Figure 11-30 illustrates the nature of the sympathetic and parasympathetic influence on SA-node function. Sympathetic stimulation increases the slope of the pacemaker potential, thus causing the pacemaker cells to reach threshold more rapidly and the heart rate to increase. Stimulation of the parasympathetics has the opposite effect on the slope of the pacemaker potential; the slope decreases, threshold is reached more slowly, and heart rate decreases. (In addition to this effect on the slope of the pacemaker potential, parasympathetic stimulation hyperpolarizes the membrane so that the pacemaker potential starts from a lower value.)

How do the neurotransmitters released by these autonomic neurons cause these changes in the pacemaker potential? Acetylcholine, the parasympathetic neurotransmitter, does so by increasing the permeability of the membrane to potassium; norepinephrine, the sympathetic neurotransmitter, enhances the rate of pacemaker depolarization mainly by increasing the flow of calcium ions into the SA-nodal cells (as described in Chapter 8, calcium-ion movements have effects on membrane potentials similar to those of sodium movements).

Factors other than the cardiac nerves also can

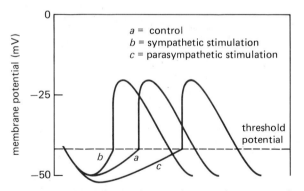

FIGURE 11-30. Effects of sympathetic and parasympathetic nerve stimulation on the slope of the pacemaker potential of an SA-node cell. Note that parasympathetic stimulation not only reduces the slope of the pacemaker potential but also lowers the "take-off" point. *(Adapted from Hoffman and Cranefield.)*

alter heart rate. Epinephrine, the main hormone liberated from the adrenal medulla, speeds the heart; this is not surprising since epinephrine acts on the same beta-adrenergic receptors as does norepinephrine (Chaps. 8 and 9). The heart rate is also sensitive to changes in temperature, plasma electrolyte concentrations, and hormones other than epinephrine. However, these are generally of lesser importance, and the heart rate is primarily regulated by balancing the slowing effects of parasympathetic discharge against the accelerating effects of sympathetic discharge, both operating on the SA node (Fig. 11-31).

As stated earlier, sympathetic and parasympathetic neurons terminate not only on the SA node but on other parts of the conducting system as well. Thus, parasympathetic stimulation not only slows the SA node but decreases conduction through the AV node; sympathetic stimulation, in contrast, accelerates the spread of excitation along almost the entire specialized conducting pathway.

Control of stroke volume

The second variable which determines cardiac output is stroke volume, the volume of blood ejected during each ventricular contraction. As emphasized earlier, the ventricles never completely empty themselves of blood during contraction. Therefore,

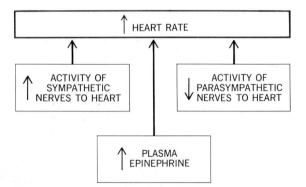

FIGURE 11-31. Summary of major factors which influence heart rate. All effects are exerted upon the SA node. The figure, as drawn, shows how heart rate is increased; reversal of all the arrows in the boxes would illustrate how heart rate is decreased.

a more forceful contraction can produce an increase in stroke volume. Changes in the force of contraction can be produced by a variety of factors, but two are dominant under most physiological conditions: (1) changes in the end-diastolic volume, i.e., the volume of blood in the ventricles just prior to contraction; and (2) alterations in the magnitude of sympathetic nervous system input to the ventricles.

Relationship between end-diastolic volume and stroke volume: Starling's law of the heart. It is possible to study the completely intrinsic adaptability of the heart by means of the so-called heart-lung preparation (Fig. 11-32). Tubes are placed in the heart and vessels of an anesthetized animal so that blood flows from the very first part of the aorta (just above the exit of the coronary arteries) into a blood-filled reservoir and from there into the right atrium. The blood then is pumped by the right heart as usual via the lungs into the left heart from which it is returned to the reservoir. A key feature of this preparation is that the pressure causing blood flow into the heart can be altered simply by raising or lowering the reservoir. This is analogous to altering the venous and right atrial pressures and causes changes in the quantity of blood entering the right ventricle during diastole. Thus when the reservoir is raised, ventricular filling increases, thereby increasing end-diastolic volume. The important finding is that the more-distended ventricle responds with a more-forceful contraction. The net result is a new steady state, in which the ventricle is distended and both diastolic filling and stroke volume are increased.

The mechanism underlying this completely intrinsic adaptation is that cardiac muscle, like other muscle, increases its strength of contraction when it is stretched over a certain range. The British physiologist Starling observed that there is a direct proportion between the diastolic volume of the heart, i.e., the length of its muscle fibers, and the force of contraction of the following systole, and this relationship is now referred to as **Starling's law of the heart**. Thus, in the experiment above, the increased diastolic volume stretches the ventricular muscle fibers and causes them to contract more

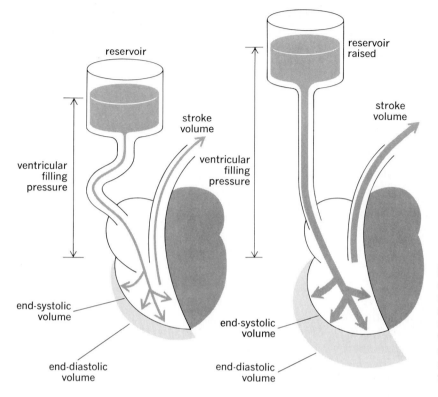

FIGURE 11-32. Experiment for demonstrating the relationship between ventricular end-diastolic volume and stroke volume (Starling's law of the heart). When the reservoir is raised, the pressure causing ventricular filling is increased. The increased filling distends the ventricle, which responds with an increased strength of contraction.

forcefully. A typical response curve obtained by progressively increasing end-diastolic volume is shown in Fig. 11-33. Thus, heart muscle manifests a length-tension relationship very similar to that described earlier for skeletal muscle. However, unlike skeletal muscle, cardiac-muscle length in the resting state is less than that which yields maximal tension during contraction, so that an increase in length produces an increase in contractile tension.

The significance of this intrinsic mechanism should be apparent; an increased flow of blood from the veins into the heart (**venous return**) automatically forces an increase in cardiac output by distending the ventricle and increasing stroke volume, just as the heart-lung apparatus did when we increased venous return by elevating the reservoir. One important function of this relationship is maintaining equality of right and left cardiac outputs. Should the right heart, for example, suddenly begin to pump more blood than the left, the increased blood flow to the left ventricle would automatically produce an equivalent increase in left ventricular output, and blood would not be allowed to accumulate in the lungs.

The sympathetic nerves. Sympathetic nerves are distributed not only to the SA node and conducting system but to all myocardial cells. The effect of the sympathetic mediator, norepinephrine, is to increase ventricular (and atrial) strength of contraction at any given initial end-diastolic volume; i.e., the increased force of contraction secondary to sympathetic-nerve stimulation is independent of a change in end-diastolic ventricular muscle-fiber length. It is termed increased **contractility**. Note that a change in force of contraction due to increased end-diastolic volume (Starling's law) does *not* reflect increased contractility, since increased contractility is specifically defined as an increased force of contraction *not* due to altered end-diastolic volume. Circulating epinephrine also increases myocardial contractility.

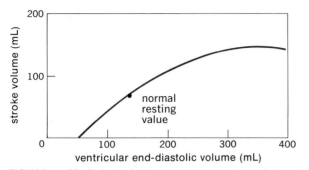

FIGURE 11-33. Relationship between ventricular end-diastolic volume and stroke volume (Starling's law of the heart). The data were obtained by progressively increasing ventricular filling pressure, as in Fig. 11-32. The horizontal axis could have been labeled "sarcomere length," and the vertical "contractile force"; i.e., this is a length-tension curve.

Not only does enhanced sympathetic nerve activity to the myocardium cause the contraction to be more powerful, it causes both the contraction and relaxation of the ventricles to occur more quickly (Fig. 11-34). These latter effects are quite important since, as described earlier, increased sympathetic activity to the heart also increases heart rate. As heart rate increases, the time available for diastolic filling decreases, but the quicker contraction and relaxation induced simultaneously by the sympathetic neurons partially compensate for this problem by permitting a larger fraction of the cardiac cycle to be available for filling.

Norepinephrine and epinephrine increase contractility by inducing (via the cyclic AMP second-messenger system) an increased movement of calcium into the cytosol during excitation. They speed relaxation by enhancing the rate at which this calcium is transported out of the cytosol into the sarcoplasmic reticulum. In contrast to the sympathetic nerves, the parasympathetic nerves to the heart have only a small effect on ventricular contractility, and we shall ignore it.

The interrelationship between Starling's law and the cardiac sympathetic nerves as measured in a heart-lung preparation is illustrated in Fig. 11-35. The dashed line is the same as the line shown in Fig. 11-33 and was obtained by slowly raising ventricular pressure while measuring end-diastolic volume and stroke volume; the solid line was obtained similarly for the same heart, during sympathetic-nerve stimulation. Starling's law still applies, but during nerve stimulation the stroke volume is greater at any given end-diastolic volume. In other words, the increased contractility leads to a more

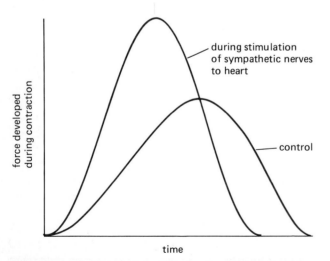

FIGURE 11-34. Effects of sympathetic stimulation on ventricular contraction. Note that both the rate of force development and the rate of relaxation are increased, as is the maximal force developed.

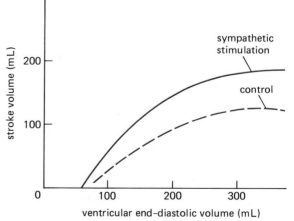

FIGURE 11-35. Effects on stroke volume of stimulating the sympathetic nerves to the heart.

complete ejection of the end-diastolic ventricular volume.

In summary (Fig. 11-36), stroke volume is controlled both by an intrinsic cardiac mechanism dependent upon changes in end-diastolic volume and by an extrinsic mechanism mediated by the cardiac sympathetic nerves (and circulating epinephrine), which causes increased ventricular contractility.

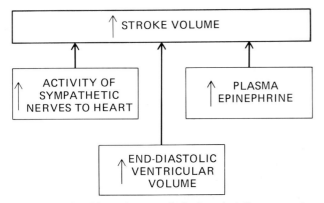

FIGURE 11-36. Major factors which directly influence stroke volume. The figure as drawn shows how stroke volume is increased; a reversal of all arrows in the boxes would illustrate how stroke volume is decreased. Refer to the text for details.

The contribution of each mechanism in specific physiological situations is described in later sections.

Summary of cardiac output control

A summary of the major factors which determine cardiac output is presented in Fig. 11-37, which combines the information of Fig. 11-31 (factors influencing heart rate) and Fig. 11-36 (factors influencing stroke volume).

Cardiac energetics

During contraction, the heart, like other kinds of muscle, converts chemical energy stored in ATP into mechanical work and heat. The total energy utilized by the heart per unit time is determined by a large number of factors, but the most important are the arterial pressure against which the ventricles must pump and the heart rate (i.e., the number of times per minute this pressure must be developed). This means that the energy expended by the heart may vary markedly in different situations having the same cardiac output. For example, if the heart

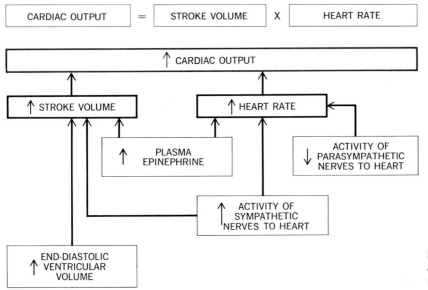

FIGURE 11-37. Major factors determining cardiac output. (This is an amalgamation of Figs. 11-31 and 11-36.)

rate is doubled and the stroke volume halved, cardiac output does not change but the energy required to pump this output increases markedly. To take another example, the hearts of persons with hypertension (high arterial blood pressure) consume more energy in pumping blood even though their cardiac output is normal. From these facts it should be clear why people with impaired coronary blood flow (i.e., decreased supply of oxygen and nutrients to the myocardium) are subjected to the greatest danger when their heart rate and arterial blood pressure increase, as during emotional excitement.

SECTION D.
THE VASCULAR SYSTEM

Figure 11-28 illustrates the pattern of pressure changes in the aorta during the cardiac cycle. Figure 11-38 illustrates the pressure changes which occur along the rest of the systemic and pulmonary vascular systems. The sections dealing with the individual vascular segments will describe the reasons for these changes in pressure, and you should refer back to this figure frequently. For the moment, note only that by the time the blood has completed its journey back to the atrium, virtually all the pressure originally generated by the ventricles' contraction has been dissipated. Note also that the pulmonary vascular pressures are considerably smaller than the systemic pressures.

The basic reason pressure decreases as blood flows through the vascular system is that the blood vessels offer resistance to this flow. We begin, then, by describing in general terms the determinants of resistance.

Resistance is a measure of the friction resulting from collisions between the tube wall and moving fluid and between the fluid molecules themselves. Resistance thus depends upon the type of fluid and the dimensions of the tube.

Syrup flows less readily than water because there is greater friction between the molecules of syrup as they slide past each other. This property of fluids is called **viscosity**; the more friction, the higher the viscosity. In abnormal states, changes in blood viscosity due to changes in hematocrit can have important effects on the resistance to flow, but under most physiological conditions, the viscosity of blood is relatively constant.

Both the length and radius of a tube affect its resistance by determining the surface area of the tube in contact with the fluid. Resistance to flow is directly proportional to the length of the tube (Fig. 11-39), but since the lengths of the blood vessels remain constant in the body, length is not a factor in the *control* of resistance. The resistance increases markedly as tube radius decreases; the exact relationship is given by the following formula, which states that the resistance (R) is inversely proportional to the fourth power of the radius (r) (the radius multiplied by itself four times):

$$R \propto 1/r^4$$

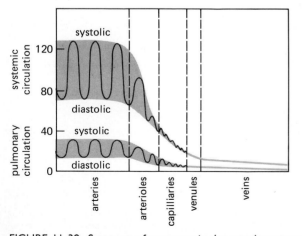

FIGURE 11-38. Summary of pressures in the vascular system.

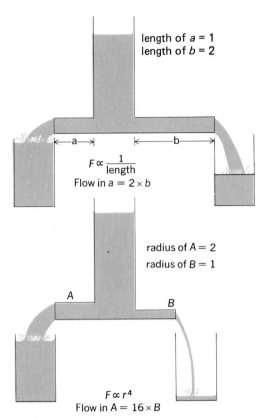

length of *a* = 1
length of *b* = 2

$$F \propto \frac{1}{\text{length}}$$

Flow in *a* = 2 × *b*

radius of A = 2
radius of B = 1

$$F \propto r^4$$

Flow in A = 16 × B

FIGURE 11-39. Effects of tube length and radius on resistance to flow.

The dependence of resistance upon radius can be seen in Fig. 11-39; doubling the radius of a tube decreases its resistance sixteenfold and therefore increases flow through it sixteenfold.

Arteries

The aorta and other systemic arteries have thick walls containing large quantities of elastic tissue. Although they also have smooth muscle, they can be viewed most conveniently as elastic tubes. Because the arteries have large radii, they serve as low-resistance conduits conducting blood to the various organs. Their second major function, related to their elasticity, is to act as a "pressure reservoir" for maintaining blood flow through the tissues during diastole; ventricular contraction produces a pressure within the arteries, and this arte-

rial pressure then constitutes the driving force for blood flow through the tissues.

Arterial blood pressure

What are the factors determining the pressure within an elastic container, like a balloon filled with water? The pressure inside the balloon depends upon the volume of water within it and the distensibility of the balloon, i.e., how easily its walls can be stretched. If the walls are very stretchable, large quantities of water can go in with only a small rise in pressure; conversely, a small quantity of water causes a large pressure rise in a balloon with low distensibility.

These principles can be applied to an analysis of arterial blood pressure. The contraction of the ventricles ejects blood into the pulmonary and systemic arteries during systole. If a precisely equal quantity of blood were to flow simultaneously out of the arteries, the total volume of blood in the arteries would remain constant and arterial pressure would not change. Such, however, is not the case. As shown in Fig. 11-40, a volume of blood equal to

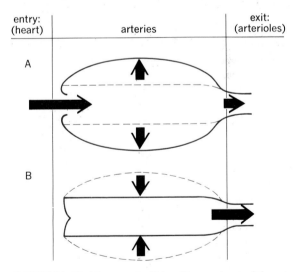

FIGURE 11-40. Movement of blood into and out of the arteries during the cardiac cycle. The lengths of the arrows denote relative quantities. During systole (A) less blood leaves the arteries than enters, and the arterial walls stretch. During diastole (B) the walls recoil passively, driving blood out of the arteries.

only about one-third the stroke volume leaves the arteries during systole. The rest of the stroke volume raises the arterial pressure and distends the arteries. When ventricular contraction ends, the stretched arterial walls recoil passively (like a stretched rubber band upon release) and the arterial pressure continues to drive blood into the arterioles during diastole. As blood leaves the arteries, the arterial volume and, therefore, the pressure slowly fall, but the next ventricular contraction occurs while there is still adequate blood in the arteries to stretch them partially. Therefore the arterial pressure does not fall to zero.

The aortic pressure pattern shown in Fig. 11-41 is typical of the pressure changes which occur in all the large systemic arteries. The maximum pressure reached during peak ventricular ejection is called **systolic pressure**. The minimum pressure occurs just before ventricular ejection begins and is called **diastolic pressure**. Arterial blood pressure is generally recorded as systolic/diastolic, that is, 125/75 mmHg in our example (see Fig. 11-41B for average values in people of different ages). The pulse, which can be felt in an artery, is caused by the difference between systolic and diastolic pressure. This difference (125 − 75 = 50 in the example) is called the **pulse pressure**.

There are three factors which determine the magnitude of the pulse pressure (i.e., how much greater systolic pressure is than diastolic). These are: (1) stroke volume, (2) speed at which the stroke volume is ejected, and (3) arterial distensibility. In other words, the increment in pressure produced by a ventricular ejection will be greater if the volume of blood ejected during the beat is increased, if the volume is ejected more quickly (as, for example, when ventricular contractility is enhanced by sympathetic nerve stimulation), or if the arteries are more resistant to expansion as the blood enters them (i.e., if they are "stiffer" or less distensible). This last phenomenon occurs in atherosclerosis (a form of hardening of the arteries) and accounts for the large pulse pressure typical of this disorder.

It is evident from Fig. 11-41A that arterial pressure (and, therefore, flow into the tissues) is constantly changing throughout the cardiac cycle. The

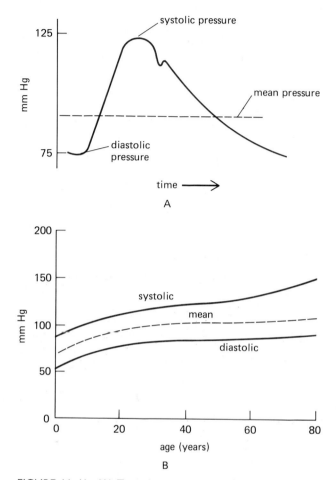

FIGURE 11-41. (A) Typical aortic pressure fluctuations during the cardiac cycle. (B) Changes in arterial pressure with age. *(Adapted from Guyton.)*

average pressure (**mean arterial pressure, MAP**) throughout the cycle is not merely the value halfway between systolic and diastolic pressure, because diastole usually lasts longer than systole. The true mean arterial pressure can be obtained by complex methods, but for most purposes it is approximately equal to the diastolic pressure plus one-third of the pulse pressure. Thus, in our example,

$$MAP = 75 + (\tfrac{1}{3} \times 50) = 92 \text{ mmHg}$$

The mean arterial pressure is the most important of the pressures described because it is the *average* pressure driving blood into the tissues throughout the cardiac cycle. In other words, if the pulsatile pressure changes were eliminated and the pressure throughout the cardiac cycle were always equal to the mean pressure, the total flow would be unchanged. Mean arterial pressure is a closely regulated quantity; indeed, the reflexes which accomplish this regulation constitute the basic cardiovascular control mechanisms and will be described in detail in a later section.

One last point: We can say ''arterial'' pressure without specifying to which artery we are referring because the aorta and other arteries have such large diameters that they offer only negligible resistance to flow, and the pressures are therefore similar everywhere in the arterial tree.

Measurement of arterial pressure

Both systolic and diastolic blood pressure are readily measured in human beings with the use of a **sphygmomanometer** (Fig. 11-42). A hollow cuff is wrapped around the arm and inflated with air to a pressure greater than systolic blood pressure (Fig. 11-42A). The high pressure in the cuff is transmitted through the tissue of the arm and completely collapses the arteries under the cuff, thereby preventing blood flow to the lower arm. The air in the cuff is then slowly released, causing the pressure in the cuff and arm to drop. When cuff pressure has fallen to a value just below the systolic pressure (Fig. 11-42B), the arterial blood pressure at the peak of systole is greater than the cuff pressure, causing the artery to open slightly and allow blood flow for this brief time. During this interval, the blood flow through the partially occluded artery occurs at a very high velocity because of the small opening for blood passage and the large pressure gradient. The high-velocity blood flow produces turbulence and vibration, which can be heard through a stethoscope placed over the artery just below the cuff. Thus, the pressure (measured on the manometer attached to the cuff) at which sounds are first heard

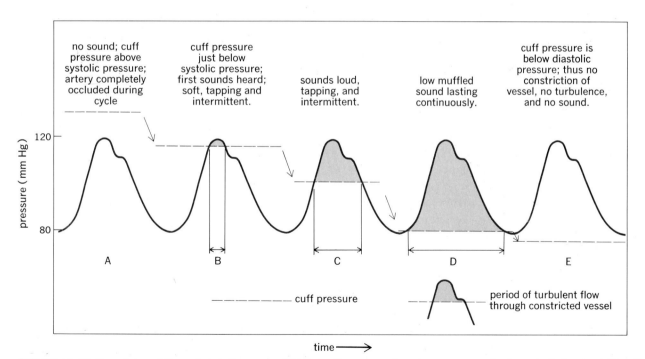

FIGURE 11-42. Sounds heard through a stethoscope while the cuff pressure of a sphygmomanometer is gradually lowered. Systolic pressure is recorded at point B and diastolic pressure is recorded at the point of sound disappearance.

as the cuff pressure is lowered is identified as the systolic blood pressure. As the pressure in the cuff is lowered further, the time of blood flow through the artery during each cycle becomes longer (Fig. 11-42C). When the cuff pressure reaches the diastolic blood pressure, the sounds become dull and muffled, as the artery remains open throughout the cycle, allowing continuous turbulent flow (Fig. 11-42D). Just below diastolic pressure all sound stops, as flow is now continuous and nonturbulent through the completely open artery. Thus, diastolic pressure is identified as the cuff pressure at which sounds disappear.

Arterioles

The arterioles are primarily responsible for determining the relative blood-flow distribution to different organs. Figure 11-43 illustrates the major principles of blood-flow distribution in terms of a simple model, a fluid-filled tank with a series of compressible outflow tubes. What determines the rate of flow through each exit tube? As stated earlier

$$\text{Flow} = \frac{\Delta P}{R}$$

Since the driving pressure is identical for each tube, differences in flow are completely determined by differences in the resistance to flow offered by each tube. The lengths of the tubes are approximately the same, and the viscosity of the fluid is a constant; therefore, differences in resistance offered by the tubes are due solely to differences in their radii. Obviously, the widest tube has the greatest flow. If

we equip each outflow tube with an adjustable cuff, we can obtain any combination of flows we wish.

This analysis can now be applied to the cardiovascular system. The tank is analogous to the arteries, which serve as a pressure reservoir, the major arteries themselves being so large that they contribute little resistance to flow. The smaller arteries begin to offer some resistance, but it is slight. Therefore, all the arteries of the body can be considered a single pressure reservoir.

The arteries branch within each organ into the next series of smaller vessels, the arterioles, which are now narrow enough to offer considerable resistance. The arterioles are the major site of resistance in the vascular tree and are therefore analogous to the outflow tubes in the model. This explains the large decrease in mean pressure (from about 90 mmHg to 35 mmHg, as illustrated in Fig. 11-38) as blood flows through the arterioles; indeed the resistance offered by the arterioles is so great that it completely dampens the pulse present upstream in the arteries and, by so doing, converts the pulsatile arterial flow into a continuous capillary flow.

Like the model's outflow tubes, the radii of arterioles of individual organs are subject to independent adjustment. The blood flow through any organ is given by the following equation, assuming for simplicity that the venous pressure, the other end of the pressure gradient, is zero:

$$F_{\text{organ}} = \frac{\text{MAP}}{R_{\text{organ}}}$$

Since the mean arterial pressure (MAP), i.e., the driving force for flow through each organ, is identical throughout the body, differences in flows be-

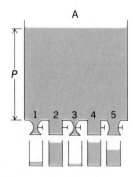

A

P

1 2 3 4 5

pressure reservoir ("arteries")

variable-resistance outflow tubes ("arterioles")

flow to different "organs"

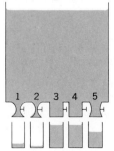

B

1 2 3 4 5

FIGURE 11-43. Physical model of the relationship between arterial pressure, arteriolar radius in different organs, and blood-flow distribution. Blood has been shifted from organ 2 to organ 3 (in going from A to B) by constricting the "arterioles" of 2 and dilating those of 3.

tween organs depend entirely on the relative resistances (R) offered by the arterioles of each organ. Arterioles contain smooth muscle which can relax or contract, thereby changing the radius of the inside (lumen) of the arteriole. Thus, the pattern of blood-flow distribution depends upon the degree of arteriolar smooth muscle contraction within each organ and tissue.

The smooth muscles surrounding arterioles are mainly single-unit smooth muscles and possess a large degree of inherent myogenic activity, i.e., "spontaneous" contraction. This tone is responsible for a large portion of the basal resistance offered by the arterioles. However, what is more important is that a variety of physiological factors play upon these smooth muscles [by altering cytosolic calcium concentration (see Chap. 10)] to either increase or decrease their degree of contraction, thereby altering the vessels' resistances. The mechanisms controlling these changes in arteriolar resistance fall into two general categories: (1) local controls and (2) extrinsic (reflex) controls.

Local controls

The term "local controls" denotes the fact that individual organs possess completely inherent mechanisms (i.e. mechanisms independent of nerves or hormones) for altering their own arteriolar resistance. This thereby confers upon the organs the capacity to self-regulate their blood flows.

Active hyperemia. Most organs and tissues manifest an increased blood flow (**hyperemia**) any time their metabolic activity is increased (Fig. 11-44). For example, the blood flow to exercising skeletal muscle increases in direct proportion to the increased activity of the muscle. This phenomenon, known as **active** (or **functional**) **hyperemia**, is the direct result of dilation of arterioles within the active organ. This vasodilation does not depend upon the presence of nerves or hormones, but is a locally mediated response. The adaptive value of the phenomenon should be readily apparent; an increased rate of activity in an organ automatically produces an increased blood flow to that organ by dilating arterioles.

The arterioles dilate because their smooth muscle relaxes, i.e., loses a fraction of its basal contractile tone. The factors acting upon smooth muscle in active hyperemia to cause it to relax are local chemical changes resulting from the increased metabolic cellular activity. The relative contributions of the various factors implicated seem to vary, depending upon the organs involved and the duration of the increased activity. At present, therefore, we will only name but not quantify some of the chemical changes which occur within organs when metabolic activity increases, and which appear to be involved in eliciting the relaxation of vascular smooth muscles. These are: decreased oxygen concentration; increased concentrations of carbon dioxide, hydrogen ion and metabolites such as adenosine; increased concentration of potassium (perhaps as a result of enhanced movement out of muscle cells during the more-frequent action potentials); increased osmolarity (i.e., decreased water concentration) as a result of the increased breakdown of high-molecular-weight substances; increased concentrations of prostaglandins; and increased generation (within sweat glands and several other glands) of a peptide known as bradykinin. Changes in all these variables—decreased oxygen, increased carbon dioxide, hydrogen ion, potassium, osmolarity, adenosine, prostaglandins, and bradykinin—have been shown to cause arteriolar dilation under controlled experimental conditions, and they all may contribute to the active-hyperemia response in one or more organs. It must be emphasized once more that all these chemical changes act locally upon the arteriolar smooth muscle causing it to relax, no nerves or homones being involved.

It should not be too surprising that the phenomenon of active hyperemia is most highly developed in heart and skeletal muscle, which show the widest range of normal metabolic activities of any organs or tissues in the body. It is highly efficient, therefore, that their supply of blood be primarily determined locally by their rates of activity.

Pressure autoregulation. The discussion of local metabolic controls has thus far emphasized increased metabolic activity of the tissue or organ as the initial event leading to vasodilation. However,

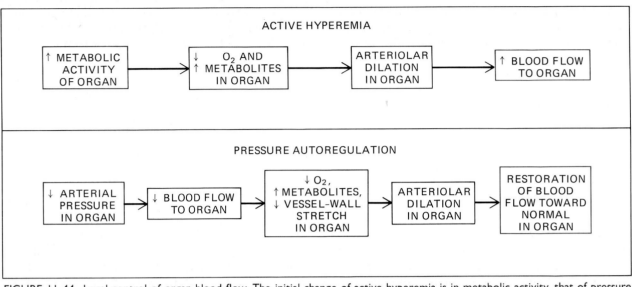

FIGURE 11-44. Local control of organ blood flow. The initial change of active hyperemia is in metabolic activity, that of pressure autoregulation is a change in the arterial blood pressure. Decreases in metabolic activity or increases in blood pressure would produce changes opposite to those shown in the figure.

similar changes in the local concentrations of some of the implicated metabolic factors may occur under a very different set of conditions, namely when a tissue or organ suffers a reduction in its blood supply due to a reduction in blood pressure (Fig. 11-44). The supply of oxygen to the tissue is diminished and the local oxygen concentration decreases. Simultaneously, the concentrations of carbon dioxide, hydrogen ion, and metabolites all increase because they are not removed by the blood as fast as they are produced, and prostaglandin synthesis is increased. In other words, the local metabolic changes which occur during increased metabolic activity are similar (although not identical) to those during decreased blood supply at constant metabolic activity because both situations reflect an imbalance of blood supply and level of cellular metabolic activity. The adaptive value of vasodilation in response to, say, partial occlusion of the artery supplying the tissue is that this response automatically tends to maintain the blood supply to the tissue (Fig. 11-44B).

Pressure autoregulation is not limited to circumstances in which arterial pressure goes down.

The opposite events may occur when, for various reasons, arterial pressure increases; the arterioles constrict in response to the increased arterial pressure, thereby maintaining local flow relatively constant in the face of the increased pressure.

Although our description has emphasized the role of *chemical* factors in the local resistance changes induced by changes in arterial pressure, it should be noted that other mechanisms may also participate. In particular, arteriolar smooth muscle may respond directly to an increased stretch (secondary to the increased arterial pressure) by increasing its basal myogenic tone. Conversely, decreased stretch (due to decreased arterial pressure) causes the vascular smooth muscle to decrease its tone. These inherent **myogenic responses** are mediated by stretch-dependent changes in calcium movement into the smooth muscle cells.

Reactive hyperemia. When an organ or tissue has had its blood supply completely occluded, there occurs a profound transient increase in its blood flow as soon as the occlusion is released. This phenomenon, known as **reactive hyperemia,** is essentially an extreme form of pressure autoregulation; i.e., during the period of no blood supply, the arterioles in the affected organ or tissue dilate, owing to the local factors described above, and blood flow is therefore very great through these wide-open arterioles as soon as the occlusion to arterial inflow is removed. It is very easy to observe this phenomenon in the skin simply by blowing up a blood-pressure cuff around the upper arm for a minute or so and then releasing it.

Response to injury. Tissue injury causes local release from cells or generation from plasma precursors of a variety of chemical substances which make arteriolar smooth muscle relax and are the causes of vasodilation in an injured area. This phenomenon, a part of the general process known as inflammation, will be described in detail in Chap. 17.

Extrinsic (reflex) controls

Sympathetic nerves. Most arterioles in the body receive a rich supply of sympathetic postganglionic nerve fibers. These neurons (with one possible exception) release norepinephrine, which combines with alpha-adrenergic receptors on the vascular smooth muscle to cause vasoconstriction. If almost all the nerves to arterioles are constrictor in action, how can reflex arteriolar dilation be achieved? Since the sympathetic nerves are seldom completely quiescent but discharge at some finite rate, which varies from organ to organ, the nerves always cause some degree of tonic constriction; from this position, further constriction is produced by increased sympathetic activity, whereas dilation can be achieved by decreasing the rate of sympathetic activity below the basal level. The skin offers an excellent example of the processes. Skin arterioles of a normal, unexcited person at room temperature are already under the influence of a moderate rate of sympathetic discharge; an appropriate stimulus (fear, loss of blood, etc.) causes reflex enhancement of this activity; the arterioles constrict further, and the skin pales. In contrast, an increased body temperature reflexly inhibits the sympathetic nerves to the skin, the arterioles dilate, and the skin flushes. This generalization cannot be stressed too strongly: Control of the sympathetic constrictor nerves to arteriolar smooth muscle can accomplish either dilation or constriction.

In contrast to the processes of active hyperemia and pressure autoregulation, the primary functions of these nerves are concerned not with the coordination of local metabolic needs and blood flow but with reflexes which redistribute blood flow throughout the body to achieve a specific function (for example, to increase heat loss from the skin). The most common reflex employing these nerves, as we shall see, is that which regulates arterial blood pressure by influencing arteriolar resistance throughout the body.

There is, as we said, one possible exception to the generalization that sympathetic nerves to arterioles release norepinephrine. In some mammalian species, a second group of sympathetic (not parasympathetic) nerves to the arterioles in skeletal muscle releases acetylcholine, which causes arteriolar dilation and increased blood flow; the only known function of these vasodilator fibers is in the response to exercise or stress. Because their existence in human beings is questionable, we shall not deal with them further.

Parasympathetic nerves. With but few major exceptions (notably the blood vessels supplying the external genitals), there is no significant parasympathetic innervation of arterioles. It is true that stimulation of the parasympathetic nerves to certain glands is associated with an increased blood flow, but this is secondary to the increased metabolic activity induced in the gland by the nerves, with resultant local active hyperemia.

Hormones. Several hormones cause constriction or dilation of arteriolar smooth muscle. One of these is **epinephrine**, the major hormone released from the adrenal medulla. Epinephrine, like norepinephrine released from sympathetic nerves, combines with alpha-adrenergic receptors on vascular smooth muscle and causes vasoconstriction. However, the story is more complex (Fig. 11-45): Some vascular

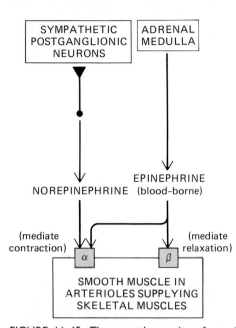

FIGURE 11-45. The smooth muscles of arterioles in skeletal muscles possess both alpha-adrenergic and beta-adrenergic receptors. Norepinephrine, released from sympathetic neurons, combines only with the alpha (α) receptors and causes only vasoconstriction. Circulating epinephrine combines with both types of receptors and hence can elicit vasodilation or vasoconstriction, depending upon this hormone's concentration. The number of beta (β) receptors in vascular beds other than those in skeletal muscle is generally quite small, so that epinephrine normally causes only vasoconstriction in them.

smooth muscle cells possess beta-adrenergic receptors as well as alpha-adrenergic receptors, and combination of epinephrine with these beta-adrenergic receptors on vascular smooth muscle causes the muscle cells to relax rather than contract. For most vascular beds, the existence of these beta-adrenergic receptors is of little, if any, importance, since they are greatly outnumbered by the alpha-adrenergic receptors. However, the arterioles in skeletal muscle are an exception; they have large numbers of beta-adrenergic receptors, and so physiological concentrations of circulating epinephrine generally cause dilation in this vascular bed. It must be emphasized that we are speaking here only of the response to epinephrine, the circulating hormone released by the adrenal medulla; stimulation of the sympathetic nerves to skeletal muscle arterioles causes only vasoconstriction.

Another hormone which exerts a potent vasoconstrictor action on most arterioles is **angiotensin**; this peptide is generated in the blood as the result of the action of an enzyme, **renin**, on a plasma protein, **angiotensinogen**. The control and multiple functions of the **renin-angiotensin system** will be described in Chap. 13; suffice it here to emphasize that an increase in plasma angiotensin will produce a widespread increase in the degree of arteriolar vasoconstriction.

Figure 11-46 summarizes the factors which determine arteriolar radius. Of course, the local metabolic factors and extrinsic reflex controls do not exist in isolation from one another but manifest important interactions.

Capillaries

At any given moment, approximately 5 percent of the total circulating blood is flowing through the capillaries. Yet it is this 5 percent which is performing the ultimate function of the entire system, namely, the exchange of nutrients and metabolic end products. (Some exchange also occurs in the very small venules, which can be viewed as extensions of capillaries.) All other segments of the vascular tree subserve the overall aim of getting ade-

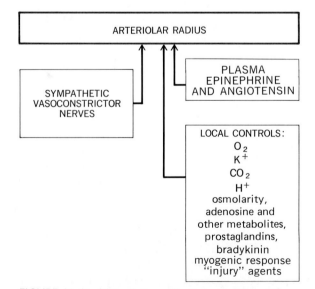

FIGURE 11-46. Major factors affecting arteriolar radius.

quate blood flow through the capillaries. The capillaries permeate almost every tissue of the body; no cell is more than 0.01 cm from a capillary. Therefore, diffusion distances are very small, and exchange is highly efficient. There are an estimated 25,000 miles of capillaries in an adult person, each individual capillary being only about 1 mm long.

Anatomy of the capillary network

Figure 11-47 illustrates diagrammatically the general anatomy of the small vessels which constitute the so-called microcirculation. Depending on the tissue or organ, blood enters the capillary network either directly from the arterioles or from vessels known as **metarterioles** which connect arterioles and venules directly. The site at which a capillary exits from a metarteriole is protected by a ring of smooth muscle, the **precapillary sphincter**, which continually opens and closes (due to local metabolic factors) so that flow through any given capillary is usually intermittent. Generally, the more active the tissue, the more precapillary sphincters are open at any moment (for example, only one-tenth of the

precapillary sphincters in skeletal muscle are open during rest). The sphincters are best visualized as functioning in concert with arteriolar smooth muscle to regulate not only the total flow of blood through the tissue capillaries but the number of functioning capillaries as well (accordingly, when we refer subsequently to arteriolar dilation or constriction, we tacitly include precapillary sphincters as well).

The structure of the capillaries themselves varies considerably from organ to organ, but the typical capillary (Fig. 11-48) is a thin-walled tube of endothelial cells one-layer thick without any surrounding smooth muscle or elastic tissue. The flat cells which constitute the endothelial tube interlock like pieces of a jigsaw puzzle with narrow, water-filled clefts (or ''pores'') between them. The width of these clefts varies in different organs and tissues. The endothelial cells generally contain large numbers of endocytotic and exocytotic vesicles, and sometimes these vesicles fuse to form a continuous channel across the cell (Fig. 11-48). The significance of these anatomical features will be discussed below. Finally, some capillaries (for example, those

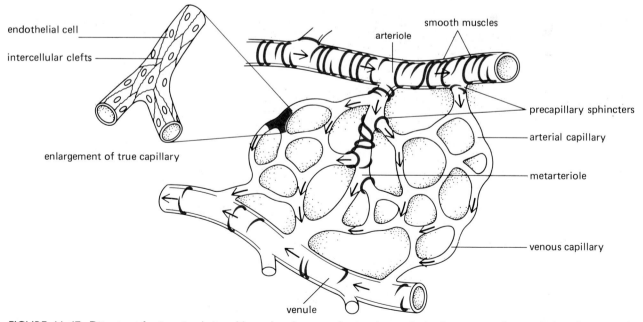

FIGURE 11-47. Diagram of microcirculation. Note the absence of smooth muscle in the true capillaries. (*Adapted from B. B. Chaffee and I. M. Lytle.*)

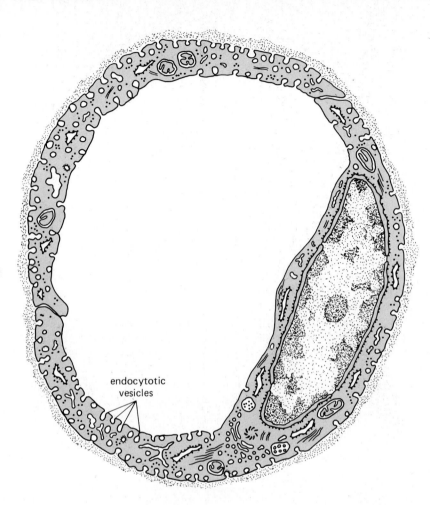

endocytotic
vesicles

FIGURE 11-48. Drawing of an electron micrograph of a capillary. (Adapted from Lentz.)

in the brain and certain parts of the kidneys) have a second set of cells adhering to the endothelial cells, and these importantly influence the ability of substances to penetrate the capillary wall.

Resistance of the capillaries

Since capillaries are very narrow, they offer a considerable resistance to flow (note the decrease in pressure along the lengths of the capillaries in Fig. 11-38), but for two reasons their resistance is not nearly as important as that of the arterioles: (1) Despite the fact that a single capillary is narrower than an arteriole, the huge total number of capillaries provides such a great cross-sectional area for flow that the *total* resistance of *all* the capillaries is

considerably less than that of the arterioles; and (2) because the capillaries have no smooth muscle, their radius (and, therefore, their resistance) is not subject to *active* control and simply reflects the volume of blood delivered to them via the arterioles (and the volume leaving via the venules).

Velocity of capillary blood flow

Figure 11-49 illustrates a simple mechanical model of a series of 1-cm-diameter balls being pushed down a single tube which branches into narrower tubes. Although each tributary tube has a smaller cross section than the wide tube, the sum of the tributary cross sections is much greater than the area of the wide tube. Let us assume that in the

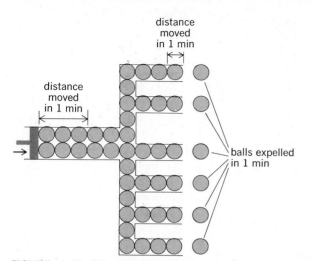

FIGURE 11-49. Relationship between cross-sectional area and velocity of flow. The total cross-sectional area of the small tubes is three times greater than that of the large tube. Accordingly, velocity of flow is one-third as great in the small tubes.

wide tube each ball moves 3 cm/min. If the balls are 1 cm in diameter and they move two abreast, six balls leave the wide tube per minute and enter the narrow tubes, and six balls leave the narrow tubes per minute. At what speed does each ball move in the small tubes? The answer is *1 cm/min.* This example illustrates the following important generalization: When a continuous stream moves through consecutive sets of tubes, the velocity of flow decreases as the sum of the cross-sectional areas of the tubes increases. This is precisely the case in the cardiovascular system (see Fig. 11-50); the blood velocity is very great in the aorta, progressively slows in the arteries and arterioles, and then markedly slows as the blood passes through the huge cross-sectional area of the capillaries (600 times the cross-sectional area of the aorta). The speed then progressively increases in the venules and veins because the cross-sectional area decreases. Because blood flows through the capillaries so slowly there is adequate time for exchange of nutrients and metabolic end products between the blood and tissues.

Diffusion across the capillary wall: Exchanges of nutrients and metabolic end products

There is no active transport of solute across the capillary wall, and materials cross primarily by diffusion. As described in the next section, there is

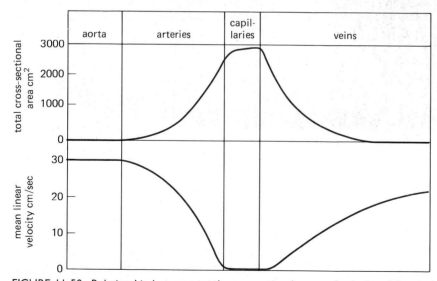

FIGURE 11-50. Relationship between total cross-sectional area and velocity of flow in each vessel type in the systemic circulation. (*Adapted from Lytle.*)

some movement of fluid by bulk flow, but it is of negligible importance for the exchange of nutrients and metabolic end products. The factors determining diffusion rates were described in Chap. 6. Because fat-soluble substances (including oxygen and carbon dioxide) penetrate cell membranes easily, they pass directly through the capillary endothelial cells. In contrast, ions and many molecules are poorly soluble in fat and must pass through water-filled ''pores'' in the endothelial lining. The most likely location of these pores is the clefts between adjacent endothelial cells. Finally, small amounts of still larger molecules, mainly proteins, pass through most capillary walls either by the processes of endocytosis and exocytosis or through a second set of larger pores, perhaps the channels formed by fused vesicles (see Chap. 6).

Because the water-filled intercellular clefts constitute less than 1 percent of the total area of the capillary endothelium, more than 100 times more area is available for diffusion of lipid-soluble substances than of water-soluble substances. Moreover, the variations in size of the clefts from tissue to tissue account for the great differences in the ''leakiness'' of capillaries in different tissues and organs. At one extreme are the extremely ''tight'' capillaries of the brain, whereas at the other end of the spectrum are liver capillaries, which have such large intercellular clefts that even protein molecules can readily pass; the latter is adaptive in that among the major functions of the liver are the synthesis of plasma proteins and metabolism of protein-bound substances. Most capillaries between these extremes allow the passage of small amounts of protein.

What is the sequence of events involved in capillary-cell transfers (Fig. 11-51)? Tissue cells do not exchange material *directly* with blood; the interstitial fluid always acts as middleman. Thus, nutrients diffuse across the capillary wall into the interstitial fluid, from which they gain entry to cells. Conversely, metabolic end products move across the tissue cells' plasma membranes into interstitial fluid, from which they diffuse into the plasma. Thus, two membrane-transfer processes must always be considered, that across the capillary wall and that across the tissue cell's plasma membrane.

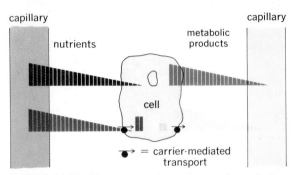

FIGURE 11-51. Movements of nutrients and metabolic products between plasma and tissue cells. The substance may enter or leave the cell by diffusion or carrier-mediated transport, but it moves across the capillary wall only by diffusion, dependent upon capillary–interstitial-fluid concentration gradients. (Two capillaries are shown for convenience; actually, of course, nutrients and products leave and enter the same capillary.)

The plasma membrane step may be by diffusion or by carrier-mediated transport, but as stated earlier, the transcapillary movement is always by diffusion (except in the brain).

How do transcapillary diffusion gradients for oxygen, nutrients, and metabolic end-products arise? Let us take two examples, those of glucose and carbon dioxide in muscle. Glucose is continuously consumed after being transported from interstitial fluid into the muscle cells by carrier-mediated transport mechanisms; this removal from interstitial fluid lowers the interstitial-fluid glucose concentration below that of plasma and creates a gradient for glucose diffusion out of the capillary. In contrast, carbon dioxide is continuously produced by muscle cells, thereby creating an increased intracellular carbon dioxide concentration, which causes diffusion of carbon dioxide into the interstitial fluid; in turn, this causes the interstitial carbon dioxide concentration to be greater than that of plasma and produces cardon doxide diffusion into the capillary. Note that in both cases net movement of the substance across the plasma membranes of the tissue cells, i.e., between interstitial fluid and cells, is the event which establishes the transcapillary (plasma-interstitial) diffusion gradients.

When a tissue increases its rate of metabolism, it must obtain more nutrients from the blood and eliminate more metabolic end products. An important mechanism for achieving this is an increase in diffusion gradients between plasma and cell (Fig. 11-52): Increased cellular consumption of oxygen and nutrients lowers their intracellular concentrations while increased production of carbon dioxide and other end products raises their intracellular concentrations. Simultaneously, blood flow to the area increases (due to the action of local metabolic factors on arterioles) and this additional blood flow helps maintain larger diffusion gradients by minimizing the fall in plasma concentration of oxygen and the rise in those of carbon dioxide and other end products as diffusion proceeds.

The vasodilation caused by the local metabolic factors facilitates diffusion by another mechanism as well—it dilates precapillary sphincters, and the resulting increase in the number of open capillaries increases the total surface area available for diffusion. Having more capillaries open also decreases the average distance from a cell to an open capillary.

Bulk flow across the capillary wall: Distribution of the extracellular fluid

Since the capillary wall is highly permeable to water and to almost all the solutes of the plasma with the exception of most plasma proteins, it behaves like a porous filter through which protein-free plasma (ultrafiltrate) moves by bulk flow through the clefts

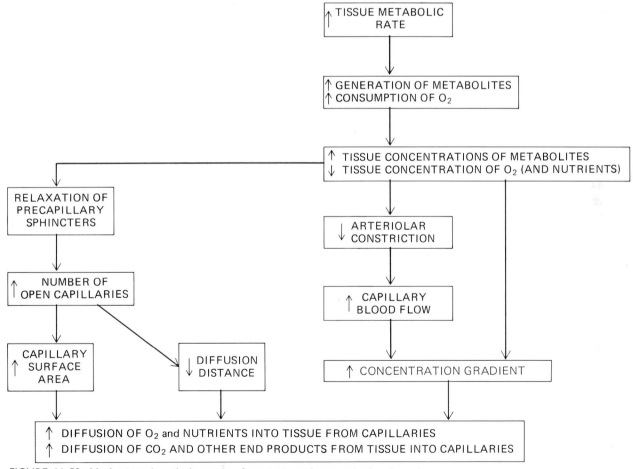

FIGURE 11-52. Mechanisms by which supply of nutrients and removal of end products is increased when a tissue increases its activity.

between the cells under the influence of a hydro-static pressure gradient. The magnitude of the bulk flow is directly proportional to the fluid (hydro-static) pressure difference between the inside and outside of the capillary, i.e., between the capillary blood pressure and the interstitial-fluid pressure. Normally, the former is much larger than the latter, so that a considerable hydrostatic pressure gradient exists to drive the filtration of protein-free plasma out of the capillaries into the interstitial fluid. Why then does all the plasma not filter out into the in-terstitial space? The explanation was first eluci-dated by Starling (the same scientist who ex-pounded the law of the heart which bears his name) and depends upon the principles of osmosis.

In Chap. 6 we described how a net movement of water occurs across a semipermeable membrane from a solution of high water concentration to a solution of low water concentration. Moreover, this osmotic flow of water "drags" along with it any dissolved solutes to which the membrane is highly permeable. Thus, a difference in water concentra-tion can result in the movement of a solution con-taining both water and permeating solute in a man-ner similar to the bulk flow produced by a hydrostatic pressure difference, and units of pres-sure (millimeters of mercury) are used in expressing the osmotic flow across a membrane.

This analysis can now be applied to capillary fluid movements (Fig. 11-53). The plasma within the capillary and the interstitial fluid outside it con-tain large quantities of low-molecular-weight sol-utes (crystalloids), e.g., sodium, chloride, or glu-cose. Since the capillary lining is highly permeable to all these crystalloids, they all have almost iden-tical concentrations in the two solutions. As we have seen, there are small concentration differences occurring for substances consumed or produced by the cells, but these tend to cancel each other, and accordingly, no significant water-concentration dif-ference is caused by the presence of the crystal-loids. In contrast, the plasma proteins cross most capillary walls only very slightly and therefore have a very low concentration in the interstitial fluid. This difference in protein concentration between plasma and interstitial fluid means that the water

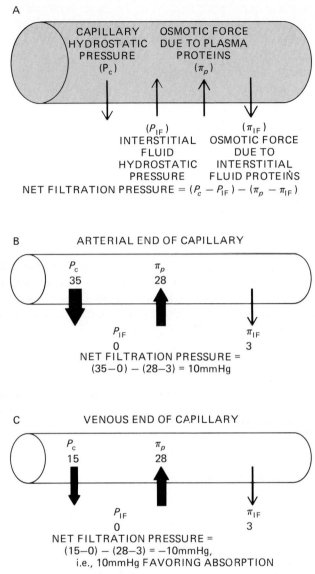

FIGURE 11-53. Forces determining fluid movement across cap-illaries. Filtration occurs at the arteriolar end of the capillary and absorption at the venous end.

concentration of the plasma is lower than that of interstitial fluid (high protein concentration means low water concentration), inducing an osmotic flow of water from the interstitial compartment into the capillary. Since all the different types of crystalloids dissolved in the interstitial fluid are carried along with the water, osmotic flow of fluid, like flow driven by a hydrostatic pressure difference, does not alter the concentrations of the low-molecular-weight substances of plasma or interstitial fluid.

In summary, two opposing forces act to move fluid across the capillary wall: (1) The pressure difference between capillary blood pressure and interstitial-fluid pressure favors the filtration of a protein-free plasma out of the capillary, and (2) the water-concentration differences between plasma and interstitial fluid, which results from differences in protein concentration, favors the flow of interstitial fluid into the capillary. Accordingly, the movements of fluid depend directly upon four variables: capillary hydrostatic pressure, interstitial hydrostatic pressure, plasma protein concentration, and interstitial-fluid protein concentration.

We may now consider quantitatively how these variables act to move fluid across the capillary wall in the systemic circulation (Fig. 11-53). Much of the arterial blood pressure has already been dissipated as the blood flows through the arterioles, so that pressure at the beginning of the capillary is about 35 mmHg. Since the capillary also offers resistance to flow, the pressure continuously decreases to approximately 15 mmHg at the end of the capillary. The interstitial pressure is essentially zero.[1] The difference in protein concentration between plasma and interstitial fluid causes a difference in water concentration, which induces a flow of fluid into the capillary equivalent to that produced by a hydrostatic pressure difference of 25 mmHg. Thus, in the first portion of the capillary the hydrostatic pressure difference across the capillary wall (35 mmHg) is greater than the opposing osmotic forces (25 mmHg), and a net movement of fluid out of the capillary occurs; in the last portion of the capillary, however, the osmotic force (25 mmHg) is greater than the hydrostatic pressure difference (15 mmHg), and fluid moves into the capillary (this is termed **absorption**). The result is that the early and

late capillary events tend to cancel each other out. For the aggregate of capillaries, in a normal person, there is a small net filtration; as we shall see, this is returned to the blood by lymphatics.

The analysis of capillary fluid dynamics in terms of different events occurring at the arterial and venous ends of the capillary is oversimplified. It is very likely that many individual capillaries (and postcapillary venules) manifest only net filtration or net absorption along their entire lengths because the arterioles supplying them are either so dilated or so constricted as to yield a capillary hydrostatic pressure above or below 25 mmHg along the entire length of the capillary.

In our quantitations of capillary filtration, we used average values for capillary pressure. Actually, capillary pressure in any vascular bed is subject to physiological regulation mediated mainly by changes in the resistance of the arterioles in that bed. As illustrated in Fig. 11-54 dilating the arterioles in a particular vascular bed raises capillary pressure because less pressure is lost between the arteries and the capillaries. Because of the increased capillary pressure, filtration is increased, and more protein-free fluid is lost to the interstitial fluid. In contrast, marked arteriolar constriction (Fig. 11-54) produces decreased capillary pressure

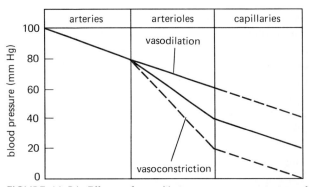

FIGURE 11-54. Effects of vasodilation or vasoconstriction of the arterioles in an organ on the blood pressure in the capillaries of that organ. Arteriolar constriction decreases capillary pressure, whereas arteriolar dilation increases it. (This applies only to those capillaries in the same organ or tissue in which the arterioles are constricting or dilating.)

---

[1] The exact value of interstitial hydrostatic pressure remains highly controversial. Many physiologists believe that it is not zero but is actually subatmospheric. The outcome of this controversy will not, however, alter the basic concepts presented here.

and net movement of interstitial fluid into the vascular compartment.

So far in these examples we have changed capillary pressure by dilating or constricting the arterioles leading to the capillaries. There are, however, other determinants of capillary pressure, and one of them—the pressure in the veins—is extremely important clinically. The capillaries, of course, drain into the veins, and so when venous pressure increases (as, for example, when a vein becomes occluded or, as we shall see, when the heart fails as a pump) the capillary pressure will also increase, and this will cause increased filtration. If great enough, this may cause accumulation of fluid in the interstitial space (**edema**).

Increases in net absorption or filtration can also be produced in the absence of capillary pressure changes whenever plasma protein concentration is altered (usually lowered). For example, in liver disease, protein synthesis decreases, plasma protein concentration is reduced, plasma water concentration is increased, net filtration is increased, and edema occurs.

The transcapillary protein-concentration difference can also be decreased by a quite different event, namely, the leakage of protein across the capillary wall into the interstitium whenever the capillary lining is damaged. This reduces the protein-concentration difference, and edema occurs as a result of the unchecked hydrostatic-pressure difference still acting across the capillary. The fluid accumulation in a blister is an excellent example.

The major function of this capillary filtration-absorption equilibrium should now be evident: It determines the distribution of the extracellular-fluid volume between the vascular and interstitial compartments. Obviously, the ability of the heart to pump blood depends upon the presence of an adequate volume of blood within the system. Recall that the interstitial-fluid volume is three to four times larger than the plasma volume; therefore, the interstitial fluid serves as a reservoir which can supply additional fluid to the circulatory system or draw off excess. The important role this equilibrium plays in the physiologial response to many situations, such as hemorrhage, will be described in a later section.

It should be stressed again that capillary filtration and absorption play no significant role in the exchange of nutrients and metabolic end products between capillary and tissues; the reason is that the total quantity of a substance (such as glucose or oxygen) moving into or out of the capillary during bulk flow is extremely small in comparison with the quantities moving by diffusion.

This analysis of capillary fluid dynamics has been in terms of the systemic circulation. Precisely the same Starling factors apply to the capillaries in the pulmonary circulation. As we shall see in the next chapter, the pulmonary capillaries lie in the thin walls between the lungs' air sacs (alveoli), and it is essential for normal gas exchange that these walls and the alveoli themselves not accumulate excess fluid. Recall that the pulmonary circulation is a low-resistance, low-pressure circuit. The normal pulmonary capillary pressure, which is the major force favoring movement of fluid out of the pulmonary capillaries into the interstitium, is only 15 mmHg (compare this with the average systemic capillary pressure of 25 mmHg). Despite this low value, there is normally a slight excess of filtration over absorption across pulmonary capillaries (see below), mainly because the interstitial protein concentration is fairly high in the lungs and so there is a relatively low osmotic difference to cause absorption. Normally the lymphatics carry away this excess fluid as fast as it filters.

Veins

Blood flows from capillaries into venules. As stated earlier, some exchange of materials with the interstitial fluid occurs across walls of the venules as well as of the capillaries. Venules also have some smooth muscle, the contractions of which can influence capillary pressure. The veins are the last set of tubes through which blood must flow on its trip back to the heart, and in the systemic circulation the force driving this venous return is the pressure difference between the peripheral veins and the right atrium. (Because the blood in the right atrium and large veins of the thorax constitutes a homogenous "pool," the **central venous pool**, it is customary to speak of **central venous pressure** rather than right atrial pressure; in order to minimize confusion, we have not adopted this convention.) Most of the pressure imparted to the blood by the heart is dissipated as blood flows through the arterioles and capillaries, so that pressure in the first portion of the peripheral veins is only approximately 15 mmHg. The right atrial pressure is close to 0 mmHg, so that the total driving pressure for flow from the peripheral veins to the right atrium is only about 15 mmHg. This is adequate because of the low resistance to flow offered by the veins owing to their large diameters. Thus, a major function of the veins is to act as low-resistance conduits for blood flow from the tissues to the heart.

In addition to acting as low-resistance conduits, the peripheral veins (i.e. those outside the thorax) perform a second important function: Their total **capacity** is altered in the face of variations in blood volume so as to maintain venous pressure and, thereby, venous return to the heart. In a previous section, we emphasized that the rate of venous return to the heart is a major determinant of end-diastolic ventricular volume and, thereby, stroke volume. Thus, we now see that venous pressure is an important determinant of stroke volume.

Determinants of venous pressure

The factors determining pressure in any elastic tube, as we know, are the volume of fluid within it and the distensibility of its wall. Accordingly, total blood volume is one important determinant of venous pressure. The veins differ from the arteries in that their walls are thinner and much more distensible and can accommodate large volumes of blood with a relatively small increase of internal pressure. This is illustrated by comparing Fig. 11-55 with Fig. 11-38; approximately 60 percent of the total blood volume is present in the systemic veins at any given moment, but the venous pressure averages less than 10 mmHg. In contrast, the systemic arteries contain less than 15 percent of the blood at a pressure of approximately 100 mmHg.

The walls of the veins contain smooth muscle innervated by sympathetic neurons. Stimulation of these neurons causes contraction of the venous smooth muscle, thereby increasing the stiffness of

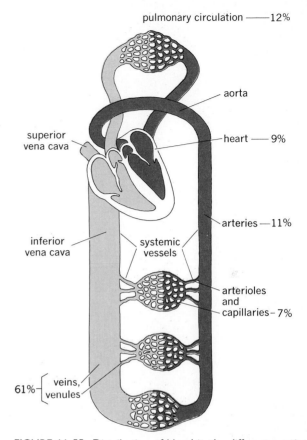

FIGURE 11-55. Distribution of blood in the different portions of the cardiovascular system. Compare this distribution of blood volumes with the pressures for the relevant areas shown in Fig. 11-38. *(Adapted from Guyton.)*

the wall, i.e., making it less distensible, and raising the pressure of the blood within the veins. Increased venous pressure drives more blood out of the veins into the right heart. The veins have such large diameters that a slight decrease in size (which has a great effect on venous capacity) produces little increase in resistance to flow. Thus, venous constriction exerts precisely the same effect on venous return as giving a transfusion.

The great importance of this effect can be visualized by the example in Fig. 11-56. A large decrease in total blood volume initially reduces the pressures everywhere in the circulatory system, including the veins; venous return to the heart decreases, and cardiac output decreases. However,

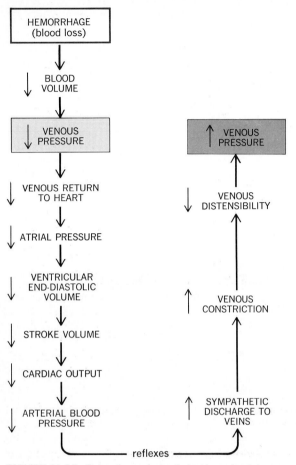

FIGURE 11-56. Role of venoconstriction in maintaining venous pressure during blood loss. The increased venous smooth-muscle constriction returns the decreased venous pressure toward, but not to, normal. The reflexes involved in this response are described later in the chapter.

reflexes to be described cause increased sympathetic discharge to the venous smooth muscle, which contracts; venous pressure is thereby returned toward normal, restoring venous return and cardiac output.

Two other mechanisms can decrease venous capacity, increase venous pressure, and facilitate venous return; these are the "skeletal muscle pump" and the effects of respiration upon thoracic and abdominal veins (the "respiratory pump"). During skeletal muscle contraction, the veins running through the muscle are partially compressed, which reduces their diameter, decreases venous ca-

pacity, and forces more blood back to the heart. As will be described in Chap. 12, during inspiration, the diaphragm descends, pushes on the abdominal contents, and increases abdominal pressure; simultaneously, the pressure in the chest (thorax) decreases. These pressure changes are transmitted passively to the intraabdominal and intrathoracic veins and heart. The net effect is to increase the pressure gradient between the veins outside the thorax and the heart; accordingly, venous return is enhanced during inspiration.

In summary (Fig. 11-57), the effects of venous

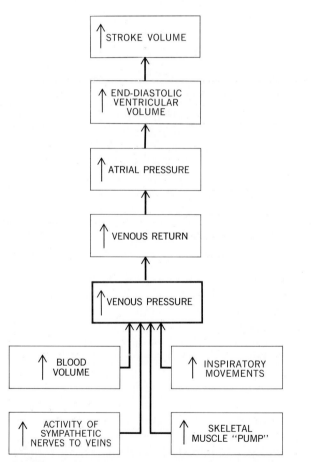

FIGURE 11-57. Major factors determining venous pressure and, hence, stroke volume. The figure shows how venous pressure and stroke volume are increased; reversing the arrows in the boxes would indicate how these can be decreased. The effects of altered inspiratory movements on end-diastolic ventricular volume are actually quite complex, and for the sake of simplicity, they are simply shown as altering venous pressure.

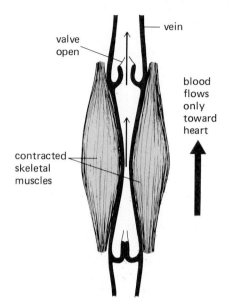

FIGURE 11-58. The skeletal muscle pump. During muscle contraction, venous diameter decreases and venous pressure rises. The resulting increase in blood flow can occur only toward the heart because of the valves in the veins.

smooth muscle contraction, the skeletal muscle pump, and the respiratory pump are to facilitate return of blood to the heart, and thereby to enhance cardiac output.

The venous valves

The veins in the limbs have valves which close so as to allow flow only toward the heart (Fig. 11-58). Why are these valves necessary if the pressure gradient created by cardiac contraction always moves blood toward the heart? We have seen how two other forces, the muscle pump and inspiratory movements, facilitate flow of venous blood. When these forces raise venous pressure locally, blood would be forced in both directions if the valves were not there to prevent backward flow. As we shall see, valves also play a role in counteracting the effects of upright posture.

Lymphatics

The lymphatics are not part of the circulatory system per se but constitute a one-way route for the movement of interstitial fluid to blood. The lym-phatic vessels arise as a group of blind-end lymph capillaries which are present in almost all organs of the body. These very thin-walled capillaries have large pores and are permeable to all interstitial-fluid constituents, including protein. Interstitial fluid enters these capillaries and (now known as **lymph**) moves through the **lymphatic vessels**, which converge to form larger and larger vessels. Ultimately, the two largest of these lymphatics drain into veins in the lower neck (valves here permit only one-way flow). Thus, the lymphatics carry fluid from the interstitial fluid to the cardiovascular system. At points during its flow through the lymphatic system, the lymph flows through **lymph nodes**.

Functions of the lymphatic system

Return of excess filtered fluid. In a normal person, the fluid filtered out of the capillaries (excluding those in the kidneys) each day exceeds that reabsorbed by approximately 3 L. This excess is returned to the blood via lymphatics (Fig. 11-59). Partly for this reason, obstruction of lymphatics leads to increased interstitial fluid, i.e., edema.

Return of protein to the blood. Most capillaries in the body have a slight permeability to protein, and

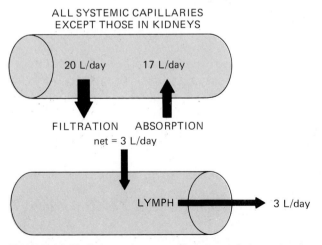

FIGURE 11-59. Relation between filtration and absorption in systemic capillaries (excluding those in the kidneys). The excess of filtration over absorption is returned to the circulation via the lymph.

accordingly, there is a small, steady movement of protein from the blood into the interstitial fluid. This protein is returned to the circulatory system via the lymphatics. The breakdown of this cycle is the most important cause of the marked edema seen in patients with lymphatic malfunction. Because some protein is normally leaked into the interstitial fluid, failure of the lymphatics to remove it by carrying away the interstitial fluid containing it allows the interstitial protein concentration to increase. This reduces or eliminates the protein-concentration difference and thus the water-concentration difference across the capillary wall and permits net movement of increased quantities of fluid out of the capillary into the interstitial space.

Specific transport functions. In addition to these nonspecific transport functions, the lymphatics also provide the pathway by which certain specific substances reach the blood. The most important is fat absorbed from the gastrointestinal tract.

Defense mechanisms. Besides its transport functions, the lymphatic system plays critical roles in the body's defenses against microbes and other foreign matter. These functions, which are mediated by the lymph nodes located along the larger lymphatic vessels, are described in Chap. 17.

Mechanism of lymph flow

How does the lymph move with no heart to push it? Lymph flow depends, in large part, upon forces external to the vessels, e.g., the pumping action of the skeletal muscles through which the lymphatics flow and the effects of respiration on thoracic-cage pressures; since the lymphatics have valves similar to those in veins, pressures cause only flow toward the points at which the lymphatics enter the circulatory system. In addition, the smooth muscle surrounding the lymphatic vessels beyond the lymphatic capillaries exerts a pump-like "sucking" action by their rhythmic contractions.

SECTION E.
INTEGRATION OF CARDIOVASCULAR FUNCTION:
REGULATION OF SYSTEMIC ARTERIAL PRESSURE

In Chap. 7 we described the fundamental ingredients of all reflex control systems: (1) the internal environmental variable being maintained relatively constant; (2) receptors sensitive to changes in this variable; (3) afferent pathways from the receptors; (4) a control center which receives and integrates the afferent inputs; (5) efferent pathways from the control center; (6) effectors which are "directed" by the efferent pathways to alter their activities, the result being a change in the level of the regulated variable. The control and integration of cardiovascular function will be described in these terms.

The major variable being regulated is the systemic arterial blood pressure. This should not be surprising, since the arterial pressure is the driving force for blood flow through all the organs; maintaining it is, therefore, a prerequisite for assuring adequate organ blood flows. *The mean arterial pressure is directly determined by only two factors—the cardiac output and the resistance to flow offered by the vascular system.* This relationship can be derived from the basic pressure-flow equation used repeatedly in this chapter.

$$F = \frac{\Delta P}{R}$$

Rearranging terms algebraically

$$\Delta P = F \times R$$

This form of the equation clearly shows the dependence of pressure upon flow and resistance. Because the systemic vascular tree is a continuous closed series of tubes, this equation holds for the entire system, i.e., from the very first portion of the aorta to the last portion of the vena cava just at the entrance to the heart. Therefore

ΔP = mean aortic pressure −
       late vena cava pressure
Flow = cardiac output (CO)
R = total peripheral resistance (TPR)

where **total peripheral resistance** means the sum of the resistances of all the vessels in the systemic vascular tree.

Since the late vena cava pressure is very close to 0 mmHg,

$$\Delta P = \text{mean aortic pressure} - O$$
or, $$\Delta P = \text{mean aortic pressure}$$

and since the mean pressure is essentially the same in the aorta and all other large arteries, ΔP is simply equal to mean arterial pressure (MAP). Thus the pressure-flow equation for the entire vascular tree now becomes

$$\begin{array}{c}\text{Mean arterial pressure (MAP)} = \\ \text{cardiac output} \times \text{total peripheral resistance} \\ \text{(CO)} \qquad\qquad \text{(TPR)}\end{array}$$

That the volume of blood pumped into the arteries per unit time (the cardiac output) is one of the two direct determinants of mean arterial pressure should come as no surprise; that the resistance of the blood vessels to flow (the total peripheral resistance) is the other determinant may not be so intuitively obvious but can be illustrated by using the model introduced in Fig. 11-43. As set up in Fig. 11-60, a pump pushes fluid into a cylinder at the rate of 1 L/min; at steady state, fluid leaves the cylinder via the outflow tubes at a rate of 1 L/min, and the height of the fluid column, which is the driving pressure for outflow, remains stable. We disturb the steady state by loosening the cuff on outflow tube 1, thereby increasing its radius, reducing its resistance, and increasing its flow. The total outflow for the system immediately becomes greater than 1 L/min, more fluid leaves the reservoir than enters via the pump, and the height of the fluid column begins to decrease. In other words, at any given pump input, a change in *total* outflow resistance must produce changes in the pressure of the reservoir.

This analysis can be applied, as previously, to the cardiovascular system by equating the pump with the heart, the reservoir with the arteries, and the outflow tubes with various arteriolar beds. An

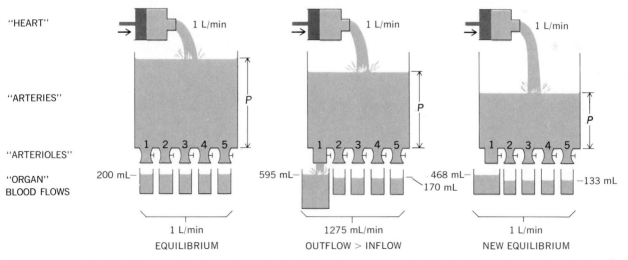

FIGURE 11-60. Model to illustrate the dependence of arterial blood pressure upon total arteriolar resistance, showing the effects of dilating one arteriolar bed upon arterial pressure and organ blood flow if no compensatory adjustments occur. The middle panel is a transient state before the new equilibrium occurs. In one respect, the illustration of the model is misleading in that the arterial reservoir is shown containing very large quantities of blood. In fact, as we have seen, the volume of blood in the arteries is quite small.

analogy to opening outflow tube 1 is exercise; during exercise, the skeletal muscle arterioles dilate, thereby decreasing resistance. If the cardiac output and the arteriolar diameters of all other vascular beds were to remain unchanged, the increased runoff through the skeletal muscle arterioles would cause a decrease in arterial pressure. (As we shall see, this does not happen because the cardiac output and the other arterioles do not remain unchanged.)

It must be reemphasized that it is the *total* arteriolar resistance which influences arterial blood pressure; the *distribution* of resistances among organs is irrelevant in this regard. Fig. 11-61 illustrates this point. Outflow tube 1 has been opened, as in the previous example, whereas tubes 2 to 4 have been simultaneously tightened. The increased resistance offered by tubes 2 to 4 compensates for the decreased resistance offered by tube 1; therefore total resistance remains unchanged from that of Fig. 11-60, and reservoir pressure is unchanged. Total outflow remains 1 L/min, although the distribution of flow is such that flow in tube 1 is increased, that in tubes 2 to 4 is decreased, and that in tube 5 is unchanged.

Applied to the body, this process is analogous to altering the distribution of vascular resistances. When the skeletal muscle arterioles (tube 1) dilate during exercise, the *total* resistance of all vascular beds could still be maintained if arterioles were to constrict in other organs, such as the kidneys, gastrointestinal tract, and skin (tubes 2 to 4), which can readily suffer moderate flow reductions for at least short periods of time. In contrast, the brain arterioles (tube 5) remain unchanged, thereby assuring constant brain blood supply. This type of resistance juggling, however, can maintain total resistance only within limits. Obviously if tube 1 opens very wide, even total closure of the other tubes could not prevent total outflow resistance from falling; we shall see that this is, in fact, the case during exercise.

The equation, MAP = CO × TPR, is the fundamental equation of cardiovascular physiology. It provides the means for integrating almost all the information presented in previous sections of this chapter. For example, we can now explain why pulmonary arterial pressure is much lower than the systemic arterial pressure (Fig. 11-38). The blood flow per minute, i.e., cardiac output through the pulmonary and systemic arteries, is, of course, the same; therefore, the pressures can differ only if the resistances differ. Thus, the pulmonary vessels must offer much less resistance to flow than the systemic arterioles. In other words, the total pulmonary vascular resistance is lower than the total systemic vascular resistance.

Figure 11-62 presents the grand scheme of fac-

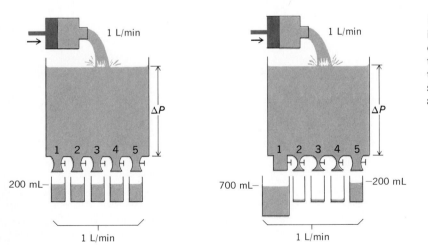

FIGURE 11-61. Compensation for dilation in one bed by constriction in others. When outflow tube 1 is opened, outflow tubes 2 to 4 are simultaneously tightened so that the total outflow resistance remains constant, total rate of runoff remains constant, and reservoir pressure remains constant.

tors which determine systemic arterial pressure.[1] None of this information is new; all of it has been presented in previous figures. A change in any single variable shown in the figure will, all others remaining constant, produce a change in mean arterial pressure by altering either cardiac output or total peripheral resistance.

But this figure also includes all the effector mechanisms and efferent pathways which rapidly *regulate* arterial pressure. Any deviation in arterial pressure from its normal operating point will call forth a reflex alteration of one or more of the variables in this figure such that cardiac output and/or total peripheral resistance will be changed in the direction required to minimize the initial change in arterial pressure. For example, a decrease in arterial pressure, for whatever reason, will reflexly induce (via changes in the autonomic nervous system and angiotensin) an increase in cardiac output and/or peripheral resistance, thus tending to restore arterial pressure toward normal. (Not all the variables shown in this figure are controlled by such blood-pressure-regulating reflexes—the major exceptions are the local controls of arteriolar radius.)

In one sense, we have approached arterial pressure regulation backward, in that the past sections have described the effectors (heart, arterioles, veins) and efferent pathways (mainly the autonomic nervous system). Now we must complete the reflexes (Fig. 11-63) by describing the monitoring systems (receptors and afferent pathways to the brain) and the control centers in the brain. Before doing so, however, we must emphasize that, despite the preeminent role of the autonomic nervous system in the reflexes controlling arterial pressure, it certainly is not the only factor, as shown in Fig. 11-62. Blood volume is particularly crucial because of its effect on venous pressure, ventricular filling, and stroke volume (Fig. 11-62). Indeed the regulation of blood volume is considered by many to be the single most critical component of blood-pressure control, particularly over long periods of time. Short-term regulation of blood volume is achieved mainly by reflex alteration of the forces which determine the distribution of extracellular fluid between the vascular and interstitial compartments; these were described earlier in the section dealing with capillaries, and an example of their potent role will be given in the section on hemorrhage appearing later in this chapter. Long-term control of blood volume is achieved principally by homeostatic regulation of red-cell volume (through the erythropoietin system described earlier in this chapter) and plasma volume (through the control of salt and water excretion by the kidneys, as will be described in Chap. 13).

Receptors and Afferent Pathways

The first link in the reflexes which regulate arterial pressure are the receptors and afferent pathways bringing information into the central nervous system (Fig. 11-63). The most important of these are the **arterial baroreceptors**.

Arterial baroreceptors

It is only logical that the reflexes which homeostatically regulate arterial pressure originate primarily with arterial receptors which respond to changes in pressure. High in the neck, each of the major arteries (carotids) supplying the head divides into two smaller arteries. At this bifurcation, the wall of the artery is thinner than usual and contains a large number of branching, vinelike nerve endings (Fig. 11-64). This portion of the artery is called the **carotid sinus**, and its nerve endings are highly sensitive to stretch or distortion; since the degree of wall

---

[1] Any model of a system as complex as the circulatory system must, of necessity, be an oversimplification. One basic deficiency in our model is that its chain of causal links is entirely unidirectional (from bottom to top in Fig. 11-62) whereas, in fact, there are also important causal interactions in the reverse direction. For example, as the figure shows, a change in cardiac output influences arterial pressure, but it is also true that a change in arterial pressure has an important effect on cardiac output. To take another example, atrial pressure is shown influencing cardiac output, but it is also true that cardiac output influences atrial pressure. A reader interested in delving further into these complex interactions and feedback loops should consult the more advanced works listed at the back of the book.

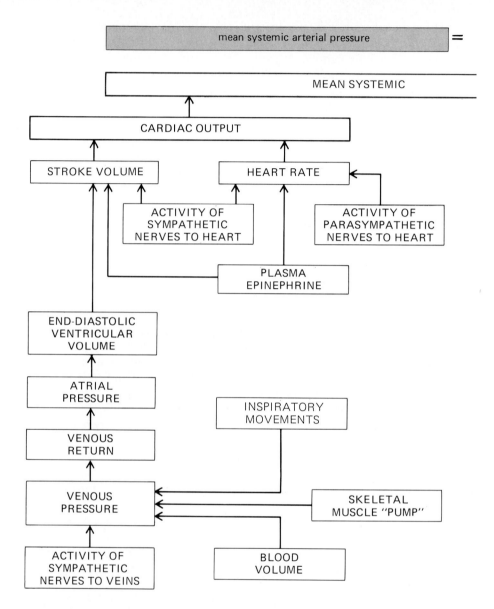

stretching is directly related to the pressure within the artery, the carotid sinus actually serves as a pressure receptor (**baroreceptor**). From the nerve endings come afferent neurons which travel to the brain, where they eventually synapse upon the neurons of cardiovascular control centers. A single area functionally similar to the carotid sinuses is found in the **arch of the aorta**, which constitutes a second important arterial baroreceptor.

Action potentials recorded in single afferent fibers from the carotid sinus demonstrate the pattern of response by these receptors. When the arterial pressure in the carotid sinus is artificially controlled at a steady nonpulsatile pressure of 100 mmHg, there is a tonic rate of discharge by the nerve. This rate of firing can be increased by raising the arterial pressure; it can be decreased by lowering the pressure. Figure 11-65 illustrates the same type

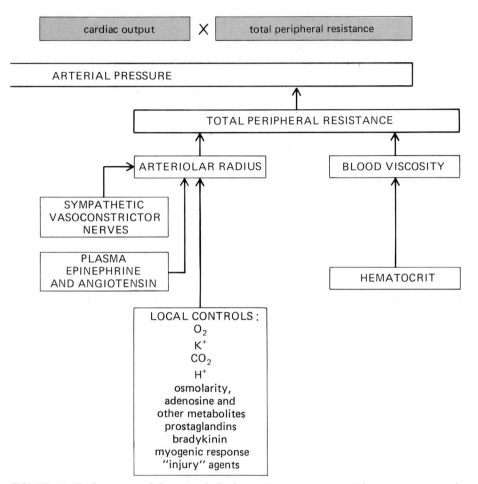

FIGURE 11-62. Summary of factors which determine systemic arterial pressure, an amalgamation of Figs. 11-37, 11-46, and 11-57, with the addition of the effect of hematocrit on resistance.

of experiment, except that pulsatile perfusion is used. Note that the arterial baroreceptors increase their rate of firing not only when *mean* arterial pressure increases but when *pulse pressure* increases as well, a responsiveness that adds a further degree of sensitivity to blood pressure regulation, since small changes in factors such as blood volume cause changes in pulse pressure before they become serious enough to affect mean pressure.

Other baroreceptors

Other portions of the vascular system contain nerve endings sensitive to stretch; these are other large arteries, the large veins, the pulmonary vessels, and the cardiac walls themselves. By means of these receptors, brain cardiovascular control centers are kept constantly informed about the venous, atrial, and ventricular pressures, and a further degree of sensitivity is gained. Thus, a slight decrease in atrial pressure may begin, reflexly, to facilitate the sympathetic nervous system even before the change lowers cardiac output and arterial pressure far enough to be detected by the arterial baroreceptors. As we shall see in Chap. 13, the atrial baroreceptors are particularly important for the control of body sodium and water, and therefore, blood volume.

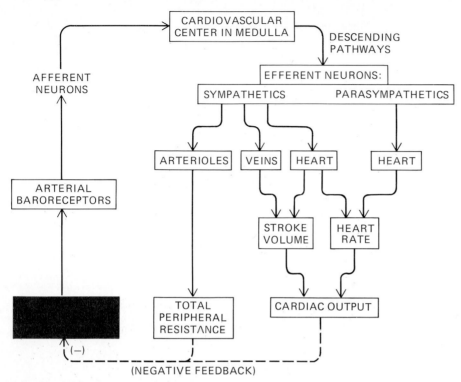

FIGURE 11-63. Basic reflex for regulation of arterial blood pressure. Start this figure at the lower left: Any change in arterial blood pressure is detected by the arterial baroreceptors (increased arterial pressure causes increased firing of the baroreceptors); the result, via the medullary cardiovascular center and the autonomic neurons to the cardiovascular system, is reflex alteration of cardiac output and total peripheral resistance so as to minimize the original change in the blood pressure.

Cardiovascular Control Centers in the Brain

The medullary cardiovascular control center

The primary cardiovascular control center for the baroreceptor reflex is in the medulla, the first segment of brain above the spinal cord. The relevant medullary neurons are sometimes divided into cardiac and vasomotor centers, which are then further subdivided and classified, but because these areas actually constitute diffuse networks of highly interconnected neurons, we prefer to call the entire complex the medullary cardiovascular center. The neurons of this center receive input from the arterial baroreceptors (as well as the other baroreceptors described above). This input determines the outflow from the center along axons which terminate upon the cell bodies and dendrites of both the vagus

(parasympathetic) neurons to the heart and the sympathetic neurons to the heart, arterioles, and veins (Fig. 11-63).[1]

The synaptic connections within the medullary cardiovascular center are such that when the arterial baroreceptors *increase* their rate of discharge, the result is a *decrease* in sympathetic outflow to the heart, arterioles, and veins and an *increase* in parasympathetic outflow to the heart (Fig. 11-66). A decrease in firing rate of the baroreceptors results in just the opposite pattern.

Our description of the major blood-pressure-regulating reflex is now complete (Fig. 11-67). If,

[1] For the sake of simplicity, we focus on the sympathetic nervous system when discussing reflex control of arterioles. Hormones, particularly angiotensin, also participate importantly in these reflexes (see Chap. 13).

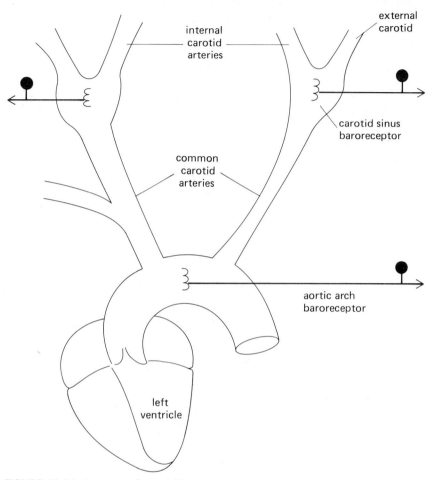

FIGURE 11-64. Location of arterial baroreceptors.

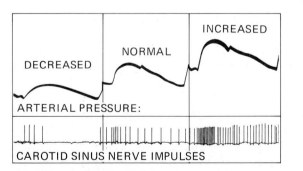

FIGURE 11-65. Effects of altering arterial pressure on firing of afferent neurons from the carotid sinus.

for some reason, arterial pressure decreases, this causes the rate of discharge of the carotid sinus and aortic arch baroreceptors to decrease; fewer impulses travel up the afferent nerves to the medullary cardiovascular center and, via appropriate synaptic connections with the neurons of this center, this induces (1) speeding of the heart because of increased sympathetic discharge and decreased parasympathetic discharge, (2) increased myocardial force of contraction because of increased sympathetic activity, (3) arteriolar constriction because of increased sympathetic activity (and increased plasma epinephrine and angiotensin), and (4) venous constriction because of increased sympathetic

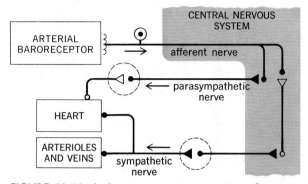

▶—● neurons which stimulate

▷—○ neurons which inhibit

FIGURE 11-66. A diagrammatic representation of reciprocal innervation in the control of the cardiovascular system. Baroreceptor input, which stimulates the parasympathetic nerves to the heart, simultaneously inhibits the sympathetic nerves to the heart, arterioles, and veins.

discharge to venous smooth muscle. The net result is an increased cardiac output (increased heart rate and stroke volume), increased total peripheral resistance (arteriolar constriction), and return of blood pressure toward normal.

In contrast, when arterial blood pressure increases for whatever reason, this causes increased firing of the arterial baroreceptors, which reflexly induces a compensatory decrease in cardiac output and total peripheral resistance.

Having emphasized the great importance of this baroreceptor reflex, we must now add an equally important qualification. The baroreceptor reflex functions primarily as a short-term regulator of arterial blood pressure; it is activated instantly by any blood-pressure change and restores blood pressure rapidly toward normal to the degree possible in the situation. Yet, the surprising fact is that if the factors impinging on the cardiovascular system to drive arterial pressure away from its normal operating point succeed in doing so for more than a few days, the arterial baroreceptors somehow come to "accept" this new pressure as normal; i.e., these receptors adapt to high pressure. Thus, in patients who have chronically elevated blood pressure (as described in the subsequent section on

hypertension), the baroreceptors continue to oppose minute-to-minute changes in blood pressure, but at the higher level. It is as though they have been "reset," i.e., had their operating point changed.

Other cardiovascular control centers

We have thus far restricted our description of blood pressure control to the baroreceptor reflex since it is the one which functions to keep arterial pressure relatively constant. However, other stimuli (for example, pain or changes in arterial oxygen concentration) and many physiological states (for example, eating and sexual activity) are associated with changes in arterial blood pressure. In addition to the specific pathways from the medullary cardiovascular center to the parasympathetic and sympathetic preganglionic neurons there are other pathways to these autonomic preganglionics, pathways which arise in diverse, interconnected nuclei in various brain levels. Thus, certain neurons in the cerebral cortex and hypothalamus synapse directly on the sympathetic neurons in the spinal cord, bypassing the medullary center altogether. Other neurons in the brain importantly influence the output of the medullary cardiovascular center under certain circumstances, as well as being influenced themselves by ascending pathways from the medulla. Thus, there is an incredible degree of flexibility and integration to the control of blood pressure. For example, all the usually observed neurally mediated cardiovascular responses to an acute emotional situation can be elicited in unanesthetized baboons by electrically stimulating a discrete area in the hypothalamus. Stimulation at other sites elicits cardiovascular changes appropriate to the maintenance of body temperature, feeding, or sleeping. It seems that such outputs are "preprogrammed," and that the complete pattern can be released by a natural stimulus which initiates the flow of afferent information to the appropriate controlling brain center. This is not to say that a given pattern will always be precisely the same, since the complex synaptic interlinking of the various centers and pathways permits neuronal integration at all levels. Thus, the

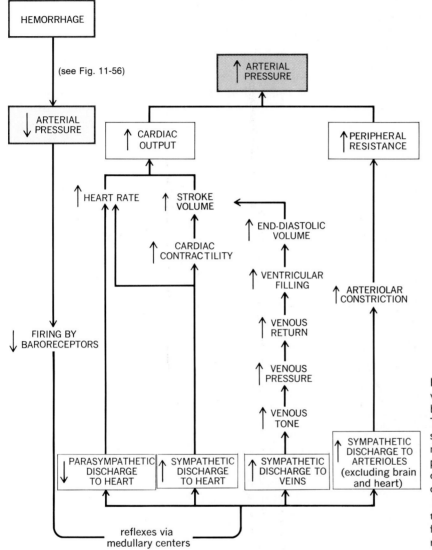

FIGURE 11-67. Reflex mechanisms by which lower arterial pressure following blood loss is brought back toward normal. The compensatory mechanisms do not restore arterial pressure completely to normal. Beyond the initial decrease in arterial pressure, all arrows signifying increases or decreases are relative to the state immediately following the hemorrhage (see Fig. 11-68). For simplicity, we have not shown the fact that plasma angiotensin is also reflexly increased and helps constrict arterioles.

cardiovascular response to warming of the body (controlled by a hypothalamic site) will be quite different in a normal person and in a person who has suffered a hemorrhage, because of interactions between the baroreceptor and temperature-regulating reflexes. In all this it is very easy to lose track of the fact that, regardless of the central nervous system sites and pathways involved in any given response, the major final common pathways to the heart and blood vessels are always the same—the parasympathetic and sympathetic neurons (and circulating epinephrine and other hormones).

SECTION F.
CARDIOVASCULAR PATTERNS IN HEALTH AND DISEASE

Hemorrhage and Hypotension

The decrease in blood volume caused by bleeding (**hemorrhage**) produces a drop in blood pressure (**hypotension**) by the sequence of events previously shown in Fig. 11-56. The most serious consequences of the lowered blood pressure are the reduced blood flow to the brain and cardiac muscle. The baroreceptor reflex restoring pressure toward normal is summarized in Fig. 11-67. Figure 11-68 adds a further degree of clarification by presenting the pattern of changes over time. The important generalization to be gained from this latter figure is that the values of factors changed by the hemorrhage, per se (these include stroke volume, cardiac output, and mean arterial pressure), are restored by the baroreceptor reflex *toward,* but not *to* normal; in contrast, values not altered by the hemorrhage itself (heart rate and peripheral resistance) are actually increased above their normal (prehemorrhage) values. (As emphasized in Chap. 7, analogous generalizations apply to many of the flow diagrams used in this book to illustrate homeostatic compensations.) The increased peripheral resistance, it should be reemphasized, is the result of increases in sympathetic outflow to the arterioles in many vascular beds but not those of the heart and brain; thus, skin blood flow may decrease markedly because of vasoconstriction of its arterioles (this is why the skin becomes cold and pale) as does kidney and intestinal blood flow, whereas brain and heart blood flow remain relatively unchanged.

A second important type of compensatory mechanism not shown in Fig. 11-67 involves capillary fluid exchange, namely, the movement of interstitial fluid into capillaries. This occurs because both the drop in blood pressure and the increase in arteriolar constriction decrease capillary hydrostatic pressure, thereby favoring absorption of interstitial fluid (Fig. 11-69). Thus, the initial event—blood loss and decreased blood volume—is in large

part compensated for by the movement of interstitial fluid into the vascular system. Indeed, as shown in Table 11-4, 12 to 24 h after a moderate hemorrhage the blood volume may be restored virtually to normal by this mechanism. At this time the entire restoration of blood volume is due to expansion of the plasma volume, for replacement of the lost erythrocytes requires many days.

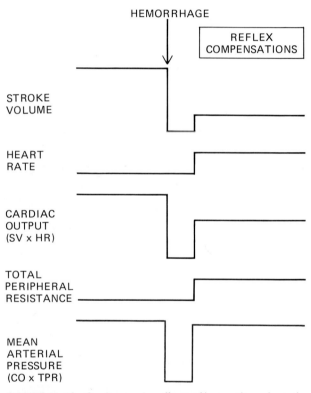

FIGURE 11-68. Cardiovascular effects of hemorrhage. In studying this figure, recall two facts: (1) Cardiac output is the product of heart rate and stroke volume; and (2) mean arterial pressure is the product of cardiac output and total peripheral resistance. Note that the entire initial decrease in arterial pressure following hemorrhage is secondary to the decrease in stroke volume (and, hence, cardiac output). This figure emphasizes the relativeness of the "increase" and "decrease" arrows of Fig. 11-67; all variables shown are increased relative to the state immediately following hemorrhage, not necessarily to the state prior to the hemorrhage.

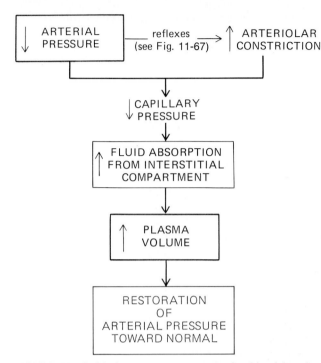

FIGURE 11-69. Mechanisms compensating for blood loss by movement of interstitial fluid into the capillaries.

TABLE 11-4. Fluid shifts after hemorrhage

	Normal	Immediately after hemorrhage	18 h after hemorrhage
Total blood volume, mL	5000	4000 ($\downarrow$ 20%)	4900
Erythrocyte volume, mL	2300	1840 ($\downarrow$ 20%)	1840
Plasma volume, mL	2700	2160 ($\downarrow$ 20%)	3060
Plasma albumin, g	135	108 ($\downarrow$ 20%)	125

At 18 h (Table 11-4) much of the plasma protein albumin lost in the hemorrhage has been replaced. This phenomenon is of great importance for restoration and expansion of plasma volume, as can be seen by considering the following discussion. As capillary hydrostatic pressure decreases as a result of the hemorrhage, interstitial fluid enters the plasma; this fluid, however, contains virtually no protein, so that its entrance dilutes the plasma proteins, which increases the plasma water concentration. The resulting reduction of the water-concentration difference between the capillary plasma and the interstitial fluid hinders further fluid absorption and would prevent the full compensatory expansion if it were not that replacement of the plasma protein minimizes this fall in protein concentration. Movement of interstitial fluid into the plasma can therefore continue. Surprisingly, synthesis of new albumin contributes to this replacement to only a minor degree during the first 24 h; most of the replacement albumin which appears in the plasma during this early stage is interstitial protein carried by the lymphatics to the vascular compartment.

We must emphasize that absorption of interstitial fluid has only *redistributed* the extracellular fluid; ultimate replacement of the plasma lost from the body involves the control of fluid ingestion and kidney function, both described in Chap. 13. Similarly, replacement of the lost red cells requires stimulation of erythropoiesis (by erythropoietin). Both these replacement processes require days to weeks in contrast to the rapidly occurring reflex compensations described in Fig. 11-69.

Loss from the body of large quantities of cell-free extracellular fluid rather than whole blood can also cause hypotension. This may occur via the skin, as in severe sweating or burns, via the gastrointestinal tract, as in diarrhea or vomiting, or via unusually large urinary losses. Regardless of the route, the loss decreases circulating blood volume and produces symptoms and compensatory phenomena similar to those seen in hemorrhage.

Hypotension may be caused by events other than blood or fluid loss. One such form is due to strong emotion and can result in fainting. Somehow, the higher brain centers involved with emotions act to inhibit sympathetic activity and enhance parasympathetic activity, resulting in a markedly decreased arterial pressure and brain blood flow. This whole process is usually transient.

Other important causes of hypotension seem to have a common denominator in the liberation within the body of chemicals which relax arteriolar smooth muscle. The cause of hypotension in these cases is reduction of peripheral resistance. An important example is the hypotension which occurs during severe allergic responses.

It may be of interest to point out the physiological reasons for not treating patients with hypotension in ways commonly favored by the uninformed, namely, administering alcohol and covering the person with mounds of blankets. Both alcohol and excessive body heat, by actions on the central nervous system, cause profound dilation of skin arterioles, thus lowering peripheral resistance and decreasing arterial blood pressure still further. As described in a later section, the worst thing is to try to get the person to stand up.

Shock

The compensatory mechanisms described above for hemorrhage are highly efficient, so that losses of as much as 1 to 1.5 L of blood (approximately 20 percent of total blood volume) can be sustained with only slight reductions of mean arterial pressure or cardiac output. In contrast, when much greater losses occur or when compensatory adjustments are inadequate, the decrease in cardiac output may be large enough to damage various tissues and organs seriously. This is known as **circulatory shock**. The cardiovascular system, itself, suffers damage if shock is prolonged, and as it deteriorates, cardiac output declines even more and shock becomes progressively worse and ultimately irreversible—the person dies even after blood transfusions and other appropriate therapy.

Just as hemorrhage is not the only cause of hypotension and decreased cardiac output, so it is not the only cause of shock. Low volume due to loss of fluid other than blood, excessive release of vasodilators as in allergy and infection, loss of neural tone to the cardiovascular system, trauma— all these can lead to severe reductions of cardiac

output and the positive-feedback cycles culminating in irreversible shock.

The Upright Posture

The simple act of getting out of bed and standing up is equivalent to a mild hemorrhage, because the changes in the circulatory system in going from a lying, horizontal position to a standing, vertical one result in a decrease in the effective circulating blood volume. Why this is so requires an understanding of the action of gravity upon the long continuous columns of blood in the vessels between the heart and the feet. All the pressures we have given in previous sections of this chapter were for the horizontal position, in which all blood vessels are at approximately the same level as the heart, and the weight of the blood produces negligible pressure. In the vertical position (Fig. 11-70), the intravascular pressure everywhere becomes equal to the usual pressure resulting from cardiac contraction plus an additional pressure equal to the weight of a column of blood from the heart to the point of

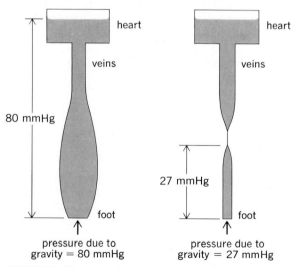

FIGURE 11-70. Role of contraction of the leg skeletal muscles in reducing capillary pressure and filtration in the upright position. The skeletal muscle contraction compresses the veins, causing intermittent complete emptying so that the column of blood is interrupted.

measurement. In a foot capillary, for example, the pressure increases from 25 (the pressure resulting from cardiac contraction) to 105 mmHg, the extra 80 mmHg being due to the weight of the column of blood (Fig. 11-70). The increased hydrostatic pressure in the veins of the legs which occurs upon standing pushes outward upon the highly distensible vein walls, causing marked distension with resultant pooling of blood in the veins; i.e., much of the blood emerging from the capillaries simply remains in the expanding veins rather than returning to the heart. Simultaneously, the marked increase in capillary pressure caused by the gravitational force produces increased filtration of fluid out of the capillaries into the interstitial space (this accounts for the fact that our feet swell). The combined effects of venous pooling and increased capillary filtration are a significant reduction in the effective circulating blood volume very similar to a mild hemorrhage. The ensuing decrease in arterial pressure causes reflex compensatory adjustments similar to those shown in Fig. 11-67 for hemorrhage.

Perhaps the most effective compensation in an awake, conscious person is contraction of skeletal muscles of the legs, which produces intermittent, complete emptying of veins within the upper leg, so that uninterrupted columns of venous blood from the heart to the feet no longer exist. The result is a decrease in venous distension and pooling and a marked reduction in capillary hydrostatic pressure and fluid filtration out of the capillaries (Fig. 11-70). An example of the importance of this compensation (really its absence) occurs when soldiers faint after standing very still, i.e., with minimal contraction of the abdominal and leg muscles, for long periods of time. Here the fainting may be considered adaptive in that venous and capillary pressure changes induced by gravity are eliminated once the person is prone; the pooled venous blood is mobilized, and the previously filtered fluid is reabsorbed into the capillaries. Thus, the wrong thing to do to anyone in a faint is to hold him or her upright.

Varicose veins (irregularly swollen superficial veins of the lower extremities) are also explainable in terms of hydrostatic pressures. They are the result of abnormal structural properties of the wall such that they are overdistensible. Therefore, they become so distended at normal hydrostatic pressures that the valves become incompetent and there is a backflow of blood from the deep veins through connections to the superficial veins. As described in a subsequent section, a major complication of venous stasis in varicose veins is the possibility of clot formation.

Exercise

During exercise, cardiac output may increase from a resting value of 5 L/min to the maximal values of 35 L/min obtained by trained athletes. The distribution of this cardiac output during strenuous exercise is illustrated in Fig. 11-71. As expected, most of the increase in cardiac output goes to the exercising muscles, but there are also increases in flow to skin (required for loss of the heat produced) and to heart (required for the additional work performed by the heart in pumping the increased cardiac output). The increases in flow through these three vascular beds are the result of arteriolar dilation in them; in both skeletal and cardiac muscle, the dilation is mediated by local metabolic factors, whereas the dilation in skin is achieved by decreasing the firing of the sympathetic neurons to the skin. At the same time arteriolar dilation is occurring in those beds, arteriolar constriction (manifested as decreased flow in Fig. 11-71) is occurring in the kidneys and gastrointestinal organs, secondary to increased activity of the sympathetic neurons supplying them.

Thus, the decrease in peripheral resistance resulting from dilation of arterioles in skeletal muscle, cardiac muscle, and skin arterioles is partially offset by constriction of arterioles in other organs, but this "resistance juggling" is quite incapable of compensating for the huge dilation of the muscle arterioles, and the net result is a marked decrease in total peripheral resistance.

What happens to arterial blood pressure during exercise? As always, the mean arterial pressure is simply the product of the cardiac output and the peripheral resistance. During most forms of exer-

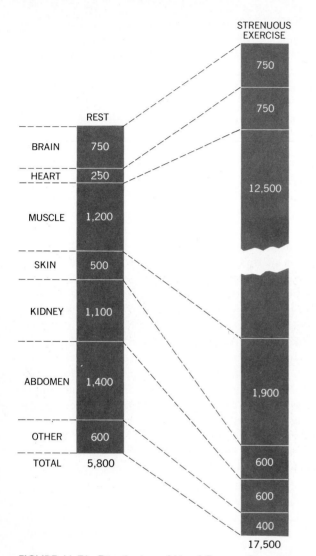

FIGURE II-71. Distribution of blood flow to the various organs and tissues of the body at rest and during strenuous exercise. The numbers show blood flow in milliliters per minute. (*Adapted from Chapman and Mitchell.*)

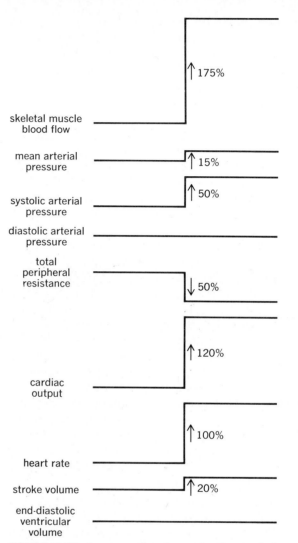

FIGURE II-72. Summary of cardiovascular changes during mild exercise.

cise (Fig. 11-72 illustrates the case for mild exercise), the cardiac output tends to increase somewhat more than the peripheral resistance decreases so that mean arterial pressure usually increases slightly. However, regardless of what happens to mean pressure, the pulse pressure shows a marked increase, mainly because of the quicker ejection of blood by the sympathetically driven heart.

The cardiac output increase during exercise is associated with greater sympathetic activity and less parasympathetic activity to the heart. Thus, heart rate and stroke volume both rise, causing an increased cardiac output. The heart-rate changes are usually much greater than stroke-volume changes. Note (Fig. 11-72) that, in our example, the increased stroke volume occurs without change in end-diastolic ventricular volume; accordingly, the increased stroke volume cannot be ascribed to Starling's law but rather is due completely to the increased contractility induced by the cardiac sympathetic nerves.

However, it would be incorrect to leave the impression that enhanced sympathetic activity to the heart completely accounts for the elevated cardiac output which occurs in exercise, for such is not the case. The fact is that cardiac output can be increased to high levels only if venous return to the heart is simultaneously facilitated to the same degree; otherwise end-diastolic volume would fall and stroke volume would decrease (because of Starling's law). Therefore, factors promoting venous return during exercise are extremely important. They are (1) marked activity of the skeletal muscle pump, (2) increased respiratory movements, (3) sympathetically mediated increase in venous tone, and (4) the greater ease of blood flow from arteries to veins through the dilated skeletal muscle arterioles. (For the sake of completeness, it should be noted that there is at least one change during exercise which opposes increased venous return, namely, a decrease in plasma volume. This occurs for two reasons: (1) the increased capillary pressure in the exercising muscles drives increased filtration across the capillary walls into the interstitial compartment; and (2) the salt and water lost during sweating come from the plasma.)

During vigorous exercise these factors may be so powerful that venous return is enhanced enough to cause an increase in end-diastolic ventricular volume unlike the situation for the mild level of exercise illustrated in Fig. 11-72. Under such conditions, stroke volume is further enhanced above what it would have been from increased contractility alone.

What are the control mechanisms by which the cardiovascular changes in exercise are elicited? As described previously, the dilation of arterioles in skeletal muscle (and cardiac muscle) once exercise is under way represents active hyperemia secondary to local metabolic factors within the muscle. But what drives the enhanced sympathetic outflow to most other arterioles, the heart, and the veins and the decreased parasympathetic outflow to the heart? The control of autonomic outflow during exercise offers an excellent example of what we earlier referred to as a "preprogrammed" pattern,

modified by continuous afferent input. One or more discrete control centers in the brain are activated during exercise and, via descending pathways to the appropriate autonomic preganglionic neurons, elicit the firing pattern typical of exercise (Fig. 11-73). (Indeed, these centers begin to "direct traffic" even before the exercise begins, since a person just about to begin exercising already manifests many of the changes in cardiac and vascular function.)

At the same time, certain of the chemical changes occurring in the exercising muscle activate chemoreceptors in the muscle, thereby causing the flow of information from the muscles along afferent neurons into the central nervous system. This afferent input, which is proportional to the degree of work being performed by the muscle, acts via the medullary cardiovascular center to facilitate the input reaching the autonomic neurons from the higher brain centers. The result is a further increase in heart rate, myocardial contractility, and vascular resistance in the nonactive organs. Such a system permits a fine degree of matching between cardiac pumping and total oxygen and nutrients required by the exercising muscles.

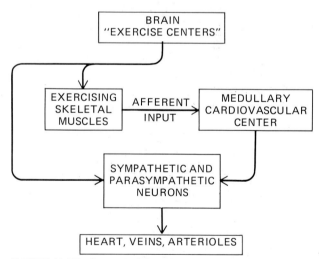

FIGURE 11-73. Control of the autonomic nervous system during exercise. The primary outflow to the sympathetic and parasympathetic neurons is via pathways from "exercise" centers in the brain, pathways parallel to those going to the exercising muscles. Afferent input from chemoreceptors in these muscles also influences the autonomic neurons by way of the medullary cardiovascular center.

What about the arterial baroreceptors (carotid sinuses and aortic arch)? Do they help drive the appropriate autonomic outflow during exercise? The answer is *no*, for the arterial baroreceptors actually oppose the observed changes in sympathetic and parasympathetic outflow. Recall that mean and pulsatile arterial pressure are elevated during exercise; therefore the arterial baroreceptors are firing at an increased frequency, which would signal the medullary cardiovascular center to decrease sympathetic outflow to the heart and vessels and increase parasympathetic outflow to the heart. Thus, we see how during exercise new groups of inputs (those from specific higher brain centers and chemoreceptors in the exercising muscle) can override the input (from arterial baroreceptors) which is preeminent at rest.

Cardiac output and training

The limiting factor in an endurance type of exercise is cardiac output, and it is stroke volume which ultimately limits cardiac output (Fig. 11-74). As we have seen, stroke volume at first increases with work load (although not to the same degree as does heart rate), but after reaching maximal values, then

it decreases at extremely high work loads (Fig. 11-74). The major factors responsible for this decrease are the very rapid heart rate (which decreases diastolic filling time) and failure of the peripheral factors favoring venous return (muscle pump, respiratory pump, venoconstriction, arteriolar dilation) to elevate venous pressure high enough to maintain adequate ventricular filling during the very short time available. The net result is a decrease in end-diastolic volume which causes, by Starling's law, a decrease in stroke volume.

A person's maximal endurance work load and oxygen consumption is not fixed at any given value but can be altered (in either direction) by the habitual level of physical activity. For example, prolonged bed rest may decrease them by 25 percent, whereas intense long-term physical training may increase them by a similar amount. It must be emphasized that, to be effective, the training must be of an endurance type i.e., it must involve large muscle groups for an extended period of time. We are still uncertain of the relative combinations of intensity and duration which are most effective but, as an example, jogging 10 to 15 min three times weekly at 5 to 8 mi/h definitely produces some conditioning effect in most people. This is manifest as

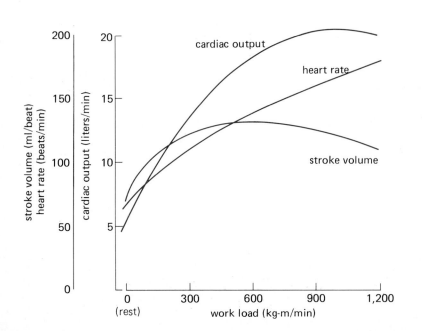

FIGURE 11-74. Changes in heart rate, stroke volume, and cardiac output with increasing work loads (measured as kg·m/ min). Note that cardiac output plateaus because stroke volume decreases at maximal levels of oxygen uptake.

an increased stroke volume and decreased heart rate (with no change in cardiac output) at rest and at any given submaximal work load, as well as an increased cardiac output at maximal work loads; the latter is due to an increased maximal stroke volume, since maximal heart rate is not altered by training. How much of this increased maximal stroke volume is due to the observed changes in the heart itself (increased cardiac muscle mass), as opposed to increases in the number of blood vessels in skeletal muscle (which would facilitate oxygen extraction, muscle blood flow, and venous return), remains unclear. Blood volume also tends to increase, as do the concentrations of oxidative enzymes and mitochondria in the affected muscles.

Hypertension

Hypertension (high blood pressure) is defined as a chronically increased arterial pressure. In general, the dividing line between normal pressure and hypertension is taken to be 140/90 mmHg. Heart attack and heart failure (see below), brain stroke (occlusion or rupture of a cerebral blood vessel), and kidney damage are all caused by prolonged hypertension and its attendant strain on the various organs.

Theoretically, hypertension could result from an increase in cardiac output or in peripheral resistance, or both. In fact, at least in most cases of well-established hypertension, the major abnormality is increased peripheral resistance due to abnormally reduced arteriolar diameter. What causes the arteriolar constriction? In only a small fraction of cases is the cause of hypertension known. For example, diseases which damage the kidneys or decrease their blood supply are often associated with hypertension; increased release of renin from the involved kidney(s) with subsequent increased generation of angiotensin almost certainly plays a role in this so-called **renal hypertension**. However, for more than 95 percent of the persons with hypertension, the cause of the arteriolar constriction is unknown, and hypertension of unknown cause is called **primary hypertension** (formerly "essential hypertension").

Many hypotheses have been proposed to explain the increased arteriolar constriction of primary hypertension. At present, much evidence seems to point to excessive sodium ingestion or retention within the body as a contributing factor in genetically predisposed persons, and certainly many persons with hypertension show a drop in blood pressure when treated with low-sodium diets or drugs (diuretics) which cause increased sodium loss via the urine. Weight reduction and exercise are also frequently effective in causing some reduction of blood pressure in overweight, sedentary persons with hypertension.

Many forms of therapy for primary hypertension involve drugs which act upon some aspect of autonomic function to produce arteriolar dilation. This does not mean that excessive sympathetic tone was the original cause of the hypertension (it probably is, in certain people) but only that, whatever the cause anything which dilates the arterioles reduces the blood pressure.

One group of drugs recently developed and now being extensively used for the treatment of primary hypertension blocks beta-adrenergic receptors. How these drugs end up reducing peripheral resistance is presently unknown.

Congestive Heart Failure

The heart may become weakened for many reasons; regardless of cause, however, the failing heart induces a similar procession of signs and symptoms grouped under the category of **congestive heart failure**. The basic defect in heart failure is a decreased contractility of the heart. As shown in Fig. 11-75, the failing heart shifts downward to a lower Starling curve. How can this be compensated for? Increased sympathetic stimulation would help to increase contractility, and this may occur. However, an even more striking compensation is increased ventricular end-diastolic volume. The heart in severe failure is generally engorged with blood, as are the veins and capillaries, the major cause being an increase (sometimes massive) in plasma volume. The sequence of events is as follows: Decreased cardiac output causes a decrease in mean and pulsatile ar-

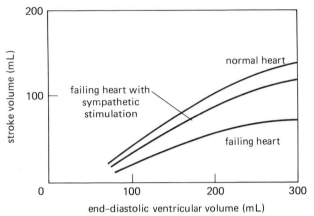

FIGURE 11-75. Relationship between end-diastolic ventricular volume and stroke volume in normal and failing hearts. The normal curves are those shown previously in Figs. 11-33 and 11-35. The failing heart can still eject an adequate stroke volume if the sympathetic activity to it is increased or if the end-diastolic volume increases, i.e., if the ventricle becomes more distended.

terial pressure; this triggers reflexes (to be described in Chap. 13) which induce the kidneys to reduce their excretion of sodium and water; the retained fluid then causes expansion of the extracellular volume, increasing venous pressure, venous return, and end-diastolic ventricular volume and thus tending to restore stroke volume toward normal.

Another (undesirable) result of this elevated venous and capillary pressure is increased filtration out of the capillaries, i.e., formation of edema, and this accumulation of tissue fluid may be the chief feature of ventricular failure. If the right ventricle fails, the legs and feet are usually most prominently involved (because of the additional effects of gravity), but the same engorgement is occurring in other organs and may cause malfunction. When the left ventricle fails, the result can be even more serious—**pulmonary edema**, which is the accumulation of fluid in the interstitial spaces of the lung or in the air spaces themselves; this impairs gas exchange. The reason for such accumulation is that the relatively ineffective left ventricle fails to pump blood, at least transiently, to the same extent as the right ventricle, and so blood in all the pulmonary

vessels increases; the resulting engorgement of pulmonary capillaries raises the capillary hydrostatic pressure above its normally very low value, which causes increased filtration out of the capillaries. This situation usually worsens at night; during the day, because of the patient's upright posture, fluid accumulates in the legs, but it is slowly absorbed when the patient lies down at night, the plasma volume expands, and an attack of pulmonary edema is precipitated.

The treatment for congestive heart failure is easily understood in these terms. The precipitating cause should be corrected if possible; ventricular contractility can be increased by a drug known as digitalis (which acts by increasing cytosolic calcium in the myocardial cells); excess fluid should be eliminated by the use of drugs which increase excretion of sodium and water by the kidneys; and the patient's activity level should be adjusted so as to reduce the cardiac output required to fulfill the body's metabolic needs.

"Heart Attacks" and Atherosclerosis

We have seen that the myocardium does not extract oxygen and nutrients from the blood within the atria and ventricles but depends upon its own blood supply via the coronary vessels. The coronary arteries exit from the aorta just above the aortic valves and lead to a branching network of small arteries, arterioles, capillaries, venules, and veins similar to those in all other organs. Insufficient coronary blood flow leads to myocardial damage in the affected region and, if severe enough, to death of that portion of the myocardium (**infarction**). This is a so-called **heart attack**. (Because damaged heart cells leak certain enzymes, the finding of these enzymes in plasma is a useful diagnostic tool.) The severity of a myocardial infarction and the likelihood of the person's survival depends on the amount of tissue damaged and the region of the heart involved. For example, damage to a portion of the conducting system tends to be more serious. Insufficient coronary flow may occur as a result of decreased arterial pressure but is more commonly due to increased coronary-artery resistance secondary to coronary atherosclerosis. **Atherosclerosis** is a disease characterized by a thickening of the arterial wall with large numbers of abnormal smooth muscle cells and deposits of cholesterol and other substances in the portion of the vessel wall closest to the lumen. The mechanisms which initiate this thickening are not clear, but it is known that cigarette smoking, a high plasma cholesterol, hypertension, diabetes, and a variety of other factors are associated with increased incidence of this disease. The suspected relationship between atherosclerosis and blood concentrations of cholesterol has probably received the most widespread attention, and many studies have documented that high blood concentrations of this lipid increase the rate and the severity of the atherosclerotic process. (Cholesterol metabolism is described in Chap. 15.)

The mechanism by which atherosclerosis reduces coronary blood flow is quite simple; the extra muscle cells and various deposits in the wall bulge into the lumen of the vessel and increase resistance to flow. This is usually progressive, leading often ultimately to complete occlusion. Acute coronary occlusion may occur because of sudden formation of a blood clot on the roughened vessel surface, the breaking off of a fragment of blood clot or fatty deposit (which then lodges downstream, completely blocking a smaller vessel), or a profound spasm of the vessel's smooth muscle. If, on the other hand, the occlusion is gradual, the heart may remain uninjured because of the development, over time, of new accessory vessels supplying the same area of myocardium. Many patients experience recurrent transient episodes of inadequate coronary blood flow, usually during exertion or emotional tension. The pain associated with this is termed **angina pectoris**.

It has recently been found that beta-adrenergic blocking drugs are effective not only in lowering blood pressure in many persons but in reducing significantly the incidence of recurrent heart attacks; the mechanism of this effect (at least in persons who are not hypertensive) is not yet known. Another new class of drugs, called calcium-entry blockers because they seem to ''block'' calcium channels, is being tested for prevention of angina, particularly the type of angina caused by acute excessive contraction of the coronary vessel's smooth muscle. Since calcium entry through membrane channels which open during membrane excitation elicits the sequence of events leading to contraction of vascular smooth muscle, drugs which block these channels prevent the contraction.

Another present treatment for angina is the surgical technique known as coronary bypass. Discrete areas of coronary narrowing can be detected (often by x-rays after injection of dye into the heart) and bypassed by implantation of a graft, usually a vein taken from elsewhere in the patient's body. This operation often produces marked relief of angina, and may also prolong life in some persons.

Finally, the question of whether regular exercise is protective against heart attacks is still controversial, although more and more circumstantial evidence favors this view. Certainly, modest exercise programs induce a variety of changes which are consistent with a protective effect: (1) increased diameter of coronary arteries; (2) decreased severity of two diseases—hypertension and diabetes (see Chap. 15)—which are risk factors for atherosclerosis; (3) decreased plasma cholesterol concentration (yet another risk factor) with a simultaneous increase in the plasma concentration of a cholesterol-carrying lipoprotein thought to be protective against atherosclerosis (see Chap. 15); and (4) improved ability to dissolve blood clots (see section on hemostasis). Finally, the results of long-term studies trying to evaluate directly the effects of exercise on the incidence of atherosclerosis are also suggestive of some degree of protection.

We do not wish to leave the impression that atherosclerosis attacks only the coronary vessels, for such is not the case. Most arteries of the body are subject to this same occluding process, and wherever the atherosclerosis becomes severe, the resulting symptoms always reflect the decrease in blood flow to the specific area. For example, cerebral occlusions (**strokes**) due to atherosclerosis are common in the aged and constitute an important cause of sickness and death.

SECTION G.
HEMOSTASIS: THE PREVENTION OF BLOOD LOSS

The basic prerequisites for bleeding hemorrhage) are (1) loss of vessel continuity so that cells and fluid can leak out and (2) a pressure inside the vessel greater than that outside. Accordingly, cessation of bleeding (**hemostasis**) occurs if at least one of two requirements is met: The pressure difference favoring blood loss is eliminated, or the damaged portion of the vessel is sealed. All hemostatic processes accomplish one of these two requirements. We shall discuss the probable sequence of events in damage to small vessels—arterioles, capillaries, venules—because they are the most common source of bleeding in everyday life and because the hemostatic mechanisms are most effective in dealing with such injuries. In contrast, the bleeding from a severed artery of medium or large size is not usually controllable by the body and requires radical aids such as application of pressure and ligatures. Venous bleeding is less dangerous because veins have low blood pressure; indeed, the drop in pressure induced by simple elevation of the bleeding part may stop the hemorrhage. In addition, if the venous bleeding is into the tissues, the accumulation of blood (**hematoma**) may increase interstitial pressure enough to eliminate the pressure gradient required for continued blood loss.

When a blood vessel is severed or injured, its immediate inherent response is to constrict. This transient response, which slows the flow of blood in the affected area, occurs in all vessels. In addition, this initial constriction presses the opposed endothelial surfaces of the vessel together, and this contact induces a stickiness capable of keeping them "glued" together against high pressures. However, this process is effective only in the very smallest vessels of the microcirculation, and the

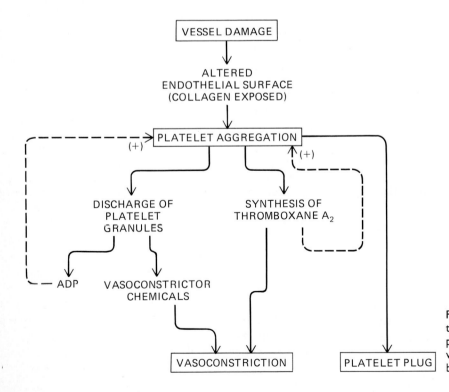

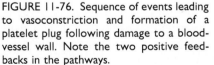

FIGURE 11-76. Sequence of events leading to vasoconstriction and formation of a platelet plug following damage to a blood-vessel wall. Note the two positive feedbacks in the pathways.

staunching of bleeding ultimately is dependent upon two other processes—formation of a platelet plug and blood coagulation.

Formation of a Platelet Plug

The involvement of platelets in hemostasis requires their adhesion to a surface. Although platelets have a propensity for adhering to any foreign or rough surfaces, they do not adhere to the normal endothelial cells lining the blood vessels. However, injury to a vessel disrupts the endothelium and exposes the underlying connective-tissue collagen molecules, fibrous proteins which constitute a major component of the interstitial matrix. Platelets adhere to collagen, and this attachment triggers the release of the platelets' granules containing potent chemical agents, including adenosine diphosphate (ADP). This ADP then causes changes in the surface of the platelets such that new platelets adhere to the old ones, and by this self-perpetuating (positive-feedback) process an aggregate or plug of platelets is rapidly built up (Fig. 11-76).

ADP is not the only stimulator of platelet aggregation. In addition, adherence of the platelets to collagen causes the conversion of arachidonic acid in the platelet membrane to thromboxane A_2 (see Chap. 7). This agent is a powerful stimulant of platelet aggregation and of secretion of platelet granules.

The platelet plug can completely seal breaks in blood-vessel walls. Its effectiveness is further enhanced by another remarkable property of platelets—contraction. Platelets contain a very high concentration of actomyosin-like contractile proteins, which are stimulated to contract in the aggregated platelets; this results in a compression and strengthening of the platelet plug.

While this plug is being built up and compacted, the vascular smooth muscle in the vessels around it is being stimulated to contract, thereby decreasing the blood flow to the area and the pressure within the damaged vessel (Fig. 11-76). This vasoconstriction is also the result of platelet activity, for it is mediated by thromboxane A_2 and by several chemicals contained in the secreted platelet granules.

Why does the platelet plug not continuously expand, spreading away from the damaged endothelium along normal endothelium in both directons? One important reason involves the ability of the adjacent vessel wall to make the prostaglandin known as prostacyclin (PGI_2). Recall that the stimulated platelets are releasing thromboxane A_2; normal vessel wall contains a different enzyme, one which converts intermediates formed from arachidonic acid not into thromboxane A_2 but into PGI_2, and PGI_2 is a profound inhibitor of platelet aggregation.

To reiterate, the platelet plug is built up extremely rapidly and is the primary sealer of breaks in vessel walls. In the next section, we shall see that platelet aggregation is also absolutely essential for the next, more slowly occurring, hemostatic event, blood coagulation; in turn, certain of the chemicals liberated in the coagulation process act as yet additional stimulators of platelet aggregation and granule release.

Blood Coagulation: Clot Formation

Blood coagulation (clotting) is the dominant hemostatic defense, as attested by the fact that, with few exceptions, the cause of abnormal bleeding is a defect in the clotting system. The event transforming blood into a solid gel around the original platelet plug is the conversion of the plasma protein **fibrinogen** to **fibrin**. Fibrinogen is a soluble, large, rod-shaped protein produced by the liver and always present in the plasma of normal persons. At the site of vessel damage, its conversion to fibrin is catalyzed by the enzyme **thrombin**:

$$\text{Fibrinogen} \xrightarrow{\text{thrombin}} \text{fibrin}$$

In this reaction, several polypeptides are split from the fibrinogen molecules, conferring upon the remaining still-quite-large molecules the ability to bind to each other to form the polymer known as fibrin (Fig. 11-77). The fibrin is initially a loose mesh

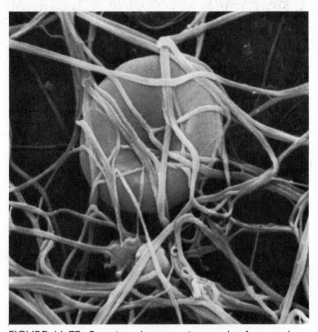

FIGURE 11-77. Scanning electron micrograph of an erythrocyte enmeshed in fibrin. *(Courtesy of Eila Kairinen, Gillette Research Institute.)*

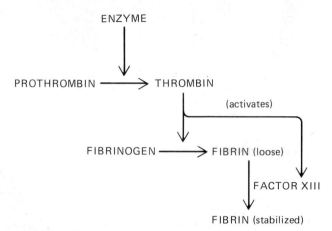

FIGURE 11-78. Generation of thrombin and its effects on fibrinogen and factor XIII.

of interlacing strands, which is rapidly stabilized and strengthened by enzymatically mediated formation of covalent cross-linkages. (This chemical linking is catalyzed by an enzyme in plasma known as **factor XIII.**) In the process many erythrocytes and other cells are trapped in the fibrin meshwork, but it must be emphasized that the clot is due to fibrin and can occur in the absence of all cells (except platelets, as we shall see). The function of the clot is to support and reinforce the platelet plug described in the previous section and to solidify blood that remains in the wound channel.

Since fibrinogen is always present in the blood, but the blood does not clot, thrombin must normally be absent. Circulating in the blood is an inactive precursor of thrombin called **prothrombin**; prothrombin is enzymatically converted to thrombin at the site of blood vessel damage and this thrombin then enzymatically converts fibrinogen to fibrin (Fig. 11-78). [Moreover, thrombin also catalyzes conversion of the inactive form of factor XIII to active factor XIII, which then acts to stabilize the fibrin, as described above (Fig. 11-78)].

We have now only pushed the essential question one step further back: Where does the enzyme come from which catalyzes the conversion of the prothrombin to thrombin? The answer is that this enzyme is always present in the plasma in an inactive form and is converted to its active form by yet another enzyme, which itself was activated by another enzyme, which . . . . Thus, we are dealing largely with a **cascade** of plasma proteins, each normally an inactive proteolytic enzyme until activated by the previous one in the sequence. Ultimately, the final enzyme in the sequence is activated and, in turn, catalyzes the conversion of prothrombin to thrombin. The first factor in the sequence is called **factor XII** or **Hageman factor** (after the patient in whom it was first discovered) and, as will be described in subsequent sections and Chap. 17, it has several other important functions in addition to initiating clotting.

But we still have not answered the basic question of what *initiates* clotting, i.e., what activates the first enzyme (factor XII) in the catalytic sequence? The answer is contact of factor XII with a damaged vessel surface, specifically with the collagen fibers underlying the damaged endothelium. Recall that this is precisely the same stimulus that triggers platelet aggregation (Fig. 11-79). This fact is important because several of the sequential enzyme-activation steps in the clotting pathway require as a cofactor a phospholipid (PF3) secreted

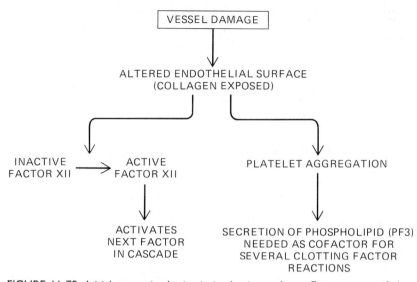

FIGURE 11-79. Initial events in the intrinsic clotting pathway. For purposes of clarity, we have not shown the fact that thrombin facilitates both platelet aggregation and several of the clotting-factor reactions in a positive-feedback manner.

by the aggregated platelets. Thus, platelets are essential for normal blood clotting.

Platelet phospholipid is not the only cofactor required for several of the clotting-sequence reactions. Calcium is also required at various steps; however, unlike the case for platelets, calcium concentration in the blood can never get so low (without killing the person from some nonbleeding cause) as to cause clotting defects.

One final concept completes this description of the basic clotting pathway. Thrombin not only catalyzes the conversion of fibrinogen to fibrin but exerts a variety of other effects within the clotting pathway: It profoundly stimulates platelet aggregation (and thereby the release of platelet phospholipid), and it facilitates several of the reactions in the sequence leading to its own formation. Therefore, once thrombin formation has begun, the overall reaction then progresses very rapidly owing to the positive-feedback effects of thrombin on its own generation.

Figure 11-80 summarizes the basic, wholly intravascular sequence of events leading to the blood clotting. The value of such a cascade system lies in the amplification gained at many steps, i.e., in the manyfold increase in the number of active molecules produced (recall from Chap. 7 that an analogous cascade of enzyme activations was described for the cyclic AMP–protein-kinase system). However, there is at least one disadvantage in such a cascade in that a single defect, either hereditary or disease-induced, anywhere in the system can block the entire cascade, thereby interfering with clot formation. For example, the classical hereditary defect of **hemophilia** results from the absence of factor VIII.

It must be emphasized that the entire sequence of reactions leading to clotting occurs only locally at the site of vessel damage. Each active component is formed, functions, and then is rapidly inactivated by enzymes in the plasma or local tissue, without spilling over into the rest of the circulation. Otherwise, because of the chain-reaction nature of the response, the appearance of both platelet phospholipid and any single activated plasma clotting factor in the overall circulation would induce massive widespread clotting throughout the body.

Why does blood coagulate when it is taken from the body and put in a glass tube? This has nothing whatever to do with exposure to air, as popularly supposed, but happens because the glass surface induces the same activation of factor XII

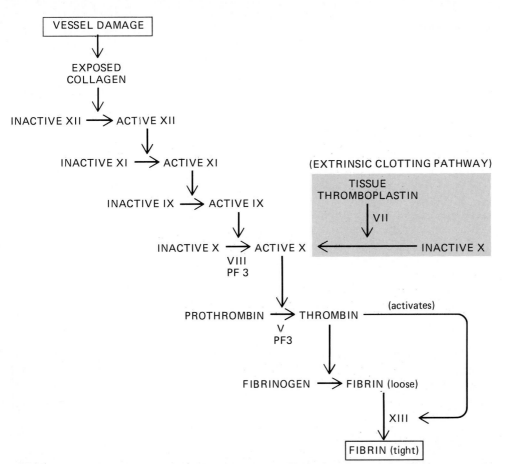

FIGURE 11-80. Intrinsic and extrinsic clotting pathways. The two pathways use different ways of getting to active factor X, but from that point on they are identical. For the sake of clarity, neither the roles of calcium nor the positive-feedback effects of thrombin are shown.

and aggregation of platelets as does a damaged blood-vessel surface. A silicone coating markedly delays clotting by reducing the activating effects of the glass surface.

The liver plays several important indirect roles in the overall functioning of the clotting mechanism. First, it is the site of production for many of the plasma clotting factors, including prothrombin and fibrinogen. Second, the bile salts produced by the liver are required for normal gastrointestinal absorption of the fat-soluble **vitamin K**, which the liver requires to produce the normal structures of some of the plasma clotting factors. Thus, persons

with liver disease or defective gastrointestinal fat absorption frequently have serious bleeding problems.

The entire sequence of events beginning with activation of factor XII is known as the **intrinsic clotting pathway**, (because everything necessary for it is in the blood). There is, however, an **extrinsic clotting pathway** leading to activation of the enzyme which converts prothrombin to thrombin (Fig. 11-80). Tissues contain a protein known as **tissue thromboplastin** which can (in concert with several cofactors) activate this enzyme. Thus, exposure of blood to tissue thromboplastin bypasses the first steps in the intrinsic clotting sequence. Since such exposure usually occurs when a blood vessel is damaged, this pathway goes into action right along with the intrinsic pathway.

The clotting pathways are also frequently triggered during infection (by chemicals secreted locally from leukocytes and macrophages, Chap. 17); they initiate interstitial fibrin clots which may block further spread of the bacteria.

The Anticlotting System

A fibrin clot is not designed to last forever; it is a transitory device until permanent repair of the vessel occurs. Clots dissolve or fail to form in several normal circumstances. Usually this is due to a proteolytic enzyme called **plasmin**, which is able to break down bonds between fibrin molecules, therefore dissolving a clot. The physiology of plasmin bears some striking similarities to that of the coagulation factors. Plasmin circulates in blood in an inactive form (**plasminogen**) which can be converted to active plasmin by an enzyme, itself activated as the last product of a cascade series of reactions. To complete the similarity, one initiator (among many) of this cascade is factor XII. Thus, the same substance, factor XII, triggers off simultaneously the clotting and anticlotting systems. This makes sense, since the generated plasmin becomes trapped within the newly formed clot and very slowly dissolves it, thereby contributing to tissue repair at a time when the danger of hemorrhage is past.

The anticlotting system no doubt has other functions as well. It may be that small amounts of fibrin are constantly being laid down throughout the vascular tree and that plasmin acts on this fibrin to prevent clotting. The lung tissue, for example, contains a substance which activates plasmin; this probably explains the lung's ability to dissolve the fibrin clumps which its capillaries filter from the blood. The uterine wall is also extremely rich in a similar activator, and thus normal menstrual blood generally does not clot.

A second naturally occurring anticoagulant is **heparin**. This substance, found in various cells of the body, acts by interfering with the activation of several of the clotting factors and with the ability of thrombin to split fibrinogen. Despite its presence in the body and the fact that it is the most powerful anticoagulant known, it is not clear whether heparin plays a normal physiological role in clot prevention. On the other hand, heparin is widely used as an anticoagulant drug in medicine.

Finally, still other naturally occurring anticoagulants are being discovered in the body. Some of these oppose the action of thrombin, others that of tissue thromboplastin, and so on. However, as with heparin, their actual contributions remain to be determined.

Excessive Clotting: Intravascular Thrombosis

Formation of a clot in a damaged vessel is obviously a homeostatic physiological response, but the formation of clots in intact vessels is pathological. Such an intravascular clot is known as a **thrombus**. It may occur in the veins, the microcirculation, or the arteries.

Numerous theories have been proposed to explain the formation of a thrombus. One of them postulates that the clotting mechanism in persons prone to thrombosis is hyperactive, as manifested by the reduced time it takes for withdrawn blood to clot in a test tube. Perhaps one of the plasma factors is present in excessive amounts, or a normally occurring anticoagulant is deficient. This theory emphasizes that the blood itself is the cause of excessive clotting. However, there seems little question that hypercoagulability is not always essential for thrombosis, since hemophiliacs have been known to suffer from coronary thrombosis.

A second category of theories puts the blame on the blood vessels. Since initiation of blood clotting is primarily dependent upon the state of the blood-vessel lining, even minor transient alterations in the endothelial surface could trigger the cascade sequence leading to clot formation. It should be noted that this concept and the hypercoagulability theory are not mutually exclusive. Both probably are valid, depending upon the circumstances.

Regardless of the initiating event, there is no question that a clot, no matter how small, provides a suitable surface upon which more clot can form. Thus the thrombus grows and may eventually occlude the entire vessel, thereby leading to damage of the tissue supplied or drained by the vessel. (Heart attacks due to occlusion of a coronary artery by a clot were described earlier in this chapter.) A second important factor in vessel closure during clot growth is the release of vasoconstrictors from freshly adhered platelets. Finally, the clot fragments may break off and be carried to the lungs (if from a vein) or other organs (if from an artery); these fragments are called **emboli**.

Considering the interrelationship of excessive clotting and atherosclerosis in the production of heart attacks, it is interesting that exercise programs induce increased activity of the plasmin anticlotting system.

RESPIRATION

Respiration has two quite different meanings: (1) utilization of oxygen in the metabolism of carbohydrate and other organic molecules (as described in Chap. 5) and (2) the processes involved in the exchanges of oxygen and carbon dioxide between the cells of an organism and the external environment. These latter processes make up the subject matter of this chapter.

Most of our cells obtain most of their energy from chemical reactions involving oxygen. In addition, cells must be able to eliminate carbon dioxide, the major end product of oxidative metabolism. A unicellular organism can exchange oxygen and carbon dioxide directly with the external environment, but this is obviously impossible for most cells of a complex organism like a human being. Therefore, the evolution of large animals required the development of a specialized system—the respiratory system—to exchange oxygen and carbon dioxide for the entire animal with the external environment.

Organization of the Respiratory System

The term **respiratory system** has a more limited compass than the term respiration, for it refers only to those structures which are involved in the exchange of gases between the blood and external environment; it comprises the lungs, the series of airways leading to the lungs, and the chest structures responsible for movement of air in and out of the lungs (Fig. 12-1).

There are two **lungs**, the right and left, each divided into several lobes. The lungs, together with the heart, great vessels, esophagus, thymus, and certain nerves, completely fill the thoracic cavity. The lungs are not simply hollow balloons but have a highly organized structure consisting of air-containing sacs and tubes, blood vessels, and elastic connective tissue. The **airways** in the lungs (Fig. 12-2) are the continuation of those which connect the lungs to the nose and mouth (Fig. 12-1); together these airways are termed the **conducting portion** of the respiratory system and constitute a series of highly branched hollow tubes becoming smaller in diameter and more numerous at each branching. The smallest of these tubes end in tiny blind sacs, the **alveoli**, which number approximately 300 mil-

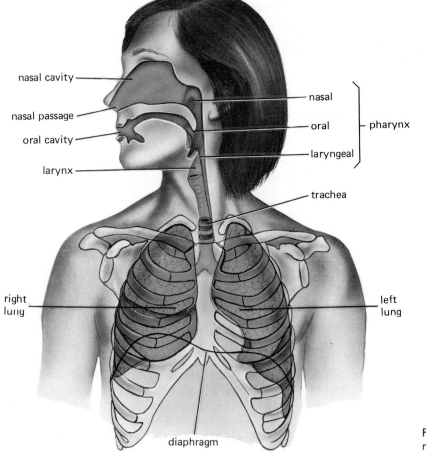

FIGURE 12-1. Organization of the respiratory system.

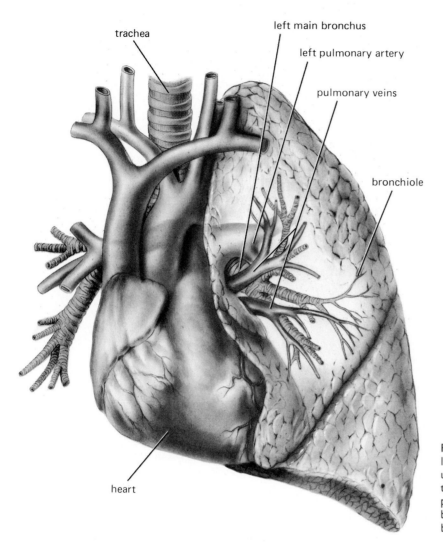

trachea

left main bronchus

left pulmonary artery

pulmonary veins

bronchiole

heart

FIGURE 12-2. Large airways of the human lung, starting with the trachea and continuing with its many branches in the lungs. In this figure the substance of the lungs appears transparent so that the relationships between the airways and blood vessels can be seen.

lion and are the actual sites of gas exchange within the lungs (Fig. 12-3). All portions of these air passageways and alveoli receive a rich supply of blood by way of blood vessels, which constitute a large portion of the total lung substance (Fig. 12-2).

Conducting portion of the respiratory system

Air can enter the respiratory passages either by nose or by mouth, although the nose is the normal route. It then passes into the **pharynx** (throat), a passage common to the routes followed by air and food. The pharynx branches into two tubes, one (the esophagus) through which food passes to the stomach and one through which air passes to the lungs. The first portion of the air passage, called the **larynx**, houses the **vocal cords**, two strong bands of elastic tissue stretched across its lumen. The larynx opens into a long tube (the **trachea**), which, in turn, branches into the two **bronchi**, one of which enters each lung. Within the lungs, these major bronchi branch many times into progressively smaller bronchi, thence into **bronchioles**, and finally into the alveolar portions of the lungs.

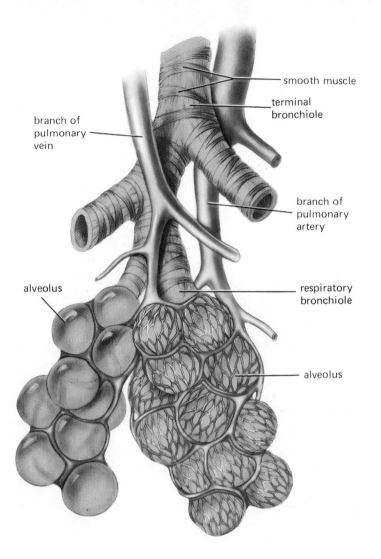

branch of
pulmonary
vein

smooth muscle

terminal
bronchiole

branch of
pulmonary
artery

respiratory
bronchiole

alveolus

alveolus

FIGURE 12-3. Clusters of alveoli are connected to the air passageways of the lung. The intimate relationship between the respiratory system and blood vessels at the level of the alveoli can be seen.

This conducting system serves several important functions:

1. The epithelial linings contain **cilia**, which constantly beat toward the pharynx (Fig. 12-4). These cilia line the respiratory airways to the end of the bronchioles; in the same regions are glands and individual epithelial cells which secrete mucus. Particulate matter, such as dust contained in the inspired air, sticks to the mucus which is continually and slowly moved by the cilia to the pharynx and then is swallowed. Besides keeping the lungs clean, this mucus escalator is important in the body's total defenses against bacterial infection, since many bacteria enter the body on dust particles. A major cause of lung infection is reduction of ciliary activity by noxious agents (for example, smoking a single cigarette can immobilize the cilia for several hours). The reduction in ciliary activity, coupled with the stimulation of mucus secretion also induced by noxious agents, may result in partial or complete airway obstruction by stationary mucus. (A smoker's early-morning cough is the

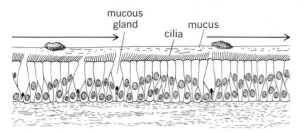

FIGURE 12-4. Epithelial lining of the airways. The arrows indicate the upward direction in which the cilia move the overriding layer of mucus, to which foreign particles are stuck.

attempt to clear this obstructive mucus from the airways.) A second protective mechanism is provided by phagocytic cells, which are present in the respiratory-tract lining in great numbers. These cells, which engulf dust, bacteria, and debris, are also injured by cigarette smoke and other air pollutants.

2. As air flows through the respiratory passages, it is warmed and moistened by contact with the epithelial lining.

3. The movement of air past the vocal cords causes them to vibrate, providing the tones of phonation (speech).

4. The walls of the airways contain smooth muscle cells which are innervated by autonomic neurons; they are also sensitive to certain hormones, notably epinephrine. Contraction or relaxation of this smooth muscle alters resistance to air flow, as we shall describe later.

Site of gas exchange in the lungs: The alveoli

The **alveoli** are tiny, cup-shaped hollow sacs whose open ends are continuous with the lumens of the smallest bronchioles and the alveolar ducts (Fig. 12-5). The alveoli (Fig. 12-5) are lined by a continuous, single, thin layer of epithelial cells resting on a very loose mesh of connective tissue elements which constitutes the interstitial space of the alveolar walls. Most of the alveolar wall consists of capillaries, the endothelial lining of which is separated from the alveolar epithelial lining only by the very thin interstitial space. Indeed, in places, the interstitial space may be absent altogether; i.e., epithe-

lium and endothelium may actually fuse. Thus, the blood within a capillary is separated from the air within an alveolus only by an extremely thin barrier (0.2 μm, compared with the 7 μm diameter of an average red blood cell). The total area of alveoli in contact with capillaries is approximately 135 m², roughly 80 times greater than the external body surface area. This immense area, combined with the thinness of the barrier, permits the rapid exchange of large quantities of oxygen and carbon dioxide by diffusion between capillary blood and alveolar air.

In addition to the thin cells, the alveolar epithelium contains smaller numbers of thicker, specialized cells (type II cells) which produce a detergentlike substance to be discussed below. The interstitial space contains phagocytic cells and other connective-tissue cells which function in the lung's defense mechanisms. Finally, there are pores in the alveolar membranes which permit some flow of air between alveoli. This **collateral ventilation** can be very important when the duct leading to an alveolus is occluded by disease, since some air can still enter this alveolus by way of the pores between it and adjacent alveoli.

Relation of the lungs to the thoracic cage

The **thoracic cage** is a closed compartment. It is bounded at the neck by muscles and connective tissue, and it is completely separated from the abdomen by a large, dome-shaped sheet of skeletal muscle, the **diaphragm**. The outer walls of the thoracic cage are formed by the breastbone (**sternum**), 12 pairs of **ribs**, and the muscles which lie between the ribs (the **intercostal muscles**). The walls also contain large amounts of elastic connective tissue.

Firmly attached to the entire interior of the thoracic cage is a thin sheet of cells, the **pleura**, which folds back upon itself to form two completely enclosed (and noncommunicating) sacs within the thoracic cage, one on each side of the midline. The relationship between the lungs and pleura in each half of the thorax can be visualized by imagining what happens when one pushes a fist into a fluid-filled balloon (Fig. 12-6): The arm represents the

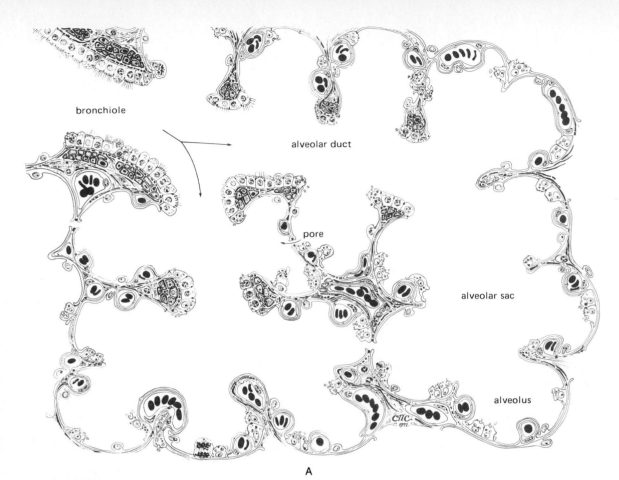

bronchiole

alveolar duct

pore

alveolar sac

alveolus

A

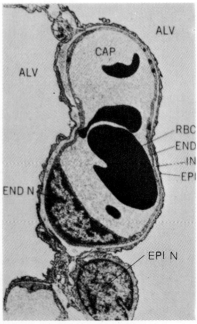

ALV

CAP

ALV

RBC
END
IN
EPI

END N

EPI N

B

FIGURE 12-5. (A) Diagram of the respiratory subdivisions in the lung. *(From R. O. Greep and L. Weiss, "Histology," 3d ed., McGraw-Hill Book Company, New York, 1973.)* (B) Higher magnification of a portion of an alveolar wall, showing a single capillary (CAP) surrounded by an alveolar epithelial cell. The nucleus (END N) and cytoplasm (END) of a capillary endothelial cell are visible, as are the nucleus (EPI N) and cytoplasm of the alveolar epithelial cell. Note that the blood in the capillary is separated from air in the alveoli (ALV) only by the thin layer consisting of endothelium, interstitial fluid (IN), and epithelium. *[From E. R. Weibel, Physiol. Rev., **53**:424 (1973)].*

major bronchus leading to the lung, the fist is the lung, and the balloon is the pleural sac. The outer portion of the fist becomes coated by one surface of the balloon. In addition, the balloon is pushed back upon itself so that its surfaces lie close together. This is the relation between the lung and pleura except that the pleural surface coating the lung is firmly attached to the lung surface. This layer of pleura and the outer layer, which is attached to and lines the interior thoracic wall, are so close to each other that they are virtually in

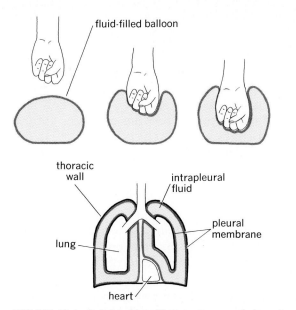

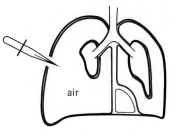

FIGURE 12-7. Lung collapse caused by cutting the thoracic cage. Note that the air in the pleural space did not come from the lungs since the lung wall is still intact.

FIGURE 12-6. Relationship of lungs, pleura, and thoracic cage, shown as analogous to pushing a fist into a fluid-filled balloon. Note that there is no communication between the right and left intrapleural fluids. The volume of intrapleural fluid is greatly exaggerated; normally it consists of an extremely thin layer of fluid between the pleural membrane lining the inner surface of the thoracic cage and the pleural membrane lining the surface of the lungs.

contact, being separated only by a very thin layer of **intrapleural fluid.**

The concept of intrapleural pressure

If the chest is opened during surgery, care being taken to cut only the thoracic cage wall but not the lung, the lung on the side of the incision partially collapses immediately (Fig. 12-7); simultaneously, the chest wall on that side moves outward. This shows that the intrinsic unstretched size of the elastic lung is less than the size it assumes inside the intact chest; it also shows that the elastic intact chest wall is normally partially pulled inward (compressed) by the presence of the lung within it. In other words, purely on the basis of their differing unstretched dimensions and their elasticity, the lungs and chest wall are tending to move away from each other, the lungs to collapse inward and the chest wall to ''pop'' outward. Yet normally, neither does (except to the slightest degree) and the situa-

tion remains stable. The reason is the counteracting pressure set up by these tendencies in the intrapleural fluid.

Visualize the situation between breaths, when no respiratory muscle contraction is occurring and no air is flowing. The pressure within the lungs (the **alveolar pressure** P_{alv}) is 0 mmHg, i.e., atmospheric pressure (all pressures in the respiratory system, as in the cardiovascular system, are given relative to atmospheric pressure—the pressure of the air surrounding the body). The pressure in the intrapleural fluid surrounding the lungs, the **intrapleural pressure** (P_{ip}), is approximately 4 mmHg less than atmospheric pressure, i.e., -4 mmHg. There is therefore a pressure difference of 4 mmHg $[0 - (-4) = 4]$ across the lung wall, and this is the force acting to hold the lungs open (there is an identical force, but opposite in direction, keeping the chest wall from moving out). Thus, the key element here is the subatmospheric intrapleural pressure. How did it become subatmospheric? Water is highly indistensible, and as the lungs and chest wall move ever so slightly away from each other, the resulting infinitesimal expansion of the water-filled intrapleural space drops the intrapleural pressure below atmospheric pressure. In this way, the elastic recoil of the lung and chest wall creates the subatmospheric intrapleural pressure which keeps them from moving any further apart.

Why the lung collapses when the chest wall is opened should now be apparent. As we have said, the intrapleural pressure is normally less than atmospheric pressure, so when the chest wall is pierced, atmospheric air rushes into the intrapleural space, the pressure difference across the lung wall is eliminated, and the stretched lung collapses. (Air in the intrapleural space is known as a **pneumothorax**.)

The intrapleural fluid pressure is normally subatmospheric between breaths, during inspiration, and during quiet expiration. (Only during a forced expiration does intrapleural pressure exceed atmospheric pressure, but even then the intrapleural pressure is always lower than the alveolar pressure.) However, as we shall see in the next sections, the magnitude of the difference between intrapleural and alveolar pressures varies during breathing and directly causes the changes in lung size which occur during inspiration and expiration.

Since intrapleural pressure is transmitted throughout the intrathoracic fluid surrounding not only the lungs but the heart and other intrathoracic structures as well, it is frequently termed the **intrathoracic pressure**.

Inventory of Steps Involved in Respiration

The following inventory is provided for orientation before we begin the detailed descriptions of each step.

1. Exchange of air between the atmosphere (external environment) and alveoli. This process, known as **ventilation**, includes the bulk flow of air in and out of the lungs and the distribution of air within the lungs.
2. Exchange of oxygen and carbon dioxide between alveolar air and lung capillaries by diffusion.
3. Transportation of oxygen and carbon dioxide by the bulk flow of blood through the circulatory system.
4. Exchange of oxygen and carbon dioxide between the blood and tissues of the body by diffusion as blood flows through tissue capillaries.

Ventilation and Lung Mechanics

Like blood, air moves by bulk flow from a high pressure to a low pressure. We have seen (Chap. 11) that bulk low can be described by the equation

$$F = \frac{\Delta P}{R}$$

That is, flow (F) is proportional to the pressure difference, (ΔP), between two points and inversely proportional to the resistance (R). For air flow into or out of the lungs, the relevant pressures are the **atmospheric pressure** (P_{atm}) and the pressure in the alveoli (**alveolar pressure**, P_{alv}):

$$F = \frac{(P_{atm} - P_{alv})}{R}$$

Air moves into (inspiration) and out of (expiration) the lungs because the alveolar pressure is made, alternately, less than and greater than atmospheric pressure (Fig. 12-8).

These alveolar pressure changes are caused by changes in the dimensions of the lungs. The relationship between the pressure exerted by a fixed number of molecules of gas in a closed container and the dimensions of the container is as follows: An increase in the dimensions of the container will decrease the pressure of the gas, whereas decreasing the dimensions of the container increases the pressure. During inspiration, alveolar pressure becomes transiently less than atmospheric pressure because the contraction of certain skeletal muscles

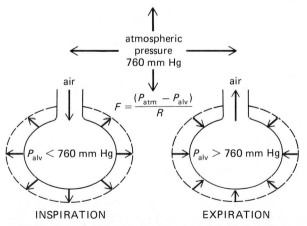

FIGURE 12-8. Relationships required for breathing. When the alveolar pressure (P_{alv}) is less than atmospheric pressure, air enters the lungs. Flow (F) is directly proportional to the pressure difference and inversely proportional to airway resistance (R).

results in an increase in the dimensions of the lungs. Then, during expiration, alveolar pressure becomes transiently greater than atmospheric pressure when the dimensions of the lungs decrease as these same muscles relax (and, sometimes, in addition, when another set of skeletal muscles contracts).

The story would be easy to understand if these skeletal muscles inserted directly on the lung surface and pulled them out or pushed them in. But such is not the case. The respiratory muscles do not insert on the lungs or act directly upon them; rather, they act on the tissues forming the boundaries of the thoracic cage. Thus their activity directly alters the dimensions of the thoracic cage, and this indirectly causes the lungs either to expand or grow smaller. The link between changes in thoracic-cage dimensions and changes in lung dimensions is provided, as we shall see, by pressure changes in the intrapleural fluid surrounding the lungs.

Throughout these discussions we will be making continual references to two *differences* in pressure, which must be distinguished from each other (Fig. 12-9): (1) The difference in pressure between the alveolar air and the intrapleural fluid ($P_{alv} - P_{ip}$); as we shall see, this pressure difference is a determinant of lung size. (2) The difference in pressure between alveolar air and atmospheric air ($P_{atm} - P_{alv}$); as stated earlier, this pressure is a determinant of air flow between the atmosphere and the alveoli. The sequence of events (Fig. 12-9) will always be that a change in $P_{alv} - P_{ip}$ causes a change in lung size, which then causes a change in $P_{atm} - P_{alv}$, which then causes the air flow into or out of the lungs.

Inspiration

Figure 12-10 and the left half of Fig. 12-11 summarize events which occur during inspiration. Just before the inspiration begins, i.e., at the conclusion of the previous expiration, the respiratory muscles are relaxed and no air is flowing. As described earlier, the intrapleural pressure is subatmospheric and the alveolar pressure is atmospheric. Inspiration is initiated by the contraction of the diaphragm and inspiratory intercostal muscles (Fig. 12-12), the dia-

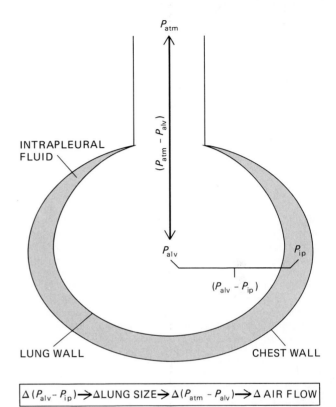

FIGURE 12-9. The difference in pressure between alveolar air and intrapleural fluid ($P_{alv} - P_{ip}$) is a determinant of lung size. The difference in pressure between alveolar air and the atmosphere ($P_{atm} - P_{alv}$) is a determinant of air flow. The sequence of events during breathing is shown in the box (Δ means "changes in"). The space occupied by the intrapleural fluid is markedly exaggerated for the sake of clarity; it is actually an extremely thin layer of fluid.

phragm being the most important during normal quiet breathing. When the diaphragm contracts, its dome moves downward into the abdomen, thus enlarging the dimensions of the thoracic cage. Simultaneously, the inspiratory intercostal muscles, which insert on the ribs, contract, leading to an upward and outward movement of the ribs and a further increase in thoracic cage size. As the thoracic cage enlarges, it moves ever so slightly away from the lung surface, and the intrapleural fluid pressure becomes even more subatmospheric. This increases the difference between the alveolar and intrapleural pressures, $P_{alv} - P_{ip}$, and the lung wall is pushed out. Thus, when the inspiratory muscles

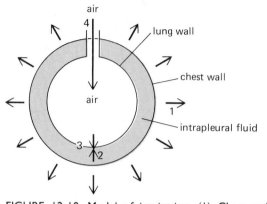

FIGURE 12-10. Model of inspiration. (1) Chest wall is expanded by contraction of the diaphragm and inspiratory intercostal muscles (as denoted by colored arrows). (2) As a result, intrapleural pressure decreases; i.e., it becomes more subatmospheric than it is between breaths. (3) Therefore, the pressure difference across the lung wall ($P_{alv} - P_{ip}$) becomes larger, and the lung wall is pushed out; i.e., the lungs begin to expand. (4) Therefore, the alveolar pressure becomes subatmospheric, a pressure difference between alveoli and atmosphere is created, and air moves in from the atmosphere.

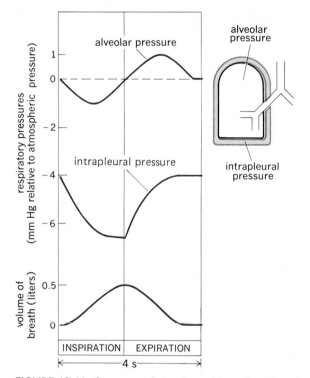

FIGURE 12-11. Summary of alveolar and intrapleural pressure changes and air flow during inspiration and expiration of 500 mL of air. Note that, on the pressure scale at the left, normal atmospheric pressure (760 mmHg) has a scale value of zero.

increase the thoracic dimensions, the lungs are also forced to enlarge virtually to the same degree because of the changes in intrapleural pressure. The lung enlargement is associated with an increase in the sizes of the alveoli; therefore, the air pressure within them drops to less than atmospheric. This produces the difference in pressure ($P_{atm} - P_{alv}$) which causes bulk flow of air from the atmosphere through the airways into the alveoli until, by the end of the inspiration, their pressure again equals atmospheric.

Expiration

When the inspiratory-muscle contraction expanding the dimensions of the thoracic cage ceases, i.e., when these muscles relax, the lungs recoil, causing a reversal of the inspiratory process, as shown in the right side of Fig. 12-11. As the lungs spring back toward their original size, alveolar air becomes temporarily compressed so that its pressure exceeds atmospheric, and air flows from the alveoli through the airways out into the atmosphere. Thus, expiration at rest is completely passive, depending only

upon the relaxation of the inspiratory muscles and recoil of the stretched lungs. Under certain conditions (during heavy exercise, for example) expiration of larger volumes is achieved by the contraction of the expiratory intercostal muscles, which actively decrease thoracic dimensions, and of abdominal muscles; contraction of these latter muscles increases intraabdominal pressure and forces the diaphragm up into the thorax.

It should be noted that the analysis of Fig. 12-11 treats the lungs as a single alveolus. The fact is that there are significant regional differences in both alveolar and intrapleural pressures throughout the lungs and thoracic cavity. These differences are due, in part, to the effects of gravity and to local differences in the elasticity of the chest structures. They are of great importance in determining the distribution of ventilation throughout the lung.

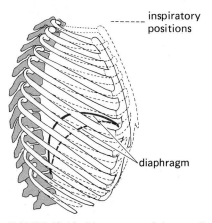

inspiratory
positions

diaphragm

FIGURE 12-12. Movements of chest wall and diaphragm during breathing. The contracting intercostal muscles move the ribs upward and outward during inspiration, while the contracting diaphragm moves downward. The dashed lines indicate the inspiratory positions.

Lung compliance

To reiterate, the degree of lung expansion at any instant is proportional to the pressure difference across the lung wall, i.e., between the alveolar air and the intrapleural fluid, $P_{alv} - P_{ip}$. But just how much any given pressure difference across the lung wall expands the lung depends upon the stretchability or compliance of the lung tissue. **Lung compliance** is defined as the magnitude of the change in lung volume produced by a given change in the pressure difference across the lung wall. The larger this ratio (lung volume change/pressure difference change), the greater the compliance (i.e., the greater the stretchability). What is the significance of having a low pulmonary compliance (due say to thickening of the lung tissue)? A low compliance means that a greater-than-normal pressure difference must be developed across the lung wall to produce a normal lung expansion; i.e., intrapleural pressure must be made more subatmospheric than usual during inspiration, and this requires more vigorous contraction of the diaphragm and inspiratory intercostal muscles. To generalize, the less compliant the lung, the more energy is required for a given amount of expansion.

You might think from this discussion of low lung compliance that a very high lung compliance might be a good thing to have. Such is not the case. In the disease **emphysema**, the internal structures of the lung become very stretchable; i.e., lung compliance increases; this results in a marked increase in resting lung (and chest) size, since the resting alveolar-intrapleural pressure difference distends the lung more than usual. The deleterious effects of emphysema for gas exchange will be described later.

Surfactant. We have certainly given the impression thus far that the elastic and other connective tissues of the lungs determine lung compliance. This is only partly true, for the single most important determinant of lung compliance is the surface tension at the air-water interfaces within the alveoli. The alveoli may be viewed as air-filled bubbles lined with water. At an air-water interface, the attractive forces between water molecules cause them to squeeze in upon the air within the bubble. This force, known as **surface tension**, makes the water lining very like highly stretched rubber which constantly tries to shorten and resists further stretching. Thus, inspiration requires considerable energy to expand the lungs because of the difficulty of distending these alveolar bubbles. Indeed, the surface tension of pure water is so great that lung expansion would require exhausting muscular effort and the lungs would tend to collapse. It is extremely important, therefore, that the type II alveolar cells produce a phospholipid known as pulmonary **surfactant**, which markedly reduces the cohesive force between water molecules on the alveolar surface, thereby lowering the surface tension and increasing total lung compliance.

A striking example of what occurs when insufficient surfactant is present is provided by the disease known as **respiratory-distress syndrome of the newborn**, which frequently afflicts premature infants in whom the surfactant-synthesizing cells are too immature to function adequately. The infant is able to inspire only by the most strenuous efforts, which may ultimately cause complete exhaustion, inability to breathe, lung collapse, and death. Normal maturation of the surfactant-synthesizing apparatus is facilitated by the hormone cortisol, the secretion of which is increased late in pregnancy. Accordingly, it has been found that administration of cortisol to a pregnant woman who is likely to deliver prematurely provides an important means of combating this disease in premature infants.

Quantitative relationship between atmosphere-alveolar pressure gradients and air flow: Airway resistance

The volume of air which flows in or out of the alveoli per unit time is directly proportional to the pressure difference between the atmosphere and alveoli ($P_{atm} - P_{alv}$) and inversely proportional to the resistance to flow offered by the airways. Normally the magnitude of this pressure gradient is increased by increasing the strength of contraction of the inspiratory muscles. This causes, in order, a more rapid expansion of the thoracic cage, a more subatmospheric intrapleural pressure, a greater pressure difference across the lung wall, increased expansion of the lungs, a more subatmospheric alveolar pressure, and increased flow of air into the lungs (Fig. 12-13).

What factors determine airway resistance? Resistance is (1) directly proportional to the magnitude of the interactions between the flowing gas molecules, (2) directly proportional to the length of the airway, and (3) inversely proportional to the fourth power of the airway radius. These factors, of course, are counterparts of the factors determining resistance in the circulatory system; in the respiratory tree, just as in the circulatory tree, resistance is largely controlled by the radius of the airways.

The airway diameters are normally so large that they offer little resistance to air flow. Interaction between gas molecules (viscosity) is also usually negligible, and so is the contribution of airway length. Therefore, the total resistance is normally so small that very small pressure differences suffice to produce large volumes of air flow. As we have seen (Fig. 12-11), the average pressure gradient during a normal breath at rest is less than 1 mmHg; yet, approximately 500 mL of air is moved by this tiny difference.

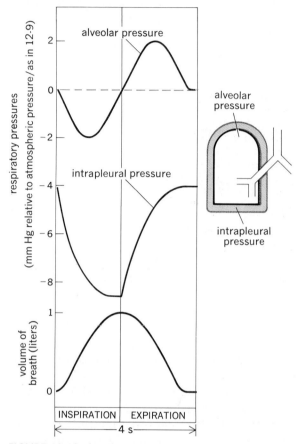

FIGURE 12-13. Summary of alveolar and intrapleural pressure changes and air flow during inspiration of 1000 mL of air. Compare these values with those given in Fig. 12-11.

The diameter of the airways becomes critically important in certain conditions such as **asthma**, which is characterized both by airway smooth muscle contraction and by plugging of the airways by accumulated mucus. Airway resistance may become great enough to prevent air flow completely, regardless of the atmosphere-alveolar pressure gradient.

Airway size and resistance may be altered by physical, nervous, or chemical factors (Table 12-1). The most important normal physical factor is simply expansion of the lungs; during inspiration, airway resistance decreases because as the lungs expand, the airways within the lung are subject to the

TABLE 12-1. Major neural and chemical factors which control airway resistance

Airways constricted by:	Histamine Parasympathetic nerves $\downarrow CO_2$ Some Prostaglandins
Airways dilated by:	Epinephrine Sympathetic nerves $\uparrow CO_2$ Some Prostaglandins

same forces as the alveoli and thus become widened. Conversely, during expiration, airway resistance increases as the airways become narrower. For this reason, persons with abnormal airway resistance, as in asthma, have much less difficulty inhaling than exhaling, with the result that more air is inspired than expired and the lung volume progressively increases.

Neural regulation of airway size is mediated mainly by parasympathetic neurons, stimulation of which causes airway smooth muscle contraction (increased resistance). Reflex parasympathetic input is important in causing airway constriction (as well as increased mucus secretion) upon inhalation of chemical irritants.

There is little, if any, sympathetic innervation of the airways, but epinephrine, the hormone from the adrenal medulla, causes airway dilation. This is a major reason for administering epinephrine or epinephrine-like drugs to patients suffering from airway constriction, as in an asthmatic attack. In contrast, histamine causes airway constriction (and increased mucus secretion) and is one of the causes of the airway constriction observed in allergic attacks. This explains the use of antihistamines to relieve the respiratory symptoms of allergies. The effects of prostaglandins on pulmonary airways in asthma and other diseases may be of particular importance since the lungs take up, metabolize, and release various members of the prostaglandin family, some of which are airway constrictors, some dilators. Finally, the airway smooth muscle is highly responsive to carbon dioxide, high carbon dioxide

producing bronchodilation and low carbon dioxide bronchoconstriction; the physiological significance of this responsiveness is discussed later.

The Heimlich maneuver. A sudden, sharp increase in alveolar pressure caused by a forceful elevation of the diaphragm forms the basis of the Heimlich maneuver, which is used to aid persons who are choking on foreign matter caught in the respiratory tract. The sudden decrease in thoracic volume, which is responsible for the increased alveolar pressure, is produced as the rescuer's fist, which has been placed against the victim's abdomen slightly above the navel and well below the tip of the sternum, is pressed into the abdomen with a quick upward thrust (Fig. 12-14). The forceful expiration thus produced expels the object caught in the respiratory tract. The Heimlich maneuver can be performed on persons who are standing, sitting, or lying down.

Lung-volume changes during breathing

The volume of air entering or leaving the lungs during a single breath is called the **tidal volume** (Fig. 12-15). Under resting conditions, this is approximately 500 mL. We are all aware that the resting thoracic excursion is small compared with a maximal breathing effort. The volume of air which can be inspired over and above the resting tidal volume is called the **inspiratory reserve volume** and amounts to 2500 to 3500 mL of air. At the end of a normal expiration, the lungs still contain a large volume of air, part of which can be exhaled by active contraction of the expiratory muscles; it is called the **expiratory reserve volume** and measures approximately 1000 mL of air. Even after a maximal expiration, some air (approximately 1000 mL) still remains in the lungs and is termed the **residual volume**.

What, then, is the maximum amount of air which can be moved in and out during a single breath? It is the sum of the normal tidal, inspiratory-reserve, and expiratory-reserve volumes. This total volume is called the **vital capacity**. Even during exercise, a person rarely uses more than 50 percent

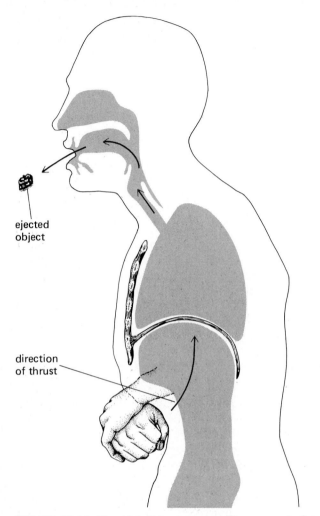

ejected
object

direction
of thrust

FIGURE 12-14. The Heimlich maneuver. The rescuer's fist forms a knob, which is placed against the victim's abdomen. A quick upward thrust causes elevation of the diaphragm and a forceful expiration. As air is forced through the trachea and larynx, the foreign object in the airway is expelled.

of the total vital capacity, because deeper breaths than this require exhausting activity of the inspiratory and expiratory muscles.

Alveolar ventilation

The **total pulmonary ventilation** per minute (also known as minute volume) is determined by the tidal volume multiplied by the respiratory rate (expressed as breaths per minute). For example, at rest, a normal person moves approximately 500 mL of air in and out of the lungs with each breath, and takes 10 breaths each minute; the total pulmonary ventilation is therefore 500 mL × 10 breaths per minute = 5000 mL of air per minute. However, because of dead space, not all of this air is available for exchange with the blood.

Dead space. The respiratory tract, as we have seen, is composed of conducting airways and the alveoli. Within the lungs, exchanges of gases with the blood occur only in alveoli and not in the conducting airways, the total volume of which is approximately 150 mL. Picture, then, what occurs during expiration: 500 mL of air is forced out of the alveoli and through the airways. Approximately 350 mL of this air is exhaled at the nose or mouth, but approximately 150 mL still remains in the airways at the end of expiration. During the next inspiration, 500 mL of air flows into the alveoli, but the first 150 mL of entering air is not atmospheric air but the 150 mL of alveolar air left behind. Thus, only 350 mL of *new* atmospheric air enters the alveoli during the inspiration. At the end of inspiration, 150 mL of fresh air also fills the conducting airways, but no gas exchange with the blood can occur there. At the next expiration, this fresh air will be washed out and again replaced by old alveolar air, thus completing the cycle. The end result is that 150 mL of the 500 mL of atmospheric air entering the respiratory system during each inspiration never reaches the alveoli but is merely moved in and out of the airways. Because these airways do not permit gas exchange with the blood, the space within them is termed the **anatomic dead space**.

Thus, the volume of fresh atmospheric air entering the alveoli during each inspiration equals the total tidal volume minus the volume of air in the anatomic dead space. For a single normal breath

$$
\begin{aligned}
\text{Tidal volume} &= 500 \text{ mL} \\
\text{Anatomic dead space} &= \underline{150 \text{ mL}} \\
\text{Fresh air entering alveoli} &= 350 \text{ mL}
\end{aligned}
$$

To determine how much fresh air enters the alveoli per minute, we simply multiply 350 mL by the breathing frequency (10 in our example), which

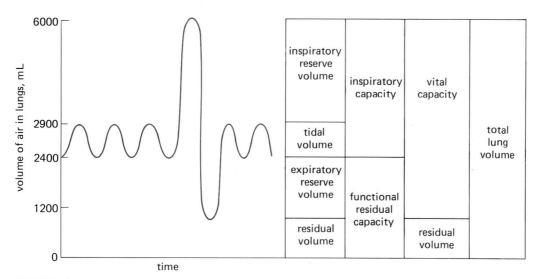

FIGURE 12-15. Lung volumes and capacities as measured on a spirograph. When the subject inspires, the pen moves up; with expiration, it moves down.

gives 3500 mL/min. This total is called the **alveolar ventilation**:

$$\text{Alveolar ventilation (mL/min)} = \text{frequency (breaths/min)} \times (\text{tidal volume} - \text{dead space}) \text{ (mL/breath)}$$

The term "alveolar ventilation" is somewhat confusing, because it seems to indicate that only 3500 mL air enters and leaves the alveoli each minute. This is not true—the total is 5000 mL of air, but only 3500 mL of this is fresh atmospheric air.

What is the significance of the anatomic dead space and alveolar ventilation? As we have men-

tioned, total pulmonary ventilation is equal to the tidal volume of each breath multiplied by the number of breaths per minute. But since only that portion of inspired atmospheric air which enters the alveoli, i.e., the alveolar ventilation, is useful for gas exchange with the blood the magnitude of the alveolar ventilation is of much greater significance than is the magnitude of the total pulmonary ventilation, as can be demonstrated readily by the highly idealized data in Table 12-2.

In this experiment subject A breathes rapidly and shallowly, B normally, and C slowly and deeply. Each subject has exactly the same total pulmonary ventilation; i.e., each is moving the same amount

TABLE 12-2. Effect of breathing patterns on alveolar ventilation

Subject	Tidal volume, mL /breath	× Frequency, breaths/min	= Total pulmonary ventilation, mL /min	Anatomic dead space ventilation, mL /min	Alveolar ventilation,* mL /min
A	150	40	6000	150 × 40 = 6000	0
B	500	12	6000	150 × 12 = 1800	4200
C	1000	6	6000	150 × 6 = 900	5100

* In the calculations of alveolar ventilation, alveolar dead space has been ignored.

of air in and out of the lungs each minute. Yet, when we subtract the anatomic-dead-space ventilation from the total pulmonary ventilation, we find marked differences in alveolar ventilation. Subject A has no alveolar ventilation (and would become unconscious in several minutes), whereas C has a considerably greater alveolar ventilation than B, who is breathing normally. The important deduction to be drawn from this example is that increased *depth* of breathing is far more effective in elevating alveolar ventilation than is an equivalent increase of breathing *rate*. Conversely a decrease in depth can lead to a critical reduction of alveolar ventilation. This is because a fraction of *each* tidal volume represents anatomic-dead-space ventilation. If the tidal volume decreases, this fraction increases until, as in subject A, it may represent the entire tidal volume. On the other hand, any increase in tidal volume goes entirely toward increasing alveolar ventilation.

These concepts have important physiological implications. Most situations that produce an increased ventilation, such as exercise, reflexly call forth a relatively greater increase in breathing depth than rate. Indeed, well-trained athletes can perform moderate exercise with very little increase, if any, in respiratory rate. The mechanisms by which rate and depth of respiration are controlled will be described in a later section of this chapter.

The anatomic dead space is not the only type of dead space. In addition, some fresh inspired air is not used for gas exchange with the blood even though it reaches the alveoli, because some alveoli, for various reasons, have deficient blood supply. This volume of air is known as **alveolar dead space**. It is quite small in normal persons but may reach lethal proportions in several kinds of lung disease. As we shall see, it is minimized by local mechanisms which match air and blood flows. The sum of the anatomic and alveolar dead space is known as the **total** (or, formerly, **physiological**) **dead space**.

The work of breathing

During inspiration, active muscular contraction provides the energy required to expand the thorax and lungs. What determines how much work these muscles must perform in order to provide a given amount of ventilation? First, there is simply the stretchability of the thorax and lungs. To expand these structures they must be stretched. As pointed out earlier, the more easily they stretch, i.e., the more compliant the lung, the less energy is required for a given amount of expansion.

The second factor determining the degree of muscular work required for a certain amount of ventilation is the magnitude of the airway resistance. When airway resistance is increased by smooth muscle contraction or by secretions, the usual pressure gradient does not suffice for adequate air inflow, and a stronger respiratory effort is required to create a larger pressure gradient.

One might imagine from this discussion (and from observing an athlete exercising) that the work of breathing uses up a major portion of the energy spent by the body. Not so; in a normal person, even during heavy exercise, the energy needed for breathing is only about 3 percent of the total expenditure. It is only in disease, when the work of breathing is markedly increased by structural changes in the lung or thorax, by loss of surfactant, or by an increased airway resistance, that breathing itself becomes an exhausting form of exercise.

Exchange of Gases in Alveoli and Tissues

We have completed our discussion of alveolar ventilation, but this is only the first step in the total respiratory process. Oxygen must move across the alveolar membranes into the pulmonary capillaries, be transported by the blood to the tissues, leave the tissue capillaries, and finally cross cell membranes to gain entry into cells (Fig. 12-16). Carbon dioxide must follow a similar path in reverse (Fig. 12-16). At rest, during each minute, body cells consume approximately 200 mL of oxygen and produce approximately the same amount of carbon dioxide. The relative amounts of these two gases depend primarily upon what nutrients are being used for energy; e.g., when glucose is utilized, one molecule of carbon dioxide is produced for every molecule of oxygen consumed:

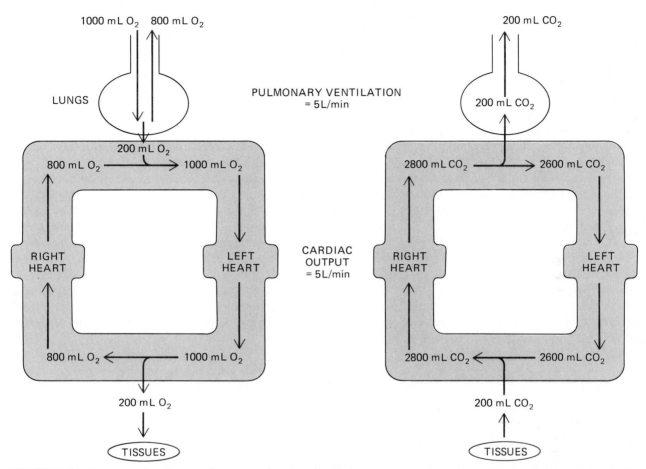

FIGURE 12-16. Summary of exchanges of oxygen and carbon dioxide between atmosphere, lungs, blood, and tissues during 1 min. It is assumed that RQ = 1.

$$C_6H_{12}O_6 + 6O_2 \rightarrow 6H_2O + 6CO_2 + energy$$

The ratio (CO_2 produced)/(O_2 consumed) is known as the **respiratory quotient (RQ)**; accordingly, for glucose RQ = 1. When fat is utilized, only 7 molecules of carbon dioxide are produced for every 10 molecules of oxygen consumed, and RQ = 0.7. On a mixed diet, the RQ is between these values, but for simplicity, Fig. 12-16 assumes that the carbon dioxide and oxygen amounts are equal and the total volumes of air inspired and expired are therefore identical.

At rest, the total pulmonary ventilation equals 5000 mL of air per minute. Since only 20 percent

of atmospheric air is oxygen, the total oxygen input is 20 percent of 5000 mL, or 1000 mL per minute. Of this inspired oxygen, 200 mL crosses the alveoli into the pulmonary capillaries, and the remaining 800 mL is exhaled. This 200 mL of oxygen is carried by 5 L of blood, which is the pulmonary blood flow (cardiac output) per minute. However, note in Fig. 12-16 that blood entering the lungs already contains large quantities of oxygen, to which this 200 mL is added. This blood is then pumped by the left ventricle through the tissue capillaries of the body, and 200 mL of oxygen leaves the blood to be taken up and utilized by cells. Because only a fraction of the total blood oxygen actually leaves the blood, a good deal of oxygen remains in the blood when it returns to the heart and lungs. The quantities of oxygen added to the blood in the lungs and removed in the tissues are identical in the steady state.

As shown by Fig. 12-16, the story reads in reverse for carbon dioxide. There is already a good deal of carbon dioxide in arterial blood; to this is added an additional amount as blood flows through tissue capillaries. This additional amount is then eliminated as blood flows through the lungs. As we shall see, most of the blood carbon dioxide is actually in the form of bicarbonate ion (HCO_3^-), but we have shown it as CO_2 for simplicity.

The pumping of blood by the heart propels oxygen and carbon dioxide between the lungs and tissues by bulk flow, but it is diffusion which is responsible for the net movement of these molecules between alveoli and blood and between blood and the cells of the body. Understanding the mechanisms involved in these diffusional exchanges depends upon familiarity with some basic chemical and physical properties of gases, to which we now turn.

Basic properties of gases

A gas consists of individual molecules constantly moving at great speeds. Since rapidly moving molecules bombard the walls of any vessel containing them, they exert a pressure against the walls. The magnitude of the pressure is increased by anything which increases the bombardment. The pressure a gas exerts is proportional to (1) the temperature (because heat increases the speed at which molecules move) and (2) the concentration of the gas, i.e., the number of molecules per unit volume. In other words, when a certain number of molecules are compressed into a smaller volume, there are more collisions with the walls. The pressure of a gas is therefore a measure of the concentration of the gas at any given temperature.

In a mixture of gases, the pressure exerted by each gas is independent of the pressure exerted by the others because gas molecules are normally so far apart that they do not interfere with each other. Since each gas behaves as though the other gases were not present, the total pressure of a mixture of gases is simply the sum of the individual pressures. These individual pressures, termed **partial pressures**, are denoted by a P in front of the symbol for the gas, the partial pressure of oxygen thus being represented by P_{O_2}. The partial pressure of a gas is a measure of the concentration of that gas in a mixture of gases. Net diffusion of a gas will occur from a region where its partial pressure (concentration) is high to a region where it is low.

Atmospheric air consists primarily of nitrogen and oxygen with very small quantities of water vapor, carbon dioxide, and inert gases such as argon. The sum of the partial pressures of all these gases is termed atmospheric pressure or barometric pressure. It varies in different parts of the world as a result of differences in altitude, but at sea level it is about 760 mmHg. Since approximately 20 percent of the molecules in air are oxygen, O_2, the P_{O_2} of inspired air is $0.20 \times 760 = 152$ mmHg at sea level.

Behavior of gases in liquids. When a gas comes into contact with a liquid, the number of gas molecules which dissolve in the liquid is directly proportional to the partial pressure of the gas. This phenomenon stems from the basic definition of pressure. Suppose, for example, that gaseous oxygen is placed in a closed container half full of water. Oxygen molecules constantly bombard the surface of the water, some entering the water and dissolving. Since the number of molecules striking the surface is directly proportional to the partial pressure of the gas, P_{O_2}, the number of molecules entering the water is also directly proportional to P_{O_2}. How many entering molecules actually stay in the water? Since the dissolved oxygen molecules are also constantly moving, some of them strike the water surface from below and escape into the free oxygen above. Eventually the rate of escape from the water and the rate of entry into the water become equal. By definition, therefore, at equilibrium the partial pressure of the oxygen in the liquid is equal to its partial pressure in the gas phase. Thus, we come back to our earlier statement: The number of gas molecules which will dissolve in a liquid is directly proportional to the partial pressure of the gas.

When the partial pressure in the gas phase is higher than the partial pressure in the liquid, there will be a net movement into the liquid. Conversely, if a liquid containing a dissolved gas at high partial pressure is exposed to that same gas at lower partial pressure, gas molecules will move from the liquid into the gas phase until the partial pressures in the two phases become equal. These are precisely the phenomena occurring between alveolar air and pulmonary capillary blood.

Dissolved gas molecules also diffuse within the liquid from a region of higher partial pressure to a region of lower partial pressure. This effect underlies the exchange of gases between cells, extracellular fluid, and capillary blood throughout the body.

This discussion has been in terms of proportionalities rather than absolute amounts. The number of gas molecules which will dissolve in a liquid is proportional to the partial pressure, but the absolute number also depends upon the solubility of the gas in the liquid. Thus, if a liquid is exposed to two different gases at the same partial pressures,

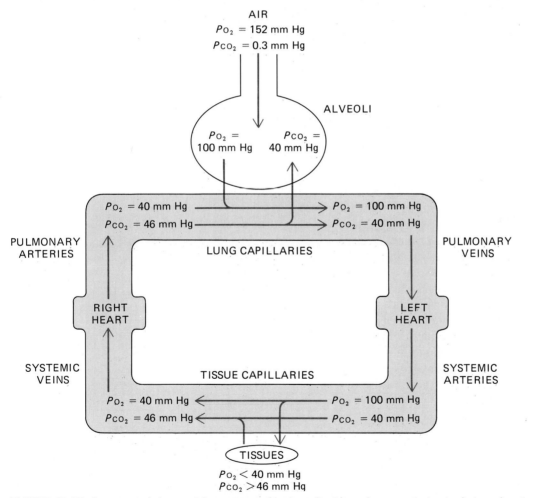

FIGURE 12-17. Summary of the partial pressures of carbon dioxide and oxygen in inspired air and various places in the body. Note that the partial pressures in pulmonary veins, left heart, and systemic arteries are shown as identical to those in the alveoli; actually there are normally slight differences, which we have ignored for the sake of clarity.

the numbers of molecules of each gas which are dissolved at equilibrium are not identical but reflect the solubilities of the two gases. (For example, carbon dioxide is more than 20 times as soluble as oxygen in water). Nevertheless, doubling the partial pressures doubles the number of gas molecules dissolved.

With these basic gas properties as foundation, we can now discuss the diffusion of oxygen and carbon dioxide across alveolar, capillary, and plasma membranes. The pressures of these gases in atmospheric air and in various sites of the body are given in Fig. 12-17 for a resting person at sea level. We start our discussion with the alveolar gas pressures because their values determine those of systemic arterial blood.

Alveolar gas pressures

One might logically reason that the alveolar gas pressures must vary considerably during the respiratory cycle, since new atmospheric air enters the alveoli only during inspiration, whereas movements of oxygen and carbon dioxide between alveoli and blood occur continuously. In fact, however, the variations in alveolar P_{O_2} and P_{CO_2} during the respiratory cycle are so small as to be negligible, because (as explained in the section on lung volumes) a large volume of gas is always left in the lungs after expiration. This remaining alveolar gas contains large quantities of oxygen and carbon dioxide, and when the relatively small volume (350 mL) of new air enters, it mixes with the large volume (2500 mL) of alveolar air already present, lowering its P_{CO_2} and raising its P_{O_2}, but only by a small amount. For this reason, the alveolar-gas partial pressures remain relatively constant throughout the respiratory cycle, and we may use the single alveolar pressures shown in Fig. 12-17 in our subsequent analysis of alveolar-capillary exchange.

The values for alveolar gas pressures, $P_{O_2} = 100$ mmHg and $P_{CO_2} = 40$ mmHg, are normal values for a healthy person breathing air at sea level. They are basically the resultant of only two variables: (1) the relative magnitudes of the alveolar ventilation, on the one hand, and (2) the oxygen consumption and carbon dioxide production, on the other. This should make intuitive sense: If the alveolar ventilation is very great (say, because a resting person is deliberately breathing rapidly and deeply) but the consumption of oxygen is unchanged, then a smaller fraction of the oxygen moving in and out of the alveoli is lost to the blood, and the steady-state alveolar P_{O_2} is relatively high (Fig. 12-18); similarly, the alveolar P_{CO_2} would be low, because the normal amount of carbon dioxide entering the alveoli would be diluted by a larger alveolar ventilation (Fig. 12-18). This is known as **hyperventilation.** Conversely, a reduced alveolar ventilation in a person with a normal oxygen consumption and carbon-dioxide production will result in a low alveolar P_{O_2} and a high alveolar P_{CO_2} (Fig. 12-18). Some causes of reduced alveolar ventilation (**hypoventilation**) are muscle weakness, decreased

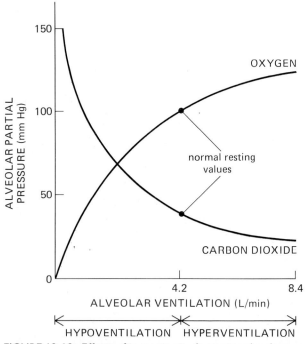

FIGURE 12-18. Effects of increasing or decreasing alveolar ventilation on alveolar partial pressures in a person breathing room air and having a constant metabolic rate (oxygen consumption).

lung compliance, and increased airway resistance. Another cause of low alveolar P_{O_2} would be breathing air containing a low P_{O_2}, as at high altitude.

Alveolar-blood gas exchange

The blood which enters the pulmonary capillaries is, of course, systemic venous blood pumped to the lungs via the pulmonary arteries. Having come from the tissues, it has a high P_{CO_2} (46 mmHg) and a low P_{O_2} (40 mmHg). As it flows through the pulmonary capillaries, it is separated from the alveolar air by only an extremely thin layer of tissue. The differences in the partial pressures of oxygen and carbon dioxide on the two sides of this alveolar-capillary membrane result in the net diffusion of oxygen into the blood and of carbon dioxide into the alveoli. As this diffusion occurs, the capillary blood P_{O_2} rises and the P_{CO_2} falls. The net diffusion of these gases ceases when the alveolar and capillary partial pressures become equal. In a normal person, the rates at which oxygen and carbon dioxide diffuse are so rapid and the blood flow through the capillaries so slow that complete equilibrium is usually reached before the end of the capillaries. Thus, the blood that leaves the pulmonary capillaries to return to the heart and be pumped into the systemic arteries has essentially the same P_{O_2} and P_{CO_2} as alveolar air. In other words, the alveolar gas pressures set the systemic arterial gas pressures.

One of the normal factors important for determining the rate of diffusion of oxygen and carbon dioxide between alveolar air and pulmonary capillary blood is the surface area available for diffusion. Many of the pulmonary capillaries are normally closed at rest. During exercise, these capillaries open and are perfused with blood, thereby increasing the surface area for diffusion and enhancing gas exchange. The mechanism by which this occurs is a simple physical one: The pulmonary circulation at rest is such a low-pressure system that the pressure in many capillaries is inadequate to keep the capillaries open, but the increased cardiac output of exercise raises pulmonary vascular pressures, which opens these capillaries.

The diffusion of gases between alveoli and capillaries may be impaired in a number of ways, resulting in inadequate gas exchange. The disease emphysema, for example, is characterized by the breakdown of the alveolar walls with the formation of fewer but larger alveoli. The result is a reduction in the total area available for diffusion. Loss of air spaces in the alveoli, as in pneumonia, also diminishes the surface area available. Decreased surface area is not the only cause of impaired diffusion; the alveolar walls may become denser and less permeable, as for example when beryllium is inhaled and deposited on the walls. But the major disease-induced cause of inadequate gas exchange between alveoli and pulmonary capillary blood is simply the mismatching of the air supply and blood supply in individual alveoli.

Matching of ventilation and blood flow in alveoli. We previously touched on the question of alveolar dead space. The lungs are composed of approximately 300 million discrete alveoli, each receiving carbon dioxide from and supplying oxygen to the pulmonary capillary blood. To be most efficient, the right proportion of alveolar air and capillary blood should be available to each alveolus, a pattern local changes in bronchiolar smooth muscle tone help maintain. For example, if an alveolus is receiving too much air for its blood supply, the concentration of carbon dioxide in it and in the surrounding tissue will be low; the airway supplying the alveolus is exposed to this low tissue carbon dioxide concentration and constricts (this effect of carbon dioxide on airway smooth muscle was stated earlier). By this completely local mechanism, ventilation can be matched to blood supply (Fig. 12-19).

The pulmonary vessels control the distribution of blood to different alveolar capillaries, thus providing a second mechanism for matching air flow and blood flow. A decreased oxygen concentration causes pulmonary vessel constriction, whereas an increased oxygen concentration causes vasodilation (note that this purely local effect of oxygen on pulmonary vessels is precisely the opposite of that exerted on systemic arterioles). How does this pro-

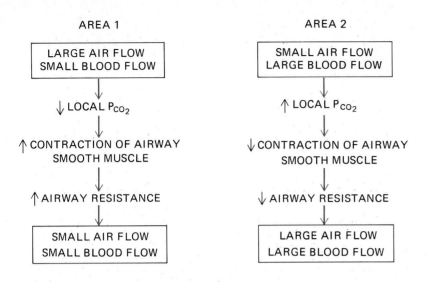

FIGURE 12-19. Local matching of ventilation and blood flow through effects of P_{CO_2} on airway resistance.

vide matching of air and blood supplies? Recall the example above in which an alveolus has a large ventilation and small blood flow. The area around the alveolus will therefore have a high oxygen concentration. This causes vasodilation of the blood vessel supplying it and an increased blood supply to match the high air supply (Fig. 12-20).

These two mechanisms for matching the ventilation and blood supply of individual alveoli are quite effective in normal persons. Even so, there is enough mismatching that gas exchange is affected, and this largely accounts for the fact that systemic arterial gas pressures are not really precisely the same as alveolar gas pressures. But in disease states, such as emphysema, the mismatching may be so great as to cause systemic arterial P_{O_2} to be markedly lower than alveolar P_{O_2}.

Gas exchange in the tissues

As the arterial blood enters capillaries throughout the body, it is separated from the interstitial fluid only by the thin capillary wall, which is highly permeable to both oxygen and carbon dioxide. The interstitial fluid, in turn, is separated from intracellular fluid by cells' plasma membranes which are also quite permeable to oxygen and carbon dioxide. Metabolic reactions occurring within these cells are constantly consuming oxygen and producing car-

bon dioxide. Therefore, as shown in Fig. 12-17, intracellular P_{O_2} is lower and P_{CO_2} higher than in blood. As a result, a net diffusion of oxygen occurs from blood to cells, and a net diffusion of carbon dioxide from cells to blood. In this manner, as blood flows through capillaries, its P_{O_2} decreases and its P_{CO_2} increases. This accounts for the venous blood values shown in Fig. 12-17. The mechanisms which enhance diffusion of oxygen and carbon dioxide between cells and blood when a tissue increases its metabolic activity were discussed in Chap. 11.

In summary, the consumption of oxygen in the cells and the supply of new oxygen to the alveoli create P_{O_2} gradients which produce net diffusion of oxygen from alveoli to blood in the lungs and from blood to cells in the rest of the body. Conversely, the production of carbon dioxide by cells and its elimination from the alveoli via expiration create P_{CO_2} gradients which produce net diffusion of carbon dioxide from blood to alveoli in the lungs and from cells to blood in the rest of the body.

Transport of Gases in the Blood

Transport of oxygen: The role of hemoglobin

Table 12-3 summarizes the oxygen content of arterial blood. Each liter of arterial blood contains the number of oxygen molecules equivalent to 200

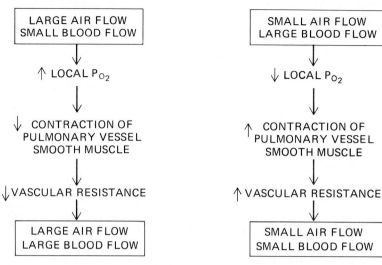

AREA 1

LARGE AIR FLOW
SMALL BLOOD FLOW

↓

↑ LOCAL P_{O_2}

↓

↓ CONTRACTION OF
PULMONARY VESSEL
SMOOTH MUSCLE

↓

↓ VASCULAR RESISTANCE

↓

LARGE AIR FLOW
LARGE BLOOD FLOW

AREA 2

SMALL AIR FLOW
LARGE BLOOD FLOW

↓

↓ LOCAL P_{O_2}

↓

↑ CONTRACTION OF
PULMONARY VESSEL
SMOOTH MUSCLE

↓

↑ VASCULAR RESISTANCE

↓

SMALL AIR FLOW
SMALL BLOOD FLOW

FIGURE 12-20. Local matching of ventilation and blood flow through effects of P_{O_2} on pulmonary vascular resistance. Events of this figure and those of Fig. 12-19 occur simultaneously; i.e., local mechanisms control the distribution of both ventilation and blood flow.

TABLE 12-3. Oxygen content of arterial blood

I liter (L) arterial blood contains:

         3 mL O_2 physically dissolved
     197 mL O_2 chemically bound to hemoglobin
Total   200 mL O_2

Cardiac output = 5 L/min

O_2 carried to tissues/min = 5 × 200 = 1000 mL

mL of pure gaseous oxygen. The oxygen is present in two forms: (1) dissolved in the blood water and (2) reversibly combined with hemoglobin molecules. The amount of oxygen which can be dissolved in blood is directly proportional to the P_{O_2} of the blood, but because oxygen is relatively insoluble in water, only 3 mL of oxygen can be dissolved in 1 L of blood at the normal arterial P_{O_2} of 100 mmHg. In contrast, in 1 L of blood, 197 mL of oxygen, more than 98 percent of the total, is carried within the red blood cells bound to hemoglobin.

As described in Chap. 11, hemoglobin is a protein composed of four polypeptide chains, each of which contains a single atom of iron. It is with the iron atom that a molecule of oxygen combines; thus each hemoglobin molecule can combine with four molecules of oxygen. However, the equation for the reaction between oxygen and hemoglobin is usually written in terms of a single polypeptide chain:

$$O_2 + Hb \rightleftharpoons HbO_2 \qquad [12\text{-}1]$$

Hemoglobin combined with oxygen (**HbO₂**) is called **oxyhemoglobin**; not combined (**Hb**), it is called reduced hemoglobin or **deoxyhemoglobin**. When all the iron atoms in hemoglobin are combined with oxygen, the hemoglobin solution is said to be fully saturated. The **percentage saturation** of a group of hemoglobin molecules therefore can vary from zero to 100 percent.

What factors determine the extent to which oxygen will combine with hemoglobin? By far the most important is the P_{O_2} of the blood, i.e., the concentration of dissolved oxygen. The blood hydrogen-ion concentration and temperature also affect the amount of oxygen bound to hemoglobin, as does a chemical—DPG—produced by the red blood cells themselves.

Effect of P_{O_2} on hemoglobin saturation. From inspection of Eq. 12-1 and the law of mass action, it is obvious that raising the P_{O_2} of the blood should increase the combination of oxygen with hemoglobin. The quantitative relationship between these variables is shown in Fig. 12-21, which is called an **oxygen-hemoglobin dissociation** (or **saturation**) **curve.** It is an S-shaped curve with a steep slope between 10 and 60 mmHg P_{O_2} and a flat portion between 70 and 100 mmHg P_{O_2}. In other words, the extent to which hemoglobin combines with oxygen increases very rapidly from 10 to 60 mmHg so that, at a P_{O_2} of 60 mmHg, 90 percent of the total hemoglobin is combined with oxygen. From this point on, a further increase in P_{O_2} produces only a small increase in oxygen binding. The adaptive importance of this plateau at higher P_{O_2} values is very great for the following reason: Many situations (including high altitudes and cardiac or pulmonary disease) are characterized by a moderate reduction of alveolar, and therefore arterial, P_{O_2}. Even if the P_{O_2} fell from the normal value of 100 to 60 mmHg, the total quantity of oxygen carried by hemoglobin would decrease by only 10 percent, since hemoglobin saturation is still close to 90 percent at a P_{O_2} of 60 mmHg. The plateau therefore provides an excellent safety factor in the supply of oxygen to the tissues.

The flatness of the upper portion of the curve also explains another fact: In a normal person at sea level, raising the alveolar (and therefore the arterial) P_{O_2}, either by hyperventilating or by breathing pure oxygen, adds very little additional oxygen to the blood. A small extra amount dissolves, but because hemoglobin is already almost completely saturated with oxygen at the normal arterial P_{O_2} of 100 mmHg, it simply cannot pick up any more when the P_{O_2} is elevated beyond this point. This applies only to normal people at sea level. If the person has an initially low arterial P_{O_2} (because of lung disease or high altitude), then there would be a good deal of unsaturated hemoglobin initially present and raising the arterial P_{O_2} would result in significantly more oxygen carriage.

We now retrace our steps and reconsider the movement of oxygen across the various mem-

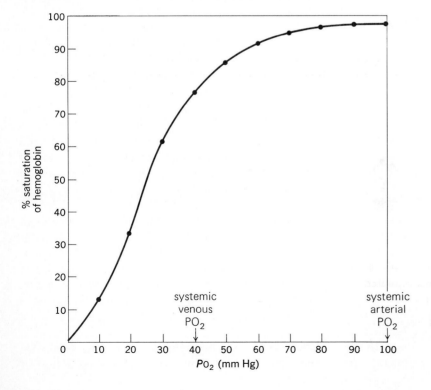

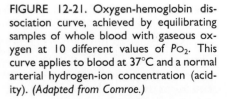

FIGURE 12-21. Oxygen-hemoglobin dissociation curve, achieved by equilibrating samples of whole blood with gaseous oxygen at 10 different values of P_{O_2}. This curve applies to blood at 37°C and a normal arterial hydrogen-ion concentration (acidity). *(Adapted from Comroe.)*

branes, this time including hemoglobin in our analysis. It is essential to recognize that the oxygen which is bound to hemoglobin does not contribute directly to the P_{O_2} of the blood. Only gas moleules which are free in solution, i.e., dissolved, do so. Therefore, the diffusion of oxygen is governed only by that portion which is dissolved, a fact which permitted us to ignore hemoglobin in discussing transmembrane pressure gradients. However, the presence of hemoglobin plays a critical role in determining the total amount of oxygen which will diffuse, as illustrated by a simple example (Fig. 12-22). Two solutions separated by a semipermeable membrane contain equal quantities of oxygen, the gas pressures are equal, and no net diffusion occurs. Addition of hemoglobin to compartment B destroys this equilibrium because much of the oxygen combines with hemoglobin. Despite the fact that the total quantity of oxygen in compartment B is still the same, the number of oxygen molecules dissolved has decreased; therefore, the P_{O_2} of compartment B is less than that of A, and net diffusion of oxygen occurs from A to B. At the new equilibrium, the oxygen pressures are once again equal, but almost all the oxygen is in compartment B and is combined with hemoglobin.

Let us now apply this analysis to the lung and tissue capillaries (Fig. 12-23). The plasma and

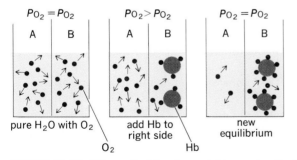

$P_{O_2} = P_{O_2}$ $P_{O_2} > P_{O_2}$ $P_{O_2} = P_{O_2}$

A   B   A   B   A   B

pure H$_2$O with O$_2$ add Hb to right side   new equilibrium

O$_2$ Hb

FIGURE 12-22. Effects of adding hemoglobin on the distribution of oxygen between two compartments separated by a semipermeable membrane. At the new equilibrium, the P_{O_2} values are equal to each other but lower than before the hemoglobin was added; however, the total oxygen, i.e., both dissolved and chemically combined with hemoglobin, is much higher on the right side of the membrane. (*Adapted from Comroe.*)

erythrocytes entering the lungs have a P_{O_2} of 40 mmHg, and the hemoglobin is 75 percent saturated. Oxygen diffuses from the alveoli, because of its higher P_{O_2} (100 mmHg), into the plasma; this increases plasma P_{O_2} and induces diffusion of oxygen into the erythrocytes, elevating erythrocyte P_{O_2} and causing increased combination of oxygen and hemoglobin. Thus, the vast preponderance of the oxygen diffusing into the blood from the alveoli does not remain dissolved but combines with hemoglobin. In this manner, the blood P_{O_2} remains less than the alveolar P_{O_2} until hemoglobin is virtually completely saturated. Thus, the diffusion gradient favoring oxygen movement into the blood is maintained despite the very large transfer of oxygen.

In the tissue capillaries, the procedure is reversed: As the blood enters the capillaries, plasma P_{O_2} is greater than interstitial fluid P_{O_2} and net oxygen diffusion occurs across the capillary wall; plasma P_{O_2} is now lower than erythrocyte P_{O_2}, and oxygen diffuses out of the erythrocyte into the plasma. The lowering of erythrocyte P_{O_2} causes the dissociation of oxyhemoglobin, thereby liberating oxygen; simultaneously, the oxygen which diffused into the interstitial fluid is moving into cells along the concentration gradient generated by cell utilization of oxygen. The net result is a transfer of large quantities of oxygen from oxyhemoglobin into cells purely by passive diffusion.

The fact that hemoglobin is still 75 percent saturated at the end of tissue capillaries under resting conditions underlies an important automatic mechanism by which cells can obtain more oxygen whenever they increase their activity. An exercising muscle consumes more oxygen, thereby lowering its tissue P_{O_2}; this increases the overall blood-to-cell P_{O_2} gradient and the diffusion of oxygen out of the blood. In turn, the resulting reduction of erythrocyte P_{O_2} causes additional dissociation of hemoglobin and oxygen. An exercising muscle can thus extract virtually all the oxygen from its blood supply. This process is so effective because it takes place in the steep portion of the hemoglobin dissociation curve. Of course, an increased blood flow to the muscles also contributes greatly to the increased oxygen supply.

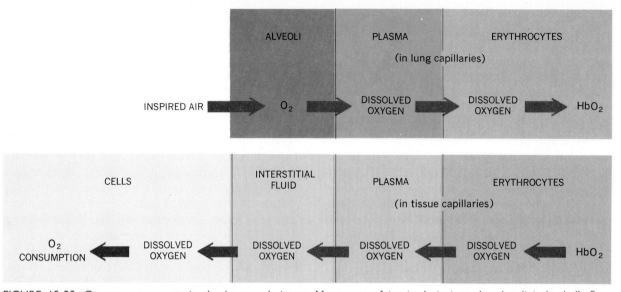

FIGURE 12-23. Oxygen movements in the lungs and tissues. Movement of inspired air into the alveoli is by bulk flow; all movements across membranes are by diffusion, driven by the differences in P_{O_2} across the membranes (prior to achievement of equilibrium).

Effect of acidity on hemoglobin saturation. The oxygen-hemoglobin dissociation curve illustrated in Fig. 12-21 is for blood having a specific hydrogen-ion concentration equal to that found in arterial blood. When the same experiments are performed at a different level of acidity, the curve changes significantly. A group of such curves is shown in Fig. 12-24 for different hydrogen-ion concentrations. It is evident that, regardless of the existing acidity, the percentage saturation of hemoglobin is still determined by the P_{O_2}. However, a change in acidity causes highly significant shifts in the entire curve. An increased hydrogen-ion concentration moves the curve to the right, which means that, at any given P_{O_2}, hemoglobin has less affinity for oxygen when the acidity is high. For reasons to be described later, the hydrogen-ion concentration in the tissue capillaries is greater than in arterial blood. Blood flowing through tissue capillaries becomes exposed to this elevated hydrogen-ion concentration and therefore loses even more oxygen than it would if the decreased P_{O_2} had been the only factor involved. Conversely, the hydrogen-ion concentration is lower in the lung capillaries than

in the systemic venous blood, so that hemoglobin picks up more oxygen in the lungs than it would if only the P_{O_2} were involved. Finally, the more active a tissue is, the greater its hydrogen-ion concentration; accordingly, hemoglobin releases even more oxygen (at any given P_{O_2}) during passage through these tissue capillaries, thereby providing the more active cells with additional oxygen. The hydrogen ion exerts its effect on the affinity of hemoglobin for oxygen by combining with the hemoglobin and altering its molecular structure.

Effect of temperature on hemoglobin saturation. The effect of temperature on the oxygen-hemoglobin dissociation curve (Fig. 12-24) resembles that of an increase in acidity. The implication is similar: Actively metabolizing tissue, e.g., exercising muscle, has an elevated temperature, which facilitates the release of oxygen from hemoglobin as blood flows through the muscle capillaries.

Effect of DPG on hemoglobin saturation. Red cells contain large quantities of the substance **2,3-diphosphoglycerate (DPG)** which is present in only

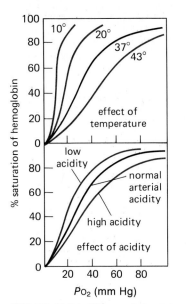

FIGURE 12-24. Effects of altering temperature and acidity on the relationship between P_{O_2} and hemoglobin saturation with oxygen. (Adapted from Comroe.)

trace amounts in other mammalian cells. DPG, which is produced by the red cells during glycolysis, binds reversibly with hemoglobin, causing it to have a lower affinity for oxygen. Therefore, the effect of increased DPG is to shift the curve to the right (just as does an increased temperature or hydrogen-ion concentration). The net result is that whenever DPG is increased there is enhanced unloading of oxygen as blood flows through the tissues. Such an increase in DPG is triggered by a variety of conditions associated with decreased oxygen supply to the tissues and helps to maintain oxygen delivery.

In summary, we have seen that oxygen is transported in the blood primarily in combination with hemoglobin. The extent to which hemoglobin binds oxygen is dependent upon P_{O_2}, hydrogen-ion concentration, temperature, and DPG concentration. These factors cause the release of large quantities of oxygen from hemoglobin during blood flow through tissue capillaries and virtually complete conversion of reduced hemoglobin to oxyhemoglobin during blood flow through lung capillaries. An active tissue increases its extraction of oxygen from

the blood because of its lower P_{O_2} and higher hydrogen-ion concentration and temperature.

Transport of carbon dioxide

The transport of carbon dioxide is achieved mainly by conversion of most of the carbon dioxide molecules to bicarbonate ions or by the binding of carbon dioxide to hemoglobin.

Carbon dioxide can undergo the reaction

$$CO_2 + H_2O \rightleftharpoons \underset{\text{Carbonic acid}}{H_2CO_3}$$

which is catalyzed by the enzyme **carbonic anhydrase**. The actual amount of carbonic acid in blood is very small because most of the carbonic acid ionizes according to the equation

$$H_2CO_3 \rightleftharpoons \underset{\text{Bicarbonate ion}}{HCO_3^-} + H^+$$

This reaction proceeds very rapidly and requires no enzyme. Combining these two equations we find

$$CO_2 + H_2O \xrightarrow{\text{carbonic anhydrase}}$$
$$H_2CO_3 \rightleftharpoons HCO_3^- + H^+ \qquad [12\text{-}2]$$

Thus, the addition of carbon dioxide to blood results in the generation of large numbers of bicarbonate and hydrogen ions.

Carbon dioxide can also react directly with proteins, particularly hemoglobin, to form **carbamino** compounds. (This reaction occurs at a site on hemoglobin different from the oxygen-binding site.)

$$CO_2 + Hb \rightleftharpoons HbCO_2 \qquad [12\text{-}3]$$

We can now apply Eqs. 12-2 and 12-3 to carbon dioxide transport by the blood. When arterial blood flows through tissue capillaries, oxyhemoglobin gives up oxygen to the tissues, and carbon dioxide diffuses from the tissues into the blood, where the following processes occur (Fig. 12-25):

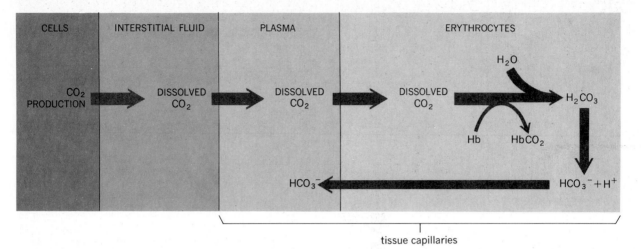

FIGURE 12-25. Summary of CO_2 movements and reactions as blood flows through tissue capillaries. All movements across membranes are by passive diffusion, driven by differences in partial pressures of CO_2 prior to attainment of any equilibrium. Note that most of the CO_2 ultimately is converted to HCO_3^-; this occurs almost entirely in the erythrocytes (because the carbonic anhydrase is located there), but most of the HCO_3^- then moves out of the erythrocytes into the plasma.

1. A small fraction (10 percent) of the carbon dioxide remains physically dissolved in the plasma and red blood cells.
2. The largest fraction (60 percent) of the carbon dioxide undergoes the reactions described in Eq. 12-2 and is converted into bicarbonate and hydrogen ions. This occurs primarily in the red blood cells because they, but not the plasma, contain large quantities of carbonic anhydrase, the enzyme required for the reaction. Once formed, most of the bicarbonate moves out of the red cells and is carried in the plasma (bicarbonate, in contrast to carbon dioxide, is extremely soluble in water). The reaction described in Eq. 12-2 also explains why tissue capillary hydrogen-ion concentration is higher than that of the arterial blood and increases as metabolic activity increases. The fate of these hydrogen ions will be discussed in the next section.
3. The remaining fraction (30 percent) of the carbon dioxide reacts directly with hemoglobin to form the carbamino compound ($HbCO_2$), as in Eq. 12-3. This process is facilitated because reduced hemoglobin has a greater affinity for carbon dioxide than does oxyhemoglobin, and a fraction of oxyhemoglobin has lost its oxygen and is transformed into reduced hemoglobin as blood flows through the tissues.

Since these are all reversible reactions, i.e., they can proceed in either direction, depending upon the prevailing conditions, why do they all proceed primarily to the right, toward generation of HCO_3^- and $HbCO_2$, as blood flows through the tissues? Once again, the answer is provided by the law of mass action: It is the increase in CO_2 concentration which drives these reactions to the right as blood flows through the tissues.

Obviously, a sudden lowering of blood P_{CO_2} has just the opposite effect. HCO_3^- and H^+ combine to give H_2CO_3, which generates CO_2 and H_2O. Similarly, $HbCO_2$ generates Hb and free CO_2. This is precisely what happens as venous blood flows through the lung capillaries (Fig. 12-26). Because the blood P_{CO_2} is higher than alveolar P_{CO_2}, a net diffusion of CO_2 from blood into alveoli occurs. This loss of CO_2 from the blood lowers the blood P_{CO_2} and drives the above chemical reactions to the left, thus generating more dissolved CO_2. Normally, as fast as this CO_2 is generated from HCO_3^- and H^+ and from $HbCO_2$, it diffuses into the al-

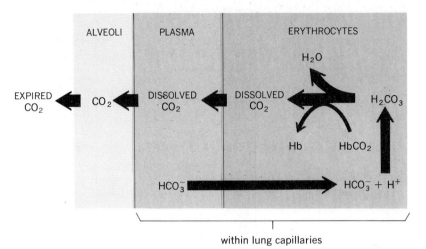

within lung capillaries

FIGURE 12-26. Summary of CO_2 movements and reactions as blood flows through the lung capillaries. All movements across membranes are by diffusion due to differences in the partial pressure of CO_2 prior to attainment of equilibrium. The plasma-erythrocyte phenomena are simply the reverse of those occurring during blood flow through the tissue capillaries, as shown in Fig. 12-25. The breakdown of H_2CO_3 to water and CO_2 is catalyzed by carbonic anhydrase.

veoli. In this manner, all the CO_2 delivered into the blood in the tissues now is delivered into the alveoli, from which it is expired and eliminated from the body.

Transport of hydrogen ions between tissues and lungs

In the previous section, we pointed out (Eq. 12-2) that hydrogen ions and bicarbonate ions are generated by the dissociation of carbonic acid after its formation by the hydration of carbon dioxide. As blood flows through the lungs, the reaction is reversed, and the hydrogen and bicarbonate ions recombine. However, most of the hydrogen ions do not remain free in the blood during transit from tissues to lungs (if they did, the high acidity would be toxic); rather, they become bound, mainly to hemoglobin. Reduced hemoglobin has a much greater affinity for hydrogen ion (just as it does for carbon dioxide) than oxyhemoglobin does. As blood flows through the tissues, a fraction of oxyhemoglobin is transformed into reduced hemoglobin, while simultaneously a large quantity of carbon dioxide enters the blood and undergoes (primarily in the red blood cells) the reactions which ultimately generate bicarbonate and hydrogen ions. Because the reduced hemoglobin has a strong affinity for hydrogen ion, most of these hydrogen ions become bound to hemoglobin (Fig. 12-27). In this

manner only a small number of hydrogen ions remain free, and the acidity of venous blood is only slightly greater than that of arterial blood. As the venous blood passes through the lungs, all these reactions are reversed. Hemoglobin becomes saturated with oxygen, and its ability to bind hydrogen ions decreases. The hydrogen ions are released, whereupon they react with bicarbonate ion to give carbon dioxide, which diffuses into the alveoli and is expired.

In a previous section, we described how the hydrogen-ion concentration of the blood is an important determinant of the ability of hemoglobin to bind oxygen. Now we have shown how the pres-

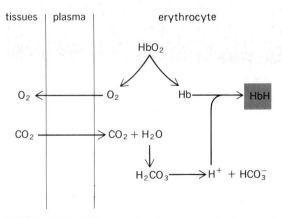

FIGURE 12-27. Buffering of hydrogen ions by hemoglobin as blood flows through tissue capillaries.

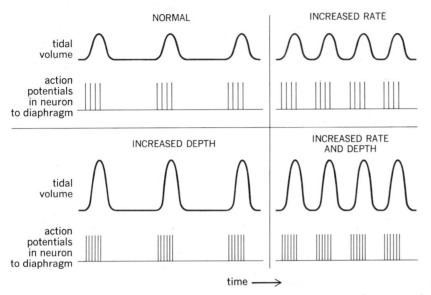

FIGURE 12-28. Recordings of depths of respiration (tidal volume) and frequency of action potentials in motor neurons to the diaphragm. Note that the rate of discharge during the inspiratory "burst" determines the tidal volume, whereas the time between bursts determines the respiratory rate.

ence of oxygen is an important determinant of hemoglobin's ability to bind both carbon dioxide and hydrogen ion.

Contol of Respiration

Neural generation of rhythmic breathing

Like cardiac muscles, the inspiratory muscles normally contract rhythmically, i.e., alternately contract and relax; however, the origins of these contractions are quite different. Certain of the cardiac muscle cells have automaticity; i.e., they are capable of self-excitation. The nerves to the heart merely alter this basic inherent rate and are not actually required for cardiac contraction. On the other hand, the diaphragm and intercostal muscles consist of skeletal muscle, which cannot contract unless stimulated by nerves. Thus, breathing depends entirely upon cyclical respiratory muscle excitation by the nerves to the diaphragm and the intercostal muscles (Fig. 12-28). Destruction of these nerves or the areas from which they originate (as in poliomyelitis, for example) results in paralysis of the respiratory muscles and death, unless some

form of artificial respiration can be rapidly instituted.

At the end of expiration, when the chest is at rest, a few impulses are still passing down these nerves. Like other skeletal muscles, therefore, the respiratory muscles have a certain degree of resting activity. This muscular contraction is too slight to move the chest wall. Inspiration is initiated both by an increased rate of firing of these motor units and by recruitment of more and more new motor units. Thus, the force of inspiration increases as it proceeds. Then, abruptly, almost all these units stop firing, the inspiratory muscles relax, and expiration occurs as the elastic lungs recoil. In addition, in situations when expiration is facilitated by contraction of expiratory muscles, the nerves to these muscles, having been quiescent during inspiration, begin firing during expiration.

By what mechanisms are nerve impulses to the respiratory muscles increased and decreased? Automatic control of this neural activity resides primarily in neurons in the **medulla**, the lower portion of the brainstem (which also contains cardiovascular control centers). By means of tiny electrodes placed in various parts of the medulla to record electrical activity, neurons have been found which discharge in synchrony with inspiration and cease discharging during expiration. These neurons are called **inspiratory neurons**. They provide, via either direct or interneuronal connections, the rhythmic input to the motor neurons innervating the inspiratory muscles.

What are the factors which determine the discharge pattern of the medullary inspiratory neurons? There are really two distinct components of this problem: (1) What factors are essential for the generation of rhythmical bursts of firing? (2) What factors modify the timing of the bursts and their intensity? This section deals only with the first question, the subsequent section with the second.

Just how the brain generates alternating cycles of firing and quiescence in the medullary inspiratory neurons remains controversial. One long-standing theory postulates that the inspiratory neurons (or neurons which synapse upon them) are pacemaker neurons; i.e., neurons that have inherent automaticity and rhythmicity. However, most experts believe that even though such pacemakers might exist, a much more complex model is required to explain the *on-off* cycle of neuronal output. Present thinking suggests the likelihood of self-cycling circuits in the medulla. For example, the medullary inspiratory neurons give off, in addition to their descending fibers, processes which synapse upon and stimulate other medullary neurons. These neurons, in turn, send processes back to the inspiratory neurons and inhibit the latter. Thus the firing of the medullary inspiratory neurons generates, via this loop, an inhibitory input to themselves, thereby helping to terminate their own firing. It is also likely that positive-feedback neuronal loops slightly more rapid than the negative-feedback loops exist within the medulla. Processes from the medullary inspiratory neurons may excite other neurons which, in turn, have excitatory synapses back upon the medullary inspiratory neurons. This positive-feedback loop could help explain the rapid buildup of discharge by the inspiratory neurons, whereas the inhibitory loop described earlier could explain the abrupt cutoff of activity.

The medullary neurons involved in all this activity also receive a rich synaptic input from neurons in various areas in the pons. This input modulates the output of the medullary neurons and may help shorten inspiration. Another potentially important "cut-off" signal comes from **pulmonary stretch receptors**. These receptors lie in the airway smooth muscle layer and are activated by lung inflation. Action potentials in the afferent nerve fibers from them travel to the brain and inhibit the medullary inspiratory neurons. Thus, feedback from the lungs themselves helps to terminate inspiration. However, the threshold of these receptors in human beings is very high, and this pulmonary stretch receptor reflex plays a role in setting respiratory rhythm only under conditions of very large tidal volumes, as in exercise.

It must be emphasized that cessation of inspiratory neuronal activity is all that is required for expiration to occur, since expiration is normally a passive process dependent only upon relaxation of the inspiratory muscles. In contrast, during active expiration, cessation of firing of the medullary inspiratory neurons is accompanied by the activation of other descending pathways which synapse upon and stimulate the motor neurons to those muscles (the expiratory intercostal and abdominal muscles) whose contractions produce active expiration.

Control of ventilation by P_{O_2}, P_{CO_2}, and hydrogen ion

In the previous section, we were concerned with the mechanisms that generate rhythmical breathing. It is obvious that the actual respiratory rate is not fixed but can be altered over a wide range. Similarly, the depth and force of breathing movements can also be altered. As we have seen, these two factors, rate and depth, determine the alveolar ventilatory volume. Depth of respiration depends upon the number of motor units firing and their frequency of discharge, whereas respiratory rate depends upon the length of time elapsing between the bursts of motor-unit activity (Fig. 12-28). Generally, rate and depth change in the same direction, although there may be important quantitative differences. For simplicity, we shall describe the control of total ventilation without attempting to discuss whether rate or depth makes the greatest contribution to the change.

There are many afferent inputs to the medullary inspiratory neurons (that from the pulmonary stretch receptors, for example), but the most important of these inputs for the involuntary control of ventilatory volume are from receptors (**peripheral chemoreceptors**) which monitor arterial P_{O_2}, P_{CO_2} and hydrogen-ion concentration, and from receptors (**central chemoreceptors**) which monitor the hydrogen-ion concentration of brain interstitial fluid. Thus, respiratory control is, in large part, designed to homeostatically regulate these chemical variables.

Control of ventilation by oxygen. Figure 12-29 illustrates an experiment in which healthy subjects breathed low P_{O_2} gas mixtures for several minutes (the experiment was performed in a way which keeps arterial P_{CO_2} constant so that the pure effects of changing only P_{O_2} could be studied). Little increase in ventilation was observed until the oxygen content of the inspired air was reduced enough to lower arterial P_{O_2} to 60 mmHg; beyond this point, any further reduction in arterial P_{O_2} caused a marked reflex increase in ventilation. Thus, large reductions in arterial P_{O_2} clearly act as a stimulus for reflexly increasing ventilation, thereby providing more oxygen to the alveoli and minimizing the drop in alveolar and arterial P_{O_2} produced by the low P_{O_2} gas mixture (Fig. 12-30). It may seem surprising that we are so insensitive to lesser reductions of arterial P_{O_2}, but look again at the hemoglobin dissociation curve (Fig. 12-21); total oxygen carriage by the blood is not really reduced very much until the arterial P_{O_2} falls below 60 mmHg, so that increased ventilation is not really "needed" up to that point.

The receptors stimulated by low arterial P_{O_2} are located high in the neck at the bifurcation of the common carotid arteries and in the thorax, on the aortic arch, quite close to, but distinct from, the arterial baroreceptors described in Chap. 11 (Fig. 12-31). These so-called **peripheral chemoreceptors** are the **carotid and aortic bodies**. Of the two groups, the carotid bodies are by far the more important. They are composed of epithelial-like cells and neuron terminals in intimate contact with the arterial blood. Afferent nerve fibers arising from these terminals pass to the medulla where they syn-

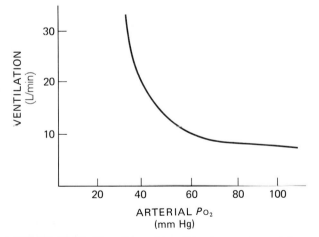

FIGURE 12-29. The effect of breathing low-oxygen mixtures on ventilation. The arterial P_{CO_2} was maintained at 40 mmHg throughout the experiment.

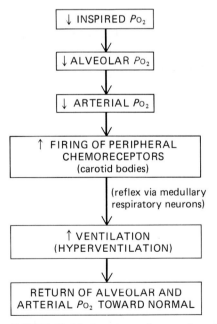

FIGURE 12-30. Sequence of events by which hypoxia causes hyperventilation, which maintains alveolar (and, hence, arterial) P_{O_2} at a value higher than would exist if the ventilation had remained unchanged.

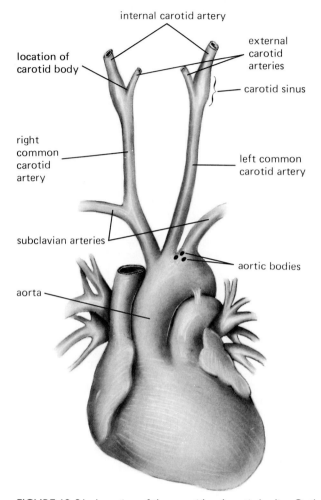

internal carotid artery

external carotid arteries

location of carotid body

carotid sinus

right common carotid artery

left common carotid artery

subclavian arteries

aortic bodies

aorta

FIGURE 12-31. Location of the carotid and aortic bodies. Both right and left common carotid bifurcations contain a carotid sinus and a carotid body.

apse ultimately with the respiratory neurons. Low P_{O_2} increases the rate at which the receptors discharge, resulting in an increased number of action potentials traveling up the afferent nerve fibers and a stimulation of the medullary inspiratory neurons.

These chemoreceptors respond to changes in arterial P_{O_2}. Accordingly, any time arterial P_{O_2} is reduced by lung disease, hypoventilation, or high altitude, the chemoreceptors will be stimulated and will initiate a compensatory increase in ventilation. Because the chemoreceptors respond to P_{O_2} (dis-

solved oxygen), not to oxygen bound to hemoglobin, they will *not* be stimulated in situations in which there are modest reductions in total oxygen content but no change in arterial P_{O_2} (e.g., moderate anemia). This same analysis holds true when total oxygen delivery is reduced by carbon monoxide, a gas which reacts with the same iron-binding sites on the hemoglobin molecule as does oxygen and has such a high affinity for these sites that even small amounts reduce the amount of oxygen combined with hemoglobin. Since carbon monoxide does not affect the amount of oxygen which can dissolve in blood, the arterial P_{O_2} is unaltered and no increase in peripheral chemoreceptor output occurs.

Control of ventilation by carbon dioxide. The important role that carbon dioxide plays in the control of ventilation can be demonstrated by a simple experiment. Figure 12-32 illustrates the effects on respiratory volume of increasing the P_{CO_2} of inspired air (normally, atmospheric air contains virtually no carbon dioxide). In the experiment illustrated, the sub-

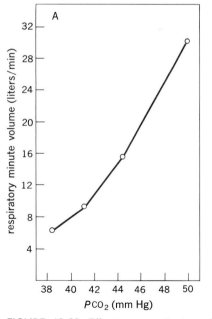

FIGURE 12-32. Effects on respiration of increasing arterial P_{CO_2}, achieved by adding carbon dioxide to inspired air.

ject breathed from bags of air containing variable quantities of carbon dioxide. The presence of carbon dioxide in the inspired air caused an elevation of alveolar P_{CO_2} and thereby an elevation of arterial P_{CO_2}. As can be seen in Fig. 12-32, this increased P_{CO_2} resulted in a markedly stimulated ventilation, an increase of 2 to 5 mmHg in alveolar P_{CO_2} causing a 100 percent increase in ventilation.

Thus, unlike the case for altered P_{O_2}, even the slightest change in arterial P_{CO_2} from its normal value is associated with a significant reflex change in ventilation. Experiments like this have documented that, at least over short periods of time, small changes in arterial P_{CO_2} are resisted by the reflex mechanisms controlling ventilation to a greater degree than are equivalent changes in arterial P_{O_2}. An increase in arterial P_{CO_2} results in stimulation of ventilation and thereby promotes the excretion of additional carbon dioxide (Fig. 12-33). Conversely, a decrease in arterial P_{CO_2} below normal removes some of the stimulus for ventilation, thereby reducing ventilation and allowing metabolically produced carbon dioxide to accumulate and return the P_{CO_2} to normal. In this manner, the arterial P_{CO_2} is stabilized at the normal value of 40 mmHg.

What are the receptor mechanisms by which changes in the arterial P_{CO_2} result in changes in ventilation? The logical prediction, of course, is that in a manner analogous to the case for oxygen, arterial P_{CO_2} receptors should exist. The surprising fact is that there are no really important receptors for carbon dioxide, per se; rather, the ability of changes in arterial P_{CO_2} to control ventilation reflexly is mainly due to associated changes in hydrogen-ion concentration, as detected by receptors sensitive to hydrogen-ion concentration. Recall the equilibrium between H^+, HCO_3^-, and CO_2

$$H_2O + CO_2 \rightleftharpoons H_2CO_3 \rightleftharpoons HCO_3^- + H^+$$

As described by the law of mass action, any increase in P_{CO_2} drives this reaction to the right, thereby liberating additional hydrogen ions. Thus, any increase or decrease in P_{CO_2} in the body fluids will cause (all other factors being equal) a similar directional change in hydrogen-ion concentration.

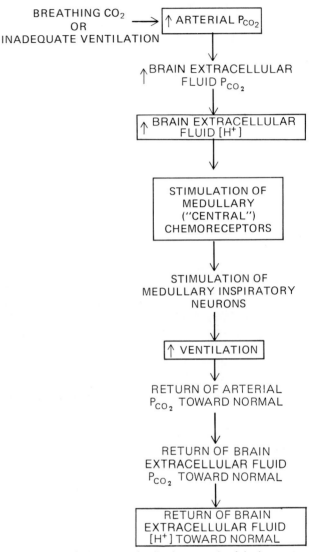

FIGURE 12-33. Negative-feedback control of hydrogen-ion concentration of extracellular fluid via regulation of ventilation.

There are two groups of hydrogen-ion receptors which participate in the reflex response. One is the same peripheral chemoreceptors—the carotid bodies—which initiate the hyperventilatory response to low P_{O_2}. Thus, these receptors are strongly stimulated by either low P_{O_2} or high hydrogen-ion concentration (the peripheral chemoreceptors also monitor P_{CO_2} directly to a very small degree, but this pathway is relatively unimportant). However, whereas the peripheral chemoreceptors are the only important receptors in the low-oxygen reflex, they play a minor role in the carbon dioxide–dependent hydrogen-ion reflex. The major receptors for this latter reflex are the so-called **central chemoreceptors**, located in the medulla of the brain; they monitor the hydrogen-ion concentration of the brain's extracellular fluid. The sequence of events is illustrated in Fig. 12-33: Carbon dioxide diffuses rapidly across the membranes separating capillary blood and brain tissue so that any increase in arterial P_{CO_2} causes a rapid, identical increase in brain extracellular-fluid P_{CO_2}; this increased P_{CO_2}, by mass action, causes an increased hydrogen-ion concentration; the latter stimulates the central chemoreceptors which, via synaptic connections, stimulate the medullary respiratory neurons to increase ventilation; the end result is a return of arterial and brain extracellular-fluid P_{CO_2} and hydrogen-ion concentration toward normal. We are left with the rather startling conclusion that the control of breathing (at least during rest) is aimed primarily at the regulation of the hydrogen-ion concentration of the brain's extracellular fluid.

Recognition of this preeminence of brain hydrogen-ion concentration, rather than P_{CO_2} itself, helps to explain an otherwise unexpected phenomenon associated with chronic elevations of blood P_{CO_2}. Many patients with chronic lung disease have a low arterial P_{O_2} and an elevated arterial P_{CO_2}. Ventilation is increased in an attempt to minimize these changes, and from our previous descriptions it would be logical to predict that the high-P_{CO_2} stimulus (via brain hydrogen-ion concentration) is more important in this regard than the low-P_{O_2} stimulus. But such is not the case, for if the arterial P_{O_2} of these patients is rapidly restored to normal, many of them literally stop breathing, an indication that the low P_{O_2} was the stimulus for their increased ventilation, with little or no contribution from the high P_{CO_2}. Thus, chronically, a low P_{O_2} is often more important in controlling ventilation than is an elevated P_{CO_2}. The reason for this loss of sensitivity to a chronically elevated P_{CO_2} is still unclear; one theory is that slowly operating transport processes in the blood-brain barrier adjust the hydrogen-ion concentration of brain extracellular fluid to normal despite the continued elevation of P_{CO_2}.

Throughout this section we have described the *stimulatory* effects of carbon dioxide on ventilation. It should also be noted that very high levels of carbon dioxide have just the opposite effect; they depress the entire central nervous system, including the respiratory neurons, and may therefore be lethal. Closed environments, such as submarines and space capsules, must be designed so that carbon dioxide is removed as well as oxygen supplied.

Finally, it should be noted that the effects of increased P_{CO_2} (via changes in hydrogen-ion concentration), on the one hand, and decreased P_{O_2}, on the other, not only exist as independent inputs to the medulla but manifest synergistic interactions as well; acute ventilatory response to combined hypoxia and increased P_{CO_2} is considerably greater than the sum of the individual responses.

Control of ventilation by changes in arterial hydrogen-ion concentration not due to altered carbon dioxide. In the previous section we described how changes in arterial and brain extracellular-fluid hydrogen-ion concentration, secondary to altered P_{CO_2}, influence ventilation. However there are many situations in which a change in arterial hydrogen-ion concentration occurs due to some cause other than a primary change in P_{CO_2}. In contrast to their lesser importance in mediating the reflex response to carbon dioxide–induced hydrogen-ion changes, the peripheral chemoreceptors play the major role in stimulating ventilation whenever arterial hydrogen-ion concentration is changed by any means other than by elevated P_{CO_2}. For example, increased addition of lactic acid to the blood causes hyperventilation (Figs. 12-34 and 12-35) almost entirely by way of the peripheral chemoreceptors. The central chemoreceptors do not respond in this case because hydrogen ions penetrate the blood-brain barrier so poorly that brain hydrogen-ion concentration is not increased (at least early on) by the administration of the lactic acid; this is in contrast to the ease with which changes in arterial P_{CO_2} produce equivalent changes in brain hydrogen-ion concentration.

The converse of the above situation is also true; i.e., when arterial hydrogen-ion concentration is lowered by any means other than by a reduction

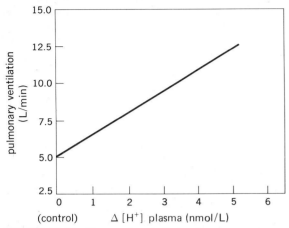

FIGURE 12-34. Changes in ventilation in response to an elevation of plasma hydrogen-ion concentration, produced by the administration of lactic acid. *(Adapted from Lambertsen.)*

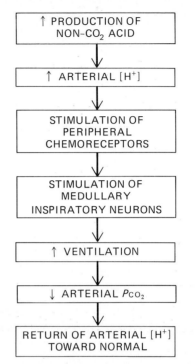

FIGURE 12-35. Reflexly induced hyperventilation minimizes the change in arterial hydrogen-ion concentration when acids such as lactic acid are produced in excess. Note that under such conditions, arterial P_{CO_2} is reflexly reduced below its normal value.

of P_{CO_2} (for example, by loss of hydrogen ions from the stomach in vomiting), ventilation is depressed because of decreased peripheral chemoreceptor output.

The adaptive value such reflexes have in regulating arterial hydrogen-ion concentration should be evident (Fig. 12-35). The hyperventilation induced by an elevated arterial hydrogen-ion concentration lowers arterial P_{CO_2}, which by mass action, lowers arterial hydrogen-ion concentration. Similarly hypoventilation induced when arterial hydrogen-ion concentration decreases results in an elevated arterial P_{CO_2} and a restoration of hydrogen-ion concentration toward normal. Notice that when a change in arterial hydrogen-ion concentration due to some acid unrelated to carbon dioxide influences ventilation via the peripheral chemoreceptors, it displaces arterial P_{CO_2} from normal; i.e., this is a reflex which regulates arterial hydrogen-ion concentration at the expense of changes in arterial P_{CO_2}.

Figure 12-36 summarizes the control of ventilation by P_{O_2}, P_{CO_2}, and hydrogen-ion concentration.

Control of ventilation during exercise

During heavy exercise, the alveolar ventilation may increase ten- to twentyfold. On the basis of our three variables—P_{O_2}, P_{CO_2}, and hydrogen-ion concentration—it might seem easy to explain the mechanism which induces this increased ventilation. Unhappily, such is not the case.

Decreased P_{O_2} as the stimulus? It would seem logical that, as the exercising muscles consume more oxygen, blood P_{O_2} would decrease and stimulate respiration. But in fact, arterial P_{O_2} is not significantly reduced during exercise (Fig. 12-37). The alveolar ventilation increases in exact proportion to the oxygen consumption; therefore, P_{O_2} remains constant. Indeed, in exhausting exercise, the alveolar ventilation may actually increase relatively more than oxygen consumption, resulting in an increased arterial P_{O_2}. Thus, in normal persons, ventilation is never rate-limiting during exercise; rather (as emphasized in Chap. 11), cardiac output is the factor limiting maximal exercise.

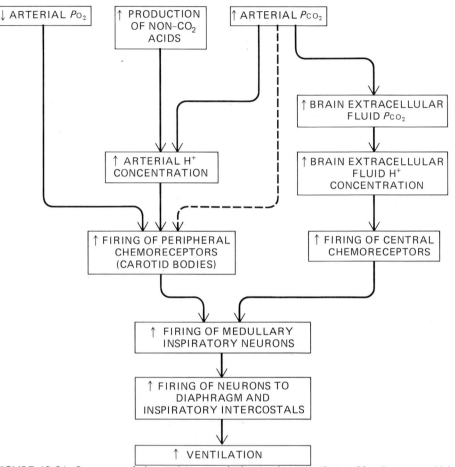

FIGURE 12-36. Summary of chemical inputs which stimulate ventilation. Ventilation would be reflexly decreased when arterial P_{O_2} increases or when P_{CO_2} or hydrogen-ion concentration decreases. The dashed line from arterial P_{CO_2} denotes the fact that the direct effect of P_{CO_2} on the peripheral chemoreceptors is minor (in contrast to its effect via hydrogen ion). This is an amalgamation of Figs. 12-30, 12-33, and 12-35.

Increased P_{CO_2} as the stimulus? This is virtually the same story. Despite the marked increase in carbon dioxide production, the increase in alveolar ventilation excretes the carbon dioxide as rapidly as it is produced, and arterial P_{CO_2} remains constant (Fig. 12-37). Indeed, for the same reasons given above for oxygen going up, the arterial P_{CO_2} may actually decrease during exhausting exercise.

Increased hydrogen ion as the stimulus? Since the arterial P_{CO_2} does not change (or decreases) during exercise, there is no accumulation of excess hydrogen ion as a result of carbon dioxide accumulation.

However, during severe exercise there actually is an increase in arterial hydrogen-ion concentration, but for quite a different reason, namely, generation and release into the blood of lactic acid. But the changes in hydrogen-ion concentration are not nearly great enough, particularly in only moderate exercise, to account for the increased ventilation.

We are left with the fact that, despite intensive study for more than 100 years by many respiratory physiologists, we do not know what input stimulates ventilation during exercise. Our big three—P_{O_2}, P_{CO_2}, and hydrogen ion—appear presently to be inadequate, but many physiologists still believe

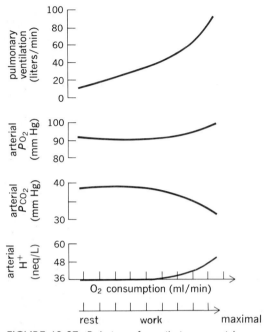

FIGURE 12-37. Relation of ventilation, arterial gas pressures, and hydrogen-ion concentration to the magnitude of muscular exercise. *(Adapted from Comroe.)*

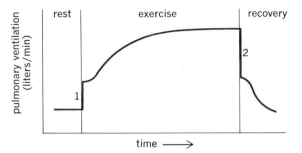

FIGURE 12-38. Ventilation changes during exercise. Note the abrupt increase (1) at the onset of exercise and the equally abrupt but larger decrease (2) at the end of exercise.

that they will ultimately be shown to be the critical inputs. They reason that the fact that P_{CO_2} remains constant during moderate exercise is very strong evidence that ventilation is actually controlled by P_{CO_2}. In other words, if P_{CO_2} were not the major controller, how else could it remain unchanged in the face of the marked increase in its production? They hypothesize: (1) there may be a change in sensitivity of the chemoreceptors during exercise; or (2) the increase in the oscillations of arterial H^+, P_{CO_2}, or P_{O_2} which occur during exercise may drive the chemoreceptors to a greater degree.

Moreover, the problem is even more complicated; as shown in Fig. 12-38, there is an abrupt increase (within seconds) in ventilation at the onset of exercise and an equally abrupt decrease at the end. Clearly, these changes occur too rapidly to be explained by alteration of chemical constituents of the blood; they probably represent a conditioned (i.e., learned) response.

Control of respiration by other factors during exercise. Many receptors in joints and muscles can be

stimulated by the physical movements which accompany muscle contraction. It is quite likely that afferent pathways from these receptors play a significant role in stimulating respiration during exercise. Thus, the mechanical events of exercise help to coordinate alveolar ventilation with the metabolic requirements of the tissues.

An increase in body temperature frequently occurs as a result of increased physical activity and contributes to stimulation of alveolar ventilation. Also, epinephrine is a respiratory stimulant and the increased plasma epinephrine associated with heavy exercise probably contributes to the increased ventilation. It should be emphasized that body temperature and plasma epinephrine increase in many situations other than exercise; when either does it will stimulate ventilation. Thus, people with fever show increased ventilation.

Finally, as seems to be the case for control of the cardiovascular system during exercise, branches may come off the axons descending from the brain to those motor neurons supplying the exercising muscles, and these branches may influence the neurons which control the respiratory muscles.

Other controls of ventilation

Protective reflexes. A group of responses protect the respiratory tract against irritant materials, the most familiar being the **cough** and the **sneeze**. These reflexes originate in receptors located between airway epithelial cells. When they are excited, the result is stimulation of the medullary respiratory neurons in such a manner as to produce a deep inspiration and a violent expiration. In this manner, particles can be literally exploded out of the respiratory tract. The cough reflex is inhibited by alcohol, which may contribute to the susceptibility of alcoholics to choking and pneumonia. Another example of a protective reflex is the immediate cessation of respiration which is frequently triggered when noxious agents are inhaled.

Pain and emotion. Painful stimuli anywhere in the body can produce reflex stimulation of the respiratory neurons. Emotional states are also often accompanied by marked alterations of respiration, as evidenced by the rapid breathing rate which characterizes fright, anxiety, and many similar emotions. In addition, the movement of air in or out of the lungs is required for such involuntary expressions of emotion as laughing and crying. In such situations, the descending pathways from higher brain centers are the controlling input.

Voluntary control of breathing. Although we have discussed in detail the involuntary nature of most respiratory reflexes, it is quite obvious that we retain considerable voluntary control of respiratory movements. This is accomplished by descending pathways from the cerebral cortex to the motor neurons of the respiratory muscles. This voluntary control of respiration cannot be maintained when the involuntary stimuli, such as an elevated P_{CO_2} or hydrogen-ion concentration, become intense (as, for example, in breath holding). Besides the obvious forms of voluntary control, respiration must also be controlled during the production of complex actions such as speaking and singing.

Breathing during sleep. Ventilation normally decreases during sleep to a degree larger than the decrease in the body's utilization of energy. Accordingly, the arterial P_{O_2} decreases and the arterial P_{CO_2} increases. This normal tendency for ventilation to decrease during sleep is profoundly and dangerously exaggerated in some persons, who suffer from **sleep apnea**, recurrent periods (up to 500 per night) during which breathing ceases completely for long periods of time (often 1 to 2 min or more).

Some of this may represent a reduced sensitivity of the chemoreceptors occurring cyclically during certain phases of sleep (see Chap. 20). Sleep apnea is being studied intensively, at present, because it may be one of the causes of so-called **sudden-infant-death syndrome**, the disease in which an infant who had seemed to be in good health is found dead in its crib. More than half of such infants have been found, on autopsy, to have underdeveloped carotid bodies.

Hypoxia

Hypoxia is defined as a deficiency of oxygen at the tissue level. There are many potential causes of hypoxia, but they can be classed in four general categories: (1) **hypoxic hypoxia**, in which the arterial P_{O_2} is reduced; (2) **anemic hypoxia**, in which the arterial P_{O_2} is normal but the total oxygen content of the blood is reduced because of inadequate numbers of red cells, deficient or abnormal hemoglobin, or competition for the hemoglobin molecule by carbon monoxide; (3) **ischemic hypoxia**, in which the basic defect is too little blood flow to the tissues; and (4) **histotoxic hypoxia**, in which the quantity of oxygen reaching the tissue is normal but the cells are unable to utilize the oxygen because a toxic agent (cyanide, for example) has interfered with their metabolic machinery.

Clearly, anemic hypoxia is mainly the result of abnormal blood formation or destruction, ischemic hypoxia is caused by failure of the cardiovascular system, and histotoxic hypoxia represents malfunction of the tissue cells, themselves. In contrast to these three categories, hypoxic hypoxia is due either to malfunction of the respiratory system or to high-altitude exposure. The causes are many, and a list of certain of the major ones provides an excellent review of the interplay between the various components of the respiratory system described in this chapter (Table 12-4). We shall describe the response to high altitude as a representative example of the range of reflex adjustments which can be brought into play to offset the reduced arterial P_{O_2} characteristic of hypoxic hypoxia, regardless of cause.

TABLE 12-4. Some causes of hypoxic hypoxia

1. Decreased P_{O_2} in inspired air (high altitude)*

2. Hypoventilation
   Increased airway resistance (foreign body, asthma)
   Paralysis of respiratory muscles (poliomyelitis)
   Skeletal deformities
   Inhibition of medullary respiratory centers (morphine)
   Decreased "stretchability" of the lungs and thorax (deficient surfactant, thickened lung or chest tissues)
   Lung collapse (pneumothorax)

3. Deficient alveolar-capillary diffusion
   Decreased area for diffusion (pneumonia)
   Thickening of alveolar-capillary membranes (berylliosis)

4. Abnormal matching of ventilation and blood flow (emphysema)

* The phrases in parentheses are specific examples in each category.

Response to high altitude

Barometric pressure progressively decreases as altitude increases. Thus, at the top of Mt. Everest (approximately 9000 m), the atmospheric pressure is 245 mmHg. The air is still 20 percent oxygen, which means that the P_{O_2} is 49 mmHg. Obviously, the alveolar and arterial P_{O_2} must decrease as one ascends unless pure oxygen is breathed. The effects of oxygen lack vary with individuals, but most persons who ascend rapidly to altitudes above 3000 m experience some degree of **mountain sickness**, consisting of breathlessness, palpitations, headache, nausea, fatigue, and impairment of mental processes (vision, judgment, etc.). Over the course of several days, these symptoms diminish and ultimately disappear, although maximal physical capacity remains reduced. The acclimatization is achieved by the compensatory mechanisms listed below. The highest villages permanently inhabited by people are in the Andes at 6000 m. These villagers work quite normally, and apparently the only major precaution they take is that the women come down to lower altitudes during late pregnancy.

In response to hypoxia, oxygen supply to the tissues is maintained in four ways: (1) by increased alveolar ventilation, (2) by increased erythrocyte and hemoglobin content of blood, (3) by tissue changes which facilitate diffusion, and (4) by a shift in the hemoglobin dissociation curve to the right. These responses have the net effect of supplying normal amounts of oxygen to the tissues with minimal deviation of *tissue* P_{O_2} from normal.

The first acute response to high altitude is reflexly increased ventilation, mediated via the carotid bodies. As we have seen, the arterial P_{O_2} must generally fall by approximately 50 percent in order to elicit increased ventilation. The arterial P_{O_2} reaches this low level when the altitude is 3000 m or greater. However, the hypoxic drive to ventilation is not unopposed. The increased ventilation produced by low arterial P_{O_2} induces *decreases* in the P_{CO_2}, and, therefore, in hydrogen-ion concentration of blood and brain extracellular fluid, which acts as a brake on ventilation (Fig. 12-39). Over several days kidney and brain transport mechanisms are brought into play which return arterial and brain extracellular-fluid hydrogen-ion concentration back toward normal despite reduced P_{CO_2}. The net result is that the hypoxic drive is less offset by changes in hydrogen-ion concentration, and ventilation progressively increases.

Over time, erythrocyte and hemoglobin syntheses are stimulated by the hormone erythropoietin (see Chap. 11), and the person's total circulating red-cell mass and hematocrit increase considerably.

Another slowly developing compensation is the growth of many new tissue capillaries. This has the important effects of increasing the surface area for diffusion and decreasing the distance which oxygen must diffuse from blood to cell. Tissue mitochondria and muscle myoglobin are also increased.

The right-shift in the hemoglobin dissociation curve occurs despite the fact that, as described above, the blood hydrogen-ion concentration is decreased at high altitude, an event which would, itself, cause a left shift of the curve; the explanation is that the hypoxia and low hydrogen-ion concentration stimulate the production of DPG by the red cells and this substance causes the right shift. However, if the decrease in blood hydrogen-ion concen-

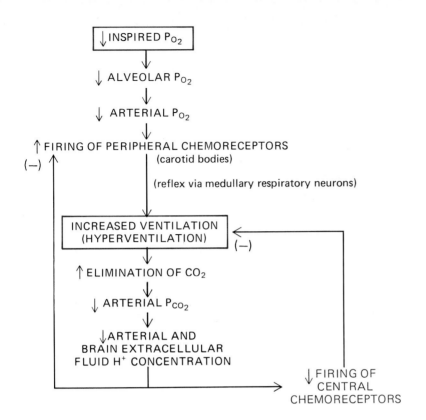

FIGURE 12-39. Sequence of events by which hypoxia causes hyperventilation, which in turn lowers P_{CO_2} and hydrogen-ion concentration of arterial blood and brain extracellular fluid. These latter changes, via both the peripheral and central chemoreceptors, limit the hyperventilation.

tration is severe enough, the increased DPG cannot compensate and the hemoglobin curve shifts to the left.

Acclimatization to high altitude is not an unmixed blessing. The change in hematocrit (due to the increased volume of red cells) raises the viscosity of the blood and thereby increases the resistance to blood flow throughout the vascular system. This emphasizes the point that acclimatization to any environmental challenge often involves trade-offs, in this case better oxygen carriage versus greater difficulty in pumping blood. There is another long-term change in response to chronic hypoxia which is frankly maladaptive. Recall that pulmonary blood vessels respond to low oxygen by constricting; that response, we saw, was useful for local matching of ventilation and blood flow. But when the entire lung manifests such vasoconstriction, the result is a marked increase in pulmonary vascular resistance and pulmonary arterial pres-

sure, work of the right ventricle, and the likelihood of heart failure.

Nonrespiratory Functions of the Lungs

The lungs have a variety of functions in addition to their roles in gas exchange and regulation of hydrogen-ion concentration. Most notable are their contributions to influencing the concentrations in arterial blood of a large number of biologically active substances. As described in Chap. 8, hormones are added to the blood by the various endocrine glands of the body, but in addition many substances (neurotransmitters and paracrines, for example) which are released locally into the interstitial fluids of their respective organs may diffuse into capillaries and thus make their way into the systemic venous system. The lungs partially or completely remove some of these substances from the blood and thereby prevent them from reaching other locations in the body via the arteries (recall that the entire cardiac output passes through the lungs). The cells which perform this function are the endothelial cells lining the pulmonary capillaries. An important example of this function is the removal of certain prostaglandins from the blood during passage through the lungs.

In contrast, the lung may *add* new substances to the blood. Some of these substances play local roles within the lungs, but if produced in large enough quantity, they may diffuse into the pulmonary capillaries and be carried to the rest of the body. For example, inflammatory responses (Chap. 17) in the lung may lead, via excessive release of potent chemicals, such as histamine, to profound alterations of systemic blood pressure or flow. In at least one case, the lungs contribute a hormone (angiotensin II) to the blood not by secreting it themselves but by enzymatically activating a hormone precursor during passage of the blood through the pulmonary capillaries (see the renin-angiotensin system in Chap. 13).

Finally, the lungs also act as a "sieve" which traps and dissolves small blood clots generated in the systemic circulation, thereby preventing them from reaching the systemic arterial blood where they could occlude blood vessels in other organs.

REGULATION OF WATER AND ELECTROLYTE BALANCE

13

One of the major themes of this book is the regulation of the internal environment. A cell's function depends not only upon receiving a continuous supply of organic nutrients and eliminating its metabolic end products but also upon the existence of stable ionic conditions in the extracellular fluid bathing it. Among the most important substances contributing to these conditions are water, sodium, potassium, calcium, and hydrogen ion. This chapter is devoted to a discussion of the mechanisms by which the total amounts of these substances in the body and their concentrations in the extracellular fluid are maintained relatively constant, mainly as a result of the regulation of their excretion by the kidneys.

Since the extracellular fluid is located between the external environment and the cells, the concentration of any substance within it can be altered by exchange in either direction. Exchanges with cells are called **internal exchanges**. For example, a decrease in extracellular potassium concentration is followed by a counteracting movement of potassium out of cells into the extracellular fluid, thereby minimizing the decrease in extracellular concentration. Each type of ion is present in cells or in bone in significant amounts, which can be partially depleted or expanded without damage to the storage site. But these stores are limited, and in the long run any deficit or excess of total body water or total body electrolyte must be compensated by changes in total body input or output.

A substance appears in the body either as a result of ingestion or as a product of metabolism. Conversely, a substance can be excreted from the body or can be consumed in a metabolic reaction. Therefore, if the quantity of any substance in the body is to be maintained at a constant level over a period of time, the total amounts ingested and produced must equal the total amounts excreted and consumed. This is a general statement of the balance concept described in Chap. 7. For water and hydrogen ion, all four possible pathways apply. However, for the mineral electrolytes such as sodium and potassium, which are neither synthesized nor consumed, total-body balance reflects only ingestion versus excretion.

As an example, let us describe the balance for total body water (Table 13-1). It should be recognized that these are average values, which are subject to considerable normal variation. The two sources of body water are metabolically produced water (resulting largely from the oxidation of organic nutrients) and water ingested in liquids and so-called solid food (a rare steak is approximately 70 percent water). There are four sites from which water is lost every day to the external environment:

TABLE 13-1. Routes of water gain and loss in adults

	Average volumes, mL/day
Intake:	
Drunk	1200
In food	1000
Metabolically produced	350
Total	2550
Output:	
Insensible loss (skin and lungs)	900
Sweat	50
In feces	100
Urine	1500
Total	2550

skin, respiratory passageways, gastrointestinal tract, and urinary tract. (Menstrual flow constitutes a fifth source of water loss in women.) The loss of water by evaporation from the cells of the skin and lining of respiratory passageways is a continuous process, often referred to as **insensible loss** because the person is unaware of its occurrence. Additional water can be made available for evaporation from the skin by the production of sweat.

Under normal conditions, as can be seen from Table 13-1, water loss exactly equals water gain, and no net change of body water occurs. This is no accident but is the result of precise regulatory mechanisms. The question then is: Which processes involved in water balance are controlled to make the gains and losses balance? The answer, as we shall see, is voluntary intake (**thirst**) and urinary loss. This does not mean that none of the other processes is controlled but that their control is not primarily oriented toward water balance. Catabolism of organic nutrients, the source of the water of oxidation, is controlled by mechanisms directed toward regulation of energy balance. Sweat production is controlled by mechanisms directed toward temperature regulation. Insensible loss is truly uncontrolled, and normal gastrointestinal loss of water (in feces) is generally quite small (but can be severe in vomiting or diarrhea).

The mechanism of thirst is certainly of great importance, since body deficits of water, regardless of cause, can be made up only by ingestion of water, but it is also true that the amount of fluid intake is often influenced more by habit and sociological factors than by the necessity of balancing water input and output. Thus the control of urinary water loss is the single most important mechanism by which body water balance is maintained.

By similar analyses, we find that the body balances of many of the ions in the extracellular fluid are regulated primarily by the kidneys. The amounts of these substances must be held within very narrow limits, regardless of large variations in intake and abnormal losses resulting from disease (hemorrhage, diarrhea, vomiting, etc.). By what mechanism does urine production rapidly increase when a person ingests several glasses of liquid? How is it that the person on an extremely low salt intake and the heavy salt eater both urinate precisely the amounts of salt required to maintain their sodium balance? What mechanisms decrease the urinary calcium excretion of children deprived of milk?

This regulatory role is obviously quite different from the popular conception of the kidneys as glorified garbage disposal units which rid the body of assorted wastes and "poisons." It is true that the chemical reactions which occur in cells result ultimately in end products collectively called waste products. The most abundant metabolic waste product is carbon dioxide, which is eliminated by the lungs, not the kidneys. The catabolism of protein produces approximately 30 g of urea per day. Other waste substances produced in relatively large quantities are uric acid (from nucleic acids), creatinine (from muscle creatine), and the end products of hemoglobin breakdown. There are many others, and most of these substances are eliminated from the body as rapidly as they are produced, primarily by way of the kidneys. Some of these waste products (for example, urea) are relatively harmless, but the accumulation of others during periods of renal malfunction accounts for certain of the disordered body functions in severe kidney disease. However, many of the problems which occur in renal disease are due simply to disordered water and electrolyte metabolism.

The kidneys have another excretory function, namely, the elimination from the body of foreign chemicals (drugs, pesticides, food additives, etc.). Finally, in addition to their excretory functions, the kidneys act as endocrine glands, secreting at least three substances which are components of hormonal systems: erythropoietin (Chap. 11), renin, and the active form of vitamin D (the last two to be discussed later in this chapter).

SECTION A.
BASIC PRINCIPLES OF RENAL PHYSIOLOGY

Structure of the Kidney and Urinary System

The kidneys are paired organs which lie in the back of the abdominal wall, one on each side of the vertebral column (Fig 13-1). Each kidney is composed of approximately 1 million functional units bound together by small amounts of connective tissue containing blood vessels, nerves, and lymphatics. One such unit, or **nephron**, is shown in Fig. 13-2. Each nephron consists of a "filtering component" called the **glomerulus** and a **tubule** extending out from the glomerulus.

The glomerulus consists of a compact tuft of interconnected capillaries (the **glomerular capillaries**) and a balloonlike hollow capsule (**Bowman's capsule**) into which the capillary tuft protrudes (Fig. 13-3). It is into the space within Bowman's capsule that fluid filters from the glomerular capillaries

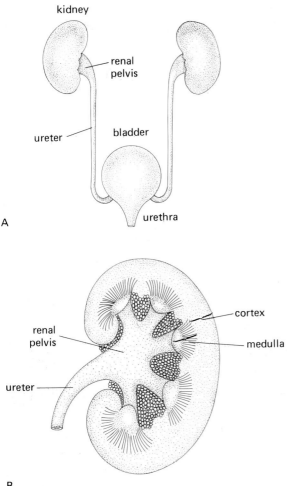

A

B

FIGURE 13-1. (A) Urinary system. The urine, formed by the kidneys, collects in the renal pelvis and then flows through the ureters into the bladder, from which it is eliminated via the urethra. (B) Section of a human kidney. Half the kidney has been sliced away. Note that the structure shows regional differences. The outer portion (cortex), which has a granular appearance, contains all the glomeruli. The collecting ducts form a large portion of the inner kidney (medulla), giving it a striped appearance; these ducts drain into the renal pelvis.

tinuous with the space in Bowman's capsule. Throughout its course, the tubule is composed of a single layer of epithelial cells, which differ in structure and function along the tubule's length. The segment of the tubule which drains Bowman's capsule is the **proximal tubule**. The next portion of the tubule is a sharp hairpinlike loop, called **Henle's loop**. Next comes the **distal tubule** and finally the **collecting duct**. From Bowman's capsule to an early portion of the collecting system, each of the million tubules is completely separate from its neighbors. The separate tubules then join to form collecting ducts, which in turn form even larger ducts, which finally empty into a large central cavity, the **renal pelvis**, at the base of each kidney (Fig. 13-1). The renal pelvis is continuous with the **ureter**, the tube which passes from the renal pelvis to the **urinary bladder**, where urine is temporarily stored and from which it is eliminated via the **urethra** during urination. The composition of the urine is not significantly altered after it leaves the collecting ducts, and from the renal pelvis on, the remainder of the urinary system simply serves as plumbing.

To return to the other component of the nephron: What is the origin of the glomerular capillaries? Blood enters the kidney via the renal artery, which then divides into progressively smaller branches. Each of the smallest arteries gives off, at right angles to itself, a series of arterioles, **afferent arterioles** (Fig. 13-3), each of which leads to a tuft of glomerular capillaries.

The glomerular capillaries, instead of leading to veins, recombine to form another set of arterioles, the **efferent arterioles**, through which blood leaves the glomerulus. Each efferent arteriole soon divides into a second set of capillaries (Fig. 13-3), the **peritubular capillaries**, which branch profusely to form a network surrounding the remaining portions of the tubule. They eventually join to form veins by which blood ultimately leaves the kidney.

Basic Renal Processes

Urine formation begins with the filtration of plasma through the glomerular capillaries into Bowman's capsule. This **glomerular filtrate** contains all low-

across the combined membranes of the capillaries and Bowman's capsule. The thin filtration barrier consists of the single-celled capillary endothelium, the single-celled epithelial lining of Bowman's capsule, and a proteinaceous membrane between them.

Extending out from Bowman's capsule is the tubule of the nephron, the lumen of which is con-

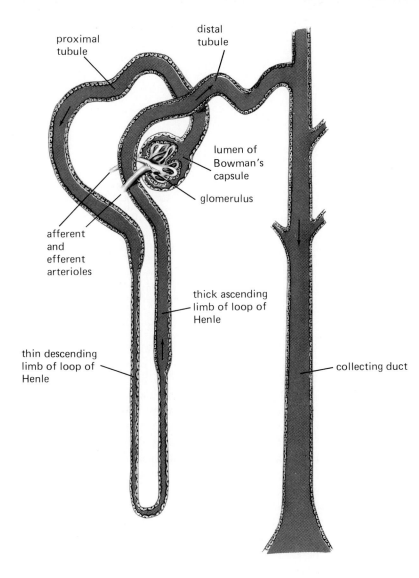

proximal
tubule

distal
tubule

lumen of
Bowman's
capsule

glomerulus

afferent
and
efferent
arterioles

thick ascending
limb of loop of
Henle

thin descending
limb of loop of
Henle

collecting duct

FIGURE 13-2. Basic structure of a nephron.

molecular-weight substances in virtually the same concentrations as in plasma; it contains almost no protein since the glomerular barrier restricts the movement of high-molecular-weight substances like the plasma proteins. The final urine which enters the renal pelvis is quite different from the glomerular filtrate because, as the filtered fluid flows from Bowman's capsule through the remaining portions of the tubule, its composition is altered. This change occurs by two general processes, **tubular reabsorption** and **tubular secretion**. The tubule is at all points intimately associated with the peritubular capillary network, a relationship that permits transfer of materials between peritubular blood and the lumen (inside) of the tubule. When the direction of transfer is from tubular lumen to peritubular capillary plasma, the process is called tubular reabsorption. Movement in the opposite direction, i.e., from peritubular plasma to tubular lumen, is called tubular secretion. (This term must not be confused with **excretion**; to say that a substance has been excreted is to say only that it appears in the final urine.) These relationships are illustrated in Fig. 13-4.

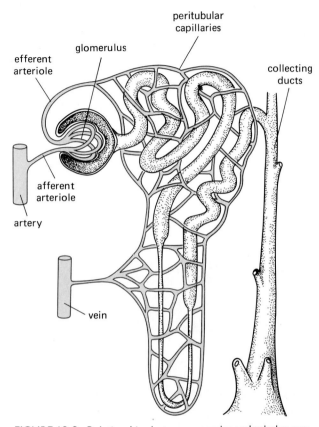

FIGURE 13-3. Relationships between vascular and tubular components of the nephron. The peritubular capillaries also supply the collecting ducts (not shown for clarity). *(Adapted from Smith.)*

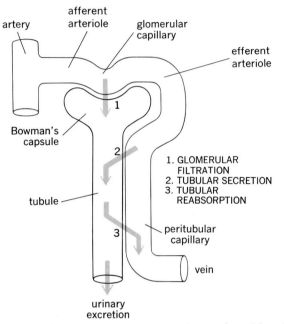

FIGURE 13-4. The three basic components of renal function.

The most common relationships between the basic renal processes—glomerular filtration, tubular reabsorption, and tubular secretion—are shown in Fig. 13-5. Plasma containing substances X, Y, and Z enters the glomerular capillaries. A certain quantity of protein-free plasma containing these substances is filtered into Bowman's capsule, enters the proximal tubule, and begins its flow through the rest of the tubule. The remainder of the plasma, also containing X, Y, and Z, leaves the glomerular capillaries via the efferent arteriole and enters the peritubular capillaries. The cells composing the epithelium of the tubule can actively transport X (not Y or Z) from the peritubular plasma into the tubular lumen (secretion), but not in the opposite direction (reabsorption). By this combination of filtration and tubular secretion all the plasma which originally entered the renal artery is cleared of substance X, which leaves the body via the urine, thus reducing the amount of X remaining in the body.

If the tubule were incapable of reabsorption, the Y and Z originally filtered at the glomerulus would also leave the body via the urine, but the tubule can transport Y and Z from the tubular lumen back into the peritubular plasma (reabsorption). The amount of reabsorption of Y is small, so that most of the filtered material does escape from the body, but for Z the reabsorptive mechanism is more efficient, and virtually all the filtered material is transported back into the plasma. Therefore no Z is lost from the body. Hence the processes of filtration and reabsorption have canceled each other out, and the net result is as though Z had never entered the kidney at all.

Each substance in plasma is handled in a characteristic manner by the nephron, i.e., by a particular combination of filtration, reabsorption, and secretion. The critical point is that the rates at which the relevant processes proceed for many of these substances are subject to physiological control.

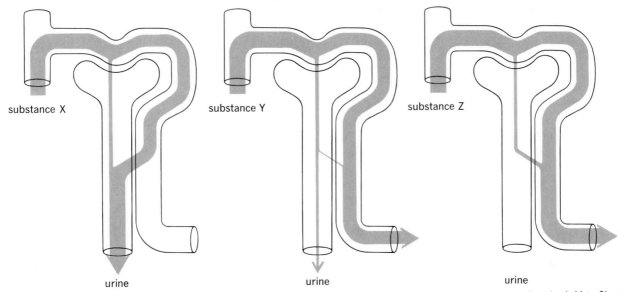

FIGURE 13-5. Renal manipulation of three substances X, Y, and Z. X is filtered and secreted but not reabsorbed. Y is filtered, and a fraction is then reabsorbed. Z is filtered and is completely reabsorbed.

What is the effect, for example, if the Y filtration rate is increased or its reabsorption rate decreased? Either change means that more Y is lost from the body via the urine. By triggering such changes in filtration or reabsorption whenever the plasma concentration of Y rises above normal, homeostatic mechanisms could regulate plasma Y.

Glomerular filtration

As described in Chap. 11, capillaries are freely permeable to water and small solutes. Most are relatively impermeable to plasma proteins (and totally prevent passage of erythrocytes). These characteristics apply to the glomerular capillaries. Accordingly, the glomerular filtrate, i.e., the fluid within Bowman's capsule, is essentially protein-free and contains all low-molecular-weight substances in virtually the same concentrations as in the plasma. (The only exceptions to this latter generalization are low-molecular-weight substances which are bound to proteins and therefore are not filtered; for example, half of the plasma calcium and virtually all of the plasma fatty acids are bound to protein.) These facts have been demonstrated by the method of **micropuncture**, in which tiny pointed tubes (micropipets) are inserted into Bowman's capsule, and extremely small quantities of fluid (approximately 1/100 of an average-size drop) are withdrawn and analyzed. Micropuncture has also been used to withdraw fluid from various portions of the tubule.

The pressure of the blood in the glomerular capillaries averages 55 mmHg. This is about half of mean arterial pressure and is higher than in other capillaries of the body because the afferent arterioles are wider than most arterioles. They therefore have a lower resistance, resulting in a lesser fall in pressure between the arteries and glomerular capillaries. This glomerular capillary hydrostatic pressure is the force favoring filtration, but it is not completely unopposed. There is, of course, fluid in Bowman's capsule which exerts a capsular hydrostatic pressure of 15 mmHg, and this opposes filtration into the capsule. A second opposing force results from the presence of protein in the plasma and its absence in Bowman's capsule. As in other capillaries, this unequal distribution of protein causes the water concentration of the plasma to be less than that of the fluid in Bowman's capsule. (Again, as in other capillaries, the water-concentration difference is due completely to the plasma protein since all the low-molecular-weight solutes have virtually identical concentrations in the plasma and Bowman's capsule.) By itself, this difference in water concentration would induce an osmotic flow of fluid (water plus all the low-molecular-weight solutes) from Bowman's capsule into the capillary, a flow the magnitude of which would be equivalent to the flow produced by a pressure difference of 30 mmHg.

As can be seen from Table 13-2, the net filtration pressure (algebraic sum of the three relevant forces) is approximately 10 mmHg. This pressure initiates urine formation by forcing an essentially protein-free filtrate of plasma through glomerular pores into Bowman's capsule and thence down the tubule into the renal pelvis. The glomerular membranes serve only as a filtration barrier and have no active, i.e., energy-requiring, function. The energy producing glomerular filtration is the energy transmitted to the blood (as hydrostatic pressure) when the heart contracts.

Before leaving this topic, we must point out the reason for use of the term "essentially protein-free" in describing the glomerular filtrate. In reality, there is a very small amount of protein in the filtrate, since the glomerular membranes are not perfect sieves for protein. This filtered protein is normally completely reabsorbed so that no protein appears in the final urine. (The reabsorbed protein

is digested in the tubular cells.) However, in diseased kidneys, the glomerular membranes may become much more leaky to protein. Moreover, the tubules sometimes may lose their ability to reabsorb protein normally. In either case, protein will appear in the urine.

Rate of glomerular filtration. A pressure of 10 mmHg may not seem to be a large net pressure favoring filtration. However, the glomerular capillaries are so much more permeable to fluid than, say, a muscle or skin capillary is, that this small pressure causes massive filtration. In a 70-kg person, the average volume of fluid filtered from the plasma into Bowman's capsule is 180 L/day (approximately 45 gal, or 450 lb of fluid). (Contrast this figure of 180 L/day net filtration at the glomeruli with the net filtration of fluid across all the other capillaries in the body—3 L/day, as described in Chap. 11.) The implications of this remarkable degree of filtration are extremely important. When we recall that the average total volume of plasma is approximately 3 L, it follows that the entire plasma volume is filtered by the kidneys some 60 times a day. It is, in part, this ability to process such huge volumes of plasma that enables the kidneys to regulate the constituents of the internal environment so precisely and to excrete large quantities of waste products.

The volume of fluid filtered into Bowman's capsule per unit time is known as the **glomerular filtration rate (GFR)**. For any low-molecular-weight substance (not bound to plasma protein), it is possible to measure the total amount of the substance filtered into Bowman's capsule (the **filtered load**) by multiplying the GFR and the plasma concentration of the substance. For example, if the GFR is 180 L/day and plasma glucose concentration is 1 g/L, then 180 g of glucose is filtered each day (180 L/day × 1 g/L). Once we know the filtered load of the substance we can compare it to the amount of the substance excreted and tell whether the substance undergoes net tubular reabsorption or secretion. The generalization is that whenever the quantity of a substance excreted in the urine is less than the amount filtered during the same period of time,

---

TABLE 13-2. Forces involved in glomerular filtration

Forces	Millimeters of mercury
Favoring filtration:	
Glomerular capillary blood pressure	55
Opposing filtration:	
Fluid pressure in Bowman's capsule	15
Osmotic force (due to protein concentration difference)	30
Net filtration pressure	10

tubular reabsorption must have occurred. Conversely, if the amount excreted in the urine is greater than the amount filtered during the same period of time, tubular secretion must have occurred.

To complete this discussion of glomerular filtration we must consider the magnitude of the total renal blood flow (Fig. 13-6). We have seen that none of the red blood cells and only a portion of the plasma which enters the glomerular capillaries is filtered into Bowman's capsule, the remainder passing via the efferent arterioles into the peritubular capillaries. The total renal plasma flow is equal to about 900 L/day; thus, the glomerular filtrate (180 L/day) constitutes approximately one-fifth of the total plasma entering the kidney. Since plasma constitutes approximately 55 percent of whole blood, the total renal blood flow, i.e., erythrocytes plus plasma, must be approximately 1640 L/day (1.1 L/min). Thus, the kidneys receive one-fifth to one-fourth of the total cardiac output (5 L/min) although

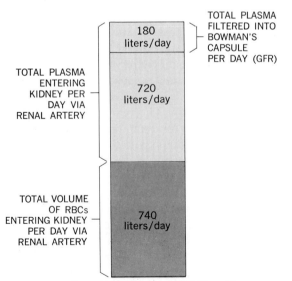

total renal blood flow = 1,640 liters/day

FIGURE 13-6. Magnitude of glomerular filtration rate, total renal plasma flow, and total renal blood flow. Only 20 percent of the plasma entering the kidneys is filtered from the glomeruli into Bowman's capsule. The remaining 80 percent flows through the glomeruli into the efferent arterioles and thence into the peritubular capillaries.

their combined weight is less than 1 percent of the total body weight! This enormous blood flow relative to tissue weight is much more than required for nutrition of the kidneys, but is required to provide the plasma for excretion of substances such as urea.

Tubular reabsorption

Many filterable plasma components are either completely absent from the urine or present in smaller quantities than were originally filtered at the glomerulus. This fact alone is sufficient to prove that these substances undergo tubular reabsorption. An idea of the magnitude and importance of these reabsorptive mechanisms can be gained from Table 13-3, which summarizes data for a few plasma components which are handled by filtration and reabsorption.

The values in Table 13-3 are typical for a normal person on an average diet. There are at least three important conclusions to be drawn from this table: (1) The quantities of material entering the nephrons via the glomerular filtrate are enormous, generally larger than their total body stores; e.g., if reabsorption of water ceased but filtration continued, the total plasma water would be urinated within 30 min. (2) The quantities of waste products, such as urea, which are excreted in the urine are generally sizable fractions of the filtered amounts; thus, in mammals, coupling a large glomerular filtration rate with incomplete urea reabsorption permits rapid excretion of the large quantities of this

TABLE 13-3. Average values for several components handled by filtration and reabsorption.

Substance	Amount filtered per day	Amount excreted per day	Percent reabsorbed
Water, L	180	1.8	99.0
Sodium, g	630	3.2	99.5
Glucose, g	180	0	100
Urea, g	54	30	44

substance produced constantly as a result of protein breakdown. (3) In contrast to urea and other waste products, the amounts of most useful plasma components (e.g., water, electrolytes, and glucose) excreted in the urine represent much smaller fractions of the filtered amounts, because more complete reabsorption has occurred. For this reason one sometimes hears the generalization that the kidney performs its regulatory function by completely reabsorbing *all* these biologically important materials and thereby preventing their loss from the body. This is a misleading half-truth, the refutation of which serves as an excellent opportunity for reviewing the essential features of renal function (and regulatory processes in general).

Let us begin by pointing out the part of the generalization that is true. Certain substances, notably glucose, are not normally excreted in the urine, because the amounts filtered are completely reabsorbed by the tubules. But does such a system permit the kidneys to regulate the plasma concentration of glucose, i.e., minimize changes from its operating point? The following example will point out why the answer is *no*. Suppose the plasma glucose concentration is 100 mg/100 mL of plasma. Since reabsorption of this carbohydrate is complete, no glucose is lost from the body via the urine, and the plasma concentration remains at 100 mg/100 mL. If, instead of 100 mg/100 mL, we set our hypothetical plasma glucose concentration at 60 mg/100 mL, the analysis does not change; no glucose is lost in the urine, and the plasma glucose stays at 60 mg/100 mL. Obviously the kidney is merely maintaining whatever plasma glucose concentration happens to exist and is not involved in the regulatory mechanisms by which the original "setting" of the plasma glucose was accomplished. It is not the kidney but primarily the liver and the endocrine system which set and regulate the plasma glucose concentration.

For comparison, consider what happens when a person drinks a lot of water; within 1 to 2 h, all the excess water has been excreted in the urine, chiefly, as we shall see, as the result of decreased renal tubular reabsorption of water, and the water concentration of the body fluids is restored to normal. The critical point is that for many plasma components, particularly certain of the inorganic ions and water, the kidney does not always completely reabsorb the total amounts filtered. The rates at which these substances are reabsorbed (and therefore the rates at which they are excreted) are constantly subject to physiological control.

Types of reabsorption. A bewildering variety of ions and molecules are found in the plasma. With the exception of the proteins (and a few ions, fatty acids, and other lipids bound to protein), these materials are all present in the glomerular filtrate, and most are reabsorbed to varying extents. It is essential to realize that tubular reabsorption is a process fundamentally different from glomerular filtration. The latter occurs completely by bulk flow in which water and all low-molecular-weight solutes move together. In contrast, there is relatively little bulk flow across the epithelial cells of the tubules from lumen to interstitial fluid, mainly because the epithelium of the tubules is relatively "nonporous" (compared to that of the glomeruli), the cells being connected to one another by tight junctions.

Given that bulk flow is not the major mechanism for transport of substances across the epithelial cells of the tubules, it follows that tubular reabsorption of various substances is not by mass movement but rather by mediated transport mechanisms and/or diffusion. The reabsorption of many substances (glucose, for example) by active mediated transport is coupled to the primary active reabsorption of sodium and is, therefore, secondary active transport (see Chap. 6).

Most, if not all, of the active reabsorptive systems in the renal tubule have a limit to the amounts of material they can transport per unit time, because the membrane proteins responsible for the transport become saturated. The classic example is the active tubular transport process for glucose (recall from Chap. 6 that glucose crosses most membranes by facilitated diffusion; however, in the kidney and intestinal tract, the transport is active). Normal persons do not excrete glucose in their urine because all filtered glucose is reabsorbed. However, it is possible to produce urinary excretion of glucose in a normal person merely by administering large quantities of glucose directly into the blood. As we have pointed out, the filtered load of any freely filterable plasma component such as glucose is equal to its plasma concentration multiplied by the glomerular filtration rate. If we assume that, as we give our subject intravenous glucose, the glomerular filtration rate remains constant, then the filtered load of glucose will be directly proportional to the plasma glucose concentration. We would find that, even after the plasma glucose concentration has doubled, the urine will still be glucose-free, indicating that the **maximal tubular capacity (T_m)** for reabsorbing glucose has not yet been reached (Fig. 13-7). But as the plasma glucose and the filtered load continue to rise, glucose finally appears in the urine. From this point on, any further increase in plasma glucose is accompanied by an increase in excreted glucose, because the T_m has been exceeded. The tubules are reabsorbing all the glucose they can, and any amount filtered in excess of this quantity cannot be reabsorbed and so appears in the urine.

This is precisely what occurs in many people with diabetes mellitus. In this disease the hormonal control of plasma glucose is defective and it may rise to extremely high values. The filtered load becomes great enough to exceed the T_m for glucose,

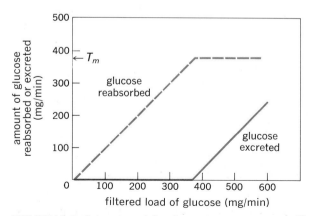

FIGURE 13-7. Saturation of the glucose transport system. Glucose is administered intravenously to a person so that plasma glucose and, thereby, filtered glucose are increased. The curves have been idealized, for the sake of clarity.

and glucose appears in the urine. There is nothing wrong with the tubular transport mechanism for glucose, it is simply unable to reabsorb the huge filtered load.

In normal persons the plasma glucose never becomes high enough to cause urinary excretion of glucose, because the reabsorptive capacity for glucose is much greater than necessary for normal filtered loads. Accordingly, as mentioned earlier, reabsorption normally prevents loss of glucose from the body via the urine, and the kidneys do not help set its plasma concentration. This pattern is true for a large number of organic substances other than glucose. For example, most amino acids and water-soluble vitamins are filtered in large amounts each day, but almost all these filtered molecules are reabsorbed, little or none being excreted in the urine. However, as is the case for glucose, if the plasma concentration of one of these substances becomes high enough, reabsorption of the filtered load will not be as complete, and the substance will appear in larger amounts in the urine. Thus, persons ingesting large quantities of vitamin C manifest progressive increases in their plasma concentrations of vitamin C until the filtered load of vitamin C exceeds the tubular reabsorptive T_m for this substance by enough to permit excretion of the amount ingested; at that point, a new steady state is reached at which plasma vitamin C concentration is elevated but not increasing any further.

Another characteristic of T_m-limited active transport systems for organic substances is that they are affected by a variety of drugs and diseases. For example, there are persons with genetic defects manifested as a deficit in the ability to reabsorb glucose or specific amino acids; the result is that large amounts of these substances are excreted in the urine, leading to their bodily depletion.

Just as glucose reabsorption provides an excellent example of an actively transported solute, urea reabsorption provides an example of passive transport. Since urea is filtered at the glomerulus, its concentration in the very first portion of the tubule is identical to its concentration in peritubular capillary plasma. Then, as the fluid flows along the tubule, water reabsorption occurs (the mechanisms will be described later), increasing the concentration of any intratubular solute (like urea) not being reabsorbed at the same rate as the water, with the result that intratubular urea concentration becomes greater than the peritubular plasma urea concentration. Accordingly, urea is able to diffuse passively down this concentration gradient from tubular lumen to peritubular capillary. Urea reabsorption is thus a passive process and completely dependent upon the reabsorption of water (itself passive, as we shall see), which establishes the diffusion gradient.

Passive reabsorption is also of considerable importance for many foreign chemicals. The renal tubular epithelium acts in many respects as a lipid barrier; accordingly, lipid-soluble substances, like urea, can penetrate it fairly readily. Recall that one of the major determinants of lipid solubility is the polarity of a molecule; the less polar, the more lipid-soluble. Many drugs and environmental pollutants are nonpolar and, therefore, highly lipid-soluble. This makes their excretion from the body via the urine difficult since they are filtered at the glomerulus and then reabsorbed, as water reabsorption causes their intratubular concentrations to increase. But evolution has provided a way out of this dilemma: The liver transforms most of these substances to more polar metabolites which, because of their reduced lipid solubility, are poorly reabsorbed by the tubules and can therefore be excreted.

Tubular secretion

Tubular secretory processes, by which substances move from peritubular capillaries into the tubular lumen, i.e., in the direction opposite to tubular reabsorption, constitute a second pathway into the tubule, the first being glomerular filtration. Of the large number of different substances which can be transported into the tubules by tubular secretion, only a few are normally found in the body, the most important being hydrogen ion and potassium. We shall see that most of the hydrogen ion and potassium appearing in the urine enters the tubules by secretion rather than by filtration. Thus, renal regulation of these two important substances is accomplished primarily by mechanisms which control the rates of their tubular secretion. The kidney is also able to secrete a large number of foreign chemicals, thereby facilitating their excretion from the body; penicillin is an example.

Metabolism by the tubules

Although renal physiologists have traditionally listed glomerular filtration, tubular reabsorption, and tubular secretion as the three basic renal processes, a fourth process—metabolism by the cells of the tubule—is of considerable importance for many substances. The cells of the renal tubules are able to synthesize certain substances (ammonia, for example), which are then added to the lumen fluid and excreted. The cells are also capable of catabolizing certain organic substances taken up either from the tubular lumen or peritubular capillaries, thus eliminating them from the body as surely as if they had been excreted in the urine.

This completes our survey of the general principles of renal physiology. In the remainder of this chapter we shall see how the kidney functions in a variety of homeostatic processes and how renal function is coordinated with that of other organs to regulate the composition of the extracellular fluid. Before turning to the individual variables being regulated, we describe the mechanisms for eliminating urine from the body.

Micturition (Urination)

From the kidneys, urine flows to the bladder through the ureters (Fig. 13-1), propelled by peristaltic contractions of the smooth muscle which makes up most of the ureteral wall. The bladder is a balloonlike chamber, with walls of thick layers of smooth muscle. As the bladder fills with urine, its highly folded wall flattens out to accommodate the increased volume. The smooth muscle at the base of the bladder is sometimes called the internal urethral spincter; however, it is not a distinct muscle but the last portion of the bladder (and the first portion of the urethra). When the bladder is relaxed, this smooth muscle closes the outlet of the bladder; when the bladder contracts, this outlet is pulled open by the changes in bladder shape occurring during contraction. As we shall see, passively produced changes in bladder shape also pull open the outlet.

In an infant, micturition is basically a local spinal reflex, and the place and time of urination are determined solely by the volume of urine in the bladder, acting via this reflex. The bladder muscle receives a rich supply of parasympathetic fibers, the stimulation of which causes bladder contraction. The bladder wall also contains stretch receptors whose afferent fibers enter the spinal cord and eventually synapse with these parasympathetic pathways and stimulate them. When the bladder contains only small amounts of urine, its internal pressure is low, there is little stimulation of the bladder stretch receptors, and the parasympathetics are relatively quiescent. As the bladder fills with urine, it becomes distended, and the stretch receptors are gradually stimulated until the reflexly elicited output of the parasympathetic neurons becomes great enough to contract the bladder. When bladder pressure exceeds urethral resistance, urination occurs.

Voluntary control of micturition, learned during childhood, is by way of skeletal muscles, particularly the muscles of the **pelvic diaphragm,** which constitutes the floor of the pelvis and helps support some of the pelvic viscera, including that part of the bladder that opens into the urethra. Micturition can be initiated voluntarily in the following way: Voluntary relaxation of the pelvic diaphragm allows the neck of the bladder to move downward; this tends to open the outlet of the bladder while simultaneously stretching the walls of the bladder; the latter leads to reflex contraction of the bladder via the parasympathetic nerves. If the increased pressure resulting from contraction of the bladder wall is insufficient to force urine into the urethra, the pressure can be further increased by voluntary contraction of the abdominal muscles and the respiratory diaphragm.

In contrast, micturition can be stopped voluntarily simply by contracting the pelvic diaphragm. The **external urethral sphincter,** or circular layer of skeletal muscle which surrounds the urethra and relaxes and contracts with the pelvic diaphragm, is not thought to play a major role in voluntary control of micturition. In an adult with damage to the spinal cord interrupting the descending pathways which mediate voluntary control of the pelvic diaphragm, micturition once again becomes the pure spinal reflex it was in infancy.

SECTION B.
REGULATION OF SODIUM, CHLORIDE, AND WATER BALANCE

Table 13-1 is a typical balance sheet for water; Table 13-4, for sodium chloride (salt). As with water, the excretion of sodium and chloride ions via the skin and gastrointestinal tract is normally quite small but may increase markedly during severe sweating, vomiting, or diarrhea. Hemorrhage, of course, can result in loss of large quantities of both salt and water.

Control of the renal excretion of sodium, chloride, and water constitutes the most important mechanism for the homeostatic regulation of these ions and water. Their excretory rates can be varied over an extremely wide range; e.g., a gross consumer of salt may ingest 20 to 25 g of sodium chlo-

TABLE 13-4. Normal routes of sodium chloride intake and loss

	Grams per day
Intake:	
Food	10.5
Output:	
Sweat	0.25
Feces	0.25
Urine	10.0
Total output	10.5

ride per day, whereas a person on a low-salt diet may ingest only 50 mg (0.05 g). The normal kidney can readily alter its excretion of salt over this range. Similarly, urinary water excretion can vary physiologically from approximately 0.4 L/day to 25 L/day, depending upon whether one is lost in the desert or participating in a beer-drinking contest.

Basic Renal Processes for Sodium, Chloride, and Water

Sodium, chloride, and water do not undergo tubular secretion; each is freely filterable at the glomerulus, and approximately 99 percent is reabsorbed as it passes down the tubules. Indeed, most renal energy utilization goes to accomplish this enormous reabsorptive task. The tubular mechanisms for reabsorption of these substances can be summarized by three generalizations:

1. The reabsorption of sodium is by primary active transport; i.e., it is carrier-mediated, requires a metabolic energy supply, and can occur against an electrical and chemical gradient (see Chap. 6 for the mechanism of sodium transport across an epithelium).
2. The reabsorption of chloride is coupled to the active reabsorption of sodium by several mechanisms, both active and passive.
3. The reabsorption of water is by diffusion (osmosis) and depends mainly upon the reabsorption of sodium and chloride.

How does the active reabsorption of sodium lead to the passive reabsorption of water? Since the concentrations of all low-molecular-weight solutes are virtually identical in plasma and the fluid of Bowman's capsule, no significant transtubular concentration gradients exist for sodium or water in the very first portion of the proximal tubule. As the fluid flows down the tubule, sodium is actively reabsorbed. What happens to the water concentrations in the tubular fluid and interstitial fluid during this reabsorption? The removal of sodium lowers the total osmolarity (raises water concentration) of the luminal fluid and simultaneously raises the osmo-

larity (lowers the water concentration) in the interstitial fluid adjacent to the epithelial cells. This difference in water concentration causes net diffusion (osmosis) of water from the lumen across the tubular cell's plasma membranes (and/or tight junctions) into the interstitial fluid, from which the reabsorbed sodium and water move together by bulk flow into peritubular capillaries.

If the water permeability of the tubular epithelium is very high, water molecules are reabsorbed by diffusion almost as rapidly as the actively transported sodium ions. However, this reabsorption of water can occur only if the tubular epithelium is permeable to water. No matter how great the water-concentration gradient, water cannot cross an epithelium impermeable to it. The permeability of the last portions of the tubules (the late distal tubules and collecting ducts) to water is subject to physiological control. The major determinant of this permeability is a hormone known as **antidiuretic hormone (ADH)** or **vasopressin**.[1]

In the absence of ADH (Fig. 13-8) the water permeability of the late distal tubule and collecting duct is very low; sodium reabsorption proceeds normally (because ADH has little or no effect on sodium reabsorption), but water is unable to follow and thus remains in the tubule to be excreted in the urine. Thus, in the absence of ADH, a large volume of urine is formed. On the other hand, in the presence of a high plasma concentration of ADH, the water permeability of the last nephron segments is very great, water reabsorption is able to keep up with sodium reabsorption, and the final urine volume is small.

ADH is a peptide closely related in structure to another hormone, oxytocin, the function of which is discussed in Chap. 16. ADH stimulates production of cyclic AMP in the tubular cells, which leads to changes in membrane proteins; it it not known how these protein alterations cause increased membrane permeability to water. How

---

[1] Throughout this chapter, we use the name "antidiuretic hormone" in keeping with the action of this hormone on the kidney; its other name, "vasopressin," which is becoming its "official" label, reflects the fact that this hormone can also exert a direct vasoconstrictor action on arterioles, resulting in elevated total peripheral resistance and, thereby, increased arterial blood pressure.

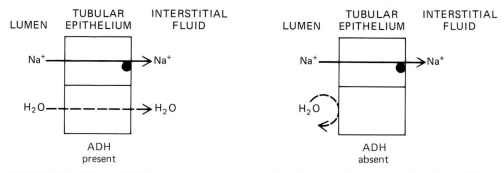

FIGURE 13-8. Effect of ADH on water reabsorption in late distal tubules and collecting ducts. ADH causes these nephron segments to be highly permeable to water, and water movement can passively accompany sodium reabsorption. Without ADH these nephron segments become quite impermeable to water, nd water reabsorption does not occur. Note that ADH does not alter reabsorption of sodium.

ADH secretion is regulated will be discussed in subsequent sections.

Several crucial aspects of sodium and water reabsorption should be emphasized:

1. The tubular response to ADH is not all-or-none but shows graded changes as the plasma concentration of ADH is altered, thus permitting fine adjustments of water permeability and excretion.
2. Excretion of large quantities of sodium or other unreabsorbed solute (such as glucose in diabetes) always results in the excretion of large quantities of water, since solute must be reabsorbed from the tubular lumen in order to create the concentration gradient for water reabsorption. If solute is not reabsorbed, water will not be reabsorbed. As we shall see, this relationship has considerable importance for the regulation of extracellular volume.
3. In contrast, large quantities of water can be excreted (when ADH concentration is very low) even though the urine is virtually free of sodium. This process we shall find critical for the renal regulation of extracellular osmolarity.

Control of Sodium Excretion: Regulation of Extracellular Volume

The renal compensation for increased intake of sodium is excretion of the excess sodium. Conversely, any initial decrease in body sodium is minimized by reducing urinary sodium, thus retaining within the body the amount which would otherwise have been lost via the urine.

Since sodium is freely filterable at the glomerulus and actively reabsorbed but not secreted by the tubules, the amount of sodium excreted in the final urine represents the resultant of two processes, glomerular filtration and tubular reabsorption:

Sodium excretion =
    sodium filtered − sodium reabsorbed

It is possible, therefore, to adjust sodium excretion by controlling either or both of these two variables (Fig. 13-9). For example, what happens if the quantity of filtered sodium increases (as a result of a higher GFR) but the rate of reabsorption remains constant? Clearly, sodium excretion increases. The same final result could be achieved by lowering sodium reabsorption while holding the GFR constant. Finally, sodium excretion could be raised greatly by elevating the GFR and simultaneously reducing reabsorption. Conversely, sodium excretion could be decreased below normal levels by lowering the GFR or raising sodium reabsorption or both.

The major reflex pathways by which changes in body sodium balance lead to changes in GFR and sodium reabsorption include: (1) cardiovascular baroreceptors and the afferent pathways leading from them to the central nervous system and to

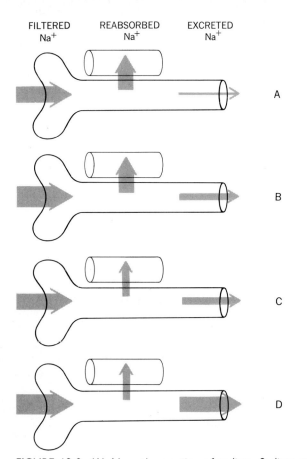

FILTERED Na⁺   REABSORBED Na⁺   EXCRETED Na⁺

FIGURE 13-9. (A) Normal excretion of sodium. Sodium excretion is increased by increasing the GFR (B), by decreasing reabsorption (C), or by a combination of both (D). The arrows indicate relative magnitudes of filtration, reabsorption, and excretion.

endocrine glands, (2) efferent neural and hormonal pathways to the kidneys, and (3) renal effector sites—the renal arterioles and tubules.

That cardiovascular baroreceptors are appropriate receptors for regulating total body sodium is shown by the following analysis. What happens when a person ingests 1 L of isotonic sodium chloride, i.e., a solution of salt with exactly the same osmolarity as the body fluids? It is absorbed from the gastrointestinal tract, all the salt and water remain in the extracellular fluid (plasma and interstitial fluid), and none enters the cells. Because of the active sodium "pumps" in cell membranes, sodium is effectively barred from cells. The water, too, remains in the extracellular fluid, since only the volume and not the osmolarity of the extracellular compartment has been changed; i.e., no osmotic gradient exists to drive the ingested water into cells. This example is one of many which could be used to illustrate the important generalization that total extracellular fluid volume depends primarily upon the mass of extracellular sodium, which in turn correlates directly with total body sodium. However, rather than extracellular volume, per se, the variables monitored in sodium-regulating reflexes are those that are altered as a result of changes in this volume, specifically intravascular and intracardiac pressures (Fig. 13-10). For example, a decrease in plasma volume tends to lower the hydrostatic pressures in the veins, cardiac chambers, and arteries. These pressure changes are detected by baroreceptors (e.g., the carotid sinus baroreceptor) in these structures and initiate the reflexes leading to renal sodium retention, which helps restore the plasma volume toward normal.

Control of GFR

Figure 13-11 summarizes the major neural reflex pathway by which a lower body sodium (as caused, for example, by diarrhea) elicits a decrease in GFR. Note that this is simply the basic baroreceptor reflex described in Chap. 11, where it was pointed out that a decrease in cardiovascular pressures causes reflex vasoconstriction in many areas of the body. Conversely, an increased GFR can result from increased plasma volume and contribute to the increased renal sodium loss which returns extracellular volume to normal.

Control of tubular sodium reabsorption

So far as long-term regulation of sodium excretion is concerned, the control of tubular sodium reabsorption is more important than that of GFR. The major factor determining the rate of tubular sodium reabsorption is aldosterone.

Aldosterone and the renin-angiotensin system. An early clue to the control of sodium reabsorption

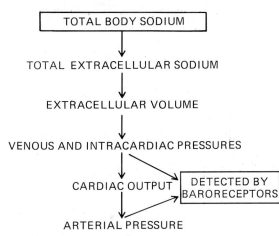

FIGURE 13-10. Dependence of cardiovascular parameters on total body sodium. Receptors which detect changes in pressure or distension supply the information required for regulation of body sodium. The magnitude of each variable in the flow sheet is determined, in large part, by the magnitude of the one above it.

was the observation that persons whose adrenal glands are diseased or missing excrete large quantities of sodium in the urine owing to decreased tubular reabsorption; indeed, if untreated, they may die because of low blood pressure resulting from depletion of plasma volume. The adrenal cortex produces a hormone, called **aldosterone,** which stimulates sodium reabsorption (specifically by the distal tubules and collecting ducts). In the complete absence of this hormone, a person may excrete 35 g of salt per day, whereas excretion may be virtually zero when aldosterone is present in large quantities. In a normal person, the amounts of aldosterone produced and salt excreted lie somewhere between these extremes, varying with the amount of salt ingested. (It is interesting that aldosterone also stimulates sodium transport by other epithelia in the body, namely, by sweat and salivary glands and the intestine; the net effect is the same as that exerted on the renal tubules—a movement of sodium out of the luminal fluid into the blood.) Aldosterone acts by inducing the synthesis of proteins involved in sodium reabsorption, but which proteins is still unclear.

Aldosterone secretion (and, thereby, tubular

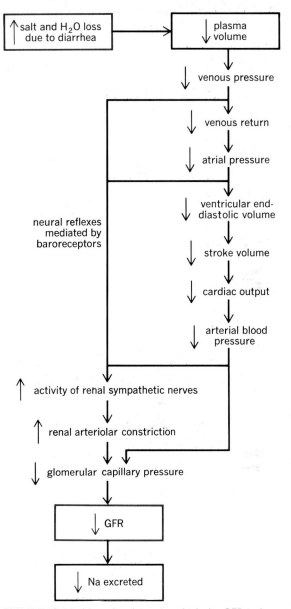

FIGURE 13-11. Neural pathway by which the GFR is decreased when plasma volume decreases. The baroreceptors which initiate the sympathetic reflex are probably located in large veins and the walls of the heart, as well as in the carotid sinuses and aortic arch. This neural pathway is much more effective in lowering GFR than is the direct effect of the lowered arterial pressure on the kidneys.

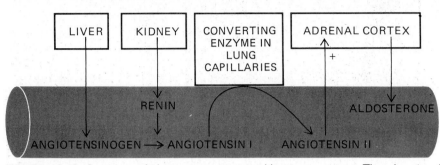

FIGURE 13-12. Summary of the renin-angiotensin-aldosterone system. The plus sign (+) denotes the stimulatory effect of angiotensin II on aldosterone secretion.

sodium reabsorption) is controlled by reflexes involving the kidneys themselves. Specialized cells lining the afferent arterioles in the kidneys synthesize and secrete into the blood an enzyme known as **renin** which splits off a small polypeptide, **angiotensin I,** from a large plasma protein, **angiotensinogen** (Fig. 13-12). Angiotensin I then undergoes further enzymatically mediated cleavage to form **angiotensin II;** this conversion is mediated by an enzyme known as **converting enzyme,** which is found in very high concentration on the luminal surface of the lung capillary endothelium. Thus, most conversion of angiotensin I to angiotensin II occurs as blood flows through the lungs. Angiotensin II is a potent stimulator of aldosterone secretion and constitutes the major input to the adrenal gland controlling the production and release of this hormone (for simplicity, we shall refer to angiotensin II simply as angiotensin).

Angiotensinogen is synthesized by the liver and is always present in the blood, and converting enzyme is also always present in the lung capillary endothelium. Therefore, the rate-limiting factor in angiotensin formation is the concentration of plasma renin, which, in turn, depends upon the rate of renin secretion by the kidneys. The critical question now becomes: What controls renin secretion?

There are multiple inputs to the renin-secreting cells, and it is not yet possible to assign quantitative roles to each of them. The renal sympathetic nerves constitute one important input (Fig. 13-13); this makes excellent sense, teleologically, since a reduction in body sodium and extracellular volume

lowers blood pressure and triggers increased sympathetic discharge to the kidneys (as shown in Fig. 13-11), thereby setting off the hormonal chain of events which restores sodium balance and extracellular volume toward normal (Fig. 13-13).

Regulation of renin secretion is not solely dependent upon sympathetic input or, for that matter, on any neuroendocrine input external to the kidneys. This is because there are several pathways totally contained within the kidneys for altering renin secretion. For example, the renin-secreting cells, themselves, are pressure-sensitive; thus, when renal arterial pressure decreases (as would occur when extracellular volume is down), these cells secrete more renin (Fig. 13-13). Clearly, this inherent response is in the same direction as that elicited by the sympathetic input.

By helping to regulate sodium balance and, thereby, plasma volume, the renin-angiotensin system contributes to the control of arterial blood pressure. However, this is not the only way in which it influences arterial pressure; recall from Chap. 11 that angiotensin is a potent constrictor of arterioles, and this effect on peripheral resistance increases arterial pressure. Abnormal increases in the activity of the renin-angiotensin system can contribute to the development of hypertension (high blood pressure) via both the sodium-retaining and arteriole-constricting effects. It should not be surprising, therefore, that attempts are being made to treat certain persons suffering from hypertension with drugs which block the renin-angiotensin system. One such group of drugs blocks the action of con-

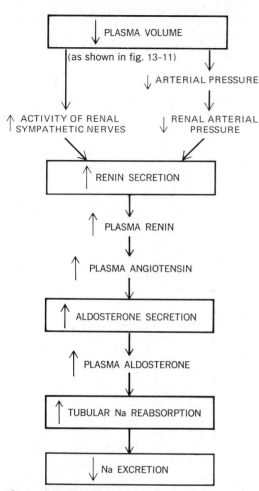

FIGURE 13-13. Pathway by which aldosterone secretion (and, hence, tubular sodium reabsorption) is increased when plasma volume is decreased.

verting enzyme, thus preventing the generation of angiotensin II (angiotensin I is relatively inactive).

Factors other than aldosterone. There are a large number of factors other than aldosterone which are capable of altering tubular sodium reabsorption, and renal physiologists are presently attempting to quantitate their relative contributions to the control of sodium excretion in both health and disease. This task is far from complete and we present here only the most exciting single line of investigation. There is much indirect evidence suggesting the existence

of a hormone, as yet unidentified, which inhibits the reabsorption of sodium and is therefore referred to as **natriuretic hormone.** This means that the tubules are probably under the influence of opposing hormonal inputs—aldosterone enhancing sodium reabsorptive rate, and the hypothesized natriuretic hormone reducing it. The site of production of natriuretic hormone is not known, but prime candidates are the cardiac atria (from which can be extracted so-called **atrial natriuretic factors**) and the brain. The pathways that might control the release of natriuretic hormone are not known.

Conclusion

The control of sodium excretion depends upon the control of two variables of renal function—the GFR and sodium reabsorption. The latter is controlled, at least in part, by the renin-angiotensin-aldosterone hormone system and, in part, by other less well-defined factors. The reflexes which control both GFR and sodium reabsorption are essentially blood-pressure-regulating reflexes, since they are probably most frequently initiated by changes in arterial or venous pressures. This is only fitting, since cardiovascular function depends upon an adequate plasma volume, which as a component of the extracellular fluid volume normally reflects the amount of sodium in the body.

In normal persons, these regulatory mechanisms are so precise that sodium balance does not vary by more than 2 percent despite marked changes in dietary intake or in losses due to sweating, vomiting, or diarrhea. In several types of diseases, however, sodium balance becomes deranged by the failure of the kidneys to excrete sodium normally. Sodium excretion may fall virtually to zero despite continued sodium ingestion, and the person may retain huge quantities of sodium and water, leading to abnormal expansion of the extracellular fluid and edema. The most important example of this phenomenon is congestive heart failure (Chap. 11). A person with a failing heart manifests a decreased GFR (due to low blood pressure) and increased aldosterone secretion rate, both of which contribute to the virtual absence of sodium from the urine. In addition to aldosterone, other less well-defined factors also cause decreased sodium excretion by enhancing tubular sodium reabsorption. The net result is expansion of plasma volume, increased capillary pressure, and filtration of fluid into the interstitial space, i.e., edema. Why are these sodium-retaining reflexes all stimulated despite the fact that the expanded extracellular volume should call forth sodium-losing responses (increased GFR and decreased aldosterone)? Because they are triggered by the lower cardiac output (a result of cardiac failure) and the resulting decrease in arterial pressure. In addition to treating the basic heart disease, physicians use **diuretics** (drugs which inhibit tubular sodium and chloride reabsorption), thereby leading to greater sodium, chloride, and water excretion.

ADH Secretion and Extracellular Volume

Although we have spoken of extracellular-volume regulation only in terms of the control of sodium excretion, the changes in sodium excretion must be accompanied by equivalent changes in water excretion to be maximally effective in altering extracellular volume. We have already pointed out that the ability of water to be reabsorbed in late distal tubules and collecting ducts depends upon ADH; the new fact is that a decreased extracellular volume reflexly calls forth increased ADH secretion as well as increased aldosterone secretion. What is the nature of this reflex? As described in Chap. 9, ADH is produced by a discrete group of hypothalamic neurons whose axons terminate in the posterior pituitary, from which ADH is released into the blood. These hypothalamic cells receive input from several vascular baroreceptors, particularly a group located in the left atrium (Fig. 13-14). The baroreceptors are stimulated by the increased atrial blood pressure which occurs when blood volume increases, and the impulses resulting from this stimulation are transmitted via afferent neurons and ascending pathways to the hypothalamus, where they inhibit the ADH-producing cells. Conversely, decreased atrial pressure causes less firing by the baroreceptors, resulting in an increase of ADH secretion (Fig.

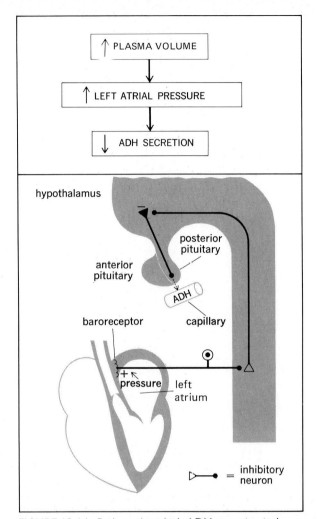

FIGURE 13-14. Pathway by which ADH secretion is decreased when plasma volume is increased. The greater plasma volume raises left-atrial pressure, which stimulates the atrial baroreceptors and inhibits ADH secretion.

13-15). The adaptive value of this baroreceptor reflex, one more in our expanding list, should require no comment.

Renal Regulation of Extracellular Osmolarity

We turn now to the renal compensation for pure water losses or gains, e.g., a person drinking 2 L of water, where no change in total salt content of

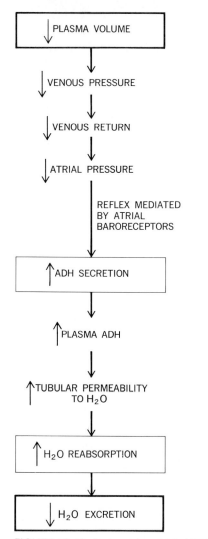

FIGURE 13-15. Pathway by which ADH secretion (and, hence, tubular permeability to water) is increased when plasma volume decreases.

the body occurs, only a change in total water. The most efficient compensatory mechanism is for the kidneys to excrete the excess water without altering their usual excretion of salt, and this is precisely what they do. ADH secretion is reflexly inhibited, as will be described below, water permeability of the late distal tubules and collecting ducts becomes very low, sodium reabsorption proceeds normally but water is not reabsorbed, and a large volume of

extremely dilute (hypoosmotic) urine is excreted. In this manner, the excess water is eliminated.

At the other end of the spectrum, when the body is in negative water balance, the final urine may reach a concentration considerably greater than that of the plasma. How can this happen if water reabsorption is always secondary to solute (particularly sodium chloride) reabsorption? To this question we now turn.

Urine concentration: The countercurrent system

The ability of the kidneys to produce concentrated (hyperosmotic) urine is not merely an academic problem. It is a major determinant of one's ability to survive without water. The human kidney can produce a maximal urinary concentration of 1400 milliosmol/L. The urea, sulfate, phosphate, and other waste products (plus the smaller number of other ions) which are excreted each day amount to approximately 600 milliosmols; therefore, the water required for their excretion constitutes an obligatory water loss and equals:

$$\frac{600 \text{ milliosmols/day}}{1400 \text{ milliosmol/L}} = 0.444 \text{ L/day}$$

As long as the kidneys are functioning, excretion of this volume of urine will occur, despite the absence of water intake.

The kidneys produce concentrated urine by a complex process involving the so-called **countercurrent multiplier system** residing in the loop of Henle. Recall that the loop of Henle, which is interposed between the proximal and distal tubules, is a hairpin loop extending into the inner portions of the kidney (the renal medulla). The tubular fluid flows in opposite directions in the two limbs of the loop; thus the name "countercurrent." Let us list the critical characteristics of this loop.

1. The ascending limb of the loop of Henle (i.e., the limb leading to the distal tubule) actively transports sodium out of the tubular lumen into the surrounding interstitium; chloride movement (also active) is coupled to that of sodium. This

limb is fairly impermeable to water, so that water cannot follow the sodium chloride.

2. The descending limb of the loop of Henle (i.e., the limb into which drains fluid from the proximal tubule) does not actively transport sodium (it is the only tubular segment that does not). Moreover, it is highly permeable to water but relatively impermeable to sodium and chloride.

Given these characteristics, imagine the loop of Henle filled with a stationary column of fluid supplied by the proximal tubule. At first, the concentration everywhere would be 300 milliosmol/L (Fig. 13-16, upper left), since fluid leaving the proximal tubule is isosmotic to plasma, equivalent amounts of sodium chloride and water having been reabsorbed by the proximal tubule.

Now let the active pump in the ascending limb transport sodium chloride into the interstitium until a steady-state limiting gradient (say 200 milliosmol/L) is established between ascending-limb fluid and interstitium (Fig. 13-16, upper right).

Given the relatively high permeability of the descending limb to water, there is a net diffusion of water out of the descending limb into the more concentrated interstitial fluid until the osmolarities are equal (Fig. 13-16, upper right). The interstitial osmolarity is maintained at 400 milliosmol/L during this equilibrium because of continued active sodium chloride transport out of the ascending limb. Thus, the osmolarities of the descending limb and interstitial fluid become equal, and both are higher than that of the ascending limb.

So far we have held the fluid stationary in the loop, but of course it is actually continuously flowing. Let us look at what occurs under conditions of flow (Fig. 13-16), simplifying the analysis by assuming the flow along the loop, on the one hand, and ion and water movements, on the other, occur in discontinuous out-of-phase steps. During the stationary phase, as described above, sodium chloride is actively transported out of the ascending limb to establish a gradient of 200 milliosmol/L across each portion of the ascending limb, and water diffuses out of the descending limb until each portion of the

descending limb has the same osmolarity as the interstitial fluid adjacent to it. During the "flow" phase, fluid leaves the loop via the distal tubule, and new fluid enters the loop from the proximal tubule.

Note that the fluid is progressively concentrated as it flows down the descending limb and then is progressively diluted as it flows up the ascending limb. Whereas only a 200 milliosmol/L gradient is maintained across the ascending limb at any given horizontal level in the medulla, there exists a much larger osmotic gradient from the top of the medulla to the bottom (312 milliosmol/L versus 700 milliosmol/L). In other words, the 200 milliosmol/L gradient established by active sodium chloride transport has been "multiplied" because of the countercurrent flow within the loop. It should be emphasized that the active sodium chloride transport mechanism in the ascending limb is the essential component of the entire system; without it, the countercurrent flow would have no effect whatever on concentrations.

But what has this system really accomplished? Certainly it concentrates the loop fluid, but then it immediately redilutes it so that the fluid entering the distal tubule is actually more dilute than the plasma. Where is the final urine concentrated and how? The site of final concentration is the collecting ducts which pass through the renal medulla parallel to the loops of Henle and are bathed by the interstitial fluid of the medulla. The real function of the loop countercurrent multiplier system is to concentrate the medullary interstitium. In the presence of maximal concentrations of ADH, fluid in the late distal tubules reequilibrates with peritubular plasma and is isosmotic to plasma (i.e., is at 300 milliosmol/L) when it enters the collecting ducts. As this fluid then flows through the collecting ducts it comes into diffusion equilibrium with the ever-increasing osmolarity of the interstitial fluid. Under the influence of ADH, the collecting ducts are highly permeable to water, which diffuses out of the collecting ducts into the interstitial fluid as a result of the osmotic gradient (Fig. 13-17). The net result is that the fluid at the end of the collecting duct has the

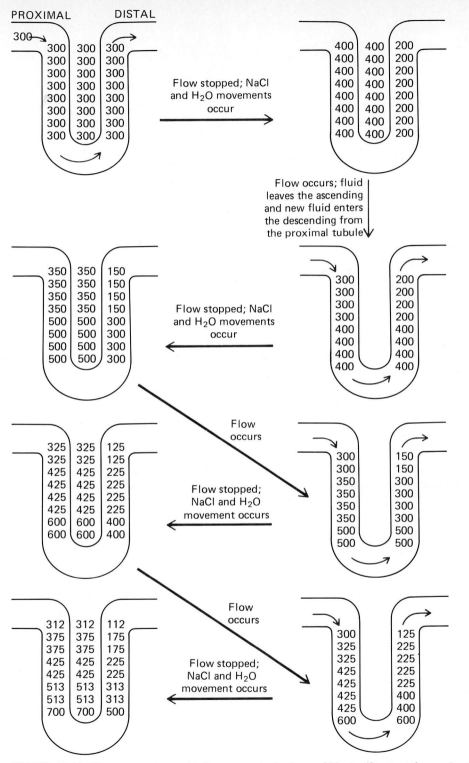

FIGURE 13-16. Countercurrent multiplier system in the loop of Henle. (See text for explanation.)

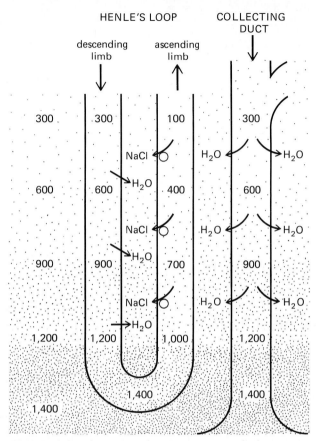

HENLE'S LOOP

descending limb

ascending limb

COLLECTING DUCT

FIGURE 13-17. Operation of countercurrent multiplier system in the formation of hypertonic urine. (Adapted from R. F. Pitts.)

same osmolarity as the interstitial fluid at the tip of the medulla. By this means, the final hyperosmotic urine contains relatively less of the filtered water than solute; this is the same as adding pure water to the extracellular fluid, and thereby compensates for a pure water deficit.

In contrast, in the presence of low plasma-ADH concentration, the late distal tubules and the collecting ducts become relatively impermeable to water. Therefore, tubular fluid in the late distal tubules does not reequilibrate with peritubular plasma, and, in addition, the medullary interstitial osmotic gradient is ineffective in inducing water movement out of the collecting ducts; therefore a large volume of hypoosmotic urine is excreted, thereby compensating for a pure water excess.

Osmoreceptor control of ADH secretion

To reiterate, pure water deficits or gains are compensated for by partially dissociating excretion of water from that of solute through changes in ADH secretion. What controls the secretion of ADH under such conditions? The answer is changes in extracellular osmolarity. This makes sense, since osmolarity is the variable most affected by pure water gains or deficits. **Osmoreceptors** are located in the hypothalamus, and information from them is transmitted by neurons to the hypothalamic cells which secrete ADH. Via these connections, an increase in osmolarity (decrease in water concentration) increases the rate of ADH secretion, and conversely, a decrease in osmolarity (increase in water concentration) inhibits ADH secretion (Fig. 13-18).

We have now described two different afferent pathways controlling the ADH-secreting hypothalamic cells, one from baroreceptors and one from osmoreceptors. These hypothalamic cells are therefore true integrators whose rate of activity is determined by the balance of excitatory and inhibitory input. To add to the complexity, these cells receive synaptic input from many other brain areas, so that ADH secretion (and therefore urine flow) can be altered by pain, fear, and a variety of other factors. For example, alcohol is a powerful inhibitor of ADH release and probably accounts for much of the large urine flow accompanying the ingestion of alcohol. However, all these effects are usually short-lived and should not obscure the generalization that ADH secretion is determined primarily by the states of extracellular volume and osmolarity.

The disease **diabetes insipidus** (which is distinct from diabetes mellitus or "sugar diabetes") illustrates what happens when the ADH system is disrupted. This disease is characterized by the constant formation of a large volume of highly dilute urine (as much as 25 L/day). In most cases, the flow can be restored to normal by the administration of ADH. These persons have lost the ability to produce their own ADH, usually as a result of damage to the hypothalamus. Thus, late-renal-tubular permeability to water is low and unchanging regardless of extracellular osmolarity or volume. The

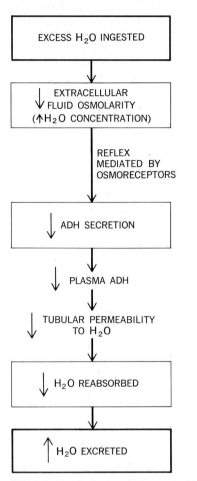

FIGURE 13-18. Pathway by which ADH secretion is lowered and water excretion raised when excess water is ingested.

very thought of having to urinate (and therefore to drink) 25 L of water per day underscores the importance of ADH in the control of renal function and body water balance.

A Summary Example: The Response to Sweating

Figure 13-19 shows the factors that control renal sodium and water excretion in response to severe sweating, as in exercise. The renal retention of fluid helps to compensate for the water and salt lost in the sweat.

Thirst and Salt Appetite

Now we must turn to the other component of any balance, control of intake. It must be emphasized that large deficits of salt and water can be only partly compensated by renal conservation, and that ingestion is the ultimate compensatory mechanism. The subjective feeling of thirst, which drives one to obtain and ingest water, is stimulated both by a lower extracellular volume and a higher plasma osmolarity, the adaptive significance of both being self-evident (Fig. 13-20). Note that these are precisely the same changes which stimulate ADH production, and the receptors (osmoreceptors and atrial baroreceptors) which control ADH secretion are probably identical to those for thirst. The brain centers which receive input from these (and other receptors) and mediate thirst are located in the hypothalamus very close to those areas which produce ADH. (These water-intake control centers are also very close to, but distinct from, the food-intake centers to be described in Chap. 15.)

Another factor influencing thirst is angiotensin, which stimulates thirst by a direct effect on the brain. Thus, the renin-angiotensin system is an important regulator not only of sodium balance but of water balance as well and constitutes one of the pathways by which thirst is stimulated when extracellular volume is decreased.

There are still other pathways controlling thirst. For example, dryness of the mouth and throat causes profound thirst, which is relieved by merely moistening them. It is facinating that animals such as the camel (and people, to a lesser extent) which have been markedly dehydrated will rapidly drink just enough water to replace their previous losses and then stop; what is amazing is that when they stop, the water has not yet had time to be absorbed from the gastrointestinal tract into the blood. Some kind of "metering" of the water intake by the gastrointestinal tract has occurred, but its nature remains a mystery.

The analog of thirst for sodium, **salt appetite,** is also an extremly important component of sodium homeostasis in most mammals, but the contribution of regulatory salt appetite to everyday sodium ho-

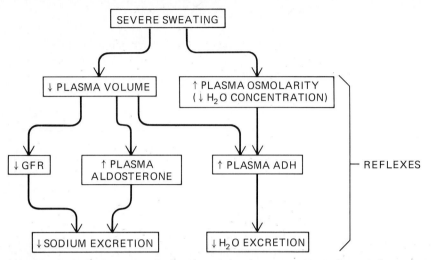

FIGURE 13-19. Pathways by which sodium and water excretion are decreased in response to severe sweating. (Sweat is a hyptonic salt solution, approximately half as concentrated as plasma.) This figure is an amalgamation of Figs. 13-12, 13-14, and 13-15, and the converse of Fig. 13-18. Nonaldosterone factors controlling sodium reabsorption are not shown in the figure.

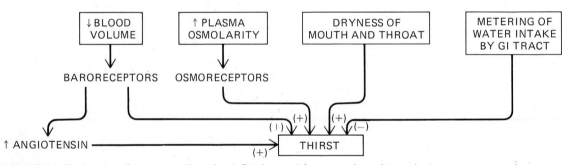

FIGURE 13-20. Inputs reflexly controlling thirst. Psychosocial factors and conditioned responses are not shown.

meostasis in normal persons is probably slight. On the other hand, people certainly like salt, as manifested by almost universally large intakes of sodium whenever it is cheap and readily available. Thus, the average American intake of salt is 10 to 15 g/ day, despite the fact that human beings can survive quite normally on less than 0.5 g/day. Present evidence suggests that a large salt intake may be an important contributor to the pathogenesis of hypertension in certain persons.

SECTION C.
REGULATION OF POTASSIUM, CALCIUM, AND HYDROGEN-ION CONCENTRATIONS

Potassium Regulation

The potassium concentration of extracellular fluid is a closely regulated quantity. The importance of controlling this concentration in the internal environment stems primarily from the role of potassium in the excitability of nerve and muscle. Recall that the resting-membrane potentials of these tissues are directly related to the relative intracellular and extracellular potassium concentrations. Raising the external potassium concentration depolarizes membranes, leading to cardiac arrhythmias. Conversely, lowering the external potassium hyperpolarizes cell membranes, and early manifestations of potassium depletion are weakness of skeletal muscles and abnormalities of cardiac muscle conduction and rhythm.

Since most of the body's potassium is found in cells, as a result of active transport of potassium into cells (Chap. 6), even a slight alteration of the rate of potassium transport across plasma membranes can produce a large change in the amount of extracellular potassium. Unfortunately, relatively little is known about the physiological control of these transport mechanisms, but they clearly play an important role in protecting against marked changes in extracellular potassium concentration.

The normal person remains in total body potassium balance by daily excreting an amount of potassium in the urine equal to the amount ingested minus the amounts eliminated in the feces and sweat. Normally, losses via the sweat and gastrointestinal tract are quite small, although large quantities can be lost during vomiting or diarrhea. Again, the control of renal function is the major mechanism by which body potassium is regulated.

Renal regulation of potassium

Potassium is freely filterable at the glomerulus. The amounts of potassium excreted in the urine are generally a small fraction (10 to 15 percent) of the quantity filtered, thus establishing the existence of tubular potassium reabsorption. However, under certain conditions, the excreted quantity may actually exceed that filtered, thus establishing the existence of tubular potassium secretion. The subject is therefore complicated by the fact that potassium can be both reabsorbed and secreted by the tubule. However, normally most of the filtered potassium is reabsorbed regardless of changes in body potassium balance. In other words, the reabsorption of potassium does not seem to be controlled so as to achieve potassium homeostasis. The important result of this phenomenon is that changes in potassium *excretion* are due mainly to changes in potassium *secretion* in the more distal portions of the nephron (Fig. 13-21). Thus, during potassium depletion, when the homeostatic response is to reduce potassium excretion to a minimal level, there is no significant potassium secretion in the distal portions of the nephron, and only the small amount of potassium escaping reabsorption is excreted. In all other situations, to this same small amount of unreabsorbed potassium is added a variable amount of secreted potassium necessary to maintain potassium balance. Therefore, in describing the homeostatic control of potassium excretion, we may focus on only the factors which alter the rate of tubular potassium secretion by the distal portions of the nephron.

One of the most important of these factors is the potassium concentration of the renal tubular cells themselves. When a high-potassium diet is ingested, potassium concentration in most of the body's cells, including the cells of the renal tubule, increases. This higher concentration leads to a greater potassium diffusion from the cells of the tubule into the lumen and raises potassium excretion. Conversely, a low-potassium diet or a negative potassium balance, e.g., from diarrhea, lowers potassium concentration in the cells of the renal tubule; this reduces potassium entry into the lumen

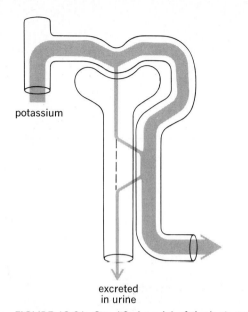

potassium

excreted
in urine

FIGURE 13-21. Simplified model of the basic renal handling of potassium. Net reabsorption occurs in the proximal portions of the nephron and net secretion in the more distal portions of the nephron. (However, during potassium depletion, the distal nephron reabsorbs potassium rather than secreting it.)

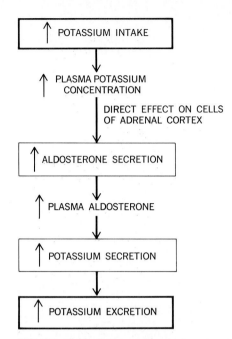

FIGURE 13-22. Pathway by which an increased potassium intake induces greater potassium excretion mediated by aldosterone.

and decreases potassium excretion, thereby helping to reestablish potassium balance.

A second important factor controlling potassium secretion is the hormone aldosterone which, besides stimulating tubular sodium reabsorption, increases tubular potassium secretion. The reflex by which changes in extracellular volume control aldosterone production is completely different from the reflex initiated by an excess or deficit of potassium. The former constitutes a complex pathway (discussed earlier), involving renin and angiotensin; the latter is much simpler (Fig. 13-22): The aldosterone-secreting cells of the adrenal cortex are, themselves, sensitive to the potassium concentration of the extracellular fluid bathing them. For example, an increased intake of potassium leads to an increased extracellular potassium concentration, which in turn directly stimulates aldosterone production by the adrenal cortex. This extra aldosterone circulates to the kidney, where it increases tubular potassium secretion and thereby eliminates the excess potassium from the body. Conversely, a lowered extracellular potassium concentration would decrease aldosterone production and thereby inhibit tubular potassium secretion. Less potassium than usual would be excreted in the urine, thus helping to restore the normal extracellular potassium concentration. Again, complete compensation depends upon the ingestion of additional potassium.

The control and renal tubular effects of aldosterone are summarized in Fig. 13-23.

Calcium Regulation

Extracellular calcium concentration is also normally held relatively constant, the requirement for precise regulation stemming primarily from the profound effects of calcium on neuromuscular excitability. A low calcium concentration increases the excitability of nerve and muscle cell membranes so that persons with diseases in which low calcium occurs suffer from **hypocalcemic tetany,** characterized by skeletal muscle spasms. Hypercalcemia is also life-threatening in that it causes cardiac ar-

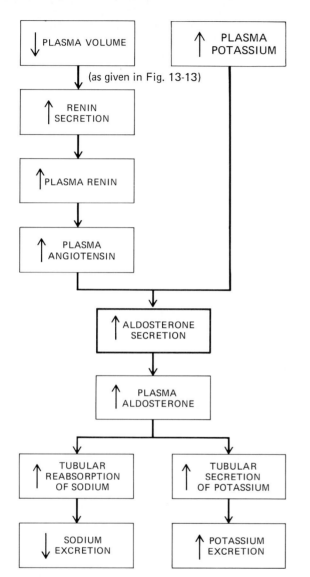

FIGURE 13-23. Summary of both the control of aldosterone and its functions.

rhythmias as well as depressed neuromuscular excitability.

Effector sites for calcium homeostasis

Earlier sections on sodium, potassium, and water homeostasis were concerned almost entirely with the renal handling of these substances. It was pos-

sible to do so for several reasons: (1) Although internal exchanges (between extracellular fluid on the one hand, and storage sites such as the matrix of bone and intracellular fluid, on the other) are of some importance for these substances, the major homeostatic controls act via the kidneys; and (2) absorption of these substances from the gut approximates 100 percent under normal circumstances and is not a major controlled variable. Neither of these statements holds true for calcium homeostasis, and so this section must deal not only with the renal handling of calcium but with the other two major effector sites for calcium homeostasis— bone and the gastrointestinal tract.

Bone. Approximately 99 percent of total body calcium is contained in the extracellular matrix of bone, which consists primarily of a framework of organic molecules (produced by bone cells) upon which calcium phosphate crystals are deposited. Contrary to popular opinion, bone is not a fixed, unchanging tissue but is constantly being remolded and, what is more important, is available for either the withdrawal or deposit of calcium from extracellular fluid.

Gastrointestinal tract. Under normal conditions a considerable amount of ingested calcium is not absorbed from the intestine and simply leaves the body along with the feces. Accordingly, control of the active transport system which moves this ion from gut lumen to blood can result in large increases or decreases in the amount of calcium absorbed.

Kidney. The kidney handles calcium by filtration and reabsorption. In addition, as we shall see, the renal handling of phosphate also plays an important role in the regulation of extracellular calcium.

Parathyroid hormone

All three of the effector sites described above are subject, directly or indirectly, to control by a hormone called **parathyroid hormone,** produced by the parathyroid glands. Parathyroid hormone production is controlled by the calcium concentration of the extracellular fluid bathing the cells of these glands. Lower calcium concentration stimulates parathyroid hormone production and release, and a higher concentration does just the opposite. It should be emphasized that extracellular calcium concentration acts directly upon the parathyroid glands (just as is true of the relation between extracellular potassium and aldosterone production) without any intermediary hormones or nerves.

Parathyroid hormone exerts at least four distinct effects influencing calcium homeostasis:

1. It increases the movement of calcium (and phosphate) from bone into extracellular fluid, making available this immense store of calcium for the regulation of extracellular calcium concentration.
2. It stimulates the activation of vitamin D (see below), and this latter hormone then increases intestinal absorption of calcium (and phosphate).
3. It increases renal tubular calcium reabsorption, thus decreasing urinary calcium excretion.
4. It reduces the renal tubular reabsorption of phospate, thus raising its urinary excretion and lowering extracellular phosphate concentration.

The adaptive value of the first three should be obvious (Fig. 13-24): They all result in a higher extracellular calcium concentration, thus compensating for the lower concentration that originally stimulated parathyroid hormone production. Conversely, an increase in extracellular calcium concentration inhibits normal parathyroid hormone production, thereby producing increased urinary and fecal calcium loss and net movement of calcium from extracellular fluid into bone.

The adaptive value of the fourth effect requires further explanation. Because of the solubility characteristics of undissociated calcium phosphate, the extracellular concentrations of ionic calcium and phosphate bear the following relationship to each other: The product of their concentrations, i.e., calcium times phosphate, is approximately a constant. In other words, if the extracellular concentration of phosphate increases, it forces the deposition of some extracellular calcium in bone, lowering the

calcium concentration and keeping the calcium phosphate product a constant. The converse is also true. Recall now that parathyroid hormone causes both calcium *and* phosphate to be released from bone into the extracellular fluid. If the phosphate concentration is allowed to increase, further movement of calcium from bone would be retarded; but in addition to this effect on bone, we have seen that parathyroid hormone also decreases tubular reabsorption of phosphate, thus permitting the excess phosphate to be eliminated in the urine. Indeed, extracellular phosphate may actually be reduced by this mechanism, which would allow even more calcium to be mobilized from bone.

Parathyroid hormone has other functions, notably a role in milk production, but the four effects discussed above constitute the major mechanisms by which it integrates the activities of various organs and tissues in the regulation of extracellular calcium concentrations.

Vitamin D

The term "vitamin D" denotes a group of closely related chemicals. One of these, called **vitamin D_3** is formed by the action of ultraviolet radiation on a substance (7-dehydrocholesterol) found in the skin (Fig. 13-25). Vitamin D_3, however, is inactive and must undergo metabolic changes (addition of hydroxyl groups), first in the liver and then in the kidneys, before it can influence its target cells (Fig. 13-25). The end result of these changes is **1,25-dihydroxyvitamin D_3**, abbreviated **1,25-$(OH)_2D_3$**. Dietary vitamin D must undergo the same metabolic changes to 1,25-$(OH)_2D_3$ as does the vitamin D_3 produced in the skin; thus, 1,25-$(OH)_2D_3$ is not a vitamin but is a hormone, since it is made in the body.

The major action of 1,25-$(OH)_2D_3$ is to stimulate active calcium absorption by the intestine. Thus, the major event in vitamin D deficiency is decreased gut calcium absorption, resulting in decreased plasma calcium. In children, the newly formed bone protein matrix fails to be calcified normally because of the low plasma calcium, leading to the disease **rickets.**

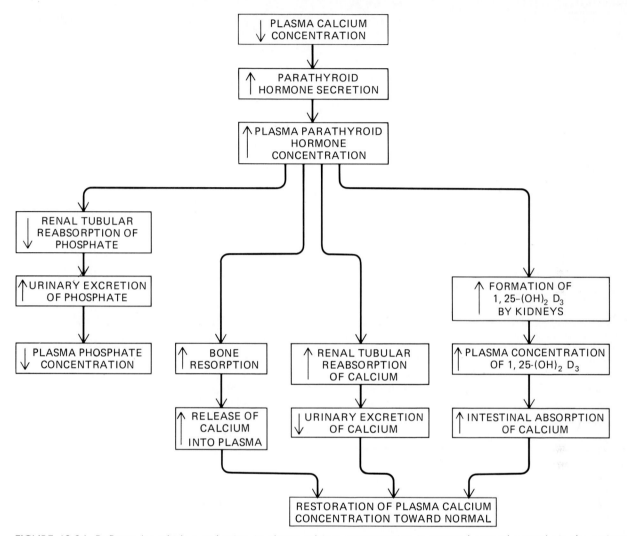

FIGURE 13-24. Reflexes by which a reduction in plasma calcium concentration is restored toward normal via the actions of parathyroid hormone and vitamin D [the active form of which is 1,25-$(OH)_2D_3$].

The blood concentration of 1,25-$(OH)_2D_3$ is subject to physiological control. The major control point is the final chemical change (addition of a hydroxyl group) which occurs in the kidneys. This step is stimulated by parathyroid hormone; thus a low plasma calcium concentration stimulates the secretion of parathyroid hormone which, in turn, enhances the production of 1,25-$(OH)_2D_3$, and both hormones contribute to the restoration of the plasma calcium toward normal.

Calcitonin

Yet a third hormone, **calcitonin**, has significant effects on plasma calcium. Calcitonin is secreted by cells in the thyroid gland which surround but are completely distinct from the cells which secrete thyroxine and triiodothyronine (Chap. 9). Calcitonin lowers plasma calcium primarily by inhibiting the release of calcium from bone. Its secretion is controlled directly by the calcium concentration of

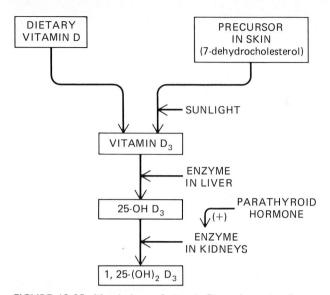

FIGURE 13-25. Metabolism of vitamin D to the active form: 1,25-$(OH)_2D_3$. The kidney enzyme which mediates the final step is activated by parathyroid hormone. The major action of 1,25-$(OH)_2D_3$ is to stimulate the absorption of calcium from the gastrointestinal tract.

the plasma supplying the thyroid gland; increased calcium causes increased calcitonin secretion. Thus, this system constitutes another feedback control over plasma calcium concentration, one that is opposed to the parathyroid hormone system. However, its overall contribution to calcium homeostasis is very minor compared with that of parathyroid hormone.

Hydrogen-Ion Regulation

Most metabolic reactions are highly sensitive to the hydrogen-ion concentration of the fluid in which they occur. This sensitivity is due to the influence on enzyme function exerted by the hydrogen ion, which changes the shapes of these molecules (Chap. 4). Accordingly, the hydrogen-ion concentration of the extracellular fluid is closely regulated.

Recall that the **hydrogen ion** is an atom of hydrogen which has lost its only electron. When dissolved in water, many compounds dissociate reversibly to produce negatively charged ions (anions) and hydrogen ions. It must be understood that the hydrogen-ion concentration of a solution refers only to the hydrogen ions which are free in solution, i.e., not bound to anions. Strong acids dissociate completely when they dissolve in water. For example, hydrochloric acid added to water gives

$$\text{HCl} \xrightarrow{\hspace{1cm}} \text{H}^+ + \text{Cl}^-$$
$$\text{\small Hydrochloric acid}$$

Virtually no hydrochloric acid molecules exist in the solution, only free hydrogen ions and chloride ions. On the other hand, weak acids are those which do not dissociate completely when dissolved in water. For example, when dissolved, only a fraction of lactic acid molecules at any instant are dissociated into lactate and hydrogen ions; these ions are in equilibrium with the undissociated lactic acid molecules. This characteristic of weak acids underlies an important chemical and physiological phenomenon, buffering.

Figure 13-26 pictures a solution made by dissolving lactic acid and sodium lactate in water. The sodium lactate dissociates completely into sodium ions and lactate ions, but only a very small fraction of the lactic acid molecules dissociate to form hydrogen ions and lactate ions. Accordingly, the solution has a relatively low concentration of hydrogen ions and relatively high concentrations of undissociated lactic acid molecules, sodium ions, and lactate ions. Lactic acid, hydrogen ion, and lactate are in equilibrium with each other:

$$\text{Lactic acid} \rightleftharpoons \text{lactate}^- + \text{H}^+$$

By the mass-action law, an increase in the concentration of any substance on one side of the arrows forces the reaction in the opposite direction. Conversely, a decrease in the concentration of any substance on one side of the arrows forces the reaction toward that side, i.e., in the direction which generates more of that substance. What happens if we add hydrochloric acid to this solution? Hydrochloric acid, a strong acid, completely dissociates and

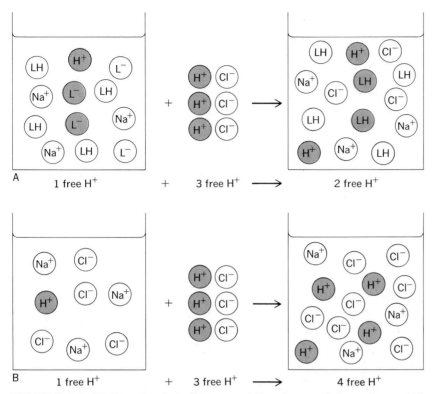

FIGURE 13-26. (A) Example of a buffer system. LH = lactic acid, L⁻ = lactate. When hydrochloric acid is added to the beaker, two of the hydrogen ions react with lactate to give lactic acid; therefore, only one of the three added hydrogen ions remains free in solution. (B) In contrast, when no buffer is present, all three of the added hydrogen ions remain free in solution.

liberates hydrogen ions. This increase in hydrogen ion concentration drives the lactic acid reaction to the left, and many of the hydrogen ions liberated by the dissociation of hydrochloric acid thus combine with lactate to give undissociated lactic acid. As a result, many of the hydrogen ions generated by the dissociation of hydrochloric acid do not remain free in solution but become incorporated into lactic acid molecules. The final hydrogen-ion concentration is therefore smaller than if the hydrochloric acid had been added to pure water or to a solution of sodium chloride (Fig. 13-26).

Conversely, if instead of adding hydrochloric acid we add a chemical which removes hydrogen ions (a **base**), the lactic acid reaction is driven to the right, lactic acid molecules dissociate to generate more hydrogen ions, and the original fall in hydrogen-ion concentration is minimized. This pro-

cess preventing large changes in hydrogen-ion concentration when hydrogen ion is added to or removed from a solution is termed **buffering**, and the chemicals (in this case lactate and lactic acid) are called **buffers** or **buffer systems**.

Body balance and homeostasis of hydrogen-ion can be analyzed in the same way as that for sodium, water, etc—i.e., in terms of input and output. The sources of input are diet and metabolism; the pathways for output are the kidneys and the gastrointestinal tract. As is the case with the other substances we have studied in this chapter (with the exception of calcium), the loss of hydrogen ion via the gastrointestinal tract is normally very small and is not regulated so as to achieve hydrogen-ion homeostasis (on the other hand, abnormalities of gastrointestinal function can lead to profound bodily losses or gains of hydrogen ion). Once again, it is the kidneys which are controlled so as to achieve a balance between gains and losses of hydrogen ion.

Generation of hydrogen ions by metabolism in the body

We ingest only a small quantity of acids (substances that can directly dissociate to release hydrogen ions) but we also ingest many organic nutrients which are not themselves acids but which yield acids within the body during their metabolism. There are three general types of such reactions:

1. Phosphorus and sulfur are present in large quantities in many proteins and other biologically important molecules. The catabolism of these molecules releases phosphoric and sulfuric acids into the extracellular fluid. These acids, to a large extent, dissociate into hydrogen ions and anions (phosphate and sulfate). For example,

$$H_2PO_4^- \longrightarrow H^+ + HPO_4^{2-}$$
$$H_2SO_4 \longrightarrow 2\ H^+ + SO_4^{2-}$$

2. Many organic acids, e.g., fatty acids and lactic acid, are produced in metabolic reactions and liberate hydrogen ion by dissociation.
3. We have already described (Chap. 12) the major source of hydrogen ion, namely, liberation of hydrogen ion from metabolically produced carbon dioxide via the reactions

$$CO_2 + H_2O \rightarrow H_2CO_3 \rightarrow H^+ + HCO_3^-$$
$$\text{Bicarbonate}$$

As described in Chap. 12, the lungs normally eliminate carbon dioxide from the body as rapidly as it is produced in the tissues. As blood flows through the lung capillaries and carbon dioxide diffuses into the alveoli, the chemical reactions which originally generated bicarbonate ion and hydrogen ion from carbon dioxide and water in the venous blood are reversed:

$$H^+ + HCO_3^- \rightarrow H_2CO_3 \rightarrow CO_2 + H_2O$$

As a result, all the hydrogen ion generated from carbonic acid is reincorporated into water molecules. Therefore, there is normally no net gain or loss of hydrogen ion in the body from this source, but what happens when lung disease prevents adequate elimination of carbon dioxide? The retention of carbon dioxide means an elevated extracellular P_{CO_2}. Some of the hydrogen ions generated from the reaction of this carbon dioxide with water are also retained and raise the extracellular hydrogen-ion concentration. This hydrogen ion must be eliminated from the body (via the kidneys) in order for hydrogen-ion balance to be maintained.

Buffering of hydrogen ions in the body

Between the generation of hydrogen ions in the body and their excretion by the kidneys, what happens? The hydrogen-ion concentration of the extracellular fluid is extremely small, approximately 0.00004 mmol/L (pH = 7.4). What would happen to this concentration if just 2 mmol of hydrogen ion remained free in solution following dissociation from an acid? Since the total extracellular fluid is approximately 12 L, the hydrogen-ion concentration would increase by 2 mmol/12 L, to a concentration of 0.167 mmol/L. Since the original concentration was only 0.00004 mmol/L, this would represent more than a 4000-fold increase. Obviously, such a rise does not occur, for we have already observed that the extracellular hydrogen ion is kept remarkably constant. Therefore, of the 2 mmol of hydrogen ion liberated in our example, only an extremely small portion can have remained free in solution. The vast majority has been bound (buffered) by other ions. The kidneys ultimately eliminate excess hydrogen ions, but it is buffering which minimizes changes in hydrogen-ion concentration until excretion occurs.

The most important body buffers are bicarbonate-CO_2, proteins, and intracellular phosphates. Recall that only free hydrogen ions contribute to the acidity of a solution. These buffers are all anions, which act by binding hydrogen ions according to the general reaction

$$\text{Buffer}^- + H^+ \rightleftharpoons H\text{—buffer}$$

It is evident that H—buffer is a weak acid in that it can exist as the undissociated molecule or can dissociate to buffer$^-$ + H$^+$. When H$^+$ concentration increases, the reaction is forced to the right and more H$^+$ is bound. Conversely, when H$^+$ concentration decreases, the reaction proceeds to the left, and H$^+$ is released. In this manner, the body buffers stabilize H$^+$ concentration against changes in either direction.

Bicarbonate-CO$_2$. In this section and Chap. 12 we have already described the relationships between HCO$_3^-$, H$^+$, and CO$_2$. Let us once more write the pertinent equations in their true forms as reversible reactions:

$$H^+ + HCO_3^- \rightleftharpoons H_2CO_3 \rightleftharpoons H_2O + CO_2$$

The basic mechanism by which this system acts as a buffer should be evident: An increased extracellular H$^+$ concentration drives the reaction to the right, H$^+$ and HCO$_3^-$ combine, H$^+$ is thereby removed from solution, and the H$^+$ concentration returns toward normal. Conversely, a decreased extracellular H$^+$ concentration drives the reaction to the left, CO$_2$ and H$_2$O combine to generate H$^+$, and this additional H$^+$ returns H$^+$ concentration toward normal. One reason for the importance of this buffer system is that the extracellular HCO$_3^-$ concentration is normally quite high and is closely regulated by the kidneys. A second reason stems from the relationship between extracellular H$^+$ concentration and CO$_2$ elimination from the body.

When additional H$^+$ is added to the extracel-lular fluid, i.e., when the H$^+$ combines with HCO$_3^-$, the extent to which this reaction can restore H$^+$ concentration to normal depends upon precisely how much additional H$^+$ actually combines with HCO$_3^-$ and is thereby removed from solution. Complete compensation (which never actually occurs) can be obtained only if the reaction proceeds to the right until all the additional H$^+$ has combined with HCO$_3^-$. However, this reaction obviously generates CO$_2$, which seriously hinders the further buffering ability of HCO$_3^-$ because any increase in the concentration of CO$_2$, by the mass-action law, tends to drive the reaction back to the left, thus preventing further net combination of H$^+$ + HCO$_3^-$. In reality, however, this expected increase in extracellular CO$_2$ does not occur. Indeed, during periods of increased body H$^+$ production from organic, phosphoric, or sulfuric acids, extracellular CO$_2$ concentration is actually *decreased*!

What causes this? We have already studied the mechanism, in Chap. 12: A greater extracellular H$^+$ concentration stimulates (via the carotid bodies) the respiratory centers to increase alveolar ventilation and thereby causes greater elimination of CO$_2$ from the body (Fig. 13-27). Thus, although an increased combination of H$^+$ and HCO$_3^-$ generates more CO$_2$, respiratory stimulation produced by the higher extracellular H$^+$ concentration results in the elimination of CO$_2$ even faster than it is generated. As a result, the extracellular CO$_2$ concentration decreases, and, by the mass-action law, the further combination of H$^+$ and HCO$_3^-$ is actually facilitated.

A lower extracellular H$^+$ concentration result-

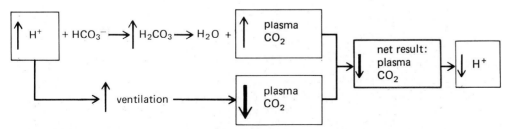

FIGURE 13-27. Effects of excess hydrogen ions on plasma carbon dioxide. The direct effect, by mass action, is to increase the production of carbon dioxide, but the indirect effect is to lower carbon dioxide by reflexly stimulating breathing. Since the latter effect predominates, the net effect is a reduction of plasma carbon dioxide.

ing from either decreased H^+ production or increased H^+ loss from the body is compensated for by just the opposite buffer reactions: (1) The lower H^+ concentration drives the HCO_3^--CO_2 reaction to the left, CO_2 and H_2O combine to generate H^+ + HCO_3^-, and this additional H^+ returns H^+ concentration toward normal; and (2) the lower H^+ concentration reflexly decreases alveolar ventilation and CO_2 elimination. The elevated extracellular CO_2 resulting from this process also serves to drive the HCO_3^--CO_2 reaction to the left, thus allowing further generation of H^+.

Renal regulation of extracellular hydrogen-ion concentration

It should be reemphasized that the buffers do not eliminate hydrogen ion from the body; instead they combine with the hydrogen ion. Binding thus removes the hydrogen ion from solution, preventing it from contributing to the free hydrogen-ion concentration, but the actual elimination of hydrogen ion from the body is normally performed only by the kidneys.

In a normal person the quantity of phosphoric, sulfuric, and organic acids formed within the body depends primarily upon the type and quantity of food ingested. A high-protein diet, for example, results in increased release of large quantities of sulfuric acid as the sulfur-containing amino acids are metabolized. The average American diet results in the liberation of 40 to 80 mmol of hydrogen ion each day. If hydrogen-ion balance is to be maintained, the same quantity must be eliminated from the body. This loss occurs via the kidneys. In addition, the kidneys must be capable of altering their hydrogen-ion excretion in response to changes in hyrogen-ion production, regardless of whether the source is carbon dioxide (in lung disease) or phosphoric, sulfuric, or organic acids. The kidneys must also be able to compensate for any gastrointestinal loss of hydrogen ion or of bicarbonate ion resulting from extensive vomiting or diarrhea, respectively.

Virtually all the hydrogen ion excreted in the urine enters the tubules via tubular secretion, the mechanism and its controls being quite complex. Suffice it to say that the presence of excess acid in the body induces an increased urinary hydrogen-ion excretion. Conversely, the kidney responds to a decreased amount of acid in the body by lowering urinary hydrogen-ion excretion. The controlling effect of the acid appears to be primarily exerted directly upon the cells of the tubules.

The ability of the kidneys to excrete hydrogen ion depends not only upon tubular hydrogen-ion secretion but upon the presence of buffers in the urine to combine with the secreted hydrogen ions (otherwise the urine acidity would become so great that secretion of hydrogen ions would cease). The major urinary buffers are HPO_4^{2-} and ammonia (NH_3):

$$HPO_4^{2-} + H^+ \longrightarrow H_2PO_4^-$$
$$NH_3 + H^+ \longrightarrow NH_4^+$$

The HPO_4^{2-} in the tubular fluid has been filtered and not reabsorbed. In contrast, the NH_3 of the tubular fluid is formed in the cells of the tubule by the deamination of certain amino acids supplied by the blood. From the cells the NH_3 diffuses into the lumen, where it combines with H^+ to form NH_4^+. Since the tubular-cell plasma membranes are quite permeable to NH_3 (a nonpolar substance) but not to NH_4^+, the NH_4^+, once formed, is trapped in the lumen and excreted. Another important feature of this system is that the rate of NH_3 production by the cells of the renal tubule increases whenever extracellular H^+ concentration remains elevated for more than 1 to 2 days; this extra NH_3 provides the additional buffers required for combination with the increased H^+ secreted.

Classification of disordered hydrogen-ion concentration

Acidosis refers to any situation in which the hydrogen-ion concentration of arterial blood is elevated; **alkalosis** denotes a reduction. It should be evident that acidosis and alkalosis are the results of an imbalance between hydrogen-ion gain and loss, and

TABLE 13-5. Changes in the arterial concentrations of hydrogen ion and carbon dioxide in acid-base disorders.

Primary disorder	H^+	CO_2	Cause of CO_2 change
Respiratory acidosis	↑	↑	Primary abnormality
Respiratory alkalosis	↓	↓	Primary abnormality
Metabolic acidosis	↑	↓	Reflex ventilatory compensations
Metabolic alkalosis	↓	↑	Reflex ventilatory compensations

all such situations fit into two distinct categories (Table 13-5): (1) **respiratory acidosis or alkalosis**; and (2) **metabolic acidosis or alkalosis**.

As its name implies, the first category results either from failure of the lungs to eliminate carbon dioxide as fast as it is produced (respiratory acidosis) or from elimination of carbon dioxide faster than it is produced (respiratory alkalosis). As described earlier, the imbalance of arterial hydrogen-ion concentrations in such cases is completely explainable in terms of mass action. Thus, the hallmark of a respiratory acidosis is elevated arterial concentrations of carbon dioxide and hydrogen ion, that of respiratory alkalosis is a reduction in both values.

The second category—metabolic acidosis or alkalosis—includes all situations other than those in which the primary problem is respiratory. Some common causes of metabolic acidosis are excessive production of lactic acid (during severe exercise or hypoxia) or of so-called ketone bodies (in uncontrolled diabetes mellitus or in fasting, Chap. 15). Metabolic acidosis can also result from excessive loss of bicarbonate, as in diarrhea. A frequent cause of metabolic alkalosis is persistent vomiting, with its associated loss of hydrogen ions (as HCl) from the stomach.

What is the arterial carbon dioxide in metabolic, as opposed to respiratory, acidosis or alkalosis? Since, by definition, metabolic acidosis and alkalosis must be due to something other than excess retention or loss of carbon dioxide, one might have predicted that arterial carbon dioxide would

be unchanged, but such is not the case. As described earlier in this chapter (and in Chap. 12) the elevated hydrogen-ion concentration associated with the metabolic acidosis, say due to retention of lactic acid, reflexly stimulates ventilation and lowers arterial carbon dioxide; by mass action this helps restore the hydrogen-ion concentration toward normal. Conversely, a person with metabolic alkalosis, say due to vomiting, will reflexly have ventilation inhibited; the result is a rise in arterial carbon dioxide and, by mass action, an associated restoration of hydrogen-ion concentration toward normal. To reiterate, the carbon dioxide changes in metabolic acidosis and alkalosis are not the cause of the acidosis or alkalosis but rather are compensatory reflex responses to primary nonrespiratory abnormalities.

Kidney Disease

The term "kidney disease" is no more specific than "car trouble," since many diseases affect the kidneys. Bacteria, allergies, congenital defects, stones, tumors, and toxic chemicals are some possible sources of kidney damage. Obstruction of the urethra or a ureter may cause injury due to a buildup of pressure and may predispose the kidneys to bacterial infection. Disease can attack the kidneys at any age. Experts estimate that there are at present more than 3 million undetected cases of kidney infection in the United States and that 25,000 to 75,000 Americans die of kidney failure each year.

Early symptoms of kidney disease depend

greatly upon the type of disease involved and the specific part of the kidney affected. Although many diseases are self-limited and produce no permanent damage, others progress if untreated. The end stage of progressive diseases, regardless of the nature of the damaging agent (bacteria, toxic chemical, etc.), is a shrunken, nonfunctioning kidney. Similarly, the symptoms of profound renal malfunction are relatively independent of the damaging agent and are collectively known as **uremia,** literally "urine in the blood."

The severity of uremia depends upon how well the impaired kidneys are able to preserve the constancy of the internal environment. Assuming that the person continues to ingest a normal diet containing the usual quantities of nutrients and electrolytes, what problems arise? The key fact to keep in mind is that the kidney destruction has markedly reduced the number of functioning nephrons. Accordingly, the many substances which gain entry to the tubule primarily by filtration are filtered in diminished amounts. Of the substances described in this chapter, this category includes sodium and chloride ions, water, and a number of waste products. In addition, the excretion of potassium and hydrogen ions is impaired because there are too few nephrons capable of normal tubular secretion. The buildup of all these substances in the blood causes many of the problems of uremia.

Other problems arise in uremia because of abnormal secretion of the hormones produced by the kidneys. Thus, decreased secretion of erythropoietin (Chap. 11) results in anemia. Decreased ability to form $1,25\text{-}(OH)_2D_3$ results in deficient absorption of calcium from the gastrointestinal tract, with a resulting decrease in plasma calcium and inadequate bone calcification. The problem with renin, the third of the renal hormones, is rarely too little secretion but rather too much by the damaged kidneys; the result is increased plasma concentration of angiotensin and the development of hypertension.

The artificial kidney and peritoneal dialysis

The artificial kidney is an apparatus that eliminates the excess ions and wastes which accumulate in the blood when the kidneys fail. Blood is pumped from one of the patient's arteries through tubing which is bathed by a large volume of fluid. The tubing then conducts the blood back into the patient by way of a vein. The tubing is generally made of a cellophane which is highly permeable to most solutes but relatively impermeable to protein—characteristics quite similar to those of capillaries. The bath fluid, which is constantly replaced, is a salt solution similar in ionic concentrations to normal plasma. The basic principle is simply that of **dialysis,** or diffusion. Because the cellophane is permeable to most solutes, as blood flows through the tubing, solute concentrations tend to reach diffusion equilibrium in the blood and bath fluid. Thus, if the plasma potassium concentration of the patient is above normal, potassium diffuses out of the blood into the bath fluid. Similarly, waste products and excesses of other substances diffuse across the cellophane tubing and thus are eliminated from the body. Many patients utilize an artificial kidney several times a week for years.

Another way of dialyzing the blood is to use not cellophane, but the lining of the person's own abdominal cavity (peritoneum). Fluid is injected into this cavity, allowed to remain there for a time during which solutes diffuse into the fluid from the person's blood, and then removed. This procedure can be performed repeatedly over long periods of time.

The major hope for patients with permanent renal failure is kidney transplantation. Although great strides have been made in assuring the success of the transplant in many patients, a major problem remains the frequent rejection of the transplanted kidney by the recipient's body (see Chap. 17).

THE DIGESTION AND ABSORPTION OF FOOD

The **gastrointestinal system** (Fig. 14-1) includes the gastrointestinal tract (mouth, pharynx, esophagus, stomach, small intestine, large intestine, and rectum), and glandular organs (salivary glands, liver, gallbladder, and pancreas) which secrete substances into the gastrointestinal tract. The function of the gastrointestinal system is to transfer organic molecules, salts, and water from the external environment to the internal environment, where they can be distributed to cells by the circulatory system.

The adult gastrointestinal tract is a tube approximately 4.5 m (15 ft) long, running through the body from mouth to anus. The lumen of the gastrointestinal tract, like the hole in a doughnut, is continuous with the external environment, which means that its contents are technically outside of the body. This fact is relevant to an understanding of some of the tract's properties. For example, the lower portion of the intestinal tract is inhabited by millions of bacteria, most of which are harmless and even beneficial in this location; however, if the same bacteria enter the body, as may happen, for example, in the case of a ruptured appendix, they may be extremely harmful and even lethal.

Most food is taken into the mouth as large particles, containing many high-molecular-weight substances such as proteins and polysaccharides,

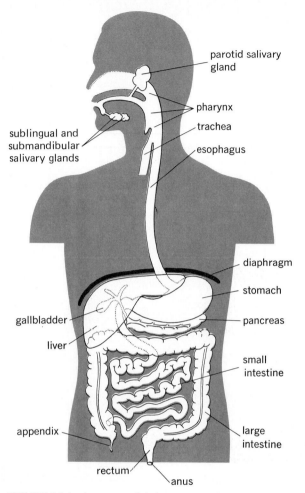

FIGURE 14-1. Anatomy of the gastrointestinal system.

enter the blood or lymph—**absorption**. While these processes are taking place, contractions of the smooth muscles in the wall of the gastrointestinal tract mix the contents of the lumen and move it through the tract from mouth to anus—**motility**. The functions of the gastrointestinal system can be described in terms of the four processes mentioned above—digestion, secretion, absorption, and motility (Fig. 14-2)—and the mechanisms controlling them.

Unlike most other organ systems, the gastrointestinal system does not have as its primary function the maintenance of some aspect of the internal environment (homeostasis) and, with several important exceptions, does not actually *regulate* the concentrations of nutrients, salts, and water in the internal environment by controlling the amounts absorbed. This regulation is primarily the function both of the kidneys, as we have already seen, and of a number of endocrine systems, as will be described in Chap. 15. The gastrointestinal system is designed to maximize the absorption of the substances in food and drink, and within certain very wide limits, it will absorb as much as is ingested.

Contrary to popular belief, the elimination of waste products from the body is only a minor function of the gastrointestinal system. Small amounts of certain metabolic end products are normally excreted via the gastrointestinal tract, but the elimination of most waste products from the internal environment is achieved by the lungs and kidneys. The material—**feces**—leaving the system at the end of the gastrointestinal tract consists mostly of bacteria and ingested material which was not digested and absorbed during its passage along the gastrointestinal tract, i.e., material that was never actually part of the internal environment.

Overview: Functions of the Gastrointestinal Organs

Digestion begins in the mouth with chewing, which breaks up large food particles into smaller particles that can be swallowed without choking. The salivary glands (Fig. 14-3) secrete into the mouth a mucus solution which moistens and lubricates the

which are unable to cross the wall of the gastrointestinal tract. Before these particles of food can be absorbed, they must be broken down into individual molecules, and many of these larger molecules must be further broken down into smaller molecules such as amino acids and monosaccharides. This breaking-down process—**digestion**—is accomplished mainly by the action of hydrochloric acid, bile, and a variety of digestive enzymes, many of which are released from the various exocrine glands of the system—by **secretion.** The molecules resulting from digestion move from the lumen of the gastrointestinal tract, across a layer of epithelial cells, and

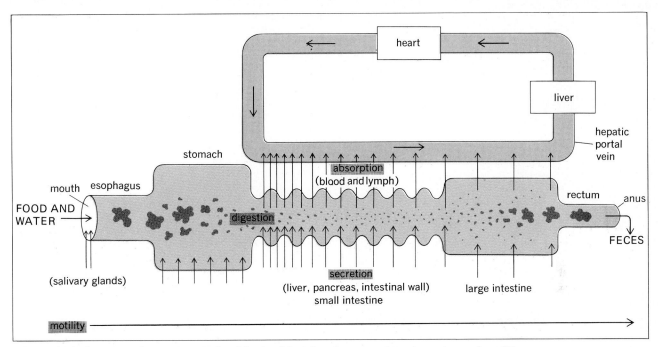

FIGURE 14-2. Summary of gastrointestinal activity involving digestion, secretion, absorption, and motility.

food particles prior to swallowing. (Mucus is also secreted by other gland cells throughout the gastrointestinal tract.) Saliva also contains a carbohydrate-digesting enzyme **amylase**, which breaks down large polysaccharides into smaller molecular fragments. A third function of saliva is to dissolve some of the molecules in the food particles; only in this dissolved state can they react with the chemoreceptors in the mouth, giving rise to the sensation of taste.

The next segments of the gastrointestinal tract, the pharynx and esophagus, contribute nothing to digestion but simply provide the pathway by which ingested food and drink reach the stomach. The motility of this region governs the process of swallowing.

The functions of the stomach are to store, dissolve, and partially digest the contents of a meal and to deliver fluid and partially digested food to the small intestine in amounts optimal for maximal digestion and absorption. Ingested food is stored in the stomach while it is being partially digested. The

glands lining the wall of the stomach secrete a very strong acid, **hydrochloric acid**, which breaks up the particles of food, forming a solution of molecules known as **chyme**. The high acidity of the chyme alters the ionization of carboxyl and amino groups in protein, thereby changing protein structure so as to break up connective tissue and cells in the ingested food. Hydrochloric acid thus continues the process begun by chewing, namely, reducing large particles of food to smaller particles and individual molecules; however, it has little ability to break down proteins and polysaccharides and it has virtually no digestive action on fats. A second function of gastric acid is to kill most of the bacteria that enter along with food. This process is not 100 percent effective, and some bacteria survive to take up residence and multiply in the intestinal tract, particularly the large intestine.

Also secreted by the stomach glands are several protein-digesting enzymes; these enzymes are collectively known as **pepsin;** they split proteins into small peptide fragments. Although polysaccha-

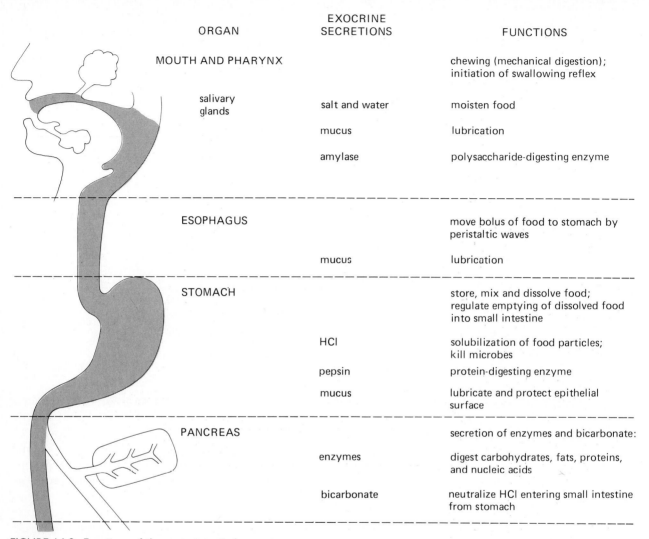

ORGAN	EXOCRINE SECRETIONS	FUNCTIONS
MOUTH AND PHARYNX		chewing (mechanical digestion); initiation of swallowing reflex
salivary glands	salt and water	moisten food
	mucus	lubrication
	amylase	polysaccharide-digesting enzyme
ESOPHAGUS		move bolus of food to stomach by peristaltic waves
	mucus	lubrication
STOMACH		store, mix and dissolve food; regulate emptying of dissolved food into small intestine
	HCl	solubilization of food particles; kill microbes
	pepsin	protein-digesting enzyme
	mucus	lubricate and protect epithelial surface
PANCREAS		secretion of enzymes and bicarbonate:
	enzymes	digest carbohydrates, fats, proteins, and nucleic acids
	bicarbonate	neutralize HCl entering small intestine from stomach

FIGURE 14-3. Functions of the gastrointestinal organs.

rides and proteins are partially digested while in the stomach, the end products are large, electrically charged molecules that still cannot cross the epithelium of the stomach wall. The fats in the gastric chyme, which are not soluble in water, aggregate into large fat droplets which also are not absorbed. Thus, little absorption of organic nutrients occurs across the wall of the stomach.

The final stages of digestion and most absorption occur in the small intestine; large molecules of intact or partially digested carbohydrate, fat, and protein are broken down by enzymes into monosaccharides, fatty acids, and amino acids. These digestive products are then able to cross the layer of epithelial cells which lines the intestinal wall and to enter the blood and/or lymph. Other organic nutrients (such as vitamins) as well as minerals (such as sodium and potassium) and water are also absorbed in the small intestine.

The exocrine portion of the pancreas secretes

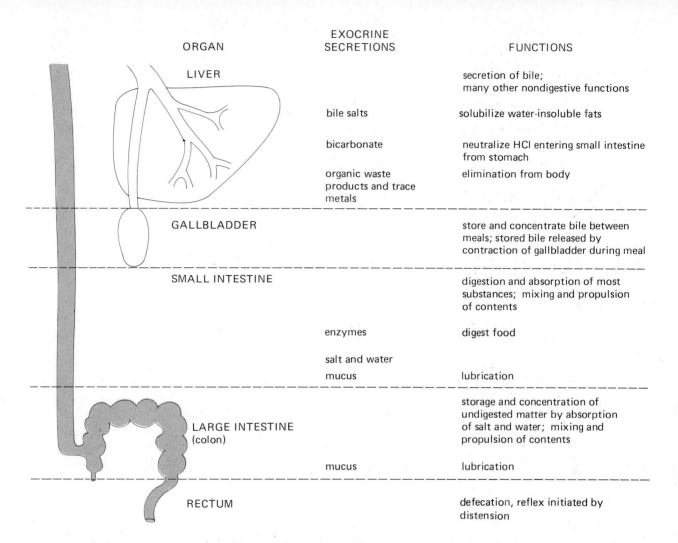

ORGAN	EXOCRINE SECRETIONS	FUNCTIONS
LIVER		secretion of bile; many other nondigestive functions
	bile salts	solubilize water-insoluble fats
	bicarbonate	neutralize HCl entering small intestine from stomach
	organic waste products and trace metals	elimination from body
GALLBLADDER		store and concentrate bile between meals; stored bile released by contraction of gallbladder during meal
SMALL INTESTINE		digestion and absorption of most substances; mixing and propulsion of contents
	enzymes	digest food
	salt and water	
	mucus	lubrication
LARGE INTESTINE (colon)		storage and concentration of undigested matter by absorption of salt and water; mixing and propulsion of contents
	mucus	lubrication
RECTUM		defecation, reflex initiated by distension

digestive enzymes specific for each of the major classes of organic molecules. These enzymes enter the small intestine through a duct leading from the pancreas to the first portion of the small intestine, the **duodenum**. The high acidity of the chyme coming from the stomach would inactivate these enzymes if the hydrochloric acid were not neutralized by bicarbonate ions also secreted in large amounts by the exocrine pancreas.

Because fat is insoluble in water, the digestion of fat in the small intestine requires special processes to solubilize these molecules. This is brought about by a group of detergent molecules, known as **bile salts**, which are secreted by the liver into bile ducts, which eventually join the pancreatic duct and empty into the duodenum. In addition to bile salts, bile contains bicarbonate ions which help to neutralize acid from the stomach. Between meals, secreted bile is stored in a small sac underneath the liver, the **gallbladder**, which concentrates the bile by absorbing salts and water. During a meal, the gallbladder contracts, causing a concentrated solution of bile to be injected into the small intestine.

Monosaccharides and amino acids are absorbed across the wall of the small intestine, mainly by specific carrier-mediated active-transport processes in the epithelial membranes, while fatty acids enter the epithelial cells by diffusion. Minerals are also actively absorbed, and water follows passively down osmotic gradients. Most digestion and absorption has been completed by the middle portion of the small intestine.

The motility of the small intestine mixes the contents of the lumen with the various secretions, brings the contents into contact with the epithelial surface where absorption takes place, and slowly advances the luminal material toward the large intestine. Since most substances are absorbed in the early portion of the small intestine, only a small volume of minerals, water, and undigested material is passed on to the large intestine, which temporarily stores the undigested material (some of which is acted upon by bacteria) and concentrates it by absorbing water. Defecation is initiated by the distension of the last segment of the gastrointestinal tract, the rectum. Contractile activities of the rectum and associated sphincter muscles result in defecation.

It must be recognized that although the average adult consumes about 800 g of food and 1200 mL of water per day, this is only a fraction of the total material entering the gastrointestinal tract. To the approximately 2000 mL of ingested food and drink are added an additional 7000 mL of fluid from the salivary glands, stomach, pancreas, liver, and intestinal tract (Fig. 14-4). Only about 100 mL of the 9000 mL of fluid that enters the gastrointestinal tract each day is lost in the feces, the rest being absorbed into the blood. Almost all of the secreted salts and enzymes in this fluid (the latter after themselves being digested) are also absorbed into the circulation. Thus, very little of the huge volume of secreted fluid is normally lost from the body. However, vomiting and diarrhea can lead to significant losses of secreted fluid.

Because of the large number of organs and the variety of activities taking place in the gastrointestinal system (Fig. 14-3) it is easy to lose sight of the fact that the overall function of this system is to get ingested nutrients digested and absorbed. The basic purpose of the many substances secreted, the variety of smooth muscle contractile patterns, and the complex controls over these processes is the achievement of optimal conditions within the tract for digestion and absorption.

This completes our overview of the basic functions of the individual organs of the gastrointestinal system. Since the major task of the system is the digestion and absorption of ingested as well as secreted substances, we shall begin our more detailed descriptions with these processes. Subsequent sections will then describe, organ by organ, the regulation of secretions and motility which produce the optimal conditions for digestion and absorption. However, a prerequisite for all this physiology is a knowledge of the general structure of the gastrointestinal tract.

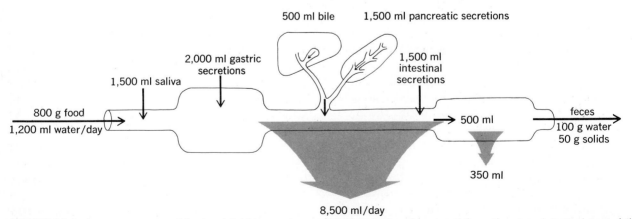

FIGURE 14-4. Average amounts of food and fluid ingested, secreted, absorbed, and excreted from the gastrointestinal tract daily.

General Structure of the Gastrointestinal Tract

Throughout its entire length, the wall of the gastrointestinal tract has the general structural organization illustrated in Fig. 14-5. The luminal surface of the tube is generally not flat and smooth, but highly convoluted, with many ridges and valleys which greatly increase the total surface area available for absorption. From the stomach on, this surface is covered by a single layer of epithelial cells, across which absorption takes place. Included in this epithelial layer are cells which secrete mucus and other cells which release hormones into the blood. Invaginations of this epithelial layer into the underlying tissue form tubular exocrine glands which secrete mucus, acid, enzymes, water, and ions into the lumen. Just below the epithelial surface is a layer of connective tissue, the **lamina propria**, through which pass small blood vessels, nerve fibers, and lymphatic ducts. The lamina propria is separated from the underlying tissues by a thin layer of smooth muscle, the **muscularis mucosa**. The combination of these three layers—the epithelium, lamina propria, and muscularis mucosa—is known as the **mucosa**.

Beneath the mucosa is a second connective tissue layer, the **submucosa**, through which pass the larger blood vessels (and lymphatics) whose branches penetrate into the overlying mucosal layer and the underlying muscular layer. Also located in this layer is a network of nerve cells known as the **submucus plexus**.

Beneath (i.e., surrounding) the submucosa is a layer of smooth muscle tissue, the **muscularis externa**, contractions of which provide the forces for moving the gastrointestinal contents. The muscularis externa consists of two layers—a relatively thick inner layer of **circular muscle**, whose fibers are oriented in a circular pattern around the tube such that contraction produces a narrowing of the lumen, and a thinner, outer layer of **longitudinal muscle**, whose contraction shortens the tube. Between these two layers of smooth muscle is another, more extensive, plexus of nerve cells, the **myenteric plexus**.

Finally, surrounding the outer surface of the tube is a layer of connective tissue, the **serosa**. Thin sheets of connective tissue, the **mesenteries**, connect the serosa to the abdominal wall to support segments of the gastrointestinal tract in the abdominal cavity. At various points along the tube, the secretions of exocrine glands lying outside the gastrointestinal tract (the salivary glands, liver, and pancreas) are delivered to the lumen of the tract via ducts.

Structure of the absorptive surface of the small intestine

Most absorption from the gastrointestinal tract occurs across the walls of the small intestine, whose mucosal surface is highly specialized for this absorptive process. The small intestine is a tube approximately 4 cm (1½ in) in diameter, leading from the stomach to the large intestine. This tube, 275 cm (9 ft) in length, is divided into three segments: an initial short 20-cm (8-in) segment, the **duodenum**, followed by the **jejunum**, and ending with the longest segment, the **ileum**. Normally, most absorption occurs in the first quarter of the small intestine, i.e., in the duodenum and jejunum. Thus, the intestine has a considerable functional reserve, making it almost impossible to exceed its absorptive capacity even when very large quantities of food are ingested.

The mucosa of the small intestine is highly folded, and the surface of these folds is further convoluted by fingerlike projections known as **villi**. The surface of each villus is covered with a single layer of epithelial cells whose surface membranes form small projections known as **microvilli** (Fig. 14-6). The combination of folded mucosa, villi, and microvilli increases the total surface area of the small intestine available for absorption about 600-fold over that of a flat-surfaced tube having the same length and diameter. The total surface area of the human small intestine is estimated to be about 300 m², or equivalent to the area of a tennis court. The folds of the intestinal mucosa are not permanent structures but change their pattern with contractions of the underlying muscularis mucosa.

A

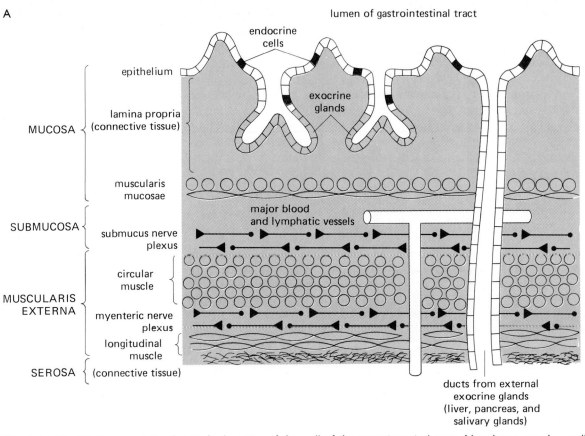

FIGURE 14-5. (A) Representative longitudinal section of the wall of the gastrointestinal tract. Not shown are the smaller blood vessels, neural connections between the two nerve plexuses, and neural terminations on muscles and glands. (B) A generalized view of the small intestine.

The structure of a villus is illustrated in Fig. 14-7. The center of each villus is occupied by a single blind-ended lymph vessel, a **lacteal**, and by a capillary network which branches from an arteriole and drains into a venule. The villi of the intestine move back and forth independently of each other, propelled by a small number of smooth muscle fibers in each of them. The motion of the villi is increased following a meal, and associated with their increased motion is a greater flow of lymph as the lacteals are "pumped," i.e., alternately compressed and released.

The plasma membranes of adjacent epithelial cells in the intestinal tract are joined near their luminal surfaces by tight junctions (Fig. 14-6) which reduce the diffusion of substances through the extracellular spaces between the cells. The leakiness of these tight junctions varies in different regions of the gastrointestinal tract, the tight junctions in the stomach and large intestine forming much more complete seals than those in the small intestine. When the tight junctions are permeable to a particular substance, a net diffusion of molecules in either direction by way of the extracellular spaces between cells will occur whenever there is a concentration difference, an electrical difference (in the case of charged particles), or a pressure difference across the epithelium.

The epithelial cells are continually being replaced by new cells which arise from the mitotic

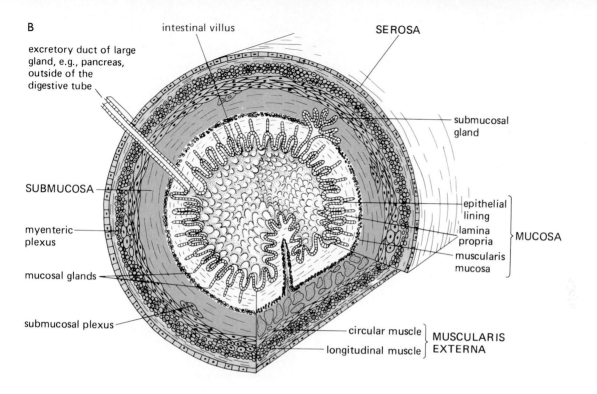

B

excretory duct of large gland, e.g., pancreas, outside of the digestive tube

intestinal villus

SEROSA

submucosal gland

SUBMUCOSA

myenteric plexus

mucosal glands

submucosal plexus

epithelial lining

lamina propria

muscularis mucosa

} MUCOSA

circular muscle

longitudinal muscle

} MUSCULARIS EXTERNA

activity of the cells at the base of the villi. These new cells differentiate and migrate to the top of the villi, replacing older cells, which disintegrate and are discharged into the lumen of the intestine. These disintegrating epithelial cells release into the lumen intracellular enzymes which contribute to the digestive process. The entire epithelium is replaced approximately every 5 days, and this continuous discharge of cells amounts to about 250 g/day. [It is because of this rapid cell turnover that the lining of the intestinal tract is so susceptible to damage by agents which inhibit cell division (such as radiation or anti-cancer drugs.)]

The blood flow to the small intestine at rest averages about 1 L/min (one-fifth the resting cardiac output), but it increases during periods of digestive activity as a result both of local autoregulation (produced by the increased metabolic activity of the intestinal cells) and of reflexes triggered by mechanical distension of the lumen by chyme.

Finally, as described in Chap. 11, the venous drainage from the intestine (as well as from the pancreas and portions of the stomach) is unusual in that the blood does not empty directly into the vena cava but passes first to the liver via the **hepatic portal vein**; there it flows through a second capillary network before leaving the liver to return to the heart. Thus, material absorbed into the capillaries of the intestine may be processed by the liver before entering the general circulation.

Digestion and Absorption

Carbohydrate

The daily intake of carbohydrate varies considerably, ranging from 250 to 800 g/day in a typical American diet. About two-thirds of this carbohydrate (Table 14-1) is in the form of the plant polysaccharide **starch**; the remainder consists mostly of the disaccharides **sucrose** (table sugar) and **lactose** (milk sugar). Only small amounts of monosaccharides are normally present. The polysaccharide **cellulose**, from vegetable matter, cannot be broken

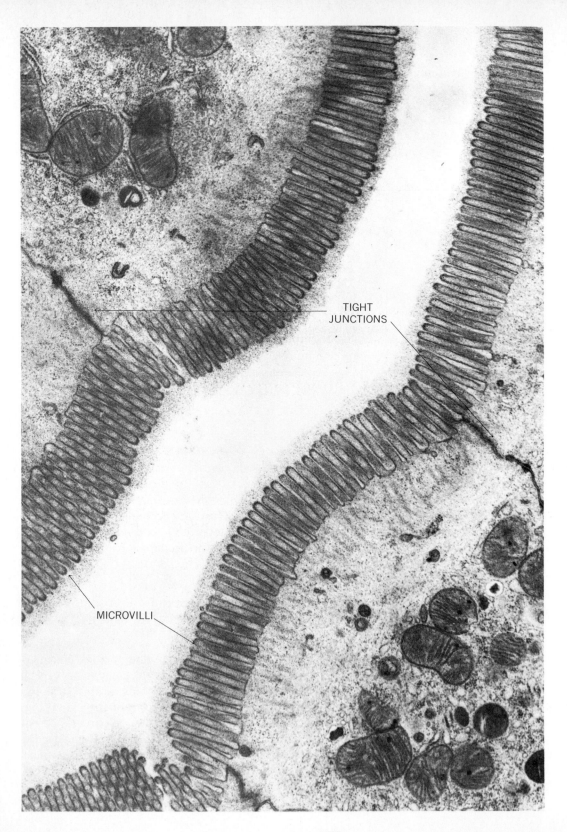

TIGHT
JUNCTIONS

MICROVILLI

down by the enzymes in the small intestine; hence it is passed on to the large intestine, where it is partially digested by the bacteria that reside there. In infants, lactose from milk makes up the greater part of the carbohydrate in the diet.

Starch is partially digested by salivary amylase, and this digestion is continued in the small intestine by pancreatic amylase. The products formed by these amylases are the disaccharide **maltose** and short chains composed of about eight glucose molecules. The enzymes that then split these molecules, as well as the disaccharides sucrose and lactose, into monosaccharides are integral membrane proteins located in the plasma membranes of the microvilli of the epithelial cells lining the small intestine.

The monosaccharides—glucose and galactose—liberated from this breakdown of polysaccharides and disaccharides are actively transported across the intestinal epithelium into the blood. For most sugars, this is a secondary active-transport system requiring a sodium gradient for the cotransport of the sugar into the epithelial cells, followed by the movement of the sugar from the cytosol to the interstitial fluid by facilitated diffusion.

Following a normal meal, most of the carbohydrate ingested is digested and absorbed within the first 20 percent of the small intestine. *Unlike* that of the kidney, the carbohydrate absorptive capacity of the small intestine is practically impossible to saturate.

Protein

An intake of about 50 g of protein each day is required by a normal adult to supply essential amino acids and replace the amino acid nitrogen converted to urea. (A typical American diet contains about 125 g of protein.) In addition to the protein in the diet, a large amount of protein (mostly enzymes) is secreted into the gastrointestinal tract by the various glands or enters it via the disintegra-

tion of epithelial cells. Most of this protein is broken down into amino acids and absorbed by the small intestine.

Proteins are broken down to peptide fragments in the stomach by pepsin and in the small intestine by **trypsin** and **chymotrypsin** which are secreted by the pancreas. The peptide fragments generated are further digested to free amino acids by **carboxypeptidase** from the pancreas and **aminopeptidase**, which is located in the epithelial membranes of the small intestine. (These enzymes' names designate the fact that they split off amino acids from the carboxyl and amino ends of the peptide chains, respectively.) The free amino acids are actively transported across the walls of the intestine, which is followed by their diffusion into the capillaries in the intestinal villi. Several different carrier systems are available for transporting different classes of amino acids. These transport carriers are sodium-dependent secondary-transport systems which operate on the same principles used to absorb carbohydrates. In addition to single amino acids, short chains of two or three amino acids are also actively absorbed. (This is in contrast to carbohydrate absorption, where disaccharides are not absorbed.) As for carbohydrates, most of the digestion and absorption of proteins is completed in the early portion of the small intestine.

Small amounts of intact proteins are able to cross the epithelium and gain access to the interstitial fluid without being digested. In this process, proteins are engulfed by the plasma membrane of the epithelial cells (endocytosis), move through the cytoplasm, and are released on the opposite side of the cell by the reverse process (exocytosis). The capacity to absorb intact proteins is greater in a newborn infant than in adults, and antibodies (proteins involved in the immunological defense system of the body, Chap. 17) secreted in the mother's milk may be absorbed by the infant in this manner, providing a short-term passive immunity until the child begins to produce its own antibodies.

FIGURE 14-6. Microvilli on the surface of intestinal epithelial cells. *[From D. W. Fawcett,* J. Histochem. Cytochem. *13:75–91 (1965). Courtesy of Susumo Ito.]*

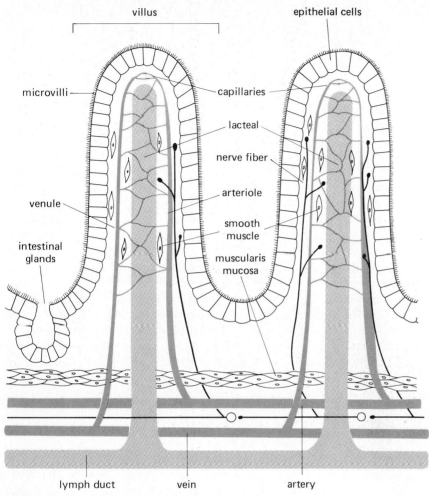

FIGURE 14-7. Structure of intestinal villi.

Fat

The amount of fat in the diet varies from about 25 to 160 g/day. Most of this fat is in the form of triacylglycerols, and its digestion occurs entirely within the small intestine. The major digestive enzyme in this process is pancreatic **lipase**, which catalyzes the splitting of the bonds linking the fatty acids to the first and third carbon atoms of glycerol, producing free fatty acids and monoglycerides as products.

$$\text{Triacylglycerols} \xrightarrow{\text{lipase}} \text{monoglycerides} + \text{fatty acids}$$

TABLE 14-1. Carbohydrates

Class	Examples	Made up of
Polysaccharides	Starch	Glucose
	Cellulose	Glucose
	Glycogen	Glucose
Disaccharides	Sucrose	Glucose-fructose
	Lactose	Glucose-galactose
	Maltose	Glucose-glucose
Monosaccharides	Glucose	
	Fructose	
	Galactose	

The triacylglycerols entering the small intestine are insoluble in water and are aggregated into large lipid droplets. Since only the lipids at the surface of the droplet are accessible to the water-soluble lipase, digestion in this state would proceed very slowly; furthermore, the products of lipase action (fatty acids and monoglycerides) are themselves insoluble in water. The role of bile salts, supplied by the liver, is to eliminate both of these problems: Bile salts break up the large lipid droplets into a number of smaller lipid droplets, a fat-solubilizing process known as **emulsification**, and they combine with the fatty acids and monoglycerides (produced by the action of lipase) to form water-soluble particles known as **micelles**.

Bile salts are formed from the nonpolar steroid molecule cholesterol, to which are attached several polar hydroxyl groups and a short carbon chain having a terminal ionized carboxyl group (Fig. 14-8). All of the polar and ionized groups are located on one side of the flat steroid ring, resulting in an amphipathic molecule having a polar and a nonpolar surface. The several bile salts secreted by the liver differ in the number of hydroxyl groups and the type of ionized side chain present, but all interact with lipids in a similar way; the nonpolar side of the steroid ring dissolves in the surface of a nonpolar lipid droplet, leaving the polar side exposed at the surface.

As mechanical agitation in the intestine breaks up the large fat droplets, the resulting smaller droplets become coated with bile salts. Because the polar and ionized groups on the bile salts at the surface of these droplets repel each other, the droplets are prevented from coalescing back into larger droplets. The resulting suspension of small lipid droplets, each about 1 μm in diameter, is known as an **emulsion** (Fig. 14-9). Because the surface area of lipid accessible to lipase in an emulsion is much greater than it is in a single large droplet, the rate of lipid digestion is increased by emulsification.

Although digestion is speeded up by emulsification, absorption of the insoluble products of the lipase reaction would be very slow if it were not for the second action of bile salts, the formation of micelles. Micelles are similar to the emulsion drop-

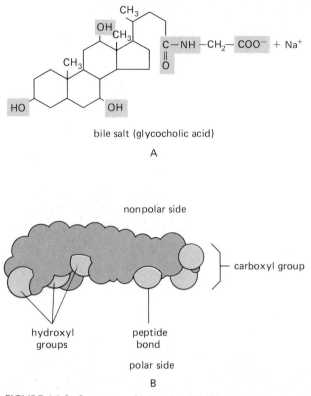

bile salt (glycocholic acid)

A

nonpolar side

carboxyl group

hydroxyl groups

peptide bond

polar side

B

FIGURE 14-8. Structure of bile salts. (A) Chemical formula of glycocholic acid, one of several bile salts secreted by the liver (polar groups in color). (B) Three-dimensional structure of a bile salt, showing the polar and nonpolar surfaces of the molecule.

lets, but they are much smaller. (Whereas a lipid emulsion appears cloudy because of the relatively large size of its emulsion droplets, a solution of micelles is perfectly clear.) Micelles consist of bile salts, fatty acids, phospholipids, and monoglycerides, all clustered together with the polar ends of each molecule oriented toward the micelle's surface and the nonpolar portions forming the core of the micelle.

How do these micelles facilitate absorption? Although free fatty acids and monoglycerides have an extremely low solubility in water, a few individual molecules do exist free in solution, and it is in this form that they are absorbed across the plasma membrane by diffusion (because of their high degree of solubility in the phospholipid bilayer of the

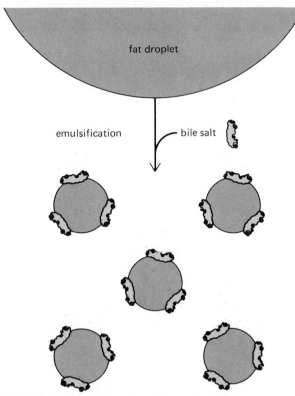

FIGURE 14-9. Emulsification of fat by bile salts.

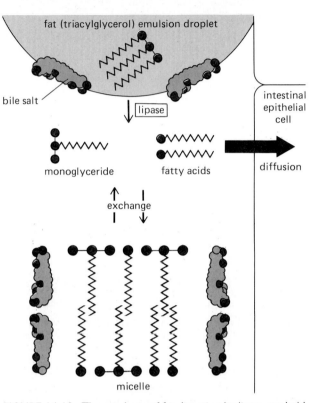

FIGURE 14-10. The products of fat digestion by lipase are held in solution in the micellar state, combined with bile salts. They rapidly exchange with the free products in aqueous solution, which are able to diffuse into intestinal epithelial cells.

plasma membrane). The fatty acids and monoglycerides in the micelles are in equilibrium with those in free solution, molecules being continuously exchanged between the two states. As the concentration of free lipids falls because of their diffusion into epithelial cells, there is a shift of more lipids out of the micelles into the free phase (Fig. 14-10). Thus, the micelles provide a means of storing large quantities of the insoluble fat digestion products in such a way that they can readily become available (through equilibrium with the free lipids) for absorption. In the absence of bile salts and, hence, without the resulting lipid emulsification and micelle formation, fat absorption occurs so slowly that much of the ingested fat is passed on to the large intestine and excreted in the feces.

Although fatty acids and monoglycerides enter the epithelial cells from the intestinal lumen, it is triacylglycerol that is released on the other side of

the cell into the interstitial fluid. During their passage through the epithelial cells, the products of luminal fat digestion are resynthesized into triacylglycerols. This occurs in the agranular endoplasmic reticulum, where the enzymes for triacylglycerol synthesis are located. Within the lumen of the endoplasmic reticulum the resynthesized fat aggregates into small droplets that are coated with an amphipathic protein which performs an emulsifying function similar to that of a bile salt. The release of this fat droplet from the cell follows the same pathway as does a secreted protein; vesicles containing the droplet pinch off the endoplasmic reticulum, are processed through the Golgi apparatus, and eventually fuse with the plasma membrane, releasing the fat droplet into the interstitial fluid. These small fat droplets are known as **chylomicrons**. In addition to triacylglycerols, chylomicrons contain other lipids (including phospholipids, cholesterol, and fat-soluble vitamins) that have been absorbed by the fat-absorption pathway.

The chylomicrons pass into the lacteals (lymph vessels) located in each villus rather than into the capillaries. The chylomicrons are unable to enter the capillaries because a **basement membrane** (an extracellular layer composed of polysaccharides) surrounds the outer surface of the capillary and provides a barrier to the chylomicrons. The lacteals do not have such a basement membrane, and thus, fat is absorbed into the lymphatic system, which eventually empties into systemic veins which carry it directly to the heart. (In contrast, most other substances absorbed from the intestinal tract enter the capillaries and pass through the liver by way of the hepatic portal vein before returning to the heart.) The chylomicrons circulating in the blood following a meal form a lipid emulsion. In the next chapter we shall describe how the lipids in the chylomicrons are made available to the cells of the body. Figure 14-11 summarizes the pathway taken by fat in moving from the lumen into the lymphatic system.

Vitamins

The fat-soluble vitamins, A, D, E, and K, are released from food by the digestive juices of the stomach and become dissolved in the fat droplets that pass into the small intestine, where they follow the pathway for fat absorption just described. Thus, any interference with the secretion of bile or the action of bile salts in the intestine decreases the absorption of the fat-soluble vitamins.

Most water-soluble vitamins are absorbed by diffusion or carrier-mediated transport across the wall of the intestinal tract. However, one—vitamin B_{12}—is a very large, charged molecule. In order to be absorbed, vitamin B_{12} must combine with a special protein, known as **intrinsic factor**. This protein, secreted by the acid-secreting cells of the stomach, binds vitamin B_{12}, and the bound complex in turn is bound to specific sites on the luminal surface of epithelial cells in the lower portion of the ileum, where vitamin B_{12} is absorbed into the blood. Vitamin B_{12} is required for red blood cell formation (see Chap. 11), and when it is deficient, a form of anemia known as **pernicious anemia** develops. This may occur when the stomach has been removed (as for example, to treat ulcers or gastric cancer) or when intrinsic factor is not secreted by the stomach. Since the absorption of vitamin B_{12} occurs specifically in the lower part of the ileum, removal of this segment of the intestine can also result in pernicious anemia.

Nucleic acids

Small amounts of nucleic acids—DNA and RNA—are present in food and are broken down to nucleotides by enzymes secreted by the pancreas. Enzymes in the intestinal epithelial membranes act upon these nucleotides, releasing free bases and monosaccharides, which are then actively transported across the epithelium by specific carrier systems.

Water and minerals

Water is the most abundant substance in the intestinal chyme. Approximately 9000 mL of ingested and secreted fluid enters the small intestine each day, but only 500 mL is passed on to the large intestine, since 95 percent of the fluid is absorbed in the small intestine. The membranes of the epithelial cells, as well as the tight junctions between the cells, are very permeable to water. Therefore, a net diffusion of water (osmosis) occurs across the epithelium whenever a water concentration gradient is established as a result of differences in the total solute concentration (osmolarity) on the two sides. This difference in osmolarity is produced by the active transport of solutes. In other words, active solute absorption indirectly causes passive water absorption. (This mechanism of coupling water movement to the transport of solutes has already been described in Chap. 6 and illustrated in the case of the kidney in Chap. 13.) Although, as described above, many organic solutes are actively absorbed across the epithelium, sodium ions account for most of the total solute actively trans-

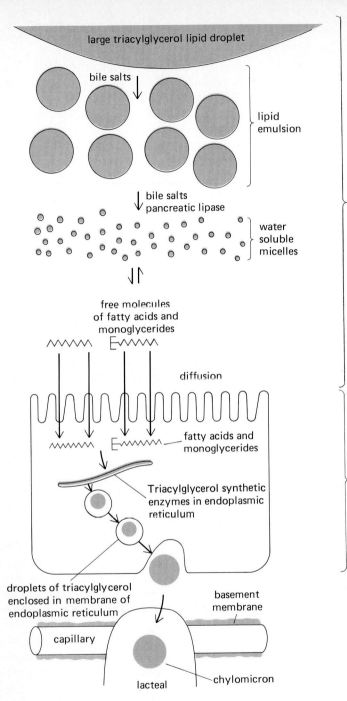

large triacylglycerol lipid droplet

bile salts

lipid emulsion

bile salts
pancreatic lipase

water soluble micelles

lumen of intestinal tract

free molecules of fatty acids and monoglycerides

diffusion

fatty acids and monoglycerides

Triacylglycerol synthetic enzymes in endoplasmic reticulum

epithelial cell

droplets of triacylglycerol enclosed in membrane of endoplasmic reticulum

basement membrane

capillary

lacteal   chylomicron

FIGURE 14-11. Summary of fat absorption across the walls of the small intestine.

ported because they constitute the majority of the total solute particles in the intestinal chyme.

The net absorption of water has an important effect upon the absorption of other substances which cross the epithelium by diffusion. As water is absorbed, the volume of the luminal contents decreases, thereby concentrating any solutes not absorbed at the same rate. This rise in concentration secondary to water absorption provides the concentration gradient for the net diffusion of many substances from intestinal lumen into the blood.

In addition to sodium and chloride ions, other

minerals present in smaller concentrations, such as potassium, magnesium, and calcium, are also absorbed, as are trace elements such as iron, zinc, and iodide. Consideration of the transport processes associated with all of these is beyond the scope of this book, and we shall briefly consider as examples the absorption of only three of them—potassium, calcium, and iron.

Potassium. Potassium moves across the intestinal epithelium by simple diffusion. As water is absorbed, the resulting rise in luminal potassium concentration produces a concentration gradient for potassium absorption. Interference with the intestine's ability to absorb water (as occurs in certain forms of diarrhea) diminishes potassium absorption and, in fact, can result in net movement of potassium from the interstitial fluid into the lumen. This secretion of potassium is explained by a small electric potential across the epithelial cells such that the luminal surface is negatively charged relative to the interstitial surface, and this electric force moves potassium into the lumen. If the driving force of this electrical gradient is greater than the concentration gradient which produces absorption, a net secretion of potassium into the lumen will occur.

Calcium. We have repeatedly emphasized that for most substances, especially the major organic nutrients, the total amount of the substance delivered to the small intestine is absorbed; i.e., the amount of nutrient absorbed is not regulated by the nutritional state of the body. The story for calcium is quite different in that the amount absorbed is controlled by reflexes that respond to changes in plasma calcium concentration. As described in Chap. 13, this regulation is mediated by the hormone 1,25-dihydroxyvitamin D_3; decreases in the plasma concentration of calcium lead to increased production of this hormone and to enhancement of calcium active transport across the intestinal epithelium.

Iron. Ferrous (Fe^{2+}) and, to a lesser extent, ferric (Fe^{3+}) ions are actively transported into intestinal epithelial cells, where most are incorporated into **ferritin,** a protein-iron complex which functions as an intracellular store of iron. Some of the absorbed iron is released from the cell and becomes bound to a specific protein in the plasma, **transferrin,** in which form it circulates throughout the body. The absorption of iron, like that of calcium, varies, depending upon the iron content of the body. When body stores are ample, the amount of iron bound to ferritin is increased and this somehow reduces further iron absorption. When the body stores of iron drop, as a result of hemorrhage, for example, so does the amount of iron bound to intestinal ferritin and this increases iron absorption. The absorption of iron is typical of that of most other trace metals in several respects: Cellular storage proteins and plasma carrier proteins are involved, and the control of absorption is the major mechanism for the homeostatic control of the body's content of the trace metal.

Iron absorption also depends on the type of food in which it is contained and the foods ingested along with it. This is because the iron becomes bound to negatively charged ions or complexes, which can either retard or enhance its transport. For example, iron in liver is much more absorbable than is iron in egg yolk, since the latter contains phosphates which bind the iron to form an insoluble complex.

Regulation of Gastrointestinal Processes

Basic principles

A special environment must be created in the lumen of the gastrointestinal tract in order for digestion and absorption to occur. Accordingly, glandular secretions and muscle contractions must be controlled in a manner that will provide this optimal environment. Unlike other control systems, which regulate some variable in the extracellular fluid, the control mechanisms of the gastrointestinal system regulate conditions in the lumen of the tract. In other words, these control mechanisms are not governed to any great extent by the nutritional state of the body but rather by the volume and composition of the luminal contents.

These control systems must be able to detect the state of the luminal environment, and so it is appropriate that the majority of the receptors for these systems are located in the wall of the gastrointestinal tract, itself. Most gastrointestinal reflexes are initiated by a relatively small number of luminal stimuli: (1) distension of the wall by the luminal contents; (2) osmolarity (total solute concentration) of the chyme; (3) acidity (H^+ concentration) of the chyme; and (4) products formed by the digestion of carbohydrates, fats, and proteins—monosaccharides, fatty acids, peptides, and amino acids. Signals initiated by these stimuli act on mechanoreceptors, osmoreceptors, and chemoreceptors and trigger reflexes which influence the effectors—the muscle layers in the wall of the tract and the exocrine glands which secrete substances into its lumen. Both the receptors and effectors of these reflexes are in the gastrointestinal system.

These reflexes, like other negative-feedback systems, prevent large changes in the variables which are the stimuli for the reflexes; in so doing they maintain optimal conditions for digestion and absorption. For example, increased acidity in the duodenum, resulting from acid emptying into it from the stomach, reflexly inhibits the secretion of acid by the stomach, thereby preventing acidification of the duodenum (the adaptive value of this reflex is that digestive enzymes in the duodenum are inhibited by acid). To take another example, distension of the small intestine, due to the presence of material not yet absorbed, reflexly increases the contractile activity of the intestinal smooth muscle, thereby facilitating mixing of the chyme with the intestinal secretions and bringing it into contact with the epithelium. At the same time, the contractions of the stomach are reflexly inhibited, thereby decreasing the rate at which additional material is emptied into the intestine.

Let us now examine the general types of neural and hormonal pathways that mediate these reflexes.

Neural regulation. The gastrointestinal tract can be considered to have its own local nervous system, in the form of two major nerve plexuses, the myenteric plexus and the submucous nerve plexus (Fig. 14-5). These two plexuses are found throughout the length of the gastrointestinal tract. They are com-

posed of neurons that either form synaptic junctions with other neurons within the plexus or end near smooth muscles and glands. Moreover, many axons leave the myenteric plexus and synapse with neurons in the submucous plexus, and vice versa, so that neural activity in one plexus influences the activity in the other. The axons in both plexuses branch profusely, and stimulation at one point in the plexus leads to impulses conducted both up and down the tract. Thus, activity initiated in the plexus in the upper part of the small intestine may affect smooth muscle and gland activity in the stomach as well as in the lower part of the intestinal tract. Many of the receptors mentioned in the previous section are, in fact, the endings of plexus neurons.

The neural connections in the plexuses permit intratract neural reflexes that are independent of the central nervous system. This is not to say that the tract is not subject to neural control via the central nervous system; nerve fibers from both the sympathetic and parasympathetic branches of the autonomic nervous system enter the intestinal tract and synapse with neurons in the plexuses. Via these pathways, the central nervous system can influence the motor and secretory activity of the gastrointestinal tract. Thus, two types of neural reflex pathways linking a stimulus to a response are found (Fig. 14-12): so-called **short reflexes** from receptors through the nerve plexuses to effector cells, and **long reflexes** traveling from receptors by way of external nerves to the central nervous system and back to the nerve plexuses and effector cells by way of the autonomic nerve fibers. Some controls are mediated solely by short pathways or long pathways, whereas others use both simultaneously.

Finally, it should be noted that not all neural reflexes controlling the gastrointestinal system are initiated by signals within the tract; whenever the reflexes are initiated by other receptors, e.g., by the sight of food, the central nervous system must be involved in the response. Complex behavioral influences, e.g., emotions, also operate, in large part, through the central nervous system.

Hormonal regulation. The first hormone to be discovered (in 1902) was a substance extracted from

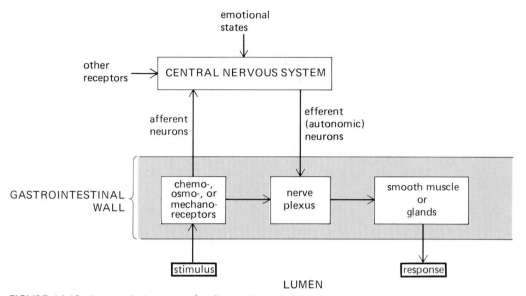

FIGURE 14-12. Long and short neural reflex pathways initiated by stimuli in the gastrointestinal tract. The long reflexes utilize neurons that link the central nervous system to the gastrointestinal tract.

the wall of the small intestine which, when injected into the blood of a dog, caused the pancreas to secrete a solution containing a high concentration of bicarbonate. This gastrointestinal hormone was named **secretin**, and with its discovery the science of endocrinology was born. Unlike the endocrine glands of the gonads, adrenals, pituitary, or thyroid, the hormone-secreting cells of the gastrointestinal tract consist of individual cells scattered throughout the epithelium of the stomach and small intestine; i.e., they are not clustered into discrete organs. One surface of the cell is exposed to the lumen of the gastrointestinal tract; at this surface, various chemical substances in the chyme stimulate the cell, causing hormonal release from the other surface, the basal portion of the cell, from which point the released hormone diffuses into capillaries. Although some "hormones" can be detected in the lumen of the gastrointestinal tract and may therefore be acting locally as paracrines, the circulation provides the route by which most of the hormones get from their site of secretion to their target cells.

Over the years substances obtained from various regions of the gastrointestinal tract have been injected into animals and found to produce changes in gastrointestinal secretory and contractile activity, as well as producing effects on other organ systems. A dozen or more substances are currently being investigated as possible gastrointestinal hormones, but only three—**secretin, cholecystokinin (CCK),** and **gastrin**—have met all the criteria for true hormones. These hormones, as well as several of the "candidate" hormones have also been found in nerve cells both within the central nervous system and the gastrointestinal plexuses, where they may be functioning as neurotransmitters or neuromodulators.

Table 14-2, which summarizes the major characteristics of the three established hormones, not only serves as a reference for future discussions but can be used to illustrate several characteristics of gastrointestinal hormones. Table 14-3 lists a number of the "candidate" hormones extracted from the gastrointestinal tract and currently under investigation. Some of the effects they are suspected of producing in gastrointestinal tissues are listed but will not be described further because it is not known whether they occur physiologically.

As described in Chap. 9, most of these peptide hormones are initially synthesized as larger proteins

TABLE 14-2. Properties of gastrointestinal hormones

	Gastrin	CCK	Secretin
Structure	Peptide	Peptide	Peptide
Location of endocrine cells	Antrum of stomach	Small intestine	Small intestine
Stimuli for hormone release	Amino acids or peptides in stomach; parasympathetic nerves (ACh)	Amino acids or fatty acids in small intestine	Acid in small intestine
Stimuli for hormone inhibition	Acid in lumen of stomach; secretin; somatostatin		
Target-cell responses*			
Stomach			
Acid secretion	Stimulates		Inhibits
Antral contraction	Stimulates		Inhibits
Pancreas			
Bicarbonate secretion		Potentiates secretin actions	Stimulates
Enzyme secretion		Stimulates	Potentiates CCK actions
Liver			
Bicarbonate secretion		Potentiates secretin action	Stimulates
Gallbladder contraction		Stimulates	
Sphincter of Oddi		Relaxes	
Small intestine motility	Stimulates ileum; inhibits ileocecal sphincter		
Large intestine motility	Stimulates mass movement		
Growth (tropic) of	Stomach and small intestine	Exocrine pancreas	Exocrine pancreas

* It should be emphasized that only the target-cell responses that are known to occur at physiological concentrations of the hormone are listed.

which undergo posttranslational processing that reduces their size. As a result, there may be several peptides of different sizes which possess the same sequences of amino acids within their structure, sequences necessary for biological activity. For example, the last five amino acids in CCK are identical to the last five amino acids in gastrin. As a result, both gastrin and CCK will bind to some of the same receptors. However, the remaining amino acids in the peptides result in different specificities and affinities of the binding of the two peptides to specific receptors.

There are two major generalizations concerning the functioning of these gastrointestinal hor-

TABLE 14-3. Some peptides found in the gastrointestinal tract and suspected of being hormones (candidate hormones)

Candidate hormone	Effects
Bombesin	Stimulates acid secretion; inhibits gastric motility
Gastric inhibitory peptide	Stimulates insulin release by pancreas; inhibits gastric motility
Motilin	Stimulates gastric and intestinal motility
Somatostatin	Inhibits acid secretion, gastric and intestinal motility, contraction of gallbladder, and enzyme and bicarbonate secretion by the pancreas
Vasoactive intestinal peptide	Stimulates salt and water secretion in small intestine; inhibits acid secretion, gastric and intestinal motility, and vascular smooth muscle contraction
Villikinin	Stimulates contraction of intestinal villi

mones that we wish to emphasize: (1) Each hormone participates in a feedback control system that regulates some aspect of the luminal environment, and (2) each hormone affects more than one target cell.

These two generalizations can be illustrated by CCK (cholecystokinin): The presence of fatty acids in the duodenum triggers the release of CCK from the duodenal wall, and CCK then stimulates the secretion of digestive enzymes from the pancreas (including lipase, which digests fat); CCK also causes the gallbladder to contract (delivering to the intestine bile salts, which are required for fat digestion and absorption). As the fat is digested and absorbed, the stimulus for CCK release (fat in the lumen) is removed.

In many cases a single target cell may contain receptors for more than one hormone (in addition to receptors for neurotransmitters and paracrines), with the result that a variety of inputs can affect the target cell response. One aspect of such interactions can be illustrated with the phenomenon known as **potentiation**. CCK by itself strongly stimulates pancreatic enzyme secretion, and secretin by itself only weakly stimulates enzyme secretion. However, the important point for the present discussion is that in the presence of a very low concentration of secretin, CCK stimulates the secretion of pancreatic enzymes more strongly than would be predicted by the sum of the stimulatory effects of CCK and secretin alone. Thus, secretin has potentiated the effect of CCK. An analogous potentiating effect of CCK on secretin stimulation of bicarbonate secretion is also seen. One of the consequences of the potentiating effects of gastrointestinal hormones is that a very small change in the plasma concentration of a particular hormone may have considerable physiological effects on the actions of other gastrointestinal hormones.

In addition to their stimulation (or in some cases inhibition) of specific target-cell functions, such as secretory and contractile activities, the gastrointestinal hormones also have tropic (growth promoting) effects on various tissues, including the gastric and intestinal mucosa and the exocrine portions of the pancreas.

Not listed in Table 14-2 are the effects of these gastrointestinal hormones on the release of other hormones, especially the release of insulin, glucagon, and somatostatin from the pancreas (see Chap. 15).

Phases of gastrointestinal control. The neural and hormonal control of the gastrointestinal system is, in large part, divisible into three phases—the cephalic, gastric, and intestinal phases—according to the location of the stimuli which initiate a particular reflex.

The **cephalic phase** is initiated by stimulation of receptors in the head (*cephalic*, head). These are sight, smell, taste, and chewing, as well as various emotional states. The efferent pathways for the reflex changes elicited by these stimuli involve both parasympathetic fibers (most of which are in the

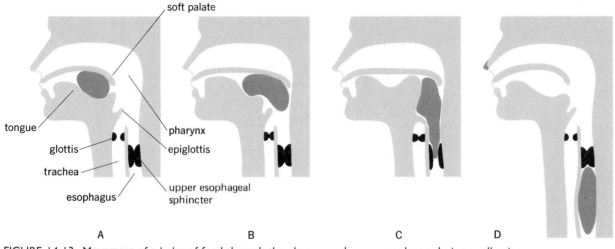

FIGURE 14-13. Movement of a bolus of food through the pharynx and upper esophagus during swallowing.

vagus nerves) and sympathetic fibers. These fibers activate neurons in the nerve plexuses, which in turn affect secretory and contractile activity.

Stimuli applied to the wall of the stomach initiate reflexes which constitute the **gastric phase** of regulation. These stimuli are distension, acid, and peptides formed during the digestion of protein. The responses to these stimuli are mediated by the nerve plexuses (short reflexes), by pathways involving the external nerves to the gastrointestinal tract (long reflexes) and by the release of gastrin.

Finally, the **intestinal phase** is initiated by stimuli in the lumen of the intestinal tract—distension, acidity, osmolarity, and the various digestive products of carbohydrates, fats, and proteins. Like the gastric phase, the intestinal phase is mediated by both long and short neural reflexes and by the release of gastrointestinal hormones, in this case secretin and CCK.

It must be emphasized that each of these phases is named for the site at which the various stimuli initiate the reflex and not for the sites of effector activity, since each phase is characterized by efferent output to virtually all organs in the gastrointestinal tract. These phases do not occur in temporal sequence except at the very beginning of a meal. Rather, during ingestion and the much longer absorptive period, reflexes characteristic of all three phases occur simultaneously.

Keeping in mind the neural and hormonal mechanisms available for regulating gastrointestinal activity, we can now examine the specific contractile and secretory processes that occur in each segment of the gastrointestinal system.

Mouth, pharynx, and esophagus

Chewing. The rhythmic act of chewing is controlled by the somatic nerves to the skeletal muscles of the mouth and jaw. In addition to the voluntary control of these skeletal muscles, rhythmic chewing motions are reflexly activated by the pressure of food against the gums, teeth, hard palate, and tongue. Activation of these pressure receptors leads to inhibition of the muscles holding the jaw closed; the resulting relaxation of the jaw reduces the pressure, allowing a new cycle of contractile activity.

Prolonged chewing of food is not essential to the digestive process (many animals, such as the dog and cat, swallow their food almost immediately). Although chewing prolongs the subjective pleasure of taste, it does not appreciably alter the rate at which the food is digested and absorbed from the small intestine. On the other hand, attempting to swallow a particle of food too large to enter the esophagus may lead to choking if the particle lodges over the trachea, blocking the entry of air into the lungs. A number of preventable deaths occur each year from choking, the symptoms of which are often confused with the sudden onset of a heart attack, so that no attempt is made to remove the obstruction from the airway. (The Heimlich maneuver, described in Chap. 12, can often be used to dislodge the obstructing particle from the airway.)

Saliva. Saliva is secreted by three pairs of exocrine glands: the **parotid,** the **submandibular**, and the **sublingual** (Fig. 14-1). Water accounts for 99 percent of the weight of the secreted fluid, the remaining 1 percent consisting of various salts and proteins. As described earlier, the major proteins of saliva are the enzyme amylase and the **mucins**, which when mixed with water form the highly viscous solution known as **mucus.**

The secretion of saliva is controlled by both the sympathetic and parasympathetic divisions of the autonomic nervous system; however, unlike their antagonistic activity in most organs, both systems stimulate salivary secretion, the parasympathetics producing the greatest secretory response. In the awake individual a basal secretory rate of about 0.5 mL/min keeps the mouth moist. The presence of food in the mouth increases the rate of salivary secretion; this reflex response is initiated by chemoreceptors and pressure receptors in the walls of the mouth and tongue. Afferent fibers from the receptors enter the brainstem medulla, which contains the integrating center controlling the autonomic output to the salivary glands. The most potent stimuli for salivary secretion are acid solutions, e.g., fruit juices and lemons, which can lead to a maximal secretion of 4 mL of saliva per minute.

Swallowing. Swallowing is a complex reflex initiated when pressure receptors in the walls of the pharynx are stimulated by a bolus of food forced into the rear of the mouth by the tongue. These receptors send afferent impulses to the swallowing center in the medulla, which coordinates the sequence of events during the swallowing process via efferent impulses to the muscles in the pharynx, larynx, esophagus, and respiratory muscles.

As the bolus of food moves into the pharynx, the soft palate is elevated and lodges against the back wall of the pharynx, preventing food from entering the nasal cavity (Fig. 14-13). The swallowing center inhibits respiration, raises the larynx, and closes the glottis (the area around the vocal cords at the beginning of the trachea), keeping food from getting into the trachea. As the tongue forces the food farther back into the pharynx, the bolus tilts a flap of tissue, the epiglottis, backward to cover the closed glottis. This pharyngeal period of swallowing lasts about 1 s.

The esophagus is a 25-cm-long tube connecting the pharynx to the stomach. It passes through the thoracic cavity, penetrates the diaphragm muscle (which separates the thoracic cavity from the abdominal cavity), and joins the stomach a few centimeters below the diaphragm. Skeletal muscles surround the upper third of the esophagus, smooth muscles the lower two-thirds. As was described in Chap. 12, the pressure in the thoracic cavity is 5 to 10 mmHg less than atmospheric, and this sub-atmospheric pressure is transmitted across the thin wall of the esophagus to the lumen. In contrast, the pressure at the beginning of the esophagus (which is continuous with the mouth and nasal passages) is equal to atmospheric pressure, and the pressure at the lower end of the esophagus (the pressure in the stomach) is greater than atmospheric. Thus, at both ends of the esophagus the pressure is greater than the pressure in the thoracic region of the esophagus. These pressure differences would tend to force air (from the top) and stomach contents (from below) into the esophagus. However, this does not normally occur, because both ends of the esophagus are closed (in the absence of swallowing) by the contraction of sphincter muscles. Skeletal muscles surround the esophagus just below the pharynx, and form the **upper esophageal sphincter**, whereas the smooth muscles in the last portion of the esophagus form the **lower esophageal sphincter** (Fig. 14-14).

The esophageal phase of swallowing begins with the relaxation of the upper esophageal sphincter. Immediately after the bolus has passed, the sphincter closes, the glottis opens, and breathing resumes. Once in the esophagus, the bolus is moved toward the stomach by a progressive wave of mus-

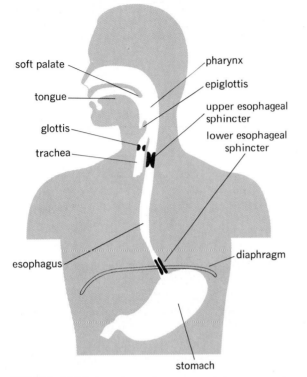

soft palate

tongue

glottis

trachea

esophagus

pharynx

epiglottis

upper esophageal
sphincter

lower esophageal
sphincter

diaphragm

stomach

FIGURE 14-14. Location of upper and lower esophageal sphincters.

cle contractions that proceeds along the esophagus, compressing the lumen and forcing the bolus ahead of it. Such waves of contraction in the muscle layers surrounding a tube are known as **peristaltic waves**. An esophageal peristaltic wave takes about 9 s to reach the stomach. Swallowing can occur while a person is upside down since it is not primarily gravity but the peristaltic wave which moves the bolus to the stomach. The lower esophageal sphincter opens and remains relaxed throughout the period of swallowing, allowing the arriving bolus to enter the stomach. After the bolus has passed, the sphincter closes, resealing the junction between the esophagus and the stomach.

Swallowing is an example of a reflex in which multiple responses occur in a regular temporal sequence that is predetermined by the pattern of synaptic connections between neurons in the coordinating center. Since both skeletal and smooth muscles are involved, the swallowing center must direct efferent activity in both somatic nerves (to skeletal muscle) and autonomic nerves (to smooth muscle), the latter nerves being the parasympathetic fibers which reach the plexuses in the esophageal wall by way of the vagus nerves. Simultaneously, afferent fibers from receptors in the wall of the esophagus send to the swallowing center information which can alter the efferent activity. For example, if a large particle of food does not reach the stomach during the initial peristaltic wave, the distension of the esophagus by the particle activates receptors which initiate reflexes causing repeated waves of peristaltic activity (**secondary peristalsis**).

The ability of the lower esophageal sphincter to maintain a barrier between the stomach and the esophagus is aided by the fact that its last portion lies below the diaphragm and is subject to the same pressures in the abdominal cavity as is the stomach. Thus, if the pressure in the abdominal cavity is raised, e.g., during cycles of respiration or by contraction of the abdominal muscles, the pressures of both the stomach contents and this terminal segment of the esophagus are raised together, preventing the formation of a pressure difference that could force the stomach contents into the esophagus.

During pregnancy, however, the growth of the fetus not only increases the pressure on the abdominal contents, it can push the terminal segment of the esophagus through the diaphragm into the thoracic cavity. The sphincter is therefore no longer assisted by changes in abdominal pressure, and so during the last 5 months of pregnancy there is a tendency for increased pressures in the abdominal cavity to force some of the contents of the stomach up into the esophagus. The hydrochloric acid from the stomach irritates the walls of the esophagus, causing contractile spasms of the smooth muscle which produces pain (known generally as **heartburn**, because the pain appears to be located over the heart). Heartburn often subsides in the last weeks of pregnancy as the uterus descends lower into the pelvis prior to delivery, decreasing the pressure on the abdominal organs. Another potential cause of heartburn in anyone is a large meal, which may raise the pressure in the stomach enough to produce reflux of acid into the esophagus.

Stomach

The stomach is a muscular sac located between the esophagus and the small intestine (Fig. 14-15). Glands in the thin-walled upper portions of the stomach, the **body** and **fundic** regions, secrete mucus, hydrochloric acid, and the enzyme precursor, pepsinogen. The lower portion of the stomach, the **antrum**, has a much thicker muscle wall; the glands in this region do not secrete acid, but they do contain the endocrine cells that secrete the hormone gastrin.

The layer of epithelial cells lining the stomach invaginates into the mucosa, forming numerous tubular glands (Fig. 14-16). The cells at the opening of these glands secrete mucus. In the body and fundic regions, deeper within the glands are the **parietal cells**, also known as **oxyntic cells**, which secrete hydrochloric acid. A third group of cells, **chief cells**, which secrete pepsinogen, are found primarily in the more basal portion of the glands throughout most regions of the stomach. Thus each of the three major exocrine secretions of the stomach—mucus, acid, and pepsinogen—is secreted by a distinct cell type, as is the hormone gastrin.

HCl secretion. The human stomach secretes about 2 L of hydrochloric acid per day. Hydrochloric acid is a strong acid which completely dissociates in water into hydrogen and chloride ions. The concentration of hydrogen ions in the lumen of the stomach may reach 150 mM, three million times greater than the concentration in the blood. The secretion of hydrochloric acid involves the active transport of both hydrogen and chloride ions.

A primary H-ATPase in the luminal membrane of the parietal cells pumps hydrogen ions out of the cell (Fig. 14-17). The hydrogen ions are obtained from the breakdown of water molecules, leaving hydroxyl ions (OH^-) behind. These hydroxyl ions are neutralized by combination with other hydrogen ions generated by the reaction between carbon dioxide and water, a reaction catalyzed by the enzyme carbonic anhydrase, which is present in high concentrations in parietal cells. The bicarbonate ions generated by this reaction move by a secondary active-transport carrier out of the parietal cell on the blood side, while chloride ions are actively transported into parietal cells from the blood (in part by countertransport with bicarbonate) and then leave the cell and enter the lumen of the gastric glands through a facilitated-diffusion carrier. Because a bicarbonate ion is formed and released into

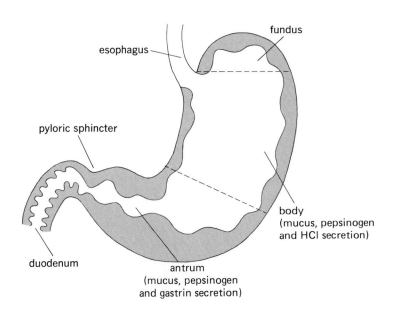

FIGURE 14-15. Structure of the stomach.

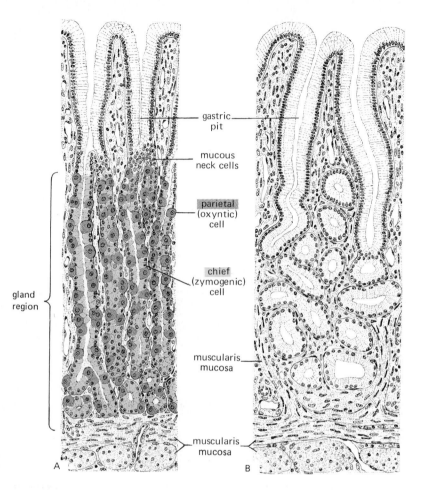

gastric pit

mucous neck cells

parietal (oxyntic) cell

chief (zymogenic) cell

gland region

muscularis mucosa

muscularis mucosa

A

B

FIGURE 14-16. Mucosa from the body region (A) and antral region (B) of the stomach. Note the relative absence of parietal (acid-secreting) and chief (pepsinogen-secreting) cells in the antrum of the stomach.

the blood for every hydrogen ion secreted into the lumen of the stomach, the venous blood leaving the stomach is more alkaline (has a higher bicarbonate concentration) than the arterial blood entering it.

During a meal, the rate of hydrochloric acid secretion increases markedly. Table 14-4 summarizes the many factors controlling HCl secretion. First, during the cephalic phase of the meal, the message for increased acid secretion reaches the stomach by way of the vagus nerves which stimulate neurons in the plexuses; fibers from these plexuses terminate near parietal cells, releasing acetylcholine, which stimulates acid secretion. Some of the plexuses' fibers also end near gastrin-releasing cells, and their stimulation releases gastrin, which in turn stimulates acid secretion by parietal cells.

Once food has reached the stomach the gastric phase of acid secretion ensues and is controlled by a variety of intragastric stimuli—distension, peptides, and changes in acidity. Distension produces its effects through neural receptors located in the wall of the stomach, receptors which activate the long and short reflex pathways leading to the stimulation of parietal-cell acid secretion. In contrast to distension, peptides and acid exert their stimulating effects directly on gastrin-secreting cells. For two reasons, the greater the protein content of a meal, the greater the acid secretion. First, peptides formed by the digestive action of pepsin on protein stimulate the release of gastrin. The second reason is more complicated but is related to the fact that a high acid concentration inhibits the release of

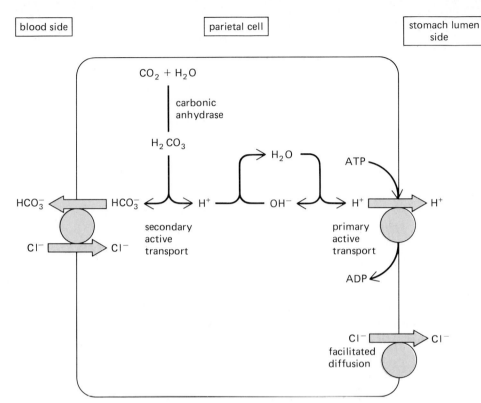

FIGURE 14-17. Secretion of hydrochloric acid by parietal cells in the stomach.

TABLE 14-4. Control of HCl secretion during a meal

Stimuli	Pathways to the parietal cells	Result
Cephalic phase Sight Smell Taste Chewing	Vagus nerves, gastrin	Increased HCl secretion
Gastric contents (gastric phase) Distension H^+ concentration Peptides	Long and short neural reflexes, gastrin	Increased HCl secretion
Intestinal contents (intestinal phase) Distension H^+ concentration Osmolarity Nutrients	Long and short neural reflexes, secretin, decreased gastrin release, other unspecified hormones	Decreased HCl secretion

gastrin and hence inhibits gastrin-stimulated acid secretion. Before food enters the stomach there is a small basal rate of acid secretion and the hydrogen-ion concentration in the lumen is high, inhibiting gastrin release. The protein in food is an excellent buffer so that as food enters the stomach, the concentration of free hydrogen ions drops as the hydrogen ions bind to the side chains of the proteins. This decrease in acidity removes the inhibition of gastrin secretion and, thereby, of acid secretion. The more protein in a meal, the greater the buffering of acid, the more gastrin secreted, and the more acid secreted—clearly this sequence provides an excellent negative-feedback control which acts to maintain a high luminal acid concentration.

Ingested substances other than protein can stimulate acid secretion when they reach the stomach. For example, caffeine, a substance found in coffee, tea, chocolate, and cola drinks, stimulates the release of gastrin and thus increases acid secretion. Alcohol, contrary to popular belief, has little effect on acid secretion in people.

We now come to the intestinal phase of the control of acid secretion, the phase in which stimuli occurring in the early portion of the small intestine reflexly influence acid secretion by the stomach. Because the digestive activity of enzymes and bile salts in the small intestine is markedly inhibited by acid solutions, it is essential that the chyme entering the small intestine from the stomach not contain so much acid that it cannot be rapidly neutralized by the bicarbonate-rich fluids secreted into the intestine by the liver and pancreas. Accordingly it is highly adaptive that excess acidity in the duodenum triggers reflexes which inhibit gastric acid secretion (other reflexes to be described later are elicited which slow gastric emptying and stimulate pancreatic secretion, two other ways of preventing a rise in acidity).

Acidity is not the only factor in duodenal chyme capable of triggering reflex inhibition of gastric acid secretion; distension, hypertonic solutions, and solutions containing amino acids, fatty acids, or monosaccharides also do so. Thus, the extent to which acid secretion is inhibited during the intestinal phase by each of the above stimuli varies, depending upon the volume and composition of the meal, but the net result is the same—balancing the secretory activity of the stomach with the digestive and absorptive functions of the small intestine.

The inhibition of gastric acid secretion during the intestinal phase is mediated by nerves (short and long reflexes) and hormones all of which inhibit gastrin secretion and/or directly inhibit parietal-cell acid secretion. Secretin appears to mediate part of the hormone-mediated inhibition of acid secretion.

We should point out that many substances and pathways other than those mentioned have been hypothesized to play a role in controlling the secretion of acid. Several of the "candidate" hormones, Table 14-3, may normally alter acid secretion, as may several paracrines, especially histamine and certain prostaglandins. **Histamine**, found in the mucosa of the stomach, can produce a marked increase in parietal-cell acid secretion (indeed, histamine is often administered to measure the acid-secreting capacity of the stomach in health and disease), but its physiological role in the control of acid secretion remains unclear.

Pepsin secretion. A group of similar proteases (enzymes which break down proteins), commonly referred to collectively as **pepsin**, are secreted by the chief cells in an inactive form known as **pepsinogen**. The high acidity in the lumen of the stomach converts pepsinogen into the active form of the enzyme, pepsin, by breaking off a small fragment of the molecule, thereby changing the shape of the protein and exposing its active site. The synthesis and secretion of inactive pepsinogen, followed by its activation to pepsin, provides an example of a process that occurs frequently when proteolytic enzymes are secreted. By synthesizing an inactive form of the enzyme the intracellular molecules that may be substrates for the enzyme are protected from digestion, thus preventing the destruction of the cell synthesizing the enzyme. Once activated by acid, pepsin can, itself, act upon pepsinogen to form even more pepsin molecules (Fig. 14-18). The activation of pepsin is thus an autocatalytic, positive-feedback process that is limited only by the amount of pepsinogen available.

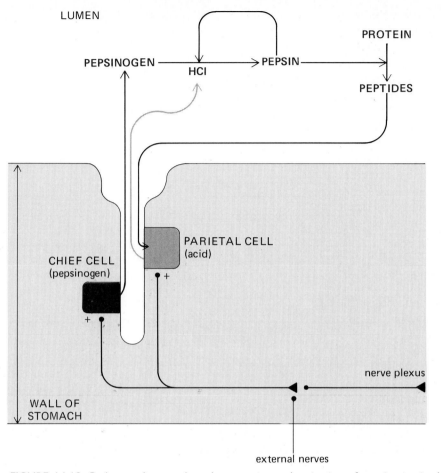

FIGURE 14-18. Pathways that regulate the secretion and activation of pepsinogen in the stomach. The peptides also affect the parietal cells via their release of gastrin.

Pepsin is active only in the presence of a high hydrogen-ion concentration (high acidity) and becomes inactive when it passes into the small intestine, where the hydrogen ions are neutralized by the bicarbonate ions secreted into the small intestine. Only a fraction of the total protein in a meal is digested by pepsin. Indeed, even in the absence of pepsin, as occurs in some pathological conditions, protein can be completely digested by the enzymes in the small intestine. (Since high acidity is required for both pepsin activation and activity, it is appropriate that, as we have seen, these peptide fragments produced by pepsin activity are a stimulus for acid secretion.)

The primary pathway for stimulating pepsinogen secretion is input to the chief cells from the nerve plexuses. In general, during the cephalic, gastric, and intestinal phases, the factors which stimulate or inhibit acid secretion exert the same effects on pepsinogen secretion; thus pepsinogen secretion parallels the secretion of acid. In addition, acid itself stimulates the secretion of pepsinogen.

Gastric motility. The empty stomach has a volume of only about 50 mL, and the diameter of its lumen is little larger than that of the small intestine. When a meal is swallowed, the smooth muscle in the fundus and body of the stomach reflexly relaxes prior to the arrival of the food, allowing the volume of the lumen of the stomach to increase to as much as 1.5 L with little increase in pressure. This **receptive relaxation** is mediated by the vagus (parasympathetic) nerves and coordinated by the swallowing center. Once distension of the stomach by a meal actually occurs, the smooth muscles in the body and fundus are reflexly relaxed by long reflexes via the vagus nerves.

Like the esophagus, the stomach manifests peristaltic waves. Each wave begins near the entry of the esophagus and produces only a weak ripple as it proceeds over the body of the stomach, a ripple too weak to produce much mixing of the luminal contents with the acid and pepsin secreted by the stomach. (As a result, the salivary amylase that is inactivated by high acidity continues to digest polysaccharides for a considerable time after food has reached the stomach.) As the wave approaches the larger mass of wall muscle surrounding the antrum, it produces a more powerful contraction that mixes the contents of the stomach and expels some of them into the duodenum. The **pyloric sphincter** is a ring of smooth muscle and connective tissue between the terminal antrum and the duodenum, and it is open most of the time. When a strong peristaltic wave arrives at the antrum, the pressure of the antral contents is increased, but the antral contraction also closes the pyloric sphincter, with the result that the pressure forces most of the antral contents back into the body of the stomach (this contributes to the mixing of the antral contents), and only small amounts pass into the duodenum with each contraction (Fig. 14-19).

What is responsible for the characteristics of the peristaltic waves? Their rhythmicity (three per minute) is due to the fact that pacemaker cells in the longitudinal muscle layer near the esophagus undergo spontaneous depolarization-repolarization cycles (slow waves) at this rate, termed the **basic electrical rhythm**. The depolarizations are propagated through gap junctions along the longitudinal muscle layer of the stomach and also induce similar slow waves in the overlying circular muscle layer (perhaps via gap junctions between the longitudinal and circular layers). However, in the absence of

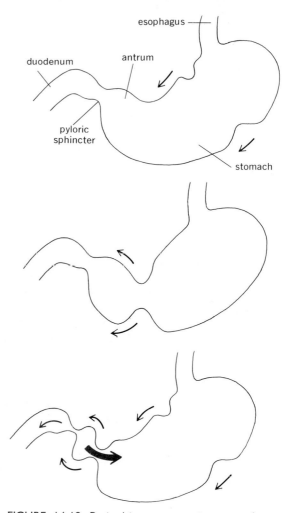

FIGURE 14-19. Peristaltic waves passing over the stomach empty a small amount of material into the duodenum. Most of the material is forced back into the antrum.

other input, these depolarizations are too small to cause the muscle membranes to reach threshold and fire the action potentials required to elicit contractions. Excitatory neurotransmitters and hormones act upon the muscle to depolarize the membrane, thereby bringing it closer to threshold. Action potentials are then generated at the peak of the slow wave cycle (Fig. 10-35); the number of spikes fired with each wave, in turn, determines the strength of the elicited smooth muscle contraction. Thus, whereas the frequency of contraction is determined by the basic electrical rhythm and remains essentially constant, the force of contraction (and therefore, the amount of gastric emptying per contraction) is determined by neural and hormonal input, i.e., by regulatory reflexes. The contraction of the gastric smooth muscle is increased by acetylcholine released by parasympathetic nerves and by gastrin, while contraction is inhibited by norepinephrine from sympathetic nerves and by secretin.

There are important gastric and intestinal phases to this reflex regulation of gastric motility. First, all of the factors previously discussed which regulate the release of gastrin alter gastric motility indirectly, since gastrin stimulates antral contraction. Distension of the stomach also stimulates antral contractions (long and short neural reflexes triggered by mechanoreceptors in the stomach mediate this response). Therefore, the larger a meal, the faster the initial rate of emptying, and as the volume of the stomach decreases, so does the rate of emptying. In contrast, distension of the duodenum or the presence of fat, acid, or hypertonic solutions in the lumen of the duodenum all produce inhibition of gastric emptying (Fig. 14-20); these responses are mediated by the release of intestinal hormones and by both long and short neural reflexes. Fat in the duodenum is the most potent of the chemical stimuli leading to this reflex inhibition of gastric motility.

In addition to the control of gastric emptying by the gastric and the duodenal contents, motility is influenced by external nerve fibers to the stom-

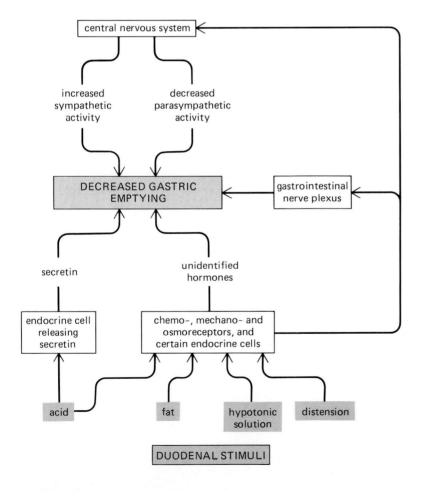

FIGURE 14-20. Intestinal-phase pathways inhibiting gastric emptying.

ach: Inhibition of motility occurs when parasympathetic activity to the stomach is decreased or sympathetic activity increased. Via these pathways, pain and emotions such as sadness, depression, and fear tend to decrease motility, whereas aggression or anger tend to increase it. These relationships are not always predictable, however, and different people show different responses to apparently similar emotional states.

Vomiting. Vomiting is the forceful expulsion of the contents of the stomach and upper intestinal tract through the mouth. Like swallowing, vomiting is a complex reflex coordinated by a region in the brainstem medulla, known as the **vomiting center**. Neural input to this center from receptors in many different regions of the body can initiate the vomiting reflex. For example, excessive distension of the stomach or small intestine, various substances acting upon chemoreceptors in the wall of the small intestine or the brain, increased pressure within the skull, rotating movements of the head (motion sickness), intense pain, and tactile stimuli applied to the back of the throat can all initiate vomiting.

What is the adaptive value of the vomiting reflex? Obviously, the removal of ingested toxic substances before they can be absorbed is of benefit. Moreover, the nausea that usually accompanies vomiting may have the adaptive value of conditioning the individual to avoid the future ingestion of foods containing the same toxic substances. Why other types of stimuli, such as those producing intense pain or motion sickness, have become linked to the vomiting center is less clear.

Vomiting is usually preceded by increased salivation, sweating, increased heart rate, pallor, and feelings of nausea—all characteristic of a general discharge of the sympathetic nervous system in response to stress. Vomiting begins with a deep inspiration, closure of the glottis, and elevation of the soft palate. The abdominal and thoracic muscles contract, raising the abdominal pressure which is transmitted to the contents of the stomach; the lower esophageal sphincter relaxes, and the high abdominal pressure forces the contents of the stomach into the esophagus. This initial sequence of events can occur repeatedly without vomiting and is known as **retching**. Vomiting occurs only when, in addition, the upper esophageal sphincter becomes reflexly relaxed. Vomiting is also accompanied by strong contractions of the upper portion of the small intestine, contractions which tend to force some of the intestinal contents back into the stomach. Thus, some bile may be present in the vomitus.

Excessive vomiting can lead to large losses of fluids and salts which normally would be absorbed in the small intestine. This can result in severe dehydration, upset the salt balance of the body, and produce circulatory problems due to a decrease in plasma volume. The loss of acid results in a lowering of body acidity. If the amount of vomiting is slight, the control mechanisms acting by way of the lungs and kidneys return the acidity to normal levels by hypoventilation and by excreting bicarbonate ions in the urine.

Pancreatic secretions

The pancreas, located below and behind the stomach, is a mixed gland containing both endocrine and exocrine portions. The endocrine cells secrete into the blood the hormones insulin, glucagon, and somatostatin (to be discussed in Chap. 15). The exocrine portion of the pancreas secretes two solutions involved in the digestive process, one containing a high concentration of bicarbonate ions and the other a number of digestive enzymes. Both exocrine solutions are secreted into ducts which converge into the single pancreatic duct, the latter joining the bile duct from the liver just before entering the duodenum (Fig. 14-21). In the intestinal lumen the bicarbonate ions neutralize the hydrochloric acid entering from the stomach, and the enzymes digest various organic nutrients. The enzyme secretions of the pancreas are released from acinar cells at the base of the exocrine glands, whereas the bicarbonate solution is secreted by the cells lining the early portions of the ducts leading from the acinar cells.

The secretion of bicarbonate ions by the pancreas is an active process. The mechanism of bicarbonate secretion is similar to the process of hy-

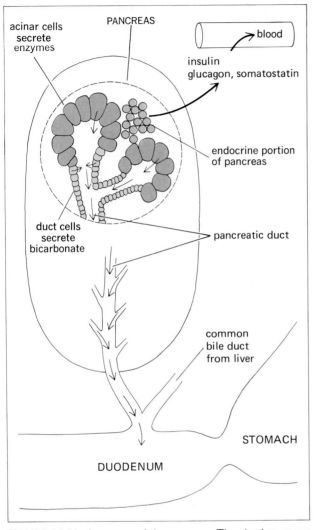

FIGURE 14-21. Structure of the pancreas. The gland areas are greatly enlarged relative to the entire pancreas and ducts.

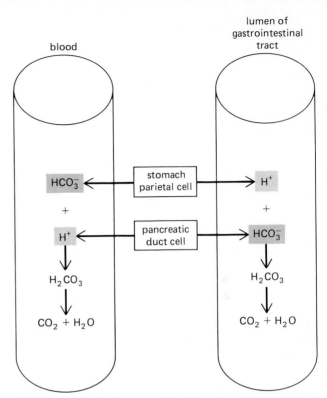

FIGURE 14-22. Acid secreted by the stomach is neutralized by the bicarbonate secreted by the pancreas (and liver); bicarbonate released into the blood by the stomach is neutralized by the acid released into the blood by the pancreas.

drochloric acid secretion by the stomach, except that the events on the two sides of the cells are reversed. Secretion of bicarbonate ions into the lumen of the pancreatic ducts is coupled to the movement of chloride ions into the cells. Simultaneously, the hydrogen ions (formed along with bicarbonate ions) are secreted across the plasma membrane on the blood side of the pancreatic cell. The hydrogen ions cause the blood leaving the pancreas to be more acid than the blood entering it.

Because the amount of bicarbonate secreted by the pancreas is reflexly regulated by the amount of acid entering the duodenum from the stomach (as will be discussed below), the amount of bicarbonate secreted by the pancreas is normally approximately equal to the amount of acid secreted by the stomach. This also means that the bicarbonate released into the blood by the stomach is normally equal to the amount of acid released into the blood by the pancreas. When the alkaline blood leaving the stomach mixes with the acidified blood from the pancreas, the overall result is no change in the acidity of the blood returning to the heart (Fig. 14-22). However, a loss of large quantities of bicarbonate ions from the intestinal tract during periods of prolonged diarrhea leads to a net accumulation of acid in the blood, just as loss of acid from the stomach by vomiting leads to a net alkalinization of the blood.

The enzymes secreted by the pancreas digest fat, polysaccharides, proteins, and nucleic acids to fatty acids, sugars, amino acids, and nucleotides, respectively. A partial list of these enzymes and their activities is given in Table 14-5. Most of the proteolytic enzymes are secreted in an inactive form, just as pepsinogen is secreted by the stomach, and are then activated in the duodenum by other enzymes. **Enterokinase**, which is embedded in the luminal plasma membrane of the intestinal epithelium, is a proteolytic enzyme which splits off a peptide from pancreatic **trypsinogen**, forming the active enzyme **trypsin**. Trypsin, like enterokinase, is also a proteolytic enzyme; once activated, it proceeds to activate the other proteolytic enzyme precursors from the pancreas (Fig. 14-23), as well as performing its major role of digesting protein. Several of the pancreatic enzymes, such as amylase and lipase, are secreted in their active form but require additional factors, such as ions and bile salts encountered in the intestinal lumen, in order to have maximal activity.

Increased pancreatic secretion during a meal is mediated mainly by the intestinal hormones secretin and CCK (Table 14-2). The major effect of secretin is to stimulate bicarbonate secretion, whereas CCK stimulates mainly the secretion of enzymes. (However, as was noted earlier, each of these hormones has a potentiating effect on the activity of the other.) Since the function of pan-

---

TABLE 14-5. Pancreatic enzymes

Enzyme	Substrate	Action
Trypsin, chymotrypsin	Proteins	Breaks amino acid bond in the interior of proteins to form peptide fragments
Carboxypeptidase	Proteins	Splits off terminal amino acid from end of protein containing a free carboxyl group
Lipase	Fat	Splits off two fatty acids from triacylglycerols, forming free fatty acids and monoglycerides
Amylase	Polysaccharide	Similar to salivary amylase; splits polysaccharides into a mixture of glucose and maltose
Ribonuclease, deoxyribonuclease	Nucleic acids	Splits nucleic acids into free mononucleotides

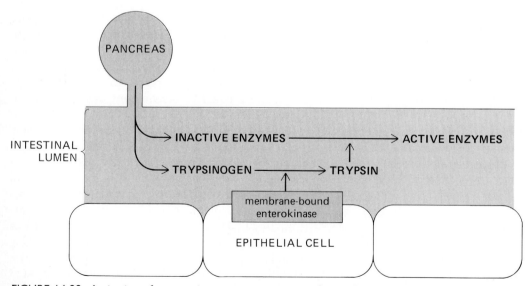

FIGURE 14-23. Activation of pancreatic enzyme precursors in the small intestine.

creatic bicarbonate is to neutralize the acid entering the duodenum from the stomach, it is appropriate that the major stimulus for secretin release from the wall of the duodenum is acid. As the acid is neutralized, the stimulus for secretin release is decreased, and less bicarbonate is secreted by the pancreas. In analogous fashion, CCK stimulates the secretion of digestive enzymes, including those for fat and protein digestion, and the stimuli for its release are, appropriately, fatty acids and amino acids. Thus, the organic nutrients in the small intestine initiate their own digestion via hormonal reflexes. Figure 14-24 summarizes the main factors controlling pancreatic secretion.

Bile secretion

Bile is secreted by the liver cells into a number of small ducts, the bile canaliculi (Fig. 14-25), which converge to form the common bile duct, which empties into the duodenum at the same site at which the pancreatic duct enters the intestine. A small sac branches from the bile duct on the underside of the liver, forming the **gallbladder**, which stores, concentrates, and releases the concentrated bile in response to various gastrointestinal stimuli. The gallbladder can be surgically removed without impairing the secretion of bile or its release into the intestinal tract; in fact, many animals (rats for example) which secrete bile do not have a gallbladder.

Bile is a solution containing six primary ingredients: (1) bile salts, (2) cholesterol, (3) lecithin (a phospholipid), (4) bile pigments and small amounts of other end products of organic metabolism, (5) certain trace metals, and (6) inorganic minerals, sodium, potassium, chloride, etc. The first three of these are synthesized in the liver and are involved in the solubilization of fat in the small intestine. The fourth and fifth are substances normally excreted from the body in the feces.

The major bile pigment is **bilirubin**, a breakdown product of the heme portion of hemoglobin, which is formed and released into the blood when macrophages destroy red blood cells, as described

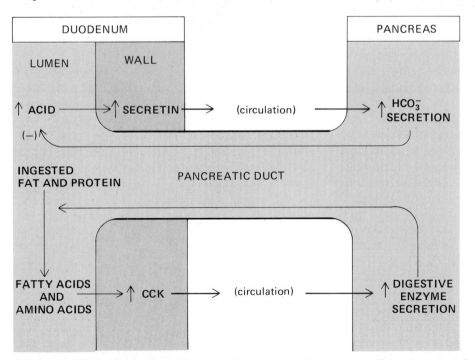

FIGURE 14-24. Feedback loops that control the pancreatic secretions of bicarbonate and enzymes. Hormonal pathways are colored. (Begin examination of this figure with events in the duodenum that lead to changes in the secretion of secretin and CCK.)

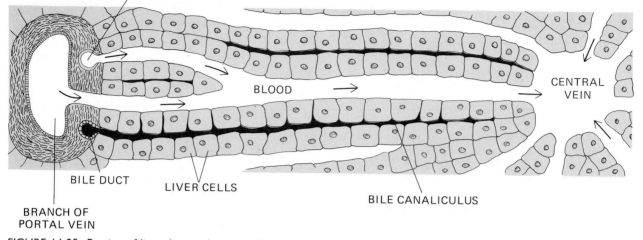

BRANCH OF HEPATIC ARTERY

BLOOD

CENTRAL VEIN

BILE DUCT   LIVER CELLS

BILE CANALICULUS

BRANCH OF
PORTAL VEIN

FIGURE 14-25. Portion of liver showing location of bile ducts with respect to liver cells. *(Adapted from Kappas and Alvares.)*

in Chap. 11. The cells of the liver extract bilirubin from the plasma and secrete it into the bile by an active process. After entering the intestinal tract via the bile, the bile pigments, which are yellow, are modified by bacterial enzymes to form the pigments which give feces their brown color. (In the absence of bile secretion, feces are grayish white.) Some of these yellow pigments are absorbed into the plasma during their passage through the intestinal tract and are eventually excreted in the urine, giving urine its yellow color.

From the standpoint of gastrointestinal function, bile salts are the most important components of bile since they are necessary to achieve an adequate digestion and absorption of fats. During the digestion of a single fatty meal, most of the bile salts entering the intestinal tract are reabsorbed in the lower part of the small intestine (ileum) and returned to the liver, which actively secretes them once again into the bile. This "recycling" pathway from the intestine to the liver (via the portal vein) and back to the intestine (via the bile duct) is known as the **enterohepatic circulation**.

Like those of pancreatic secretions, the components of the bile are secreted by two different types of cells: The bile salts, bilirubin, cholesterol and lecithin are secreted by **hepatocytes** (liver cells), whereas most of the inorganic salt solution contain-

ing sodium, chloride, and bicarbonate ions and water is secreted by the epithelial cells lining the bile ducts. Since bicarbonate ions in the bile help to neutralize acid in the duodenum, it is appropriate that the salt solution secreted by the bile ducts (just like that secreted by the pancreas) is stimulated by secretin in response to the presence of acid in the duodenum.

Although secretin stimulates the secretion of the inorganic salt solution by the bile ducts, it does not stimulate the secretion of bile salts by hepatocytes. The amount of bile salt secretion is controlled by the concentration of bile salts in the blood; the greater the plasma concentration of bile salts, the greater is the stimulus for their secretion. Thus, between meals, when there is little bile salt in the intestine and, therefore, little absorbed, the plasma concentration of bile salts is low and there is little bile salt available to be transported into the bile ducts (i.e., the enterohepatic cycling of bile salts is minimal). During a meal, the higher plasma concentration of absorbed bile salts leads to an increased rate of bile salt secretion. The liver also synthesizes new bile salts to replace the bile salts that escape absorption in the small intestine. The uptake, by hepatocytes, of bile salts from the plasma inhibits the synthesis of bile salts.

Cholesterol secretion into the bile is one of the

mechanisms by which cholesterol homeostasis is maintained (as will be discussed in Chap. 15). Cholesterol is an insoluble molecule, and its solubility in bile is achieved by its incorporation into micelles formed in the bile with the aid of bile salts, just as described for micelle formation from fat digestion products in the intestine. The presence of lecithin in the bile increases the capacity of the micelles for solubilizing cholesterol.

Although, as we have seen, the greatest volume of bile is secreted during and just after a meal, some bile is always being secreted by the liver. Surrounding the bile duct at the point where it enters the duodenum is a ring of smooth muscle known as the **sphincter of Oddi**. When this sphincter is closed, the bile secreted by the liver is shunted into the gallbladder. The cells lining the gallbladder actively transport sodium from the bile back into the plasma. As solute is pumped out of the bile, water follows by osmosis. The end result is a marked concentration of the constituents of the bile. Shortly after the beginning of a fat-containing meal, the sphincter of Oddi relaxes and the smooth muscles in the wall of the gallbladder contract, discharging concentrated bile into the duodenum. The signal for gallbladder contraction and sphincter relaxation is the intestinal hormone CCK—appropriately so, since a major stimulus for this hormone's release is the presence of fat in the duodenum. (It is from this ability to cause contraction of the gallbladder that cholecystokinin received its name: *chole,* bile; *cysto,* bladder; *kinin,* to move). Figure 14-26 summarizes the factors controlling the entry of bile into the small intestine.

Small intestine

Secretions. In addition to the fluids entering the small intestine from the stomach, liver, and pancreas, approximately 2000 mL of fluid moves across the wall of the small intestine from the blood into the lumen each day. Still larger volumes of fluid are simultaneously moving in the opposite direction—from the lumen into the blood—so that normally there is an overall net absorption of fluid from the small intestine.

One reason for water movement into the lumen is that the chyme entering the intestine from the stomach may be hypertonic, because of a high concentration of solutes in a particular meal and because digestion breaks down large molecules (such as proteins and polysaccharides) into many more small molecules (amino acids and monosaccharides). This causes the osmotic movement of water from the isotonic interstitial fluids in the wall of the intestine into the hypertonic solution in the lumen. As we have seen, the osmolarity of the duodenum constitutes one of the stimuli inhibiting gastric emptying. This prevents too much hypertonic fluid from accumulating in the duodenum; otherwise, a large net movement of water from the blood into the intestine would occur, with a resulting decrease in plasma volume and the consequences thereof.

In the description above, water movement was the result of events in the lumen. In addition, the intestine itself secretes ions (such as sodium, potassium, chloride, and bicarbonate ions) across its epithelial membranes into the lumen, and water moves with these ions.

Another source of material entering the small intestine is the continuous disintegration of the intestinal epithelium. These cells divide, differentiate, function for a few days, and then are discharged from the surface of the intestine. Their breakdown releases enzymes, other organic materials, ions, and water into the lumen of the intestine.

The secretion of mucus throughout the length of the intestine provides yet another source of organic material and fluid.

Motility. When the motion of the small intestine is observed by x-ray fluoroscopic examination shortly after a meal, the contents of the lumen are seen to move back and forth with little apparent net movement toward the large intestine. Thus, in contrast to the waves of peristaltic contraction that sweep over the surface of the stomach, the motion of the small intestine during a meal is a stationary oscillating contraction and relaxation of rings of smooth muscle (Fig. 14-27). Each contracting segment is only a few centimeters long and the contraction lasts a few seconds. The chyme in the lumen of a

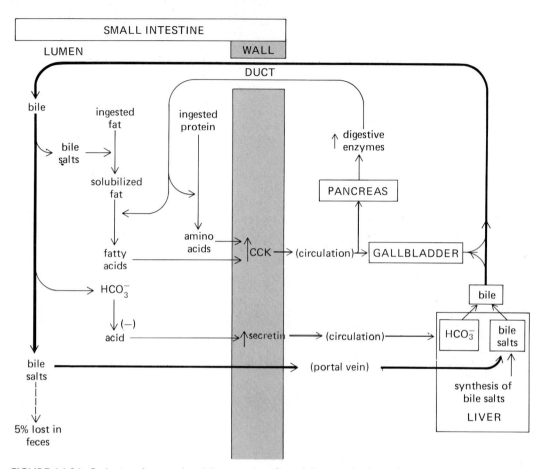

FIGURE 14-26. Pathways that regulate bile secretion. Start following the loops beginning with ingested fat, ingested protein, and acid. Note the interaction with the pancreatic loop, diagrammed in Fig. 14-24. Hormonal pathways are colored. The thick outer lines denote the enterohepatic circulation.

contracting segment is forced both up and down the intestine. This rhythmical contraction and relaxation of the intestine, known as **segmentation**, produces a continuous division and subdivision of the intestinal contents in a manner which thoroughly mixes the chyme in the lumen and brings it into contact with the intestinal wall.

These segmenting movements of the small intestine are initiated by electrical activity generated by pacemaker cells located in the longitudinal smooth muscle. Like the slow waves in the stomach, the pacemaker slow waves produce oscillations in smooth muscle membrane potential, which,

if threshold is reached, trigger action potentials causing muscle contraction. The frequency of segmentation is set by the frequency of the intestinal basic electrical rhythm, but unlike the stomach, which normally has a single basic electrical rhythm, the intestine has a rhythm that varies along the length of the intestine, each successive pacemaker having a slightly lower frequency than the one above. For example, segmentation in the duodenum occurs at a frequency of about 12 contractions per minute, whereas in the terminal portion of the ileum the rate is only 9 contractions per minute. Since the frequency of contraction is greater in the upper

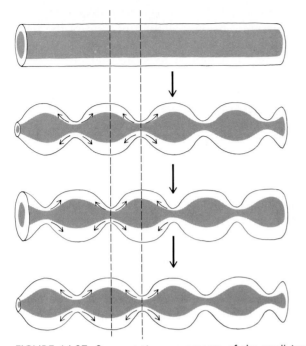

FIGURE 14-27. Segmentation movements of the small intestine. The small arrows indicate the movements of the luminal contents.

portion of the small intestine than in the lower, more chyme is forced downward, on the average, than is forced upward.

The intensity of the segmentation can be altered by the nerve plexuses, external nerves (parasympathetic activity increases contractile activity, and sympathetic stimulation decreases it), and hormones. As is true for the stomach, these reflexes produce changes in the intensity of smooth muscle contraction but do not change the natural frequency of the pacemakers. Following a meal, distension increases (by both long and short reflexes) the intensity of segmenting contractions (paradoxically, the actual propulsion achieved is decreased, because the contracting segments narrow the lumen of the intestine, producing a high resistance to flow).

After most of a meal has been absorbed, the segmenting contractions decline and are replaced by peristaltic waves that begin to sweep over seg-

ments of the intestine, and are known as the **migrating motility complex**. (The mechanism controlling the generation and progression of these waves is unknown.) These waves travel about 60 to 70 cm and then die out; however, the site at which the waves are initiated slowly migrates from the duodenum down the small intestine. By the time the wave activity reaches the end of the ileum, new waves are beginning in the duodenum and the process is repeated. These peristaltic waves sweep any material still remaining in the lumen of the small intestine on into the large intestine. Upon the arrival of a new meal in the stomach, the migrating motility complex ceases and is replaced by the segmenting type of contractile activity.

Several other specific reflexes are worth mentioning: Contractile activity in the ileum increases during periods of gastric emptying, and this is known as the **gastroileal reflex** (conversely, distension of the ileum produces decreased gastric motility, the **ileogastric reflex**). Also, large distensions of the intestine, injury to the intestinal wall, and various bacterial infections in the intestine lead to a complete cessation of motor activity, the **intestino-intestinal reflex**. Most of these reflexes appear to have as their efferent pathways the external nerves. Gastrin may play a role in the gastroileal reflex since gastrin increases the motility of the ileum and relaxes the ileocecal sphincter between the ileum and the beginning of the large intestine.

A person's emotional state can also affect the contractile activity of the intestine and, therefore, the rate of propulsion of chyme. Fear tends to decrease motility whereas hostility increases it, although these responses vary greatly in different individuals. These responses are all mediated by the external nerves.

As much as 500 mL of air may be swallowed along with a meal. Most of this air travels no farther than the esophagus, from which it is eventually expelled by belching. Some of the air, however, reaches the stomach and is passed on to the intestines, where its percolation through the chyme as the intestinal contents are mixed produces gurgling sounds which are often quite loud.

Large intestine

The large intestine (colon), a tube about 6 cm (2.5 in) in diameter, forms the last 120 cm (4 ft) of the gastrointestinal tract (Fig. 14-28). The first portion of the large intestine, the **cecum**, forms a blind-ended pouch from which the **appendix**, a small fingerlike projection, extends; the appendix has no known digestive function. The colon consists of three relatively straight segments—the ascending, transverse, and descending portions. The terminal portion of the descending colon is S-shaped, forming the sigmoid colon, which empties into a short section, the **rectum**. Although the large intestine has a greater diameter than the small intestine, it is about half as long, its mucosa lacks villi, and its surface is not convoluted; therefore, its epithelial surface area is far less than that of the small intestine. The secretions of the colon are scanty and consist mostly of mucus. The large intestine se-

cretes no digestive enzymes. It is responsible for the absorption of only about 4 percent of the total intestinal contents per day, mainly ions and water. Its primary function is to store and concentrate fecal material prior to defecation.

Chyme enters the colon through the ileocecal sphincter separating the ileum from the cecum. This sphincter is normally closed, but after a meal, when the gastroileal reflex increases the contractile activity of the ileum, the sphincter relaxes each time the terminal portion of the ileum contracts, allowing chyme to enter the large intestine. Distension of the colon, on the other hand, produces a reflex contraction of the sphincter, preventing further material from entering.

About 500 mL of chyme enters the colon from the small intestine each day. Most of this material is derived from the secretions of the lower small intestine, since most of the ingested food has been absorbed before reaching the large intestine.

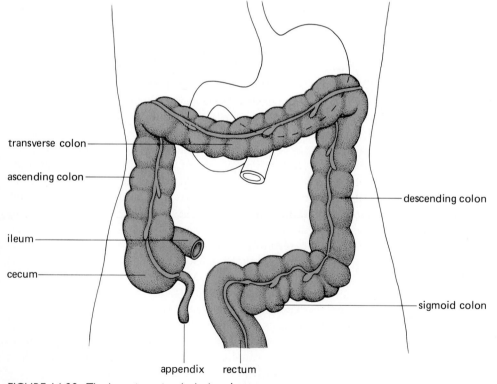

transverse colon

ascending colon

ileum

cecum

descending colon

sigmoid colon

appendix   rectum

FIGURE 14-28. The large intestine (colon) and rectum.

The primary absorptive process in the large intestine is the active transport of sodium from the lumen to blood with the accompanying osmotic reabsorption of water. If fecal material remains in the large intestine for a long time, almost all the water is absorbed, leaving behind dry fecal pellets. The cells lining the large intestine are unable to absorb either glucose or amino acids. There is a small net movement of potassium into the colon, and severe depletion of total body potassium can result when large volumes of fluid are excreted in the feces.

The large intestine also absorbs some of the products (vitamins, for example) synthesized by the bacteria inhabiting this region. Although this source of vitamins generally provides only a small part of the normal daily requirement, it may make a significant contribution when dietary intake of vitamins is low.

Other bacterial products contribute to the production of intestinal gas (**flatus**). This gas is a mixture of nitrogen and carbon dioxide, with small amounts of the flammable gases hydrogen, methane, and hydrogen sulfide. Bacterial fermentation produces gas in the colon at the rate of about 400 to 700 mL/day. Certain foods, beans for example, contain types of carbohydrates that are not digested by intestinal enzymes but are readily metabolized by bacteria in the large intestine, producing very large amounts of gas.

Motility and defecation. Contractions of the circular smooth muscle in the colon produce a segmentation motion considerably slower than that in the small intestine; a contraction may occur only once every 30 min. Because of this slow movement, material entering the colon from the small intestine remains for some 18 to 24 h, providing time for bacteria to grow and multiply. Three to four times a day, generally following a meal, a wave of intense contraction known as a **mass movement** spreads rapidly over the colon towards the rectum. This usually coincides with the gastroileal reflex, described earlier. Unlike a peristaltic wave, in which the smooth muscle at each point relaxes after the wave of contraction has passed, the smooth muscle of the colon

remains contracted for some time after a mass movement.

The **anus**, the exit from the large intestine, is normally closed by the **internal anal sphincter**, which is composed of smooth muscle, and the **external anal sphincter**, consisting of skeletal muscle under voluntary control. The sudden distension of the walls of the rectum produced by the mass movement of fecal material is the normal stimulus for defecation. It initiates the defecation reflex, which is mediated primarily by the nerves to the terminal end of the large intestine. The reflex response consists of a contraction of the rectum, relaxation of both anal sphincters, and increased peristaltic activity in the sigmoid colon. This activity is sufficient to propel the feces through the anus. Defecation is normally assisted by a deep inspiration, followed by closure of the glottis and contraction of the abdominal and chest muscles, causing a marked increase in intraabdominal pressure, which is transmitted to the contents of the large intestine and assists in the elimination of feces. This maneuver also causes a rise in intrathoracic pressure, which leads to a sudden rise in blood pressure followed by a fall in pressure as the venous return to the heart is decreased (Chap. 11). The cardiovascular stress resulting from excessive strain during defecation may precipitate a stroke or heart attack in individuals with cardiovascular disease.

Brain centers may, via descending pathways to somatic nerves, override the afferent input from the defecation reflex, thereby keeping the external sphincter closed and allowing a person to delay defecation. As with most skeletal muscle activity, voluntary control of the external anal sphincter is learned; thus a child gradually learns to control the anal sphincter. The conscious urge to defecate, mediated by stretched mechanoreceptors, accompanies the initial distension of the rectum. If defecation does not occur, the tension in the walls of the rectum decreases and the urge to defecate subsides until the next mass movement propels more feces into the rectum, increasing its volume and again initiating the defecation reflex.

About 150 g of feces, consisting of about 100 g of water and 50 g of solid material, is normally eliminated from the body each day. The solid matter is made up mostly of bacteria, undigested cellulose, debris from the turnover of the intestinal epithelium, bile pigments, and small amounts of salts.

Pathophysiology of the Gastrointestinal Tract

Since the function of the gastrointestinal system is to digest and absorb nutrients, salts, and water, most malfunctions of this organ system affect either the nutritional state of the body or its salt and water content. The effects of gastrointestinal malfunctioning on nutrition should be obvious, but the consequences for maintaining body fluid balance may require amplification. We have seen that the gastrointestinal system secretes 7000 mL into the lumen each day, the salt and water of these secretions being derived from plasma whose volume is only 3000 mL. Therefore, any significant decrease in the absorption of fluid by the intestinal tract leads to a decrease in plasma volume.

The following provide only a few examples of the most frequent disorders of gastrointestinal function.

Ulcers

Considering the high concentration of acid and pepsin secreted by the stomach, it is natural to wonder why the stomach does not digest itself. Several factors protect the walls of the stomach (and duodenum). The surface of the mucosa is lined with cells which secrete a slightly alkaline mucus, which forms a thin layer over the stomach surface. The protein content of mucus and its alkalinity both tend to neutralize hydrogen ions in the immediate area of the epithelium; i.e., mucus forms a chemical barrier between the highly acid contents of the lumen and the cell surface. In addition, the tight junctions between the epithelial cells lining the stomach have a very low permeability to hydrogen ions, preventing their diffusion into the underlying tissues. Finally, these epithelial cells are replaced every few days by new cells arising by the division of cells within the gastric pits.

Yet, in some people these protective mechanisms are inadequate, and erosions (**ulcers**) of the gastric wall occur. Ulcers may also occur in the lower part of the esophagus, as a result of the reflux of acid into the esophagus from the stomach, and in the duodenum which is bathed by acid emptied from the stomach. About 10 percent of the population of the United States have ulcers, which are about 10 times more frequent in the walls of the duodenum than in the stomach. Damage to blood vessels in the tissues underlying the ulcer may cause bleeding into the gastrointestinal lumen. On occasion, the ulcer may penetrate the entire wall, resulting in leakage of the luminal contents into the abdominal cavity.

The pain associated with ulcers probably results from irritation of exposed nerve fibers in the region of the ulcer, as well as contractile spasms of the smooth muscles irritated by acid. The pain from an ulcer often subsides following a meal, which buffers the acid secreted by the stomach. The pain is most intense during the early hours of the morning, when the acidity of the empty lumen is greatest. Ulcer formation requires that the mucosal barrier be broken, exposing the underlying tissue to the corrosive action of acid and pepsin, but it is not clear what produces the initial damage to the barrier, and it is likely that many factors are involved— genetic susceptibility, drugs, bile salts, and the excessive secretion of acid and pepsin are just some of the possibilities.

Regardless of the cause of barrier breakdown, reducing the level of acid and pepsin secretion, as for example by cutting the vagus nerves, promotes healing and tends to prevent the recurrence of ulcers. On the other hand, the secretion of large quantities of acid (due to the presence of gastrin-releasing tumors, for example) results in very severe duodenal ulcers. One should not get the impression, however, that a simple correlation exists between acid secretion and ulcer formation; many patients with ulcers have normal or even subnormal rates of acid secretion.

Despite popular notions that ulcers are due to emotional stress and despite the existence of a pathway (the vagus nerves) for mediating stress-induced increases in acid secretion, the actual role of stress in initiating ulcer formation remains uncertain.

Gallstones

As described earlier, bile secreted by the liver into the bile duct contains not only bile salts but cholesterol and phospholipids, which are water-insoluble and are maintained in soluble form in the bile fluid as micelles in association with the bile salts. When the concentration of cholesterol in the bile becomes too high in relation to the amounts of phospholipid and bile salts, cholesterol will crystallize out of solution forming **gallstones**. This occurs either when the liver secretes excessive amounts of cholesterol or when the bile becomes highly concentrated in the gallbladder. The concentration of bile by the gallbladder can cause a rise in the concentration of cholesterol to the point where it exceeds its capacity to be solubilized by the bile salts, thereby forming gallstones.

If the gallstone is small, it may pass through the bile duct into the intestine with no complications. A larger stone may become lodged in the gallbladder at its connection with the common bile duct, causing contractile spasms of the smooth muscle and pain. A more serious complication arises when a gallstone lodges in the common bile duct, thereby preventing bile from entering the intestine. The absence of bile in the intestine decreases the rate of fat digestion and absorption, so that approximately half of the ingested fat is not digested; it passes on to the large intestine and eventually appears in the feces. Furthermore, bacteria in the large intestine convert some of this fat into fatty acid derivatives which act upon the wall of the large intestine, altering salt and water movements and causing a net flow of fluid into the large intestine; the result is diarrhea and increased fluid loss from the body.

The buildup of pressure in a blocked bile duct inhibits further secretion of bile. As a result, bilirubin, which is normally secreted into the bile from the blood, accumulates in the blood and diffuses into tissues where it produces the yellowish coloration in the skin known as **jaundice**. Jaundice can also occur in the absence of a blocked bile duct, if the hepatocytes are damaged by liver disease and therefore fail to secrete bilirubin into the bile.

Since in most individuals the duct from the pancreas joins the common bile duct just before it enters the duodenum, a gallstone that becomes lodged at this point prevents both bile and pancreatic secretions from entering the intestine, resulting in failure both to neutralize acid from the stomach and to digest adequately most organic nutrients, not just fat. This results in severe nutritional deficiencies.

Why some individuals develop gallstones and others do not is still unclear. Women, for example, have about twice the incidence of gallstone formation of men, and American Indians have a very high incidence compared with other ethnic groups in America. Increased cholesterol secretion by the liver may contribute to gallstone formation in these susceptible individuals, but the factors responsible for the increased secretion are unknown.

Lactose intolerance

Lactose is the major carbohydrate of milk. It cannot be absorbed directly, but must first be digested into its components—glucose and galactose—which are readily absorbed by active-transport processes. This digestion is performed by the enzyme **lactase**, which is embedded in the plasma membrane of the intestinal epithelial cells. Lactase is present at birth (except in those rare cases where individuals have a genetic defect and are unable to form the active enzyme); its concentration then declines in some individuals between 18 and 36 months of age. (This decline does not occur in approximately 75 percent of white Americans and in most northern Europeans.) In the absence of lactase, lactose is not digested and remains in the lumen of the small intestine, where it tends to reduce water absorption (since the absorption of water requires prior absorption of solute to provide an osmotic gradient for osmosis). This unabsorbed lactose-containing fluid is passed on to the large intestine where bacteria, which do have enzymes capable of metabolizing lactose, produce from it large quantities of gas (which distends the colon, producing pain) and organic products which inhibit active-transport processes, thereby leading to an increase in the luminal osmolarity. The result is the movement of fluid into the lumen of the large intestine. Fluid loss in infants with congenital lack of lactase may be so large as to be fatal if they are not transferred to a diet without lactose. The response to milk ingestion by adults whose lactase levels have diminished during development varies from mild discomfort to severely dehydrating diarrhea, according to the volume of milk ingested and the amount of lactase present in the intestine.

Constipation and diarrhea

Many people have a mistaken belief that unless they have bowel movements every day, the absorption of toxic substances from fecal material in the large intestine will somehow poison them. Attempts to identify such toxic agents in the blood following prolonged periods of fecal retention have been unsuccessful. In unusual cases where defecation has not occurred for a year or more because of blockage of the rectum, no serious ill effects were noted (except for the discomfort of carrying around the extra weight of accumulated feces). There appears to be no physiological necessity for having bowel movements regulated by a clock; whatever maintains a person in a comfortable state is physiologically adequate, whether this means a bowel movement after every meal, once a day, or only once a week.

On the other hand, there often are some symptoms—headache, loss of appetite, nausea, and abdominal distension—which may arise when defecation has not occurred for long periods of time. These symptoms of **constipation** appear to be caused not by toxins but by the distension of the rectum, since inflating a balloon in the rectum of an individual without these symptoms produces similar sensations. In addition, the longer fecal material remains in the large intestine, the more water is absorbed, and the harder and drier the feces become, making defecation more difficult and sometimes painful. Thus, constipation tends to promote constipation.

Decreased motility of the large intestine is the primary factor causing constipation. This often occurs in old age or may result from damage to the nerve plexuses of the colon which coordinate motility. Emotional stress can also decrease the motility of the large intestine. One of the factors increasing the motility, and thus opposing the development of constipation, is distension of the large intestine. A variety of plant products, including cellulose and other complex polysaccharides collectively known as **dietary fiber**, provide a natural laxative in ingested food. Bran, cabbage, and carrots have a relatively high content of dietary fiber. These substances are not digested by the enzymes in the small intestine and are passed on to the large intestine, where their bulk produces distension and thereby increases motility. Much of the bulk of these substances is due to the large amount of water that they bind.

Laxatives, which increase the frequency of defecation, act through a variety of mechanisms. Some, such as mineral oil, simply lubricate the feces, making defecation easier and less painful. Others, containing magnesium and aluminum salts, are substances that are absorbed slowly, and thus their presence in the lumen leads to a retention of water. Still others, such as castor oil, stimulate the motility of the colon and affect membrane processes associated with ion movements across the wall (thus indirectly affecting water movements).

In some cases the excessive use of laxatives in attempting to maintain a preconceived notion of regularity leads to a decreased responsiveness of the colon to the normal defecation-promoting signals initiated by distension. In such cases, a long period without defecation may occur following cessation of laxative intake, appearing to confirm the person's belief in the necessity of taking laxatives to promote regularity.

Diarrhea is characterized by an increased water content in the feces and is usually accompanied by an increased frequency of defecation, stimulated by the distension of the colon from the increased fluid in the lumen. Diarrhea is ultimately the result of decreased fluid absorption or increased fluid secretion. The increased motility which usually accompanies diarrhea probably does not cause the diarrhea (by decreasing the time available for fluid absorption) but rather is the result of the distension produced by increased luminal fluid, which is the primary cause of the diarrhea.

A number of bacterial, protozoan, and viral diseases of the intestinal tract cause diarrhea. In many cases they do so by releasing substances which either alter various ion-transport processes in the epithelial membranes or damage the epithelial barriers by direct penetration into the mucosa. As we have described earlier, the presence of unabsorbed solutes in the lumen, as a result of decreased digestion or absorption, also results in retained fluid and diarrhea.

The major consequences of severe diarrhea are the effects upon blood volume resulting from the loss of fluid, and the effects on electrolyte and acid-base homeostasis resulting from the loss of ions such as potassium and bicarbonate.

15

REGULATION OF ORGANIC METABOLISM AND ENERGY BALANCE

SECTION A.
CONTROL AND INTEGRATION OF
CARBOHYDRATE, PROTEIN, AND FAT METABOLISM

In Chap. 5, we described the basic chemistry of living cells and their need for a continuous supply of nutrients. Although a certain fraction of these organic molecules is used in the synthesis of structural cell components, enzymes, coenzymes, hormones, antibodies, and other molecules serving specialized functions, most of the molecules in the food we eat are used by cells to provide the chemical energy required to maintain cell structure and function.

Crucial for an understanding of organic metabolism is the remarkable ability of most cells, particularly those of the liver, to convert one type of molecule into another. These interconversions permit the utilization of the wide range of molecules found in different foods, but there are limits and certain molecules must be present in the diet in adequate amounts. These are known collectively as **essential nutrients** (Chap. 5 and Appendix 2). For example, not only must enough protein be ingested to provide the nitrogen needed for synthesis of amino acids and other nitrogenous substances but this protein must also contain an adequate quantity of the specific **essential amino acids** which cannot be formed in the body by conversion from another molecule type; eight of the twenty amino acids fit this category. The other essential organic nutrients are several **fatty acids** and the **vitamins**.

The concept of a dynamic catabolic-anabolic steady state is also a critical component of organic metabolism. With few exceptions, e.g., DNA, virtually all organic molecules are continuously broken down and rebuilt, often at a rapid rate. For example, the turnover rate of body protein is approximately 100 g/day; i.e., this quantity is broken down into amino acids and resynthesized each day. Few of the molecules present in a person's skeletal muscle a month ago are still there today.

With the basic concepts of molecular interconvertibility and dynamic steady state as foundation, we can discuss organic metabolism in terms of total body interactions. Figure 15-1 summarizes the ma-

jor pathways of protein metabolism. The **amino acid pools**, which consist of the body's total free amino acids, are derived primarily from ingested protein (which is degraded to amino acids during digestion) and from the continuous breakdown of body protein. These pools are the source of amino acids for resynthesis of body protein and a host of specialized amino acid derivatives, such as nucleotides, epinephrine, etc. A very small quantity of amino acids and protein is lost from the body via the urine, skin, hair, fingernails, and, in women, the menstrual fluid.

The interactions between amino acids and the other nutrient types—carbohydrate and fat—are extremely important: (1) An amino acid may be converted into carbohydrate or fat by removal of its amino group; and (2) one type of amino acid may participate in the formation of another by passing its amino group to a carbohydrate. Both these processes were described in greater detail in Chap. 5. The ammonia, NH_3, formed during deamination is converted by the liver into **urea**, which is then excreted by the kidneys as the major end product of protein catabolism. Thus, not all the events relating to amino acid metabolism occur in all cells—urea is formed in one organ (the liver) but excreted by another (the kidneys)—but the concept of a pool is valid because all cells are interrelated by the vascular system and blood.

If any of the essential amino acids is missing from the diet, negative nitrogen balance (i.e., output greater than intake) always results. Apparently, the proteins for which that amino acid is essential cannot be synthesized, and the other amino acids which would have been incorporated into the proteins are metabolized. It should be obvious, therefore, why a dietary requirement for protein cannot be specified without regard to the amino acid composition of that protein. Protein is graded in terms of how closely its ratio of essential amino acids approximates their relative proportions in body protein. The highest-quality proteins are those found

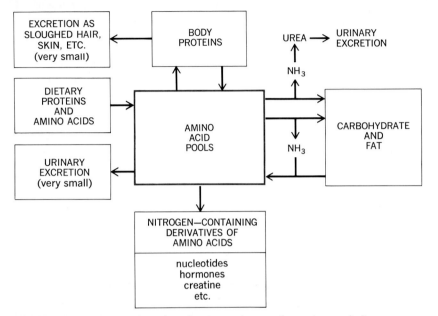

FIGURE 15-1. Amino acid pools and major pathways of protein metabolism.

in animal products, whereas the quality of most plant proteins is lower. Nevertheless, it is quite possible to obtain adequate quantities of all essential amino acids from a mixture of plant proteins alone, although the total quantity of protein ingested must be larger.

Figure 15-2 summarizes the metabolic pathways for carbohydrate and fat, which are considered together because of the high rate of conversion of carbohydrate to fat (see below and Chap. 5). The similarities between Figs. 15-1 and 15-2 are obvious, but there are several critical differences: (1) The major fate of both carbohydrate and fat is catabolism to yield energy, whereas amino acids can supply energy only after they are converted to carbohydrate or fat; and (2) excess carbohydrate and fat can be stored as such, whereas excess amino acids are not stored as protein but are converted to carbohydrate and fat.

In discussing the mechanisms which regulate the magnitude and direction of these molecular interconversions, we shall see that the liver, adipose tissue (the storage tissue for fat), and muscle are the dominant effectors, and that the major control-

ling input to them is a group of hormones and the sympathetic nerves to adipose tissue and the liver. At this point, the reader should review the biochemical pathways described in Chap. 5, particularly those dealing with glucose and the interconversions of carbohydrate, protein, and fat.

Events of the Absorptive and Postabsorptive States

When food is readily available, human beings can get along by eating small amounts of food all day long if they wish; however, this situation clearly does not hold for most other animals and did not hold for early man; indeed, food is not constantly available for most persons today. Mechanisms have evolved for survival during alternating periods of plenty and fasting. We speak of two functional states: the **absorptive state**, during which ingested nutrients are entering the blood from the gastrointestinal tract, and the **postabsorptive state**, during which the gastrointestinal tract is empty and energy must be supplied by the body's endogenous stores.

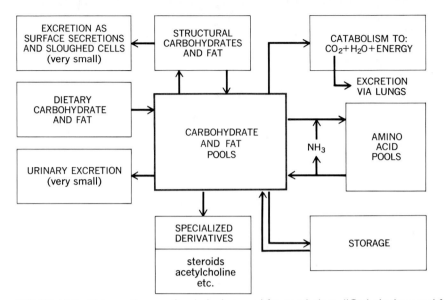

FIGURE 15-2. Major pathways of carbohydrate and fat metabolism. "Carbohydrate and fat pools" include the simple, unspecialized carbohydrates and fats dissolved in the body fluids. Note that structural carbohydrate and fat are constantly being broken down and resynthesized.

Since an average meal requires approximately 4 h for complete absorption, our usual three-meal-a-day pattern places us in the postabsorptive state during the late morning and afternoon and almost the entire night. The average person can easily withstand a fast of many weeks (so long as water is provided), and extremely obese persons have fasted (under medical supervision) for many months, being given only water and vitamins.

The absorptive state can be summarized as follows: (1) Glucose provides the major energy source, (2) only a small fraction of the absorbed amino acids and fat is utilized for energy, (3) another fraction of amino acids and fat is used to resynthesize the continuously degraded body proteins and structural fat, respectively, and (4) most of the amino acids and fat as well as carbohydrate not oxidized for energy are transformed into adipose-tissue fat.

In the postabsorptive state: (1) Carbohydrate is synthesized in the body, but its utilization for energy is greatly reduced, (2) the oxidation of endogenous fat provides most of the body's energy supply, and (3) fat and protein synthesis are curtailed and net breakdown occurs.

Figures 15-3 and 15-4 summarize the major pathways to be described. Although they may appear formidable at first glance, they should give little difficulty after we have described the component parts, and they should be referred to constantly during the following discussion.

Absorptive state

We shall assume an average meal to contain approximately 65 percent carbohydrate, 25 percent protein, and 10 percent fat. Recall from Chap. 14 that these nutrients enter the blood and lymph from the gastrointestinal tract primarily as monosaccharides, amino acids, and triacylglycerols, respectively. The first two groups enter the blood, which leaves the gastrointestinal tract to go directly to the liver by way of the hepatic portal vein, allowing this remarkable biochemical factory to alter the composition of the blood before it returns to the heart to be pumped to the rest of the body. In contrast, fat is absorbed not into the blood but into the lymph in the form of chylomicrons; the lymph drains into the systemic venous system in the neck,

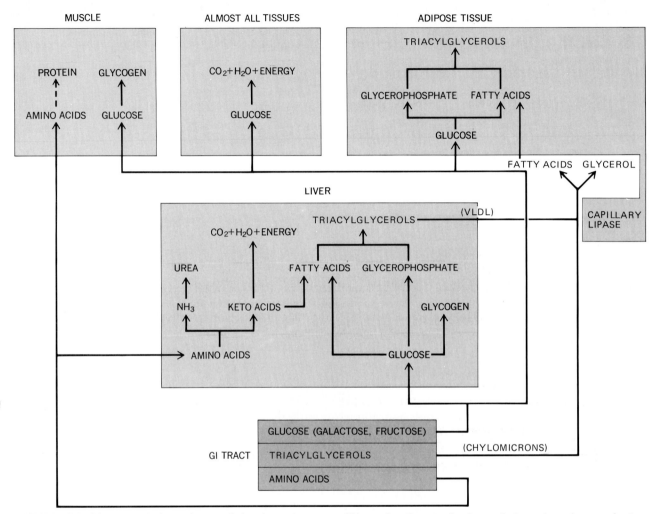

FIGURE 15-3. Major metabolic pathways of the absorptive state. Glycerophosphate is the form of glycerol used to synthesize triacylglycerol. See text for handling of triacylglycerols in adipose tissue. VLDL = very low-density lipoproteins.

and therefore the liver does not get first crack at the absorbed fat.

Glucose. Some of the absorbed carbohydrate is galactose and fructose, but since the liver converts most of these carbohydrates immediately into glucose (and because fructose enters essentially the same metabolic pathways as does glucose), we shall simply refer to these sugars as glucose. As shown in Fig. 15-3, much of the absorbed carbohydrate enters the liver cells, but little of it is oxidized for energy; instead it is stored as the polysaccharide glycogen or transformed into fat. The importance of glucose as a precursor of fat cannot be overemphasized; note that glucose provides both the glycerol (as **glycerophosphate**) and the fatty acid components of triacylglycerols. Some of this fat synthesized in the liver may be stored there, but most is packaged, along with specific proteins, into lipid-protein aggretates called **lipoproteins**, which are secreted from the liver cells and enter the blood. They are called **very low density lipoproteins**

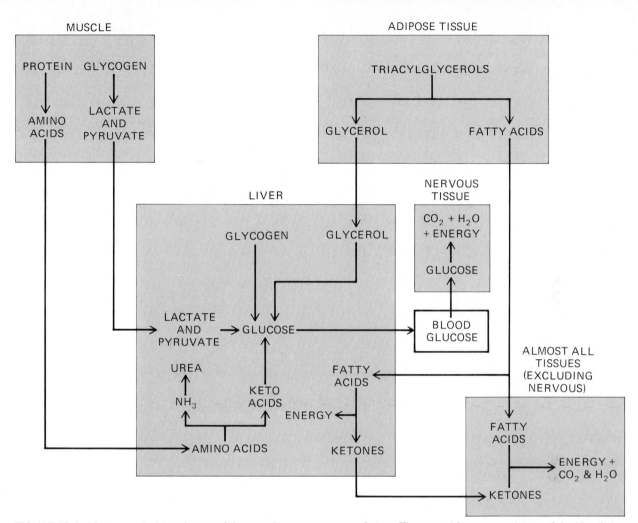

FIGURE 15-4. Major metabolic pathways of the postabsorptive state or fasting. The central focus is regulation of the blood glucose concentration.

(VLDL), because they contain much more fat (not only triacylglycerols but smaller amounts of phospholipids and cholesterol) than protein, and fat is less dense than protein. The synthesis of VLDL by liver cells occurs by processes similar to those for chylomicrons by intestinal mucosal cells, as described in Chap. 14.

Once in the bloodstream, VLDL complexes, being quite large, do not readily penetrate capillary walls, and their surface triacylglycerols are hydrolyzed into component molecules—glycerol and fatty acids—by a lipase located on the surface of capillary endothelium in contact with blood. The released fatty acids then move across the capillary wall and into the tissue cells where they combine with glycerophosphate (supplied largely by conversion from glucose) to form triacylglycerols once again. The capillaries within adipose tissue are particularly rich in this lipase, and so most of the fatty acids in the VLDL triacylglycerol synthesized by the liver end up being stored in adipose tissue.

What of the glucose absorbed from the gastrointestinal tract which does not enter liver cells? Much of it enters adipose-tissue cells, where (as in the liver) it is transformed into glycerophosphate and fatty acids, which then combine to form triacylglycerols. Another fraction is stored as glycogen in skeletal muscle and certain other tissues, and a very large fraction enters the various cells of the body and is oxidized to carbon dioxide and water, thereby providing the cells' energy requirements. Glucose is usually the body's major energy source during the absorptive state.

Triacylglycerols. As described in Chap. 14, almost all ingested fat is absorbed into the lymph as chylomicrons, containing mainly triacylglycerols with small amounts of other lipids as well as protein. The biochemical processing of these chylomicron triacylglycerols is quite similar to that just described for VLDL produced by the liver. Their fatty acids are released (mainly in adipose tissue) by action of the endothelial lipase, and the released fatty acids enter adipose-tissue cells and combine with glycerophosphate to form triacylglycerols. Thus, there are three major sources of the fatty acids found in adipose-tissue triacylglycerol: (1) fatty acids synthesized in the liver and transported via the blood in VLDL triacylglycerol to the adipose tissue, (2) fatty acids synthesized in adipose tissue, and (3) fatty acids in ingested fat transported to adipose tissue in chylomicron triacylglycerol. The use of fat to store excess calories is an excellent adaptation for mobile animals, because 1 g of triacylglycerol contains more than twice as many calories as 1 g of protein or glycogen and because there is very little water in adipose tissue. For simplicity, we have not shown in Fig. 15-3 that a fraction of fat is also oxidized during the absorptive state by various organs to provide energy. The actual amount utilized depends upon the content of the meal and the person's nutritional status.

Amino acids. Many of the absorbed amino acids enter liver cells and are converted into carbohydrate (keto acids) by removal of the amino group. The keto acids formed can enter the Krebs tricarboxylic acid cycle and be oxidized to provide energy for the liver cells. The keto acids can also be converted to fatty acids, thereby participating in fat synthesis by the liver. Most of the ingested amino acids are not taken up by the liver cells but enter other cells (Fig. 15-3). Although virtually all cells require a constant supply of amino acids for protein synthesis, we have simplified the diagram by showing only muscle because it constitutes the great preponderance of body mass and therefore contains the most important store, quantitatively, of body protein. Other organs, of course, participate, but to a lesser degree, in the amino acid exchanges occurring during the absorptive and postabsorptive states.

After entering the cells, the amino acids may be synthesized into protein. This process is represented by the dotted line in Fig. 15-3 to call attention to an important fact: Excess amino acids are not stored as protein, in the sense that glucose and fat are stored as fat and to a lesser degree as glycogen. Eating large amounts of protein does not significantly increase body protein; the excess amino acids are merely converted into carbohydrate or fat. On the other hand, a minimal supply of ingested amino acids is essential to maintain normal protein stores by preventing net protein breakdown in muscle and other tissues. During the usual alternating absorptive and postabsorptive states, the fluctuations in total body protein are relatively small. This discussion applies only to the adult; the growing child, of course, manifests a continuous increase of body protein.

Summary. During the absorptive period, anabolism exceeds catabolism, energy is provided primarily by glucose, body proteins are maintained, and excess calories (regardless of source) are stored mostly as fat. Glycogen constitutes a quantitatively less important storage form for calories.

Postabsorptive state

The essential problem during this period is that no glucose is being absorbed from the intestinal tract, yet the plasma glucose concentration must be maintained because the nervous system normally utilizes only glucose for energy. Perhaps the most convenient way of viewing the events of the postabsorptive state is in terms of how the blood glucose concentration is maintained. These events fall into two categories: (1) sources of glucose and (2) glucose sparing (fat utilization).

Sources of blood glucose. The sources of blood glucose during the postabsorptive period (Fig. 15-4) are as follows:

1. Glycogen stores in the liver are broken down to glucose, which is released into the blood. After the absorptive period is completed, the normal liver contains less than 100 g of glycogen; at 4 kcal/g, this provides 400 kcal, enough to fulfill the body's total caloric need for only 4 h.
2. Glycogen in muscle (and to a lesser extent other tissues) contains approximately the same amount of glucose as the liver. A complication arises because muscle lacks the necessary enzyme to form free glucose, and only free glucose can cross the plasma membrane and enter the blood. But glycolysis breaks the glycogen down into pyruvate and lactate, which are then liberated into the blood, circulate to the liver, and are converted into glucose, which can then enter the blood. Thus, muscle glycogen contributes to the blood glucose indirectly via the liver.
3. As shown in Fig. 15-4, the catabolism of triacylglycerols yields glycerol and fatty acids. The former can be converted into glucose by the liver, but the latter cannot. Thus, a potential source of glucose is adipose-tissue triacylglycerol breakdown, in which glycerol is liberated into the blood, circulates to the liver, and is converted into glucose.
4. The major source of blood glucose during a fast of more than a few hours is protein. Large quantities of protein in muscle and to a lesser extent other tissues are not absolutely essential for cell function; i.e., a sizable fraction of cell protein can be catabolized, as during prolonged fasting, without serious cellular malfunction. There are, of course, limits to this process, and continued protein loss ultimately means functional disin-

tegration, sickness, and death. Before this point is reached, protein breakdown can supply large quantities of amino acids, which are converted into glucose by the liver.

In conclusion, for survival of the brain (at least during the first day or so of a fast) plasma glucose concentration must be maintained. Glycogen stores, particularly in the liver, form the first line of defense, are mobilized quickly, and can supply the body's needs for several hours, but they are inadequate for longer periods. During longer fasts, protein and, to a much lesser extent, fat supply amino acids and glycerol, respectively, for production of glucose by the liver. Hepatic synthesis of glucose from pyruvate, lactate, glycerol, and amino acids is known as **gluconeogenesis**, i.e., new formation of glucose. During a 24-h fast, it amounts to approximately 180 g of glucose. The kidneys are also capable of glucose synthesis from the same sources, particularly in a prolonged fast, at the end of which they may be contributing as much glucose as the liver.

Glucose sparing (fat utilization). A simple calculation reveals that even the 180 g of glucose per day produced by the liver during fasting cannot possibly supply all the body's energy needs: 180 g/day × 4 kcal/g = 720 kcal/day, whereas normal total energy expenditure equals 1500 to 3000 kcal/day. The following essential adjustment must therefore take place during the transition from absorptive to postabsorptive state: The nervous system continues to utilize glucose, but virtually all other organs and tissues markedly reduce their oxidation of glucose and increase their utilization of fat, which becomes the major energy source, thus sparing the glucose produced by the liver for use by the nervous system. The essential step is the catabolism of adipose-tissue triacylglycerol to liberate fatty acids into the blood (in which they circulate in combination with plasma albumin). These fatty acids are picked up by virtually all tissues (excluding the nervous system), enter the Krebs cycle, and are oxidized to carbon dioxide and water, thereby providing energy. The liver, too, utilizes fatty acids for its energy, oxidizing them to acetyl CoA. However, this acetyl CoA is then processed into a group of compounds called **ketone bodies** instead of being oxidized further via the Krebs cycle. (One of the ketone bodies is acetone, some of which is exhaled and accounts for the distinctive breath odor of persons undergoing prolonged fasting or suffering from severe untreated diabetes mellitus.) These ketone bodies are released into the blood and provide an important energy source for the many tissues capable of oxidizing them via the Krebs cycle.

The net result of fatty acid utilization during fasting is provision of energy for the body and sparing glucose for the brain. The combined effects of gluconeogenesis and the switch-over to fat utilization are so efficient that, after several days of complete fasting, the plasma glucose concentration is reduced only by a few percent. After one month, it is decreased only 25 percent.

There also occurs an important change in brain metabolism with prolonged starvation. Many areas of the brain are capable of utilizing ketone bodies for energy, and so as these substances begin to build up in the blood after the first few days of a fast, the brain begins to utilize large quantities of ketone bodies, as well as glucose, for its energy source. The survival value of this phenomenon is very great; if the brain significantly reduces its glucose requirement (by utilizing ketone bodies instead of glucose), much less protein need be broken down to supply the amino acids for gluconeogenesis. Accordingly, the protein stores will last longer, and the ability to withstand a long fast without serious tissue disruption is enhanced.

Thus far, our discussion has been purely descriptive; we now turn to the endocrine and neural factors which so precisely control and integrate these metabolic pathways and transformations. As before, the reader should constantly refer to Figs. 15-3 and 15-4. We shall focus primarily on the following questions (Fig. 15-5) raised by the previous discussion: (1) What controls the shift from the net anabolism of protein, glycogen, and triacylglycerol to net catabolism? (2) What induces primarily glucose utilization during absorption and fat utilization during postabsorption; i.e., how do cells "know"

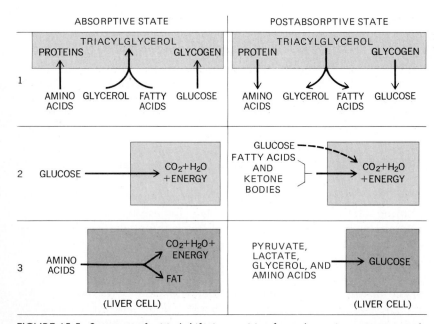

FIGURE 15-5. Summary of critical shifts in transition from absorptive state to postabsorptive state. The phrase "absorptive state" could be replaced with "actions of insulin"; "postabsorptive state" with "results of decreased insulin."

they should start oxidizing fatty acids and ketones instead of glucose? (3) What drives net liver glucose synthesis (gluconeogenesis) and release during postabsorption?

Endocrine and Neural Control of the Absorptive and Postabsorptive States

The major controls of the transitions from feasting to fasting and vice versa are a group of pancreatic hormones (insulin, glucagon, and somatostatin), the hormone epinephrine (from the adrenal medulla), and the sympathetic nerves to liver and adipose tissue. The pancreatic hormones are proteins secreted by the **islets of Langerhans**, clusters of endocrine cells in the pancreas. Appropriate histological techniques reveal four distinct types of islet cells, termed A, B, D, and F cells, each of which secretes a different hormone. The **B cells** are the source of **insulin**, the **A cells** of **glucagon**, the **D cells** of **somatostatin**, and the **F cells** of **pancreatic polypeptide**. The function of pancreatic polypeptide is not known.

Insulin

Insulin acts directly or indirectly on most tissues of the body with the notable exception of most brain areas. The effects of insulin are so important and widespread that an injection of this hormone into a fasting person duplicates the absorptive-state pattern of Fig. 15-5; conversely, persons suffering from insulin deficiency manifest the postabsorptive pattern of Fig. 15-5. From these statements alone, one might predict (correctly) that secretion of insulin is stimulated by eating and inhibited by fasting.

Insulin induces in its target cells a large number of changes which fall into two general categories: alteration of membrane transport or alteration of enzyme function (Fig. 15-6). What are the biochemical mechanisms by which these changes are brought about? Insulin combines with receptors specific for it on the outer surface of the plasma membrane, and it is this combination which triggers the sequences of events leading ultimately to the changes in membrane transport and enzyme func-

tion. Almost certainly involved in some of these sequences is at least one "second messenger," but its (or their) identity has not been established with certainty. Insulin is clearly one hormone that does not activate adenylate cyclase; indeed it inhibits this enzyme, thereby decreasing intracellular cyclic AMP.

Effects on membrane transport. Glucose enters most cells by the carrier-mediated mechanism which we described as facilitated diffusion in Chap. 6. Insulin stimulates this facilitated diffusion of glucose into many cells, particularly cells of muscle and adipose tissue (but not brain or liver), and greater glucose entry into cells increases the availability of glucose for all the reactions in which glucose participates. Insulin also stimulates the active transport of amino acids into most cells, thereby making more amino acids available for protein synthesis.

Effects on enzymes. The enhanced membrane transport of glucose and amino acids induced by insulin favors, by mass action, glucose oxidation and net synthesis of glycogen, triacylglycerol, and protein. However, in addition, insulin alters the activities or concentrations of many of the intracellular enzymes involved in the anabolic and catabolic pathways of these substances so as to achieve the same result. Glycogen synthesis provides an excellent example: Increased glucose uptake per se stimulates glycogen synthesis, but in addition, insulin increases the activity of the rate-limiting step in glycogen synthesis and inhibits the enzyme which catalyzes glycogen catabolism; thus, insulin favors glucose transformation into glycogen by a triple-barrelled effect!

The situation for triacylglycerol storage is analogous: Increased entry of glucose into adipose tissue provides the precursors for the synthesis of fatty acid and glycerophosphate, which then combine to form triacylglycerol; simultaneously, insulin increases the activity of certain of the enzymes which catalyze fatty acid synthesis, and most important, inhibits the enzyme which catalyzes triacylglycerol breakdown. Again there is a triple-barrelled effect favoring, in this case, triacylglycerol storage.

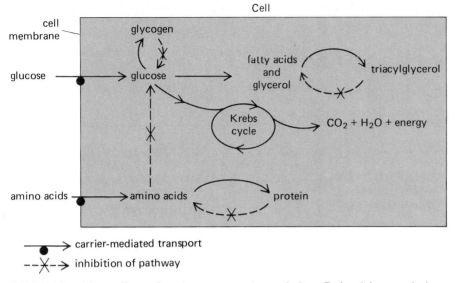

FIGURE 15-6. Major effects of insulin upon organic metabolism. Each solid arrow (→) represents a process enhanced by insulin, whereas a dashed and marked arrow (-X→) denotes a reaction inhibited by insulin. Conversely, when insulin concentration is low, the activity of the dashed pathways is enhanced and that of the solid pathways is decreased. The nontransport effects are mediated by insulin-dependent enzymes. All these effects are not necessarily exerted upon all cells.

And similarly for protein: Insulin stimulates uptake of amino acids while simultaneously increasing the activity of some of the ribosomal enzymes which mediate the synthesis of protein from these amino acids and inhibiting the enzymes which mediate protein catabolism.

Finally, insulin inhibits almost all the critical liver enzymes which catalyze gluconeogenesis; the net result is that insulin abolishes glucose release by the liver. At the same time, insulin enhances the metabolism of glucose by the liver. It does so not by stimulating the transport process for glucose (as it does in many other cells), but rather by increasing the activity of the enzyme which catalyses the first step in glucose metabolism. Other enzymes important for glucose utilization, both in liver and elsewhere, are also activated by insulin.

Effects of decreased insulin. We have thus far dealt with the positive effects of insulin. Insulin deficit will have just the opposite results; when the blood concentration of insulin decreases, the metabolic pattern is shifted toward decreased glucose entry and oxidation and a net catabolism of glycogen, protein, and triacylglycerol. In other words, these

metabolic pathways are in a dynamic state, capable of proceeding, in terms of net effect, in either direction. For this reason, energy metabolism can be shifted from the absorptive to the postabsorptive pattern merely by lowering the rate of insulin secretion: Glucose entry and oxidation decrease; glycogen breakdown increases; net protein catabolism liberates amino acids into the blood; net triacylglycerol catabolism liberates glycerol and fatty acids into the blood, and the resulting higher fatty acid concentration in blood facilitates cellular uptake of fatty acids, which in turn stimulates fatty acid oxidation; and gluconeogenesis is stimulated not only by the increased availability of precursors (amino acids and glycerol) but by enzyme changes in the liver itself. The glucose released by the liver can be utilized by the brain since the brain's glucose uptake is not insulin-dependent.

Control of insulin secretion. One direct control of insulin secretion is the glucose concentration of the blood flowing through the pancreas, a simple system requiring no participation of nerves or other hormones. An increase in blood glucose concentration stimulates insulin secretion; conversely, a reduction inhibits secretion. The feedback nature of this system is shown in Fig. 15-7. A rise in plasma glucose stimulates insulin secretion; insulin induces rapid entry of glucose into cells as well as cessation of glucose output by the liver; these effects reduce the blood concentration of glucose, thereby removing the stimulus for insulin secretion, which returns to its previous level.

Figure 15-8 illustrates typical changes in plasma glucose and insulin concentrations following glucose ingestion. Note the association between the rising blood glucose (resulting from gastrointestinal absorption) and the plasma insulin increase induced by the glucose rise. The low postabsorptive values for plasma glucose and insulin concentrations are not the lowest attainable, and prolonged fasting induces even further reductions of both variables until insulin is barely detectable in the blood.

Plasma glucose is not the sole control over insulin secretion, for insulin secretion is sensitive to numerous other types of input (Fig. 15-9). One is the plasma concentration of certain amino acids, an elevated amino acid concentration causing enhanced insulin secretion. This is easily understandable since amino acid concentrations increase after eating, particularly after a high-protein meal. The increased insulin stimulates cell uptake of these amino acids.

There are also important neural and hormonal controls over insulin secretion. For example, several of the hormones secreted by the gastrointes-

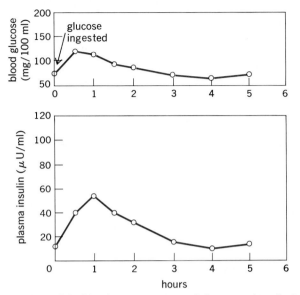

FIGURE 15-8. Blood concentrations of glucose and insulin following ingestion of 100 g of glucose (glucose tolerance test). Study performed on normal human subjects. *(Adapted from Daughaday and Kipnis.)*

tinal tract (see Chap. 14) in response to the presence of food stimulate the release of insulin; this provides an "anticipatory" (or "feedforward") component to glucose regulation, i.e., insulin secretion will rise earlier than it would if plasma glucose were the only controller. Important neural controls are the autonomic neurons to the islets. Activation of the parasympathetic neurons, which occurs during ingestion of a meal, stimulates secretion of insulin and constitutes a second type of anticipatory regulation. Activation of the sympathetic neurons to the islets, as occurs during exercise or "stress," inhibits insulin secretion, and the significance of this relationship will be described later. Finally, insulin secretion is influenced by a variety of hormones acting directly on the B cells, but the physiological significance of these inputs is still unclear.

Diabetes mellitus. The name diabetes, meaning syphon or running through, was used by the Greeks over 2000 years ago to describe the striking urinary volume excreted by certain people. Mellitus, meaning sweet, distinguishes this urine from the large

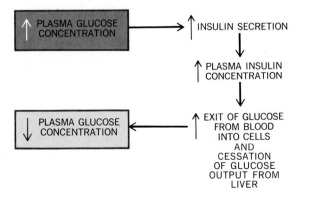

FIGURE 15-7. Negative-feedback nature of plasma glucose control over insulin secretion.

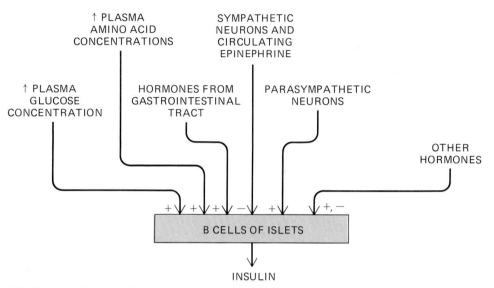

FIGURE 15-9. Control of insulin secretion.

quantities of nonsweet (insipid) urine produced by persons suffering from ADH deficiency. Currently, as described in Chap. 13, the latter disorder is known as diabetes insipidus, and the unmodified word "diabetes" is used as a synonym for diabetes mellitus. (Why there is a relationship between urine formation and organic metabolism will be described shortly.)

Diabetes is not one but several diseases with differing causes. Classification of these rests on the status of insulin secretion and dependence on insulin treatment. In **Type 1** ("insulin-dependent") **diabetes**, the hormone is completely or almost completely absent; in **Type 2** ("insulin-independent") **diabetes**, the hormone is often present at near-normal or even above-normal levels.

Type 1 is the less common, affecting 10 to 20 percent of diabetic patients. It is due to the total or near-total destruction of the pancreatic B cells, which means that the afflicted person is absolutely dependent on therapy with exogenous insulin. The cause of the B cell destruction probably varies, to some extent, from patient to patient, but seems to involve some interaction of genetic susceptibility, viral infection, and an attack on the B cells by the body's own defense mechanism (autoimmune disease, see Chap. 17).

Because of their absolute deficiency of insulin, untreated patients with Type 1 diabetes always have markedly elevated plasma glucose concentrations, both because glucose fails to enter cells normally and because the liver continues to make and release glucose. We have previously described how a low insulin concentration induces the metabolic profile characteristic of the postabsorptive state, and so the changes (other than the increased plasma glucose) seen in untreated Type 1 diabetes (Fig. 15-10) represent a gross caricature of this state. The catabolism of triacylglycerol with resultant elevation of plasma fatty acids and ketones is an appropriate response because these substances must provide energy for the body's cells, most of which are prevented from taking up adequate glucose by the insulin deficiency (only the brain is spared glucose deprivation since its uptake of glucose is not insulin-dependent). In contrast, glycogen and protein catabolism and the marked gluconeogenesis so important to maintain plasma glucose during fasting are completely inappropriate in the diabetic since plasma glucose is already high. These reactions serve only to raise the plasma glucose still higher.

The elevated plasma glucose of diabetes induces changes in renal function with serious consequences. In Chap. 13, we pointed out that a nor-

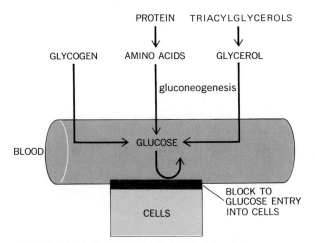

FIGURE 15-10. Factors which elevate blood glucose concentration in insulin deficiency.

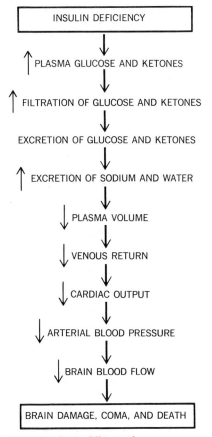

FIGURE 15-11. Effects of severe untreated insulin deficiency on renal function.

mal person does not excrete glucose, because all glucose filtered at the glomerulus is reabsorbed by the tubules. However, the elevated plasma glucose of diabetes may so increase the filtered load of glucose that the maximum tubular reabsorptive capacity is exceeded and large amounts of glucose may be excreted. For the same reasons, large amounts of ketones may also appear in the urine. These urinary losses, of course, only aggravate the situation by further depleting the body of nutrients and leading to weight loss. Far worse, however, is the effect of these solutes on sodium and water excretion. In Chap. 13, we saw how tubular water reabsorption is a passive process induced by active solute reabsorption. In diabetes, the osmotic force exerted by unreabsorbed glucose and ketones holds water in the tubule, thereby preventing its reabsorption. For several reasons (the mechanisms are beyond the scope of this book) sodium reabsorption is also retarded. The net result is marked excretion of sodium and water, which can lead, by the sequence of events shown in Fig. 15-11, to hypotension, brain damage, and death.

Another serious abnormality in diabetes is a markedly increased plasma hydrogen-ion concentration, due primarily to the accumulation of ketone bodies, several of which are moderately strong acids. The kidneys respond to this increase by excreting more hydrogen ion and are generally able

to maintain balance fairly well, at least until the volume depletion described above interferes with renal function. What effect does the increased hydrogen-ion concentration have on respiration? Hyperventilation occurs in response to stimulation of the peripheral chemoreceptors by hydrogen ion, and the resulting overexcretion of carbon dioxide further helps to regulate the hydrogen-ion concentration below lethal limits.

The major therapy for patients with Type 1 diabetes is administration of insulin, which must be given by injection since, as a protein, it is broken down by gastrointestinal enzymes. The dose must be determined carefully since an overdose excessively lowers the plasma glucose concentration, which can interfere with brain function.

The grim picture painted above is seen only in patients with untreated Type 1 diabetes, i.e., those with almost total inability to secrete insulin. The great majority of diabetics are in the Type 2 category and never develop these severe metabolic derangements. Indeed many of these persons can be treated adequately without administration of exogenous insulin (this is why Type 2 diabetes is called "insulin independent"). A further indication of the difference in severity between Type 1 and Type 2 is that many persons with Type 2 diabetes have relatively normal plasma glucose concentrations when fasting; such persons require a glucose tolerance test, i.e., administration of concentrated glucose, to bring out their defective handling of glucose.

As stated earlier, Type 2 diabetes is characterized by near-normal or even above-normal plasma levels of insulin. The problem is that insulin's target cells do not respond normally to the circulating insulin, because of alterations either in the insulin receptors or some process beyond the receptors, i.e., in one of the intracellular receptor-triggered metabolic events.

An insufficient number of insulin receptors per target cell seems to be the major problem in the great majority of Type 2 diabetics who are overweight (obese). (Given the earlier mention of progressive weight loss as a symptom of diabetes, it may seem contradictory that many diabetics are overweight, but the paradox is resolved when one recalls that only patients with severe insulin deficiency suffer so much loss of glucose in the urine as to lose weight.) The sequence of events coupling obesity to a diminished number of insulin receptors is as follows (Fig. 15-12): Because, as we have seen, insulin secretion is increased during ingestion and absorption of food, any person (diabetic or not) who chronically overeats secretes, on the average, increased amounts of insulin; the resulting elevation of plasma insulin induces, over time, a reduction in the number of insulin receptors (downregulation, Chap. 9); thus, insulin, itself, is responsible for the decrease in target-cell responsiveness, the end result of which is a rise in plasma glucose concentration. The latter is small in nondiabetic overeaters because the islet cells respond to it by secreting enough additional insulin to get the job done despite the reduction in available receptors; in contrast, the diabetes-prone person may also secrete additional

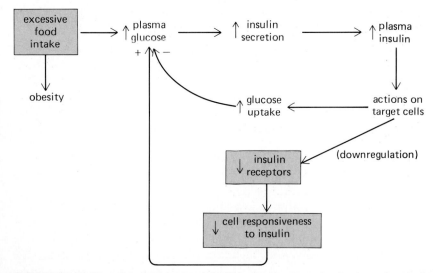

FIGURE 15-12. Postulated mechanism by which *chronic* overeating leads to chronically elevated plasma insulin and diminished cell responsiveness to insulin. Note that a chronically increased plasma insulin contributes to both a negative-feedback and (via downregulation) a positive-feedback effect on plasma glucose.

insulin but not enough to prevent significant hyperglycemia.

The alteration in receptor number secondary to overeating can usually be completely reversed if the person simply reduces his or her caloric intake. As the number of receptors returns to normal (this begins even while the person is still overweight as long as caloric intake is below expenditure), so do the responses of target cells to insulin; thus dietary control, without any other therapy, is frequently sufficient to eliminate the elevated blood glucose of Type 2 diabetes. An exercise program also is useful, since the number of insulin receptors is increased by frequent performance of endurance exercise, such as jogging, independently of changes in body weight.

Given the fact that even untreated Type 2 diabetics do not usually develop the severe metabolic and urinary derangements typical of untreated Type 1, one might wonder why therapy is so important. Unfortunately, persons with either type of diabetes tend to develop a variety of chronic abnormalities, including atherosclerosis, small-vessel and nerve disease, susceptibility to infection, and blindness. The factors responsible are still unclear, but it is quite likely that an elevated plasma glucose, itself, contributes; accordingly the assumption is that the more normal the plasma glucose, the more slowly these complications will arise.

Glucagon

Insulin unquestionably plays a central role in controlling the metabolic adjustments required for feasting or fasting, but glucagon, a protein hormone produced by the A cells of the pancreatic islets,[1] is also very important. The major effects of glucagon on organic metabolism (mediated by generation of cyclic AMP in its target cells) are all opposed to those of insulin (Fig. 15-13). They include: increased glycogen breakdown in liver; increased synthesis of glucose (gluconeogenesis) and ketone bodies by liver; and increased breakdown of adi-

pose-tissue triacylglycerol (fat mobilization). Thus, the overall results of glucagon's effects are to increase the plasma concentrations of glucose and fatty acids, all important events of the postabsorptive period.

From a knowledge of these effects, one would logically suppose that the secretion of glucagon should be increased during the postabsorptive period and prolonged fasting, and such is the case. The stimulus is a decreased plasma glucose concentration. The adaptive value of such a reflex is obvious; a decreasing plasma glucose concentration induces increased release of glucagon which, by its effects on metabolism, serves to restore normal blood glucose levels and, at the same time, supply fatty acids for cell utilization (Fig. 15-13). Conversely, an increased or increasing plasma glucose inhibits glucagon's secretion, thereby helping to return plasma glucose toward normal.

The pathway for this reflex is quite simple: The A cells in the pancreas respond directly to changes in the glucose concentration of the blood perfusing the pancreas.

Thus far, the story is quite uncomplicated; the B and A cells of the pancreatic islets constitute a push-pull system for regulating plasma glucose by producing two hormones (insulin and glucagon) whose actions and controlling input are just the opposite of each other. However, we must now point out a complicating feature: A second major control of glucagon secretion is the plasma amino acid concentration (acting directly on the A cells), and in this regard the effect is identical rather than opposite to that for insulin: Glucagon secretion, like that of insulin, is strongly stimulated by a rise in plasma amino acid concentration such as occurs following a protein-rich meal. Thus, during absorption of a carbohydrate-rich meal containing little protein, there occurs an increase in insulin secretion alone, caused by the rise in plasma glucose, but during absorption of a low-carbohydrate–high-protein meal, glucagon also increases, under the influence of the increased plasma amino acid concentration. The usual meal is somewhere between these extremes and is accompanied by a rise in insulin and relatively litle change in glucagon, since

---

[1] Glucagon and glucagon-like substances are also secreted by cells in the lining of the gastrointestinal tract morphologically similar to the A cells; the function of these substances is unclear.

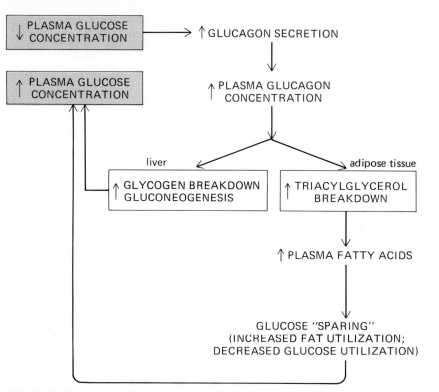

FIGURE 15-13. Negative-feedback nature of plasma glucose control over glucagon secretion (compare with Fig. 15-7).

the simultaneous increases in blood glucose and amino acids counteract each other so far as glucagon secretion is concerned. Of course, regardless of the type of meal ingested, the postabsorptive period is always accompanied by a rise in glucagon secretion (and a decrease in insulin secretion).

What is the adaptive value of the amino acid–glucagon relationship, i.e., of the fact that a high-protein meal stimulates glucagon secretion? Imagine what might occur were glucagon not part of the response to a high-protein meal: Insulin secretion would be increased by the amino acids but, since little carbohydrate was ingested and therefore available for absorption, the increase in plasma insulin could cause a marked and sudden drop in plasma glucose. In reality, the rise in glucagon secretion caused by the amino acids permits the hyperglycemic effects of this hormone to counteract the hypoglycemic actions of insulin, and the net result is a stable plasma glucose. Thus, a high-protein

meal virtually free of carbohydrate can be absorbed with little change in plasma glucose despite a marked increase in insulin secretion.

The secretion of glucagon, like that of insulin, is controlled not only by the plasma concentrations of nutrients but by neural and hormonal inputs to the islets. Perhaps the most important of these for glucagon secretion is the sympathetic nerves, activation of which stimulates glucagon secretion (just the opposite of their effect on insulin secretion). The adaptive significance of this relationship for exercise and stress will be described subsequently.

Because glucagon induces an elevation of plasma glucose, an important question is whether diabetes might involve an excess of glucagon. One approach to this question has been to measure the concentration of glucagon in the plasma of patients with diabetes. Recall that, since glucagon secretion is inhibited by an elevated plasma glucose, one would expect to find a low plasma glucagon in diabetic persons (because they have increased plasma glucose) if glucagon had nothing to do with their disease. However, the actual finding is that most diabetic patients have plasma glucagon concentrations which are increased (either absolutely or relative to their elevated plasma glucose). The failure of glucagon to decrease seems to result from a partial loss of glucose-sensing ability by the A cells. It is now clear that absolute or relative glucagon excess contributes to the carbohydrate and fat abnormalities typical of diabetes. For example, if glucagon secretion is eliminated in experimental animals with Type 1 diabetes, the massive hepatic overproduction of glucose and ketone bodies usually observed in this type of diabetes does not occur, despite the total absence of insulin.

Pancreatic somatostatin

The third pancreatic hormone, secreted by the D cells of the islets, is somatostatin. This peptide is an excellent example of an intercellular chemical messenger produced by cells in a variety of locations. As mentioned in Chap. 9, the somatostatin secreted by neuroendocrine cells in the hypothalamus functions as the hormone which inhibits secretion of growth hormone by the anterior pituitary. Somatostatin is also found in other nervous system locations where it probably serves as a neurotransmitter, and it is produced by cells lining the gastrointestinal tract.

The physiology of pancreatic somatostatin is much less well understood than that of insulin and glucagon. The secretion of somatostatin by the D cells is increased by several of the same stimuli that increase insulin secretion—glucose, amino acids, and several gastrointestinal hormones. Therefore, somatostatin is released from the pancreas during absorption of a meal. It then exerts a variety of inhibitory effects on the gastrointestinal tract, including slowed gastric emptying, decreased gastric acid secretion, and diminished contraction of the gall bladder. Thus, pancreatic somatostatin, in a negative-feedback fashion, helps prevent excessive levels of nutrients in the plasma by restraining the rates at which a meal is digested and absorbed.

In describing the factors which control the secretion of insulin, glucagon, and somatostatin, we did not mention the fact that under appropriate experimental conditions, each of these three hormones can exert potent effects on the secretion of one or both of the others (for example, administration of small amounts of somatostatin can inhibit the secretion of both insulin and glucagon). This fact, coupled with the intricate anatomical arrangement of the A, B, and D cells within the islets, has led to the hypothesis that the hormones released from their distinct islet cells may diffuse to other islet cells and influence their function. However, the role of any such paracrine communication is still unknown.

Epinephrine and sympathetic nerves to liver and adipose tissue

Circulating epinephrine, the major hormone secreted by the adrenal medulla (the endocrine gland which is an integral part of the sympathetic nervous system), inhibits insulin secretion and stimulates glucagon secretion. In addition, this hormone also exerts direct effects on nutrient metabolism. These include stimulation of: (1) glycogenolysis (by both liver and skeletal muscle, (2) gluconeogenesis (by the liver), and (3) breakdown of adipose tissue triacylglycerol (fat mobilization).

Activation of the sympathetic nerves to liver and adipose tissue elicits essentially the same responses as does circulating epinephrine. Thus, the direct overall effects on organic metabolism of enhanced activity of the sympathetic nervous system are the opposite of those of insulin and similar to those of glucagon: increased plasma concentrations of glucose, glycerol, and fatty acids (Fig. 15-14).

As might be predicted from these effects, an important stimulus leading to increased secretion of epinephrine and activity of the nerves to liver and adipose tissue is a decreased plasma glucose. This is the same stimulus which, as described above, leads to increased secretion of glucagon (although the receptors and pathways are totally different), and the adaptive value of the response is also the same: Blood glucose is restored toward normal and fatty acids are supplied for cell utilization. The

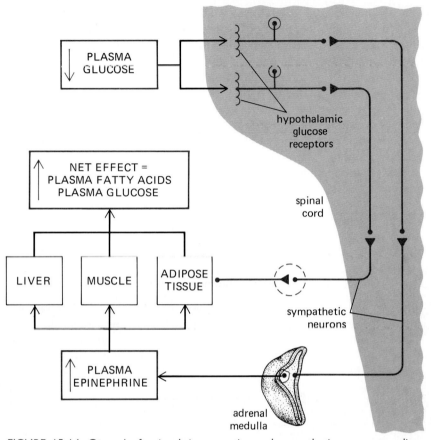

FIGURE 15-14. Control of epinephrine secretion and sympathetic nerves to adipose tissue by plasma glucose concentration. Sympathetic neurons to the liver (not shown in the figure) also participate. A decrease in plasma glucose stimulates the hypothalamic glucose receptors and, via the reflex chain shown, restores plasma glucose to normal while at the same time increasing plasma fatty acids.

pathway for this reflex is illustrated in Fig. 15-14. As described in Chap. 9, epinephrine release is controlled entirely by the preganglionic sympathetic fibers to the adrenal medulla; the receptors, glucose receptors in the brain (probably in the hypothalamus), initiate the reflexes which lead ultimately to increased activity in these neurons and in the sympathetic pathways to the liver and adipose tissue. This reflex is triggered when the plasma glucose concentration is rapidly decreasing, and many of the symptoms associated with such a state are accounted for by the actions of the large quantity of epinephrine, particularly those on the cardiovas-

cular and nervous systems—palpitations, tremor, headache, nervousness, cold sweating, dizziness, etc.

It is not yet known whether epinephrine and the sympathetic nervous system play any role in the metabolic adjustments to *prolonged* fasting (as contrasted to acute reductions in plasma glucose described above). Recent evidence suggests that the overall activity of the sympathetic nervous system, contrary to expectations, actually decreases under such conditions; the possible adaptive significance of this change is discussed later in this chapter.

Other hormones

In addition to the four hormones described in this section, there are others—growth hormone, cortisol, thyroid hormone, sex steroids, vasopressin, and angiotensin—which have various effects on organic metabolism. However, the secretion of these hormones is not primarily keyed to the transitions between the absorptive-postabsorptive states but is controlled by other factors; i.e., these hormones are, for the most part, involved in other homeostatic processes. Therefore, their effects are mainly described elsewhere. However, several of the effects of cortisol and growth hormone are so important for the maintenance of plasma glucose concentration that they do warrant brief description here.

Cortisol. Cortisol, the major glucocorticoid produced by the adrenal cortex (see Chap. 9), is essential for the metabolic adaptations to fasting. Its role in this regard offers an excellent example of "permissiveness" in hormonal actions (Chap. 9). We have described how fasting is associated with stimulation of both gluconeogenesis and lipolysis; however, neither of these critical metabolic transformations occurs in a person deficient in cortisol. In other words, plasma cortisol need not *rise* during fasting (and usually does not), but it must be present in the blood to some degree in order to "permit" other hormonal changes (decreased insulin, increased glucagon and epinephrine) to be effective in eliciting gluconeogenesis and lipolysis. Somehow, the continuous presence of even small amounts of cortisol in the blood maintains the concentrations of the key liver and adipose-tissue enzymes required for these metabolic processes. It should be evident, therefore, that persons with cortisol deficiency develop severe decreases in their plasma glucose concentrations in response to fasting (or to injection of insulin).

We have emphasized that cortisol does not normally increase during fasting and does not itself trigger the metabolic responses to fasting. On the other hand, whenever its plasma concentration does increase greatly, as occurs physiologically during stress (see Chap. 17), cortisol does, in fact, elicit many of the metabolic events ordinarily associated with fasting. In large amounts, cortisol: (1) stimulates protein catabolism, (2) stimulates the liver's uptake of amino acids and their conversion to glucose (gluconeogenesis), (3) stimulates mobilization of adipose tissue fat, and (4) inhibits glucose uptake and oxidation by many peripheral tissues. Clearly, here is another hormone, in addition to glucagon and epinephrine, which can exert actions opposite to those of insulin, and persons suffering from cortisol excess often develop symptoms of diabetes.

Growth hormone. The primary physiological effects of growth hormone are to stimulate protein anabolism and growth (as will be described later in this chapter). Compared to these effects, those it exerts on carbohydrate and lipid metabolism are quite minor under most physiological situations. Nonetheless, as is true for cortisol, either severe deficiency or excess of the hormone does produce significant derangement of the flow of lipids and carbohydrates. Indeed, growth hormone's effects on these nutrients (but not on protein) are, overall, quite similar to those of cortisol and are opposite to those of insulin. Growth hormone: (1) facilitates fatty acid and glycerol mobilization from adipose tissue, (2) increases gluconeogenesis by the liver, and (3) decreases glucose uptake by many peripheral tissues. Accordingly, as is the case with cortisol, persons with chronic growth hormone excess may manifest symptoms of diabetes.

We should qualify the statement made earlier that growth hormone secretion does not vary with the feasting-fasting pattern. Some people do show a small burst of growth hormone secretion several hours after a meal, but it is unlikely that this change in plasma growth hormone concentration contributes significantly to the rapid postabsorptive adaptations. Much larger increases generally occur in response to severe hypoglycemia, long-term fasting (more than 24 h), heavy exercise, and stress, and it is likely that, under these circumstances, the elevated growth hormone level does contribute to carbohydrate and lipid homeostasis.

Conclusion

To a great extent insulin may be viewed as the "hormone of plenty;" it is increased during the absorptive period and decreased during postabsorption. The effects on organic metabolism of glucagon and the sympathetic nervous system are, in various ways, opposed to those of insulin, and these systems are activated (the latter at least acutely) during the postabsorptive period. The influence of plasma glucose concentration is paramount in this regard, producing opposite effects on insulin secretion and on the secretion of glucagon, epinephrine, and the activity of the sympathetic nerves to liver and adipose tissue. Two other hormones—cortisol and growth hormone—also exert actions on lipid and carbohydrate metabolism opposed to those of insulin, but their rates of secretion are not usually coupled to the absorptive-postabsorptive pattern. Nevertheless, their presence in the blood at basal concentrations is necessary for normal adjustment to the postabsorptive period, and excessive amounts of these hormones may so oppose insulin as to induce the picture of diabetes.

Hypoglycemia

Hypoglycemia means "low blood glucose concentration." This term is the subject of intense controversy; many people (including some physicians) claim that hypoglycemia is responsible for a wide variety of symptoms, whereas most experts argue that the condition is relatively uncommon, and that most symptoms ascribed to it are mainly due to emotional problems. There is no question that in persons with certain diseases (insulin-producing tumors, liver disease, and cortisol deficiency, among others) and in some with no discernible explanation, blood glucose can drop to very low values during prolonged fasting and produce symptoms such as fatigue, headache, inability to concentrate, and sleepiness. But the real controversy is whether the appearance of such symptoms several hours after a carbohydrate-rich meal really is due to hypoglycemia in any significant fraction of those persons reporting these and other nonspecific symptoms. The problem is that most people who have been told (or have simply assumed) that they have so-called reactive hypoglycemia have never had their blood glucose concentrations measured at the time of the symptoms. And in those cases where measurements have been made, the blood sugar is most often within normal range.

A conservative conclusion at present is that the true incidence of hypoglycemia is not known, and that the diagnosis is usually made on the basis of insufficient data.

Fuel Homeostasis in Exercise and Stress

Exercise is another situation in which large quantities of endogenous fuels must be mobilized, in this case to provide the energy required for muscle contraction. There are at least three fuels used by exercising muscle (this description applies to moderate levels of endurance-type exercise): plasma glucose, plasma fatty acids, and the muscle's endogenous glycogen.

During the earliest phase of exercise (5 to 10 min), muscle glycogen (via glycogenolysis) is the major fuel consumed. During the middle phase (10 to 40 min) blood-borne fuels, brought to the muscle by the increased blood flow, become dominant, plasma glucose and fatty acids contributing approximately equally to the oxygen consumption of the muscle. Beyond the middle phase, fatty acids become progressively more important as glucose utilization decreases. Finally, as fatigue is approached, muscle becomes very dependent on its own glycogen, and fatigue coincides with depletion of muscle glycogen. (Why glycogen depletion causes muscle to lose its ability to contract despite the availability of ample blood-borne fuels is not known.) Thus, both fats and carbohydrates serve as fuels during exercise. The relative contribution of each depends, in part, on its availability, which can be altered by diet.

What happens to blood glucose concentration during exercise (Fig. 15-15)? It changes very little in short-term mild-to-moderate exercises (and may actually increase slightly with more severe short-term activity). However, during prolonged exercise

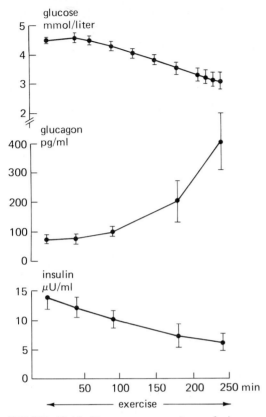

FIGURE 15-15. Plasma concentrations of glucose, glucagon, and insulin during prolonged exercise. *(Adapted from Felig and Wahren.)*

(more than 90 min) it decreases, sometimes by as much as 25 percent but rarely enough to cause symptoms of hypoglycemia. These relatively minor changes in blood glucose in the presence of marked increases in glucose utilization clearly demonstrate that glucose output by the liver must have increased approximately in proportion to the increased utilization until later stages when it begins to lag somewhat behind. The liver provides this glucose both by breakdown of its glycogen stores and by conversion of lactate, glycerol, and amino acids into glucose. The glycerol is made available by a marked increase in the catabolism of adipose-tissue triacylglycerol with resultant release of glycerol and fatty acids into the blood, the latter serving as a fuel source for the exercising muscle.

It should be evident that this combination in exercise of increased hepatic glucose production, triacylglycerol breakdown, and fatty acid utilization is similar to that seen in a fasted person, and the hormonal controls are also the same—exercise is characterized by a fall in insulin secretion, a rise in glucagon secretion, and increased activity of the sympathetic nervous system. In addition, cortisol and growth hormone are also secreted in increased amounts [as is endorphin, an endogenous opiate (see Chap. 9)].

What accounts for the changes in secretion of glucagon and insulin? One important signal during prolonged exercise is the decrease in plasma glucose, the same signal which controls these hormones in fasting. However, this cannot be the signal during less prolonged exercise since, as we have seen, plasma glucose is either unchanged or elevated. Rather, the controlling input appears to be enhanced activity of the sympathetic neurons supplying the cells of the pancreatic islets; this results in stimulation of glucagon release by the A cells and inhibition of insulin release by B cells (circulating epinephrine exerts similar effects on these cells). Thus, the increased sympathetic nervous system activity characteristic of exercise not only contributes directly to fuel mobilization but contributes indirectly by inhibiting the release of insulin and stimulating that of glucagon.

It should be noted, however, that one component of the response to exercise is quite different from fasting—in exercise, glucose uptake and utilization by the muscles is enhanced, whereas in fasting it is markedly reduced. How is it that glucose uptake can remain high in the presence of reduced plasma insulin? One hypothesis is that some local chemical change associated with muscle contraction occurs which increases the affinity of insulin receptors for insulin. An important result of this phenomenon is that patients with diabetes require less insulin to keep glucose normal when they are physically active than when they are sedentary.

Exercise is not the only situation characterized by this neuroendocrine profile (decreased insulin, with increased glucagon, sympathetic activity, cortisol, growth hormone, and endorphin). The other large category is a variety of nonspecific "stress," both physical and emotional; the adaptive value of these neuroendocrine responses is thought to be that the resulting shifts in organic metabolism prepare the body for exercise ("fight or flight") in the face of real or threatened injury. Moreover, an individual faced with an injury or threat is likely to have to forego eating, and these metabolic changes are essential for survival during fasting. Finally, the amino acids liberated by catabolism of body protein stores (because of decreased insulin and increased cortisol) not only provide energy, via gluconeogenesis, but also constitute a potential source of amino acids for tissue repair should injury occur. The subject of stress and the body's responses to it are described in greater detail in Chap. 17.

Regulation of Plasma Cholesterol

In the previous sections, we described the flow of lipids to and from adipose tissue in the form of fatty acids and triacylglycerols complexed with proteins. Another very important lipid—cholesterol—was not mentioned in the earlier sections because it, unlike the fatty acids and triacylglycerols, does not serve as a metabolic fuel but rather as a precursor for plasma membranes, bile salts, steroid hormones, and other specialized molecules. A huge number of studies have been devoted to the factors which govern the plasma concentration of cholesterol, because it is generally accepted that high plasma cholesterol predisposes to the development of atherosclerosis, the arterial thickening which leads to strokes, heart attacks, and other forms of damage (see Chap. 11 for a discussion of the etiology of this thickening).

Cholesterol circulates in lipoprotein complexes, including the chylomicrons made by the intestinal mucosa cells (during ingestion of a meal) and VLDL made by the liver. But two other plasma lipoprotein complexes are more important in the metabolism of cholesterol—**low-density lipoproteins (LDL)** and **high-density lipoproteins (HDL)**. LDL, the major cholesterol carrier in plasma, originates from the peripheral metabolism of VLDL (the loss

of fat accounts for the change in density from "very low" to "low"). The precursor form of HDL is not LDL, as one might have expected; rather, it is made by the liver and cells of the intestinal tract. The cholesterol portion of HDL comes mainly from other cells, as we shall see.

Figure 15-16 summarizes the average total plasma cholesterol concentration in American males of different ages. At birth, the value is approximately 50 mg/100 mL, but it then rises very rapidly to reach 165 mg/100 mL by the age of 2. There is relatively little change until late adolescence, at which time, in men, the mean value begins a progressive rise to reach a peak of 245 mg/100 mL when the man is in his 50s. Women show a smaller rise until after menopause (this has been hypothesized to explain why premenopausal women are relatively protected from atherosclerotic cardiovascular diseases).

A schema for cholesterol balance is illustrated in Fig. 15-17. The sources of cholesterol gain are shown on the left. Dietary cholesterol comes from animal sources, egg yolk being by far the richest in this lipid; a single egg contains about 250 mg of cholesterol. The averge daily intake of cholesterol per person by Americans is 550 mg. However, some of this ingested cholesterol is not absorbed into the blood but simply moves the length of the gastrointestinal tract to be excreted in the feces. The second

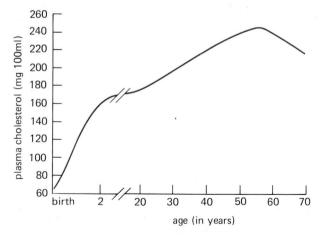

FIGURE 15-16. Average plasma cholesterol concentrations in healthy American males of different ages.

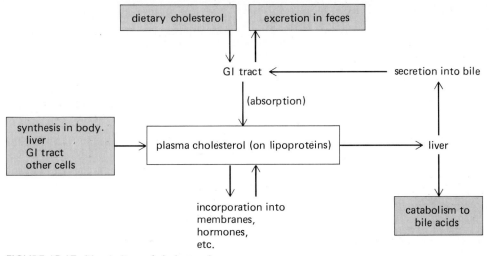

FIGURE 15-17. Metabolism of cholesterol.

source of cholesterol gain is synthesis of cholesterol within the body. Almost all cells can synthesize some of the cholesterol required for their plasma membranes, but most cannot do so in adequate amounts and depend upon receiving, via the bloodstream, cholesterol supplied by the diet or synthesized either by the liver or by cells lining the gastrointestinal tract.

An interesting aspect of cholesterol metabolism is the emerging picture of just how this plasma cholesterol is taken up by cells. As mentioned earlier, most of the cholesterol in plasma is carried by LDL. Almost all cells have, on the outer surface of their plasma membranes, receptors which specifically bind LDL. This binding of LDL to the membrane receptor then triggers the invagination of the cell membrane and the engulfment of the bound LDL (endocytosis). Once inside the cell, the LDL-containing vesicle merges with one of the cell's lysosomes, and enzymes break down the LDL, liberating cholesterol in the process. When a cell is short of cholesterol, it synthesizes more LDL receptors and deposits them on its plasma membrane. This allows the cell to bind more LDL, imbibe it, and liberate its cholesterol, which is perferentially used over the cholesterol synthesized by the cell itself. Indeed, the acquisition of cholesterol from

plasma LDL inhibits the rate-limiting enzyme in the cell's own cholesterol-synthesizing machinery.

In contrast to LDL, which provides cholesterol for cells, HDL is crucial for the removal of cholesterol from cells. Some cholesterol is always coming off the plasma membranes of cells and entering the plasma; this cholesterol is taken up by plasma HDL, a process which maintains the gradient for continued loss of cholesterol from the plasma membranes. Once it has been picked up by HDL, this cholesterol can be transferred to other plasma lipoproteins and be metabolized by the liver.

This brings us to the other side of the balance for cholesterol—the pathways for net cholesterol loss from the body. Because of recycling processes, they appear in Fig. 15-17 to be more complex than they really are. First of all, some plasma cholesterol is picked up from the plasma lipoproteins by liver cells and enters the bile, which flows into the intestinal tract. Here it is treated much like ingested cholesterol, some being absorbed back into the blood and the remainder being excreted in the feces. Second, much of the cholesterol picked up by the liver cells is metabolized into bile acids. After their production by the liver, these bile acids flow through the bile duct into the small intestine. Most are then reclaimed by absorption back into the blood across the wall of the lower small intestine, only those escaping absorption being excreted in the feces. This is an important point, because the drain on plasma cholesterol represented by its conversion to bile acids will depend upon how many of these bile acids are ultimately lost in the feces.

It should be apparent that the liver is the center of the cholesterol universe. At one and the same time, it may be synthesizing new cholesterol, secreting old cholesterol from blood into bile, and transforming it into bile acids. The homeostatic control mechanisms which operate to keep plasma cholesterol relatively constant operate on all three of these processes, but the single most important response involves cholesterol production: The synthesis of cholesterol by the liver is inhibited whenever dietary cholesterol is increased. This is because cholesterol, itself, inhibits the enzyme critical for cholesterol synthesis by the liver (an example of end-product inhibition, described in Chap. 5). Thus, as soon as plasma cholesterol starts rising because of increased cholesterol ingestion, hepatic synthesis is inhibited, and the plasma concentration remains close to its original value. Conversely, when dietary cholesterol is reduced, and plasma cholesterol begins to fall, hepatic synthesis is stimulated (released from inhibition), and this increased production opposes any further fall. This homeostatic control of synthesis is the major reason why it is difficult to alter plasma cholesterol very much in either direction by altering only dietary cholesterol.

How much can changes in dietary cholesterol alter plasma cholesterol? There is still some disagreement (because different scientists have obtained different results) over the answer to this question. However, you can assume that, on the average, reducing your cholesterol intake from 550 to 300 mg/day (equivalent to eating one less egg per day) would lower your blood cholesterol about 3 to 5 percent (conversely, raising intake from 550 to 800 mg/day would increase plasma cholesterol 3 to 5 percent); going all the way down to zero cholesterol intake would drop off another 5 to 10 percent.

Thus far, the homeostatic regulation of plasma cholesterol has been emphasized. However, there are other environmental and genetic factors which can alter plasma cholesterol. Those that have achieved the most attention are the quantity and type of dietary fatty acids. This is not because the fatty acids are precursors for cholesterol but because they influence one or more of the metabolic pathways affecting cholesterol balance. **Saturated fatty acids**, the dominant fatty acids of animal fat, tend to raise plasma cholesterol, both by stimulating cholesterol synthesis and by inhibiting cholesterol conversion to bile acids. In contrast, most **polyunsaturated fatty acids** (the dominant plant fatty acids) tend to lower plasma cholesterol, at least in part by increasing the fecal excretion of both the cholesterol and bile acids secreted by the liver into the bile (loss of the bile acids contributes to lowering of plasma cholesterol because the liver must then convert more cholesterol into bile acids to replace those lost).

Still other environmental and genetic factors are known or suspected to influence plasma cholesterol. One example is a group of genetic diseases characterized by an extremely high plasma cholesterol and an extraordinary propensity toward atherosclerosis and coronary heart disease, some of the people with these diseases dying in their 20s. The defect in at least one of these diseases has been identified: Because afflicted persons lack the gene which codes for LDL receptors on plasma membranes, their cells cannot remove LDL from the blood at a normal rate and so these lipoproteins, loaded with cholesterol, build up to very high concentrations in the blood. Genetic defects such as these account for only a small fraction of persons with atherosclerotic heart disease and do not explain the relatively high plasma cholesterol concentrations which are the norm in this country.

In contrast to persons with increased plasma LDL, people with elevated plasma concentrations of HDL have unusually low incidences of atherosclerotic cardiovascular disease. It should not be surprising that HDL is protective against such disease, since as we have seen, HDL promotes the removal of cholesterol from cells and its excretion by the liver. It is now apparent that the best single indicator of the likelihood of developing atherosclerotic heart disease is not *total* plasma cholesterol but rather the *ratio* of plasma LDL-cholesterol to plasma HDL-cholesterol; i.e., at any given plasma cholesterol the higher the HDL, the less the risk. In this regard, it is significant that cigarette smoking (a known risk factor for heart attacks) lowers plasma HDL, whereas regular exercise increases it.

Control of Growth

People manifest two periods of rapid growth (Fig. 15-18), one during the first 2 years of life, which is actually a continuation of rapid fetal growth, and the second during adolescence. Note that total body growth may be a poor indicator of the rate of growth of specific organs (Fig. 15-18). An important implication of differential growth rates is that the so-called critical periods of development vary from organ to organ.

A simple gain in body weight does not necessarily mean true growth, since it may represent retention of either excess water or fat. In contrast, true growth involves lengthening of the long bones and increased cell division and enlargement; common to all these is increased accumulation of protein. Lengthening of the long bones is perhaps the most striking component of growth. Bone is a living tissue, consisting of a protein matrix upon which calcium salts (particularly phosphates) are deposited. The cells responsible for laying down this ma-

trix are **osteoblasts**. Growth of a long bone depends upon actively proliferating layers of cartilage (**epiphyseal plates**) near the ends of the bone. The osteoblasts at the edge of the epiphyseal plates replace the cartilaginous tissue with bone while new cartilage is simultaneously formed in the plates (Fig. 15-19).

External factors influencing growth

An individual's growth capacity is genetically determined, but there is no guarantee that the maximum capacity will be attained. Adequacy of food supply and freedom from disease are the primary external factors influencing growth. Lack of sufficient amounts of any of the essential amino acids, essential fatty acids, vitamins, or minerals interferes with growth. Total protein and total calories must also be adequate. No matter how much pro-

= "MARKER" PIECE OF CARTILAGE

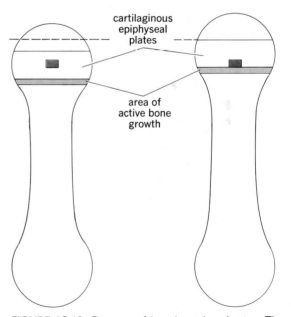

FIGURE 15-19. Diagram of long-bone lengthening. The edge of the cartilaginous plate is converted into bone, but the cells of the plate keep producing more cartilage to take its place. Thus the marker, once in the center of the plate, is now at the edge and is about to be converted into bone; yet the marker itself has not moved.

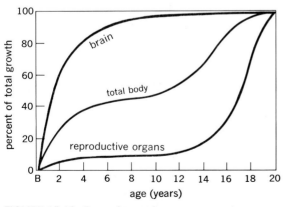

FIGURE 15-18. Rate of growth.

tein is ingested, growth cannot be normal if caloric intake is too low, since the protein is simply catabolized for energy.

Many studies have demonstrated that the growth-inhibiting effects of malnutrition are most profound when they occur very early in life. Indeed, malnutrition during infancy may cause irreversible stunting of total body growth (and brain development). The individual seems "locked in" to a younger developmental age. A related problem is that maternal malnutrition may cause retardation of growth in the fetus. One reason for the great importance of this phenomenon is that low birth weight is strongly associated with increased infant mortality; accordingly, prenatal malnutrition causes increased numbers of prenatal and early postnatal deaths. Moreover, irreversible stunting of brain development may be caused by prenatal malnutrition.

On the other hand, it is important to realize that one cannot stimulate growth beyond the genetically determined maximum by eating more than adequate vitamins, protein, or total calories; this produces obesity, not growth.

Sickness can stunt growth, but if the illness is temporary, upon recovery the child manifests a remarkable growth spurt which rapidly brings him up to his normal growth curve.

Hormonal influences on growth

The classic hormones most important to human growth are **growth hormone**, **thyroid hormone**, **insulin**, **androgen**, and **estrogen**, which exert widespread effects. In addition, ACTH, TSH, prolactin, and FSH and LH selectively influence the growth and development of their target organs, the adrenal cortex, thyroid gland, breasts, and gonads, respectively. Finally, in recent years, increasing attention has been directed toward a group of peptide **growth factors**, each of which is highly effective in stimulating multiplication and growth of a specific cell type or tissue (for example, **nerve growth factor** stimulates the differentiation and growth of sympathetic neurons). The production of these growth factors is not confined to the usual endocrine glands, although in some cases their secretion may be controlled by the more classic hormones. Many of the growth factors are produced and released in the immediate vicinity of their sites of action and so are categorized as paracrines or autocrines (Chap. 7).

Growth hormone. Pituitary deficiency in children arrests growth. Conversely, secretion of excessively large quantities of growth hormone in children causes excessive growth. (When excess growth hormone is secreted in adults after the actively growing cartilaginous areas of the long bones have disappeared, it cannot lengthen the bones further, but it does produce the disfiguring bone thickening and overgrowth of other organs known as acromegaly.) Thus growth hormone is essential for normal growth and, in abnormally large amounts, can cause excessive growth.

We earlier described how growth hormone increases plasma glucose and facilitates fat mobilization, effects which are opposed to those of insulin and which are of minor importance under most physiological conditions. In contrast, its very important growth-promoting effects are due primarily to its ability to stimulate protein synthesis in many tissues. This it does by increasing the membrane transport of amino acids into cells, the synthesis of RNA and ribosomes, and the activity of ribosomes, all events essential for protein synthesis. Growth hormone also causes large increases of mitotic activity and cell division, the other major components of growth. The effects on cartilage and bone are dramatic. Growth hormone promotes bone lengthening by stimulating protein synthesis in both the cartilaginous center and bony edge of the epiphyseal plates as well as by increasing the rate of osteoblast mitosis.

The effects on growth described in the previous paragraph represent direct actions of growth hormone on its target tissues. In addition, there is a very important indirect mechanism by which growth hormone enhances protein anabolism and mitosis: It stimulates the release from the liver (and perhaps other tissues) of several peptide growth factors known collectively as **somatomedins**, which are carried by the blood to target tissues, where they induce growth-promoting effects, particularly in cartilage and bone. Because the somatomedins are structurally and functionally similar to insulin, they are included in the category of **insulin-like growth factors**. Both the secretion and the activity of the somatomedins are controlled not only by growth hormone but by several other hormones and, very importantly, by the nutritional status of the individual (inadequate intake of protein, for example, inhibits the action of somatomedins). Interestingly, it has recently been found that pygmies have very low plasma concentrations of the most potent of the somatomedins, and this genetic deficit very likely accounts for their short stature.

Growth hormone offers an excellent example of a hormone whose secretion occurs in episodic bursts and manifests a striking diurnal rhythm (Chap. 9). During most of the day, in unstressed nonexercising people, there is little or no growth hormone secreted, although a small burst may occur occasionally. Larger bursts may be elicited by certain stimuli such as stress, severe hypoglycemia, or exercise. In contrast, 1 to 2 h after a person falls asleep, one or more larger, prolonged bursts of secretion may occur. It is hypothesized that only during these times is the plasma growth hormone concentration high enough to produce the growth-promoting effects of this hormone.

How do all these events elicit increased secretion of growth hormone? Recall from Chap. 9 that growth hormone secretion, like that of other anterior pituitary hormones, is under the control of hormones secreted by neuroendocrine cells in the hypothalamus and reaching the anterior pituitary via the hypothalamus–anterior-pituitary portal vessels. The control system for growth hormone involves two hypothalamic hormones: **somatostatin** (we are dealing here with hypothalamic, not pancreatic, somatostatin) which is inhibitory; and **growth hormone releasing hormone (GHRH)**, which is stimulatory. Therefore, as shown in Fig. 15-20, the neural input to the hypothalamus associated with each of the situations that trigger secretion of GH (for example, synaptic input from glucose receptors or the sleep centers in the brain) could theoretically act by inhibiting the firing of somatostatin-releasing neurons or by stimulating the firing of GHRH-releasing neurons. Which pathway predominates in each specific case is generally not known.

At what stages of growth is growth hormone important? It is generally accepted that growth hormone has little or no influence on the growth of the fetus or very young infant, but exerts an important influence thereafter. The total 24-h secretion rate of growth hormone (almost all during sleep) is highest during adolescence (the period of most rapid growth), next highest in children, and lowest in adults.

Thyroid hormone. Infants and children with deficient thyroid function manifest retarded growth, which can be restored to normal by administration of physiological quantities of thyroid hormone (TH) (this term refers collectively to two hormones—thyroxine and triiodothyronine—secreted by the thyroid gland). Administration of excess TH, however, does not cause excessive growth (as is true of growth hormone); rather, it causes marked catabolism of protein and other nutrients, as will be explained in the section on energy balance. The essential point is that normal amounts of TH are necessary for normal growth, the most likely explanation being that TH promotes the effects of growth hormone (or somatomedin) on protein synthesis; certainly the absence of TH significantly reduces the ability of growth hormone to stimulate amino acid uptake, ribosomal activation, and RNA synthesis. TH also plays a crucial role in the closely related area of organ development, particularly that of the central nervous system. Hypothyroid infants (cretins) are mentally retarded, a defect that can be completely repaired by adequate treatment with TH, although if the infant is untreated for more than several months, the development failure is largely irreversible.

Insulin. It should not be surprising that adequate amounts of insulin are necessary for normal growth since insulin is, in all respects, an anabolic hormone. Its stimulatory effects on amino acid uptake

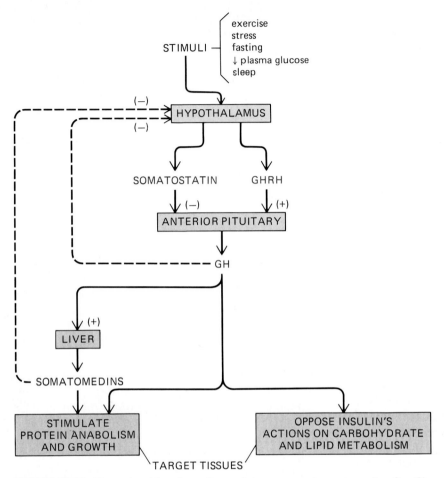

STIMULI — exercise
stress
fasting
↓ plasma glucose
sleep

(−)

HYPOTHALAMUS

(−)

SOMATOSTATIN   GHRH

(−)   (+)

ANTERIOR PITUITARY

GH

(+)

LIVER

SOMATOMEDINS

STIMULATE
PROTEIN ANABOLISM
AND GROWTH

OPPOSE INSULIN'S
ACTIONS ON CARBOHYDRATE
AND LIPID METABOLISM

TARGET TISSUES

FIGURE 15-20. Control and functions of growth hormone and somatomedins. See Chap. 9 for discussion of the feedback controls on the hypothalamus.

and protein synthesis are particularly important in favoring growth.

Androgen and estrogen. In Chap. 16 we shall describe in detail the various functions of the sex hormones in directing the growth and development of the sexual organs and the obvious physical characteristics which distinguish male from female. Here we are concerned only with the effects of these hormones on general body growth.

Sex hormone secretion begins in earnest at about the age of eight to ten and progressively increases to reach a plateau within 5 to 10 years. The testicular hormone, **testosterone**, is the major male sex hormone, but other **androgens** similar to it are

also secreted in significant amounts by the **adrenal cortex** of both sexes. Females manifest a sizable increase of adrenal androgen secretion during adolescence. However, the adrenal androgens are generally not nearly so potent as is testosterone. During adolescence large increases in secretion of **estrogens**, the dominant female sex hormones, are virtually limited to the female. Thus the relative quantities of androgen and estrogen are very different between the sexes.

Androgens strongly stimulate protein synthesis in many organs of the body, not just the reproductive organs, and the adolescent growth spurt in both sexes is due, at least in part, to these anabolic effects. Similarly, the increased muscle mass of men compared with women may reflect their greater amount of more potent androgens. Androgens stimulate bone growth but also ultimately stop bone growth by inducing complete conversion of the epiphyseal plates to bone. This accounts for the patterns seen in adolescence, i.e., rapid lengthening of the bones culminating in complete cessation of growth for life, and explains several clinical situations: (1) Unusually small children treated before puberty with large amounts of androgens may grow several inches very rapidly but then stop completely; and (2) eunuchs may be very tall because bone growth, although slower, continues much longer due to persistence of the epiphyseal plates.

Estrogen profoundly stimulates growth of the female sexual organs and sexual characteristics during adolescence (Chap. 16). However, in contrast to the androgens, the net effect of estrogens on overall somatic growth is inhibitory (the mechanism may be inhibition of somatomedin production). Like testosterone, estrogen causes complete conversion of the epiphyseal plates to bone, thereby ending growth.

Cortisol. Cortisol, one of the major hormones secreted by the adrenal cortex, has potent antigrowth effects. It stimulates protein catabolism in many organs, inhibits cartilage growth, and blocks the secretion of both growth hormone and somatomedin. In children, the elevation in plasma cortisol which accompanies infections and other stresses is, at least in part, responsible for the retarded growth that occurs with illness.

Compensatory growth

We have dealt thus far only with growth during childhood. During adult life, maintenance of the status quo is achieved by the mechanisms described earlier in this chapter. In addition, a specific type of organ growth, known as **compensatory growth**, can occur in many human organs and is actually a type of regeneration. For example, within 24 h of the surgical removal of one kidney, the cells of the other begin to manifest increased mitotic activity, ultimately growing until the total mass approaches the initial mass of the two kidneys combined. What causes this compensatory growth? It certainly does not depend upon the nerves to the organ since it still occurs after their removal or destruction. Several types of experiments indicate that as yet unidentified blood-borne agents are responsible.

SECTION B.
REGULATION OF
TOTAL BODY ENERGY BALANCE

Basic Concepts of Energy Expenditure and Caloric Balance

The breakdown of organic molecules liberates the energy locked in their intramolecular bonds (Chap. 5). This is the source of energy utilized by cells in their performance of the various forms of biological work (muscle contraction, active transport, synthesis of molecules). As described in Chap. 5, the first law of thermodynamics states that energy can be neither created nor destroyed but can be converted from one form to another. Thus, internal energy liberated (ΔE) during breakdown of an organic mol-

ecule can either appear as heat (H) or be used for performing work (W).

$$\Delta E = H + W$$

During metabolism, about 60 percent of the energy appears immediately as heat and the rest is used for work. (As described in Chap. 5, the energy used for work must first be incorporated into molecules of ATP, the subsequent breakdown of which serves as the immediate energy source for the work.) It is essential to realize that the body is not a heat engine since it is totally incapable of converting heat into work. The heat is, of course, valuable for maintaining body temperature.

It is customary to divide biological work into two general categories: (1) **external work**, i.e., movement of external objects by contracting skeletal muscles; and (2) **internal work**, which comprises all other forms of biological work, including skeletal muscle activity not used in moving external objects. As we have seen, much of the energy liberated from the catabolism of nutrients appears immediately as heat. What may not be obvious is that all internal work is ultimately transformed into heat except during periods of growth (Fig. 15-21). Several examples will illustrate this essential point:

1. Internal work is performed during cardiac contraction, but this energy appears ultimately as heat generated by the resistance to flow (friction) offered by the blood vessels.
2. Internal work is performed during secretion of HCl by the stomach and $NaHCO_3$ by the pancreas, but this work appears as heat when the H^+ and HCO_3^- react in the small intestine.
3. The internal work performed during synthesis of a plasma protein is recovered as heat during the inevitable catabolism of the protein, since with few exceptions, all bodily constituents are constantly being built up and broken down. However, during periods of net synthesis of protein, fat, etc., energy is stored in the bonds of these molecules and does not appear as heat.

Thus, the total energy liberated when organic nutrients are catabolized by cells may be trans-

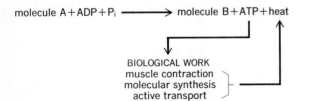

FIGURE 15-21. General pattern of energy liberation in a biological system. Most of the energy released when nutrients, such as glucose, are broken down appears immediately as heat. A smaller fraction goes to form ATP, which can be subsequently broken down and the released energy coupled to biological work. Ultimately, the energy which performs this work is also completely converted into heat.

formed into body heat, appear as external work, or be stored in the body in the form of organic molecules, the latter occurring only during periods of growth (or net fat deposition in obesity). The total energy expenditure of the body is therefore given by the equation

$$\text{Total energy expenditure} = \text{heat produced} + \text{external work} + \text{energy storage}$$

The units for energy are kilocalories,[1] and total energy expenditure per unit time is called the **metabolic rate**.

The metabolic rate of a person can be measured directly or indirectly. In either case, the measurement is much simpler if the person is fasting and at rest; total energy expenditure then becomes equal to heat production since energy storage and external work are eliminated. The direct method is simple to understand but difficult to perform; the subject is placed in a **calorimeter**, an instrument large enough to accommodate a person, and heat production is measured by the temperature changes in the calorimeter (Fig. 15-22). This is an excellent method in that it measures heat production directly, but calorimeters are found in only a few research laboratories; accordingly, a simple, indirect method has been developed for widespread use.

Using the indirect procedure, one simply measures the subject's oxygen uptake per unit time (by measuring total ventilation and the oxygen content of both inspired and expired air). From this value one calculates heat production based on the fundamental principle (Fig. 15-23) that the energy liberated by the catabolism of foods in the body must be the same as that liberated when they are catabolized to the same products outside the body. We know precisely how much heat is liberated when 1 L of oxygen is consumed in the oxidation of fat, protein, or carbohydrate outside the body; this same quantity of heat must be produced when 1 L of oxygen is consumed in the body. Fortunately, we do not need to know precisely which type of nutrient is being oxidized internally, because the

---

[1] In the field of nutrition, Calorie (with a capital C), large calorie, and kilocalorie are synonyms.

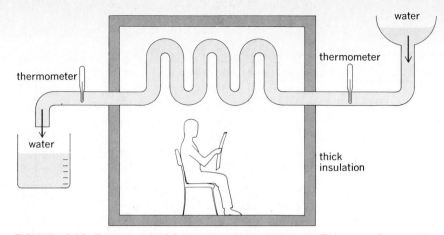

FIGURE 15-22. Direct method for measuring metabolic rate. The water flowing through the calorimeter carries away the heat produced by the person's body. The amount of heat is calculated from the total volume of water and the difference between inflow and outflow temperatures.

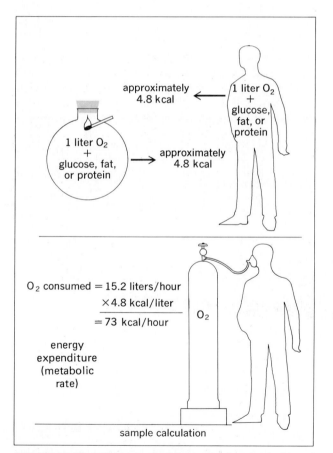

FIGURE 15-23. Indirect method for measuring metabolic rate. The calculation depends upon the basic principle that when 1 L of oxygen is utilized in the oxidation of organic nutrients, approximately 4.8 kcal is liberated.

quantities of heat produced per liter of oxygen consumed are reasonably similar for the oxidation of fat, carbohydrate, and protein, and average 4.8 kcal/L of oxygen. When more exact calculations are required, it is possible to estimate the relative quantity of each nutrient. Figure 15-23 presents values obtained by the indirect method from a normal, fasted, resting adult person. This person's rate of heat production is approximately equal to that of a single 100-W bulb.

Determinants of metabolic rate

Since many factors cause the metabolic rate to vary (Table 15-1), when one wishes to evaluate an individual's metabolism, it is essential to control as many of the variables as possible. The test used clinically and experimentally, the **basal metabolic rate (BMR)**, specifies certain standardized conditions: The subject is at mental and physical rest, in a room at comfortable temperature, and has not eaten for at least 12 h. These conditions are arbitrarily designated "basal," even though the metabolic rate during sleep is actually less than the BMR. The measured BMR is then compared with previously determined normal values for a person of the same weight, height, age, and sex.

BMR is often appropriately termed the metabolic **cost of living**. Under these conditions, most of the energy is expended, as might be imagined, by the heart, liver, kidneys, and brain. Its magnitude

535

TABLE 15-1. Some factors affecting the metabolic rate

Age

Sex

Height, weight, and surface area

Growth

Pregnancy, menstruation, lactation

Infection or other disease

Body temperature

Recent ingestion of food (SDA)

Prolonged alteration in food intake

Muscular activity

Emotional state

Sleep

Environmental temperature

Circulating levels of various hormones, especially epinephrine and thyroid hormone

is related not only to physical size but to age and sex as well. The growing child's metabolic rate, relative to size, is considerably higher than the adult's, in part because the child expends a great deal of energy in net synthesis of new tissue. On the other end of the age scale, the metabolic cost of living gradually decreases with advancing age. A woman's resting metabolic rate is generally less than that of a man (even taking into account size differences) but increases markedly, for obvious reasons, during pregnancy and lactation. The greater demands upon the body by infection or other disease generally increase total energy expenditure.

The ingestion of food also acutely increases the metabolic rate, as shown by measuring the oxygen consumption or heat production of a resting person before and shortly after eating; the metabolic rate is 10 to 20 percent higher after eating. This effect of food on metabolic rate is known as the **specific dynamic action (SDA)**. Protein gives the greatest effect, carbohydrate and fat less. One might logi-

cally expect that the cause of SDA would be the energy expended in the digestion and absorption of ingested food. However, these processes account for only a small fraction of the increased metabolic rate, as evidenced by the fact that intravenous administration of amino acids produces almost the same SDA effect as oral ingestion of the same material. Most of the increased heat production is secondary to the processing of the exogenous nutrients by the liver. To reiterate, the SDA effect is the *rapid* increase in energy expenditure in response to ingestion of a meal; as we shall see, *prolonged* alterations in food intake (either increased or decreased total calories) also have marked effects on metabolic rate.

The factor which can increase metabolic rate to the greatest extent is altered activity of skeletal muscle (Table 15-2). Even minimal increases in muscle tone significantly increase metabolic rate, and severe exercise may raise heat production more than fifteenfold. Changes in muscle activity also explain part of the effects on metabolic rate of sleep (decreased muscle tone), reduced environmental

TABLE 15-2. Energy expenditure during different types of activity for a 70-kg person

Form of activity	Energy kcal/h
Lying still, awake	77
Sitting at rest	100
Typewriting rapidly	140
Dressing or undressing	150
Walking on level, 2.6 mi/h	200
Sexual intercourse	280
Bicycling on level, 5.5 mi/h	304
Walking on 3 percent grade, 2.6 mi/h	357
Sawing wood or shoveling snow	480
Jogging, 5.3 mi/h	570
Rowing, 20 strokes/min	828
Maximal activity (untrained)	1440

temperature (increased muscle tone and shivering), and emotional state (involving changes in muscle tone).

Metabolic rate is also strongly influenced by the hormones **epinephrine** and **thyroid hormone**. A marked increase in the plasma concentration of epinephrine may promptly increase heat production by more than 30 percent. As we have seen, epinephrine has powerful effects on organic metabolism, and its **calorigenic**, i.e., heat-producing, **effect** may be related to its stimulation of glycogen and triacylglycerol catabolism, since ATP splitting and energy liberation occur in both the breakdown and the subsequent resynthesis of these molecules. Regardless of the mechanism, whenever epinephrine secretion is stimulated, the metabolic rate rises. This probably accounts for part of the greater heat production associated with emotional stress, although increased muscle tone is also contributory.

Thyroid hormone (TH) also increases the oxygen consumption and heat production of most body tissues, a notable exception being the brain. The important effects of TH relating to growth and development, described earlier, appear to be quite distinct from this calorigenic effect, the mechanism of which remains controversial. Long-term excessive TH, as in persons with hyperthyroidism, induces a host of effects secondary to the hypermetabolism, effects which well illustrate the interdependence of bodily functions. The increased metabolic demands markedly increase hunger and food intake; the greater intake frequently remains inadequate to meet the metabolic needs, and net catabolism of endogenous protein and fat stores leads to loss of body weight; excessive loss of skeletal muscle protein results in muscle weakness; catabolism of bone protein weakens the bones and liberates large quantities of calcium into the extracellular fluid, resulting in increased plasma and urinary calcium; the hypermetabolism increases the requirement for vitamins, and vitamin deficiency diseases may occur; respiration is increased to supply the required additional oxygen; cardiac output is also increased and, if prolonged, the enhanced cardiac demands may cause heart failure; the greater heat production activates heat-dissipating mechanisms, and the person suffers from marked intolerance to warm environments. These are only a few of the many results induced by the calorigenic effect of TH.

Total body energy balance

Using the basic concepts of energy expenditure and metabolic rate as a foundation, we can consider total body energy balance in much the same way as any other balance, i.e., in terms of input and output. The laws of thermodynamics dictate that, in the steady state, the total energy expenditure of the body equals total body energy intake. We have already identified the ultimate forms of energy expenditure: internal heat production, external work, and net molecular synthesis (energy storage). The source of input, of course, is the energy contained in ingested food. Therefore the energy-balance equation is

$$\text{Energy (from food intake)} =$$
$$\text{internal heat produced} + \text{external work} + \text{energy storage}$$

Our equation includes no term for loss of fuel from the body via excretion of nutrients. In a normal person, almost all the carbohydrate and amino acids filtered at the glomerulus are reabsorbed by the tubules, so that the kidneys play no significant role in the regulation of energy balance. In certain diseases, however, the most important being diabetes, urinary losses of organic molecules may be quite large and would have to be included in the equation. In all normal persons very small losses occur via the urine, feces, and as sloughed hair and skin, but we can ignore them as being negligible.

As predicted by the energy-balance equation, three states are possible, as shown in Table 15-3.

Is **total body energy content** regulated, i.e., maintained relatively constant at some operating point? The fact that, in most adults, body weight remains remarkably constant over long periods of time strongly suggests that the answer to this question is *yes,* but it should be noted that body weight and total energy content are not identical. The ma-

TABLE 15-3. Three possible states of energy balance

State	Result
1. Energy intake = (internal heat production plus external work)	Body energy content remains constant
2. Energy intake > (internal heat production plus external work)	Body energy content increases
3. Energy intake < (internal heat production plus external work)	Body energy content decreases

jor component of body weight, as we have seen, is water, which is regulated by mechanisms unrelated to energy flow. Therefore, regulation of body energy content is not equivalent to regulation of body weight but rather reflects regulation of the total energy content of the organic molecules in the body's lean tissues and fat-storage depots. Techniques more sophisticated than merely measuring body weight have demonstrated that body energy content is, indeed, a regulated quantity.

Theoretically, body energy content could be maintained constant through regulatory adjustments of metabolic energy expenditure to compensate for randomly determined food intake, or through adjustment of food intake to changing metabolic expenditures, or both. There is no question that adjustment of food intake is the major one.

Nevertheless, there is evidence that some degree of control of metabolic expenditure aimed at maintaining energy balance does exist. The most striking experiment designed to study this problem was the reduction of caloric intake in normal male volunteers from 3492 kcal/day to 1570 kcal/day for 24 weeks. These men were soldiers performing considerable physical activity, and weight loss was initially very rapid and ultimately averaged 24 percent of the men's original body weights. The important fact is that the weights did stabilize, i.e., energy balance was reestablished despite the continued caloric intake of 1570 kcal/day. Clearly, metabolic rate must have decreased by an identical amount. This was due partly to a reduction in physical activity as the men became apathetic and reluctant to engage in any activity. BMR was also decreased, due both to a decreased total body mass and to a decrease in metabolism by the liver and other organs out of proportion to their changes in weight.

Experiments have also been done at the other end of the spectrum, to see whether prolonged deliberate overfeeding of volunteers would induce an increased metabolic rate adequate to prevent very much weight gain before caloric balance was reestablished. Most of these studies have suggested that, at least in some subjects, the metabolic rate does, in fact, increase (because of increased heat production, i.e., less efficient utilization of energy), since the weight gains of these subjects were considerably less than that expected from the degree of overfeeding.

Studies of these types have led to the conclusion that, after several weeks, deviations in an individual's total body energy content trigger significant, although generally relatively small, counteracting changes in metabolism. This partially explains why some dieters lose 10 pounds or so of fat fairly easily and then become stuck at a plateau, losing additional weight at a much slower rate. It would also help explain why some very thin people have difficulty trying to gain much weight.

We now turn to the more important mechanism for keeping body energy content constant—control of food intake.

Control of food intake

The control of food intake can be analyzed in the same way as any other biological control system; i.e., in terms of its various components (regulated variable, receptor sensitive to this variable, afferent pathway, etc.). As our previous description emphasized, the variable which is being maintained relatively constant in this system is total body energy content. Here we have a problem similar to that described in Chap. 13 for total body sodium balance: What kind of receptors could possibly detect the total body content of a particular variable, in this case calories? It is very unlikely that any such receptors exist and so, instead, the system must depend on signals which are intimately related to food intake and total energy storage—the plasma concentrations of glucose, fatty acids, glycerol, amino acids, and the hormones which regulate organic metabolism. It is easy to visualize how these inputs could constitute the **satiety signals** for terminating a meal and setting the time period before hunger returns once again. For example (Fig. 15-24) plasma glucose concentration (and the rate of cellular glucose utilization) rises during eating; detection of this increase by glucose receptors (in the brain) constitutes a signal to those portions of the brain which control eating behavior and leads to cessation of eating. Conversely, after the meal has been absorbed, plasma glucose and glucose utilization decrease, the signal to the brain glucose receptors is removed, and the individual once again becomes hungry.

Similarly, meal-to-meal changes in plasma concentrations of the hormones which regulate carbohydrate and fat metabolism may be important signals to the brain's integrating centers. For example, insulin (which increases during food absorption) has been shown to suppress hunger, whereas glucagon, which increases during fasting, is known to stimulate hunger (Fig. 15-25).

A nonchemical correlate of food ingestion—body temperature—may be yet another signal. The specific dynamic action of food, i.e., the increase in metabolic rate induced by eating, tends to raise body temperature slightly and this constitutes a signal inhibitory to eating. All the satiety signals described thus far reflect events occurring as a result of nutrient absorption. But there are also important signals initiated within the gastrointestinal tract itself. These include neural signals triggered by stimulation of both stretch receptors and chemoreceptors in the stomach and duodenum, as well as by several of the hormones released from the duodenum during eating.

Thus far we have been dealing with all those inputs helping to control meal size and frequency, factors which are certainly important in determining total caloric intake. But to be effective over long periods of time in regulating total caloric intake and matching it to energy expenditure, the magnitude of these signals must reflect not merely meal-eating patterns but the total-body energy content itself. Earlier in this chapter we saw one example of this—how plasma glucose and insulin concentrations are influenced by a person's degree of obesity. Another example is glycerol, the plasma concentration of which (averaged over long periods of time), seems to be an excellent indicator of total-body fat stores, because the basal rate of glycerol release from adipose tissue is in direct proportion to how much adipose tissue there is.

It would certainly be simpler if we could quantitate the precise contributions of each of these signals (and of the many others postulated) to the control of food intake and describe the location and nature of the receptors stimulated by each, but much of this information is simply not available at present. Moreover, the confusion does not lessen when we consider the brain areas which integrate all these afferent inputs and cause the individual either to seek out and ingest food or to desist from doing so; areas of the hypothalamus are certainly involved, but there are other brain areas which play equally important roles. The ways in which these various neuronal collections interact remains to be determined.

Thus far we have described the control of food intake in a manner identical to that for any other homeostatic control system, such as those regulating sodium or potassium. Thus, the automatic ''involuntary'' nature of the system has been emphasized. However, although total energy balance

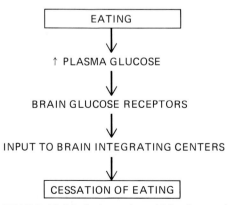

FIGURE 15-24. Pathway by which plasma glucose acts as a ''satiety'' signal to shut off eating.

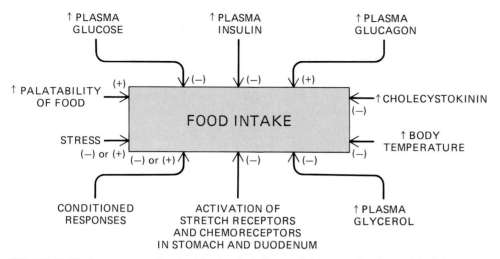

FIGURE 15-25. Inputs controlling food intake. ($-$) denotes inhibition of eating and ($+$) denotes stimulation of eating.

unquestionably reflects, in large part, the reflex input from some combination of glucose receptors, glycerol receptors, thermoreceptors, etc., it also is strongly influenced by the reinforcement (both positive and negative) of such things as smell, taste, and texture. For example, rats given a ''cafeteria-style'' choice of highly palatable human foods overeat by 70 to 80 percent and soon become obese; this obesity is rapidly reversed when the usual rat chow is restarted. [Interestingly, this overeating can be completely blocked by a drug which antagonizes the action of endogenous opiates (see Chaps. 8 and 9)]. An analogous (but reverse) experiment has shown that people who must obtain all their food from a monotonous diet delivered via a tube in response to lever pressing lose weight (and then plateau). Thus, the behavioral concepts of reinforcement, drive, and motivation to be described in a later chapter must be incorporated into any comprehensive theory of food-intake control. So must factors such as stress; for example, simple repeated tail-pinching will cause rats to overeat (and this, too, is blocked by drugs which antagonize endogenous opiates).

Obesity

Obesity is defined as an excessive content of fat (as triacylglycerols) in adipose tissue. However, the dividing line between normal and ''excessive'' fat is

an arbitrary one. All experts would agree that persons who are more than 25 percent heavier than the values on the widely used ''ideal'' or ''desired'' weight charts should be classified as obese. Such persons are clearly at greater risk of developing certain diseases, notably high blood pressure, atherosclerotic heart disease, and diabetes. However, strong differences of opinion exist as to whether lesser degrees of fat storage have negative health consequences. The truth is that the charts, which grew out of the life insurance industry's desire to identify high-risk persons, are based on what are now recognized to be faulty assumptions and data collection. More recent studies have documented that life expectancy is low not only for the clearly obese but also for the lightest-weight group of persons in the population (this health-risk associated with extreme thinness was hitherto unrecognized). Between these extremes, weight has relatively little correlation with mortality, although the data do suggest the ''best'' weight is 10 to 15 pounds higher than that given in present ''ideal'' charts; indeed, it is at or slightly above the present national averages.

The accumulation of fat obviously represents an imbalance between caloric intake and energy expenditure, but what causes this imbalance is unknown for the great majority of obese persons. It is widely assumed that the predominant cause of obesity in our society is overeating due to psychological factors, but most investigators believe that this view is not really well supported by experimental data (such as psychological profiles of obese and thin persons).

Indeed, the very concept of "overeating" is being reevaluated, since a variety of studies have shown that on the average (and there are certainly many exceptions), fat persons eat no more food than thinner persons. For example, a study of obese high school girls revealed that they ate, on the average, less than a control group of normal-weight girls. The obese girls had much less physical activity than the control group and were eating "too much" only in relationship to their physical activity. Moreover it is now apparent that very low levels of activity do not induce similar reductions of food intake but actually stimulate eating. Thus, a modest exercise program particularly in obese persons, reduces body weight not only by increasing energy expenditure but by reducing food intake. On the other hand, inactivity usually causes otherwise normal individuals to gain weight.

We certainly do not wish to leave the impression that lack of activity is the sole, or even the major, cause of obesity. As described earlier, many other factors (for example, various hormones, the plasma concentrations of nutrients, the palatability of available food, and conditioned responses reflecting habit and social custom) are involved in determining the setpoint for total-body energy stores; alteration of any of these could theoretically lead to an increased setpoint. So could altered function of the brain centers which integrate all this information.

Another important question concerns the possibility that influences early in life may induce physiological changes which predispose to adult obesity. Adipose-tissue growth is due to both an increased size and increased number of cells. The number of cells stabilizes sometime during early adulthood and does not change thereafter. Some physiologists believe that overfeeding early in life, particularly during infancy, causes the generation of an abnormally large number of adipose-tissue cells which persists throughout later life, and that the presence of these extra cells may well act as a stimulus for increased food intake (possibly through the glycerol signal described earlier). This view remains controversial, and others contend that genetic factors are more important than early-life experiences of this type. Certainly identical twins who have been separated soon after birth and raised in different households manifest strikingly similar body weights as adults.

Finally, as an exercise in energy balance, let us calculate how rapidly a person can expect to lose weight on a reducing diet. Suppose a person whose metabolic rate per 24 h is 2000 kcal goes on a 1000 kcal/day diet. How much of the person's own body fat will be required to supply this additional 1000 kcal/day? Almost all of the organic nutrients lost are fat, and fat contains 9 kcal/g; therefore

$$1000 \text{ kcal/day} \div 9 \text{ kcal/g} = 111 \text{ g/day},$$
$$\text{or } 777 \text{ g/week}$$

Approximately another 77 g of water is lost from the adipose tissue along with this fat (adipose tissue is 10 percent water), so that the grand total for 1 week's loss equals 854 g, or 1.8 lb. Thus, during the diet, the person can reasonably expect to lose approximately this amount of weight per week. Actually, the amount of weight lost during the first week might be considerably greater, since, for reasons poorly understood, a large amount of water may be lost early in the diet, particularly when the diet contains little carbohydrate. This early loss, which is really of no value so far as elimination of excess fat is concerned, often underlies the claims made for fad diets (indeed, to enhance this effect, drugs which cause the kidneys to excrete even more water are sometimes included in the diet).

Anorexia nervosa

Anorexia nervosa is a disorder which almost exclusively affects adolescent girls and young women. The typical patient becomes pathologically afraid of gaining weight and reduces her food intake so severely that she may die of starvation.

It is not known whether the cause of anorexia nervosa is primarily psychological or biological.

There are many other abnormalities associated with it—loss of menstrual periods, low blood pressure, low body temperature, altered secretion of many hormones—but it is not clear whether these are simply the result of starvation or whether they represent signs, along with the eating disturbances, of primary hypothalamic malfunction.

Regulation of Body Temperature

Animals capable of maintaining their body temperatures within very narrow limits are termed **homeothermic**. The adaptive significance of this ability stems primarily from the effects of temperature upon the rate of chemical reactions, in general, and enzyme activity, in particular. Homeothermic animals are spared the slowdown of all bodily functions which occurs when the body temperature falls. However, the advantages obtained by a relatively high body temperature impose a requirement for precise regulatory mechanisms, since large elevations of temperature cause nerve malfunction and protein denaturation. Some people suffer convulsions at a body temperature of 41°C (106°F), and 43°C is the absolute limit for survival of human cells. In contrast, most body tissues can withstand marked cooling (to less than 7°C), which has found an important place in surgery when the heart must be stopped, since the dormant, cold tissues require little nourishment.

Figure 15-26 illustrates several important generalizations about normal human body temperature:

(1) Oral temperature averages about 0.5°C less than rectal; thus, not all parts of the body have the same temperature. (2) Internal temperature is not absolutely constant but varies several degrees in perfectly normal persons in response to activity pattern and external temperature; in addition, there is a characteristic diurnal fluctuation, so that temperature is lowest during sleep and higher during the awake state even if the person remains relaxed in bed. An added variation in women is a higher temperature during the last half of the menstrual cycle (Chap. 16).

If temperature is viewed as a measure of heat "concentration," temperature regulation can be studied by our usual balance methods. In this case, the total heat content of the body is determined by the net difference between heat produced and heat lost from the body. Maintaining a constant body temperature implies that, overall, heat production must equal heat loss. Both these variables are subject to precise physiological control.

Temperature regulation offers a classic example of a biological control system; its generalized components are shown in Fig. 15-27. The balance between heat production and heat loss is continuously being disturbed, either by changes in metabolic rate (exercise being the most powerful influence) or by changes in the external environment which alter heat loss. The resulting small changes in body temperature reflexly alter the output of the effectors so that heat production or heat loss is altered and body temperature restored toward normal.

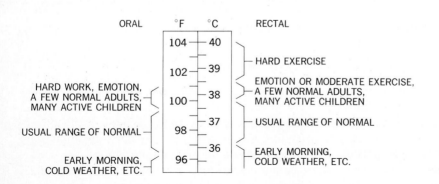

FIGURE 15-26. Ranges of body temperatures in normal persons. (*Adapted from Dubois.*)

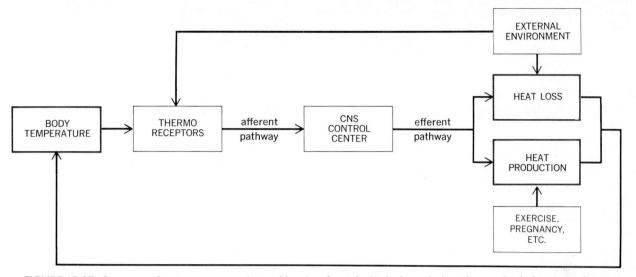

FIGURE 15-27. Summary of temperature regulation. Heat loss from the body depends directly upon both the external environment and changes controlled by temperature-regulating reflexes. In certain environments, heat gain rather than heat loss may actually occur.

Heat production

The basic concepts of heat production have already been described. Recall that heat is produced by virtually all chemical reactions occurring in the body, and that the cost-of-living metabolism by all organs sets the basal level of heat production, which can be increased as a result of skeletal muscular contraction or the action of several hormones.

Changes in muscle activity. The first muscle changes in response to cold are a gradual and general increase in skeletal muscle tone. This soon leads to **shivering**, the characteristic muscle response to cold, which consists of oscillating rhythmical muscle tremors occurring at the rate of about 10 to 20 per second. So effective are these contractions that body heat production may be increased severalfold within seconds to minutes. Because no external work is performed, all the energy liberated by the metabolic machinery appears as internal heat. As always, the contractions are directly controlled by the efferent motor neurons to the skeletal muscles. During shivering, these nerves are controlled by descending pathways under the primary control of

the hypothalamus. This "shivering pathway" can be suppressed, at least in part, by input from the cerebral cortex, since a cold person usually ceases to shiver when starting to perform voluntary activity. Besides increased muscle tone and shivering, people also use voluntary heat-production mechanisms such as foot stamping, hand clapping, etc.

Thus far, our discussion has focused primarily on the muscle response to cold; the opposite reactions occur in response to heat. Muscle tone is reflexly decreased and voluntary movement is also diminished ("It's too hot to move"). However, these attempts to reduce heat production are relatively limited in capacity both because muscle tone is already quite low normally and because the increased body temperature acts directly on cells to increase metabolic rate.

Nonshivering ("chemical") thermogenesis. In most experimental animals chronic cold exposure induces an increase in metabolic rate which is not due to increased muscle activity. Indeed, as this so-called nonshivering thermogenesis increases over time, it is associated with a decrease in the degree of shivering. The cause of nonshivering thermogenesis has been the subject of considerable controversy; present evidence suggests that it is due mainly to an increased secretion of epinephrine (with some contribution of thyroid hormone, as well). Equally controversial is the question whether nonshivering thermogenesis occurs at all in adult human beings (it does seem to occur in infants). Regardless of the outcomes of these debates, it seems clear that hormonal changes and nonshivering thermogenesis are of secondary importance in adult humans and that changes in muscle activity constitute the major control of heat production for temperature regulation, at least in the early response to temperature changes.

Heat-loss mechanisms

The surface of the body exchanges heat with the external environment by radiation, conduction, and convection (Fig. 15-28) and by water evaporation.

Radiation, conduction, convection. Radiation is the process by which the surface of the body constantly emits heat in the form of electromagnetic waves. Simultaneously, all other dense objects are radiating heat. The rate of emission is determined by the temperature of the radiating surface. Thus, if the surface of the body is warmer than the *average* of the various surfaces in the environment, net heat is lost, the rate being directly dependent upon the temperature difference. The sun, of course, is a powerful radiator, and direct exposure to it may cause heat gain.

    Conduction is the exchange of heat not by radiant energy but simply by transfer of thermal energy from atom to atom or molecule to molecule. Heat, like any other quantity, moves down a concentration gradient, and thus the body surface loses or gains heat by conduction only through direct contact with cooler or warmer substances, including, of course, the air.

    Convection is the process whereby air (or water) next to the body is heated, moves away, and is replaced by cool air (or water) which, in turn, follows the same pattern. It is always occurring because warm air is less dense and therefore rises,

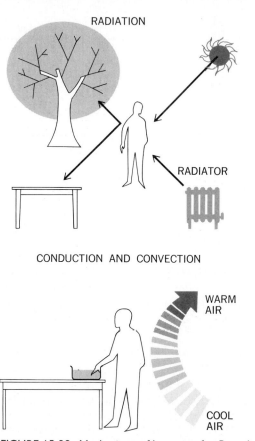

FIGURE 15-28. Mechanisms of heat transfer. By radiation, heat is transferred by electromagnetic waves (solid arrows); in conduction, heat moves by direct transfer of thermal energy from molecule to molecule (dashed arrow).

but it can be greatly facilitated by external forces such as wind or fans. Thus, convection aids conductive heat exchange by continuously maintaining a supply of cool air. In the absence of convection, negligible heat would be lost to the air, and conduction would be important only in such unusual circumstances as immersion in cold water. Henceforth we shall also imply convection when we use the term "conduction." (Because of the great importance of air movement in aiding heat loss, attempts have been made to quantitate the cooling effect of combinations of air speed and temperature; the most useful tool has been the wind-chill index.)

It should now be clear that heat loss by radiation and conduction is largely determined by the temperature difference between the body surface and the external environment. It is convenient to view the body as a central core surrounded by a shell consisting of skin and subcutaneous tissue (for convenience, we shall refer to the complex shell of tissues simply as skin) whose insulating capacity can be varied. It is the temperature of the central core which is being regulated at approximately 37°C; in contrast, as we shall see, the temperature of the outer surface of the skin changes markedly. If the skin were a perfect insulator, no heat would ever be lost from the core; the outer surface of the skin would equal the environmental temperature (except during direct exposure to the sun), and net conduction or radiation would be zero. The skin, of course, is not a perfect insulator, so that the temperature of its outer surface generally lies somewhere between that of the external environment and that of the core.

Of great importance for temperature regulation of the core is that the skin's effectiveness as an insulator is subject to physiological control by a change of the blood flow to the skin. The more blood reaching the skin from the core, the more closely the skin's temperature approaches that of the core. In effect, the blood vessels diminish the insulating capacity of the skin by carrying heat to the surface (Fig.15-29). These vessels are controlled by vasoconstrictor sympathetic nerves. Vasoconstriction may be so powerful that the skin of the finger, for example, may undergo a 99 percent reduction in blood flow during exposure to cold.

Exposure to cold increases the temperature difference between core and environment; in response, skin vasoconstriction increases skin insulation, reduces skin temperature, and lowers heat loss. Exposure to heat decreases (or may even reverse) the difference between core and environment; in order to permit the required heat loss, skin vasodilation occurs, the difference between skin and environment increases, and heat loss increases. Although we have spoken of skin temperature as if it were uniform throughout the body, certain areas participate much more than others in the vasomotor responses; accordingly, skin temperatures vary with location.

What are the limits of this type of process? The lower limit is obviously the point at which maximal skin vasoconstriction has occurred; any further drop in environmental temperature increases the environment-skin temperature difference and causes increased heat loss. At this point, the body must increase its heat production to maintain temperature. The upper limit is set by the point of maximal vasodilation, the environmental temperature, and the core temperature itself. At high environmental temperatures, even maximal vasodilation cannot establish a core-environment

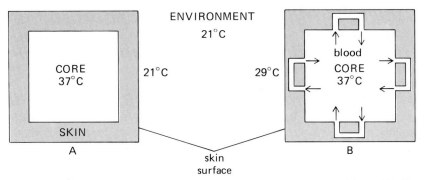

FIGURE 15-29. Relationship of skin's insulating capacity to its blood flow. (A) Skin as a perfect insulator, i.e., with zero blood flow, the temperature of the skin's outer surface equaling that of the external environment. (B) When the skin blood vessels dilate the increased flow carries heat to the body surface; i.e., the insulating capacity of the skin is reduced, and the surface temperature becomes intermediate between that of the core and that of the external environment.

temperature difference large enough to eliminate heat as fast as it is produced (of course, when environmental temperature is higher than core temperature, heat gain occurs by conduction). Another heat-loss mechanism—sweating—is therefore brought strongly into play. Thus, the skin vasomotor contribution to temperature regulation is highly effective in the midrange of environmental temperature (20 to 28°C), but the major burden is borne by increased heat production at lower temperatures and by increased heat loss via sweating at higher temperatures.

Two *behavioral* mechanisms for altering heat loss by radiation and conduction remain: changes in surface area and in clothing. Curling up into a ball, hunching the shoulders, and similar maneuvers in response to cold reduce the surface area exposed to the environment, thereby decreasing radiation and conduction. In human beings, clothing is also an important component of temperature regulation, substituting for the insulating effects of feathers in birds and fur in other mammals. The principle is similar in that the outer surface of the clothes now forms the true "exterior" of the body surface. The skin loses heat directly to the air space trapped by the clothes; the clothes in turn pick up heat from the inner air layer and transfer it to the external environment. The insulating ability of clothing is determined primarily by its type as well as by the thickness of the trapped air layer.

We have spoken thus far only of the ability of clothing to *reduce* heat loss; the converse is also desirable when the environmental temperature is greater than body temperature, since radiation and conduction then favor heat gain. People therefore insulate themselves against temperatures which are greater than body temperature by wearing clothes. The clothing, however, must be loose so as to allow adequate movement of air to permit evaporation, the only source of heat loss under such conditions. White clothing is cooler since it reflects more radiant energy, which dark colors absorb. Contrary to popular belief, loose-fitting light-colored clothes are far more cooling than going nude during direct exposure to the sun.

Evaporation. **Evaporation** of water from the skin and lining membranes of the respiratory tract is the other major process for loss of body heat. Thermal energy is required to transform water from the liquid to the gaseous state. Thus, whenever water vaporizes from the body's surface, the heat required to drive the process is conducted from the surface, thereby cooling it. Even in the absence of sweating, there is still loss of water by diffusion through the skin, which is not completely waterproof; a like amount is lost during expiration from the respiratory lining. This **insensible water loss** amounts to approximately 600 mL/day in human beings and accounts for a significant fraction of total heat loss. In contrast to this passive water loss, **sweating** requires the active secretion of fluid by **sweat glands** and its extrusion into ducts, which carry it to the skin surface.

Production of sweat is stimulated by the sympathetic nerves. Sweat is a dilute solution containing primarily sodium chloride. The loss of this salt and water during severe sweating can cause diminution of plasma volume adequate to provoke hypotension, weakness, and fainting. It has been estimated that there are over 2.5 million sweat glands spread over the adult human body, and production rates of over 4 L/h have been reported. This is 9 lb of water, the evaporation of which would eliminate almost 2400 kcal from the body! It is essential to recognize that sweat must evaporate in order to exert its cooling effect. The most important factor determining evaporation is the water vapor concentration of the air, i.e., the **relative humidity**. The discomfort suffered on humid days is due to the failure of evaporation; the sweat glands continue to secrete, but the sweat simply remains on the skin or drips off.

Most other mammals differ from human beings in lacking sweat glands. They increase their evaporative losses primarily by panting, thereby increasing pulmonary air flow and increasing water losses from the lining of the respiratory tract, and they deposit water for evaporation on their fur or skin by licking.

Heat loss by evaporation of sweat gradually

dominates as environmental temperature rises, since radiation and conduction decrease as the body-environment temperature gradient diminishes. At environmental temperatures above that of the body, heat is actually gained by radiation and conduction, and evaporation is the sole mechanism for heat loss. A person's ability to survive such temperatures is determined by the humidity and by the maximal sweating rate. For example, when the air is completely dry, a person can survive a temperature of 130°C (255°F) for 20 min or longer, whereas very moist air at 46°C (115°F) is not bearable for even a few minutes.

Summary of effector mechanisms in temperature regulation

Table 15-4 summarizes the mechanisms regulating temperature, none of which is an all-or-none response, but calls for a graded, progressive increase

TABLE 15-4. Summary of effector mechanisms in temperature regulation

Desired effect	Mechanism
Stimulated by cold	
Decrease heat loss	Vasoconstriction of skin vessels; reduction of surface area (curling up, etc.); behavioral response (put on warmer clothes, raise thermostat setting, etc.)
Increase heat production	Increased muscle tone; shivering and increased voluntary activity; increased secretion of thyroid hormone and epinephrine; increased appetite
Stimulated by heat	
Increase heat loss	Vasodilation of skin vessels; sweating; behavioral response (put on cooler clothes, turn on fan, etc.)
Decrease heat production	Decreased muscle tone and voluntary activity; decreased secretion of thyroid hormone and epinephrine; decreased appetite

or decrease in activity. As we have seen, heat production via skeletal muscle activity becomes extremely important at the cold end of the spectrum, whereas increased heat loss via sweating is critical at the hot end.

Brain centers involved in temperature regulation

Neurons in the hypothalamus and other brain areas, via descending pathways, control the output of motor neurons to skeletal muscle (muscle tone and shivering) and of sympathetic neurons to skin arterioles (vasoconstriction and dilation), sweat glands, and the adrenal medulla. When thyroid hormone is a component of the response to cold, these centers also control the output of hypothalamic TSH releasing hormone (TRH).

Afferent input to the integrating centers

The final component of temperature-regulating systems to be described is really the first component, i.e., the afferent input. Obviously these temperature-regulating reflexes require receptors capable of detecting changes in the body temperature. There are two groups of receptors, one in the skin (**peripheral thermoreceptors**) and the other in deeper body structures (**central thermoreceptors**).

Peripheral thermoreceptors. In the skin (and certain mucous membranes) are nerve endings usually categorized as **cold** and **warm receptors**. In one sense, these are misleading terms, since cold is not a separate entity but a lesser degree of warmth. Really there are two populations of temperature-sensitive skin receptors, one stimulated by a lower and the other by a higher range of temperatures. Information from these receptors is transmitted via the afferent neurons and ascending pathways to the hypothalamus and other integrating areas, which respond with appropriate efferent output; in this manner, the firing of cold receptors stimulates heat-producing and heat-conserving mechanisms.

Central thermoreceptors. The skin thermoreceptors alone would be highly inefficient regulators of body temperature for the simple reason that it is the core temperature, not the skin temperature, which is actually being regulated. On theoretical grounds alone it has been apparent that core, i.e., central, receptors have to exist somewhere in the body, and numerous experiments have localized them to the hypothalamus, spinal cord, abdominal organs, and other internal locations. In unanesthetized dogs, local warming of certain hypothalamic neurons causes the neurons to fire rapidly and reproduces the entire picture of the dog's usual response to a warm environment: He becomes sleepy, stretches out to increase his surface area, pants heavily, salivates, and licks his fur. Conversely, local cooling induces vasoconstriction, intensive shivering, fluff-

ing out of the fur, and curling up. These hypothalamic thermoreceptors have synaptic connections with hypothalamic and other integrating centers, which also receive input from other central thermoreceptors as well as from the skin thermoreceptors. The precise relative contributions of the various thermoreceptors remain the subject of considerable debate. Temperature-regulating reflexes are summarized in Fig. 15-30.

Acclimatization to heat

Changes in sweating determine people's chronic adaptation to high temperatures. A person newly arrived in a hot environment has poor ability to do

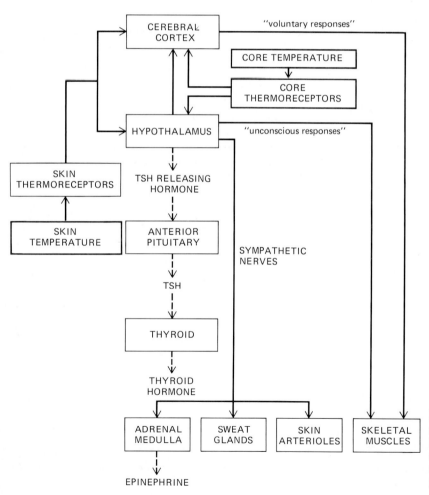

FIGURE 15-30. Summary of temperature-regulating mechanisms. The dashed lines are hormonal pathways, which are probably of minor importance in human beings. Not shown are other nonhypothalamic integrating areas.

work initially; body temperature rises and severe weakness and illness may occur. After several days, there is a great improvement in work tolerance with little increase in body temperature, and the person is said to have **acclimatized** to the heat. Body temperature is kept low because there are an earlier onset of and increased rates of sweating. There is also an important change in the composition of the sweat, namely, a marked reduction in its sodium concentration. This adaptation, which minimizes the loss of sodium from the body via the sweat, is due to **aldosterone**, which stimulates the reabsorption of sodium from the fluid originally secreted by the sweat glands as the fluid flows along the ducts toward the skin surface. The reflexes which activate the renin-angiotensin-aldosterone system during exercise were described in Chap. 13.

Fever

The elevation of body temperature so commonly induced by infection is due not to a breakdown of temperature-regulating mechanisms but to a "resetting of the thermostat" in the hypothalamus or other brain area. Thus, a person with a fever regulates body temperature in response to heat or cold but at a higher set point. The onset of fever is frequently gradual but it is most striking when it occurs rapidly in the form of a chill. It is as though the brain thermostat were suddenly raised; the person suddenly feels cold, and marked vasoconstriction and shivering occur; the person also curls up and puts on more blankets. This combination of decreased heat loss and increased heat production serves to drive body temperature up to the new set point, where it stabilizes. It will continue to be regulated at this new value. Later, the fever "breaks" as the thermostat is reset to normal; the person feels hot, throws off the covers, and manifests profound vasodilation and sweating.

What is the basis for the resetting? As is described in the chapter on resistance to infection (Chap. 17) a chemical known as **endogenous pyrogen (EP)** is released from monocytes and macrophages in the presence of infection or inflammation. EP acts upon the thermoreceptors in the hypothalamus (and perhaps other brain areas), altering their rate of firing and their input to the integrating centers. This effect of EP may be mediated via local release of **prostaglandins** which then directly alter thermoreceptor function. Consistent with this hypothesis is the fact that aspirin, which reduces fever by restoring thermoreceptor activity toward normal, inhibits the synthesis of prostaglandins.

One would expect fever, which is such a consistent concomitant of infection, to play some important protective role, and recent evidence strongly suggests that such is the case. Increased body temperature seems to stimulate a large number of the body's defensive responses to infection (to be discussed in Chap. 17) and perhaps to cancer as well. Very importantly, it is now clear that EP has many effects in addition to eliciting fever; these effects, also to be described in Chap. 17, have the common denominator of enhancing resistance to infection and promoting the healing of damaged tissue.

The likelihood that fever is a beneficial response raises questions concerning the widespread use of aspirin and other drugs to suppress fever during infection. This usually makes the person "feel better," and it will require much more experimentation before we know just which diseases really have their courses prolonged or worsened enough by fever-reducing therapy to make the trade-off not worthwhile. It must be emphasized that this discussion has been with reference to the usual modest fevers; there is no question that an extremely high fever may be quite harmful, particularly in its effects on the central nervous system, and must be vigorously opposed with drugs and other forms of therapy.

Other causes of hyperthermia

The term "fever" is often used loosely to denote any elevation of a person's body temperature above the normally accepted range. However, physiologists prefer the term **hyperthermia** for this definition and reserve "fever" for the form of hyperthermia just described—the specific elevation in body temperature (produced by endogenous pyrogen) due to an altered setpoint. There are a variety of situations other than fever in which hyperthermia occurs because of an imbalance between the rates of heat production and heat loss.

Exercise. The most common cause of hyperthermia in normal people is sustained exercise, during which body temperature rises and is maintained as long as the exercise continues (Fig. 15-31). It is possible that a small fraction of this rise represents a new setpoint (some endogenous pyrogen may be released during sustained exercise, perhaps because of slight tissue damage), but most of the hyperthermia is simply a physical consequence of the internal heat load generated by the exercising muscles. As shown in Fig. 15-31, during the initial stage of exercise, heat production rises immediately and exceeds heat loss, causing heat storage in the body and a rise in core temperature. This rise in core temperature, in turn, triggers reflexes, via the core thermoreceptors, for increased heat loss (increased skin blood flow and sweating), and the discrepancy between heat production and heat loss starts to diminish but does not disappear; therefore core temperature continues to rise. Ultimately, however, core temperature will be high enough to drive, via the core thermoreceptors, the reflexes for heat loss at a rate such that heat loss does become equal to heat production, and core temperature stabilizes despite continued exercise.

Heat exhaustion and heat stroke. Heat exhaustion is a state of collapse (often taking the form of fainting) due to hypotension brought on both by depletion of plasma volume (secondary to sweating) and by extreme dilation of skin blood vessels, i.e., by decreases in both cardiac output and peripheral resistance. Thus, heat exhaustion occurs as a direct consequence of the activity of heat-loss mechanisms; because these mechanisms have been so active, the body temperature is only modestly elevated. In a sense, heat exhaustion is a safety valve which, by forcing cessation of work in a hot environment when heat-loss mechanisms are overtaxed, prevents the larger rise in body temperature which would precipitate the far more serious condition of heat stroke.

Unlike heat exhaustion, **heat stroke** represents a complete breakdown in heat-regulating systems. It is an extremely dangerous situation, characterized by collapse, delirium, seizures, or prolonged unconsciousness—all due to marked elevation of body temperature. It almost always occurs in association with exposure to or overexertion in hot and humid environments. In some persons, particularly the elderly, heat stroke may appear with no apparent prior period of severe sweating, but in most cases, it comes on as the end-stage of prolonged untreated heat exhaustion. Exactly what triggers the transition to heat stroke is not clear, but the striking finding is that even in the face of a rapidly rising body temperature, the person fails to sweat. This sets off a positive-feedback situation in which the rising body temperature directly stimulates metabolism, i.e., heat production, which further raises body temperature. A recent finding of potential importance is that some of the commonly used tranquilizer drugs interfere with neurotransmitters in hypothalamic thermoregulatory centers, and some people using these drugs are very prone to developing heat stroke.

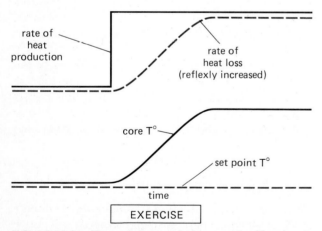

FIGURE 15-31. Thermal changes during exercise. Heat loss is reflexly increased, and when it once again equals heat production, core temperature stabilizes. The set point temperature is shown remaining constant during the exercise; actually, a small increase may occur. T° = temperature.

REPRODUCTION

16

Before beginning detailed individual descriptions of male and female reproductive systems, it is worthwhile to summarize some general terminology. The primary reproductive organs are known as the **gonads**, the **testes** in the male and the **ovaries** in the female. In both sexes, the gonads serve dual functions: (1) production of the reproductive cells, gametes—**spermatozoa (sperm)** by males and **ova** by females; and (2) secretion of the so-called **sex hormones—testosterone** by males and **estrogen** and **progesterone** by females.

At some point, the **germ cells**, which differentiate into the gametes, undergo **meiosis**, in which the chromosomal content of the germ cells is halved from 46 to 23 chromosomes. The result is that when a mature sperm and ovum come together at fertilization, the resulting fertilized ovum, the **zygote**, has a full complement of chromosomes genetically different from those of either parent. The formation of genetic diversity within a population through sexual reproduction provides a richer overall genetic pool for natural selection to act upon and, hence, a greater likelihood for resilience of at least some of the individuals in the face of environmental challenge.

The systems of ducts through which the sperm or ova are transported and the glands lining or emptying into the ducts are termed the **accessory reproductive organs** (in the female the breasts are also usually included in this category). The **secondary sexual characteristics** comprise the many external differences (hair, body contours, etc.) between male and female; these are not directly involved in reproduction.

The gonads and accessory reproductive organs are present at birth but remain relatively small and nonfunctional until the onset of puberty at about 10 to 14 years of age, at which time the secondary sexual characteristics also begin to appear. The term **puberty** signifies the attainment of sexual maturity in the sense that conception becomes possible; as commonly used, it refers to the 3 to 5 years of sexual development culminating in the attainment of sexual maturity. The term **adolescence** has a much broader meaning than does "puberty" and includes the total period of transition from childhood to adulthood in all respects, not just sexual development.

Reproductive function is largely (but not exclusively) controlled by a sequential chain of hormones (Fig. 16-1). The first hormone in the chain is **gonadotropin releasing hormone (GnRH)**, secreted by neuroendocrine cells in the hypothalamus and reaching the anterior pituitary via the hypothalamo-pituitary portal blood vessels (see Chap. 9). There, GnRH stimulates the release from specific anterior pituitary cells of the **pituitary gonadotropins—follicle-**

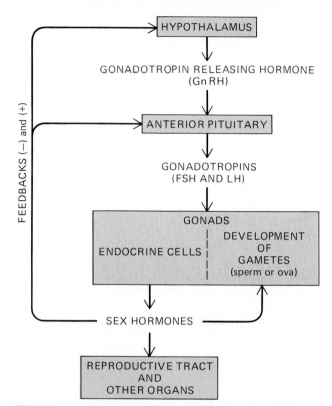

FIGURE 16-1. General pattern of the control of reproduction in both males and females. GnRH reaches the anterior pituitary via the hypothalamo-pituitary portal vessels. Not shown in this introductory figure but described later is another hormone, inhibin, which is secreted by the gonads.

stimulating hormone (FSH) and **luteinizing hormone (LH)**. These two hormones then act upon the gonads to stimulate both production of gametes (gametogenesis) and sex hormone secretion. Finally, the sex hormones exert many effects on all portions of the reproductive system (including the gonads from which they come) and other parts of the body, as well. In addition, the sex hormones (and at least one additional gonadal hormone—**inhibin**) exert feedback effects on the secretion of GnRH, FSH, and LH. This brief description is simply for orientation, and the specifics of these control processes will be dealt with in subsequent sections.

SECTION A.
MALE REPRODUCTIVE PHYSIOLOGY

The essential male reproductive functions are the formation of sperm (**spermatogenesis**) and the deposition of the sperm in the female. The organs which carry out these functions are the testes, epididymides (singular, epididymis), ductus ("vas") deferens, ejaculatory ducts, and penis, together with the accessory glands: the seminal vesicles, prostate, and bulbourethral glands (Fig. 16-2). The **testes** are suspended outside the body in a sac, the **scrotum**, in which each testis lies in a separate pouch. During embryonic development, the testes lie in the rear wall of the abdomen, but during the seventh month of intrauterine development, they descend into the scrotum. This descent is essential for normal sperm production during adulthood, since sperm formation requires a temperature lower than normal internal body temperatures.

Each testis contains many tiny, convoluted **seminiferous tubules**, the combined length of which is 250 m; they are lined with the sperm-producing **spermatogenic cells**. The testes also serve an endocrine function, the manufacture of the primary male sex hormone **testosterone**. The endocrine **interstitial cells (Leydig cells)**, which secrete the testosterone, lie in the small connective-tissue spaces between the seminiferous tubules (Fig. 16-3). Thus, the sperm-producing and testosterone-producing functions of the testes are carried out by different cells.

Spermatogenesis

In Fig. 16-3, a microscopic section of an adult human testis, it can be seen that the seminiferous tubules contain many so-called **germ cells** (or spermatogenic cells), the vast majority of which are in various stages of division. Each seminiferous tubule is surrounded by a membrane, and only the outermost layer of spermatogenic cells is in contact with this membrane; these are undifferentiated germ cells termed **spermatogonia**, which, by dividing mitotically, provide a continuous source of new germ cells. Some spermatogonia move away from the membrane and increase markedly in size. Each of these large cells, now termed a **primary spermatocyte**, undergoes a meiotic division to form two **secondary spermatocytes**, each of which in turn divides into two **spermatids**, the latter ultimately being transformed into **spermatozoa (sperm)** (Fig. 16-4).

The division of the primary and secondary spermatocytes by meiosis differs from the ordinary mitotic division. During mitosis (Chap. 4), the chromosomes are duplicated and each daughter cell receives the full number of 46 chromosomes; in meiosis, which is really two cell divisions in succession, each daughter cell receives only half of the chromosomes present in the original cell. The primary spermatocyte contains 46 chromosomes, 23 from each parent. Prior to its first meiotic division, each of the 46 chromatin threads is duplicated, and during their condensation into chromosomes, the homologous duplicated maternal and paternal chromatin threads pair with each other, forming 23 four-stranded chromosomes (Fig. 16-5). When the primary spermatocyte divides, the two maternal strands pass into one of the secondary spermatocytes and the paternal strands into the other. Since each of the 23 four-stranded chromosomes becomes attached to the spindle fibers at random, one chromosome may have its maternal strands oriented toward one pole of the cell and the adjacent chromosome have its maternal strands oriented toward the opposite pole. Thus, when the maternal and paternal strands separate, the two secondary spermatocytes receive a mixture of maternal and paternal chromosomes. It would be extremely improbable that all 23 maternal chromosomes would end up in one cell and all 23 paternal chromosomes in the other.

The random distribution of maternal and paternal chromosomes during meiosis provides the basis for much of the variability in the genetic constitution of the sperm (and, as we shall see, the ova) formed by a single individual. Over 8 million (2^{23})

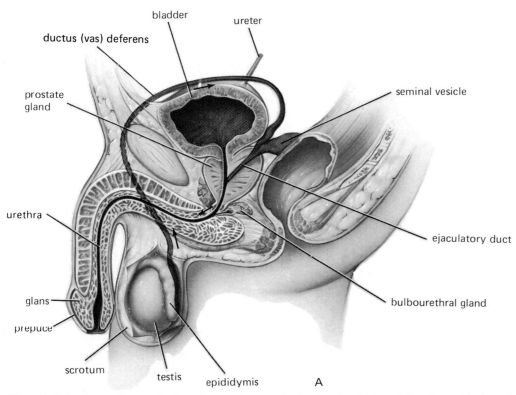

FIGURE 16-2. Anatomic organization of the male reproductive tract and interrelationships with the urinary system. (A) As seen from the side. (B) As seen from the rear. The spermatozoa are formed in the highly coiled seminiferous tubules. Part of the left testis has been cut away to reveal internal structures.

different combinations of maternal and paternal chromosomes can result from the distribution of the 23 chromosomes during meiosis. But this is only the minimum number of possible combinations of genetic material, since segments of the maternal and paternal chromosomes may exchange with each other in a process known as **crossing over**, which occurs as a result of chromosomal pairing during the first meiotic division. Crossing over results in chromosomes containing segments of both maternal and paternal genetic material in the same chromosome. Thus, all the genes initially present in a paternal or maternal chromosome do not necessarily pass into the same cell during meiosis. When the secondary spermatocytes divide (the second division of meiosis), the two-stranded chromosomes (which duplicated during the first meiotic division) separate so that each of the two resulting cells, the spermatids, have 23 nonduplicated chromosomes.

The final phase of spermatogenesis, the differentiation of the spermatids into spermatozoa, involves extensive cell remodeling, including elongation, but no further cell divisions. The head of a spermatozoon (Fig. 16-6) consists almost entirely of the nucleus, containing the DNA which bears the sperm's genetic information. The tip of the nucleus is covered by the **acrosome**, a protein-filled vesicle containing several lytic enzymes that play an important role in the sperm's penetration of the ovum. The tail contains a group of contractile filaments forming a flagellum, which produces movements (like those of a cilium) capable of propelling the sperm at a velocity of 1 to 4 mm/min. The mitochondria of the spermatid form the midpiece of the tail and probably provide the energy for its movement.

The developing germ cells remain intimately associated with another type of cell, the **Sertoli cell**

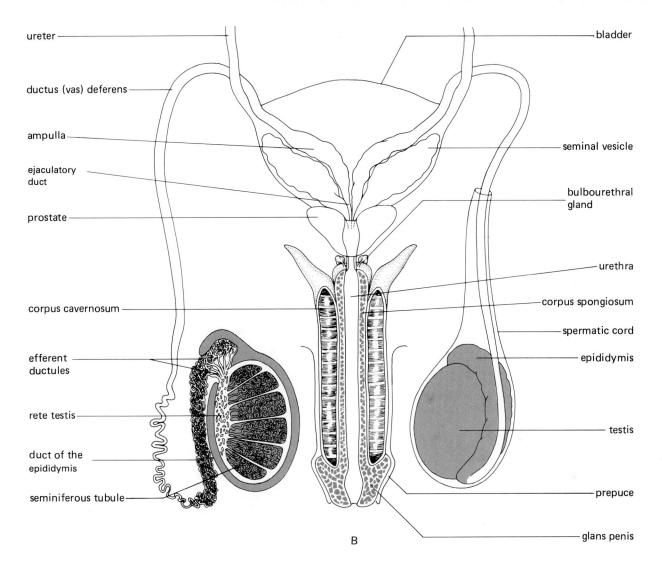

ureter

ductus (vas) deferens

ampulla

ejaculatory duct

prostate

corpus cavernosum

efferent ductules

rete testis

duct of the epididymis

seminiferous tubule

bladder

seminal vesicle

bulbourethral gland

urethra

corpus spongiosum

spermatic cord

epididymis

testis

prepuce

glans penis

B

(Fig. 16-7). Each Sertoli cell extends from the membrane of the seminiferous tubule all the way to the fluid-filled lumen and is joined to adjacent Sertoli cells by means of tight junctions. Thus, the Sertoli cells form an unbroken ring around the outer circumference of the seminiferous tubule and divide the tubule into two compartments—a basal compartment (between the basement membrane and the tight junction), and a central (adluminal) compartment including the lumen.

This arrangement has several very important results: (1) The ring of interconnected Sertoli cells forms a "barrier" which limits the movement of many chemicals from blood into the lumen of the seminiferous tubule (the membranes surrounding the entire tubule form a second component of the so-called **blood-testis barrier**); and (2) mitosis of the spermatogonia takes place entirely in the basal compartment, and the primary spermatocytes must move through the tight junctions of the Sertoli cells (which are transiently disrupted to make way for them) to gain entry into the central compartment. In this compartment, the spermatids are contained in recesses formed by invaginations of the Sertoli-cell plasma membranes until they are mature enough for release. Apparently, the Sertoli cells

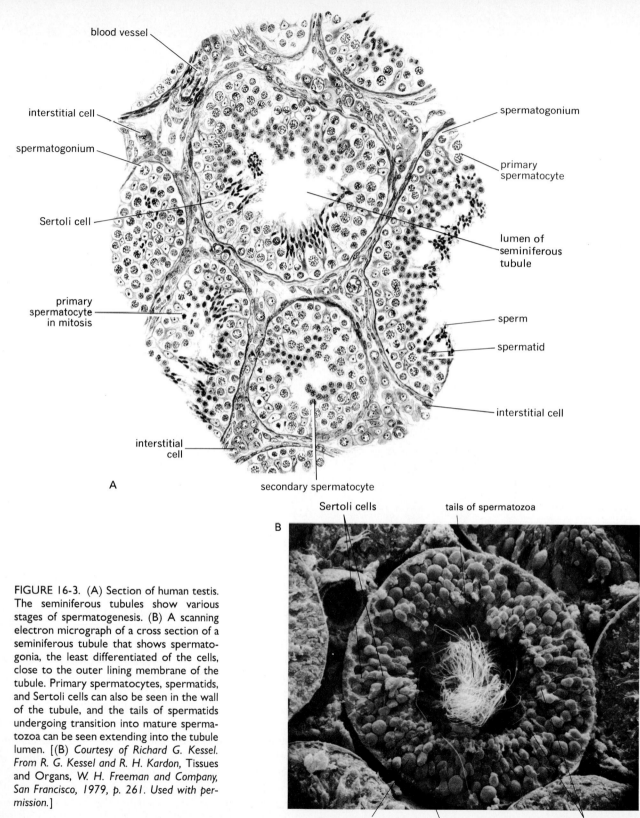

blood vessel

interstitial cell

spermatogonium

Sertoli cell

primary
spermatocyte
in mitosis

interstitial
cell

spermatogonium

primary
spermatocyte

lumen of
seminiferous
tubule

sperm

spermatid

interstitial cell

A

secondary spermatocyte

Sertoli cells

tails of spermatozoa

B

primary spermatocyte

spermatogonium

spermatids

FIGURE 16-3. (A) Section of human testis. The seminiferous tubules show various stages of spermatogenesis. (B) A scanning electron micrograph of a cross section of a seminiferous tubule that shows spermatogonia, the least differentiated of the cells, close to the outer lining membrane of the tubule. Primary spermatocytes, spermatids, and Sertoli cells can also be seen in the wall of the tubule, and the tails of spermatids undergoing transition into mature spermatozoa can be seen extending into the tubule lumen. [(B) *Courtesy of Richard G. Kessel. From R. G. Kessel and R. H. Kardon, Tissues and Organs, W. H. Freeman and Company, San Francisco, 1979, p. 261. Used with permission.*]

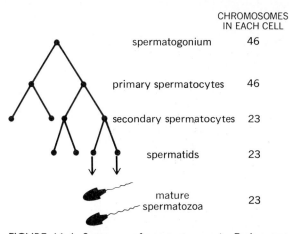

FIGURE 16-4. Summary of spermatogenesis. Each spermatogonium yields eight mature sperm, each of which contains 23 chromosomes.

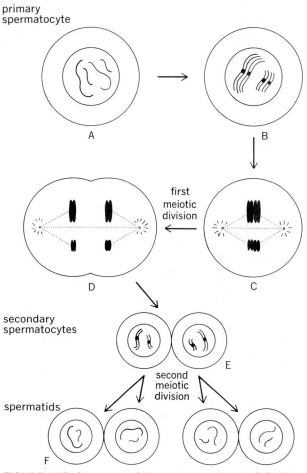

FIGURE 16-5. Separation of chromosomes during cell divisions of meiosis.

serve as the route by which nutrients and chemical signals reach these developing germ cells. As we shall see, they may also produce hormones and other chemicals important for the control of spermatogenesis.

The Sertoli cells have still other functions (Table 16-1). They secrete most of the fluid in the lumen of the seminiferous tubules, fluid which has a highly characteristic ionic composition and contains, among other organic components, androgens (see below) and an **androgen-binding protein**. Finally, as we shall see, the Sertoli cells play a central role in the hormonal control of spermatogenesis.

In any small segment of seminiferous tubules, the entire process of spermatogenesis proceeds in a regular sequence. For example, at any given time, virtually all the primary spermatocytes in one portion of the tubule are undergoing division, whereas in an adjacent segment, the secondary spermatocytes may be dividing. The entire process, from primary spermatocyte to spermatozoon, in a single area takes approximately 64 days. Perhaps the most amazing characteristic of spermatogenesis is its sheer magnitude: The normal human male may manufacture several hundred million sperm per day.

Transport of sperm

When spermatid elongation is completed, the Sertoli-cell cytoplasm around them retracts, and the

spermatids are released into the lumen of the seminiferous tubule, bathed by the luminal fluid. From the seminiferous tubules, the still-immature sperm pass through a network of interconnected highly coiled ducts which join to form a single duct within the **epididymis** (Fig. 16-8), which in turn leads to the large thick-walled **ductus deferens**.

Movement of the sperm as far as the epididymis results from the pressure created by the continuous formation of fluid by the Sertoli cells back in the seminiferous tubules (the sperm themselves are nonmotile at this time). During passage through the epididymis (which takes about 12 days), the sperm undergo a **maturation process** which confers upon them the ability to swim and the ability to fertilize

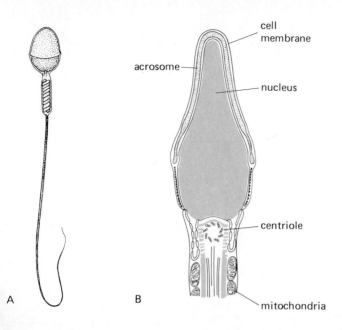

cell
membrane

acrosome

nucleus

centriole

mitochondria

A   B

FIGURE 16-6. (A) Human mature sperm seen in frontal view. (B) A close-up of the side view.

an ovum after a period in the female tract. This maturation process (accompanied by changes in sperm size, shape, and metabolic properties) is crucially dependent upon adequate stimulation of the epididymis by testosterone (most of which reaches the epididymis not via the blood but, instead, complexed with androgen-binding protein in the fluid entering the epididymis from the seminiferous tubules).

Another event which occurs during passage through the epididymis is a hundredfold concentration of the sperm by fluid absorption from the lumen. Thus, as the sperm pass from the end of the epididymis into the next segment, the ductus deferens, they are a densely packed mass whose transport is no longer a result of fluid movement but is rather the result of peristaltic-like contractions of the smooth muscle in the epididymis and ductus deferens. (The absence of a large quantity of fluid accounts for the fact that a man with a tied-off ductus deferens (**vasectomy**) does not accumulate much fluid behind the tied-off point; the sperm do build up, however, and these are removed mainly by phagocytosis.)

The ductus deferens and the last portion of the epididymis serve as the storage reservoir for sperm

prior to ejaculation. A pair of large glands, the **seminal vesicles**, drain into the two ductus deferens, which now become the **ejaculatory ducts** which pass into the **prostate gland** and join the urethra (Fig. 16-2). The prostate and seminal vesicles, as well as the **bulbourethral glands** just below the prostate, secrete the bulk of the fluid in which ejaculated sperm are suspended. This fluid, plus the sperm cells, constitutes **semen**, which contains a large number of different chemical substances, the functions of which are still generally unclear. Another function of the seminal fluid is to dilute the sperm (sperm constitutes only a few percent of total ejaculated semen); without such dilution, motility is impaired.

Erection

The primary components of the male sexual act are **erection** of the penis, which permits entry into the vagina of the female, and **ejaculation** of the semen into the vagina. Erection is a vascular phenomenon which can be understood from the structure of the penis (Fig. 16-9). This organ consists almost entirely of three cylindrical cords of **erectile tissue**, which are actually vascular spaces. Normally the vessels supplying these chambers are constricted

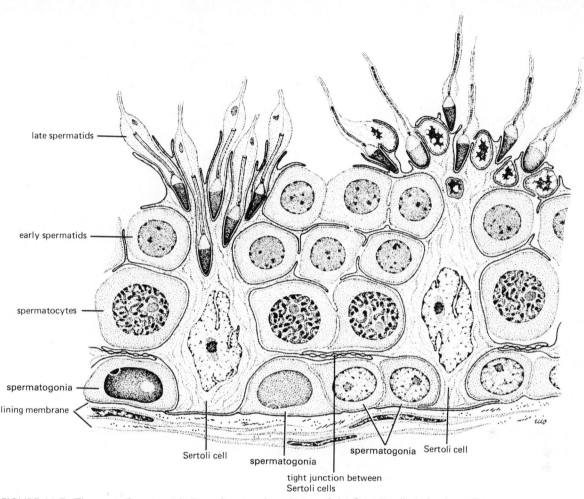

late spermatids

early spermatids

spermatocytes

spermatogonia

lining membrane

Sertoli cell spermatogonia spermatogonia Sertoli cell

tight junction between
Sertoli cells

FIGURE 16-7. The seminiferous epithelium, showing the relation of the Sertoli cells and germ cells.

TABLE 16-1. Functions of Sertoli cells

1. Provide a blood-testis barrier to chemicals

2. Nourish developing sperm

3. Secrete the luminal fluid, including androgen-binding protein, of the seminiferous tubules

4. Are the site of action for testosterone and FSH in control of spermatogenesis

5. Secrete the protein hormone inhibin

---

so that the erectile tissue contains little blood and the penis is flaccid; during sexual excitation, the inflow vessels dilate, the chambers become engorged with blood at high pressure, and the penis becomes rigid. Moreover, as the erectile tissues expand, the veins emptying them are passively compressed, thus importantly contributing to the engorgement. This entire process occurs rapidly, complete erection sometimes taking only 5 to 10 s. The vascular dilation is accomplished by stimulation of the **parasympathetic nerves** and inhibition of the **sympathetic nerves** to the blood vessels of the

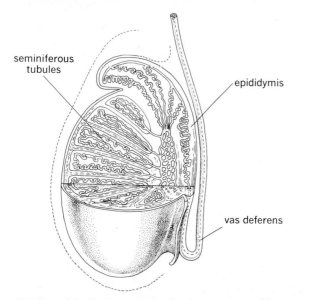

FIGURE 16-8. Cross section of a human testis. The highly coiled seminiferous tubules are the sites of sperm production.

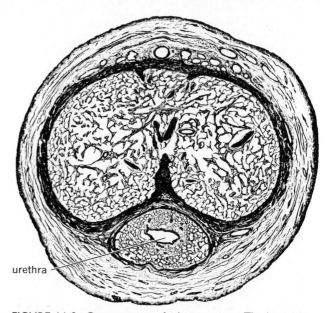

FIGURE 16-9. Cross section of a human penis. The large central area and that surrounding the urethra are vascular spaces which become filled with blood to cause erection. *(Adapted from Bloom and Fawcett.)*

penis (Fig. 16-10). This appears to be one of the few cases of direct parasympathetic control over high-resistance blood vessels. In addition to these vascular effects, the parasympathetic nerves stimulate urethral glands to secrete a mucuslike material which aids in lubrication of the head of the penis.

What receptors and afferent pathway initiate these reflexes? The primary input comes from highly sensitive mechanoreceptors in the genital regions, particularly in the head of the penis. The afferent fibers carrying the impulses synapse in the lower spinal cord and trigger the efferent outflow. It must be stressed, however, that higher brain centers, via descending pathways, may exert profound stimulatory or inhibitory effects upon the autonomic neurons to the arterioles of the penis. Thus, thoughts or emotions can induce erection in the complete absence of mechanical stimulation of the penis; conversely, failure of erection (**impotence**) may frequently be due to psychological factors acting through descending pathways. The ability of alcohol to inhibit erection is probably due to its effects on higher brain centers.

Ejaculation

This process is also basically a spinal reflex, the afferent pathways apparently being identical to those described for erection. When the level of stimulation produces sufficient summation of synaptic potentials, a patterned autonomic sequence of efferent discharge is elicited which can be divided into two phases: (1) The smooth muscles of the prostate, ductus deferens, and seminal vesicles contract as a result of sympathetic stimulation, emptying the semen into the urethra (**emission**); and (2) the semen is then expelled from the urethra by a series of rapid contractions of the urethral smooth muscle as well as the skeletal muscle at the base of the penis. During ejaculation, the sphincter at the base of the bladder is closed so that sperm cannot enter the bladder nor can urine be expelled.

The rhythmical contractions of the muscles which occur during ejaculation are associated with intense pleasure and many systemic physiological changes, the entire event being termed an **orgasm**.

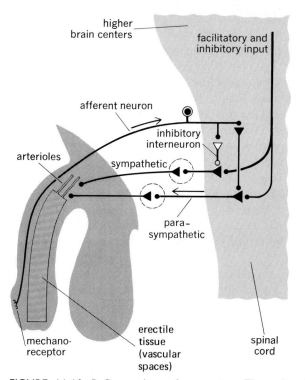

FIGURE 16-10. Reflex pathway for erection. The reflex is initiated by mechanoreceptors in the penis. Input from higher centers can facilitate or inhibit this reflex.

A marked skeletal muscle contraction throughout the body is followed by the rapid onset of muscular and psychological relaxation, and there is also a marked increase in heart rate and blood pressure. Once ejaculation has occurred, there is a so-called latent period during which a second erection is not possible; this period is quite variable but may last from minutes to hours in perfectly normal men. As is true of erection, premature ejaculation, failure to ejaculate, or lack of generalized orgasm at the time of ejaculation can be the result of influence by higher brain centers via descending pathways.

The average volume of semen ejaculated is 3 mL, containing approximately 300 million sperm. However, the range of normal values is extremely large and although quantity is important, it is obvious that the "quality" of the sperm is another critical determinant of fertility. Numerous tests are available to assess the structure, metabolism, and motility of sperm.

Hormonal Control of Male Reproductive Functions

As briefly mentioned earlier, male reproductive function is controlled by a chain of hormones (Fig. 16-11): (1) a single gonadotropin releasing hormone (GnRH) from the hypothalamus stimulates the secretion of both gonadotropins—follicle-stimulating hormone (FSH) and luteinizing hormone (LH)[1]— from the anterior pituitary; (2) FSH and LH stimulate the gonads, the result being production of mature sperm and the secretion of testosterone; and (3) testosterone exerts a wide variety of effects on the reproductive system (including the gonads themselves) and other organs or tissues.

Each link in this hormonal chain is essential, for malfunction of either the hypothalamus or the anterior pituitary can result in failure of gametogenesis or testosterone secretion as surely as if the gonads themselves were diseased.

Effects of GnRH, FSH, and LH

Recall from Chap. 9 that the hypothalamic cells which produce releasing hormones, including GnRH, are neurons which secrete their hormones into the hypothalamo-pituitary portal vessels as a result of action potentials generated in the neurons. In a normal adult man, the GnRH-secreting neuroendocrine cells fire a synchronous burst of action potentials approximately every 1.5 h and so release GnRH at these times, with essentially no secretion in between. It is not yet known what underlies the rhythmical nature of the electrical activity. It is possible that these neurons possess autorhythmicity; i.e., that they spontaneously depolarize to threshold as a result of inherent pacemaker potentials (like the SA node of the heart). Alternatively,

---

[1] These two protein hormones were named for their effects in the female, but their molecular structures are the same in both sexes. (LH, in the male, is sometimes called interstitial cell–stimulating hormone, ICSH.)

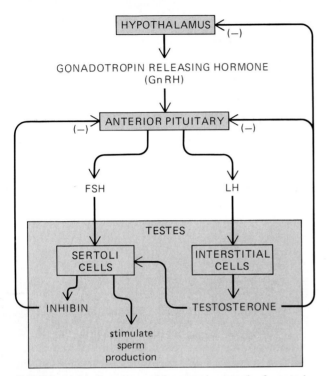

FIGURE 16-11. Summary of hormonal control of testicular function. GnRH reaches the anterior pituitary via the hypothalamo-pituitary portal vessels. The negative signs (−) denote the negative-feedback effects exerted by inhibin and testosterone on the anterior pituitary and hypothalamus. Note that FSH acts only on the Sertoli cells, and its secretion is inhibited by inhibin, a protein hormone secreted by the Sertoli cells. LH acts only on the interstitial cells, and its secretion is inhibited by testosterone, the hormone secreted by the interstitial cells.

synaptic interconnections with other neurons could constitute an oscillating circuit (analogous to that described for control of breathing) periodically activating the neuroendocrine cells.

The GnRH reaching the anterior pituitary (via the hypothalamo-pituitary portal vessels) during each periodic pulse triggers the release from the anterior pituitary of both LH and FSH (although not necessarily in equal amounts, as we shall see). Accordingly, systemic plasma concentrations of FSH and LH also show rhythmical episodic changes—rapid increases during the pulse followed by slow decreases over the next 90 min or so, during which no LH or FSH is being secreted and the two hormones are slowly removed from the plasma (Fig. 16-12). As we shall see, since LH is the stimulus for testosterone secretion by the testes, testosterone also shows similar rhythmical changes in its plasma concentration. (As described in Chap. 9, such fluctuations in plasma hormone concentrations prevent the marked down-regulation of the hormone's receptors which could occur in response to a more constant plasma concentration of the hormone.)

There is a clear separation of function of FSH and LH on the testes (Fig. 16-11). FSH acts on the Sertoli cells in the seminiferous tubules to facilitate spermatogenesis, whereas LH stimulates testosterone secretion by a direct action on the interstitial cells.

The last components of the hypothalamo-pituitary control of male reproduction which remain to be discussed are the negative feedbacks exerted by testicular hormones. That such inhibition exists is evidenced by the fact that castration results in marked increases in the secretion of both LH and FSH. Testosterone is clearly responsible for the inhibition of LH secretion but has much less effect on FSH. There are two ways, one indirect and one

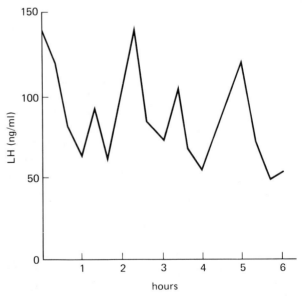

FIGURE 16-12. Secretory pattern of LH in a normal man. *(Adapted from Santen and Bardin.)*

direct, by which testosterone normally keeps LH secretion at a relatively reduced level (Fig. 16-11): (1) It acts on the hypothalamus to decrease the frequency of GnRH bursts, which thereby results in less GnRH reaching the pituitary over any given period of time; and (2) it acts directly on the anterior pituitary to cause less LH secretion in response to any given level of GnRH. This direct action on the pituitary helps explain why testosterone inhibits LH secretion so much more than FSH secretion.

How does the presence of functioning testes reduce FSH secretion, if not via testosterone? The inhibitory signal (exerted directly on the anterior pituitary) is a protein hormone, termed **inhibin**, which is secreted by the Sertoli cells of the seminiferous tubules (Fig. 16-11). That the Sertoli cells are the source of this tonic feedback inhibition of FSH secretion makes sense since, as pointed out above, the facilitatory effect of FSH on spermatogenesis is exerted via the Sertoli cells; thus, these cells are in all ways the link between FSH and spermatogenesis.

Despite these complexities, one should not lose sight of the fact that the total amounts of GnRH, LH, FSH, and testosterone secreted (and sperm produced) are relatively constant from day to day in the adult male. This is completely different from the large cyclical swings of activity so characteristic of the female reproductive processes. However, this is not to say that the interplay between GnRH, LH, FSH, testosterone, and inhibin results in absolutely unchangeable levels of these hormones; the GnRH-secreting cells in the hypothalamus receive much synaptic input from other neurons, some excitatory and some inhibitory. This input can cause changes in the frequency at which the GnRH pulses occur and, hence, changes in the average rates at which FSH, LH, and testosterone are secreted. For example, watching sexually arousing movies causes a marked increase in the secretion of all these hormones.

Effects of testosterone

Testosterone, the hormone produced by the testes, is the major male sex hormone (Table 16-2). As pointed out in Chap. 9, other steroids with actions

---

TABLE 16-2. Effects of testosterone and other androgens

1. Are necessary for spermatogenesis
2. Induce differentiation and maintain function of male accessory organs
3. Induce male secondary sex characteristics
4. Stimulate protein anabolism and bone growth
5. Maintain sex drive and (?) enhance aggressive behavior
6. Exert feedback inhibition of LH secretion

---

similar to those of testosterone are produced by the adrenal cortex and are, with testosterone, collectively known as **androgens**. Although the adrenally produced androgens constitute a large fraction of total blood androgens in men, they are much less potent than testosterone and are unable to maintain testosterone-dependent functions should testosterone secretion be decreased or eliminated by disease or castration (surgical removal of the gonads). Accordingly, we shall not further discuss adrenal androgens in our description of male reproductive function. We shall also ignore the small amounts of estrogen produced by the testes.

Spermatogenesis. As pointed out earlier, FSH is required for normal spermatogenesis. So are adequate amounts of testosterone, and sterility is an invariable result of testosterone deficiency. As mentioned earlier, the cells which secrete testosterone are the interstitial cells (Fig. 16-3), which lie scattered between the seminiferous tubules. The stimulatory effects of testosterone on spermatogenesis are exerted locally by the hormone transported from the interstitial cells into the seminiferous tubules. There, testosterone enters Sertoli cells and it is largely, if not entirely, via these cells that testosterone facilitates spermatogenesis. Some testosterone in the luminal fluid binds to androgen-binding protein (also produced by the Sertoli cells).

It must be emphasized that although testosterone is required for the process of spermatogenesis, testosterone production does not depend upon spermatogenesis. In other words, testosterone deficiency produces sterility by interrupting spermatogenesis, but interference with the function of the seminiferous tubules does not alter normal testosterone production by the interstitial cells. This relationship has relevance for the simple, effective method of sterilizing the male known as **vasectomy**—surgical ligation and removal of a segment of each of the two vas deferens, which carry sperm from the testes. This procedure prevents the delivery of sperm but does not alter secretion of testosterone. (Whether vasectomy might have other effects on health is presently controversial; more than half the men who have undergone vasectomy develop circulatory antibodies to sperm proteins and, as described in Chap. 17, it is conceivable that such antibodies might exert harmful effects on the cardiovascular system.)

Accessory reproductive organs. The morphology and function of the entire male duct system, glands, and penis all depend upon testosterone. Following removal of the testes (castration) in the adult, all the accessory reproductive organs decrease in size, the glands markedly reduce their rates of secretion, and the smooth muscle activity of the ducts is inhibited. Erection and ejaculation may be deficient. These defects disappear upon the administration of testosterone.

Secondary sex characteristics. In men, virtually all the obvious masculine secondary characteristics are testosterone-dependent. For example, a male castrated before puberty does not develop a beard or either axillary or pubic hair. An unexplained finding is that baldness, although genetically determined in part, does not occur in castrated men (castration will not, however, reverse baldness). Other testosterone-dependent secondary sexual characteristics are the deepening of the voice (resulting from growth of the larynx), thick secretion of the skin oil glands (predisposing to acne), and the masculine pattern of muscle and fat distribution.

This leads us into an area of testosterone effects usually categorized as general metabolic effects but very difficult to separate from the secondary sex characteristics. It is obvious that the bodies of men and women (even excepting the breasts and external genitals) have very different appearances; a woman's shape is due in large part to the feminine distribution of fat, particularly in the region of the hips and lower abdomen but in the limbs as well. A castrated male gradually develops this pattern, whereas a woman treated with testosterone loses it. A second very obvious difference is that of skeletal muscle mass, which is increased by testosterone. The overall relationship of testosterone to general body growth was described in Chap. 15.

Behavior. Most of our information in this area comes from experiments on animals other than human beings, but even from our fragmentary information about people, there is little doubt that the development and maintenance of normal sexual drive and behavior in men are, in large part, testosterone-dependent and may be impaired by castration. However, it is a mistake to assume that deviant male sexual behavior must therefore be due to testosterone deficiency or excess. For example, most (but not all) male homosexuals have normal rates of testosterone secretion; although administration of exogenous testosterone may sometimes increase sexual activity in these men, they remain homosexual. To date, no clear-cut correlation has been established between homosexuality or hypersexuality and hormonal status in either men or women.

A question which has recently become the subject of controversy is whether testosterone influences other human behavior in addition to sex; i.e., are there any inherent male-female differences, or are the observed differences in behavior all socially conditioned? There is little doubt that behavioral differences based on sex do exist in other mammals; for example, aggression is clearly greater in males and is testosterone-dependent. Obviously, it will be difficult to answer such questions with respect to human beings, but attempts are now being made to study them in a controlled scientific manner.

Mechanism of action of testosterone. Testosterone, like other steroid hormones, crosses plasma membranes readily and, within target cells, combines with specific cytoplasmic receptors. The hormone-receptor complex then moves into the nucleus where it influences the transcription of certain genes into messenger RNA. The result is a change in the rate at which the target cell synthesizes the proteins coded for by these genes, and it is the changes in the concentrations of these proteins which underlie the cell's overall response to the hormone. For example, testosterone induces increased synthesis of enzymes in the prostate gland which then catalyze the formation of the gland's secretions.

In Chap. 9 we mentioned that hormones sometimes must undergo transformation in their target cells in order to be most effective, and this is true of testosterone in certain of its target cells. In some of these, after its entry to the cytoplasm, testosterone undergoes an enzyme-mediated transformation into another steroid—**dihydrotestosterone**—and it is mainly this molecule which then combines with androgen receptor and moves into the nucleus to exert its effect (Fig. 16-13).

Quite startling is the fact that, in still other target cells, notably neurons in certain areas of the brain, testosterone is transformed not to dihydrotestosterone but to **estrogen**, which then combines with cytoplasmic estrogen receptor, moves into the nucleus . . . etc. (Fig. 16-13). Thus, a "male" sex hormone must first be transformed into a "female" sex hormone to be able to influence neurons, which mediate "male" behavior!

Prolactin

A third anterior pituitary hormone, prolactin, has only recently been recognized to have important functions in men (again, its name reflects its long-recognized major function in women). By itself, prolactin exerts very little effect on male reproductive function, but it significantly potentiates the effects of testosterone on many of the latter's target cells. This "permissive" action is probably mediated by prolactin's ability to increase the number of receptors for testosterone on these cells.

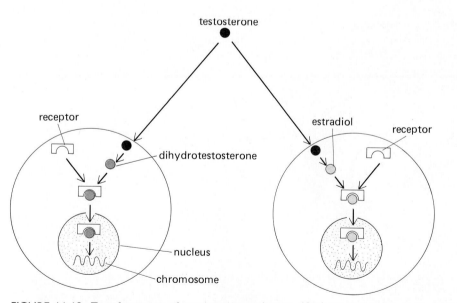

FIGURE 16-13. Transformation of testosterone either to dihydrotestosterone or to estradiol (an estrogen) is accomplished by enzymes in many target cells. In other target cells, testosterone does not undergo transformation. *(Adapted from McEwen.)*

SECTION B.
FEMALE REPRODUCTIVE PHYSIOLOGY

The ovaries and female reproductive duct system —uterine tubes, uterus, and vagina—are located in the pelvic region and constitute the **internal genitalia** (Fig. 16-14A and B). In the female, the urinary and reproductive duct systems are entirely separate.

The **uterine tubes** (also known as oviducts or Fallopian tubes) are not directly connected to the ovaries but open into the abdominal cavity close to them. This opening of each uterine tube is a trumpet-shaped expansion surrounded by long fingerlike projections (the **fimbriae**) which are lined with ciliated epithelium. The other ends of the uterine tubes are attached to the uterus and empty directly into its cavity (the womb). The **uterus** is a hollow, thick-walled, muscular organ lying between the bladder and rectum. The lower portion of the uterus is the **cervix**. A small opening in the cervix leads from the uterus to the **vagina**.

The **external genitalia** (Fig. 16-14C) include the mons pubis, labia majora and minora, the clitoris, the vestibule of the vagina, and the vestibular glands. The term **vulva** includes all these parts. The **labia majora**, the female analogue of the scrotum, are two prominent skin folds. The **labia minora** are small skin folds lying between the labia majora;

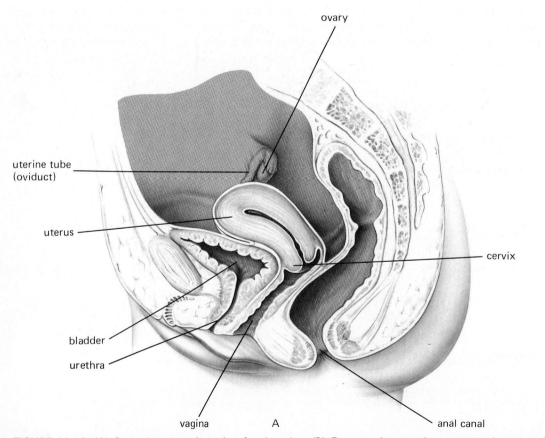

FIGURE 16-14. (A) Sagittal section through a female pelvis. (B) Diagram showing the continuity between the oviducts, uterus, and vagina. (C) Female external genitalia.

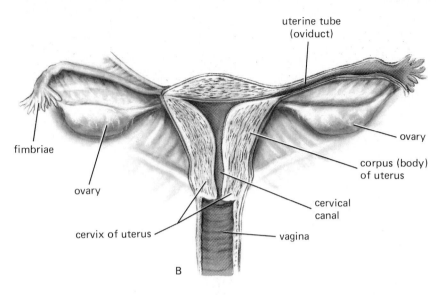

uterine tube (oviduct)

fimbriae

ovary

cervix of uterus

ovary

corpus (body) of uterus

cervical canal

vagina

B

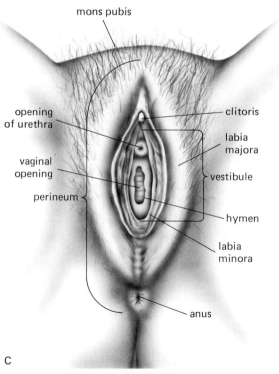

mons pubis

opening of urethra

vaginal opening

perineum

clitoris

labia majora

vestibule

hymen

labia minora

anus

C

they surround the urethral and vaginal openings, and the area thus enclosed is the **vestibule**. The vaginal opening lies behind that of the urethra. Partially overlying the vaginal opening is a thin fold of

mucous membrane, the **hymen**. The **clitoris** is an erectile structure homologous to the penis.

Unlike the continuous sperm production of the male, the maturation and release of the female gamete, the ovum, is cyclic. This pattern is true not only for ovum development but for the function and structure of virtually the entire female reproductive system. In human beings these cycles of approximately 28 days duration are called **menstrual cycles**.

Ovarian Function

The **ovary**, like the testis, serves a dual purpose: (1) production of **ova**, and (2) secretion of the female sex hormones **estrogen** and **progesterone**.

Ovum and follicle growth

In Fig. 16-15, a cross section of an ovary, note the discrete cells clusters known as **primary follicles**. Each is composed of one ovum surrounded by a single layer of **granulosa cells** (Fig. 16-16A). (As will be described in a subsequent section, the ovum, like the sperm, is referred to by different names at each stage of development, but for clarity, we shall often simply use the term "ovum.") At birth, normal human ovaries contain an estimated one million follicles, and no new ones appear after birth. Thus, in marked contrast to the male, the newborn female

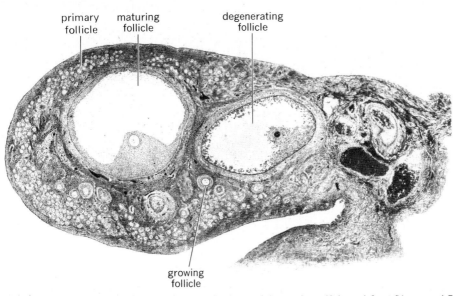

primary follicle maturing follicle degenerating follicle

growing follicle

FIGURE 16-15. Cross section of an ovary from an adult monkey. *(Adapted from Bloom and Fawcett.)*

already has all the germ cells she will ever have. Only a few, perhaps 400, are destined to reach full maturity during her active productive life. All the others degenerate at some point in their development so that few, if any, remain by the time she reaches menopause at approximately 50 years of age. One result of this is that the ova which are released (ovulated) near menopause are 30 to 35 years older than those ovulated just after puberty; it has been suggested that certain congenital defects, much commoner among children of older women, are the result of aging changes in the ovum.

The development of the follicle (Fig. 16-16) is characterized by an increase in size of the ovum and a proliferation of the surrounding granulosa cells. The ovum becomes separated from the granulosa cells by a thick layer of material, the **zona pellucida**, secreted by the granulosa cells; however, despite the presence of the zona pellucida, the inner layer of granulosa cells remains intimately associated with the ovum by means of cytoplasmic processes which traverse the zona pellucida and form gap junctions with the ovum. Whether chemical messengers cross these gap junctions from granulosa cell to ovum is unknown.

As the follicle grows, new cell layers are formed, not only from mitosis of the original follicle cells but from the growth of specialized ovarian connective-tissue cells. Thus, the follicle, originally composed of the ovum and its surrounding layers of granulosa cells, becomes invested with additional outer layers of cells known as the **theca**. When the follicle reaches a certain diameter, a fluid-filled space, the **antrum**, begins to form in the midst of the granulosa cells as a result of fluid they secrete.

By the time the antrum begins to form, the ovum has reached full size. From this point on, the follicle grows in part because of continued follicular cell proliferation but largely because of the expanding antrum. Ultimately, the ovum, surrounded by the zona pellucida and several layers of granulosa cells, occupies a little hill (the **cumulus oophorus**) projecting into the antrum. The antrum becomes so large (about 1.5 cm) that the completely mature follicle actually balloons out on the surface of the ovary.

Ovulation occurs when the ovarian wall at this site of ballooning ruptures, and the ovum, surrounded by its tightly adhering zona pellucida and cumulus, is carried out of the ovary by the antral fluid. Many women experience varying degrees of abdominal pain at approximately the midpoint of

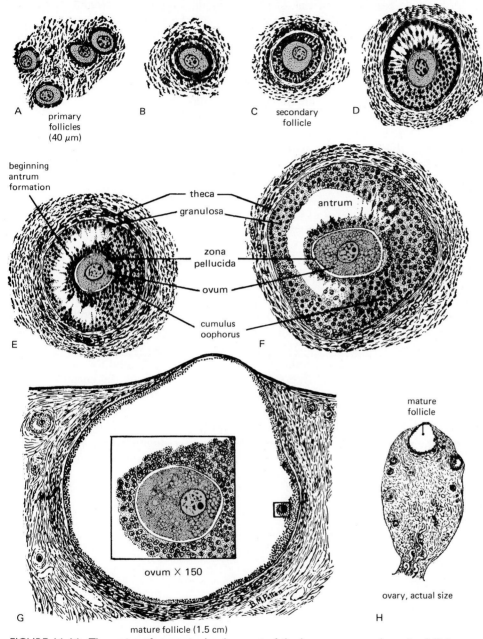

A primary follicles (40 μm)

B

C secondary follicle

D

beginning antrum formation

theca

granulosa

antrum

zona pellucida

ovum

cumulus oophorus

E

F

ovum × 150

mature follicle (1.5 cm)

G

mature follicle

ovary, actual size

H

FIGURE 16-16. The series of stages in development of the human ovum and ovarian follicle.

the menstrual cycles; this has generally been presumed to represent abdominal irritation induced by the entry of follicular contents at ovulation. However, precise timing of ovulation has indicated that this time-honored concept may be wrong, and the cause of discomfort remains unclear.

In the adult human ovary, there are, at any moment, several antrum-containing follicles of varying sizes, and during each cycle, several begin to enlarge. However, normally only one follicle reaches the complete maturity just described, the process requiring approximately 2 weeks. All the other partially matured preantral and antral follicles undergo degeneration at some stage in their growth. On occasion (1 to 2 percent of all cycles), two or more follicles reach maturity, and more than one ovum may be ovulated. This is the most common cause of multiple births; in such cases the siblings are fraternal, not identical.

Ovum division

The ova present at birth are the result of numerous mitotic divisions of the primitive ova, the **oogonia** (a term analogous to spermatogonia in the male), which had occurred during early fetal development (Fig. 16-17). At some point in fetal life the oogonia cease dividing, and from this point on no new ova are generated. The oogonia then develop into **primary oocytes** (analogous to primary spermatocytes). In the fetus, these **primary oocytes** all begin the first division of meiosis but do not complete it; accordingly, all the ova present at birth are primary oocytes containing 46 replicated DNA strands and in a state of meiotic arrest. This division is completed only just before an ovum is about to be ovulated and is analogous to the division of the primary spermatocyte, because each daughter cell receives 23 replicated chromosomes. However, in this division

one of the two daughter cells, the **secondary oocyte**, retains virtually all the cytoplasm; the other (the "polar body") is very small. In this manner, the already full-size ovum loses half of its chromosomes but almost none of its nutrient-rich cytoplasm. The second cell division of meiosis occurs in the oviduct after ovulation (indeed, only after penetration by a sperm), and the daughter cells each receive 23 single-stranded chromosomes. Once again, one daughter cell, now the fertilized ovum, retains nearly all the cytoplasm. The net result is that each primary oocyte is capable of producing only one mature fertilizable ovum (Fig. 16-17); in contrast, each primary spermatocyte produces four viable spermatozoa.

Formation of the corpus luteum

After rupture of the follicle and discharge of the antral fluid and the ovum, a transformation occurs within the remnant follicle, which collapses, and the antrum fills with partially clotted fluid, ultimately to be replaced with connective tissue. The follicular cells enlarge greatly, and the entire gland-like structure is now known as the **corpus luteum**. If the discharged ovum is not fertilized, i.e., if pregnancy does not occur, the corpus luteum reaches its maximum development within approximately 10 days and then rapidly degenerates. If pregnancy does occur, the corpus luteum grows and persists until near the end of pregnancy.

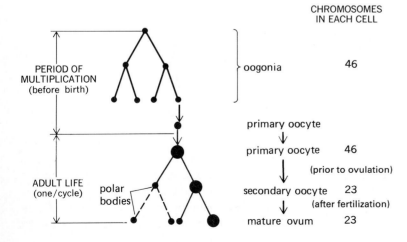

CHROMOSOMES
IN EACH CELL

PERIOD OF MULTIPLICATION (before birth)

oogonia   46

ADULT LIFE (one/cycle)   polar bodies

primary oocyte
↓
primary oocyte   46
(prior to ovulation)
↓
secondary oocyte   23
(after fertilization)
↓
mature ovum   23

FIGURE 16-17. Summary of ovum development. Compare with the male pattern of Fig. 16-4. Each primitive ovum produces only one mature ovum containing 23 chromosomes. It is a semantic oddity that the ovum is not termed "mature" until after fertilization occurs.

Cyclical nature of ovarian function

The length of a menstrual cycle varies considerably from woman to woman, averaging about 28 days. Day 1 is the first day of menstrual bleeding, and in a typical 28-day cycle ovulation occurs around day 14. In terms of ovarian function, therefore, the menstrual cycle may be divided into two approximately equal phases: (1) the **follicular phase**, during which a single follicle and ovum develop to full maturity; and (2) the **luteal phase**, during which the corpus luteum is the active ovarian structure. It must be stressed that the day of ovulation varies from woman to woman and frequently in the same woman from month to month.

Ovarian hormones

Just as ''androgen'' refers to a group of hormones with similar actions, the term **estrogen** denotes not a single specific hormone but rather a group of steroid hormones which have similar effects on the female reproductive tract. These include **estradiol**, **estrone**, and **estriol**, of which the first is the major estrogen secreted by the ovaries. It is common to refer to any of them simply as ''estrogen'' and we shall follow this practice.

Estrogen is secreted by the granulosa cells and, following ovulation, by the corpus luteum. **Progesterone**, the second ovarian steroid hormone, is secreted in very small amounts by the granulosa cells, just prior to ovulation, but its major source is the corpus luteum. The detailed physiology of these hormones will be described subsequently.

Control of Ovarian Function

The basic factors controlling ovarian function are analogous to the controls described for testicular function in that they constitute a hormonal chain made up of GnRH, the anterior pituitary gonadotropins (FSH and LH), and gonadal sex hormones (estrogen and progesterone). However, the overall schema is more complex in the female since it includes a cycling quite different from the more stable rates of hormone secretion in the male. As in the male, the entire sequence of controls depends upon the secretion of GnRH from hypothalamic neuroendocrine cells in episodic pulses, reflecting the existence of inherent autorhythmicity in the secretory cells and/or oscillating neuronal circuits. However, in the female, the frequency of these pulses (and, hence, the total amount of GnRH secreted during a 24-hour period) changes in a patterned manner over the course of the menstrual cycle. So does both the responsiveness of the anterior pituitary to GnRH and of the ovaries to FSH and LH.

For purposes of orientation, let us look first at the changes in the systemic arterial blood concentrations of four of the five participating hormones during a normal menstrual cycle (Fig. 16-18). Note that FSH is slightly elevated in the early part of the follicular phase of the menstrual cycle and then steadily decreases throughout the remainder of the period except for a small transient midcycle peak. LH is quite constant during most of the follicular phase but then shows a very large midcycle surge (peaking approximately 18 h before ovulation) followed by a progressive slow decline during the luteal phase. The estrogen pattern is more complex. After remaining fairly low and stable for the first week (as the follicle develops), it rises to reach a peak the day before LH starts off on its surge. This peak in estrogen concentration is followed by a dip, a second rise (due to secretion by the corpus luteum), and finally, a rapid decline during the last days of the cycle. The progesterone pattern is simplest of all; virtually no progesterone is secreted by the ovaries during the follicular phase (until just prior to ovulation), but very soon after ovulation the developing corpus luteum begins to secrete progesterone, and from this point the progesterone pattern is similar to that for estrogen. It is hoped that, after the following discussion, the reader will understand how these changes are all interrelated to yield a self-cycling pattern.

Control of follicle development and estrogen secretion prior to the LH surge

Ovum maturation and growth of the follicle are independent of hormonal control until just before formation of the antrum. From this point on, de-

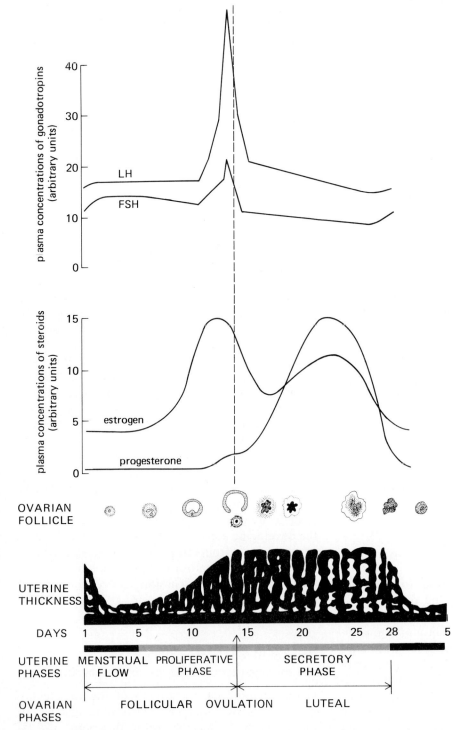

FIGURE 16-18. Summary of plasma hormone concentrations, ovarian events, and uterine changes during the menstrual cycle.

velopment depends directly upon the estrogen produced by the follicle and upon the pituitary gonadotropins—FSH and LH.

The interplay between these three hormones and its effects on ovarian cells during the early and middle portions of the follicular phase are illustrated in Fig. 16-19. During that period there is a separation of function of LH and FSH (the secretion of both being tonically stimulated by GnRH) on the thecal cells and granulosa cells, respectively. LH acts upon the thecal cells to stimulate the synthesis of androgens (note that the thecal cell is analogous to the male Leydig cell); these androgens diffuse into the granulosa cells and are converted there into estrogen by enzymes whose activity is stimulated by FSH. Thus, the overall process of estrogen production by the follicle requires the interplay of both types of follicle cells and both pituitary gonadotropins. Much of the estrogen formed by the granulosa cells diffuses into the blood (and into the antrum), but some remains in the granulosa cells and stimulates their proliferation and, hence, even more estrogen secretion. Whether estrogen exerts any effects on the ovum itself, either directly or via the granulosa cells, is not known for certain.

Throughout these portions of the follicular phase, estrogen is exerting a degree of **negative-feedback inhibition** over the entire hormonal chain (Table 16-3 and Fig. 16-20). At the plasma concentrations existing during this period, estrogen probably acts directly on the hypothalamus to decrease the frequency of GnRH pulses (and, hence, the total amount of GnRH secreted over any time period), but it also definitely acts directly on the anterior pituitary to reduce the amount of FSH and LH secreted in response to any given amount of GnRH. The follicle also secretes the protein hormone inhibin, which (as in the male) preferentially inhibits secretion of FSH; this may help explain why plasma FSH, unlike plasma LH, clearly decreases during this period (Fig. 16-18).

Control of ovulation

The follicle and ovum mature for 2 weeks under the influence of FSH, LH, and estrogen, and then ovulation is triggered by a rapid brief outpouring from the pituitary of large quantities of LH (Fig. 16-18). This LH then induces, within hours, a variety of changes in the ovum and follicle, followed by ovulation: (1) The nuclear membrane of the ovum (recall that it is a primary oocyte at this stage) breaks down, and the first meiotic division is completed; (2) simultaneously, there occurs a marked increase

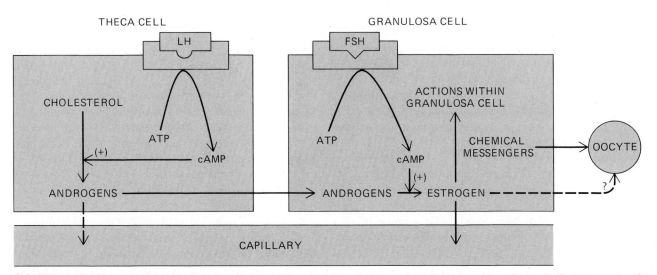

FIGURE 16-19. Control of ovarian function during early and middle portions of the follicular phase. Cyclic AMP does not itself catalyze the formation of the steroids; rather, it leads to activation of the enzymes which do so.

TABLE 16-3. Summary of major feedback effects of estrogen, progesterone, and inhibin

1. Estrogen, in *low* plasma concentrations, inhibits the hypothalamic neurons which secrete GnRH and also causes the anterior pituitary to secrete less FSH and LH in response to GnRH. Result: Negative-feedback inhibition of FSH and LH secretion.

2. Estrogen, in *high* plasma concentrations, causes the anterior pituitary cells to secrete more LH (and FSH) in response to GnRH and may also stimulate the hypothalamic neurons which secrete GnRH. Result: Positive-feedback stimulation of LH (and FSH) secretion.

3. High plasma concentrations of progesterone, in the presence of estrogen, inhibit the hypothalamic neurons which secrete GnRH. Result: Negative-feedback inhibition of FSH and LH secretion and prevention of LH surges.

4. Inhibin inhibits the secretion of FSH.

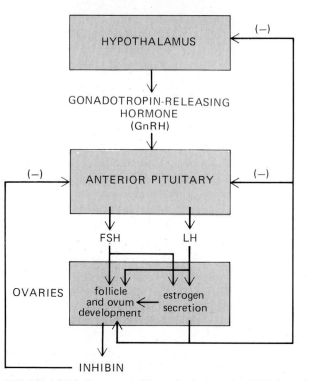

FIGURE 16-20. Summary of hormonal control of follicle and ovum development and estrogen secretion during the early and middle follicular portions of the menstrual cycle. Compare with the analogous pattern of the male (Fig. 16-11). The negative signs (−) indicate that estrogen inhibits both the hypothalamus and the anterior pituitary. Estrogen reaches the developing ovum and follicle both by local diffusion and by release into the blood and recirculation to the ovaries. Inhibin is thought to be secreted by the granulosa cells and to inhibit FSH secretion.

in follicular size (due almost entirely to increased accumulation of antral fluid), the onset of progesterone secretion by granulosa cells, and a decrease in estrogen secretion (which accounts for the midcycle drop in the plasma concentration of this latter hormone); and, finally, (3) the appearance of enzymes which break down the thin follicular-ovarian membranes at the bulge of the follicle on the ovarian surface. Most of these effects of LH reflect actions of this hormone upon the granulosa cells, which release chemical signals influencing the ovum and adjacent tissues. That LH influences the granulosa cells may come as a surprise, in light of Fig. 16-19, which shows no LH receptors on granulosa cells; however, during the period preceding the LH surge, LH receptors do appear in large numbers on the granulosa cells (induced by FSH with the assistance of estrogen).

The midcycle surge of LH emerges as perhaps the single most decisive event of the entire menstrual cycle. Indeed, it is the presence of this surge which most distinguishes the female pattern of pituitary secretion from the relatively unchanging pattern of the male.

What causes the LH surge? In the previous section we described how estrogen exerts a nega-

tive-feedback inhibition on the hypothalamus and pituitary; however, this inhibitory effect occurs only when blood estrogen concentration is relatively low, as during the first part of the follicular phase. In contrast, *high* blood concentrations of estrogen for 1 to 2 days, as occurs during the estrogen peak of the late follicular phase, act directly upon the pituitary to *enhance* the sensitivity of LH-releasing mechanisms to GnRH (Table 16-3 and Fig. 16-21); the high concentration of estrogen may also act on the hypothalamus to increase the frequency of GnRH pulses. These stimulatory effects of estrogen are often called **positive-feedback** effects.

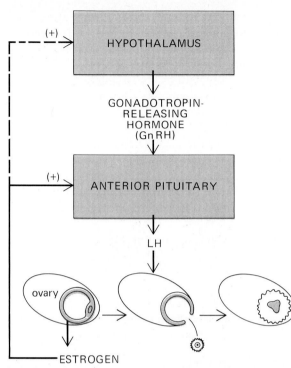

FIGURE 16-21. At the end of the follicular phase of the menstrual cycle, ovulation and corpus luteum formation are induced by the markedly increased LH, itself induced by the stimulatory effects of high levels of estrogen. The pituitary is the major site of this "positive-feedback" effect of estrogen.

The net result is that, as estrogen secretion rises rapidly during the late portion of the follicular phase, its blood concentration eventually becomes high enough for a long enough time to cause an outpouring of LH, which, in turn, induces ovulation.

Why is there only one surge per menstrual cycle; i.e., why does not the elevated estrogen existing throughout most of the luteal phase keep inducing LH surges? The answer is that, as a result of ovulation and succeeding corpus luteum formation (both induced by the LH surge), large amounts of progesterone are secreted, and progesterone (in the presence of estrogen) acts on the hypothalamus to suppress the secretion of GnRH (by reducing the frequency of GnRH pulses) (Table 16-3). (Interestingly, the much lower plasma concentrations of progesterone which occur during the late follicular phase may have just the opposite effect; i.e., they may facilitate the estrogen-induced LH surge.)

Control of the corpus luteum

The LH surge not only induces ovulation but triggers the reactions which transform the cells of the ovarian follicle from which the ovum came into a corpus luteum. Once formed, maintenance of the corpus luteum (and its secretion of hormones) requires some stimulatory support from LH, but the amount of LH needed is quite small. Recall that the corpus luteum degenerates within 2 weeks if pregnancy does not occur. What causes this? The most likely hypothesis at present is that the corpus luteum cells "self-destruct" after 10 to 14 days by producing large quantities of a **prostaglandin** which interferes locally with their own hormone secretion and cell function (Table 16-4); the stimulus for this late-luteal-phase secretion of prostaglandin is not yet known with certainty.

During its short life in the nonpregnant woman, the corpus luteum secretes large quantities of estrogen and progesterone. As mentioned above, a high blood concentration of progesterone (in the presence of estrogen) exerts a powerful inhibition of GnRH secretion and, hence, of FSH and LH secretion. Accordingly, during the luteal phase of the cycle, pituitary gonadotropin secretion is reduced, which explains the diminished rate of follicular maturation during this second half of the cycle. With degeneration of the corpus luteum, blood estrogen and progesterone concentrations decrease; GnRH, FSH, and LH secretions increase, as a result of being relieved from the inhibiting effects of progesterone and estrogen; and a new follicle is stimulated to mature.

Summary

The events described thus far in this section are summarized in Fig. 16-18, which shows the ovarian and hormonal changes during a normal nonpregnant menstrual cycle:

1. Under the influence of FSH, LH, and estrogen, a single follicle and ovum reach maturity at about 2 weeks.

TABLE 16-4. Some effects of prostaglandins* on the female reproductive system.

Site of production of prostaglandin	Action of prostaglandin	Result
Ovary (corpus luteum)	Interferes with corpus luteum's hormone secretion and function	Death of corpus luteum
Uterus (endometrium)	Constricts blood vessels in endometrium	Onset of menstruation
	Causes changes in endometrial blood vessels and cells early in pregnancy	Facilitates implantation
Uterus (myometrium)	Increases contraction of uterine smooth muscle	Helps initiate menstruation; helps initiate parturition

* The term "prostaglandins" is used here to denote not only the classical prostaglandins but closely related molecules such as the thromboxanes (see Chap. 7).

2. During the second week of the cycle, under the influence of LH and FSH, estrogen secretion by the granulosa cells progressively increases.
3. For several days near midperiod, production of LH (and FSH, to a lesser extent) increases sharply as a result of the stimulatory (positive-feedback) effects of high levels of estrogen on the pituitary and possibly also on the hypothalamus (this LH surge may be facilitated by the small amounts of progesterone present at this time).
4. The high concentration of LH induces final maturation of the follicle and ovum, followed by ovulation (it is not known what role, if any, the increased FSH plays).

5. The remaining follicle cells are transformed into a corpus luteum, which secretes large quantities of both estrogen and progesterone.
6. The high blood concentration of progesterone (in the presence of estrogen) inhibits the release of GnRH and, hence, of LH and FSH, thereby lowering their blood concentrations and preventing the development of a new follicle or ovum during the last 2 weeks of the cycle; additional LH surges are also prevented.
7. Failure of ovum fertilization is associated with the degeneration of the corpus luteum during the last days of the cycle, mediated by prostaglandin.
8. The disintegrating corpus luteum is unable to maintain its secretion of estrogen and progesterone, and their blood concentrations drop rapidly.
9. The marked decrease of estrogen and progesterone removes the inhibition of GnRH, FSH, and LH secretion.
10. The blood concentrations of FSH and LH begin to rise, follicle and ovum development are stimulated, and the cycle begins anew.

Uterine Changes in the Menstrual Cycle

Profound changes in uterine morphology occur during the menstrual cycle which are completely attributable to the effects of estrogen and progesterone (Table 16-5). Estrogen stimulates growth of the uterine smooth muscle (**myometrium**) and the glandular epithelium (**endometrium**) lining the uterine cavity. In addition, it induces the synthesis of receptors for progesterone. Then, during the luteal phase, progesterone acts upon this estrogen-primed endometrium to convert it to an actively secreting tissue: The tubular glands become coiled and filled with secreted glycogen, the blood vessels become spiral and more numerous, and various enzymes accumuluate in the glands and connective tissue. All these changes are ideally suited to provide a hospitable environment for implantation and survival of the embryo.

Estrogen and progesterone also have important effects on the **mucus** secreted by the cervix of the uterus. Under the influence of estrogen alone, this mucus is abundant, clear, and nonviscous; all these characteristics are most pronounced at the time of ovulation and facilitate movement through the mucus by sperm that have been deposited in the vagina. In contrast, progesterone causes the cervical mucus to become thick and sticky; in essence it becomes a ''plug'' which may constitute an important blockade against the entry of sperm or bacteria from the vagina—the latter a further protection for the fetus should conception occur.

Progesterone also has an important inhibitory effect on the contraction of uterine smooth muscle, in large part by opposing the stimulatory actions of estrogen and locally generated prostaglandins.

The uterine changes throughout the normal menstrual cycle can now be explained:

1. The fall in blood progesterone and estrogen, which results from regression of the corpus luteum, deprives the highly developed endometrial lining of its hormonal support; the immediate result is profound constriction of the uterine blood vessels due to production of vasoconstrictor prostaglandins (Table 16-4), which leads to diminished supply of oxygen and nutrients. Disintegration starts, and the entire lining (except for a thin, deep layer which will regenerate the endometrium in the next cycle) begins to slough. Also, the uterine smooth muscle begins to undergo rhythmical contractions, mediated by locally produced prostaglandins no longer opposed by high levels of progesterone. (The major cause of menstrual cramps—**dysmenorrhea**—suffered by a large number of normal women is overproduction of these prostaglandins, leading to excessive uterine contractions; their effects on smooth muscle elsewhere in the body also account for the systemic symptoms that sometimes accompany the cramps, such as nausea, vomiting, and headache.)
2. After the initial period of vascular constriction, the endometrial arterioles dilate, resulting in hemorrhage through the weakened capillary walls; the **menstrual flow** consists of this blood mixed with endometrial debris. Average blood

loss per menstrual period equals 50 to 150 mL of blood.
3. The entire period of menstruation is known as the **menstrual phase** of the menstrual cycle; it continues for 3 to 5 days, during which time blood estrogen levels are low.
4. The menstrual flow ceases as the endometrium repairs itself and then grows under the influence of the rising blood estrogen concentration; this period, the **proliferative phase**, lasts for the 10 days or so between cessation of menstruation and ovulation.
5. Following ovulation and formation of the corpus luteum, progesterone, acting in concert with estrogen, induces the secretory type of endometrium described above.
6. This period, the **secretory phase**, is terminated by disintegration of the corpus luteum and withdrawal of hormonal stimulation of the uterus, completing the cycle.

It is evident that the phases of the menstrual cycle can be named either in terms of the ovarian or uterine events (Fig. 16-18). Thus, the ovarian follicular phase includes the uterine menstrual and proliferative phases; the ovarian luteal phase is the same as the uterine secretory phase. The essential point is that the uterine changes simply reflect the effects of varying blood concentrations of estrogen and progesterone throughout the cycle.

Nonuterine Effects of Estrogen and Progesterone

The uterine effects of the sex hormones described above represent only one set of many exerted by estrogen and progesterone (Table 16-5); the nonuterine effects are also discussed in this chapter and are listed here together for reference. The effects of estrogen in the female are analogous to those of testosterone in the male in that estrogen influences all the accessory sex organs and secondary sex characteristics. Estrogenic stimulation maintains the entire female genital tract (uterus, uterine tubes, and vagina), the glands lining the tract, the external

TABLE 16-5. Effects of female sex steroids

Effects of estrogen

1. Stimulation of growth of ovaries and follicles

2. Stimulation of growth and maintenance of the smooth muscle and epithelial linings of the entire reproductive tract

   a. Uterine tubes: Increase of motility and ciliary activity

   b. Uterus: Increase of motility

   Secretion of abundant, clear cervical mucus

   Preparation of endometrium for progesterone's action by inducing progesterone receptors

   c. Vagina: Increase of "cornification" (layering of epithelial cells)

3. Stimulation of growth of external genitalia

4. Stimulation of growth of breasts (particularly ducts)

5. Stimulation of development of female body configuration: narrow shoulders, broad hips, converging thighs, diverging arms

6. Stimulation of more-fluid sebaceous gland secretions ("antiacne")

7. Stimulation of development of female pattern of pubic hair (actual growth of pubic and axillary hair is androgen-stimulated)

8. Closure of the epiphyses of bone

9. Vascular effects (deficiency produces "hot flashes")

10. Feedback effects on hypothalamus and anterior pituitary (see Table 16-3)

11. Retention of fluid

12. Stimulation of prolactin secretion but inhibition of prolactin's milk-inducing actions on the breasts.

Effects of progesterone

1. Stimulation of secretions by endometrial glands

2. Induction of thick, sticky cervical secretions

3. Stimulation of growth of myometrium (in pregnancy)

4. Decrease of motility of uterine tubes and uterus

5. Decrease of vaginal "cornification"

TABLE 16-5. (continued)

6. Stimulation of breast growth (particularly glandular tissue)

7. Inhibition of milk-inducing effects of prolactin on the breasts

8. Feedback effects on hypothalamus and anterior pituitary (see Table 16-3)

genitalia, and the breasts. Removal of estrogen leads to a decrease in the size and function of all these organs. Estrogen is also responsible for the female body-hair distribution and the general female body configuration: narrow shoulders, broad hips, and the characteristic female "curves" (the result of fat deposition in, for example, the hips and abdomen). Estrogen does not have testosterone's general anabolic effect, but it does cause closure of the epiphyses of bone, resulting in cessation of growth. Finally, as described above, estrogen is required for follicle and ovum maturation; its increased secretion at puberty, in concert with that of the pituitary gonadotropins, permits ovulation and the onset of menstrual cycles. For the rest of the woman's reproductive life, estrogen continues to support the ovaries, accessory organs, and secondary characteristics. Because estrogen's blood concentration varies so markedly throughout the menstrual cycle, associated changes in all these dependent functions occur, the uterine manifestations being the most striking.

As is true for testosterone, estrogen acts on the cell nucleus, and its biochemical mechanism of action is at the level of the genes themselves.

It should be reemphasized that estrogen is not a uniquely female hormone. Small and usually insignificant quantities of estrogen are secreted by the male adrenal and the testicular interstitial cells, the latter probably being responsible for the breast enlargement so commonly observed in pubescent boys; apparently, the rapidly developing interstitial cells release significant quantities of estrogen along with the much larger amounts of testosterone (also, there is some conversion of testosterone to estrogen in peripheral tissues).

Progesterone is present in significant amounts only during the luteal phase of the menstrual cycle, and its effects are less widespread than those of estrogen, the endometrial changes being the most prominent. Progesterone also exerts important effects on the breasts, the uterine tubes, and the uterine smooth muscle, the significance of which will be described later. Progesterone also causes a transformation of the cells lining the vagina (decreased cornification), and the microscopic examination of some of these cells provides an indicator that ovulation has or has not occurred. Note that in this regard, and others in Table 16-5, progesterone exerts an "antiestrogen effect" (it probably does so by inhibiting the synthesis of estrogen receptors).

Another indicator that ovulation has occurred is a small rise (approximately 0.5° C) in body temperature that usually occurs at this time and persists throughout the luteal phase. This change was previously ascribed to an action of progesterone on temperature regulatory centers in the brain, but this is probably incorrect, and the actual cause is not known.

Androgens in Women

Testosterone is not a uniquely male hormone but is found in very low concentration in the blood of normal women (as a result of production by the ovaries and adrenal glands). Of greater importance is the fact that androgens other than testosterone are found in quite significant concentrations in the blood of women. The sites of production are mainly the adrenal glands, which contribute these same nontestosterone androgens in the male; in contrast to their lack of significance in the male, adrenal androgens do play several important roles in the female, notably maintenance of sexual drive (see below). In several disease states, the female adrenals may secrete abnormally large quantities of androgen, which produce virilism; the female fat distribution disappears, a beard appears along with the male body-hair distribution, the voice lowers in pitch, the skeletal muscle mass enlarges, the clitoris (homologue of the male penis) enlarges, and the breasts diminish in size. These changes illustrate the sex-hormone dependency of secondary sex characteristics.

Female Sexual Response

The female response to sexual intercourse is very similar to that of the male in that it is characterized by marked vasocongestion and muscular contraction in many areas of the body. For example, increasing sexual excitement is associated with engorgement of the breasts and erection of the nipples, resulting from contraction of muscle fibers in them. The **clitoris** (Fig. 16-14), which is a homologue of the penis and is composed primarily of similar erectile tissue and endowed with a rich supply of sensory nerve endings, also becomes erect. During intercourse, the vaginal epithelium becomes highly congested and secretes a mucuslike lubricant.

The female has no counterpart to male ejaculation, but with this exception, as mentioned above, the physical correlates of orgasm are very similar in the sexes: There is a sudden increase in skeletal muscle activity involving almost all parts of the body; the heart rate and blood pressure increase; the female counterpart of male genital contraction is transient rhythmical contraction of the vagina and uterus. Orgasm seems to play no essential role in assuring fertilization, since conception can occur in the absence of orgasm.

A question related to the female sexual response is sex drive. Incongruous as it may seem, sexual desire in adult women is more dependent upon androgens than estrogen. Thus, sex drive is usually not altered by removal of the ovaries (or by the physiological analogue, menopause). In contrast, it is greatly reduced by adrenalectomy, since these glands are the major source of androgens in women.

This completes our survey of normal reproductive physiology in the nonpregnant female. In weaving one's way through this maze, it is all too easy to forget the prime function subserved by this entire system, namely, reproduction. Accordingly, we must now return to the mature ovum we left free in the abdominal cavity, obtain a sperm for it, and carry the fertilized ovum through pregnancy and delivery.

Pregnancy

Following their ejaculation into the vagina, the sperm live approximately 48 h; after ovulation, the ovum remains fertile for 10 to 15 h. The net result is that for pregnancy to occur, sexual intercourse must be performed no more than 48 h before or 15 h after ovulation (these are only average figures, and there is probably considerable variation in the survival time of both sperm and ovum). However, even these short time limits are probably too generous since, although fertile, less-recently ovulated ova manifest a variety of malfunctions after fertilization, frequently resulting in their death.

Ovum transport

At ovulation, the ovum is extruded from the ovary and its first mission is to gain entry into the oviduct. The end of the uterine tube has long, fingerlike projections lined with ciliated epithelium (Fig. 16-14). At ovulation, the smooth muscle of these projections causes them to pass over the ovary while the cilia beat in waves toward the interior of the duct; these ciliary motions sweep in the ovum as it emerges from the ovary.

Once in the uterine tube, the ovum moves rapidly for several minutes, propelled by cilia and by contractions of the tube's smooth muscle coating; the muscular contractions soon diminish, and ovum movement (driven almost entirely by the cilia) becomes so slow that the ovum takes about 4 days to reach the uterus. Thus, if fertilization is to occur, it must do so in the uterine tube because of the short life span of the unfertilized ovum.

Sperm transport

Within a minute or so after intercourse, some sperm can be detected in the uterus. The act of intercourse itself provides some impetus for transport out of the vagina into the uterus because of the fluid pressure of the ejaculate and the pumping action of the penis during orgasm. In addition, beating of the cilia on the inner surface of the cervix probably wafts the sperm towards the cervical canal. Passage through the cervical mucus by the swimming sperm is dependent on the estrogen-induced changes in consistency of the mucus described earlier. Transport of the sperm the length of the uterus and into the uterine tube probably is via the sperm's own propulsions and the currents set up by the beating of uterine cilia; the possible role of uterine smooth muscle contractions remains unclear.

The mortality rate of sperm during the trip is huge; of the several hundred million deposited in the vagina, only a few hundred reach the uterine tube. This is one of the major reasons that there must be so many sperm in the ejaculate to permit pregnancy.

Sperm capacitation and activation

In addition to aiding transport of sperm, the female reproductive tract exerts a second critical effect on them, namely, the conferring upon them of the capacity for fertilizing the egg. Although, as we have mentioned, sperm undergo "maturation" during their stay in the epididymis, they are still not able to fertilize the ovum until they have resided in the female tract for several hours. The first event in this two-stage process is known as **capacitation**, and it involves the stripping-off from the sperm's surface of a layer of glycoprotein molecules (originally acquired in the epididymis). Next, the capacitated sperm undergo multiple changes collectively known as **activation**: Fusions occur between the sperm's surface membrane and the outer acrosomal membrane, exposing to the exterior the enzymes within the acrosomal vesicle and attached to the inner acrosomal membranes; in addition, the previously regular wavelike beats of the sperm's tail are replaced by a more whiplashlike action, which propels the sperm forward in strong lurches; and finally, the sperm's plasma membrane becomes altered so that it is capable of fusing with the surface membrane of the ovum.

Entry of the sperm into the ovum

The sperm makes initial contact with the cells still surrounding the ovum presumably by random motion, there being no good evidence for the existence of "attracting" chemicals. Having made contact, the sperm rapidly moves between these adhering cells and through the zona pellucida; the path through these structures is created by acrosomal enzymes which break down cell connections and intermolecular bonds.

Once through the zona pellucida, the sperm plasma membrane makes contact with the ovum plasma membrane, fuses with it, and the sperm slowly passes through into the cytoplasm (frequently losing its tail in the process). Upon penetration by the sperm, the ovum completes its last division, and the one daughter cell with practically no cytoplasm—the second polar body—is extruded. The nuclei of the sperm and ovum then unite; the cell now contains 46 chromosomes, and **fertilization** is complete.

Viability of the fertilized egg (zygote) depends upon stopping the entry of additional sperm. The mechanism of this "block to polyspermy" is as follows: The initial fusing of the sperm and ovum membranes causes secretory vesicles located around the ovum's periphery to fuse with the ovum's plasma membrane and release their contents into the space between the plasma membrane and zona pellucida; some of these molecules are enzymes which break down binding sites for sperm in both the zona pellucida and the plasma membrane of the ovum itself, thereby preventing additional sperm from moving through the zona or fusing with the ovum. The immediate trigger for the release of the contents of these vesicles is a transient increase in the ovum's cytoplasmic calcium concentration, resulting from membrane changes induced by fusion of sperm and ovum; i.e., calcium is acting as a "second messenger" in this system (the sperm being the "first messenger"). This calcium also triggers activation of ovum enzymes required for the ensuing cell divisions and embryogenesis.

The fertilized egg is now ready to begin its development as it continues its passage down the uterine tube to the uterus. If fertilization had not occurred, the ovum would slowly disintegrate and be phagocytized by cells lining the uterus. Rarely, a fertilized ovum remains in the oviduct, where implantation may take place; such tubal pregnancies cannot succeed because of lack of space for the fetus to grow, and surgery may be necessary. Even more rarely a fertilized ovum may be expelled into the abdominal cavity where implantation may take place and (rarely) proceed to term, the infant being delivered surgically.

Early development, implantation, and placentation

During the leisurely 3- to 4-day passage through the uterine tube, the fertilized ovum undergoes a number of cell divisions (identical twins result when, at some point during this very early stage of development, the dividing cells become completely separated into two independently growing cell masses). After reaching the uterus, the fertilized ovum floats free in the intrauterine fluid (from which it receives nutrients) for approximately three more days, all the while undergoing cell division. The fertilized ovum has by now developed into a ball of cells surrounding a recently formed central fluid-filled cavity and is known as a **blastocyst**. This entire time span corresponds to days 14 to 21 of the typical menstrual cycle; thus, while the ovum is undergoing fertilization and early development, the uterine lining is simultaneously being prepared by estrogen and progesterone to receive it. By approximately the twenty-first day of the cycle, i.e., 7 days after ovulation, **implantation** occurs.

A section of the blastocyst is shown in Fig. 16-22; note the disappearance of the zona pellucida, an event necessary for implantation. The inner cell mass (**embryoblast**) of the blastocyst is destined to develop into the **embryo** itself, whereas the cells lining the blastocyst cavity differentiate into specialized cells, **trophoblasts**, which will form the nutrient membranes (placenta) for the fetus, as described below. Thus, both the embryo and the placenta arise from the fertilized ovum. The trophoblast cells of blastocysts recovered from the uterus

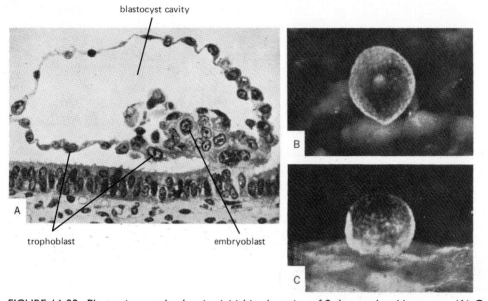

blastocyst cavity

trophoblast

embryoblast

FIGURE 16-22. Photomicrographs showing initial implantation of 9-day monkey blastocysts. (A) Cross section of the embryo and uterine lining. (B) An entire blastocyst attached to the uterus, viewed from above. (C) The same embryo viewed from the side. *[From C. H. Heuser and G. L. Streeter, Carnegie Contrib. Embryol. 29:15 (1941).]*

have been found to be quite sticky, particularly in the region overlying the embryoblast; it is this portion which adheres to the endometrium upon contact and initiates implantation.

This initial contact somehow induces rapid development of the trophoblasts, fingerlike projections of which penetrate between endometrial cells. The endometrium, too, is undergoing changes at the site of contact: increase in vascular permeability, edema, and compositional changes in the intercellular matrix and in the endometrial cells themselves. These changes, which facilitate implantation, are induced largely by **prostaglandins** released by endometrial cells (Table 16-4), but the release of these agents is, itself, triggered by some unknown chemical messenger secreted by the blastocyst. The embryo is soon completely embedded within the endometrium (Fig. 16-23), the nutrient-rich cells of which provide the metabolic fuel and raw materials for the developing embryo. The system, however, is adequate to provide for the embryo only during the first few weeks when it is very small. The system taking over after this is the fetal circulation and **placenta** (after the end of the second

month of development, the embryo is known as a **fetus**).

The placenta is a combination of interlocking fetal and maternal tissues which serves as the organ of exchange between mother and fetus. The expanding trophoblastic layer releases enzymes which break down endometrial blood vessels, allowing maternal blood to ooze into the spaces surrounding it; clotting is prevented by the presence of some anticoagulant substance produced by the trophoblasts. Soon the trophoblastic layer (remember that this layer is of fetal, not maternal, origin) completely surrounds and projects into these oozing areas (Fig. 16-23). By this time, the developing embryo has begun to send out into these trophoblastic projections blood vessels which are all branches of larger vessels, the **umbilical arteries and veins**, communicating with the main intraembryonic arteries and veins via the **umbilical cord** (Fig. 16-24). Five weeks after implantation, this system has become well established, the fetal heart has begun to pump blood, and the entire mechanism for nutrition of the fetus and excretion of its waste products is in operation. Waste products move from the fetal blood

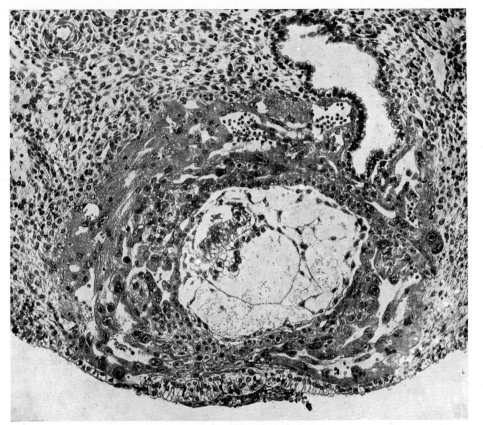

FIGURE 16-23. Eleven-day human embryo, completely embedded in the uterine lining. *[From A. T. Hertig and J. Rock,* Carnegie Contrib. Embryol. **29:**127 (1941).]

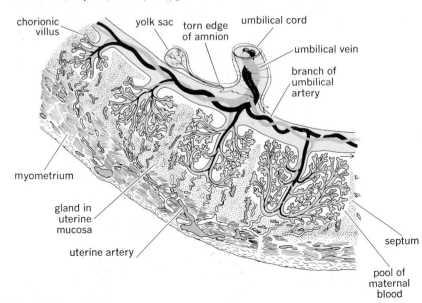

FIGURE 16-24. Interrelations of fetal and maternal tissues in formation of placenta. The placenta becomes progressively more *developed from left to right. (From B. M. Patten and B. M. Carlson "Human Embryology,"* 3d ed., McGraw-Hill Book Company, New York, 1968.)

across the placental membranes into the maternal blood; nutrients move in the opposite direction. Many substances, such as oxygen and carbon dioxide, move by simple diffusion, whereas other substances are carried by mediated-transport mechanisms in the placental membranes.

At first, as described above, the trophoblastic projections simply lie in endometrial spaces filled with blood, lymph, and some tissue debris. This basic pattern is retained throughout pregnancy, but many structural alterations have the net effect of making the system more efficient; e.g., the trophoblastic layer thins, the distance between maternal and fetal blood thereby being reduced. It must be emphasized that there is exchange of materials between the two blood streams but no actual mingling of the fetal and maternal blood. The maternal blood enters the placenta via the uterine artery, percolates through the spongelike endometrium, and then exits via the uterine veins; similarly, the fetal blood never leaves the fetal vessels.

The fetus, floating in its completely fluid-filled cavity (the **amniotic cavity**) and attached by the umbilical cord to the placenta (Fig. 16-25), develops into a viable infant during the next 9 months. Description of intrauterine development is beyond the scope of this book but can be found in any standard textbook of human embryology.

A point of great importance is that the developing embryo and fetus is subject to considerable influence by a host of factors (noise, chemicals, viruses, etc.) affecting the mother. For example, drugs taken by the mother can reach the fetus via the placenta and influence growth and development of the fetus. The thalidomide disaster is our major reminder of this fact in recent years, as is the growing number of babies who suffer heroin withdrawal symptoms after birth as a result of their mothers' drug use during pregnancy.

In this regard, it must be emphasized that aspirin, alcohol, and the chemicals in cigarette smoke are potent agents. For example, when mothers ingest aspirin within 5 days of delivery, many of their offspring have transient bleeding tendencies because of inadequate platelet aggregation (see Chap. 11). Both alcohol ingestion and cigarette smoking

by pregnant women have also been shown to have serious deleterious effects on the fetus.

Hormonal changes during pregnancy

Throughout pregnancy, the specialized uterine structures and functions depend upon high concentrations of circulating estrogen and progesterone (Fig. 16-26). During approximately the first 2 months of pregnancy, almost all these steroid hormones are supplied by the extremely active corpus luteum formed after ovulation. Recall that if pregnancy had not occurred, this glandlike structure would have degenerated within 2 weeks after its formation; in contrast, continued growth of the corpus luteum and steroid secretion occur if the woman becomes pregnant.

The persistence of the corpus luteum during pregnancy is due to a hormone secreted by the trophoblast cells and called **chorionic gonadotropin (CG)**. Almost immediately after beginning their endometrial invasion, the trophoblastic cells start to secrete CG into the maternal blood. This protein hormone is very similar to but not identical with LH, and it maintains the corpus luteum and strongly stimulates steroid secretion by it.

Recall that secretion of GnRH and, hence, of LH and FSH is powerfully inhibited by progesterone in the presence of estrogen. Since both these steroid hormones are present in high concentration throughout pregnancy, the blood concentrations of the pituitary gonadotropins remain extremely low; by this means, further follicle development, ovulation, and menstrual cycles are eliminated for the duration of the pregnancy.

The detection of CG in urine or plasma is the basis of most pregnancy tests. The secretion of CG increases rapidly during early pregnancy, reaching a peak at 60 to 80 days after the end of the last menstrual period; it then falls just as rapidly, so that by the end of the third month it has reached a low, but definitely detectable, level which remains relatively constant for the duration of the pregnancy. Associated with this falloff of CG secretion, the placenta itself begins to secrete large quantities of estrogen and progesterone, and the very marked

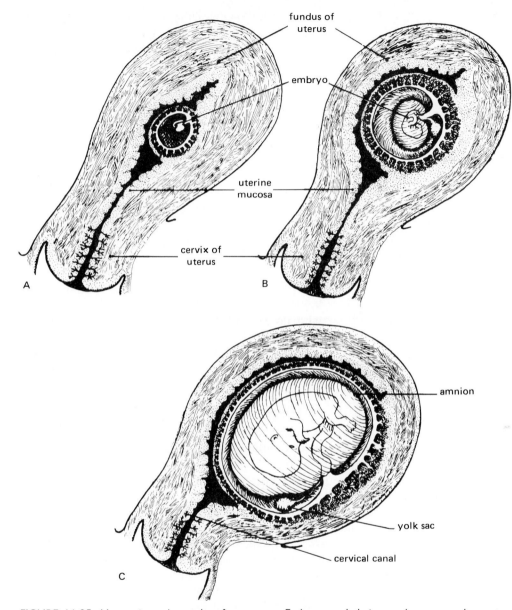

FIGURE 16-25. Uterus in early weeks of pregnancy. Embryos and their membranes are drawn to actual size. Uterus is within actual size range. (A) At fertilization age of 3 weeks; (B) at 5 weeks; (C) at 8 weeks. *(From B. M. Patten and B. M. Carlson, "Human Embryology," 3d ed., McGraw-Hill Book Company, New York, 1968.)*

increases in blood steroids during the last 6 months of pregnancy are due almost entirely to the placental secretion. The corpus luteum remains, but its contribution is dwarfed by that of the placenta; indeed, removal of the ovaries during the last 7 months has no effect at all upon the pregnancy, whereas removal during the first 2 months causes immediate loss of the fetus **(abortion)** due to disintegration of the endometrium.

An important and clinically useful aspect of

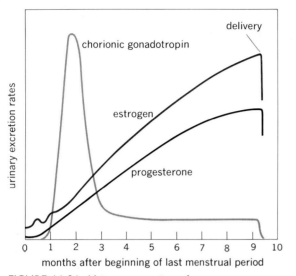

FIGURE 16-26. Urinary excretion of estrogen, progesterone, and chorionic gonadotropin during pregnancy. Urinary excretion rates are an indication of blood concentrations of these hormones.

placental steroid secretion is that the placenta does not have the enzymes required for the complete synthesis of the major estrogen of pregnancy, **estriol**, from progesterone (which it does synthesize in its entirety). However, the enzymes the placenta lacks are present in special cells in the adrenal cortex of the fetus; therefore, the placenta and fetal adrenals, working together with intermediates transported between them via the fetal circulation, produce the estriol required to maintain the pregnancy. Since the fetus is necessary for estriol production, measurement of this hormone in maternal blood provides a means for monitoring the well-being of the fetus.

Finally, the placenta not only produces steroids and CG but many other hormones as well. Some of these hormones are identical to hormones normally produced by other endocrine glands, whereas others are different. One different hormone has effects very similar to those of growth hormone and prolactin. This hormone, **placental lactogen**, may play important roles in the mother: maintaining a positive protein balance, mobilizing fats for energy, stabilizing plasma glucose at relatively high levels

to meet the needs of the fetus, and facilitating development of the breasts.

Of the numerous other physiological changes, hormonal and nonhormonal, in the mother during pregnancy, many, such as increased metabolic rate and appetite, are obvious results of the metabolic demands placed upon her by the growing fetus. Of great importance is salt and water metabolism. During a normal pregnancy, body sodium and water increase considerably, the extracellular volume alone rising by approximately 1 L. The important factors causing fluid retention are renin, aldosterone, ADH, and estrogen, all acting upon the kidneys. Some women retain abnormally great amounts of fluid and manifest both protein in the urine and hypertension. These are the symptoms of the disease known as **toxemia of pregnancy**. In the United States it is seen primarily in women in the lower socioeconomic segments of the population who do not obtain adequate medical care during pregnancy, since it can usually be well controlled by salt restriction. All attempts to determine the factors responsible for the disease have failed.

Parturition (delivery of the infant)

A normal human pregnancy lasts approximately 40 weeks, although many babies are born 1 to 2 weeks earlier or later. Delivery of the infant, followed by the placenta, is produced by strong rhythmical contractions of the uterus. Actually, beginning at approximately 30 weeks, weak and infrequent uterine contractions occur, gradually increasing in strength and frequency. During the last month, the entire uterine contents shift downward so that the baby is brought into contact with the outlet of the uterus, the cervix. In over 90 percent of births, the baby's head is downward and acts as the wedge to dilate the cervical canal. **Labor** is said to have begun when the uterine contractions become coordinated and quite strong (although usually painless at first) and occur at approximately 10- to 15-min intervals. At this time or before, the membrane surrounding the fetus ruptures and the intrauterine fluid escapes out of the vagina.

As the contractions, which begin in the upper

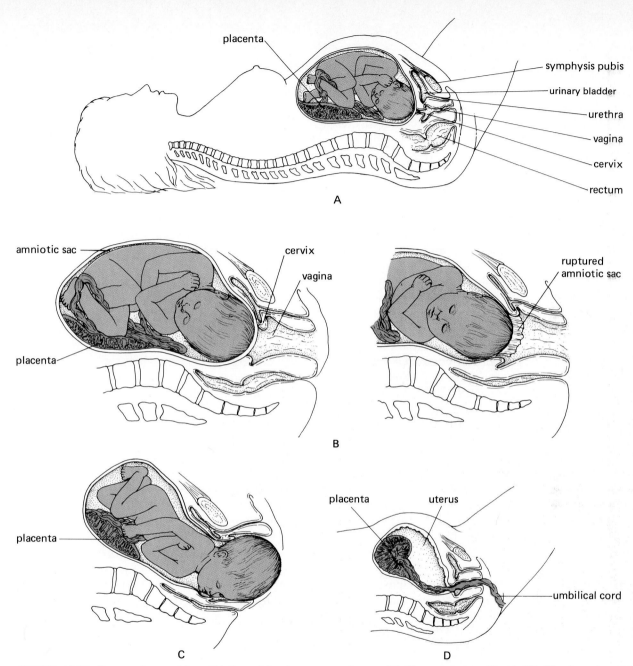

placenta

symphysis pubis

urinary bladder

urethra

vagina

cervix

rectum

A

amniotic sac

cervix

vagina

ruptured amniotic sac

placenta

B

placenta

placenta

uterus

umbilical cord

C

D

FIGURE 16-27. Stages of parturition. (A) Parturition has not yet begun. (B) The cervix has dilated. (C) The fetus is moving through the birth canal. (D) The placenta is coming loose from the uterine wall preparatory to its expulsion.

portion and sweep down the uterus, increase in intensity and frequency, the cervical canal is gradually forced open to a maximum diameter of approximately 10 cm. Until this point, the contrac- tions have not moved the fetus out of the uterus but have served only to dilate the cervix (Fig. 16-27). Now the contractions move the fetus through the cervix and vagina. At this time the mother, by bearing down to increase abdominal pressure, can help the uterine contractions to deliver the baby. The umbilical vessels and placenta are still functioning, so that the baby is not yet on its own, but within minutes of delivery both the infant's and mother's placental vessels completely constrict, stopping blood flow to the placenta, the entire placenta becomes separated from the underlying uterine wall, and a wave of uterine contractions delivers the placenta (the **afterbirth**).

Ordinarily, the entire process of parturition from beginning to end proceeds automatically and requires no real medical intervention, but in a small percentage of cases, the position of the baby or some maternal defect can interfere with normal delivery. The position is important for several reasons: (1) If the baby is not oriented head first, another portion of its body is in contact with the cervix and is generally a far less effective wedge; (2) because of its large diameter compared with the rest of the body, if the head were to go through the canal last, the cervical canal might obstruct its passage, leading to obvious problems when the baby attempts to breathe; (3) if the umbilical cord becomes caught between the birth canal and the baby, mechanical compression of the umbilical vessels can result. Despite these potential problems, however, most babies who are not oriented head first are born normally.

What mechanisms control the events of parturition? Let us consider a set of fairly well-established facts:

1. The uterus is composed of smooth muscle capable of autonomous contractions and having inherent rhythmicity, both of which are facilitated by stretching the muscle.

2. The efferent neurons to the uterus are of little importance in parturition, since anesthetizing them does not interfere with delivery.

3. Progesterone exerts a powerful inhibitory effect upon uterine contractility (apparently by decreasing its sensitivity to estrogen, oxytocin, and prostaglandins, all of which stimulate uterine contractions). Shortly before delivery, the secretion of progesterone sometimes drops, perhaps owing to "aging" changes in the placenta.

4. **Oxytocin**, one of the hormones released from the posterior pituitary, is an extremely potent uterine-muscle stimulant. It not only acts directly on uterine muscles but stimulates them to synthesize prostaglandins (see below). Oxytocin is reflexly released as a result of afferent input into the hypothalamus from receptors in the uterus, particularly the cervix.

5. The pregnant uterus near term secretes a prostaglandin (Table 16-4) which is a profound stimulator of uterine smooth muscle; an increase in the release of this substance has been demonstrated during labor.

These facts can now be put together in a unified pattern, as shown in Fig. 16-28. Once started, the uterine contractions exert a positive-feedback effect upon themselves via local facilitation of inherent uterine contractility and via reflex stimulation of oxytocin secretion. But what *starts* the contractions? A decrease in progesterone cannot be essential since it simply does not occur in most women. Nor is uterine distension a requirement, as attested

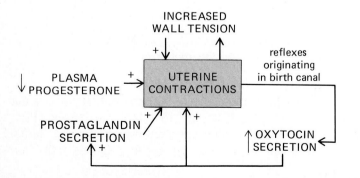

FIGURE 16-28. Factors stimulating uterine contractions during parturition. Note the positive-feedback nature of several of the inputs.

by the remarkable fact that typical "labor" begins at the expected time in some animals from which the fetus has been removed weeks previously. The most tenable hypothesis at present involves both oxytocin and prostaglandin: During pregnancy the concentration of oxytocin receptors in the uterus markedly and progressively increases (perhaps stimulated by estrogen); therefore, even through there is no dramatic change in circulating oxytocin just prior to the onset of labor, the increased number of receptors in the myometrium makes the uterus so sensitive to oxytocin that contractions become coordinated and strong; simultaneously, the increased oxytocin receptors result in an increased oxytocin-induced local synthesis of uterine prostaglandin, which further stimulates uterine contractions.

Lactation

The **breasts** are formed of epithelium-lined ducts which converge at the **nipples**. These ducts branch all through the breast tissue and terminate in saclike glands called **alveoli**. The alveoli, which secrete **milk**, look like bunches of grapes with stems terminating in the ducts. The alveoli and the ducts immediately adjacent to them are surrounded by specialized contractile cells called **myoepithelial cells**. Before puberty the breasts are small with little internal glandular structure. With the onset of puberty, the increased estrogen causes a marked enhancement of duct growth and branching but relatively little development of the alveoli, and much of the breast enlargement at this time is due to fat deposition. Progesterone secretion also commences at puberty (during the luteal phase of each cycle) and this hormone also contributes to breast growth.

During each menstrual cycle, the breasts undergo fluctuations in association with the changing blood concentrations of estrogen and progesterone, but these changes are small compared with the marked breast enlargement which occurs during pregnancy as a result of the stimulatory effects of high plasma concentrations of estrogen, progesterone, prolactin, and placental lactogen. This last hormone, as described earlier, is secreted by the placenta, whereas prolactin is secreted by the anterior

pituitary. Under the influence of all these hormones, both the ductal and the alveolar structures become fully developed.

As pointed out in Chap. 9, the secretion of **prolactin** by the anterior pituitary is controlled by several hypothalamic hormones, at least one of which is inhibitory and one stimulatory; the dominant influence most of the time is **prolactin inhibiting hormone (PIH)**, which has been identified as the chemical messenger **dopamine**. Under the tonic influence of dopamine released from the hypothalamus, prolactin secretion is low prior to puberty but increases considerably at puberty in girls (but not in boys). This increase is caused by the increased plasma concentration of estrogen which occurs at this time, for estrogen acts on the anterior pituitary to reduce the sensitivity of the prolactin-secreting cells to dopamine. During pregnancy, there is a marked increase in prolactin secretion (due to stimulation by the elevated plasma estrogen), beginning at about 8 weeks and rising throughout the remainder of the pregnancy.

Prolactin is the single most important hormone promoting milk production. Yet despite the fact that prolactin is elevated and the breasts markedly enlarged and fully developed as pregnancy progresses, there is no secretion of milk. This is because estrogen and progesterone, in large concentrations, prevent milk production by inhibiting the action of prolactin on the breasts (Table 16-5). Thus, although estrogen causes an increase in the secretion of prolactin and acts with it in promoting breast growth and differentiation, it (along with progesterone) is antagonistic to prolactin's ability to induce milk secretion. Delivery removes the sources of the large amounts of sex steroids and, thereby, the inhibition of milk production.

Following delivery, basal prolactin secretion decreases from its peak late-pregnancy levels (and after several months may actually return to prepregnancy levels even though the mother continues to nurse), but superimposed upon this basal level are large secretory bursts of prolactin during each nursing period. This occurs because afferent input from **nipple receptors** (stimulated by **suckling**) causes specific hypothalamic neurons to release a pulse of one or more **prolactin releasing hormones (PRH)**, which override the action of PIH and stimulate the anterior pituitary to secrete prolactin (Fig. 16-29). These episodic pulses of prolactin are the signals to the breasts for maintenance of milk production. Milk production ceases several days after the mother completely stops nursing her infant but continues uninterrupted for years if nursing is continued.

One final reflex process is essential for nursing. Milk is secreted into the lumen of the alveoli, but because of their structure, the infant cannot suck the milk out. It must first be moved into the ducts, from which it can be sucked. This process is called

milk **let-down** and is accomplished by contraction of the myoepithelial cells surrounding the alveoli; the contraction is directly under the control of oxytocin, which is reflexly released by suckling (Fig. 16-30), just like prolactin (but much more rapidly). Higher brain centers can also exert an important influence over oxytocin release: A nursing mother may actually leak milk when hearing her baby cry. In view of the central role of the nervous system in lactation reflexes, it is no wonder that psychological factors can interfere with a woman's ability to nurse.

Milk contains four major constituents: water, protein, fat, and the carbohydrate lactose (milk sugar). The mammary alveolar cells must be capable of extracting the raw materials—amino acids, fatty acids, glycerol, glucose, etc.—from the blood and building them into the higher-molecular-weight substances. Although prolactin is the single most important hormone controlling these synthetic processes, insulin, growth hormone, cortisol, and still other hormones also participate.

Another neuroendocrine reflex triggered by suckling (and mediated, in part, by prolactin) is

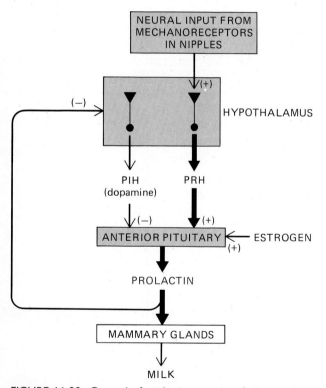

FIGURE 16-29. Control of prolactin secretion during nursing. Neural input from nipple mechanoreceptors stimulates the hypothalamic neuroendocrine cells which secrete PRH. PRH overrides any simultaneous inhibitory effect of PIH on the anterior pituitary. Estrogen also sensitizes the anterior pituitary to release prolactin. The line from prolactin to the hypothalamus denotes that this hormone exerts a negative-feedback effect over its own secretion (either via increasing PIH or decreasing PRH).

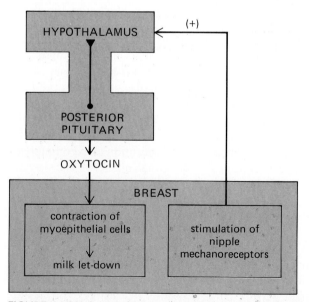

FIGURE 16-30. Suckling-reflex control of oxytocin secretion and milk let-down. As with Fig. 16-29, begin following the figure with the nipple mechanoreceptors.

inhibition of the hypothalamic-pituitary-ovarian chain at a variety of steps, with resultant block of ovulation. This inhibition apparently is relatively short-lived in many women, and approximately 50 percent begin to ovulate despite continued nursing. Pregnancy is common in women lulled into false security by the mistaken belief that nursing always prevents ovulation.

Fertility control

There are a variety of techniques of birth control (**contraception**) which prevent sperm from reaching the ovum: vaginal diaphragms, sperm-killing jellies, and condoms. Another widely used method is the so-called rhythm method, in which couples merely abstain from sexual intercourse near the time of ovulation. Unfortunately, it is difficult to time ovulation precisely even with laboratory techniques; e.g., the small rise in body temperature or change in cervical mucus and vaginal epithelium, all of which are indicators of ovulation, occur only *after* ovulation. This problem, combined with the marked variability of the time of ovulation in many women, explains why this technique is only partially effective.

Development of **oral contraceptives** was based on the knowledge that combinations of estrogen and progesterone can inhibit pituitary gonadotropin release, thereby preventing ovulation. The most commonly used agents, at least at first, were combinations of an estrogen-like and a progesterone-like (progestogen) substance. Each month, these pills are taken for 20 days, then discontinued for 5 days; this steroid withdrawal produces menstruation, and the net result is a menstrual cycle without ovulation. Another type of regimen is the so-called sequential method in which estrogen is administered alone for 15 days followed by estrogen plus progestogen for 5 days, followed by withdrawal of both steroids. As with the combination pills, this regimen interferes with the orderly secretion of gonadotropins and prevents ovulation.

It has now become clear that the oral contraceptives do not always prevent ovulation, yet still are effective because they have multiple antifertility effects. In other words, the hormonal milieu required for normal pregnancy is such that these exogenous steroids interfere with many of the steps between intercourse and implantation of the blastocyst. Taken correctly, they are almost 100 percent effective. Unfortunately, potentially lethal side effects, such as intravascular clotting, have been reported, but only in a very small number of women (particularly those over 35 who smoke); the fatality rate overall is considerably less than that associated with pregnancy.

Another type of contraceptive which is highly effective (although not 100 percent) is the **intrauterine device** (Table 16-6). Placing one of these small objects in the uterus prevents pregnancy, perhaps by somehow interferring with the endometrial preparation for acceptance of the blastocyst.

Fertility prevention is of great importance, but the other side of the coin is the problem of unwanted infertility. Approximately 10 percent of the married couples in the United States are infertile. There are many reasons—some known, some unknown—for infertility. Careful investigation of infertile couples frequently permits diagnosis and therapy of the basic problem.

TABLE 16-6. Mean effectiveness of contraceptive methods

Method	Pregnancies per 100 women per year
None	115
Douche	31
Rhythm	24
Jelly alone	20
Withdrawal	18
Condom	14
Diaphragm	12
Intrauterine device	5
Oral contraceptive	
Average	1
Correctly used	0

SECTION C.
THE CHRONOLOGY OF SEX DEVELOPMENT

Sex Determination

Sex is determined by the genetic inheritance of the individual, specifically by two chromosomes called the "sex chromosomes." All human cells (excepting the ovum and sperm) have 46 chromosomes, 22 pairs of **autosomal** (nonsex) **chromosomes**, and 1 pair of **sex chromosomes**. The larger of the sex chromosomes is called the X chromosome and the smaller the Y chromosome. Genetic males possess one X and one Y, whereas females have two X chromosomes. Thus, the genetic difference between male and female is simply the difference in one chromosome. The reason for the approximately equal sex distribution of the population should be readily apparent (Fig. 16-31): The female can contribute only an X chromosome, whereas half of the sperm produced by the male during meiosis are X and half are Y. When the sperm and ovum join, 50 percent should have XX and 50 percent XY.

Interestingly, however, sex ratios at birth are not really 1:1. Rather, there tends to be a slight preponderance of male births (for example, in England, the ratio of male to female births is 1.06). Even more surprising, the ratio at the time of conception seems to be much higher. From various types of evidence, it has been estimated that there may be 30 percent more male conceptions than female. There are several implications of these facts: (1) There must be a considerably larger in utero death rate for males; and (2) the "male"

sperm, i.e., the Y-bearing sperm, must have some advantage over the "female" or X-bearing sperm in reaching and fertilizing the egg. It has been suggested, for example, that since the Y chromosome is lighter than the X, the "male" sperm might be able to travel more rapidly. There are numerous other theories. It should be pointed out that conception and birth ratios show considerable variation in different parts of the world and, indeed, in rural and urban areas of the same country.

An easy method of identifying the sex-chromosome composition of cells exists. The cells of female tissue (scrapings from the cheek mucosa are convenient) contain a readily detected nuclear mass which is a condensed, nonfunctional X chromosome (in the normal female only one X chromosome functions in each cell). This has been called the **sex chromatin** and is not usually found in male cells. This method has proved valuable when genetic sex is in doubt. Its use and that of the more exacting tissue-culture visualization of chromosomes have revealed a group of genetic sex abnormalities characterized by such bizarre chromosomal combinations as XXX, XXY, X, and many others. The end result of such combinations is usually the failure of normal anatomical and functional sexual development.

Sex Differentiation

It is not surprising that persons with abnormal genetic endowment manifest abnormal sexual development, but careful study has also revealed people with normal chromosomal combinations but abnormal sexual appearance and function. This kind of puzzle leads us into the realm of **sex differentiation**, i.e., the process by which the fetus develops the male or female characteristics directed by its genetic makeup. The genes directly determine only gonadal sex, i.e., whether the individual will have testes or ovaries; all the rest of the sexual differentiation depends upon this genetically determined gonad.

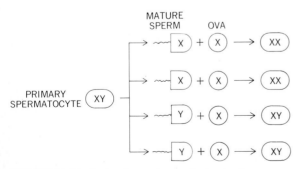

FIGURE 16-31. Basis of genetic sex determination.

Differentiation of the gonads

The male and female gonads derive embryologically from the same site in the body. Until the sixth week of life, there is no differentiation of this site. During the seventh week, in the genetic male, the testes begin to develop; in the genetic female, several weeks later, ovaries begin to develop instead. It is now clear that the presence or absence of the Y chromosome determines which gonad develops; if the Y chromosome is present, testes develop, whereas in its absence, ovaries develop. Recently, the link between presence of the Y chromosome and testis differentiation has been discovered: It is a protein called **H-Y antigen**, which is secreted by primitive cells (perhaps Sertoli cell precursors) in the male gonad and "directs" testicular development (Fig. 16-32). (H-Y antigen is not just a fetal protein but appears on the plasma membrane of all cells in males.)

Differentiation of internal and external genitalia

So far as its internal duct system and external genitalia are concerned, the very early fetus is capable of developing into either sex. The primitive reproductive tract consists, in addition to the primitive gonad, of a double genital duct system (**Wolffian and Müllerian ducts**) and a common opening for the genital ducts and urinary system to the outside. Normally, the internal reproductive organs develop from only one of these duct systems: In the male, the Wolffian ducts persist and the Müllerian ducts regress, whereas in the female, the opposite happens; i.e., the Müllerian duct system persists and the Wolffian ducts degenerate. The external genitalia in the two sexes and the vagina do not develop from these duct systems but from other structures.

Which of the two duct systems develops depends not on the genetic makeup of the duct cells but rather on the presence or absence of two substances secreted by the fetal testes—testosterone (from the interstitial cells) and **Müllerian inhibiting substance (MIS)**. MIS causes the Müllerian duct system to degenerate, and testosterone causes the Wolffian ducts to differentiate into epididymis, ductus deferens, ejaculatory duct, and seminal vesicle.

Externally (and somewhat later), under the influence of testosterone, a penis forms and the tissue near it fuses to form the scrotum, into which the testes will ultimately descend (Fig. 16-32A). In contrast, the female embryo, lacking the two secretory products of a testis, develops uterine tubes and uterus (from the Müllerian system) and (from other structures) vagina and female external genitalia. Thus, the system is basically quite simple. If a functioning male gonad is present to secrete testosterone and MIS, a male duct system and external genitalia will form, but if these "male" secretions are absent (or ineffective), a female reproductive system will develop. A female gonad need not be present for the female organs to develop (Fig. 16-32B).

To take one example of abnormal sex differentiation: In the syndrome known as **testicular feminization**, the genotype of XY (male) and testes are present, but the individual has female external genitalia, a vagina, and no internal genitalia at all. The problem causing this is a lack of androgen receptors due to a genetic defect. The fetal testes differentiate as usual under the influence of H-Y antigen, and they secrete both MIS and testosterone. MIS causes the Müllerian ducts to regress, but the inability of the Wolffian ducts to respond to testosterone also causes them to regress, and so no internal genitalia develop. It should be clear from the preceding discussion why a vagina and female external genitalia do develop in these people.

Sexual differentiation of the central nervous system

As we have seen, the male and female hypothalamus differ in that there is cyclical secretion of GnRH in the female but rather fixed continuous release in the male. In people (and other primates), it appears that this difference does not reflect any qualitative difference in hypothalamic development between male and female; rather, it is due simply to the fact, described previously, that large amounts of the dominant female sex hormone, estrogen, can stimulate LH release, whereas testosterone can only inhibit it. Thus, it has been demonstrated that the administration of estrogen to castrated male

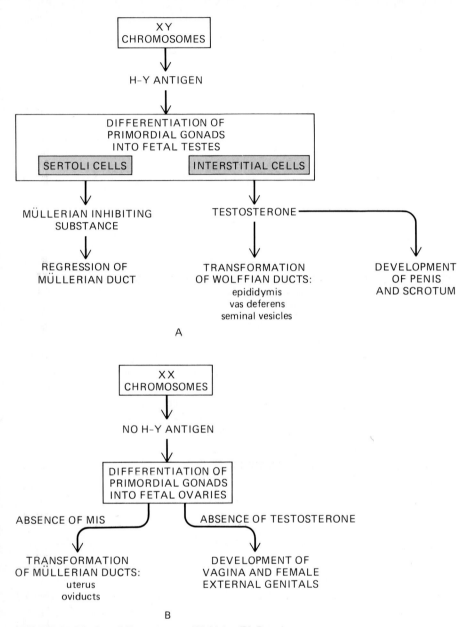

FIGURE 16-32. Sex differentiation. (A) Male. (B) Female.

monkeys elicited LH surges indistinguishable from those shown by females.

The situation may be quite different for sexual behavior in that qualitative differences in the brain may be formed during development: Genetic female monkeys given testosterone during late fetal life manifest not only masculinized external genitalia but evidence of masculine sex behavior (mounting for example) as adults. Related to this is the question of whether exposure to androgens during fetal existence is necessary for development of other behavior patterns in addition to sex. Again, for other primates, the answer seems a clear-cut *yes*; for example, the female monkey offspring given testosterone also manifest a high degree of male-type play behavior during growth.

The evidence for human beings is very scanty; perhaps the best-studied group has been women who were accidentally virilized during late fetal development because their mothers were given synthetic hormones (to prevent abortion) not recognized to be androgenic at the time. At birth they had male external genitalia (enlarged clitoris and fused empty scrotum) but normal internal genitalia; this was corrected surgically very early in life and their behavior for the next 5 to 15 years was studied carefully and compared with a control group matched to them in every possible way. Certain of the behaviors of these young girls differed from those of the girls in the control group; for example, they took part in more rough-and-tumble outdoor activities. Nonetheless, on the basis of other findings in this study and several others like it, the present tentative conclusions are that, although prenatal sex hormones influence certain forms of behavior, the most powerful factors determining gender identity are the individual's experience and socialization.

In summary (Fig. 16-32), it is evident that the presence or absence of the Y chromosome dictates whether testes or ovaries develop. In the male, once the testes have developed, the male gonadal secretions play a crucial role in determining normal sexual differentiation in the fetus and just after birth. Shortly after delivery, testosterone secretion becomes very low until puberty, when it once again increases to stimulate the organs it had previously helped to differentiate. In contrast, estrogen probably takes little active part in fetal development, female differentiation requiring only the absence of the testicular secretion. Estrogen is, of course, the stimulating agent for the female sex organs during puberty and adult life.

Puberty

Puberty is the period, usually occurring sometime between the ages of 10 and 14, during which the reproductive organs mature and reproduction becomes possible. In the male, the seminiferous tubules begin to produce sperm; the genital ducts, glands, and penis enlarge and become functional; the secondary sex characteristics develop; and sex drive is initiated. All these phenomena are effects of testosterone, and puberty in the male is the direct result of the onset of increased testosterone secretion by the testes (Fig. 16-33). Although significant testosterone secretion occurs during late fetal life, within the first days of birth testosterone secretion becomes very low and remains so until puberty.

The critical question is: What stimulates testosterone secretion at puberty? Or conversely: What inhibits it before puberty? The hypothalamus is the critical site of control. Before puberty, the hypothalamus fails to secrete sufficient quantities of GnRH to stimulate secretion of FSH and LH (Fig. 16-34); deprived of these gonadotropic hormones, the testes fail to produce sperm or large amounts of testosterone. Puberty is initiated by an unknown alteration of brain function which permits increased secretion of hypothalamic GnRH, but the mechanism of this change remains unknown. One likely hypothesis is that the hypothalamus prior to puberty is so sensitive to the negative-feedback effects of testosterone that even the extremely low blood concentration of this hormone in prepubescent boys is adequate to block secretion of the releasing hormone; of course, this simply raises the question of what brain change occurs at puberty to lower responsiveness to the negative feedback. In any case, the process is not abrupt but develops over several years, as evidenced by slowly rising plasma concentrations of the gonadotropins and testosterone. Some children with brain tumors or other lesions of the hypothalamus may undergo **precocious puberty**, i.e., sexual maturation at an unusually early age, sometimes within the first 5 years of life.

The picture for the female is analogous to that for the male. Throughout childhood, estrogen is secreted at very low levels (Fig. 16-35). Accordingly, the female accessory sex organs remain small and nonfunctional; there are minimal secondary sex characteristics, and follicle maturation does not occur. As for the male, prepubertal dormancy is probably due mainly to deficient secretion of hypothal-

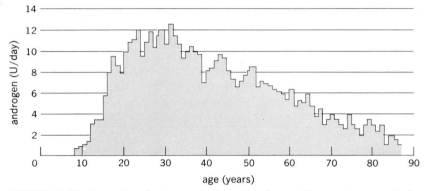

FIGURE 16-33. Excretion of androgen in the urine of normal boys and men, an indicator of the blood concentration of androgen. *(Adapted from Pedersen-Bjergaard and Tonnesen,* Acta Med. Scand.*)*

amic GnRH, and the onset of puberty is occasioned by an alteration in brain function which raises secretion of this releasing hormone, which, in turn, stimulates secretion of pituitary gonadotropins. The increased estrogen stimulated by the gonadotropins then induces many of the striking changes associated with puberty.

Precocious puberty also occurs in females. The youngest mother on record gave birth to a full-term,

healthy infant by cesarean section at 5 years, 8 months.

It should be recognized that the maturational events of puberty usually proceed in an orderly sequence, but that the ages at which they occur may vary among individuals. In boys, the first sign of puberty is acceleration of growth of testes and scrotum; pubic hair appears a trifle later, and axillary and facial hair still later. Acceleration of penis

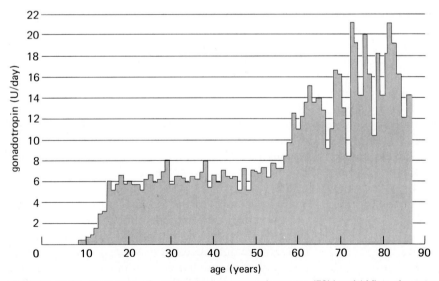

FIGURE 16-34. Excretion of anterior pituitary gonadotropins (FSH and LH) in the urine of normal boys and men, an indicator of the blood concentration of gonadotropins. *(Adapted from Pedersen-Bjergaard and Tonnesen,* Acta Med. Scand.*)*

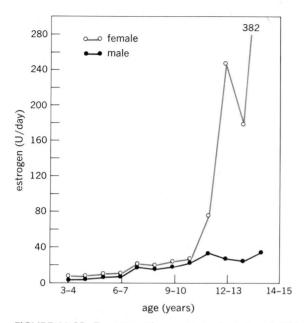

FIGURE 16-35. Excretion of estrogen in the urine of children, an indicator of the blood concentration of estrogen. *(Adapted from Nathanson et al.)*

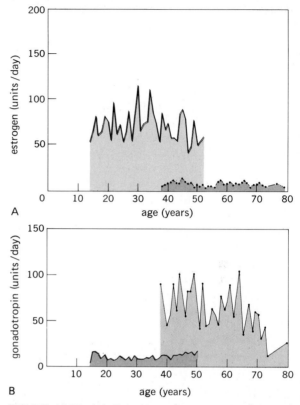

FIGURE 16-36. (A) Excretion of estrogen in the urine of women from puberty to senescence, an indicator of blood concentration of estrogen. (Color) before menopause. (Gray) after menopause. (B) Excretion of gonadotropins in the urine of nonpregnant women, an indicator of blood concentrations of gonadotropins. (Color) before menopause. (Gray) after menopause. *(Adapted from Pedersen-Bjergaard and Tonnesen.)*

growth begins on the average at 13 years (range 11 to 14.5) and is complete by 15 (13.5 to 17). But note that, because of the overlap in ranges, some boys may be completely mature, whereas others at the same age (say 13.5) may be completely prepubescent. Obviously, this can lead to social and psychological problems. In girls, appearance of "breast buds" is usually the first event (average age of 11) although pubic hair may, on occasion, appear first. **Menarche**, the first menstrual period, is a later event (average of 12.3 years) and occurs almost invariably after the peak of the total body-growth spurt has passed.

Menopause

Fertility in women peaks in their mid 20s and declines gradually after the age of 30. This is associated with a decline in ovarian function. Figure 16-36 demonstrates that the cause of the decline is decreasing ability of the aging ovaries to respond to pituitary gonadotropins (in part because of the diminished number of follicles). Estrogen secretion

drops despite the fact that the gonadotropins, partially released from the negative-feedback inhibition by estrogen, are secreted in greater amounts. Around the age of 50, on the average, menstrual periods become less regular, and most are anovulatory because not enough estrogen is produced to trigger an LH surge. Ultimately, menstrual periods cease entirely and the last period (the counterpart of menarche—the first period) is known as the **menopause**. The entire phase of life (beginning with menstrual irregularity) involving physical and emotional changes as sexual maturity gives way to cessation of reproductive function is known as the

climacteric (the counterpart of puberty). Some ovarian secretion of estrogen generally continues beyond the menopause, but it gradually diminishes until it is inadequate to maintain the estrogen-dependent tissues: The breasts and genital organs gradually atrophy to a large degree, and marked bone demineralization occurs. Sexual drive is frequently not diminished and may even be increased. The hot flashes, so typical of menopause, result from dilation of the skin arterioles, causing a feeling of warmth and marked sweating; why estrogen deficiency causes this is unknown. Many of the symptoms of the menopause can be reduced by the administration of estrogen, but the safety of such administration is controversial because of the possibility that it may facilitate development of breast or uterine cancers. Another aspect of menopause is its relationship with cardiovascular diseases.

Women have much less hypertension and atherosclerosis than men until after the menopause, when the incidence becomes similar in both sexes.

Male reproductive system changes with aging are less drastic. Once testosterone and pituitary gonadotropin secretions are initiated at puberty, they continue to some extent throughout adult life. However, as shown in Fig. 16-33, there is a steady decrease in testosterone secretion beginning at about the age of 40, which apparently reflects slow deterioration of testicular function (the mirror-image rise in gonadotropin secretion—Fig. 16-34—is due to diminishing negative-feedback inhibition by the decreasing plasma testosterone concentration). Along with the decreasing testosterone levels, both libido and potency diminish, although many men continue to enjoy active (and fertile) sex lives in their 70s and 80s.

DEFENSE MECHANISMS OF THE BODY: IMMUNOLOGY, FOREIGN CHEMICALS, AND STRESS

SECTION A.
IMMUNOLOGY: THE BODY'S DEFENSES
AGAINST FOREIGN MATTER

Immunity constitutes all the physiological mechanisms which allow the body to recognize materials as foreign or abnormal and to neutralize or eliminate them; in essence, these mechanisms maintain uniqueness of "self." Classically, immunity referred to the resistance of the body to **microbes** (viruses, bacteria, and protozoa), as well as to fungi and multicellular organisms such as parasitic worms. It is now recognized, however, that the immune system has more diverse functions than this. It is involved in the elimination of "worn-out" or damaged body cells (such as old erythrocytes) and in the destruction of abnormal or mutant cells which arise within the body. This last function, known as **immune surveillance**, constitutes a major defense against cancer.

It has also become evident that immune responses are not always beneficial and may result in serious damage to the body. In addition, the immune system seems to be involved in the process of aging. Finally, it constitutes the major obstacle to successful transplantation of organs. Because of these broad relationships, few other research areas of biology have grown so rapidly or have produced such a wealth of new and exciting, albeit sometimes bewildering, information.

Immune responses may be classified into two categories: specific and nonspecific. **Specific immune responses**, which are mediated by lymphocytes (and cells derived from lymphocytes), depend upon prior exposure to a specific foreign material, recognition of it upon subsequent exposure, and reaction to it. In contrast, the **nonspecific immune responses** do not require previous exposure to the particular foreign material, and they nonselectively protect against foreign materials without having to recognize their specific identities. The nonspecific immune responses are particularly important during the initial exposure to a foreign organism.

Immune responses can be viewed in the same way as other homeostatic processes in the body, i.e., as stimulus-response sequences of events. In such an analysis, the groups of cells which mediate the final responses are effector cells. Before we begin our description of these effector cells, let us first at least mention some of their major opponents—bacteria and viruses.

Bacteria are unicellular organisms which have not only a cell membrane but also an outer coating, the cell wall. Most bacteria are self-contained complete cells in that they have all the machinery required to sustain life and reproduce themselves. In contrast, **viruses** are essentially nucleic acids that are surrounded by a protein coat. They lack both the enzyme machinery for energy production and the ribosomes essential for protein synthesis. Thus, they cannot survive by themselves but must "live" inside other cells whose biochemical apparatus they make use of. The viral nucleic acid directs the synthesis by the host cell of the proteins required for viral replication, with nucleotides and energy sources also being supplied by the host cell. Other types of microorganisms and multicellular parasites are potentially harmful to human beings, but we shall devote most of our attention to the body's defense mechanisms against bacteria and viruses.

How microorganisms cause damage and endanger health depends upon the specific bacterium or virus involved. Some bacteria directly destroy cells by releasing enzymes which break down nearby cell membranes and organelles. Others give off toxins which disrupt the functions of organs and tissues. The effect of viral habitation and replication within a cell depends upon the type of virus. Some viral particles, after entering a cell, multiply very rapidly and kill the cell by depleting it of essential components or by directing it to produce toxic substances; with death of the host cell, the viral particles leave and move on to another cell. In contrast, other viral particles replicate inside their host cells very slowly, and the viral nucleic acid may even become associated with the cell's own DNA molecules, replicating along with them and being passed on to the daughter cells during cell division;

such a virus may remain in the cell or its offspring for many years. As we shall see, cells infected with these so-called **slow viruses** may be damaged by the body's own defense mechanisms turned against the cells because they are no longer "recognized" as "self." Finally, it is likely that certain viruses cause transformation of their host cells into cancer cells.

Effector Cells of the Immune System

The major types of effector cells of the immune system (Table 17-1) are the different types of leukocytes (white blood cells), the plasma cells, and the macrophages. (All the functions listed in Table 17-1 will be described in subsequent sections and are presented here together for orientation and reference.) The **leukocyte** types (**neutrophils, eosinophils, basophils, monocytes,** and **lymphocytes**) were described in Chap. 11 and should be reviewed at this time. **Plasma cells** are derived from lymphocytes and are the cells which secrete antibodies. **Macrophages** have multiple functions, including phagocytosis, and are found scattered throughout the tissues of the body (the structure of these cells varies from place to place). Some macrophages are capable of replication, but the major source of tis-

TABLE 17-1. Major effector-cell types of the immune system

	White blood cells (leukocytes)						
	Polymorphonuclear granulocytes						
	Neutrophils	Eosinophils	Basophils	Lymphocytes	Monocytes	Plasma cells	Macrophages
Percent of total leukocytes	50–70	1–4	0.1	20–40	2–8		
Primary sites of production	Bone marrow	Bone marrow	Bone marrow	Bone marrow, thymus, and lymphoid tissues	Bone marrow	Derived from B lymphocytes in lymphoid tissue	Most are formed from monocytes
Primary known function	Phagocytosis; release of chemicals involved in inflammation (chemotaxins, etc.)	Destruction of parasitic worms	Release of histamine and other chemicals; transformed into mast cells with similar functions	B cells: production of antibodies (after transformation into plasma cells) T cells: different subgroups responsible for cell-mediated immunity, "helping" or suppressing B cells and other T cells	Transformed into tissue macrophages; functions similar to macrophages	Production of antibodies	Phagocytosis; assist in antibody formation and T-cell sensitization; secretion of chemicals involved in inflammation, regulation of lymphocytes, and total-body response to infection or injury

sue macrophages is by influx of blood monocytes and their differentiation locally into macrophages. Monocytes and macrophages are often termed the **mononuclear phagocyte system** to distinguish them from the other major phagocytes of the body, the neutrophils (eosinophils are also phagocytic, but as shown in Table 17-1, they are few in number and are highly selective in their targets). Because of the large number of cell types and chemical mediators involved in immune responses, a miniglossary defining them is given at the end of the section on immunology.

The effector cells of the immune system are

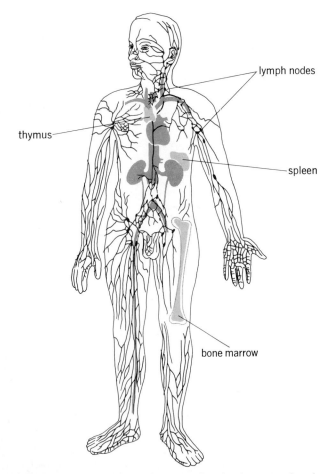

FIGURE 17-1. Location of various lymphoid organs: lymph nodes, thymus, and spleen. Not shown are the tonsils and other lymphoid patches in the various epithelial linings of the body. The bone marrow is the site of production of many of the cells which come to reside in the other lymphoid organs.

distributed throughout the organs and tissues of the body, but are found particularly in the so-called **lymphoid tissues** (Fig. 17-1 and Table 17-2)—lymph nodes, spleen, thymus, aggregates of lymphoid follicles associated with the gastrointestinal tract (for example, the tonsils), and bone marrow.

Lymph nodes function as filters along the course of the lymph vessels, lymph flowing through them before being returned to the general circulation. Lymph enters the node via afferent lymphatic vessels, trickles through the lymphatic sinuses of the node, and leaves via the efferent lymphatic vessels on the other side. The lymphatic sinuses are relatively open channels, and as the lymph flows through them, some lymphocytes are removed to be stored temporarily in the node and others are added to the lymph. Some of the lymphocytes released into the lymph are those previously stored, but others have been newly formed, after an appropriate stimulus, in the node. The lymphatic sinuses are lined with macrophages, which phagocytize particulate matter, such as dust (inhaled into the lungs), cellular debris, bacteria, and other microorganisms. Certain lymphocytes in the lymph nodes can differentiate into plasma cells which secrete antibodies.

The **spleen** is the largest of the lymphoid organs and lies in the left part of the abdominal cavity between the stomach and the diaphragm. The interior of the spleen is filled with a reticular meshwork, the **red pulp** and the **white pulp**. Blood, rather than lymph, percolates through the red pulp (thus its name), and erythrocytes as well as lymphocytes and macrophages are collected in the spaces in the meshwork. In the human fetus, the spleen is an important blood-cell-forming organ, but in the adult only lymphocytes are formed there. They are produced in the white pulp of the spleen. The macrophages of the spleen phagocytize many of the products of red cell degradation as well as various kinds of foreign matter. In essence, the spleen is to the circulating blood what the lymph nodes are to the lymph. Plasma cells derived from lymphocytes in the spleen, as in the lymph nodes, produce antibodies.

The **thymus** lies in the upper part of the chest, and its size varies with age; it is relatively large at birth and grows until puberty (10 to 15 years of age)

TABLE 17-2. Some functions of lymphoid organs

Lymph nodes (including unencapsulated lymphoid tissues)	1. Produce antibodies and sensitized T cells 2. Remove and store lymphocytes 3. Form and add new lymphocytes to lymph flow through nodes 4. Remove particulate matter through phagocytosis by macrophages
Spleen	1. Produce antibodies and sensitized T cells 2. Produce red blood cells in fetus but not adult 3. Produce new lymphocytes 4. Remove products of red-cell degradation and other foreign matter through phagocytosis by macrophages 5. Store red blood cells, which can be added to circulation by contraction of the spleen
Thymus	1. Produce T lymphocytes 2. Secrete hormones (thymosin)

when it gradually atrophies and is replaced by fatty tissue. The thymus is the organ responsible for differentiation of one of the major classes of lymphocyte (T lymphocytes); it also secretes a group of hormones known collectively as **thymosin.**

The **bone marrow** is also classified as a lymphoid organ because it produces lymphocytes. Indeed, the bone marrow is the source of virtually all the lymphocytes which ''seed'' the other lymphoid tissues (this will be described more fully in a subsequent section); thus the ancestry of almost all the lymphocytes in an adult can be traced back to the bone marrow [in fact, this is true of all effector cells of the immune response (Fig. 17-2)]. However, unlike the lymph nodes and spleen, the bone marrow is not a site in which lymphocytes become activated to secrete antibodies or participate in other specific immune responses.

Nonspecific Immune Responses

External anatomical and chemical ''barriers''

The body's first lines of defense against infection are the barriers offered by surfaces exposed to the external environment. Very few microorganisms can penetrate the intact skin, and the sweat, seba-

ceous, and lacrimal glands secrete chemical substances which are highly toxic to certain forms of bacteria. The mucous membranes also contain antimicrobial chemicals, but more important, mucus is sticky. When particles adhere to it, they can be swept away by ciliary action, as occurs in the upper respiratory tract, or engulfed by phagocytic cells. Other specialized surface ''barriers'' are the hairs at the entrance to the nose, the cough reflexes, and the acid secretion of the stomach. Finally, a major ''barrier'' to infection is the normal microbial flora of the skin and other linings exposed to the external environment; these microbes suppress the growth of other potentially more virulent ones.

Inflammatory response

Despite the effectiveness of the external barriers, small numbers of microorganisms penetrate them every day. Think of all the small breaks produced in the skin or mucous membranes by tooth-brushing, shaving, tiny scratches, etc. In addition, we now recognize that many viruses are able to penetrate seemingly intact healthy skin or mucous membranes.

Once the invader has gained entry, it triggers off **inflammation**, the response to injury. The local manifestations of the inflammatory response are a complex sequence of highly interrelated events, the overall functions of which are to bring phagocytes into the damaged area so that they can destroy (or inactivate) the foreign invaders and set the stage for tissue repair. The sequence of events which constitute the inflammatory response varies, depending upon the injurious agent (bacteria, cold, heat, trauma, etc.), the site of injury, and the state of the body, but the similarities are in many respects more striking than the differences. It should be emphasized that, in this section, we describe inflammation in its most basic form, i.e., the *nonspecific* innate response to foreign material. As we shall see, inflammation remains the basic scenario for the acting out of *specific* immune responses as well, the difference being that the entire process is amplified and made more efficient by the participation of antibodies and sensitized lymphocytes, i.e., by agents of specific immune responses.

The sequence of events in an inflammatory response is briefly as follows, using bacterial infection as our example:

1. Initial entry of bacteria
2. Vasodilation of the vessels of the microcirculation leading to increased blood flow
3. Marked increase in vascular permeability to protein
4. Filtration of fluid into the tissue with resultant swelling
5. Exit of neutrophils (and, later, monocytes) from the vessels into the tissues
6. Destruction of the bacteria either through phagocytosis and intracellular killing, or by mechanisms not requiring prior phagocytosis
7. Tissue repair

The familiar gross manifestations of this process are redness, swelling, heat, and pain, the latter being the result both of distension and of the effect of locally produced substances on afferent nerve endings.

Each of the events of inflammation is induced and regulated by chemical mediators of varying origins. Surprisingly, some are produced by the bacteria themselves. Some are released from tissue cells present in the area prior to the invasion; an example is **histamine**, present in mast cells scattered throughout the tissues of the body and a potent inducer of steps 2 to 4. Other chemical mediators are secreted by neutrophils and mononuclear phagocytes which enter the area (step 5). Still others are newly generated in the interstitial fluid of the area from inactive plasma precursors; an example is the **kinins**, peptides which are split in the inflamed area from proteins (**kininogens**) circulating in the blood (Chap. 11), and which not only stimulate steps 2 to 5 but may activate neuronal pain receptors.

There are a very large number of these chemical mediators and an even larger number of stimuli for their release from cells or generation in the interstitial fluid. Therefore, we shall limit our naming of them to a few examples (like histamine and the kinins) and the single most important group of mediators, known as **complement**, which will be discussed later.

Vasodilation and increased permeability to protein. Immediately upon microbial entry, chemical mediators dilate most of the vessels of the microcirculation in the area and somehow alter the material between the endothelial cells so as to make the capillaries quite leaky to large molecules. Tissue swelling is directly related to these changes: The arteriolar dilation increases capillary blood pressure, thereby favoring filtration of fluid out of the capillaries; more important, the protein which leaks out of the vessels as a result of increased permeability builds up locally in the interstitium, thereby diminishing the difference in protein concentration between plasma and interstitium (recall from Chap. 11 that this difference is mainly responsible for the osmotic flow of fluid from interstitium into capillaries).

The adaptive value of all these vascular changes is twofold: (1) The increased blood flow to the inflamed area increases the delivery of phagocytic leukocytes and plasma proteins crucial for immune responses; and (2) the increased capillary permeability to protein ensures that the relevant plasma proteins—all normally restrained by the

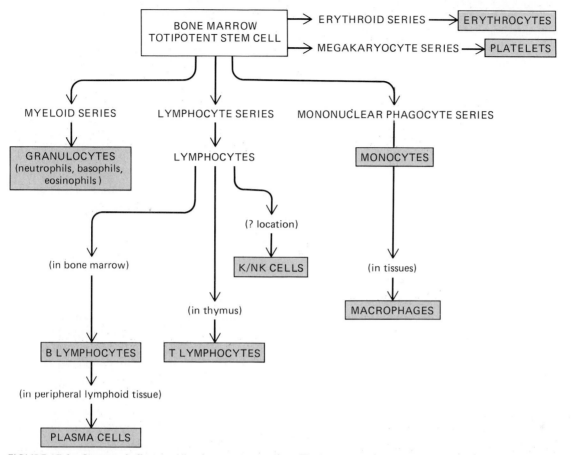

FIGURE 17-2. Origins of effector cells of immune responses. The bone marrow produces granulocytes, lymphocytes, and monocytes (as well as erythrocytes and platelets). Further transformation of the monocytes (into macrophages), B lymphocytes (into plasma cells), and lymphocytes destined to be T lymphocytes occurs in sites outside the bone marrow. K/NK cells are described later.

capillary membranes—can gain entry to the inflamed area.

Chemotaxis. Within 30 to 60 min after the onset of inflammation, a remarkable interaction occurs between the capillary endothelium and circulating neutrophils. First, the blood-borne neutrophils begin to stick to the inner surface of the endothelium. Following their surface attachment, the neutrophils manifest considerable amoebalike activity. Soon a narrow amoeboid projection is inserted into the space between two endothelial cells and the entire neutrophil then squeezes into the interstitium (**neutrophil exudation**). The neutrophil may simply pry the intercellular connections apart by the force of its amoeboid movement, or it may secrete substances which chemically disrupt the connection. In this way, huge numbers of neutrophils migrate into the inflamed areas of tissue and move toward the microbes. This entire response of the neutrophils is known as **chemotaxis** and is induced by chemical mediators (**chemotaxins**) which move out from the inflamed area by diffusion, forming a chemotaxin concentration gradient, with the area immediately around the microbes containing the greatest concentration. This permits the chemotaxins to polarize the motion of the neutrophils along this gradient. The neutrophils do not swim but rather "crawl" toward the microbe. The process is initiated by the binding of the chemotaxins to plasma-membrane receptors on the neutrophil's surface, and this binding activates increased entry of calcium into the neutrophil's cytosol; it is this calcium which activates the force-generating processes associated with actin microfilaments that lead to amoeboid motion along the chemotaxin gradient.

Movement of leukocytes into the tissue is not limited only to neutrophils. Monocytes follow, but usually later, and once in the tissue are transformed into macrophages. Meanwhile some of the macrophages normally present in the tissue may multiply by mitosis and become motile. Thus, all the major phagocytic cell types are present in the inflamed area. Usually the neutrophils predominate early in the infection but tend to die off more rapidly than the others, thereby yielding a more mononuclear picture later. In contrast, in certain types of allergies and inflammatory responses to parasites, eosinophils are in striking preponderance.

Phagocytosis. Phagocytosis is a primary function of the inflammatory response, and the increased blood flow, vascular permeability, and leukocyte exudation mentioned above serve largely to ensure the presence of adequate numbers of phagocytes and to provide the milieu required for the performance of their function.

The initial step in phagocytosis is contact between the surfaces of the phagocyte and microbe (or foreign particle or damaged native cell). But such contact is not, itself, always sufficient to cause firm attachment and trigger engulfment. This is particularly true for many virulent bacteria which have a thick polysaccharide capsule that makes attachment and engulfment very difficult. As we shall see, the presence of molecular factors (produced by the body), which actually bind the phagocyte to the object it is to engulf, markedly enhances the process of phagocytosis. (Any substance that does this is known as an **opsonin**, from the Greek word that means "to prepare for eating.")

The process of ingestion is illustrated in Fig. 17-3. The phagocyte engulfs the organism by endocytosis, i.e., membrane invagination and pouch (**phagosome**) formation. Once inside, the microbe remains in the phagosome, a layer of plasma membrane separating it from the phagocyte's cytosol. The next step is known as **degranulation** (Fig. 17-3): The membrane surrounding the phagosome makes contact with one of the phagocyte's lysosomes (these appear microscopically as "granules") which are filled with a variety of hydrolytic enzymes. Once contact between the phagosome and lysosome is made, the membranes fuse (the disappearance of the granules during this fusion leads to the term "degranulation"), and the combined vesicles are now called the **phagolysosome**,

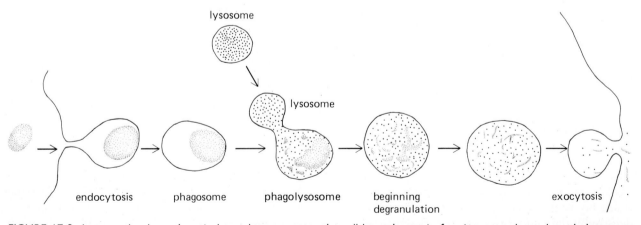

FIGURE 17-3. Large molecules and particulate substances enter the cell by endocytosis, forming a membrane-bound phagosome (left). A phagosome may then merge with a lysosome, which brings together the digestive enzymes of the lysosome and the contents of the phagosome. After digestion has taken place, the contents may be released to the outside of the cell by exocytosis.

inside which the microbe is exposed to the lysosomal enzymes capable of breaking down its macromolecules.

However, the lysosomal enzymes are not the major mechanisms within the phagolysosome for killing bacteria. The phagolysosome produces a high concentration of hydrogen peroxide and other derivatives of oxygen, which are extremely destructive to macromolecules. This production is mediated by enzymes in the phagolysosomal membrane, activated in response to the presence of the foreign particle.

The phagolysosomal enzymes and oxygen derivatives not only kill the microorganism but catabolize it into low-molecular-weight products which can then be safely released from the phagocyte or actually utilized by the cell in its own metabolic processes. This entire process need not kill the phagocyte, which may repeat its function over and over before dying. Nondegradable foreign particles (such as wood, tattoo dyes, or metal) and certain species of microorganisms may be retained indefinitely within macrophages.

Neutrophils and macrophages function in the inflammatory process not only as phagocytes but as secretory cells (Fig. 17-4). Contact with microbes stimulates them to release one type of their lysosomes not into the phagosome but into the extracellular fluid. In addition to releasing these preformed intravesicle chemicals, they are also stimulated to synthesize and release many other substances. In the extracellular fluid, some of these chemicals (both those newly synthesized and those released from vesicles) function as mediators of the inflammatory response. For example, several chemicals secreted by the neutrophil induce histamine release from nearby mast cells and generate kinins from kininogens in the extracellular fluid.

Some of the other substances released by the phagocyte into the extracellular fluid are toxic to microbes (as will be described in the next section). Still others are enzymes which can trigger both the clotting and anticlotting pathways described in Chap. 11. Still others, released specifically by macrophages (and their precursors, monocytes) enter the circulation and function as hormones to produce adaptive changes in the function of organs and tissues far removed from the inflammatory site (as will be described later).

Direct killing of microbes. Phagocytosis is not the only way of destroying microbes. As mentioned above, the phagocytes release into the extracellular fluid antimicrobial substances. For example, the enzymes which generate the antimicrobial oxygen derivatives within the phagolysosomes are also lo-

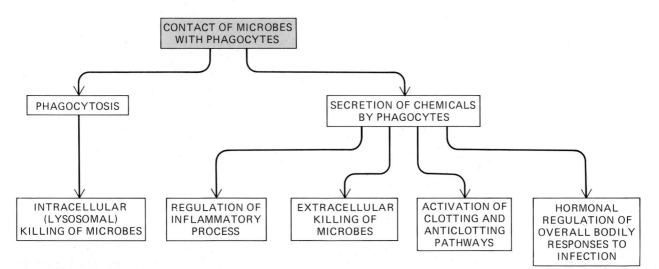

FIGURE 17-4. Role of phagocytes in nonspecific immune responses.

cated on the outer surface of the phagocyte's plasma membrane, and so these substances also gain entry to the extracellular fluid. So do certain digestive enzymes contained in the vesicles released extracellularly by the phagocyte. An example of a more subtle antimicrobial substance secreted by neutrophils into the extracellular fluid is **lactoferrin**, a protein which tightly binds iron and hence prevents the microbes, whose multiplication requires considerable iron, from obtaining it.

There are, in addition to the antimicrobial chemicals released into the extracellular fluid by phagocytes, at least two other mechanisms for destroying the cell without phagocytosis which act during the period prior to the generation of the specific immune responses. One of these involves a class of lymphocyte known as **natural killer (NK)** cells, but because its role is best established in the body's defenses against cancer and virus-infected cells, it will be described in those contexts. The remaining mechanism for early extracellular killing of microbes—the **complement** system—is the most important of all in nonspecific immune response against bacteria. However, its actions extend well beyond cell-killing, and so we will treat it in its entirety after describing the last stage in local inflammation—tissue repair.

Tissue repair. The final stage of the inflammatory process is tissue repair. Depending upon the tissue involved, regeneration of organ-specific cells may or may not occur (for example, regeneration occurs in skin and liver but not in the central nervous system). In addition, fibroblasts in the area divide rapidly and begin to secrete large quantities of collagen, forming **scar** tissue.

The end result may be complete repair (with or without a scar), **abscess** formation, or **granuloma** formation. An abscess is a bag of pus (microbes, leukocytes, and liquified debris) walled off by fibroblasts and collagen. This occurs when tissue breakdown is very severe and when the microbes cannot be eliminated but only contained. When this stage has been reached, the abscess must be drained, for it will not be absorbed spontaneously.

A granuloma occurs when the inflammation has

been caused by certain microbes (such as the bacteria causing tuberculosis) which are engulfed by phagocytes but survive within them. It also occurs when the inflammatory agent is a nonmicrobial substance which cannot be digested by the phagocytes. The granuloma consists of numerous layers of phagocytic-type cells, the central ones of which contain the offending material. The whole thing is, itself, usually surrounded by a fibrous capsule. Thus, a person may harbor live tuberculosis-producing bacteria for many years and show no ill effects as long as the microbes are contained within a granuloma and are not allowed to escape.

The complement system. The **complement** system is yet another example (the clotting, anticlotting, and kinin systems are others) of a group of plasma proteins which normally circulate in the blood in an inactive state; upon activation of the first protein of the group, there occurs a sequential cascade in which active molecules are generated from inactive precursors. In the complement system, the final active protein generated in the cascade directly kills foreign cells by attacking their plasma membranes. This protein is really a complex of five different proteins known appropriately as the **membrane attack complex**, or **MAC**. MAC imbeds itself in the microbial surface and creates channels through the membrane, making it leaky and killing the microbe. This is the major mechanism whereby microbes can be killed directly without prior phagocytosis in nonspecific inflammatory responses.

Thus far, the complement system is analogous to the other cascade systems we have described, in that the final activated molecule (fibrin in clotting, plasmin in anticlotting, kinins in the kinin system) is the direct mediator of the biological response of that system. However, in the complement system, certain of the different active proteins generated along the cascade prior to the terminal protein also function as mediators of a particular inflammatory response (vasodilation, chemotaxis, etc.). Figure 17-5 illustrates these in terms of one of the most important complement molecules, C_3. (Since this system consists of at least 20 distinct proteins, it is extremely complex and we will make no attempt to

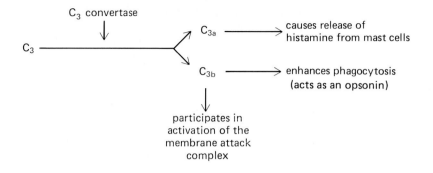

FIGURE 17-5. An example of events in the complement system. The complement protein C_3 is split into two proteins, C_{3a} and C_{3b}. C_{3a} functions as one of the mediators of the inflammatory response, in large part by stimulating mast cells to release histamine. C_{3b} attaches to the microbe where it not only facilitates phagocytosis but participates in the activation of the next complement protein in the sequence.

identify the roles of individual complement proteins except for purposes of illustration or clarity.)

As summarized in Fig. 17-6, one or more of the activated complement molecules is capable of mediating virtually every step of the inflammatory response. Certain of them enhance vasodilation and increased capillary permeability by stimulating release of histamine from mast cells and by generating kinins. Some stimulate neutrophil exudation by acting as powerful chemotaxic agents. Another complement component (C_{3b}) enhances phagocytosis, i.e., it acts as an opsonin (the only major one in nonspecific inflammatory responses), attaching the phagocyte to the object it is to engulf (Fig. 17-7). One portion of this complement molecule binds covalently and nonspecifically to the surface of the microbe, whereas another portion binds to specific

receptor sites for it on the plasma membrane of the phagocyte. In this manner, the complement molecule forms a link between the two cells, providing the surface contact needed to initiate phagocytosis.

In describing the actions of the complement system we have emphasized the sequential or cascade nature of the reactions required for the generation of active mediators, but we have so far ignored the problem of just how the sequence is initiated. As we shall see in a later section, antibody is required to activate the very first protein (C_1) in the entire complement sequence now known as the **classical complement pathway**, but antibody is not present in the nonspecific inflammatory responses we are presently describing. There is, however, an **alternate pathway** for complement activation which is not antibody-dependent and which bypasses the

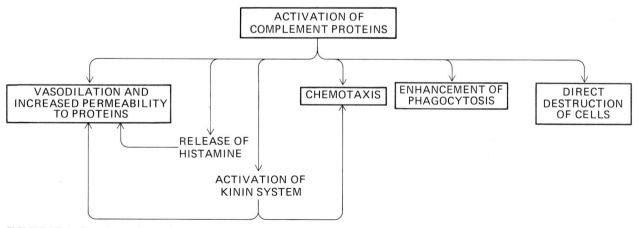

FIGURE 17-6. Functions of complement.

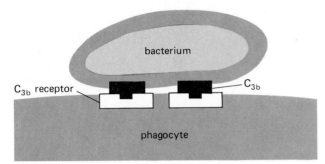

FIGURE 17-7. Function of complement C_{3b} as an opsonin. C_{3b} attaches to the bacterial surface by covalent bonds and to the phagocyte via plasma-membrane receptors for it. The structures in this figure are not drawn to scale.

first steps in the classical pathway. Relatively little biological activity is lost in bypassing the first steps, because no essential inflammatory mediators are generated by them. The first really critical step in this regard is the cleavage of C_3 to its subunits C_{3a} and C_{3b} (Fig. 17-5), and this is where the alternate pathway plugs in.

The critical question for initiation via the alternate pathway can, therefore, be stated as follows: How does the presence of microbes in the absence of antibodies lead to the generation of an enzyme (**C_3 convertase**) which can cleave C_3, and, thereby, initiate the rest of the complement pathway? Unfortunately, the answer to this question is still not completely clear, but appears to involve the participation of six alternate-pathway complement proteins. Suffice it to say that the the key element required to generate C_3 convertase from these proteins is the presence on the microbial surface of carbohydrate molecules not found on normal human cells. Moreover, as the C_{3b} proteins are formed by the action of C_3 convertase, they bind to the surface of the microbes, and this permits the remainder of the complement reactions, including the ultimate formation of the MAC, to proceed in the immediate vicinity of the microbe.

To recapitulate, the alternate pathway of complement activation provides a means of natural resistance to infectious agents and is brought into play even in the total absence of antibody. Thus, it provides an immediately available line of defense that does not require previous exposure to the invading microbe. It may also turn out that this alternate pathway plays an important role in attacking not only extracellular microbes but virus-infected host cells as well as cancer cells.

Systemic manifestations of inflammation

We have thus far described the *local* aspects of the inflammatory response. What are the *systemic* responses, i.e., responses of organs and tissues distant from the site of inflammation? Probably the single most common and striking sign of injury is fever. The substance primarily responsible for the resetting of the hypothalamic thermostat, as described in Chap. 15, is a protein known as **endogenous pyrogen**, which is secreted by monocytes and macrophages.

One would expect fever, being such a consistent concomitant of infection, to play some important protective role, and recent evidence strongly suggests that such is the case (this is a very important question in light of the widespread use of aspirin and other drugs to suppress fever). In contrast to the possible benefits of fever, there is no question that an extremely high fever may be quite harmful, particularly in its effects on the functioning of the central nervous system, and convulsions are not infrequent in young children with high fevers.

Another striking systemic change which occurs in response to infection or tissue injury is a decrease in the plasma concentrations of iron (and zinc), due to changes in the intake and/or release of iron by liver, spleen, and other tissues. This phenomenon is beneficial since, as mentioned before, bacteria require a high concentration of iron to multiply (recall that another way of reducing the iron available to microbes is release of an iron-binding protein by neutrophils in the inflamed area).

In addition, the liver is stimulated during infection to release a host of proteins known collectively as **acute phase proteins**, which exert a bewildering array of effects on the inflammatory process, immune cell function, and tissue repair. The bone marrow is stimulated to produce and release neutrophils and other granulocytes (**granulopoiesis**).

The eliciting of these last three effects (decreased plasma iron and zinc, release of acute phase proteins, and stimulation of granulopoiesis) is ascribed to a substance named **leukocyte endogenous mediator (LEM)** (Fig. 17-8), secreted by monocytes and macrophages. In fact, it is likely that LEM is the same substance as endogenous pyrogen, EP (or at least, a very close relative).

To add to the diversity of actions exerted by EP, it now appears that **interleukin 1 (IL-1)**, a protein which is secreted by macrophages and which is required for activation of lymphocytes, as we shall see, is also identical (or very similar) to EP. Thus, EP/LEM/IL-1 acts as hormone and paracrine to elicit a broad spectrum of events throughout the body which are adaptive in the response to infection or injury. (The list of effects given here is by no means complete.) Indeed, the release of EP/LEM/

IL-1 is probably elicited by other stresses as part of a nonspecific protective response. For example, some of the physiological responses to endurance exercise closely parallel the response to infection, and EP release has been reported in this situation.

Interferon

Interferon, a family of proteins with differing amino acid sequences, sites of production, and ranges of biological activity is a major nonspecific defense mechanism, particularly against viral infection. Interferon nonspecifically inhibits viral replication inside host cells. In response to the presence of a virus (its nucleic acid is the inducer), interferon is synthesized and secreted into the extracellular fluid by many different cell types (including, but not limited to, some which are effector cells of the immune

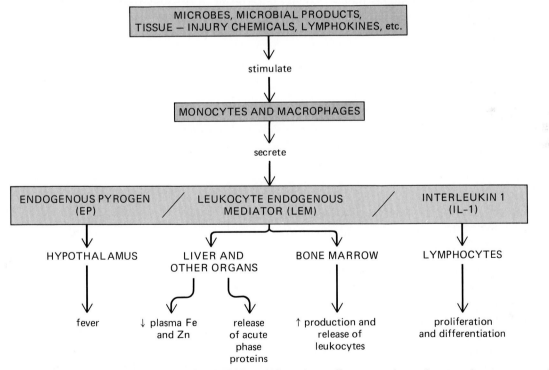

FIGURE 17-8. Origin and functions of EP/LEM/IL-1. When these effects were being discovered, it was not realized that they are probably due to the action of a single protein (or several very closely related proteins); rather, it was thought that three distinct proteins were involved (fever caused by EP, activation of the immune system by IL-1, etc.), and this accounts for the multiple names. (Adapted from W. R. Beisel.)

response, like macrophages and lymphocytes). It then binds to plasma membrane receptors on nearby bodily cells (additionally, some types of interferon may also enter the circulation and reach cells at far-removed sites).

The interferon molecule does not, itself, have direct antiviral activity; rather its binding to the plasma membrane triggers the synthesis of several enzymes by the cell to which it is bound, and these enzymes block synthesis within the cell of proteins the virus requires for its multiplication (Fig. 17-9). Importantly, the enzymes synthesized in response to interferon are inactive until the cell is actually infected by a virus (the virus's RNA conveniently serves as activator); this protects the cell's normal protein-synthesizing machinery from inhibition should the cell never be invaded. The fact that interferon itself is not antiviral but rather stimulates the cells to make their own antiviral proteins is another example of amplification—a single molecule of interferon can trigger the synthesis of many

antiviral proteins so that only a few interferon molecules are required to protect a cell.

To reiterate, interferon is not specific; many viruses induce interferon synthesis, and interferon, in turn, can inhibit the multiplication of many different viruses. This nonspecificity permits the very rapidly reacting interferon system to contain a viral infection until the more slowly reacting specific immune responses can take over.

In this regard, one of the most important interactions between interferon and the specific immune system should be mentioned here (it will be described in more detail later). Interferon stimulates the activity of those lymphocytes which directly attack and kill virus-infected cells. Our description thus far has focused entirely on interferon's antiviral role. However, it is now known that other microbes, particularly those with an intracellular phase to their growth cycle, can induce cells to make interferon. It is likely that interferon helps to protect against these invaders as well. But the most

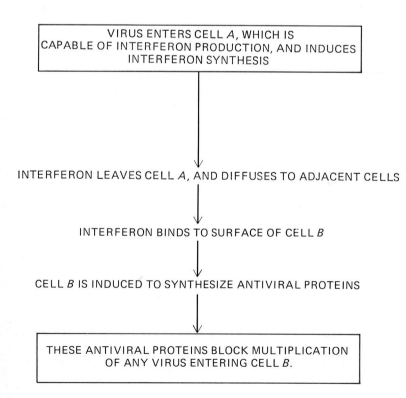

FIGURE 17-9. Role of interferon in preventing viral replication.

exciting finding in the study of interferon is that interferon exerts anticancer effects, not because cancers are caused by viruses (few, if any, human cancers are) but rather because the lymphocytes stimulated into activity by interferon attack and destroy cancer cells as well as virus-infected cells. A great deal of research is now in progress to determine whether interferon (manufactured commercially mainly by recombinant DNA techniques) can be used medically in the therapy or prevention of viral diseases and cancer.

Specific Immune Responses

The roles of lymphocytes and macrophages: An overview

Traditionally, specific immune responses are placed in one of two categories, according to the nature of the effector mechanisms employed in the response. One category has been termed **humoral** or **antibody-mediated immunity**, in recognition of the central role of circulating antibodies in the destructive process; the antibodies are secreted by **B lymphocytes** (to be more precise, by the **plasma cells** into which B lymphocytes differentiate upon being stimulated). The second category of specific immune defenses is **cell-mediated immunity**, and it is mediated not by antibodies but by intact cells, in this case largely by a second population of lymphocytes, the **T lymphocytes**, which are distinct from the B lymphocytes.

After fetal life, lymphocytes, like all leukocytes, are derived from precursor cells in bone marrow. After entering the blood, the lymphocytes take up residence in lymph nodes, thymus, spleen, and other lymphoid organs. There, upon appropriate stimulation, they undergo mitosis, so that most new lymphocytes actually arise in these lymphoid organs rather than in bone marrow. The crucial event determining whether a lymphocyte and all generations of the cells it gives rise to by mitosis will be a B cell or T cell is entry of the cell into the thymus or lack of entry.

At some point in their travels, cells destined to be T cells enter (or arise within) the thymus (thus the name T cell), which in some manner confers

upon them the ability to differentiate and mature into cells competent to act as effectors in cell-mediated immunity. The T cells then leave the thymus and take residence in the other lymphoid tissues, but the thymus continues to somehow stimulate them and their offspring by means of hormones (**thymosin**) which its epithelial cells secrete. In contrast, the B cells do not enter the thymus prior to their taking up residence in lymph nodes and other peripheral lymphoid tissues.

Unlike the nonspecific defense mechanisms, B-cell and T-cell responses depend upon the cells "recognizing" the specific foreign matter to be attacked or neutralized. This recognition is made possible by the fact that molecular components of foreign cells (and other foreign matter) combine specifically with receptor sites on the surfaces of B cells and/or T cells, triggering the attack. These foreign molecular components which stimulate a specific immune response are known as **antigens**.

Most antigens are either proteins or polysaccharides. They may exist as conjugates of each other (glycoproteins) or with other substances (lipopolysaccharides or lipoproteins). An essential determinant of a molecule's capacity to serve as an antigen is size (most antigens have molecular weights of greater than 10,000). However, many low-molecular-weight substances can become attached to an already existing antigen and thereby acquire the ability to trigger a specific immune response; in such cases the small molecule is known as a **hapten**.

The B-cell system is characterized by an ability to recognize an enormous variety of different specific antigens, and its major function is to confer, via its secreted antibodies, resistance against most bacteria (and their toxins) and some viruses. By contrast, the T cells confer major resistance against cells infected by viruses or other infectious agents, some cancer cells, and solid-tissue transplants.

B cells and T cells influence each other in a variety of ways. On the one hand, antibodies (a B-cell product) may either facilitate or decrease the ability of an attacking T cell to destroy a foreign cell. On the other hand, T cells may either enhance or suppress the secretion of antibody by B cells.

This last phenomenon led to the discovery that T cells do not constitute a homogeneous population but are of three kinds: (1) **cytotoxic T cells**, which upon activation perform the role classically ascribed to T cells—the destruction of those cells listed in the previous paragraph; (2) **helper T cells**, which enhance both antibody production and cytotoxic T-cell function; and (3) **suppressor T cells**, which suppress both antibody production and cytotoxic T-cell function. As we shall see, most of the actions of T cells are mediated by chemicals which they secrete. All nonantibody chemical messengers secreted by lymphocytes (whether B or T) are collectively termed **lymphokines**; most, if not all, lymphokines are proteins.

To reiterate, it is the ability of B and T cells to recognize specific foreign antigens that confers specificity upon the immune responses in which they participate. However, another cell type we have already dealt with extensively—the macrophage—is also crucial for specific immune responses, not because it "recognizes" specific foreign antigens (it does not) but because it "cooperates" with B and T cells in a variety of essential ways (Fig. 17-10). First, macrophages must somehow "process" and "present" the foreign antigen to B cells and T cells or else the antigen will not be recognized by or activate the lymphocyte. This processing probably occurs as a result of the initial nonspecifically mediated phagocytosis carried on by macrophages that we described earlier. A second macrophage-lymphocyte interaction is that the macrophage is stimulated to secrete proteins which have profound regulatory effects on the lymphocytes. The most important of these is **interleukin 1 (IL-1)** (recall that this substance is probably identical to endogenous pyrogen), which is required for the maturation and proliferation of both B cells and T cells. Thus, the macrophages not only get the antigen to the lymphocyte, but, via IL-1, initiate a sequence of events which stimulate the now-activated lymphocyte to undergo its differentiation and multiplication.

The third macrophage-lymphocyte interaction is in the opposite causal direction; we have so far described how the macrophages influence the lymphocytes, but once fully activated and functioning, the lymphocytes induce the macrophages, as we shall see, to participate in the destruction of the foreign cells or matter which started off the entire process.

Humoral immunity

Table 17-3 summarizes the sequence of events which results in antibody-directed destruction of bacteria.

Antibodies. An antibody is a specialized protein capable of combining with the specific antigen which stimulated its production. "Specific" is essential in the definition, since an antigenic substance reacts only with the type of antibody elicited by its own kind (or an extremely closely related kind) of molecule. Specificity is thus related to the chemical structure of the antigen and its antibody.

All antibodies are proteins, each composed of four interlinked polypeptide chains (Fig. 17-11). The two long chains are called heavy chains, and the two short ones are light chains (these chains are all joined by disulfide bridges). The amino acid sequences along the chains are constant from one antibody to the next except for relatively short sequences in portions of opposing light and heavy chains (Fig. 17-11). These latter sequences are

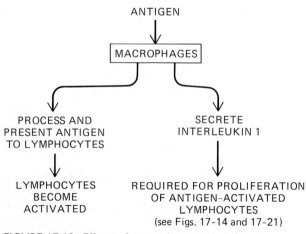

FIGURE 17-10. Effects of macrophages on lymphocytes.

TABLE 17-3. Sequence of events in humoral immunity against bacteria

1. Bacterial antigen is carried to lymphoid tissues, where it binds to surface of specific B lymphocytes. For this to occur, macrophages must first "process" and present the antigen.
2. These B lymphocytes proliferate and differentiate into plasma cells, which secrete antibodies specific for that antigen. This is facilitated by interleukin 1, released by macrophages. It may be facilitated or suppressed by helper or suppressor T cells, respectively.
3. Antibody circulates to site of infection and combines with antigen on surface of bacteria.
4. Presence of antibody bound to antigen facilitates phagocytosis, activates complement system that further enhances phagocytosis and also directly kills bacteria, directs K cells to kill the bacteria directly.
5. Certain of the B lymphocytes differentiate into memory cells capable of responding very rapidly should the bacteria be encountered again.

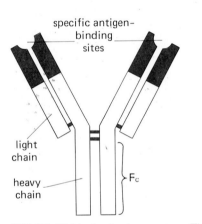

FIGURE 17-11. Antibody structure. The links represent disulfide bonds.

unique for each of the extremely large number of antibodies that an individual can produce and constitute the binding sites (two per antibody) for the antigen specific for that antibody. The constant sequences in the "stem" (or Fc portion) of the heavy chains contain nonspecific binding sites for molecules and cells which function as the effectors for antibody action (see below).

Antibodies belong to the family of proteins known as **immunoglobulins**, which may be subdivided into five classes according to the types of light and heavy chains they contain. It is very easy to misunderstand this concept, which states that there are five *classes* of immunoglobulins (antibodies), not five antibodies; each class contains many thousands of unique antibodies. These classes are designated by the letters G, A, M, D, and E after the symbol Ig (for immunoglobulin). **IgG** antibodies (also commonly called "gamma globulin"), which are the most abundant of the plasma antibodies, and **IgM** antibodies provide the bulk of specific immunity against bacteria and viruses. The other class we shall be concerned with is **IgE**, for these antibodies mediate certain allergic responses. **IgA** antibodies are secreted by the linings of the gastrointestinal, respiratory, and genitourinary tracts and exert their major activities in these secretions; they are also the major antibodies in milk. The function of the **IgD** class is presently uncertain.

Antibody production. When a foreign antigen reaches lymphoid tissues, it triggers off antibody synthesis (Table 17-3 and Fig. 17-12). The antigen may reach the spleen via the blood, but much more commonly it is carried from its site of entry via the lymphatics to lymph nodes. There it stimulates a tiny fraction of the B lymphocytes to enlarge and undergo cell division, most of the progeny of which then differentiate into **plasma cells**, which are the major antibody producers. The most striking aspect of this transformation is a marked expansion of the cytoplasm, which consists almost entirely of the granular type of endoplasmic reticulum (Fig. 17-13) found in other cells which manufacture protein for export (so all-consuming is this transformation that plasma cells die after several days of antibody production). After synthesis, the antibodies are released into the blood or lymph, which carries them

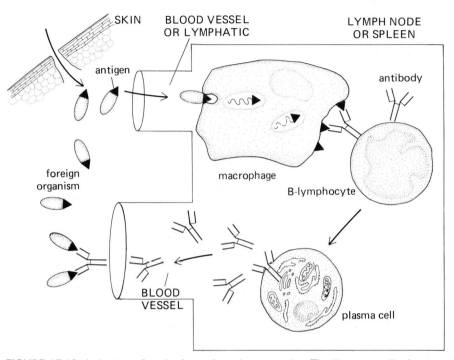

FIGURE 17-12. Induction of antibody synthesis by a microbe. The "processing" of antigen by macrophage is still not clear. The roles of helper T cells and of chemicals secreted by the macrophages are not shown (see Fig. 17-14). *(Adapted from Singer and Hilgard.)*

to the site of infection or presence of foreign matter. There the antibodies combine with the antigenic type that had elicited the antibody's production.

Some of the B-cell progeny do not fully differentiate into plasma cells but rather differentiate into "**memory**" **cells**, ready to respond rapidly should the antigen ever reappear at a future time. Thus, the presence of the memory cells will avoid much of the delay which occurs during the initial infection.

Note that we stated that only a tiny fraction of the total B cells respond to any given antigen. Different antigens stimulate entirely different populations (**clones**) of B cells. This is because the cells of any one lymphocyte clone (and the plasma cells it gives rise to) are capable of secreting only one kind of antibody. Thus, according to this clonal theory, different antigens do not direct a single cell to produce different antibodies; rather each specific antigen triggers activity in the clone of cells already

predetermined to secrete only antibody specific to that antigen. The antigen selects this particular clone and no other because each B cell has immunoglobulin binding sites on its plasma membrane similar to those of the antibodies which it is capable of producing (after differentiation into a plasma cell). These surface immunoglobulins act as receptor sites with which the antigen can combine, triggering off the entire process of division, differentiation, and antibody secretion just described. The staggering but statistically possible implication is that there must exist millions of different clones, at least one for each of the possible antigens an individual *might* encounter during life.

It must be emphasized that interaction between antigen and B-cell receptor site is considerably more complex than that just described, for macrophages also play a crucial role (Fig. 17-14). During antigen activation of the B cells, there occurs a clustering of macrophages around the relevant B-

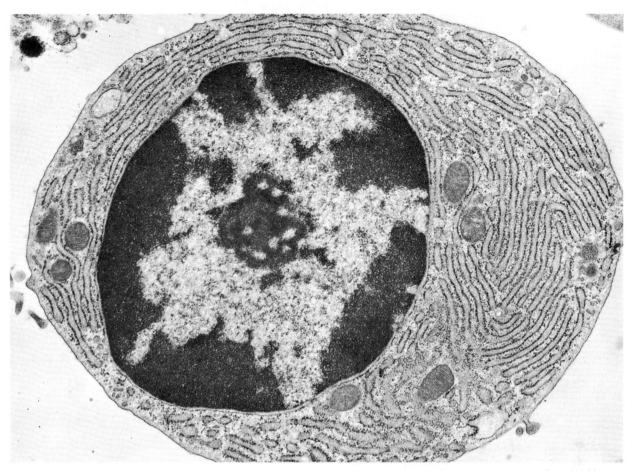

FIGURE 17-13. Electron micrograph of a guinea pig plasma cell. Note the extensive endoplasmic reticulum. *(From W. Bloom and D. W. Fawcett, "A Textbook of Histology," 9th ed., W. B. Saunders Company, Philadelphia, 1968.)*

cell clone. The macrophages "process" and "present" the antigen in some way so as to allow it to activate the B-cell receptor sites. In addition to presenting the antigen to the B cell, the macrophages secrete interleukin 1 (IL-1), which stimulates proliferation of the antigen-activated B cells.

For many antigens, only the interaction between macrophage and B lymphocyte is required to initate B-cell proliferation, differentiation, and antibody secretion. However, for others, participation of **helper T cells** is also essential. In these cases, the antigen processed by the macrophage is presented in such a way that helper T cells, as well as B cells, are activated by it. One way the acti-

vated helper T cell stimulates the B cell is via its secretion of a chemical (**B-cell growth factor**) which acts in concert with interleukin 1. In some cases, **suppressor T cells**, which inhibit B-cell function, are brought into play; these suppressor T cells provide an important dampening effect when antibody production is becoming excessive.

Functions of antibodies. In a previous section, we described how a local inflammatory response is induced nonspecifically by any tissue damage. Now we shall see that the presence of antigen-antibody complexes triggers events which profoundly *amplify* the inflammatory response. In other words,

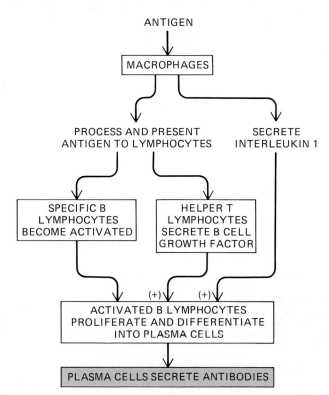

FIGURE 17-14. Activation, proliferation, and differentiation of B lymphocytes.

the major function of humoral immune mechanisms is to enhance and make more efficient the inflammatory response and cell elimination already initiated in a nonspecific way by the invaders. Thus, vasodilation, increased vascular permeability to protein, neutrophil exudation, phagocytosis, and killing of microbes without prior phagocytosis are all markedly enhanced.

Activation of complement system. There are several mechanisms by which the presence of antigen-antibody complexes enhances inflammation, but the single most important involves the complement system. As described earlier, this system, which generates not only terminal proteins of the membrane attack complex that directly kills microbes, but mediators of virtually every process in inflammation, is involved in the nonspecific inflammatory re-

sponse. However, the presence of antibody attached to antigen increases its participation because the antibody-antigen complex is a powerful activator of the first step in the so-called **classical complement pathway** (Fig. 17-15). The first molecule in this pathway, C_1, binds to complement receptors on the Fc (stem) portions of the antibody to which antigen has become complexed (Fig. 17-16), and this binding activates the enzymatic portions of C_1, thereby initiating the entire complement pathway leading through cleavage of C_3 and on to the formation of the membrane attack complex (Fig. 17-5).

It is important to note that C_1 binds not to the unique antigen-specific binding sites in the antibody's "prongs" but rather to binding sites in the antibody's stem (Fig. 17 11). Since the latter are the same in virtually all antibodies of the IgG and

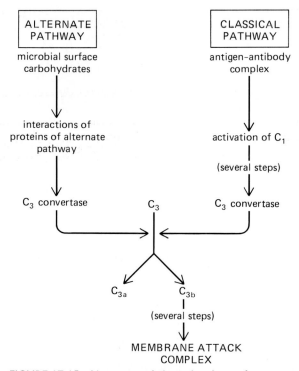

FIGURE 17-15. Alternate and classical pathways for generating C_3 convertase (the C_3 convertases in the two pathways have different structures, but both cleave C_3). Once C_3 is cleaved, the pathways are identical.

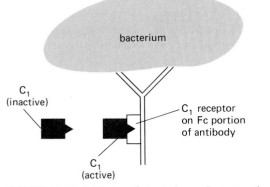

FIGURE 17-16. Initiation of classical complement pathway. The enzymatic ability of C_1 is activated by its binding to the Fc (stem) portion of an antibody, itself bound to specific antigen. Active C_1 then activates the next complement molecule in the pathway while attached to the antibody. The structures in the figure are not drawn to scale.

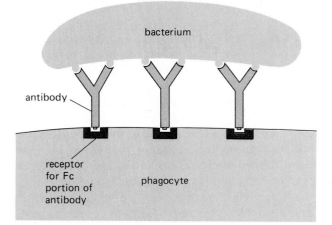

FIGURE 17-17. Enhancement of phagocytosis by antibody. The antibody links the phagocyte to the bacterium.

IgM classes, the complement molecule will bind to *any* antibodies belonging to these classes. In other words, there is only one set of complement molecules, and once activated, they do essentially the same thing, regardless of the specific identity of the invader. In contrast, the formation of antibodies to antigens on the invader and their subsequent combination are highly specific.

To reiterate, the function of the antibodies is to "identify" the invading cells as foreign by combining with antibody-specific antigens on the cell's surface. The complement system is subsequently activated when the first complement molecules in the sequence combine with this antigen-bound antibody, and the cascade of activated complement molecules then mediates the actual attack.

Direct enhancement of phagocytosis. Activation of complement is not the only mechanism by which antibodies enhance phagocytosis. Merely the presence of antibody attached to antigen on the microbe's surface has some enhancing effect, i.e., antibody acts as an opsonin.

The mechanism is analogous to that for complement in that the antibody links the phagocyte to the antigen. As shown in Fig. 17-17, the phagocyte has receptors on its surface to which bind the Fc (stem) portion of antibodies. This linkage promotes attachment of the antigen to the phagocyte and the triggering of phagocytosis.

Activation of K cells. The presence of antibody attached to specific antigens on a foreign cell brings into play not only phagocytes but a little-understood population of lymphocytes known as **K cells** (for **killer cells**). These cells also have receptors for the Fc portion of antibodies and so are brought into intimate contact with foreign cells via this linkage (in Fig. 17-17, all one would have to do is change "phagocyte" to "K cell"). K cells, however, do not phagocytize the foreign cells but rather kill them directly by lysing their outer membranes.

Direct neutralization of bacterial toxins and viruses. Bacterial toxins and certain viral components act as antigens to induce antibody production. The antibodies then combine with the toxins and viruses to "neutralize" them. Neutralization in reference to a virus means that the combined antibody prevents attachment of the virus to host cell membranes, thereby preventing virus entry into the cell. Similarly, antibodies neutralize bacterial toxins by combining with them, thus preventing the interaction of the toxin with susceptible cell-membrane sites. In both cases, since each antibody has two binding sites for combination with antigen, chains of antibody-antigen complexes are formed (Fig. 17-18) which are then phagocytized.

Active and passive humoral immunity. We have been discussing antibody formation without regard to the course of events in time. The response of the antibody-producing machinery to invasion by a foreign antigen varies enormously, depending upon whether it has previously been exposed to that antigen. Antibody response to the first contact with a microbial antigen occurs slowly over several days, with some circulating antibody remaining for long periods of time, but a subsequent infection elicits an immediate and marked outpouring of additional antibody (Fig. 17-19). This represents a type of "memory" which converts a greatly enhanced resistance toward subsequent infection with that particular microorganism. This resistance, built up as a result of actual contact with microorganisms and

their toxins or other antigenic components, is known as **active immunity**. Until modern times, the only way to develop active immunity was actually to suffer an infection, but now a variety of other medical techniques are used, i.e., the injection of **vaccines**, or microbial derivatives. The actual material injected may be small quantities of living or dead microbes, e.g., polio vaccine, small quantities of toxins, or harmless antigenic materials derived from the microorganism or its toxin. The general principle is always the same: Exposure of the body to the agent results in the induction of the antibody-synthesizing machinery required for rapid, effective response to possible future infection by that particular organism. However, for many microorganisms (the common cold virus, for example), the memory component of the antibody response does not occur, and antibody formation follows the same time course regardless of how often the body has been infected with the particular microorganism.

A second kind of immunity, known as **passive immunity**, is simply the direct transfer of actively

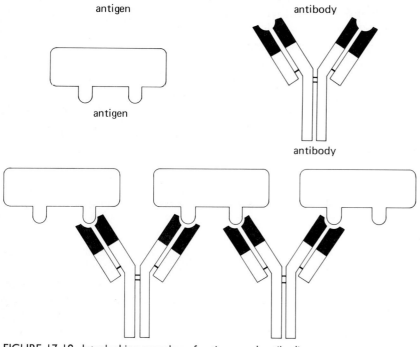

FIGURE 17-18. Interlocking complex of antigens and antibodies.

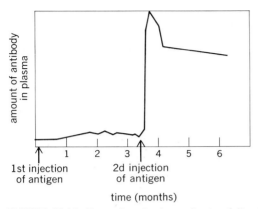

FIGURE 17-19. Rate of antibody production following the initial contact with an antigen and a subsequent contact with the same antigen.

formed antibodies from one person (or animal) to another, the recipient thereby receiving preformed antibodies. Such an exchange (for IgG) normally occurs between fetus and mother across the placenta. Breast-fed children also receive antibodies (of the IgA class) in the mother's milk. These are important sources of protection for the infant during the first months of life, when the antibody-synthesizing capacity is relatively poor.

The same principle is used clinically when specific antibodies or pooled gamma globulin is given a person exposed to or actually suffering from certain infections, such as measles, hepatitis, or tetanus. The protection afforded by this transfer of antibodies is relatively short-lived, usually lasting only a few weeks. The procedure is not without danger since the injected antibodies (often of nonhuman origin) may themselves serve as antigens, eliciting antibody production by the recipient and possibly severe allergic responses.

Summary. We may now summarize the interplay between nonspecific and specific humoral immune mechanisms in resisting a bacterial infection. When a bacterium is encountered for the first time, *nonspecific* defense mechanisms resist its entry and, if entry is gained, attempt to eliminate it both by phagocytosis and, to some extent, by nonphagocytic killing. Simultaneously, the bacterial antigens induce the differentiation of specific B-cell clones into plasma cells capable of antibody production. If the nonspecific defenses are rapidly successful, these slowly developing *specific* immune responses may never play an important role. If the nonspecific defenses are only partly successful, the infection may persist long enough for significant amounts of antibody to reach the scene; the presence of antibody, in a variety of ways, leads to both enhanced phagocytosis and direct destruction of the foreign cells. In either case, all subsequent encounters with that type of bacteria will be associated with the same sequence of events, with the crucial difference that the specific humoral immune responses may be brought into play much sooner and with greater force; i.e., the person might have active immunity against that type of bacteria.

Cell-mediated immunity

Cytotoxic T cells. The cytotoxic T lymphocytes are responsible for highly specific cell-mediated immunity. Individual T cells, like B cells, are clonal in that each T cell bears on its plasma membrane genetically determined receptor sites for a specific foreign antigen. These specific surface receptors are similar to immunoglobulins. In contrast to B cells, cytotoxic T cells will (with one major exception) ignore a foreign antigen unless it is present on a cell surface which also contains an antigen identical to one of the T cell's own "self-antigens."

Each of us has a group of genes (the **major histocompatibility complex** or **MHC**) which codes for many proteins important for immune function. Among these are a group of proteins located on the plasma membranes of all nucleated cells in an individual's body and called **human leukocyte associated antigens (HLA antigens)** (the name reflects the fact that these proteins were first discovered in leukocytes; they are also called **histocompatibility antigens**). Since no two persons (other than identical twins) have the same MHC genes, no two persons have the same HLA antigens. To restate the earlier generalization: T cells will be activated to attack a cell which bears on its surface a foreign antigen *plus* an antigen identical to one of the T cell's own HLA antigens. As we shall see, this is the situation which exists for several categories of the cytotoxic T cell's favorite targets—viral-infected host cells and cancer cells—and explains why the cytotoxic T cells have a much more restricted set of targets than do the B cells.

Upon initial exposure to the appropriate antigen, which must first be "processed" by macrophages, the cytotoxic T cell becomes "sensitized"; i.e., it undergoes, over 1 to 2 weeks, enlargement and mitosis, with differentiation of the daughter cells. So far, the story sounds similar to that described earlier for B cells, but the crucial difference has to do with the activities of these differentiated daughter cells: Instead of secreting antibodies (as do the plasma-cell progeny of B cells), these sensitized cytotoxic T cells actually combine with the foreign antigen on the surface of the cell being attacked (Fig. 17-20), and this triggers the release by the cytotoxic T cell of one or more chemicals which lyse the attached cell's plasma membrane, thus kill-

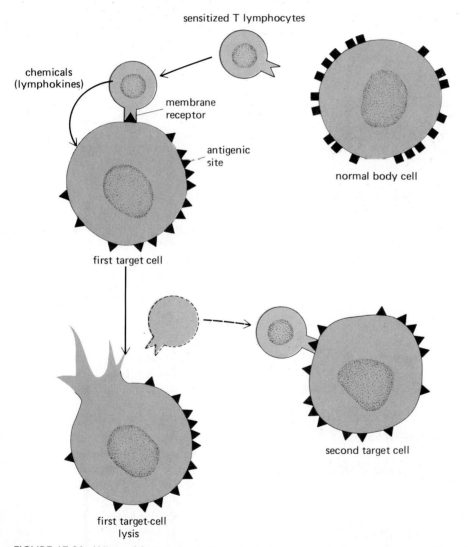

FIGURE 17-20. Killing of foreign (target) cells by sensitized cytotoxic T lymphocytes. The killing of the target cell is achieved by chemicals (lymphokines) secreted by the sensitized T cell. (Adapted from Cerottini.)

ing it directly (i.e., without prior phagocytosis). We must emphasize the important geographic difference between plasma cell and sensitized T-cell function: Antibodies are secreted by plasma cells located in lymph nodes and other lymphoid organs often far removed from the invasion site and reach the site via the blood; in contrast, a sensitized cytotoxic T cell travels to the invasion site where it combines with foreign antigen on the cell under attack and kills it.

As is true for the B system, some of the sensitized T cells do not actually participate in the immune response but serve as a "memory bank" which greatly speeds up and enhances the immune response if the person is ever exposed to the specific antigen again. Thus, active immunity exists for cell-mediated immune responses just as for antibody responses. Passive immunity also exists in this system and can be conferred by administering sensitized lymphocytes taken from a previously infected person (or animal).

Cytotoxic T cells are not the only class of T cells which participate in cell-mediated immunity. Helper and suppressor T cells importantly regulate the cytotoxic T cells, just as they do B cells in humoral immunity. For one thing, helper T cells secrete a lymphokine known as **interleukin 2 (IL-2)**, which stimulates the proliferation of antigen-activated T cells (thus, IL-2 does for T cells what B-cell growth factor—also secreted by helper T cells—does for B cells). This secretion of IL-2 is, itself, stimulated by interleukin 1 released from macrophages (Fig. 17-21).

But IL-2 is not the only chemical messenger secreted by helper T cells in cell-mediated immune responses. They also secrete proteins which act as an amplification system for the facilitation of the inflammatory response and killing by macrophages (Fig. 17-21). Thus, cell-mediated immunity is analogous to humoral immunity in that it serves, in large part, to enhance and make more efficient the nonspecific defense mechanisms already elicited by the foreign material. The major difference is that humoral immunity utilizes a circulating group of plasma proteins (the complement system) as its major amplification system, whereas the T cells literally produce and secrete their own chemical amplification system.

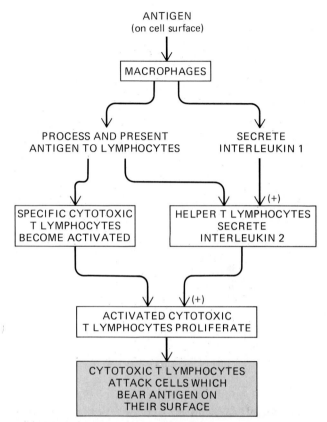

FIGURE 17-21. Activation and proliferation of cytotoxic T cells. Compare to Fig. 17-14. Not shown in the figure is the fact that interleukin 2 also causes helper T cells to proliferate, and secrete many lymphokines which facilitate inflammation.

ally produce and secrete their own chemical amplification system.

As might be predicted, some of these chemicals are chemotaxins. These serve to attract some neutrophils but many more monocytes to the area. The monocytes are converted to macrophages and begin their job of phagocytosis. The T cells not only secrete chemotaxins to attract the macrophages-to-be but they also secrete another substance which keeps the macrophages in the area and stimulates them to greater activity (indeed, such revved-up macrophages are known as "angry" macrophages). Finally, interferon is also released in large quantity from activated T cells; as mentioned earlier, this substance not only exerts inhibitory effects on viral replication but markedly enhances the killing action of cytotoxic T cells.

Natural killer (NK) cells. Our description of specific cell-mediated immunity has been restricted thus far to T cells. Recently, another population of cells, known as **natural killer (NK) cells** has been discovered which have certain properties similar to those of cytotoxic T cells, namely, they combine with foreign antigens on their target cells and kill them directly (by lysing their membranes). Thus, to visualize NK cell action, just substitute "NK cell" for "sensitized T cell" in Fig. 17-20. Moreover, as is true for cytotoxic T cells, major targets for NK cells are tumor cells and virus-infected host cells.

But NK cells (which may be yet another subset of lymphocytes) are definitely not cytotoxic T cells, from which they differ in important ways. First, they are not specific for only one kind of foreign antigen; rather there are a relatively few clones of NK cells, each of which can recognize and bind to a spectrum of different foreign antigens (neither the binding sites on NK cells nor the nature of the antigens on the target cells are known). Second, and very important, NK cells do not need prior exposure to antigen followed by a long maturation period before they can function (this is why they are called "natural" killers). Thus NK cells provide (along with the completely nonspecific mechanisms described earlier in this chapter) an immediately responding first line of defense against cancer cells, virus-infected host cells, and perhaps other microbes as well.

A central role in the function of NK cells is played by interferon. Activated NK cells (like activated T cells) secrete large quantities of interferon. This interferon then not only inhibits viral replication but also markedly enhances the killing actions of the NK cells (just as it does those of cytotoxic T cells and macrophages). Thus interferon and NK cells operate in a positive-feedback manner (Fig. 17-22).

One more point about NK cells concerns their relationship to K cells, the lymphocyte subset which kills cells directly when guided to do so by antibodies. NK cells do not require antibodies to kill their target cells, and so it has been presumed that K and NK cells are distinct cells. It now seems much more likely that they are, in fact, one and the same cell, and whether antibody influences their activity depends upon the milieu in which the attack occurs.

Cell-mediated defenses against viral infection. The body's defenses against viral infection (summarized in Table 17-4) include virtually all the processes we have been describing. It should be clear from Table 17-4 that the host defense mechanisms against viruses operate in two spheres: (1) when the virus is still in the extracellular fluid; and (2) after the virus has taken up residence within the body's cells (something it must do to survive and replicate; replicated virus may leave the original host cell and once again enter the extracellular fluid to seek out new host cells).

The specific immune defense against intracell-

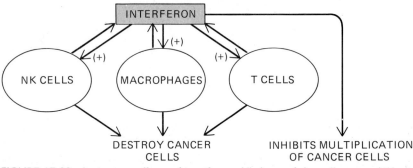

FIGURE 17-22. Anticancer effects of interferon. All three of the cell types which directly attack cancer cells secrete interferon, which then enhances the ability of the cells to destroy cancer cells and also, itself, inhibits multiplication of cancer cells.

TABLE 17-4. Summary of host responses to viruses

	Main cells involved	Comment on action
Nonspecific responses		
Anatomic barriers	Surface linings of body	Simple barrier; antiviral chemicals
Inflammation	Tissue macrophages	Phagocytosis of extracellular virus
Interferon	Multiple cell types	Prevention of viral replication inside host cells
Specific responses		
Humoral immunity	Plasma cells derived from B lymphocytes secrete antibodies that perform the actions listed to the right	Neutralize virus and thus prevent entry to cell. Activate complement that leads to both enhanced phagocytosis and direct destruction of extracellular virus
Cell-mediated immunity	Sensitized T lymphocytes and NK cells secrete chemicals that perform the actions listed to the right	Destroy host cell and thus induce release of virus so that it can be phagocytized. Prevent viral replication (the chemical mediating this effect is interferon.)

ular virus is cell-mediated, by cytotoxic T lymphocytes and by NK cells. In order to attack a virus once it has gained cellular entry, the T cell or NK cell must destroy the cell itself. Such destruction occurs because the viral nucleic acid codes for a plasma-membrane protein foreign to the body, and once this protein has been synthesized by the host cell, it can act as an antigen to elicit a T-cell or NK cell attack. (Recall that cytotoxic T cells "recognize" and attack a cell bearing foreign antigen on its surface when the surface also contains "self-antigens" of the HLA group—clearly the case in viral-infected host cells.) The result is destruction of the host cell with release of the viruses, which can then be directly attacked. Generally, only a few host cells must be sacrificed in this way, but once viruses have had a chance to replicate and spread from cell to cell, so many host cells may be attacked by the body's own defenses that serious malfunction may result.

Immune surveillance: Defense against cancer. Another function of cell-mediated immunity is to recognize and destroy cancer cells. This is made possible by the fact that virtually all cancer cells have some surface antigens different from those of other body cells and can, therefore, be recognized as "foreign." It is likely that cancer arises as a result of genetic alteration (by chemicals, radiation, etc.) in previously normal body cells. One manifestation of the genetic change is the appearance of the new surface antigens (side by side with the cell's own HLA antigens). Circulating T cells encounter and become sensitized to these cells, combine with the antigens on their surface, and release the chemicals which destroy the cells directly; they also call in attacking macrophages.

As mentioned earlier, NK cells also participate in the attack on cancer cells. Indeed, because they do not require a long period of sensitization, they are the single most important effector cell in immune surveillance. Note also (Fig. 17-22) that all three of the cell types which directly attack cancer cells secrete interferon, which, in turn, stimulates the cytotoxicity of these cells.

There occur important interactions between the humoral and cell-mediated systems, one of which actually protects the cancer cell. We have pointed out that the cancer-specific antigens stimulate the development of sensitized T cells against the cancer cells. Simultaneously, they may also stimulate the production by B cells of circulating antibodies (of the IgG class). These antibodies combine with the cell-surface antigen sites, and the next events may be quite variable; in some cases, the antibody may facilitate the destruction of the cancer cell, whereas in most cases it may actually protect the cell by preventing T cells from combining with the antigen. In the latter cases, the antibodies are called "blocking" antibodies and the process is called **immune enhancement**. Whether facilitation or blocking occurs seems to depend upon precisely how the antibody interacts with the antigen, but many unanswered questions remain (for example, why do the blocking antibodies not activate complement which would destroy the cancer cells?).

Rejection of tissue transplants. The cell-mediated immune system is also mainly responsible for the recognition and destruction, i.e., **rejection**, of tissue transplants (Fig. 17-23). As described earlier, on the surfaces of all nucleated cells of an individual's body are genetically determined antigenic protein molecules known as HLA or histocompatibility antigens. When tissue is transplanted from one individual to another, those surface antigens which differ from the recipient's are recognized as foreign and are destroyed by sensitized cytotoxic T cells. (This constitutes the one exception to the generalization that cytotoxic T cells will attack only cells

which bear on their surface one of the T cell's own HLA antigens side by side with foreign antigen.) As is true for the response to cancer cells, the foreign cells may also stimulate the secretion of circulating "blocking" antibodies; if this occurs, the chances for graft survival are enhanced. Note that, in immunology, whether a phenomenon is desirable or undesirable depends on the point of view; cancer enhancement is undesirable whereas graft enhancement is desirable.

Some of the most valuable tools aimed at reducing graft rejection are radiation and drugs which kill actively dividing lymphocytes and, thereby, decrease the T-cell population. Unfortunately, this also results in depletion of B cells as well, so that antibody production is diminished and the patient becomes highly susceptible to infection. A more discriminating method presently being tried is to prepare and inject into the recipient antibodies against the T cells; by this means, the T cells would

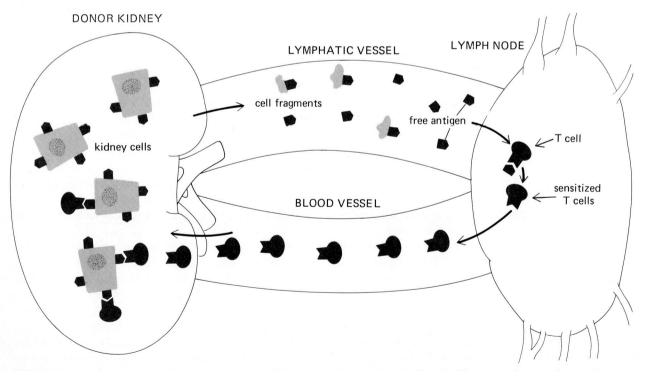

FIGURE 17-23. Mechanism of rejection of renal transplant by sensitized cytotoxic T cells. The roles of macrophages in antigen processing, T-cell sensitization, and kidney-cell destruction are not shown. Nor is the role of helper T cells. *(Adapted from Hume.)*

be destroyed but not the B cells. Recently, an extremely effective drug (cyclosporin) has come into use which blocks either the production or action of interleukin 2, the substance produced by helper T cells which is required for proliferation of antigen-activated cytotoxic T cells; this drug leaves the B-cell system essentially untouched.

Transfusion Reactions and Blood Types

Transfusion reactions are a special example of tissue rejection, one which illustrates the fact that antibodies rather than sensitized T cells can sometimes be the major factor in leading to the destruction of nonmicrobial cells. Among the large number of erythrocyte membrane antigens, we still recognize those designated A, B, and O as most important. These antigens are inherited, A and B being dominant. Thus, an individual with the genes for either A and O or B and O will develop only the A or B antigen. Accordingly, the possible blood types are A, B, O, and AB. If the typical pattern of antibody induction were followed, one would expect that a type A person would develop antibodies against type B cells only if the B cells were introduced into the body. However, what is atypical of this system is that even without initial exposure, the type A person always has a high plasma concentration of anti-B antibody. The sequence of events during early life which lead to the presence of the so-called **natural antibodies** in all type A persons is unknown. Similarly, type B persons have high levels of anti-A antibodies; type AB persons obviously have neither anti-A nor anti-B antibody; type O persons have both; anti-O antibodies are usually not present in anyone.

With this information as background, what will happen if a type A person is given type B blood? There are two incompatibilities: (1) The recipient's anti-B antibody causes the transfused cells to be attacked; and (2) the anti-A antibody in the transfused plasma causes the recipient's cells to be attacked. The latter is generally of little consequence, however, because the transfused antibodies become so diluted in the recipient's plasma that they are

ineffective. It is the destruction of the transfused cells which produces the problems. The range of possibilities is shown in Table 17-5. It should be evident why type O people are frequently called "universal donors" whereas type AB people are "universal recipients." These terms, however, are misleading and dangerous since there are a host of other erythrocyte antigens and plasma antibodies besides those of the ABO type. Therefore, except in dire emergency, the blood of donor and recipient must be carefully matched.

Another antigen of medical importance is the so-called **Rh factor** (because it was first studied in rhesus monkeys) now known to be a group of erythrocyte membrane antigens. The Rh system follows the classic immunity pattern in that no one develops anti-Rh antibodies unless exposed to Rh-type cells (usually termed Rh-positive cells) from another person. Although this can be a problem in an Rh-negative person, i.e., one whose cells have no Rh antigen, subjected to multiple transfusions with Rh-positive blood, its major importance is in the mother-fetus relationship. When an Rh-negative mother carries an Rh-positive fetus, some of the fetal erythrocytes may cross the placental barriers into the maternal circulation, inducing her to synthesize anti-Rh antibodies. Because the movement of fetal erythrocytes into the maternal circulation occurs mainly during separation of the placenta at delivery, a first Rh-positive pregnancy rarely offers any danger to the fetus, since delivery occurs before the antibodies can be made. In future pregnan-

TABLE 17-5. Summary of ABO blood-type interactions

Recipient	Donor	Compatible ?
A	O, A	yes
B	O, B	yes
AB	O, A, B, AB	yes
O	O	yes
A	AB, B	no
B	AB, A	no
AB	— — —	— — —
O	AB, A, B	no

cies, however, these anti-Rh antibodies will already be present in the mother and can cross the placenta to attack the erythrocytes of an Rh-positive fetus. The risk increases with each Rh-positive pregnancy as the mother becomes more and more sensitized. Fortunately, Rh disease can be prevented by giving any Rh-negative mother gamma globulin against Rh erythrocytes within 72 h after she has delivered an Rh-positive infant. These exogenous antibodies bind to the antigenic sites on any Rh erythrocytes which might have entered the mother's blood during delivery and prevent them from inducing antibody synthesis by the mother (both the cells and exogenous antibody are soon destroyed).

Factors Which Alter the Body's Resistance to Infection

Let us examine two seemingly opposed statements: (1) Tuberculosis is caused by the tubercle bacillus; (2) tuberculosis is caused by malnutrition. The first statement seems the more accurate in the sense that the disease, tuberculosis, will not occur in the absence of infection by tubercle bacilli. Yet we also know that many people harbor these bacteria, but never develop tuberculosis. There must exist, therefore, other factors which, by upsetting the balance between host and microbe, permit invasion, multiplication, and production of symptoms. In a sense, then, malnutrition "causes" tuberculosis by doing just this. In fact, there need be no quibbling over semantics once one realizes that the presence of the microbe is the necessary but frequently not sufficient cause of the disease. Therefore, it becomes very important to define those influences which determine the body's capacity to resist infection (as well as cancer cells and transplants). We offer here only a few examples.

Nutritional status is extremely important, and **protein-calorie malnutrition** is, worldwide, the single greatest contributor to decreased resistance to infection. The role that excesses or deficiencies of individual nutrients play in altering resistance to infection is now receiving much study, and no simple conclusions can be made. For example, severe iron deficiency predisposes to infection, but so does an excess of iron.

A preexisting disease (infectious or noninfectious) can also predispose the body to infection. Diabetics, for example, suffer from a propensity to numerous infections, at least partially explainable on the basis of defective leukocyte function. Moreover, any injury to a tissue lowers its resistance, perhaps by altering the chemical environment or interfering with blood supply.

It is also likely (although this point remains controversial) that a person's state of mind (whether the person is stressed, happy, depressed, etc.) can influence his resistance to infection; the physiological links between "state of mind" and resistance remain to be determined (one possibility will be presented in the section of this chapter dealing with "stress").

In numerous examples one of the basic resistance mechanisms itself is deficient. A striking case is that of congenital deficiency of plasma gamma globulin, i.e., failure to synthesize antibodies; these patients cannot survive without either frequent intravenous injections of gamma globulin or isolation from microbes. Similarly, a variety of complement deficiencies exist. A decrease in the production of leukocytes is also an important cause of lowered resistance, as, for example, in patients given drugs specifically to inhibit rejection of tissue or organ transplants. The total quantity of leukocytes circulating is not necessarily critical. Patients with leukemia, for example, may have tremendous numbers of blood neutrophils or monocytes, but these cells are almost all immature or otherwise incapable of normal function; such patients are extremely prone to infection. Reduced functional activity of lymphocytes is also seen in the elderly and may account for their decreased resistance.

One of the most striking examples of the lack of a basic resistance mechanism is the disease called **acquired immune deficiency syndrome (AIDS)**. This disease is characterized by profound inability to resist many infections. The major defect is a lack of helper T cells, caused by destruction or alteration of these cells by certain viruses. Clinical trials are underway to determine whether the administration to these patients of interleukin 2, a major lymphokine secreted by helper T cells, can result in improvement.

Finally, we must mention the most important of the external agents we employ in altering resistance to infection—the **antibiotics**, such as penicillin. Antibiotics exert a wide variety of effects, but the common denominator is interference with the synthesis of one or more of the bacteria's essential macromolecules. Use of these antibacterial agents is made possible because they are harmful to microbes but relatively innocuous to the body's cells. This characteristic distinguishes them from the **disinfectants**, which are highly effective antibacterial agents but equally toxic to the body. The word "relatively" is quite important, since all antibiotics are toxic to a lesser or greater degree and must not be used indiscriminately. A second reason for judicious use is the problem of drug resistance. Most large bacterial populations contain mutants which are not sensitive to the drug and which are thus selected out by the drug. These few are capable of multiplying into large populations resistant to the effects of that particular antibiotic. Perhaps even more important, resistance can be transferred from one microbe directly to another previously nonresistant microbe by means of chemical agents ("**resistance factors**") passed between them. A third reason for the judicious use of antibiotics is that these agents may actually contribute to a new infection by altering the normal flora so that overgrowth of an antibiotic-resistant species occurs.

Allergy (Hypersensitivity Reactions)

A certain portion of the population is capable of acquiring specific-immune reactivity to environmental antigens such as dusts, pollens, food constituents, etc. Despite the relative harmlessness of most of these antigens, subsequent exposure to them elicits an immune attack which produces distressing symptoms and, often, outright bodily damage. This phenomenon, known as **allergy** or **hypersensitivity**, is in essence immunity gone wrong, for the response is really inappropriate to the stimulus.

Depending upon the antigen, allergic responses may be due to activation of either the humoral (antibody) or cell-mediated systems. In the latter cases (the skin rash after contact with poison ivy is an example), the inflammatory response is simply a T-cell–mediated affair of the kind we have discussed; because it takes several days to develop, it is known as **delayed hypersensitivity**. In contrast, antibody-mediated hypersensitivity responses are usually *immediate* in onset and involve a group of antibodies and a sequence of events different from the classical antibody-mediated response to bacteria.

Immediate hypersensitivity

Initial exposure to the antigen in the immediate-hypersensitivity type of allergy leads to some antibody synthesis but, more important, to the memory storage which characterizes active immunity. Upon reexposure, the antigen elicits a more powerful antibody response. So far, none of this is unusual, but the fact is that these particular antigens stimulate the production of the **IgE** class of antibodies. Upon their release from plasma cells, these antibodies circulate to various parts of the body and attach themselves to **mast cells**. When the antigen then combines with the IgE attached to the mast cell, this triggers release of the mast cell's secretory vesicles which contain **histamine**, other vasoactive chemicals, and chemotaxins specific for **eosinophils**. In addition to releasing these preformed inflammatory mediators stored in mast-cell vesicles, the binding of IgE to receptors triggers the mast cell to synthesize **prostaglandins** and related substances (Chap. 7), which exert many inflammation-stimulating effects. All these chemicals then initiate a local inflammatory response. Thus, the symptoms of antibody-mediated allergy are due to the various effects of these chemicals and the body site in which the antigen–IgE–mast-cell combination occurs. For example, when a previously sensitized person inhales ragweed pollen, the antigen combines with IgE-mast cells in the respiratory passages. The chemicals released cause increased mucus secretion, increased blood flow, leakage of protein, and contraction of the smooth muscle lining the airways. (**Leukotrienes** are particularly potent in inducing this last effect.) Thus, there follow the symptoms of congestion, running nose, sneezing, and difficulty in breathing which characterize hayfever.

In this manner, the allergic symptoms may be localized to the site of entry of the antigen. However, if very large amounts of the chemicals released enter the circulation, systemic symptoms may result and cause severe hypotension and bronchiolar constriction. This sequence of events (**anaphylactic shock**) can actually cause death (due to circulatory and respiratory failure) and can be elicited in some sensitized people by the antigen in a single bee sting.

As may well be imagined, **antihistamines** offer some relief in allergies; it is usually incomplete, however, since other inflammatory agents are also released from the mast cells. In severe cases, the anti-inflammatory powers of large doses of **cortisol** are employed.

Given the inappropriateness of most immediate hypersensitivity responses, why should such a system have evolved? The normal physiological function of the IgE–mast-cell–eosinophil pathways is to repel invasion by parasitic worms; the eosinphils in such cases serve as the major killer cells against the worms.

Autoimmune Disease

Given the huge diversity of recognition sites for antigens on lymphocytes, why does the immune system not attack the body's cells? Put in a different way, how does the immune system discriminate between self and nonself antigens? In some cases, the self molecules are located deep within cells and tissues and never make contact with lymphocytes. In other cases, the self molecules are not associated with HLA antigens and so fail to stimulate lymphocytes. These two cases offer no problem, but many other self antigens are accessible to lymphocytes and do associate with HLA antigens. It is thought that during in utero and early postnatal life, the clones of lymphocytes that would recognize these self molecules are either destroyed outright or sur-

vive but are repressed by suppressor T cells. How these events occur is not understood, although there is general agreement that they occur in the thymus.

Unfortunately, the body does, all too often, produce antibodies or sensitized T cells against its own tissues, the result being cell damage or alteration of function. A growing number of human diseases are being recognized as **autoimmune** in origin.

There are multiple potential causes for the body's failure to recognize its own cells: (1) normal antigens may be altered by combination with drugs or environmental chemicals; (2) the cell may be infected by a virus whose nucleic acid codes for a new protein (antigen); (3) genetic mutations may yield new antigens; (4) the body may encounter microbes whose antigens are so close in structure to certain self antigens that the antibodies or sensitized lymphocytes produced against these antigens cross-react with the self antigens; and (5) suppressor T cells may normally prevent the production of autoantibodies or activated, cytotoxic T cells, and a deficiency of these suppressor T cells could contribute to the development of autoimmunity. This list of possibilities is by no means complete, but whatever the cause, a breakdown in self-recognition results in turning the body's immune mechanisms against its own tissues.

The above description centers on the production of antibodies or sensitized lymphocytes against the body's own cells. However, autoimmune damage may also be brought about in several other quite different ways. An overzealous response (too much generation of complement or release of chemicals from platelets, neutrophils, or sensitized lymphocytes) may cause damage not only to invading foreign cells but to neighboring normal cells or membranes as well. For example, were a circulating antigen-antibody complex to be trapped within capillary membranes, the generation of complement or release of chemicals into the area might cause damage to the adjacent membranes. As might be predicted, the kidney glomeruli, with their large filtering surface, are prime targets for such so-called **immune-complex disease**.

Another important question in the field of autoimmunity concerns the relationship between mother and fetus: Why does the mother's immune system not reject the fetus (half of the fetal antigens are, of course, paternal and therefore foreign to the mother)? No generally accepted explanation is available at present.

Summary

Table 17-6 presents a summary of immune mechanisms in the form of a miniglossary of cells and chemical mediators involved in immune functions. All the material in this table has been covered in this chapter.

TABLE 17-6. Miniglossary of cells and chemical mediators involved in immune functions

---

Cells

B cells (B lymphocytes). lymphocytes which, upon activation, proliferate and differentiate into antibody-secreting plasma cells

cytotoxic T cells. class of T lymphocytes which, upon activation by specific antigen, directly attack the cell bearing that type of antigen

eosinophils. polymorphonuclear granulocytic leukocytes involved in allergic responses and destruction of parasitic worms

helper T cells. class of T cells which enhance antibody production and cytotoxic T-cell function

killer (K) cells. class of cells (probably lymphocytes) which are brought into action by the presence of antibody bound to antigen on the surface of a foreign cell and which kill the cell directly without prior phagocytosis

lymphocytes. type of leukocyte responsible for specific immune defenses; categorized mainly as B cells and T cells

macrophages. cell type which functions as a phagocyte, processes and presents antigen to lymphocytes, and secretes chemicals involved in inflammation, proliferation of activated lymphocytes, and total-body responses to infection or injury

memory cells. B and T cells produced during an initial infection and which respond rapidly during a subsequent exposure to the same antigen

monocytes. type of leukocyte; leaves the bloodstream and is transformed into a macrophage

natural killer (NK) cells. class of cells (probably lymphocytes) which bind relatively nonspecifically to cells bearing foreign antigens and kill them directly; no prior exposure to the antigen is required

neutrophils. polymorphonuclear granulocytic leukocytes which function as phagocytes and also release chemicals involved in inflammation

plasma cells. cells which are derived from activated B lymphocytes and which secrete antibodies

suppressor T cells. class of T cells which inhibit antibody production and cytotoxic T-cell function

T cells (T lymphocytes). lymphocytes derived from precursors that at one time were in the thymus; see cytotoxic T cells, helper T cells, and suppressor T cells

---

Chemical Mediators

---

antibody. specialized protein secreted by plasma cells and capable of combining with the specific antigen which stimulated its production

"blocking" antibodies. antibodies whose production is induced by cancer cells or tissue transplants and which block the killing of these cells by cytotoxic T cells

C_1. The first protein in the classical complement pathway

C_3 convertase. enzyme in the complement pathway which cleaves complement C_3 and initiates the rest of the cascade

chemotaxin. general name given to any chemical mediator which stimulates chemotaxis of neutrophils or other phagocytes

complement. group of plasma proteins which, upon activation, kill microbes directly and facilitate every step of the inflammatory process, including phagocytosis; the classical

complement pathway is triggered by antigen-antibody complexes, whereas the alternate pathway operates independently of antibody

endogenous pyrogen (EP). protein secreted by monocytes and macrophages, and which acts in the brain to cause fever; probably the same molecule as LEM and IL-1

histamine. inflammatory mediator secreted mainly by mast cells; acts on microcirculation to cause vasodilation and increased permeability to protein

IgA. class of antibodies secreted by the lining of the body's various "tracts"

IgD. class of antibodies whose function is unknown

IgG. most abundant class of plasma antibodies

IgE. class of antibodies which mediates immediate hypersensitivity ("allergy")

TABLE 17-6. (continued)

Chemical Mediators

IgM. class of antibodies that along with IgG provide the bulk of specific humoral immunity against bacteria and viruses

immunoglobulin (Ig). synonym for "antibody"; the five classes are IgG, IgM, IgA, IgE, and IgD

interferon. family of proteins which nonspecifically inhibit viral replication and also stimulate the activity of cytotoxic T cells, NK cells, and macrophages

interleukin 1 (IL-1). protein secreted by macrophages which stimulates activated B cells to proliferate and helper T cells to secrete interleukin 2; probably the same molecule as LEM and EP

interleukin 2 (IL-2). protein secreted by helper T cells which causes activated T cells to proliferate; its secretion is stimulated by interleukin 1

kinins. peptides split from kininogens in inflamed areas and which facilitate the vascular changes associated with inflammation; may also activate neuronal pain receptors

lactoferrin. protein secreted by neutrophils and which tightly binds iron

leukocyte endogenous mediator (LEM). protein secreted by monocytes and macrophages which induces decreases in plasma zinc and iron, release of acute-phase proteins, and stimulation of granulocyte production; probably the same molecule as EP and IL-1

lymphokines. general term denoting all nonantibody chemical messengers secreted by lymphocytes

membrane attack complex (MAC). group of complement proteins which imbed in the surface of a microbe and kill it

natural antibodies. antibodies to the erythrocyte antigens not present in a person's body; these antibodies are present without prior exposure to the antigen

opsonin. general name given to any chemical mediator which promotes phagocytosis

prostaglandins. mediators produced by many cells, including mast cells, and which facilitate inflammation

SECTION B.
METABOLISM OF FOREIGN CHEMICALS

The body is exposed to a huge number of environmental chemicals, including inorganic nonnutrient elements, naturally occurring fungal and plant toxins, and synthetic chemicals. This last category is by far the largest, since there are now more than 10,000 foreign chemicals being commercially synthesized (over 1 million have been synthesized at one time or another); these are "foreign" in the sense that they are not normally found in nature. These foreign chemicals inevitably find their way into the body, either because they are purposely administered, as drugs (medical or "recreational"), or simply because they are in the air, water, and food we use.

As described in section A of this chapter, foreign materials can induce inflammation and specific immune responses. However, these defenses are directed mainly against foreign cells, and although noncellular foreign chemicals can also elicit certain of them (as in allergy, for example), such immune responses do not constitute the major defense mechanisms against most foreign chemicals. Rather, molecular alteration (**biotransformation**) and **excretion** constitute the primary mechanisms.

A central focus for the body's handling of any foreign chemical is those factors which determine the effective concentration of the chemical at its sites of action (Fig. 17-24). First, the chemical must gain entry to the body through the gastrointestinal tract, lungs, or skin (or placenta in the case of a fetus). Accordingly, its ability to move across these barriers will have an important influence on its blood concentration. But as Fig. 17-24 illustrates, the rate of entry into the body is only one of many factors determining the concentration of the chemical at its site of action. Once in the blood, the chemical may become bound reversibly to plasma proteins or to erythrocytes; this lowers its free concentration and, thereby, its ability to alter cell function. It may accumulate in storage depots (for example, DDT in fat tissue) or it may undergo biotransformation. The metabolites resulting from

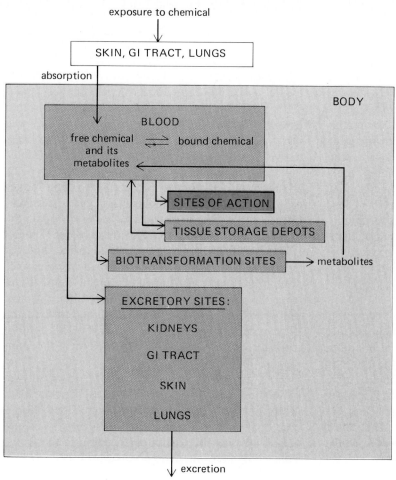

FIGURE 17-24. Metabolic pathways for foreign chemicals.

these latter enzyme-mediated reactions themselves enter the blood and are subject to the same fates as the parent molecules. Finally, the foreign chemical and its metabolites may be eliminated from the body in the urine, expired air, skin secretions, or feces (having entered the feces by biliary secretion).

Absorption

In practice, most organic molecules move through the lining of some portion of the gastrointestinal tract fairly readily, either by simple diffusion or by carrier-mediated transport. This should not be sur-

prising, since the gastrointestinal tract evolved to favor absorption of the wide variety of nutrient molecules in the environment; the nonnutrient chemicals are the beneficiaries of these relatively nondiscriminating transport mechanisms.

The lung alveoli are highly permeable to most organic chemicals and therefore offer an easy entrance route for airborne chemicals. They are also an important entry site for airborne metals, which generally penetrate the gastrointestinal tract very poorly. One important aspect of absorption through the lungs is that the liver does not get first crack at the chemical (as it does when entry is via the gastrointestinal tract); by the same token, any chemical which is toxic to the liver is not as dangerous when it enters via the lungs.

Lipid solubility is all-important for entry through the skin, so that this route is of little importance for charged molecules but can be used by oils, steroids, and other lipids.

The penetration of the placental membranes by foreign chemicals is one of the most important fields in toxicology, since the effects of environmental agents on the fetus during critical periods of development may be quite marked and, in many cases, irreversible. Diffusion is an important mechanism for lipid-soluble substances, and carrier-mediated systems (which evolved for the carriage of endogenous nutrients) may be usurped by foreign chemicals to gain entry into the fetus.

Storage Sites

The major storage sites for foreign chemicals are cell proteins, bone, and fat. The chemical bound to cell or bone proteins or dissolved in the fat is in equilibrium with the free chemical in the blood, so that an increase in blood concentration causes more movement into storage (up to the point of saturation). Conversely, as the chemical is eliminated from the body and its blood concentration falls, movement occurs out of storage sites.

These storage sites obviously are a source of protection, but it sometimes happens that the storage sites accumulate so much chemical that they become damaged themselves.

Excretion

To appear in the urine, a chemical must either be filtered through the glomerulus or secreted across the tubular epithelium (Chap. 13). Glomerular filtration is, as emphasized in Chap. 13, a bulk-flow process so that all low-molecular-weight substances in plasma undergo filtration; accordingly, there is considerable filtration of most environmental chemicals except for those which are bound to plasma proteins or erythrocytes. (Note that such binding

is, therefore, a mixed blessing in that it reduces toxicity but impedes excretion). In contrast, tubular secretion is by discrete transport processes, and many environmental chemicals (penicillin is a good example) utilize the mediated-transport systems available for naturally occurring substances.

Once in the tubular lumen, either via filtration or tubular secretion, the foreign chemical may still not be excreted, for it may be reabsorbed back across the tubular epithelium into the blood. This is a major problem, since so many foreign chemicals are highly lipid-soluble; as the filtered fluid moves along the renal tubules, these molecules passively diffuse along with reabsorbed water through the tubular epithelium and back into the blood. The net result is that little is excreted in the urine, and the chemical is retained in the body. If these chemicals could be transformed into more polar (and, therefore, less lipid-soluble) molecules, their passive reabsorption from the tubule would be retarded and they would be excreted more readily. This type of transformation is precisely what occurs in the liver, as described in the next section.

An analogous problem exists for those foreign molecules (and trace metals) secreted in the bile. Many of these substances, having reached the lumen of the small intestine, are absorbed back into the blood, thereby escaping excretion in the feces. This cyclic enterohepatic circulation was described in Chap. 14.

Biotransformation

The metabolic alteration of foreign molecules occurs mainly in the liver (but to some extent also in kidney, skin, placenta, and other organs). A large number of distinct enzymes and pathways are involved, but the common denominator of most of them is that they transform chemicals into more polar, less lipid-soluble substances. One consequence of this transformation is that the chemical may be rendered less toxic, but this is not always so. The second, more important, consequence is that its tubular reabsorption is diminished and urinary excretion facilitated. Similarly, for substances handled by biliary excretion, gut absorption of the metabolite is less likely so that fecal excretion is also enhanced.

The hepatic enzymes which perform these transformations are called the **microsomal enzyme system (MES)** and are located mainly in the smooth endoplasmic reticulum. One of the most important facts about this enzyme system is that it is easily inducible, i.e., the number of these enzymes can be greatly increased by exposure to a chemical which acts as a substrate for the system.

However, all is not really so rosy, for the hepatic biotransformation mechanisms vividly demonstrate how an adaptive response may, under some circumstances, turn out to be maladaptive. These enzymes all too frequently "toxify" rather than "detoxify" a drug or pollutant; in fact many foreign chemicals are quite nontoxic until the liver enzymes biotransform them. Of particular importance is the likelihood that many chemicals which cause cancer do so only after biotransformation. For example, a major component of cigarette smoke and charcoal-broiled foods is transformed by the MES into a carcinogenic compound.

The MES can also cause problems in another way, because they evolved primarily not to defend against foreign chemicals (which were much less prevalent during our evolution) but rather to metabolize endogenous substrates, particularly steroids and other lipid-soluble molecules. Therefore, their induction by a drug or pollutant increases metabolism not only of that drug or pollutant, but of the endogenous substrates as well. The result is a decreased concentration in the body of that normal substrate.

Another fact of great importance concerning the microsomal enzyme system is that just as certain chemicals induce it, others inhibit it. The presence of such chemicals in the environment could have deleterious effects on the system's capacity to protect against those chemicals it transforms. (Just to illustrate how complex this picture can be, note that any chemical which inhibits the microsomal enzyme system may actually confer protection against those other chemicals which must undergo transformation in order to become toxic.)

Alcohol: An example

Ethyl alcohol offers an excellent example of the role biotransformation plays in determining a substance's toxicity and its influence on the body's responses to other chemicals. The overuse of alcohol is associated with liver damage, and for many years it was thought that this damage was due to the malnutrition so frequently accompanying alcoholism. It is now clear that, although severe malnutrition may play some role, the toxic damage to the liver is caused mainly by the metabolites of alcohol, produced by the liver cells, themselves.

Alcohol is initially broken down by liver cells to hydrogen and acetaldehyde, and these seem to be the major culprits. Acetaldehyde damages mitochondria. The excess hydrogen exerts its damaging effect more subtly by influencing metabolism so as to cause the accumulation of fat in liver cells, which damages them. (We have restricted our discussion to the toxic effects of hydrogen and acetaldehyde on liver cells; other organs and tissues are also damaged by them in a variety of ways.)

The metabolism of alcohol not only damages the liver cells but leads to marked changes in the metabolism of other drugs and foreign chemicals, all explainable by the fact that these other agents share certain of alcohol's metabolic pathways. First, when alcohol is taken at the same time as another drug, such as a barbiturate, enhanced effects of the two drugs are observed. This is because the alcohol and barbiturate compete for the same hepatic microsomal enzyme system, resulting in a decreased rate of catabolism of both (and, therefore, increased blood concentrations).

The situation just described was for simultaneous administration of the two drugs and is independent of whether the individual is a chronic overuser of alcohol or not. Let us look now at the long-term effects of chronic overuse. Alcohol is a powerful inducer of the microsomal enzyme system, so that its chronic presence causes an increase in the system's activity. The result is an increase in the rate of catabolism of other chemicals, such as barbiturates, which share the enzymes. Thus, the chronic use of alcohol decreases the potency of any given dose of barbiturate by causing its blood concentration to fall more rapidly.

Putting these last two paragraphs together, one can see that the effect of alcohol on the blood concentrations of other chemicals depends both on whether the two drugs are taken simultaneously and whether the person is a chronic overuser of alcohol. But the situation is even more complex, for if the person has suffered serious liver damage due to chronic alcoholism, then the microsomal enzyme system, instead of being induced, may be inadequate simply because the cells have been damaged. At this stage, then, the metabolism of the other agents will be diminished and their effects will be exaggerated.

Finally, we have presented the chronic effects of alcohol only in terms of the effects on the metabolism of other chemicals. It should be clear, however, that similar effects are exerted on the metabolism of alcohol itself. Thus, chronic overuse, before significant damage has occurred, results in increased catabolism (because of the induction of the microsomal enzymes), and this accounts for much of the "tolerance" to alcohol, i.e., the fact that increasing doses must be taken to achieve a given magnitude of effect. Once severe damage has occurred the decline in rate of catabolism may lead to increased sensitivity to a given dose, just the opposite of tolerance.

Inorganic Elements

Thus far our discussion of environmental chemicals has dealt mainly with organic molecules. Many potentially dangerous substances are not organic chemicals but are inorganic elements (like mercury and lead) present in excess in the environment because of human activity. As is true for organic drugs and pollutants, the concentrations of inorganic elements at their sites of action depend on rates of entry, excretion, and storage. The gastrointestinal tract offers an important first line of defense since absorption of them is usually quite limited. In contrast, airborne inorganic elements gain entry to the blood more readily through the lungs. Little is known of the mechanisms by which the kidneys and liver handle potentially harmful trace elements; specifically, it is not known whether adaptive increases in excretion are induced by exposure to the element.

SECTION C.
RESISTANCE TO STRESS

Much of this book has been concerned with the body's response to stress in its broadest meaning of an environmental change which must be adapted to if health and life are to be maintained. Thus, any change in external temperature, water intake, etc., sets into motion mechanisms designed to prevent a significant change in some physiological variable. In this section, however, we describe the basic stereotyped response to **stress** in the more limited sense of noxious or potentially noxious stimuli. These comprise an immense number of situations, including physical trauma, prolonged heavy exercise, infection, shock, decreased oxygen supply, prolonged exposure to cold, pain, fright, and other emotional stresses. It is obvious that the overall response to cold exposure is very different from that to infection, but in one respect the response to all these situations is the same: Invariably, secretion of **cortisol** is increased; indeed, the term "stress" has come to mean to physiologists any event which elicits increased cortisol secretion (by the **adrenal cortex**). Also, **sympathetic nervous activity** is usually increased.

Historically, activation of the sympathetic nervous system was the first overall response to stress to be recognized and was labeled the **fight-or-flight** response. Only later did further work clearly establish the contribution of the adrenal cortical response. The increased cortisol secretion is mediated entirely by the hypothalamus–anterior-pituitary system described in Chap. 9. As illustrated in Fig. 17-25, afferent input to the hypothalamus induces secretion of **corticotropin releasing hormone,** which is carried by the hypothalamo-pituitary portal vessels to the anterior pituitary and stimulates **ACTH** release. The ACTH, in turn, circulates to the adrenal and stimulates cortisol release. As described in Chap. 9, the hypothalamus receives input from virtually all areas of the brain and receptors of the body, and the pathway involved in any given situation depends upon the nature of the stress, e.g., pathways from other brain centers mediate the response to emotional stress. The destination is always the same, namely, synaptic connection with the hypothalamic neurons which secrete corticotropin releasing hormone. These same pathways also converge on the brain areas which control sympathetic nervous activity (including release of **epinephrine** from the **adrenal medulla**).

Functions of Cortisol in Stress

The effects of increased cortisol on organic metabolism were described in Chap. 15. To reiterate, cortisol (1) stimulates protein catabolism; (2) stimulates liver uptake of amino acids and their conversion to glucose (gluconeogenesis); and (3) inhibits glucose uptake and oxidation by many body cells ("insulin antagonism") but not by the brain. These effects are ideally suited to meet a stressful situation. First, an animal faced with a potential threat is usually forced to forego eating, and these metabolic changes are essential for survival during fasting. Second, the amino acids liberated by catabolism of body protein stores not only provide energy, via gluconeogenesis, but also constitute a potential source of amino acids for tissue repair should injury occur.

A few of the many medically important impli-

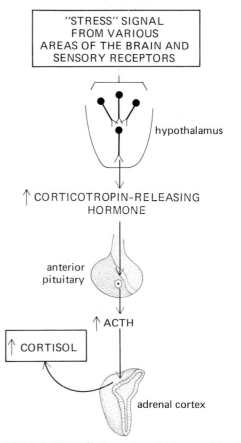

FIGURE 17-25. Pathway by which stressful stimuli elicit increased cortisol secretion.

cations of these cortisol-induced metabolic effects associated with stress are as follows: (1) Any patient ill or subject to surgery catabolizes considerable quantities of body protein; (2) a diabetic who suffers an infection requires much more insulin than usual; and (3) a child subject to severe stress of any kind manifests retarded growth. The explanations for these phenomena should be evident.

Cortisol has important effects other than those on organic metabolism. One of the most important is that of enhancing **vascular reactivity**. A patient lacking cortisol faced with even a moderate stress may develop hypotension and die if untreated. This is due primarily to a marked decrease in total peripheral resistance. For unknown reasons, stress would induce widespread arteriolar dilation, despite massive sympathetic nervous system discharge, if large amounts of cortisol were not present. A large part of its counteracting effect is ascribable to the fact that moderate amounts of cortisol permit norepinephrine to induce vasoconstriction, but this can be only part of the story since considerably larger amounts of cortisol are required to prevent stress-induced hypotension completely. In other words, the normal cardiovascular response to stress requires *increased* cortisol secretion, not just permissive quantities.

Thus far we have presented the adaptive value of the stress-induced cortisol increase mainly in terms of its role in preparing the body physically for fight or flight, and there is no doubt that cortisol does function importantly in this way. However, in recent years, it has become apparent that cortisol may have other important functions. Table 17-7 is a partial list of the large variety of psychosocial situations demonstrated to be associated with increased cortisol secretion. Common denominators of many of them are novelty and challenge. Of great interest, therefore, are recent experiments which

---

TABLE 17-7. Psychosocial situations shown to be associated with increased plasma concentration or urinary excretion of adrenal cortical steroids

---

Experimental animals

---

1. Any "first experience" characterized by novelty, uncertainty, or unpredictability
2. Conditioned emotional responses; anticipation of something previously experienced as unpleasant
3. Involvement in situations in which the animal must master a difficult task in order to avoid or forestall aversive stimuli (the animal must really be "trying")
4. Situations in which long-standing rules are suddenly changed so that previous behavior is no longer effective in achieving a goal
5. Socially subordinate animals (dominant animals have decreased cortisol)
6. Crowding (increased social interactions)
7. Fighting or merely observing other animals fighting

Human beings

---

1. Normal persons
   a. Acute situations
      (1) Aircraft flight
      (2) Awaiting surgical operation
      (3) Final exams (college students)
      (4) Novel situations
      (5) Competitive athletics
      (6) Anticipation of exposure to cold
      (7) Workdays, compared to weekends
      (8) Many job experiences
   b. Chronic life situations
      (1) Predictable personality-behavior profile: aggressive, ambitious, time-urgency
      (2) Discrepancy between levels of aspiration and achievement
   c. Experimental techniques
      (1) "Stress" or "shame" interview
      (2) Many motion pictures
2. Psychiatric patients
   a. Acute anxiety
   b. Depression, but only when patient is aware of and involved in a struggle with it

---

suggest that cortisol affects memory in experimental animals. Even more striking, ACTH (independent of its stimulation of cortisol secretion) is one of the peptides which facilitates learning and memory (Chap. 20). Thus, it may well be that the rise in ACTH secretion induced by psychosocial stress helps one to cope with the stress by facilitating the learning of appropriate responses.

Cortisol's pharmacological effects and disease

There are several situations in which adrenal corticosteroid levels in human beings become abnormally elevated. Patients with excessively hyperactive adrenals (there are several causes of this disease) represent one such situation, but the common occurrence is that of steroid administration for medical purposes. When corticosteroids are present in very high concentration, the previously described effects on organic metabolism are all magnified, but in addition there may appear one or more new effects, collectively known as the **pharmacological effects** of cortisol and closely related steroids. The most obvious is a profound reduction in the inflammatory response to injury or infection (indeed, reducing the inflammatory response in allergy, arthritis, other diseases, and transplantation rejection is the major reason for administering the cortisol to patients). Large amounts of cortisol inhibit almost every step of inflammation (vasodilation, increased vascular permeability, phagocytosis), destroy lymphocytes in lymphoid tissues, and decrease antibody production. Unfortunately, such therapy also decreases the ability of the person to resist infections. In addition, large amounts of cortisol may accelerate development of hypertension, atherosclerosis, and gastric ulcers, and may interfere with normal menstrual cycles.

As emphasized above, these pharmacological effects are known to be elicited when cortisol levels are extremely elevated. An unsettled question of great importance is whether long-standing lesser elevations of cortisol may do the same thing, albeit more slowly and less perceptibly. Put in a different way, do the psychosocial stresses, noise, etc., of everyday life contribute to disease production via increased cortisol?

Functions of the Sympathetic Nervous System in Stress

A list of the major effects of increased general sympathetic activity, including secretion of adrenal medullary hormones, almost constitutes a guide on how to meet emergencies. Since all these actions have been discussed in other sections of the book, they are listed here with little or no comment:

1. Increased hepatic and muscle glycogenolysis (provides a quick source of glucose)
2. Increased breakdown of adipose tissue triacylglycerol (provides a supply of glycerol for gluconeogenesis and of fatty acids for oxidation)
3. Decreased fatigue of skeletal muscle
4. Increased cardiac output secondary to increased cardiac contractility and heart rate
5. Shunting of blood from viscera to skeletal muscles by means of vasoconstriction in the former beds and vasodilation in the latter
6. Increased ventilation
7. Increased coagulability of blood

The adaptive value of these responses in a fight-or-flight situation is obvious. But what purpose do they serve in the psychosocial stresses so common to modern life when neither fight nor flight is appropriate? As for cortisol, a question yet to be answered is whether certain of these effects, if prolonged, might not enhance the development of certain diseases, particularly atherosclerosis and hypertension. For example, one can easily imagine the increased blood lipid concentration and cardiac work contributing to the former disease. Considerable work remains to be done to evaluate such possibilities.

Other Hormones Released During Stress

Other hormones which are usually released during many kinds of stress are aldosterone, vasopressin (ADH), growth hormone, and glucagon (insulin is usually decreased). The increases in vasopressin and aldosterone ensure the retention of water and sodium within the body, an important adaptation in the face of potential losses by hemorrhage or sweating (vasopressin may also influence learning). As described in Chap. 15, the overall effects of the changes in growth hormone, glucagon, and insulin are, like those of cortisol and epinephrine, to mobilize energy stores.

This list of hormones whose secretion rates are altered by stress is by no means complete. It is likely that the secretion of almost every known hormone may be influenced by stress. For example, prolactin and thyroid hormone are often increased, whereas the pituitary gonadotropins (LH and FSH) and the sex steroids (testosterone or estrogen) are decreased. The adaptive significance of many of these changes is unclear.

The secretion of endorphin and B-lipotropin is also increased during stress. As described in Chap.

9, these substances are secreted from the anterior pituitary along with ACTH. Endorphin is a potent endogenous opiate, and its possible role in mediating analgesia and mood alterations in stress is a subject of great interest. Moreover, endorphin has many other effects as well; for example, it decreases appetite.

Finally, we would like to raise once again the potential role of the EP/LEM/IL-1 hormone(s) secreted by stimulated macrophages. To the extent that any particular stress actually causes significant tissue damage, it will trigger release of these proteins, which exert the far-reaching protective effects described earlier. A question for future study is whether psychological stresses or physical stresses that cause little, if any, tissue damage can also trigger this system, perhaps via hormonal input to the macrophages.

THE SENSORY SYSTEMS

A human being's awareness of the world is determined by the physiological mechanisms involved in the processing of afferent information by the nervous system. The initial step of this processing is the conversion of stimulus energy into action potentials in nerve fibers. This code represents information from the external world even though, as is frequently the case with symbols, it differs vastly from what it represents.

Intuitively, it might seem that sensory systems operate like familiar electrical equipment, but this is true only up to a point. As an example, let us compare telephone transmission with our auditory sensory system. The telephone changes sound waves into electric impulses, which are then transmitted along wires to the receiver; thus far the analogy holds. (Of course, the mechanisms by which electric currents and action potentials are transmitted are quite different, but this does not affect our argument.) The telephone then changes the coded electric impulses *back into sound waves*. Here is the crucial difference, for our brain does not physically translate the code into sound; rather the coded information itself or some correlate of it is what we perceive as sound.

Afferent information may or may not have a conscious correlate; i.e., it may or may not give rise to a conscious awareness of the physical world. Afferent information which does have a conscious correlate is called **sensory information**, and the conscious experience of objects and events of the external world which we acquire from the neural processing of sensory information is called **perception**. At present there is absolutely no understanding of how coded action potentials or composites of them become conscious sensations.

Neural Pathways in Sensory Systems

We mentioned in Chap. 8 that afferent information from receptors travels over discrete neural pathways beginning with afferent neurons, passing to second-order neurons, on to third-order neurons, and so on, along the synaptically interconnected chains of interneurons which make up the neural pathways in the sensory systems.

Some of the interneurons at each stage along these neural pathways are influenced by information originating at only a single receptor type. Thus, one interneuron may only be influenced by information from afferent neurons having mechanoreceptors at their peripheral endings, whereas another may be influenced only by information from temperature-sensitive afferent neurons. The ascending spinal cord and brain pathways influenced by such interneurons retain the same specificity in that they carry information that pertains to only one sensory modality. These are known as the **specific pathways** ("labeled lines," as it were) for the different modalities (Fig. 18-1A). Many parallel chains of interneurons are grouped together in each specific pathway as it ascends through the central nervous system. Each chain consists of three to five synaptically connected neuronal links.

The specific pathways (except for the olfactory pathways) pass to the brainstem and thalamus, the final neurons in the pathways going to different areas of the cerebral cortex (Fig. 18-2), collectively known as **primary sensory cortex**. For example, the

fibers transmitting information from somatic receptors (touch, temperature, etc.) synapse at cortical levels in **somatosensory cortex** (the primary somatic sensory area), a strip of cortex which lies in the parietal lobe just behind the junction between the parietal and frontal lobes. (The word "somatic" in this context refers to the framework or outer walls of the body, including skin, skeletal muscle, tendons, and joints, as opposed to the viscera, i.e., the organs in the thoracic and abdominal cavities.)

The specific pathways which originate in receptors of the taste buds, after synapsing in the brainstem and thalamus, probably pass to cortical areas adjacent to the face region of the somatosensory strip. The specific pathways from the eyes, ears, and nose do not pass to the somatosensory cortex but go instead to other primary cortical receiving areas (Fig. 18-2). The pathways subserving olfaction are different from all the others in that

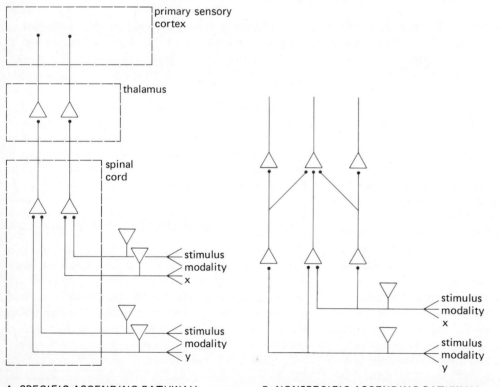

A SPECIFIC ASCENDING PATHWAY   B NONSPECIFIC ASCENDING PATHWAY

FIGURE 18-1. Diagrammatic representation of (A) specific and (B) nonspecific ascending pathways.

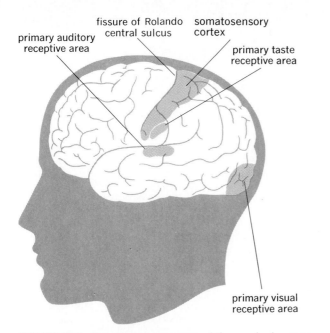

fissure of Rolando central sulcus
primary auditory receptive area
somatosensory cortex
primary taste receptive area
primary visual receptive area

FIGURE 18-2. Primary sensory areas of the cerebral cortex.

they do not pass through thalamus and have no representation in cerebral cortex; they pass instead into parts of the limbic system (see Fig. 8-12).

However, these specific pathways are not the only ascending pathways. In contrast to the specific pathways, neurons in the **nonspecific pathways** are activated by sensory units of several different types and therefore signal only general information about the level of excitability; i.e., they indicate that *something* is happening, usually without specifying just what (or where). A given neuron in the non-specific pathways may respond, for example, to maintained skin pressure, heating, cooling, skin stretch, and other stimuli applied to afferent nerves (Fig. 18-1B). Such neurons, which respond to different kinds of stimuli, are called **polymodal**. The nonspecific pathways feed into the brainstem reticular formation and regions of the thalamus and cortex which are not highly discriminative.

Finally, the processing of afferent information does not end in primary sensory cortex or lower way stations. Information goes from these areas to so-called association areas of cerebral cortex, as will be discussed at the end of this chapter.

Somatic Sensation

The different kinds of somatic receptors respond to mechanical stimulation of the skin or hairs and underlying tissues, rotation or bending of joints, temperature changes, and some chemical changes. Their activation gives rise to the sensations of touch, pressure, heat, cold, the awareness of the position and movement of the parts of the body, or pain. Each of these sensations is associated with a specific type of receptor; i.e., there are distinct receptors for heat, cold, touch, pressure, joint position, and pain. (Recall that by receptor we mean the specialized peripheral ending of an afferent nerve fiber or the specialized receptor cell associated with it.)

After entering the central nervous system, the afferent nerve fibers from the somatic receptors synapse on interneurons which form the specific pathways going to somatosensory cortex via the brainstem and thalamus as well as upon interneurons which form the nonspecific pathways. The specific pathways for somatic sensation cross in the spinal cord or brainstem from the side of afferent-neuron entry to the opposite side of the central nervous system; thus, the somatic sensory pathways from receptors on the left side of the body go to the somatosensory cortex of the right cerebral hemisphere and vice versa.

In somatosensory cortex, the endings of the individual fibers of the specific somatic pathways are grouped according to the location of the receptors. The pathways which originate in the foot terminate nearest the longitudinal dividing line between the right and left cerebral hemispheres. Passing laterally over the surface of the brain, one finds the terminations of the pathways from leg, trunk, arm, hand, face, tongue, throat, and viscera (Fig. 18-3). Within somatosensory cortex there are several distinct regions lying next to one another. We assume that these subdivisions of somatosensory cortex serve different functions because their neurons have different response characteristics. For example, the neurons in one region respond to stimulation of rapidly adapting receptors, whereas those in another region respond to slowly adapting

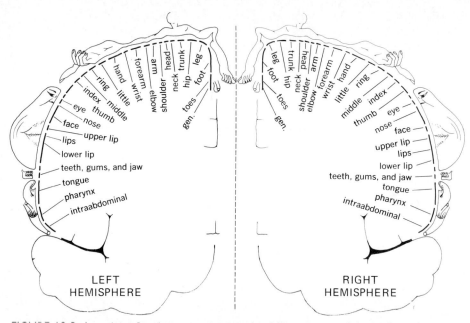

FIGURE 18-3. Location of pathway terminations for different parts of the body in the somatosensory cortex. The left half of the body is represented on the right hemisphere of the brain, and the right half of the body is represented on the left cerebral hemisphere.

ones; one region responds to stimulation of the surface of the skin, another to stimulation of deeper tissues; one responds to stationary stimuli, another responds better to moving stimuli, particularly to stimuli moving across the surface of the skin in one certain direction. The parts of the body with the greatest sensitivity (fingers, thumb, and lips) are represented by the largest areas of somatosensory cortex.

Touch-pressure

The mechanoreceptors which are sensitive to touch-pressure are classified into two major categories according to the nature of their response to sustained pressure. About half the receptors are **rapidly adapting** and respond with a burst of action potentials only at the onset and removal of the stimulus, i.e., when the stimulus is moving. The other receptors are **slowly adapting** and respond with a sustained discharge throughout the duration of the stimulus. Within each of these two categories

are two other categories: Some of both the rapidly and slowly adapting receptors have small, well-defined receptive fields and are able to provide precise information about the contours of objects indenting the skin. As might be expected, these receptors are concentrated at the finger tips, which are used for the exploration of objects by touch. (In fact, a single impulse from such a receptor on the finger tips can reach consciousness.) In contrast, the other receptors (both slowly and rapidly adapting) have large receptive fields, sometimes covering a whole finger or a large part of the palm, with obscure boundaries; these cells are not involved in detailed spatial discrimination but signal information about vibration, stretch of the skin, and movement of the joints.

Specificity of receptors is often determined by the characteristics of the surrounding tissue, and the **pacinian corpuscle**, discussed in Chap. 8 is an example of this. The nerve terminal of the pacinian corpuscle, which is a rapidly adapting receptor, is surrounded by alternating layers of cells and extracellular fluid in such a way that a mechanical stimulus reaches the enclosed nerve terminal only after displacing the layers of the surrounding capsule (see Fig. 8-48). Rapidly applied pressures are transmitted to the nerve terminal without delay and give rise to a receptor potential by the mechanisms described in Chap. 8. In contrast, the energy of slowly administered or sustained forces is, in part, absorbed by the elastic-tissue components of the capsule and is therefore partially dissipated before it reaches the nerve terminal at the core. Thus the receptor fires only at the fast onset—and perhaps again at the release—of a mechanical stimulus but not under sustained pressure. What the pacinian corpuscles signal is not pressure, but *changes* in pressure with time. They can effectively discriminate stimuli vibrating up to frequencies of 300 Hz (cycles per sound).

Slowly adapting mechanoreceptors provide information about both the rate of pressure application and the stimulus intensity. A person's experience of the intensity of a stimulus, however, is not related only to the activity from individual slowly adapting receptors; it is also influenced by the recruitment of other receptors and by processing in the central nervous system.

Proprioception and kinesthesia

Proprioception is the sense of the position of the body in space, whereas **kinesthesia** commonly refers to sensations associated with joint position and movement. Both types of information arise, at least in part, from the stimulation of mechanoreceptors in the joints, tendons, and muscles. Slowly adapting receptors are particularly responsive to movements of the limbs, probably because of their great sensitivity to the changing patterns of tension in the skin which occur with such movements. Figure 18-4 shows the response of a single receptor in a joint during flexion and extension of the joint. Different joint receptors respond oppositely, firing faster during extension and slower during flexion.

Other receptors implicated in proprioception and kinesthesia will be discussed in Chap. 19 (the muscle spindle and the Golgi tendon organ) or later

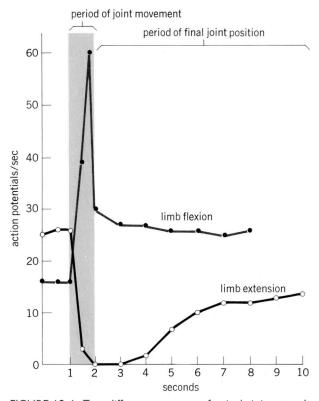

FIGURE 18-4. Two different responses of a single joint stretch receptor to movements in opposite directions, the upper curve during flexion and the lower curve during extension. *(Redrawn from Boyd and Roberts.)*

in this chapter (the receptors of the eyes and vestibular apparatus).

Temperature

The thermoreceptors in the skin are classified according to their responses as cold and warm receptors, even though distinguishing anatomical structures have not been identified for either type. Warm receptors respond to temperatures between 30 and 43°C, increase their discharge upon warming, and show a temporary inhibition upon cooling, whereas cold receptors are stimulated by temperatures between 35 and 20°C and increase their discharge upon cooling. The way temperatures alter the endings of the thermosensitive afferent neurons to cause action potentials to be generated is not known.

Pain

A stimulus which causes (or is on the verge of causing) tissue damage usually elicits a sensation of pain, which can be identified in terms of its location, intensity, and duration. But pain differs from the other somatosensory modalities because, in addition to the perceived sensation, there occurs a reflex escape or withdrawal response as well as a gamut of physiological changes similar to those elicited during fear, rage, and aggression. These changes, mediated by the sympathetic nervous system, include faster heart rate, higher blood pressure, greater secretion of epinephrine into the bloodstream, increased blood glucose concentration, less gastric secretion and motility, decreased blood flow to the viscera and skin, dilated pupils, and sweating. In addition, a painful stimulus elicits feelings of fear, anxiety, and unpleasantness. Probably more than any other type of sensation, the experience of pain can be altered by past experiences, suggestion, emotions (particularly anxiety), and the simultaneous activation of other sensory modalities. Thus, painfulness is not a physical property of the stimulus (as, for example, temperature or shape would be); rather, painfulness depends on the subjective opinion of the person receiving the stimulus.

To reiterate, the stimuli which give rise to pain result in a sensory experience *plus* a reaction to it. The sensation of pain can be dissociated from the emotional and behavioral components by drugs, e.g., morphine, or by selective brain operations which interrupt pathways connecting the frontal lobe of the cerebrum with other parts of the brain. When the reactive components are no longer associated with the sensation, pain is still felt, but it is not necessarily disagreeble; the patient does not mind as much. Thus pain relief can be obtained even though the perception of painful stimuli is not reduced.

The receptors whose stimulation gives rise to pain are at the ends of certain small unmyelinated or lightly myelinated afferent neurons. These receptors, known as **nociceptors** all appear the same when viewed with an electron microscope, yet some endings respond preferentially to intense mechanical stimulation, others to mechanical and thermal stimulation, and still others to irritant chemicals as well. It is hypothesized that a chemical is released from the damaged tissue and depolarizes the nearby nerve endings, initiating action potentials in the afferent nerve fibers. The chemical has not been identified for certain, but histamine, bradykinin, and potassium ions are likely possibilities.

Different nociceptors have different thresholds; some respond only to severe procedures such as cutting the skin, pricking it with needles, or heating it to 50°C, whereas receptors having lower thresholds respond to firm pressure or temperatures as low as 42°C, neither of which damages the skin. These latter receptors signal impending harm, and the sensation they invoke is described as threatening or warning rather than actual pain. Both types of receptors fire more vigorously as stimulus intensity increases.

The primary afferents coming from nociceptors synapse onto several types of second-order interneurons after entering the central nervous system (one transmitter thought to be released at these synapses is **substance P**), and information about pain is transmitted to higher centers via specific and nonspecific pathways. It is hypothesized that the specific ascending pathways convey information about where, when, and how strongly the stimulus was applied and about the sharp, localized aspect of pain. The nonspecific pathways are believed to convey information about the aspect of pain which is duller, longer lasting, and less well localized. Neurons of the reticular formation and thalamus which are activated by the pain pathways connect with the hypothalamus and other areas of the brain which play major roles in integrating autonomic and endocrine stress responses and in generating the behavioral patterns of aggression and defense.

Several *descending* pathways originate in the brainstem and selectively inhibit the transmission of information originating in nociceptors. One of these pathways utilizes **serotonin** as the transmitter, whereas another releases **norepinephrine**. The descending axons of both these pathways end at spinal levels on interneurons in the pain pathways and on the terminals of the afferent neurons themselves. At least some of the terminals of these descending pathways seem to achieve their inhibitory effect by activating interneurons which, in turn, release inhibitory neurotransmitters (either **GABA** or **enkephalin**). It is not known whether the descending pathways merely modify and filter the incoming information or whether they are part of a defense mechanism to lessen stress by operating in anticipation of pain rather than in response to it.

The endocrine system has also been implicated in the inhibition of pain pathways, since removal of the pituitary or adrenal glands increases the response to nociceptor stimulation. The specific hormone (or hormones) involved has not been identified, but **endorphin** is a likely suspect.

Electric stimulation of specific areas of brainstem reticular formation produces a profound reduction of pain (**analgesia**), a phenomenon called stimulus-produced analgesia, and in this form of pain relief, the descending inhibitory pathways are possibly involved. There are other forms of pain control, such as transcutaneous nerve stimulation (in which the painful site itself or the nerves leading from it are stimulated by electrodes placed on the surface of the skin), acupuncture, and hypnosis, each of which provides relief for some people but whose mechanism of action is not well understood.

As we have mentioned, at some of the synaptic links in the pain-suppression pathways the neurotransmitter enkephalin seems to be involved, and morphine and other opiates are thought to relieve pain by their ability to activate the receptor sites for enkephalin. Since the dull, long-term, poorly localized type of pain responds better to opiates than does the sharp, well-localized type, the inhibitory pathways sensitive to these drugs presumably affect mainly the nonspecific pathways.

Vision

Light

The receptors of the eye are sensitive to only that tiny portion of the vast spectrum of electromagnetic radiation which we call light (Fig. 18-5). Radiant

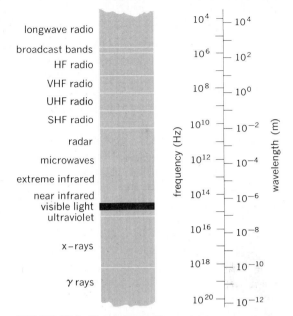

FIGURE 18-5. Electromagnetic spectrum.

energy is described in terms of wavelengths and frequencies. The **wavelength** is the distance between two successive wave peaks of the electromagnetic radiation (Fig. 18-6); it varies from several kilometers at the top of the spectrum to minute fractions of a millimeter at the bottom end. Those wavelengths (the **visible spectrum**) capable of stimulating the receptors of the eye are between 400 and 700 nm (i.e., billionths of a meter), and light of different wavelengths in this band is associated with different color sensations.

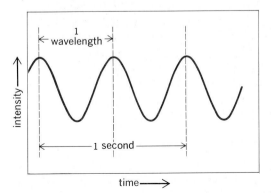

FIGURE 18-6. Properties of a wave. The frequency of this wave is 2 Hz (cycles/s).

The optics of vision

Light can be represented most simply by a ray or line drawn in the direction in which the wave is traveling. Light waves are propagated in all directions from every point of a visible object. These divergent light waves must pass through an optical system which focuses them back into a point before an accurate image of the object is achieved. In the eye itself, the image of the object being viewed must be focused upon the **retina**, a thin layer of neural tissue lining the back of the eyeball (Fig. 18-7), where the light-sensitive receptor cells of the eye are located.

The **lens** and **cornea** of the eye (Fig. 18-7) are the optical systems which focus the image of the object upon the retina. At a boundary between two substances, such as the cornea of the eye and the air outside it, the light rays are bent so that they travel in a new direction. The degree of bending depends in part upon the angle at which the light enters the second medium. The cornea plays the larger role in focusing light rays because the rays are bent more in passing from air into the cornea than into and out of the lens or other transparent structures of the eye.

The surface of the cornea is curved so that light rays coming from a single point source hit the cornea at different angles and are bent different amounts, but all in such a way that they are directed to a point after emerging from the lens (Fig. 18-8A). When the object being viewed has more than one dimension, the image on the retina is upside down relative to the original light source (Fig. 18-8B), and it is also reversed right to left.

The shape of the cornea and lens and the length of the eyeball determine the point where light rays reconverge. Moreover, light rays from objects close to the eye strike the cornea at greater angles (are more divergent) and have to be bent more in order to reconverge on the retina. Although the cornea performs the greater part quantitatively of focusing the visual image on the retina, all *adjustments* for distance are made by changing the shape of the lens. Such changes are called **accommodation**. The shape of the lens is controlled by the **ciliary muscles** and the tension they apply to ligaments of the lens. The lens is flattened when distant objects are to be focused upon the retina, and it is allowed to assume a more spherical shape to provide additional bending of the light rays when near objects are viewed (Fig. 18-9). The ciliary muscles are controlled by autonomic nerve fibers: The sympathetic fibers relax the ciliary muscles, thereby tightening the suspensory ligament and flattening the lens for far vision, whereas parasympathetic activity causes the ciliary muscle to contract, which relaxes the ligament, allowing the lens to become more convex for near vision.

Cells are added to the lens throughout life but only to the outer surface. This means that cells at the center of the lens are both the oldest and the farthest away from the nutrient fluid which bathes the outside of the lens. (If capillaries ran through the lens, they would interfere with its transpar-

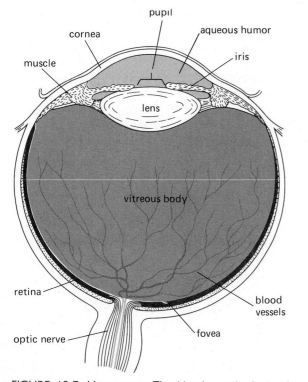

FIGURE 18-7. Human eye. The blood vessels depicted run along the back of the eye between the retina and the vitreous body.

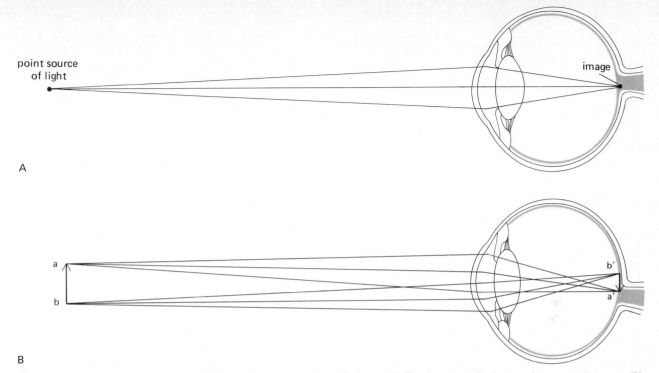

point source of light

image

A

a
b

b'
a'

B

FIGURE 18-8. Refraction (bending) of light by the lens system of the eye. (A) The focusing of light rays from a single point. (B) The focusing of light rays from more than one point to form the image of the object on the retina. In reality, the image is not focused on the place of exit of the optic nerve, which is a "blind spot"; rather, it is focused on the fovea, the area of the retina with greatest acuity, which is near the exit of the nerve.

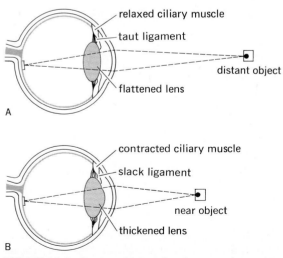

relaxed ciliary muscle
taut ligament

distant object

flattened lens

A

contracted ciliary muscle
slack ligament

near object

thickened lens

B

FIGURE 18-9. Accommodation for distant and near vision by the pliable lens. (A) The lens is stretched for distant vision so that it adds the minimum of focusing power. (B) The lens thickens for near vision to provide greater focusing power.

ency.) These central cells age and die first, and with death they become swollen and stiff, so that accommodation of the lens for near and far vision becomes more difficult. This impairment is known as **presbyopia** and is one reason why many people who never needed glasses before start wearing them in middle age.

Cells of the lens can also become opaque so that detailed vision is impaired; this is known as **cataract**. The defective lens can usually be removed surgically from persons suffering from cataract, and with the aid of compensating eyeglasses, contact lenses, or even an implanted artificial lens effective vision can be restored, although the ability to accommodate will be lost.

Defects in vision occur if the eyeball is too long in relation to the lens size, for then the images of near objects fall on the retina but the images of far objects come into focus at a point in front of the retina. This is a **nearsighted**, or **myopic**, eye, which is unable to see distant objects clearly (Fig. 18-10A). If the eye is too short for the lens, distant objects are focused on the retina while near objects come into focus behind it (Fig. 18-10B); this eye is **farsighted**, or **hyperopic**, and near vision is poor. The corrections for near- and farsighted vision are shown in Fig. 18-10. Defects in vision also occur where the lens or cornea does not have a smoothly spherical surface; this is known as **astigmatism** and can usually be compensated for by eyeglasses.

The amount of light entering the eye is controlled by a ringlike pigmented muscle known as the **iris**, the color being of no importance as long as the tissue is sufficiently opaque to prevent the passage of light. The hole in the center of the iris through which light enters the eye is the **pupil**. The iris muscle contracts reflexly in bright light, decreasing the diameter of the pupil; this not only reduces the amount of light entering the eye but also directs the light to the central and most optically accurate part of the lens. Conversely, the iris

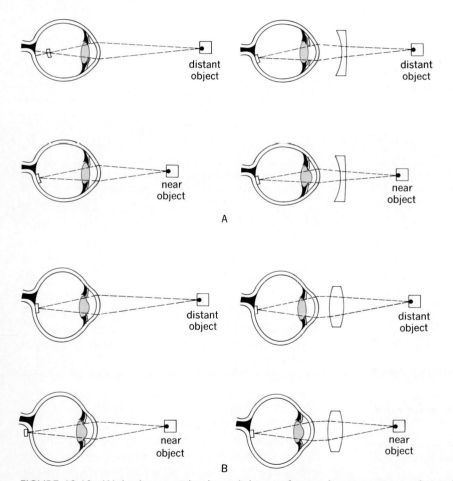

FIGURE 18-10. (A) In the nearsighted eye, light rays from a distant source are focused in front of the retina. A concave lens placed before the eye bends the light rays out sufficiently to move the focused image back onto the retina. When near objects are viewed through concave lenses, the eye accommodates to focus the image on the retina. (B) The farsighted eye must accommodate to focus the image of distant objects on the retina. (The normal eye views distant objects with a flat, stretched lens.) The accommodating power of the lens of the eye is sufficient for distant objects, and these objects are seen clearly. The lens cannot accommodate enough to keep images of near objects focused on the retina, and they are blurred. A convex lens converges light rays before they enter the eye and allows the eye's lens to work in a normal manner.

relaxes in dim light, when maximal sensitivity is needed.

Receptor cells

The receptor cells in the retina (the **photoreceptors**) are called **rods** and **cones** because of their microscopic appearance (Fig. 18-11). Photoreceptors contain light-sensitive molecules called **photopigments** which absorb light. There are four different photopigments in the retina, each of which is made up of a protein (**opsin**) bound to a **chromophore** molecule. The chromophore is always **retinal** (a slight variant of vitamin A) but the opsin differs in each of the four types. This difference causes each of the four photopigments to absorb light most ef-

fectively at a different part of the visual spectrum. For example, one photopigment absorbs wavelengths in the range of red light better than those in blue or green ranges, whereas another absorbs green light better than red or blue.

The events leading to receptor activation have been most studied for the rods. The rod-shaped ending of these receptor cells contains many stacked, flattened membrane disks parallel to the light-receiving surface of the retina (Fig. 18-12). The disk membranes contain a high concentration of the rod photopigment **rhodopsin**. The opsin component of rhodopsin is covalently bonded to the chromophore, which, upon exposure to light, changes its shape without breaking internal chemical bonds (enzymes are not required for this reaction, for it is catalyzed by light). This reaction is extremely fast, occurring in a matter of picoseconds (trillionths of a second). The initial change in molecular shape, caused by the absorption of light, leads to

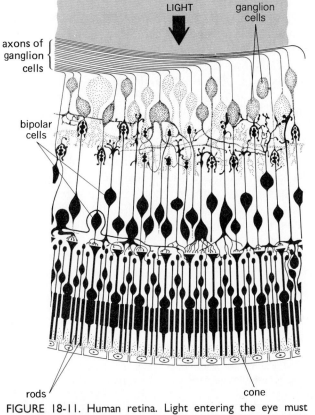

FIGURE 18-11. Human retina. Light entering the eye must pass through the fibers and cells of the retina before reaching the sensitive tips of the rods and cones. (Redrawn from Gregory.)

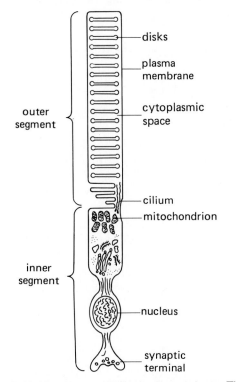

FIGURE 18-12. A rod cell of a vertebrate. The inner segment is concerned with cellular metabolism. (Redrawn from O'Brien.)

an unstable molecule which spontaneously undergoes further changes resulting in the dissociation of the chromophore from the opsin. These changes in rhodopsin lead to the closing of sodium channels in the plasma membrane of the receptor cells (Fig. 18-12), which in turn leads to a hyperpolarizing receptor potential. However, since rhodopsin is bound to the membranous disks inside the rod, which are physically separate from the plasma membrane where the ion-channel changes take place, some sort of internal transmitter (second messenger) must mediate between these two sites. The identity of this substance is not certain, but is most likely either calcium or cyclic GMP.

Note that since the closing of the sodium channels results in a *decrease* in sodium permeability, the receptor potential in this case is a hyperpolarization. Thus the receptors are relatively depolarized in the dark and hyperpolarized in the light; accordingly, transmitter release, which occurs with depolarization, increases in the dark and is reduced by illumination. Needless to say, this is still an effective signaling mechanism.

After dissociation of the photopigment, rhodopsin is regenerated by reactions which do not depend on light but are enzyme-mediated.

Visual pathways

The visual pathways begin with the retina. The rods and cones are connected via gap junctions with neighboring receptor cells and synapse upon several types of second-order neurons in the retina, one of which is the **bipolar cell** (Fig. 18-13). The bipolar cell, in turn, synapses (still within the retina) upon the **ganglion cells**. Other cells pass information horizontally from one part of the retina to another so that a great deal of information processing takes place in the retina, and its task is much more complicated than the simple point-to-point transmission of information about the light and dark areas of the visual image.

Generally, cone receptor cells have relatively direct lines to the brain; i.e., there is little convergence along the neural pathway. This relative lack of convergence provides precise information about the area of the retina that was stimulated, and cone visual acuity is very high. Because cones are concentrated in the center of the retina, it is that part that we use for finely detailed vision. However, the lack of convergence in the cone visual pathways offers little opportunity for the summation of subthreshold events to fire action potentials down the line in the ganglion cells, and high levels of illumination are needed to activate these pathways. Thus, the cone pathways are useful only in "daytime" vision. In contrast, there is much convergence in the rod visual pathways, and although acuity is poor, opportunities for spatial and temporal summation are good. Therefore, a relatively low-intensity light stimulus that would result in only a subthreshold response in a cone ganglion cell can result in an action potential in a rod ganglion cell. Thus, the difference in acuity and light sensitivity between rod and cone vision (summarized in Table 18-1) is due, at least in part, to the anatomical wiring patterns of the retina.

These differences explain why objects in a darkened room are indistinct; with such low illumination the cones do not generate effective signals, so that all vision is supplied through the more sensitive but less accurate rod vision. Moreover, rods contain only one type of photopigment (rhodopsin) and, as we shall see later, cannot give rise to color vision. Thus, objects in a darkened room appear in shades of gray.

The output from the retina is via the ganglion cells, whose axons form the **optic nerve** which leads

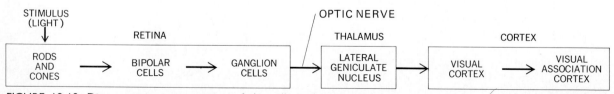

FIGURE 18-13. Diagrammatic representation of the cells in the visual pathway.

TABLE 18-1. Properties of rod and cone vision

Rod system

Much convergence in neural pathways
High sensitivity
Night vision
Low visual acuity
Black-and-white vision

Cone system

Little convergence in neural pathways
Low sensitivity
Daytime vision
High visual acuity
Color vision

from the retina into the brain. These are the first retinal neurons which respond to activation with action potentials (the receptors, bipolar cells, and horizontally placed retinal cells generally have only graded potentials). As will be discussed later, there are many different kinds of ganglion cells, each with a distinct functional role. The axons of the ganglion cells (i.e., the optic nerves) from the two eyes meet at the base of the brain to form the **optic chiasm,** where some of the fibers cross to the opposite side of the brain. This partial crossover provides both cerebral hemispheres with input from both eyes.

The optic nerve fibers project to many structures in the brain, the largest number of fibers passing to the **lateral geniculate nucleus** (Fig. 18-13), a relay nucleus in the thalamus, which transmits their information on to the visual areas of cerebral cortex. This relay of information follows a precise point-to-point pattern projection from the retina to the lateral geniculate nucleus and from this nucleus on to the **visual cortex.** Visual cortex comprises several subdivisions, each representing the complete visual field but responding to a distinct aspect of the visual stimulus. For example, different regions of visual cortex may respond specifically to color, or to direction of movement of the stimulus, or to orientation of the contours of objects in the visual field, or to binocular stimulation.

Many of the functional differences between different areas of visual cortex are determined by the input these areas receive from the distinct classes of ganglion cells, and an important aspect of lateral geniculate function is to separate and relay different kinds of retinal information to different cortical zones. Some of the information is kept separate even after it leaves visual cortex; for example, pathways that transmit information dealing with objects in the visual field (analyzing the "what" of the visual stimulus) travel from visual cortex to the association areas in the frontal lobe via the temporal lobe, whereas those dealing with spaces in the visual field (analyzing the "where" of the visual stimulus) travel to the frontal lobe via the parietal lobe. Thus, there is a growing understanding that different aspects of visual information are carried in parallel pathways and are processed simultaneously in a number of more or less independent ways in different parts of cerebral cortex before they are reintegrated to produce the conscious sensation of sight.

A substantial number of fibers of the visual pathway project to regions of the brain other than visual cortex, and part of the lateral geniculate nucleus serves as a relay point for much of this information, too. For example, visual information is transmitted to a nucleus lying just above the optic chiasm which is thought to play a major role in a variety of behaviors associated with biological clocks, e.g., eating, locomotion, sleep/wakefulness, and reproduction. In this nucleus the information about diurnal cycles of light intensity is used to synchronize the rhythmical activity of the neuronal biological clocks. Other visual information is passed to the brainstem and cerebellum, where the information plays an essential role in the coordination of eye and head movements, fixation of gaze, and constriction of the pupil. Some of this visual information in noncortical areas can be used to detect spaces and to guide the visual attention to specific areas of the visual field, although the ability to discriminate patterns with noncortical vision is quite poor.

Visual system coding

This discussion is based on research done chiefly on cats and monkeys, but it almost certainly applies to people as well. In most of the experiments dealing with coding in the visual system, simple visual shapes such as white bars against a black background were projected onto a screen in front of an anesthetized animal while the activity of single cells in the visual system was recorded. Different parts of the retina (and therefore different populations of receptor cells) could be stimulated by varying the position of the bar on the screen.

We shall refer in the following discussion to the **receptive fields** of neurons within the visual pathway and to the responses of these neurons to light. It is essential to recognize that only the rods and cones respond *directly* to light; all other components of the pathway are influenced only by the synaptic input to them. Thus when we speak of the receptive field of a neuron in the visual pathway, we really mean that area of the retina which, when stimulated, can influence the activity of that neuron. Similarly the neuron's ''response to light'' is really its response to neural activity in the visual pathway initiated by light falling upon the rods and cones. We shall follow the information processing elucidated by these experiments through the stages of the visual pathway, starting at the level of the ganglion cells (Fig. 18-13).

We have mentioned that an amazing amount of data processing occurs in the retina. The retinal ganglion cells discharge spontaneously; i.e., they fire in the absence of any light stimulus. This spontaneous activity gives the cells an important flexibility with which to work; they can either increase or decrease their rate of firing. Each converging receptor–bipolar-cell–ganglion-cell chain is synaptically connected to other similar chains by cells which conduct laterally through the retina.

The receptive fields of the ganglion cells vary, but most are circular; i.e., any light falling within a specific circular area of the retina influences the activity of a given ganglion cell (by way of the receptor–bipolar-cell chain). The responses of the ganglion cells vary markedly, depending on the type of ganglion cell and the region stimulated in its receptive field. Some ganglion cells speed up their rate of firing when a spot of light is directed at the center of the cell's receptive field and slow down their firing when the periphery is stimulated. Such

a cell is said to have an **on center**. The activity of a ganglion cell of this type is shown in Fig. 18-14A. Other ganglion cells (**off-center cells**, such as cell 2 in Fig. 18-14) have just the opposite response, decreasing activity when the center of the cell's receptive field is stimulated (Fig. 18-14B) and increasing activity when the light stimulates the periphery (Fig. 18-14C); notice the spontaneous activity of this *off*-center cell and the abrupt inhibition of its activity when the light is turned on.

The receptive field of single cells in the lateral geniculate nucleus, i.e., that area of the retina which when stimulated can influence the activity of a given lateral geniculate neuron, resembles that of a retinal ganglion cell in being concentric with either an *on*-center–*off*-periphery or vice versa. The visual processing continues in the visual cortex, and while the receptive fields of cells there vary widely in organization, the neurons are categorized as either **simple**, **complex**, or **hypercomplex** cells. All these neurons are relatively insensitive to the amount of illumination in their receptive fields (as long as there is enough light so that the objects are clearly visible), and they are insensitive to stationary images on the retina. They are, however, particularly sensitive to boundaries between light and dark. For

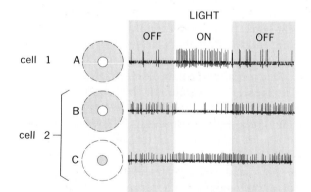

FIGURE 18-14. Recordings of the activity of a single ganglion cell. (A) Response of an *on*-center ganglion cell (cell 1) which increases its activity when light stimulates the center of its receptive field. (B) Activity of an *off*-center cell (cell 2) is suppressed when the center of its receptive field is stimulated. (C) The same *off*-center cell increases its activity when the light is restricted to the periphery. *(Redrawn from Hubel and Wiesel., J. Physiol., **154**.)*

simple cells the boundaries might take the form of straight edges or bands (Fig. 18-15), whereas for hypercomplex cells they might be corners or tongue-shaped objects. For simple cells, the stimuli have a clearly defined location on the retina as well as an optimal size and orientation (sometimes they even have a preferred direction of movement), but the complex cells are often insensitive to where in the receptive field the boundary occurs, and the receptive field can be quite wide. Note that the cells of the visual pathways are organized to handle information about line, contrast, movement, and as we shall see, color, but they do not form a picture in the brain; rather, they form a specifically coded electrical statement.

Color vision

Light is the source of all colors. Pigments, such as those mixed by a painter, serve only to reflect, absorb, or transmit different wavelengths of light, yet the nature of the pigments determines how light of different wavelengths will react. For example, an object appears red because the shorter wavelengths are absorbed by the material whereas the long wavelengths are reflected to excite the photopigment of the retina most sensitive to them. Light perceived as white is a mixture of all wavelengths, and black is the absence of all light. Sensation of any color can be obtained by the appropriate mixture of three lights, red, blue, and green (Fig. 18-16). Light and pigments are properties of the physical world, but color exists only as a sensation in the mind of the beholder.

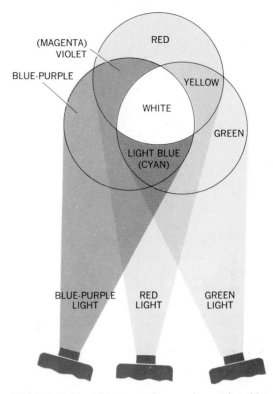

FIGURE 18-16. All known colors can be produced by different combinations of the three primary wavelengths of light, those giving rise to color sensations of red, blue, and green. *(Redrawn from Gregory.)*

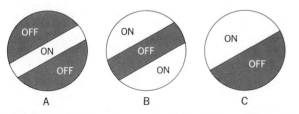

FIGURE 18-15. Retinal receptive fields of three different simple cortical cells. The receptive fields are no longer arranged concentrically but are organized to provide information about lines and borders. *(Redrawn from Hubel and Wiesel, J. Physiol., 160.)*

Color vision begins with the activation of the photopigments in the cone receptor cells. Human retinas have three kinds of cones, which contain either yellow-, green-, or blue-sensitive photopigments, responding optimally to light of 560-, 530-, and 420-nm wavelengths, respectively. (The 560-nm pigment sensitivity extends far enough to sense the long wavelengths which correspond to red, and this pigment is sometimes called the red photopigment.) Although each type of cone is excited most effectively by light of one particular wavelength, it also responds to other wavelengths; thus, for any given wavelength, the three cone types are excited to different degrees (Fig. 18-17). For example, in response to a light of 530-nm wavelength, the green cones respond maximally, the red cones less, and the blue cones hardly at all. Our sensation of color

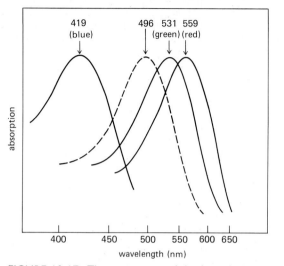

FIGURE 18-17. The sensitivities of the four photopigments of the normal human retina. The solid lines are for the three types of cones, the dotted curve for the rods. (Redrawn from Mollon.)

depends upon the ratios of these three cone outputs and their comparison by higher-order cells in the visual system.

A given cone's membrane polarization increases with the rate at which photons are absorbed by its photopigment, but the input of photons to the cell varies with both the wavelength and intensity of the light stimulus. A change in the response of a single cone type would occur with a change in either variable, and therefore vision with a single cone type, i.e. one photopigment type, by itself cannot signal color. It is for this reason that rods cannot signal color vision either; thus, vision is "color blind" at low levels of light intensity when only rods are stimulated. (Recall that there is only one type of photopigment, rhodopsin, in rods; changes in rod output are taken to mean changes in light intensity.)

At high light intensities (daylight vision), most people (over 90 percent of the male population and over 99 percent of the female population) have normal color vision; i.e., their color vision is determined by the differential activity of the three types of cones. Most color-blind (or better, color-defective) people appear to lack one of the three photo-

pigments, and their color vision is therefore formed by the diffential activity of the remaining two types of cones. For example, people with green-defective vision see as if they have only red- and blue-sensitive cones.

The pathways for color vision follow those described earlier for the line-contrast processors (Fig. 18-13); the cones synapse upon bipolar cells (and other cells of the retina as well), and the bipolar cells synapse upon ganglion cells, etc. One type of ganglion cell receives input (via bipolar cells) from cones, each of which indicates both the wavelength range and brightness of light activating them. These cells transmit to the brain information about short-, middle-, and long-wavelength brightness rather than information about specific colors, and they allow the brain to scale light intensity information across the entire visual field so that, for example, an apple seen in the shade is perceived to be the same color as an apple seen in the sunlight.

A second type of ganglion cell codes specific colors and is called the **opponent color cell.** These cells receive excitatory input from one of the three cone types and inhibitory input from another. For example, the cell in Fig. 18-18 increased its rate of firing when stimulated by a blue light but decreased

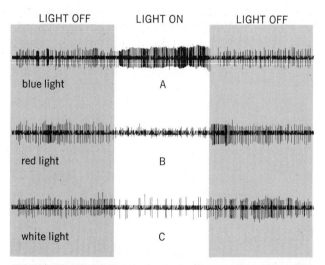

FIGURE 18-18. Response of a single ganglion cell to blue light, red light, and white light. (Redrawn from Hubel and Wiesel, J. Physiol., **154.**)

it when a red light replaced the blue. The cell gave a weak response when stimulated with a white light because the light contained both blue and red wavelengths.

Eye movement

The cones are most concentrated in a specialized area of the retina known as the **fovea**, and images focused there are seen with the greatest acuity. In order to get the visual image focused on the fovea and keep it there, the six muscles attached to the outside of each eyeball perform two basic movements, fast and slow.

The fast movements include **saccades**, small jerking movements which rapidly bring the eye from one fixation point to another to allow search of the visual field. In addition, saccades move the visual image over the receptors, thereby preventing adaptation. In fact, if saccades are prevented, all color and most detail fade away in a matter of seconds. Saccades also occur during certain periods of sleep when the eyes are closed, and perhaps they are associated with ''watching'' the visual imagery of dreams. Saccades are among the fastest movements in the body.

Slow eye movements are involved in tracking visual objects as they move throughout the visual field and during compensation for movements of the head. If the head is moved to the left, the eyes must be moved an equal distance to the right if the image of a stationary visual object is to remain focused on the fovea; if the head moves up, the eyes must move down. The control centers for these compensating movements obtain their information about the movement of the head from the semicircular canals of the vestibular system, which will be described shortly. Control systems for the other slow movements require the continual feedback of visual information about the moving object.

Hearing

Sound energy is transmitted through air by a movement of air molecules. When there are no air molecules, as in a vacuum, there can be no sound. The disturbance of air molecules that makes up a **sound wave** consists of regions of compression, in which the air molecules are close together and the pressure is high, alternating with areas of rarefaction, where the molecules are farther apart and the pressure is lower. Anything capable of creating such disturbances can serve as a sound source.

A tuning fork at rest emits no sound (Fig. 18-19), but if it is struck sharply, it vibrates. As the arms of the turning fork move, they push air molecules ahead of them, creating a zone of compression, and leave behind a zone of rarefaction (Fig. 18-19B). As they move in the opposite direction, they again create pressure waves of compression and rarefaction (Fig. 18-19C).

The molecules in an area of compression, pushed together by the vibrating prong of the tuning fork, bump into the molecules ahead of them, push them together, and create a new region of compression. Individual molecules travel only short distances, but the disturbance passed from one molecule to another can travel many miles, and it is by these disturbances (sound waves) that sound energy is transmitted. The sound dies out only when so much of the original sound energy has been dissipated that one sound wave can no longer disturb the air molecules around it. A tone emitted by the tuning fork is said to be *pure* because the tuning fork vibrates at a single fixed frequency and the waves of rarefaction and compression are regularly spaced. The waves of speech and most other common sounds are not regularly spaced but are complex waves made up of many frequencies of vibration.

The sounds heard most keenly by human ears are those from sources vibrating at frequencies between 1000 and 4000 Hz (Fig. 18-20), but the entire range of frequencies audible to human beings extends from 20 to 20,000 Hz. The **frequency** of vibration of the sound source is related to the **pitch** we hear; the faster the vibration, the higher the pitch. We can also detect loudness and tonal quality, or timbre, of a sound. The difference between the packing (or pressure) of air molecules in zones of compression and rarefaction, i.e., the **amplitude** of the sound wave, is related to the loudness of the

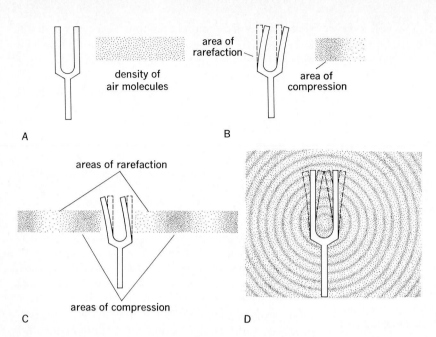

A   B

C   D   FIGURE 18-19. Formation of sound waves.

sound that we hear. The number of sound frequencies in addition to the fundamental tone, i.e., the lack of **purity** of the sound wave, is related to the timbre of the sound. We can distinguish some 400,000 different sounds; we can distinguish the note A played on a piano from the same note on a violin, and we can identify voices heard over the telephone. We can also selectively *not* hear sounds, tuning out the babel of a party to concentrate on a single voice.

The first step in hearing is the entrance of pressure waves into the **ear canal** (Fig. 18-21). The waves reverberate from the sides and end of the ear canal, filling it with the continuous vibrations of pressure waves. The **tympanic membrane** (eardrum) is stretched across the end of the ear canal. The air molecules, under higher pressure during a wave of compression, push against the membrane, causing it to bow inward. The distance the membrane moves, although always very small, is a function of the force with which the air molecules hit it and is therefore related to the loudness of the sound. During the following wave of rarefaction, the membrane returns to its original position. The exquisitely sensitive tympanic membrane responds to all the varying pressures of the sound waves, vibrating slowly in response to low-frequency sounds and rapidly in response to high tones. It is sensitive to pressures to which the most delicate touch receptors of the skin are totally insensitive.

The tympanic membrane separates the ear canal from the **middle-ear cavity** (Fig. 18-21). The pressures in these two air-filled chambers are nor-

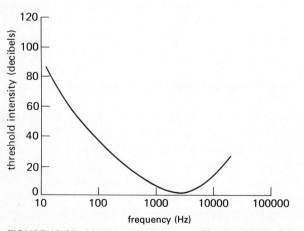

FIGURE 18-20. Human audibility curve. The threshold of hearing varies with sound frequency. The conversational voice of an average male is about 120 Hz, that of an average female about 250 Hz.

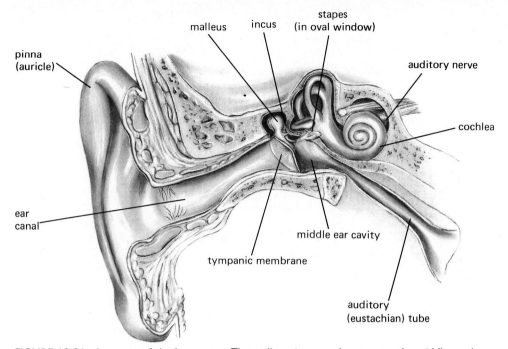

FIGURE 18-21. Anatomy of the human ear. The malleus, incus, and stapes are the middle-ear bones.

mally equal. The ear canal is at atmospheric pressure, and the middle ear is exposed to atmospheric pressure through the **auditory (eustachian) tube,** which connects the middle ear to the pharynx. The slitlike ending of this tube in the pharynx is normally closed, but during yawning, swallowing, or sneezing, when muscle movements of the pharynx open the entire passage, the pressure in the middle ear equilibrates with atmospheric pressure. A difference in pressure can be produced with sudden changes in altitude, as in an elevator or airplane, when the pressure outside the ear and in the ear canal changes while the pressure in the middle ear remains constant because of the closed auditory tube. This pressure difference excessively bends the tympanic membrane and causes pain.

The second step in hearing is the transmission of sound energy from the tympanic membrane, through the middle-ear cavity, and then to the fluid-filled chambers of the **inner ear**. Because the liquid in the inner ear is more difficult to move than air, the pressure transmitted to the inner ear must be amplified. This is achieved by a movable chain of

three small bones (the **malleus**, **incus**, and **stapes**), which couple the tympanic membrane to a membrane-covered opening (the **oval window**) separating the middle and inner ear (Fig. 18-22).

The *total* force on the tympanic membrane is

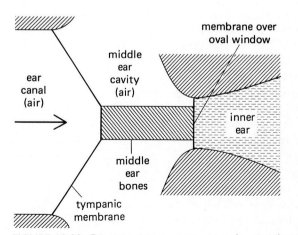

FIGURE 18-22. Diagrammatic representation showing that the middle-ear bones act as a piston against the fluid of the inner ear. (*Redrawn from von Bekesy.*)

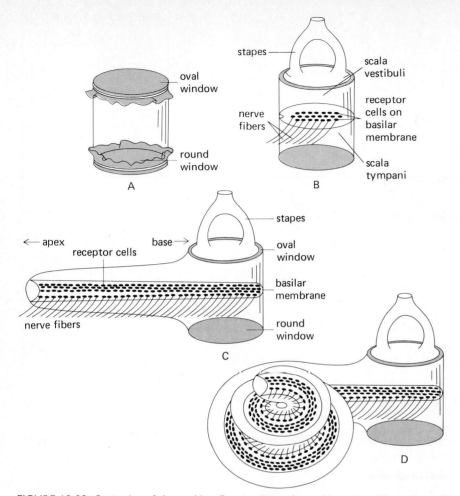

FIGURE 18-23. Basic plan of the cochlea. For simplicity, the cochlear duct (the scala media), which lies between the scala vestibuli and the tympanic membrane, is not shown, but see Fig. 18-24. (A) The cochlea is represented as a fluid-filled container with an elastic membrane (the oval and round window membranes) covering each end. The rigid wall of the container corresponds to the bony walls of the cochlea. (B) A driving piston (the stapes) and an elastic partition (the basilar membrane) are added. Sensory hair cells are placed on the basilar membrane. The nerve fibers from these cells will form the auditory nerve. A hole is placed in one end of the basilar membrane. (C) The wall of the cochlea near the hole is extended and the basilar membrane is elongated. (D) The elongated cochlea is coiled along its length. (Redrawn from Kiang.)

transferred to the oval window, but because the oval window is so much smaller, the *force per unit area* (i.e., pressure) is increased 15 to 20 times. Additional advantage is gained through the lever action of the middle-ear bones. The amount of energy transmitted to the inner ear can be modified by the contraction of two small muscles in the middle ear which alter the tension of the tympanic membrane and the position of the third middle-ear bone (the stapes) in the oval window. These muscles protect the delicate receptor apparatus of the inner ear from intense sound stimuli and possibly aid intent listening over certain frequency ranges.

Thus far, the entire system has been concerned with the transmission of the sound energy into the inner ear, where the receptors are located. The inner ear, or **cochlea**, is a spiral-shaped passage in the temporal bone. It is almost completely divided lengthwise by a fluid-filled membranous tube, the **cochlear duct**, which follows the cochlear spiral. The floor of the cochlear duct is the **basilar membrane** (Figs. 18-23 and 18-24). On either side of the

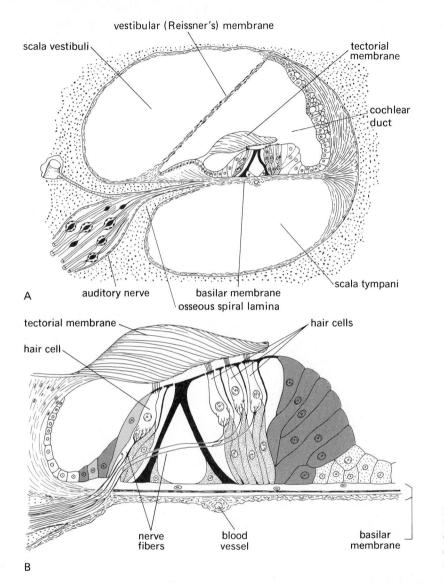

vestibular (Reissner's) membrane

scala vestibuli

tectorial
membrane

cochlear
duct

auditory nerve

basilar membrane

osseous spiral lamina

scala tympani

A

tectorial membrane

hair cells

hair cell

nerve
fibers

blood
vessel

basilar
membrane

B

FIGURE 18-24. Cross section of the membranes and compartments of the inner ear with detailed view of the hair cells and other structures on the basilar membrane. *(Redrawn from Rasmussen.)*

cochlear duct are fluid-filled compartments: the **scala vestibuli**, which is on the side of the cochlear duct which ends at the oval window, and the **scala tympani**, which is below the cochlear duct and closer to a second membrane-covered opening to the middle ear (the **round window**).

As the pressure wave in the ear canal pushes in on the tympanic membrane, the chain of bones rocks the footplate of the stapes against the membrane covering the oval window, causing it to bow

into the scala vestibuli compartment of the cochlea (Fig. 18-23) and create there a wave of pressure. The wall of the scala vestibuli is largely bone, but there are two paths by which the pressure wave can be dissipated. One path is to the end of the scala vestibuli, where the waves pass around the end of the cochlear duct into the scala tympani and back to the round-window membrane, which they bow out into the middle-ear cavity. However, most of the pressure is transmitted from the scala vestibuli to the cochlear duct and thereby to the basilar membrane, which is deflected into the scala tympani.

The pattern by which the basilar membrane is deflected is important because this membrane carries the **organ of Corti**, which consists of the sensitive receptor cells which transduce sound energy, i.e., the pressure wave in the cochlea, into action potentials, and the supporting cells which surround the receptor cells. At the end of the cochlea closest to the middle-ear cavity, the basilar membrane is narrow and relatively stiff, but it becomes wider and more elastic as it extends through the length of the cochlear spiral. The stiff end nearest the middle-ear cavity vibrates immediately in response to pressure changes transmitted to the scala vestibuli, but the responses of the more distant, more elastic parts are slower. Thus, with each change in pressure in the inner ear, a wave of vibrations is made to travel down the basilar membrane (Fig. 18-25).

The region of maximal displacement of the basilar membrane varies with the frequency of vibration of the sound source. The properties of the membrane nearest the middle ear are such that this region resonates best with high-frequency tones and undergoes the greatest amplitude of vibration when high-pitched tones are heard. The vibration of the basilar membrane in response to high-frequency sound waves soon dies out once it is past this region. Lower tones also cause the basilar membrane to vibrate near the middle-ear cavity, but the vibration wave travels out along the membrane for greater distances. The more distant regions of the basilar membrane vibrate maximally in response to

low tones. Thus the frequencies of the incoming sound waves are, in effect, sorted out along the length of the basilar membrane (Fig. 18-26). This spatial representation of frequency in the cochlea is known as the **place principle** and is preserved throughout the auditory pathway, i.e., different

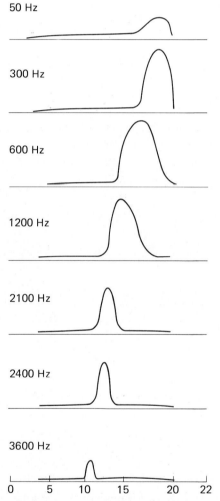

distance from stapes in mm (approx.)

FIGURE 18-26. Point along the basilar membrane where the traveling wave peaks is different with different sound frequencies. The region of maximal displacement of the basilar membrane occurs near the end of the membrane for low-pitched (low-frequency) tones and occurs near the oval window and middle ear for high-pitched tones. *(Redrawn from Kim and Molnar.)*

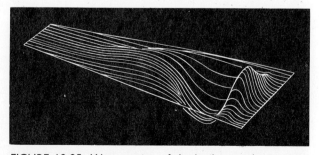

FIGURE 18-25. Wave motion of the basilar membrane in response to pressure changes in the inner ear. *(Redrawn from Alpern, Lawrence, and Wolsk.)*

neurons transmit information about different sound frequencies.

Vibration of the basilar membrane serves to stimulate the receptor cells (**hair cells**) of the organ of Corti (Fig. 18-24), which are essentially mechanoreceptors and ride upon the membrane; the greatest stimulation occurs wherever displacement of the basilar membrane is greatest. The fine hairs on top of the receptor cells are in contact with the overhanging **tectorial membrane,** which projects inward from the side of the cochlea. As the basilar membrane is displaced by pressure waves, the hair cells move in relation to the tectorial membrane and, consequently, the hairs are displaced. The resulting mechanical deformation of the hair cells, in an as yet unknown manner, opens ion channels in the receptor-cell membrane, allowing ions to flow into the cell and depolarizing it slightly. This depolarization of the hair cell—the receptor potential— causes the release of a chemical transmitter (whose identity is unknown) from the hair cell. The transmitter diffuses across an extracellular gap and activates sites on the terminals of the afferent neurons which underlie the hair cell, and the neurons are depolarized.

The hair cells are easily damaged by exposure to high-intensity noises such as the typical amplified rock-music concerts, engines of jet planes, and revved-up motorcycles. The damaged sensory hairs form giant, abnormal hair structures or are lost altogether, and in cases of long exposure to loud sounds some hair cells and their supporting cells completely degenerate (Fig. 18-27). Much lesser noise levels also cause damage if exposure is chronic.

The depolarizations that occur in the endings of the afferent neurons trigger bursts of action potentials which are transmitted into the central nervous system. The greater the energy (intensity) of the sound wave, the greater the movement of the basilar membrane, the larger the amplitude of the hair-cell receptor potential, the greater the amount of transmitter substance released from the hair cell, the greater the depolarization of the terminals of the afferent neurons, and the greater the frequency of action potentials in them.

Nerve pathways from different parts of the bas-

ilar membrane are connected to specific sites along the strips of auditory cortex in an orderly manner, according to sound frequency, in much the same way that signals from different regions of the body are represented at different sites in somatosensory cortex; it is as though the basilar membrane were unrolled and spread along a strip of cortex. And just as there are several complete representations of the body parts in somatosensory cortex, so there are several repetitions of the cochlear map in auditory cortex. Afferent fibers from both ears converge on neurons in the brainstem; therefore information carried by auditory pathways from these cells to auditory cortex is not specific for one ear or the other.

Each neuron along the auditory pathway responds at threshold levels of sound intensity (loudness) to a very small range of sound frequencies, its so-called **best frequency**. Different neurons have different best frequencies, depending on the location of their receptive fields on the basilar membrane. Although in the previous section of this chapter we described neurons in visual cortex that respond to meaningful features of the visual environment, e.g., edges and corners, very few neurons in the auditory cortex have been found to respond to specific features of sounds, such as components of speech or other biologically significant sounds.

In addition to the auditory nerve pathways that proceed from the inner ear to the cerebral cortex, there are pathways that proceed in the opposite direction. The efferent neurons that pass from the brain stem to the cochlea end both on the hair cells and on the terminals of the afferent nerve fibers. The function of these efferent fibers is not certain, but animals lacking them are not as capable of discriminating different sound frequencies or telling meaningful signals from noise.

Vestibular System

Changes in both the motion and position of the head are also detected by hair cells, which are mechanoreceptors and similar to the hair cells in the inner ear. The changes in motion that are detected include acceleration in a straight line in any direction and

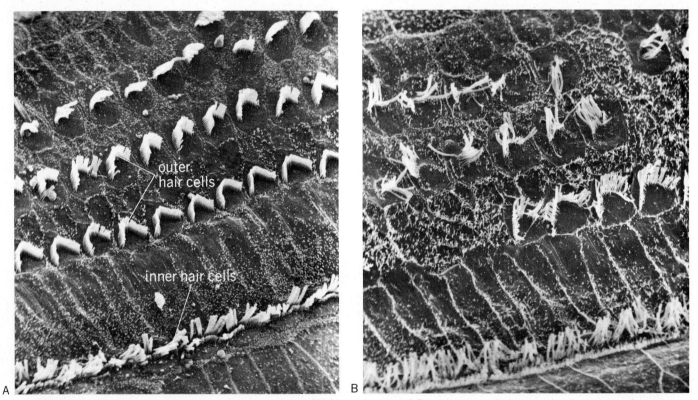

FIGURE 18-27. Injury to the inner ear by intense noise. (A) Normal organ of Corti (guinea pig) showing the three rows of outer hair cells and single row of inner hair cells. (B) Injured organ of Corti after 24-h exposure to noise levels typical of very loud rock music (2000-Hz-octave band at 120 dB). Several outer hair cells are missing, and the cilia of others no longer form the orderly pattern of the normal ear. Note also the increased number and size of small villi on the cell surfaces. *(Scanning electron micrograph by Robert E. Preston. Courtesy of Joseph E. Hawkins, Kresge Hearing Research Institute.)*

angular acceleration along three perpendicular axes. (The three axes are those activated while nodding the head up and down as in signifying "yes," shaking the head from side to side as in signifying "no," and tipping the head so the ear touches the shoulder on the same side.)

The hair cells which serve these functions are part of the **vestibular apparatus**, a series of membranous sacs which connect with the membranous cochlear duct of the inner ear and are filled with the same specialized fluid (endolymph). The vestibular apparatus consists of three **semicircular canals** and two saclike swellings, the **utricle** and **saccule** (Fig. 18-28A), which lie in tunnels in the temporal bone of the skull on each side of the head.

The three semicircular canals lie in different planes at right angles to each other (Fig. 18-28B), and they detect angular acceleration of the head in any direction in three-dimensional space. The hairs of the receptor cells of the semicircular canals are closely ensheathed by a gelatinous mass which extends into the lumen of the membranous semicircular duct at the **ampulla**, a slight bulge in the wall of each duct. Whenever the head is moved, the bony tunnel wall of the semicircular canal, its enclosed membranous sac, and the attached bodies of the hair cells all turn with it. The endolymph fluid filling the membranous sac, however, is not attached to the skull; therefore, because of inertia, the fluid tends to retain its original position, i.e., to

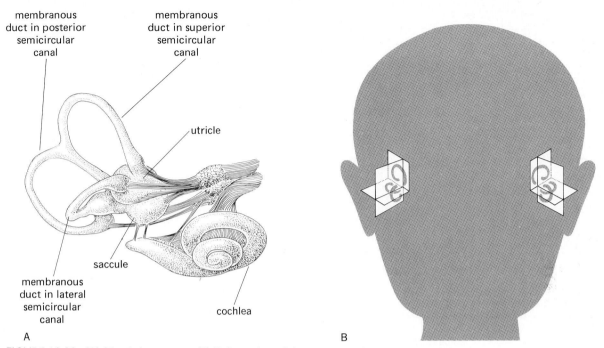

FIGURE 18-28. (A) Vestibular system. (B) Relationship of the two sets of semicircular canals.

be "left behind," and pushes against the gelatinous mass, moving the hairs within it and thereby stimulating the underlying hair cells.

As the inertia is overcome and the endolymph begins to move at the same rate as the rest of the head, the hairs slowly return to their resting position. For this reason, the hair cells are stimulated only during *changes* in the rate of motion, i.e., during acceleration, of the head. In contrast, stimulation of the hair cells ceases during motion at a constant speed.

The speed and magnitude of the movement of the head determine the way in which the hairs are bent and the hair cells stimulated. Each receptor has one direction of maximum sensitivity, and when the hairs are bent in this direction, the receptor cell depolarizes; when the hairs are bent in the opposite direction, the cell hyperpolarizes. The semicircular canals on the opposite sides of the head are paired such that at least one pair is activated during movement of the head, but the hair cells in the paired canals are polarized in opposite directions (Fig. 18-29). For example, when the head turns to the

left, the endolymph in the duct on the left side of the head will push the hair bundles in their direction of polarization, and the receptor cells there will depolarize. On the right side of the head, however, the hair bundles will be moved against their polarized direction, and the receptor cells on the right side of the head will hyperpolarize.

Similar to the hair cells of the organ of Corti, the hair cells of the semicircular canals are connected to the afferent neurons underlying them by chemically mediated synapses. Some transmitter is released from the hair cells in the absence of stimulation so that even when the head is motionless (or moving at a constant speed), the afferent neurons fire action potentials at a relatively low rate. Thus, the vestibular receptors can signal information by either increasing or decreasing the frequency of action potentials in the afferent nerve fiber. When the hairs are bent toward their polarized direction and the cells are depolarized from their resting level, an increased amount of transmitter is released from them, and the action-potential frequency in the afferent nerve fiber increases;

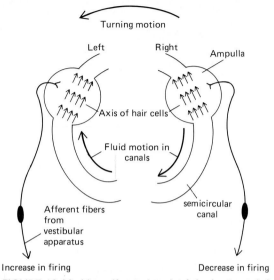

FIGURE 18-29. View (from above) of the horizontal ducts of the semicircular canals showing how the paired canals on opposite sides of the head work together to indicate head movement. *(Redrawn from Kandel and Schwartz.)*

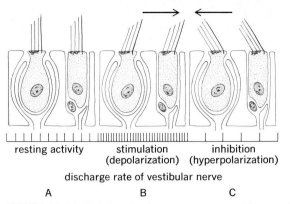

FIGURE 18-30. Relationship between position of hairs and activity in afferent neurons. (A) Resting activity. (B) Movement of hairs in one direction increases the action potential frequency in the afferent nerve fiber activated by the hair cell. (C) Movement in the opposite direction decreases the rate relative to the resting rate. *(Redrawn from Wersall, Gleisner, and Lundquist.)*

when the hairs are moved opposite to their polarized direction and the receptor cells are hyperpolarized, the amount of transmitter released is decreased, and the afferent-nerve fibers fire at a rate slower than their resting rate (Fig. 18-30). Thus, the frequency of action potentials in the afferent nerve fibers is related not only to the amount of force bending the hairs on the receptor cells but also to the direction in which this force is applied.

The semicircular canals are the receptor organs for *angular* acceleration, but the **utricle** and **saccule** provide information about linear acceleration and changes in the position of the head relative to the forces of gravity. The receptor cells here, too, are mechanoreceptors sensitive to the displacement of projecting hairs. The hair cells of the utricle and saccule are collected into groups from which the hairs protrude into a gelatinous substance. Tiny calcium carbonate crystals, or **otoliths**, are embedded in the gelatinous mass covering the hair cells, making it heavier than the surrounding endolymph. When the head is tipped, the gelatinous-mass-otolith material changes its position, pulled by gravitational forces against the hair cells so that the hairs

are bent and the receptor cells stimulated. The hairs of the utricle and saccule receptor cells have directional sensitivity, but they do not all face the same direction; thus, in any position of the head, some cells are depolarized and others are hyperpolarized. As with input from the paired semicircular canals, the central nervous system receives information in the form of increased activity in one set of nerve fibers coupled with decreased activity in another.

Information is relayed from the vestibular apparatus to a cluster of nuclei in the brain stem. Components from some of these nuclei send information higher up in the brain stem to the motor neurons controlling the muscles that move the eyes; others descend in a highly organized fashion to the spinal cord. Thus, the information from the vestibular apparatus is used for two purposes. One is to control the muscles which move the eyes so that, in spite of changes in the position of the head, the eyes remain fixed on the same point. As the head is turned to the left, for example, impulses from the vestibular nuclei activate the ocular muscles which turn the eyes to the right; the eyes therefore turn toward the right and remain fixed on the point of interest.

The second use of vestibular information is in

the reflex mechanisms for maintaining upright posture. In monkeys, cats, and dogs, for example, the vestibular apparatus plays a definite role in the postural fixation of the head, orientation of the animal in space, and reflexes accompanying locomotion. However, in people, very few postural reflexes are known to depend primarily on vestibular input, despite the fact that the vestibular organs are sometimes called the sense organs of balance.

Chemical Senses

Receptors sensitive to specific chemicals in the environment are **chemoreceptors**. Some of these respond to chemical changes in the internal environment (e.g., the oxygen and hydrogen-ion receptors in certain of the large blood vessels); others respond to external chemical changes, and in this category are the receptors for taste and smell. Taste and smell affect a person's appetite, flow of saliva and gastric secretions, and the avoidance of harmful substances.

Taste

The specialized receptor organs for the sense of taste are the 10,000 or so **taste buds** which are on the tongue, roof of the mouth, pharynx, larynx, and upper third of the esophagus. In the taste buds, the receptor cells and their adjacent supporting cells are arranged like the segments of an orange so that the multifolded upper surfaces of the receptor cells extend into a small pore at the surface of the taste bud, where they are bathed by the fluids of the mouth (Fig. 18-31).

Taste sensations are traditionally divided into four basic groups: sweet, sour, salt, and bitter, but different types of taste buds or receptor cells which would support this specificity have not been identified. In fact, a single receptor cell can respond in varying degrees to many different chemical substances which fall into more than one of these basic categories.

The mechanisms by which the taste receptors are stimulated are not known. It has been suggested that the first step is a loose binding of the individual

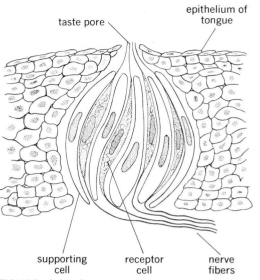

FIGURE 18-31. Structure and innervation of a taste bud.

molecules of the chemical substance (after they have dissolved in the saliva at the taste pore) with specific sites on the plasma membrane of the receptor cell. The fact that a single receptor can respond to more than one basic taste quality could be explained if the cell had several different sites, each capable of binding with a different type of molecule.

Afferent nerve fibers enter the taste buds to end on the receptor cells, separated from them by an extracellular cleft. Transmitter substances, released from the receptor cells upon stimulation, diffuse across the cleft to activate the afferent neurons. One nerve fiber generally innervates several receptor cells, and each receptor is innervated by several different neurons. There is clearly no one-to-one relationship by which each receptor cell has a direct line into the nervous system. The pathways in the central nervous system for taste project to parietal cortex, near the "mouth" region of the somatosensory cortex.

The frequency of action potentials in single nerve fibers increases in response to increasing concentrations of the chemical stimulant; therefore, frequency signals the intensity of the stimulus, but what signals the quality? How can we distinguish so many different taste sensations when the receptor cells lack specificity both in terms of the kind of chemical to which they respond and the way in which they are connected to the brain?

The afferent fibers involved in taste show different firing patterns in response to different substances; e.g., one fiber may fire very rapidly when the stimulatory substance is salt but only sporadically when it is sugar, and another fiber may have just the opposite response. Therefore, awareness of the specific taste of a substance probably depends upon the pattern of firing in a group of neurons rather than that in a specific neuron. Identification of a substance is also aided by information about its temperature and texture which is transmitted to the central nervous system from other receptors on the tongue and surface of the oral cavity. The odor of the substance clearly helps, too, as is attested by the common experience that food lacks taste when one has a stuffy head cold.

Smell

The olfactory receptors which give rise to the sense of smell lie in a small patch of mucus-secreting membrane (the **olfactory mucosa**) in the upper part of the nasal cavity (Fig. 18-32A). The olfactory receptors are not separate cells but are specialized endings of afferent neurons. The cell body of these neurons has an enlarged knob from which several long ciliary processes extend out to the surface of the olfactory mucosa (Fig. 18-32B). The knob and cilia of the afferent neurons contain the receptor sites, and the axons of these neurons project to the brain as the olfactory nerve.

Before an odorous substance can be detected, it must release molecules which diffuse into the air and pass into the nose to the region of the olfactory mucosa. Once there, the molecules dissolve in a layer of mucus covering the receptor cells, interact with receptor sites on the cells, and depolarize the membrane enough to initiate action potentials in the afferent-nerve fiber. Upon stimulation of the olfactory mucosa with odorous substances, one can record receptor potentials which change with stimulus quality and intensity.

The physiological basis for discrimination be-

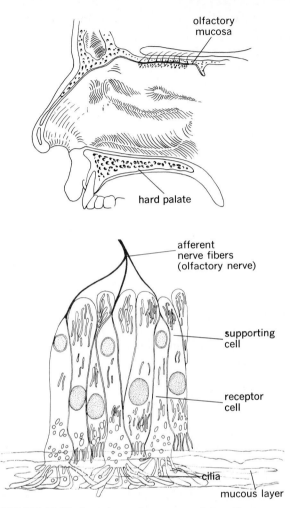

FIGURE 18-32. Location and structure of the olfactory receptors.

tween the tens of thousands of different odor qualities is speculative. There are no apparent differences in receptor cells, at least on a microscopic level, but differences between receptors presumably exist at molecular levels. It is known, however, that olfactory discrimination depends only partially upon the action-potential pattern generated in the different olfactory neurons; it also varies with attentiveness, state of the olfactory mucosa (acuity decreases when the mucosa is congested, as in a head cold), hunger (sensitivity is greater in hungry subjects), sex (women in general have keener olfactory sensitivities than men), and smoking (decreased sensitivity has been repeatedly associated with smoking). And just as the awareness of the taste of an object is aided by other senses, so the knowledge of the odor of a substance is aided by the stimulation of other receptors. This is responsible for the description of odors as pungent, acrid, cool, or irritating.

Olfactory information is handled in the central nervous system by pathways which pass to a region of cerebral cortex that lies on the undersurface of the frontal lobe and is considered to be part of the limbic system. It is not surprising that olfactory information is relayed directly to the limbic system, i.e., that part of the brain most intimately associated with neuroendocrine regulation and both food-getting and sexual behavior.

Further Perceptual Processing

Our actual perception of the events around us often involves areas of the brain other than primary sensory cortex. Information from the primary sensory areas achieves further elaboration through the neural activity in the **cortical association areas** (Fig. 18-33), particularly the association area in parietal lobe. These brain areas lie outside the classic primary cortical sensory or motor areas but are connected to them. Although it has not often been possible to elucidate the specific role performed by the association areas, they are assumed to serve complex integrative functions and are implicated in many different forms of complex behavior.

The parietal association cortex can be divided into two regions, the outer portion lying next to the primary cortical receiving areas for somatosensory, auditory, and visual information. Neurons in this region receive input directly from the primary cortical receiving area adjacent to them (e.g., the neurons next to somatosensory cortex receive information about touch, temperature, or pain, whereas cells next to visual cortex receive visual information). This portion of association cortex therefore might serve relatively simple sensory-related functions, such as the recognition of patterns in the incoming information.

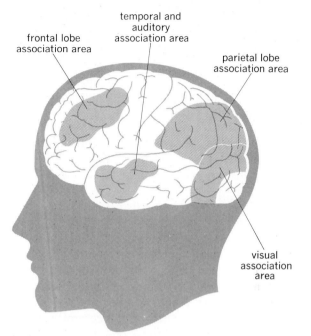

FIGURE 18-33. Areas of association cortex.

The inner region of the parietal association cortex receives information only after it has undergone initial processing by the outer region, and some of the neurons in this region receive input concerning two or even three of the different modalities. This region of association cortex is assumed to subserve more complex functions. Thus, a neuron receiving input from the "head" region of somatosensory cortex and from the visual cortex as well might be concerned with integrating sensory information about head position with visual information so that, for example, a tree is understood to be vertical even though the viewer's head is tipped sideways.

Neurons of this inner region connect with still other association areas which are part of the limbic system, and through these connections highly processed sensory information can be invested with emotional and motivational significance. For example, removal of part of the limbic association areas may abolish the emotional response to painful stimuli.

Further perceptual processing involves not only arousal, attention, learning, memory, language, and emotions, but also comparing the information presented via one sensory modality with that of another. For example, we may hear a growling dog, but our perception of the actual event taking place varies markedly, depending upon whether our visual system detects that the sound source is an angry animal or a loudspeaker.

We put great trust in our sensory-perceptual processes despite the inevitable modifications we know to exist. Some factors known to distort our perceptions of the real world are as follows:

1. Afferent information is distorted by receptor mechanisms and by its processing along afferent pathways, e.g., by adaptation.
2. Such factors as emotions, personality, experience, and social background can influence perceptions so that two people can witness the same events and yet perceive them differently.
3. Not all information entering the central nervous system gives rise to conscious sensation. Actually, this is a very good thing because many unwanted signals, generated by the extreme sensitivity of our receptors, are cancelled out. The afferent systems are very sensitive. Under ideal conditions the rods of the eye can detect the flame of a candle 17 mi away. The hair cells of the ear can detect vibrations of an amplitude much lower than that caused by the flow of blood through the blood vessels of the ear and can even detect molecules in random motion bumping against the tympanic membrane. Olfactory receptors respond to the presence of only four to eight odorous molecules. It is possible to detect one action potential generated by a pacinian corpuscle. If no mechanisms existed to select, restrain, and organize the barrage of impulses from the periphery, life would be unbearable. Information in some receptors' afferent pathways is not cancelled out; it simply does not give rise to a conscious sensation. For example, stretch receptors in the walls of some of the largest blood vessels effectively monitor both absolute blood pressure and its rate of change, but people have no conscious awareness of their blood pressure.
4. We lack suitable receptors for many energy forms. For example, we can have no direct information about radiation and radio or television waves until they are converted to an energy form to which we are sensitive. Many regions of the body are insensitive to certain forms of energy because they lack the appropriate receptors. The brain itself has no pain or pressure receptors, and brain operations can be performed painlessly on patients who are still awake, provided that the cut edges of the sensitive brain coverings are infused with local anesthetic.

However, the most dramatic examples of a clear difference between the real world and our perceptual world can be found in illusions and drug- and disease-induced hallucinations, when whole worlds can be created.

Any sense organ can give false information; e.g., pressure on the closed eye is perceived as light in darkness, and electric stimulation of the afferent nerve fibers from any sense organ produces the sensory experience normally arising from the activation of that receptor. Why do such illusions appear? The two bits of false information just mentioned arise because most afferent fibers transmit information about only one modality, and action potentials in a given afferent pathway going to a certain area of the brain signal the kind of information normally carried in that pathway.

In conclusion, the two processes—transmitting data through the nervous system and interpreting it—cannot be separated. Information is processed at each synaptic level of the afferent pathways. There is no one point along the afferent pathways and no one particular level of the central nervous system below which activity cannot be a conscious sensation and above which it is a recognizable, definable sensory exerience. Perception has many levels, and it seems that the many separate stages are arranged in a hierarchy, with the more complex stages of sensory processing receiving input only after it has been processed by the more elementary systems. Every synapse along the afferent pathways adds an element of organization and contributes to the sensory experience.

Overview of the Motor Systems

Carrying out a coordinated movement is a complicated process. Consider the events associated with reaching out and grasping an object. The fingers are extended (straightened) and then flexed (bent), the degree of extension depending upon the size of the object to be grasped, and the force of flexion depending upon the weight and consistency of the object. Simultaneously, the wrist, elbow, and shoulder are extended, and the trunk is inclined forward, the exact movements depending upon the distance of the object and the direction in which it lies. The shoulder is stabilized to support first the weight of the arm and then the added weight of the object. Upright posture is maintained, despite the body's continually shifting center of gravity.

The building blocks for this simple action—as for all movements—are active motor units, each comprising one motor neuron together with all the skeletal muscle fibers the motor neuron innervates (Chap. 10). The only way that the nervous system controls the contraction of skeletal muscle is by means of synaptic input to the motor neurons of the motor units.

All the motor neurons for a given muscle comprise the **motor neuron pool** for that muscle. Different motor neurons within the pool are used for different muscle movements. Recall (Chap. 10) that there are three basic types of motor units characterized by the type of muscle fibers they contain: "oxidative slow fibers," "oxidative fast fibers," or "glycolytic fast fibers." These motor unit types, as their names suggest, differ widely in their contractile and metabolic properties. The motor neurons controlling each of the three types of motor units are also different, particularly in their size and the order in which they are recruited so that, in a voluntary contraction of graded tension, for example, the smaller motor units, which have oxidative slow fibers, are recruited early during the course of the contraction. The motor units having the oxidative fast muscle fibers are recruited later as the contraction reaches higher tension levels. However, the order of recruitment of the motor neurons need not be "specified" by the synaptic input to them: Rather, the order of recruitment is determined by their excitability characteristics, which in turn are determined by their size.

Neural inputs from many sources converge

upon the motor neurons to control their activity, and the precision of a coordinated muscle movement depends upon the *balance* of this input. For example, if an inhibitory system is damaged, lessening its input to the motor neurons, the still-normal excitatory input will be unopposed; the motor neurons will fire excessively and the muscle will be hyperactive. No one source of input to a motor unit is essential for movement, but to provide the precision and speed of normally coordinated movements, a balanced input from all sources is necessary.

Each of the myriad coordinated body movements is characterized by a set of motor unit activities occurring over space and time, and the interrelating inputs which converge upon the motor neurons to control their activity are the subject of this chapter. We present first a summary of a model of motor-system functioning (Fig. 19-1) and then describe each component in detail.

First, a general "command" such as "pick up sweater" or "write signature" or "answer telephone" is generated, but it is not known how the neurons involved in this function are activated or even exactly where in the brain the "command" neurons are located. Simultaneously, neurons in certain areas of the brain are receiving information from receptors in the muscles, tendons, joints, skin, vestibular apparatus, and eyes, providing these brain areas with information about the starting position of the body part that is to be "commanded" to move. This information is integrated with the signals from the "command neurons," and the initial **program**, i.e., pattern of neuronal activity needed to perform the desired movement, is prepared. The various regions of the brain involved in

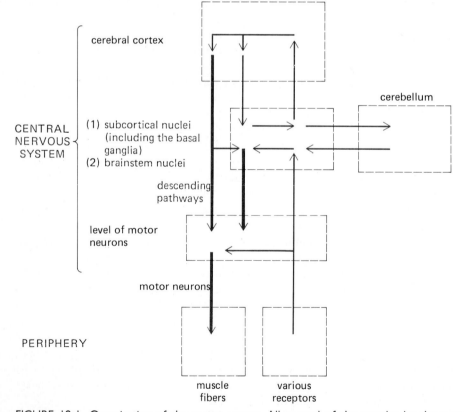

FIGURE 19-1. Organization of the motor system. All control of the muscles by the motor system is exerted via the motor neurons.

this overall process include areas of association cortex and cerebellum.

In many cases, the program is relayed to specific motor regions of cortex, where it causes changes in the firing patterns of neurons there. The axons from some of these motor-cortex neurons descend directly to the level of the motor neurons (in the lower brain stem or spinal cord); this direct descending pathway is the **corticospinal pathway**. This pathway is not the only route for conveying programs and other motor commands to the motor neurons; a second pathway, known as the **multineuronal pathway**, exists and will be described later. Thus, the execution, or performance, of a movement is achieved by brain areas different from those involved in the initiation and preparation of the program for that movement.

As the descending pathways pass through the various areas of the brain, those regions concerned with motor control [particularly the basal ganglia (Fig. 19-2) and cerebellum], receive from them a constant stream of information about what actions

are supposed to be taking place. They also simultaneously receive reports from receptors in the periphery about the actions that actually are taking place. Any discrepancies between the intended and the actual movements are detected, and program corrections are sent to the motor cortex. The pathways that loop from the motor cortex to the basal ganglia and cerebellum and back again continue to operate throughout the course of the movement. However, although activation of these loops takes only about $\frac{1}{50}$ s, very rapid movements do not provide enough time for continual correction, and the entire course of the action in such movements is completely preprogrammed.

Voluntary and involuntary actions

Given such a model, it is difficult to use the word **voluntary** with any real precision, but we shall use it to refer to those actions which have the following characteristics:

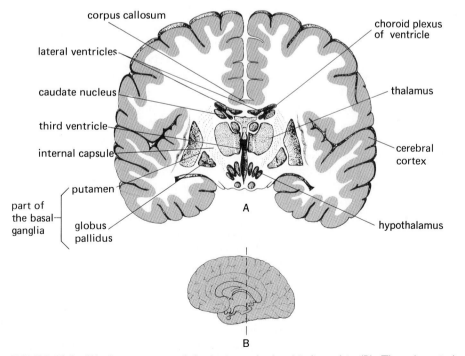

FIGURE 19-2. (A) Cross section of the brain at the level indicated in (B). The subcortical nuclei labeled caudate, putamen, and globus pallidus make up a large part of the basal ganglia.

1. These are actions we think about. The movement is accompanied by a conscious awareness of what we are doing and why we are doing it rather than the feeling that it "just happened," a feeling that often accompanies reflex responses.
2. Our attention is directed toward the action or its purpose.
3. The actions are the result of learning. Actions known to have disagreeable consequences are less likely to be performed voluntarily.

In the previous example of reaching to pick up an object, the activation of some of the motor units, such as those actually involved in grasping the object, can be classified clearly as voluntary, but most of the muscle activity associated with changing postural support during the act is initiated without any conscious, deliberate effort. In fact almost all motor behavior involves both conscious and unconscious components, and the distinction between the two cannot be made easily.

Even a highly conscious act such as threading a needle involves the unconscious postural support of the hand and arm and inhibition of the antagonistic muscles (those muscles whose activity would oppose the intended action, in this case the muscles which straighten the fingers). Moreover, unconscious basic reflexes such as dropping a hot object can be influenced by conscious effort. If the hot object is something that took a great deal of time and effort to prepare, one probably would not drop it but would try to inhibit the reflex, holding on to the object until it could be put down safely.

Most motor behavior is neither purely voluntary nor purely involuntary but falls at some point on a spectrum between these two extremes. But even this statement is of little help because patterned muscle movements shift along the spectrum according to the frequency with which they are performed. For example, when a person first learns to drive a car with standard transmission, stopping is a fairly complicated process involving the accelerator, brake, and clutch. The sequence and force of the various movements depends upon the speed of the car, and their correct implementation requires a great deal of conscious attention. With

practice the same actions become automatic. If a child darts in front of the car of an experienced driver, the driver does not have to think about the situation and decide to remove his or her foot from the accelerator and depress the brake and clutch. Upon seeing the child, the driver immediately and automatically stops the car. A complicated pattern of muscle movements is shifted from the highly conscious end of the spectrum over toward the involuntary end by the process of learning.

Such a shift presumably occurs because relevant programs have been established, possibly by altering the number or effectiveness of synaptic connections between relevant neurons (see Chap. 20). During early stages of motor learning, movements are slow and clumsy and are characterized by hesitant exploration; at this time the movements depend upon sensory feedback for guidance. With repetition of the movement, programming (i.e., learning) takes place. Thus, with later attempts at the movement, the altered synaptic connections are utilized, the dependence on sensory feedback is diminished, and the movement is carried out with greater speed and efficiency.

Whether activated voluntarily or involuntarily, given motor units are frequently called upon to serve many different functions. For example, one demand upon the muscles of the limbs, trunk, and neck is made by postural mechanisms; these muscles must support the weight of the body against gravity, control the position of the head and other parts of the body relative to each other to maintain equilibrium, and regain stable, upright posture after shifts in position. Superimposed upon these basic postural requirements are the muscle movements associated with locomotion. For these purposes, the muscles must be capable of transporting the body from one place to another under the coordinated commands of neural mechanisms for alternate stepping movements. And added to the requirements of posture and locomotion can be highly skilled movements such as those of a ballet dancer or hockey player. The motor units are activated, and the sometimes conflicting demands are settled, usually without any conscious, deliberate effort.

We now turn to an analysis of the individual components of the motor-control system, beginning with local control mechanisms, because their activity serves as a base upon which the pathways descending from the brain frequently exert their influence.

Local Control of Motor Neurons

Some of the synaptic input to motor neurons arises not from the descending pathways emphasized in the model but from neurons at the same level of the central nervous system as the motor neurons themselves. (For example, motor neurons whose axons leave the spinal cord via the ventral root of the first cervical nerve are influenced by synaptic input from interneurons at this same spinal level or by afferent fibers that have entered the cord at this level of the spinal cord.) Indeed, some of these neurons are influenced by receptors in the very muscles controlled by the motor neurons, in other nearby muscles, and in the tendons associated with these muscles. These receptors monitor muscle length and tension and pass this information via afferent nerve fibers into the central nervous system. This input forms the afferent component of purely local reflexes which provide negative-feedback control over muscle length and tension. In addition, this afferent information is transmitted to motor areas of the brain, such as the cerebral cortex, cerebellum, and basal ganglia, where it can be integrated with input from other types of receptors.

Length-monitoring systems and the stretch reflex

Muscle length and changes in muscle length are monitored by stretch receptors embedded within skeletal muscle. These receptors are made up of afferent nerve endings wrapped around modified muscle cells, both partially enclosed in a fibrous capsule. The entire structure is called a **muscle spindle**. The modified muscle fibers within the spindle are known as **spindle fibers**, whereas the skeletal muscle fibers which form the bulk of the muscle are the **skeletomotor** (or **extrafusal**) **fibers** (Fig. 19-3).

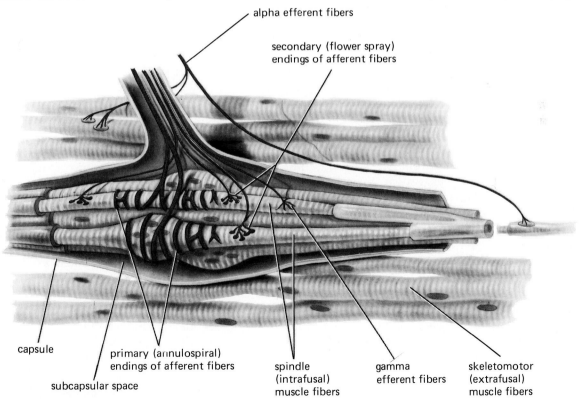

FIGURE 19-3. A muscle spindle. *(Redrawn from Merton.)*

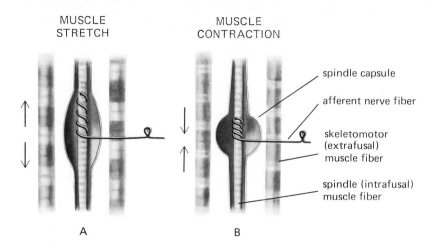

MUSCLE
STRETCH

MUSCLE
CONTRACTION

spindle capsule

afferent nerve fiber

skeletomotor
(extrafusal)
muscle fiber

spindle (intrafusal)
muscle fiber

A

B

FIGURE 19-4. (A) Passive stretch of the muscle activates the spindle stretch receptors and causes an increased rate of firing in the afferent nerve. (B) Contraction of the muscle removes tension on the spindle stretch receptors and lowers the rate of firing in the afferent nerve. Arrows indicate the direction of force on the muscle spindles.

The muscle spindles are parallel to the muscle fibers such that passive stretch of the muscle pulls on the spindle fibers, stretching them and activating their receptor endings; the greater the stretch, the faster the rate of firing. In contrast, contraction of the skeletomotor fibers and the resultant shortening of the muscle release tension on the muscle spindles and slow down the rate of firing of the stretch receptors (Fig. 19-4).

There are different kinds of spindle receptors, one responding best to how much the muscle has been stretched, another to both the absolute magnitude of the stretch and the speed with which it occurs. Although the two kinds of stretch receptors are separate entities, they will be referred to collectively as the **muscle-spindle stretch receptors.**

When the afferent neurons from the muscle spindle enter the central nervous system, they divide into branches which can take several different paths. One group of terminals (A in Fig. 19-5) directly forms excitatory synapses upon motor neurons going back to the muscle that was stretched, thereby completing a reflex arc known as the **stretch reflex.** This reflex is probably most familiar in the form of the knee jerk, which is tested as part of the routine medical examination. The physician taps the patellar tendon, which stretches over the knee and connects muscles in the thigh to a bone in the foreleg. As the tendon is depressed, the muscles to which it is attached are stretched, and the receptors within the muscle spindles are activated. Information about the change in length of the muscles is

fed back to the motor neurons controlling the same muscles. The motor units are excited, the thigh muscles shorten, and the patient's foreleg is raised to give the knee jerk. The proper performance of the knee jerk tells the physician that the afferent limb of the reflex, the balance of synaptic input to the motor neurons, the motor neurons themselves, the neuromuscular junctions, and the muscles are functioning normally. In usual circumstances, of course, the stretch receptors are activated neither simultaneously nor so strongly, and the response is not a jerk.

Because the group of afferent terminals mediating the stretch reflex synapses directly with the motor neurons without the interposition of any interneurons, the stretch reflex is called **monosynaptic.** Stretch reflexes are the only known monosynaptic reflex arcs in people; all other reflex arcs are **polysynaptic,** having at least one interneuron (and usually many) between the afferent and efferent neuron.

A second group of afferent terminals (B in Fig. 19-5) ends on interneurons which, when excited, inhibit the motor neurons controlling antagonistic muscles whose contraction would interfere with the reflex response. For example, the normal response in the knee-jerk reflex is straightening of the knee to extend the foreleg. The antagonists to these extensor muscles are a group of flexor muscles which, when activated, draw the foreleg back and up against the thigh. If both of these opposing groups of muscles are activated simultaneously, the knee on interneurons which, when excited, activate **synergistic muscles**, i.e., muscles whose contraction assists the intended motion. For example, in the knee jerk, interneurons facilitate motor neurons which control other leg extensor muscles.

A fourth group of afferent terminals (D in Fig. 19-5) synapses with interneurons which convey information about the muscle length to areas of the brain dealing with coordination of muscle movement. Information from the muscle stretch receptors that reaches cerebral cortex has a conscious correlate and, with information from the joints, ligaments, and skin, contributes to the conscious awareness of the position of a limb or joint.

Alpha-gamma coactivation. Because the muscle spindles are parallel to the large skeletomotor muscle fibers, stretch on the spindle fibers is removed when the skeletomotor fibers contract to even a small extent. If the spindle stretch receptors were permitted to go slack at this time, they would stop firing action potentials and this source of information over the very large range of shortening would be lost. To prevent this, the spindle muscle fibers at each end of the spindle themselves are frequently made to contract during the shortening of the skeletomotor fibers, thus maintaining tension in the central region of the spindle where the stretch receptors are located. The spindle fibers are not large and strong enough to shorten whole muscles and move joints; their sole job is to produce tension on the spindle stretch receptors.

The muscle fibers in the spindles shorten in response to motor neuron activity (Fig 19-6), but the motor neurons which activate the spindle fibers are usually not the same motor neurons as those which activate the skeletomotor muscle fibers. The motor neurons controlling the skeletomotor muscle fibers are larger and are classified as **alpha motor neurons**, whereas the smaller neurons whose axons innervate the spindle fibers are known as **gamma motor neurons**. Experiments have shown that alpha and gamma motor neurons are **coactivated**, i.e., fired at almost the same time, during many voluntary and involuntary movements. It appears that the alpha motor neurons are activated by synaptic input according to the strength of the intended

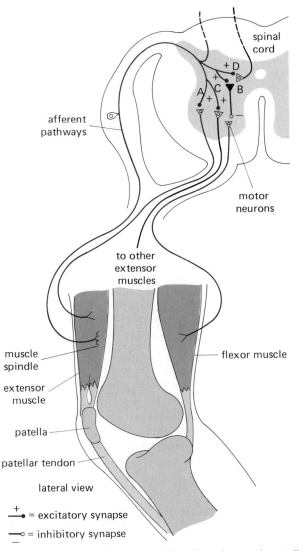

FIGURE 19-5. The afferent nerve fiber from the muscle spindle and the terminals which are involved in the knee jerk.

joint is immobilized and the leg becomes a stiff pillar. This is certainly what is required in some situations, but if the foreleg is to be extended from a flexed position, the motor neurons which activate the flexor muscles must be inhibited as the motor neurons controlling the extensor muscles are activated. The excitation of one muscle and the simultaneous inhibition of its antagonistic muscle is called **reciprocal innervation**.

A third group of terminals (C in Fig. 19-5) ends

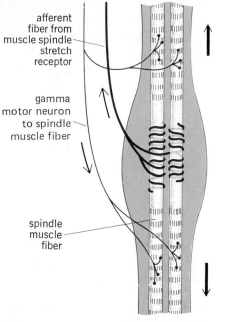

afferent
fiber from
muscle spindle
stretch
receptor

gamma
motor neuron
to spindle
muscle fiber

spindle
muscle
fiber

CONTRACTED

FIGURE 19-6. As the two striated ends of a spindle fiber contract in response to gamma motor neuron activation, they pull on the center of the fiber and stretch the receptor, which is located there.

movement, whereas the gamma motor neurons are simultaneously activated according to the distance of the intended movement. As long as there is no discrepancy between the two parameters, the firing rate of the muscle spindle afferents does not change; if a discrepancy should arise, however, the difference between the actual length of the muscle and the intended length will be detected and corrections will be made.

It seems, however, that during most voluntary actions in people, the gamma-efferent activity is of limited usefulness, since the spindle afferents provide a continuous supply of information to the central nervous system only during very slow movements, and any change in spindle afferent activity is relatively small even when an ongoing movement is physically prevented. Probably, feedback from the muscle spindles is much more important during the first performance of an unfamiliar motor act. Once the act is well learned and can be performed without conscious thought, the gamma-efferent ac-

tivity seems to be ineffective and irrelevant, and the feedback information from the muscle-spindle afferents is unnecessary for accurate performance of the movement. Thus, it is likely that the nervous system uses information from the muscle spindles in different ways for different movements under different conditions, but a more specific statement than this cannot be made at present.

Tension-monitoring systems

A second component of the local motor-control apparatus monitors tension rather than length. A given input to a muscle by a group of motor neurons does not always produce the same amount of tension. The tension developed by a contracting muscle depends on the force-velocity properties of the muscle, the muscle length, and the degree of muscle fatigue (Chap. 10). Because one set of inputs to the motor neurons can lead to a large number of different tensions, feedback of information is necessary to inform the motor control systems of the tension actually achieved.

The receptors employed in the tension-monitoring system are the **Golgi tendon organs**, which are located in the tendons near their junction with the muscle. Endings of afferent nerve fibers are wrapped around collagen bundles of the tendon, which are slightly bowed in the resting state. When the attached muscle fibers contract, they pull on the tendon, straighten the collagen bundles, and distort the receptor endings of the afferent neurons. These receptors discharge in relation to the tension generated by the contracting muscle, and the afferent neuron's activity supplies the motor control systems with continuous information about this tension. In addition, branches of the afferent neuron inhibit, via polysynaptic reflexes, the motor neurons of the contracting muscle.

Other local afferent systems

In addition to the afferent information from the spindle stretch receptors and Golgi tendon organs of the activated muscle, other input is fed into the local motor control systems. For example, stimulation of the skin causes a facilitation of the flexor motor neurons and an inhibition of the extensor motor neurons so that the body part is moved away from the stimulus. (Recall the flexion reflex described in Chap. 8, which demonstrates this point.) On the other hand, stimulation of the skin causes just the opposite response on the other side of the body; activation of the flexor motor neurons on the stimulated side of the body is accompanied by excitation of the extensor motor neurons on the opposite side (the **crossed-extensor reflex**, Chap. 8). Information from joint receptors also affects the local motor control systems. These inputs are all polysynaptic, i.e., one or more interneurons lie between the afferent fibers and motor neurons.

Interneurons

The interneurons are of several types; some are confined to the general level of the motor neuron upon which they synapse, others have processes that extend up or down through several segments of the spinal cord, whereas still others have very long processes that can extend throughout the length of the cord. The interneurons with long processes are important in movements that involve the coordinated interaction of, for example, an arm and a leg, whereas the interneurons with shorter processes are activated in movements that involve a smaller region of the body, for example, a shoulder and arm.

The interneurons of the spinal cord have emerged as integrative elements in their own right. Varied inputs converge upon them, not only from local sources but from higher centers, too (Fig. 19-7). For example, the inhibitory interneuron in the stretch reflex receives a large number of inputs from higher levels of the nervous system as well as from other local receptors, and one can conclude that such interneurons are concerned with far more than converting an excitatory input from an afferent fiber into an inhibitory output upon an alpha motor neuron.

One additional role suggested for interneurons is that they act as "switches," which enable a reflex to be turned on or off under the command of higher motor centers. Such switching has been shown to occur phasically during cyclical events such as walking, so that a reflex that is present during one phase of the cycle may be missing or altered at

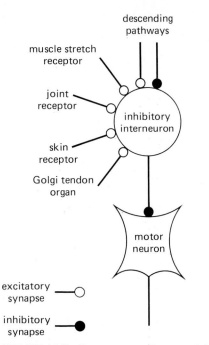

FIGURE 19-7. Convergence of axons of descending pathways and afferent nerve fibers and joint, skin, and muscle-tendon receptors onto a local interneuron.

another phase of the same cycle. Thus, local reflexes do not necessarily have the obligatory nature we commonly associate with reflex behavior but rather have permissive, modulatory influences that are active only when higher motor centers allow them to be. But just as inputs from motor centers in the brain may switch reflexes on and off, so too might a given descending motor command have different effects depending upon the state of the muscles as signalled by afferent fibers conveying information from the muscles and tendons. In other words, because of convergence of local and descending inputs onto interneurons in the motor pathways, descending commands are subject to local influence even before the commands arrive at the motor neurons.

Descending Pathways and the Brain Centers Which Control Them

The various motor centers in the brain influence the motor neurons by descending pathways, which are composed of axons that are derived from neurons in either the cerebral cortex or brainstem. There are three mechanisms by which these pathways alter the balance of excitatory and inhibitory synaptic input converging upon the alpha motor neurons:

1. By synapsing directly upon the alpha motor neurons themselves. This has the advantage of speed and specificity.
2. By synapsing on the gamma motorneurons, which produce tension on the muscle spindles and influence the alpha motor neurons indirectly. This pathway and the pathway above seem always to operate together (alpha-gamma coactivation). As described earlier, this has the advantage of maintaining output from the stretch receptors.
3. By synapsing on interneurons, often, as we have mentioned, the same ones subserving the local reflexes. Although this route is not as fast as directly influencing the motor neurons, it has the advantage of the coordination built into the interneuron network as described earlier (e.g., recruitment of synergistic muscles and reciprocal innervation).

The degree to which each of these three mechanisms is employed varies, depending upon the nature of the motor task to be performed and the descending pathway that is utilized. As we have mentioned, the descending pathways can be divided into two major categories: the **corticospinal pathway** and the **multineuronal pathways**.

Corticospinal pathway

The fibers of the corticospinal pathway, as the name implies, have their cell bodies in the cerebral cortex. The axons of these cortical neurons pass without any additional synapsing to end in the immediate vicinity of the motor neurons (Fig. 19-8). The group of fibers innervating muscles of the eye, face, tongue, and throat leave these descending pathways in the brainstem to influence motor neurons whose axons travel out with the cranial nerves; the rest descend to their various terminations in the spinal cord, traveling down the spinal cord to innervate

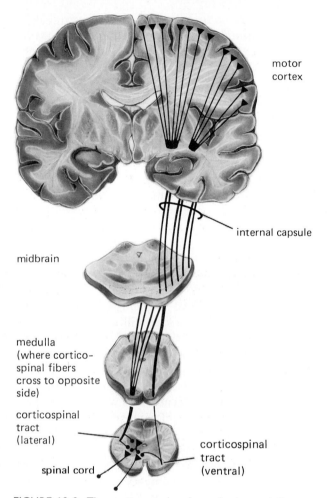

FIGURE 19-8. The corticospinal pathway. In the medulla most of its fibers cross to descend in the opposite side of the spinal cord.

motor neurons or interneurons there. Near the junction of the spinal cord and brainstem, most of these fibers cross the spinal cord to descend on the opposite side. Thus, the skeletal muscles on the left side of the body are controlled largely by neurons in the right half of the brain. The corticospinal pathway is also called the **pyramidal tract** or **pyramidal system**, perhaps because of its shape in some parts of the brain or because it was formerly thought to arise solely from the giant pyramidal cells of the cortex.

The corticospinal pathway is the major mediator of fine intricate movements, such as threading a needle. However, it is not the sole mediator of these movements, since surgical transection of the pathway in people does not completely eliminate such movements, although it does make them weaker, slower, and less well coordinated. Clearly, the multineuronal pathways (to be described next) must also contribute to the performance of delicate .movements.

The fibers of the corticospinal pathway end in all three of the ways described, i.e., on alpha motor neurons, gamma motor neurons, and interneurons. In addition, they (or their collateral branches) end presynaptically on the terminals of *afferent* neurons and on neurons of the ascending afferent pathways. The overall effect of their input to afferent systems is to limit the areas of skin, muscle, or joints allowed to influence the cortical neurons, thereby sharpening the focus of the afferent signal and improving the contrast between important and unimportant information. The collaterals also convey the information that a certain motor command is being delivered and possibly give rise to the sense of effort. Because of this descending (motor) control over ascending (sensory) information, there is clearly no real functional separation of these two systems.

Multineuronal pathways

The other axons that descend into the lower brainstem and spinal cord to influence motor neurons have their cells of origin in the upper brainstem. Yet these brainstem cells are themselves only the final link in a multineuronal pathway that consists of many neurons and involves many of the other brain regions, including the motor regions of cerebral cortex, the basal ganglia and other subcortical nuclei, and the cerebellum, that are important in motor control (Fig. 19-9). At successive neurons, the information carried by the pathways is altered according to the balance of excitatory and inhibitory input to them. Some of these complex motor pathways in the brain loop back to earlier way stations, including the cerebral cortex. These loops constitute the paths by which the programs and program corrections described in our original model modify the activity of the motor cortex, including the neurons of the corticospinal pathway and the cortical neurons which are part of the multineuronal pathways feeding onto the brainstem descending neurons. These multineuronal pathways and the structures which they connect are also called the **extrapyramidal system** to distinguish them from the corticospinal (pyramidal) pathways.

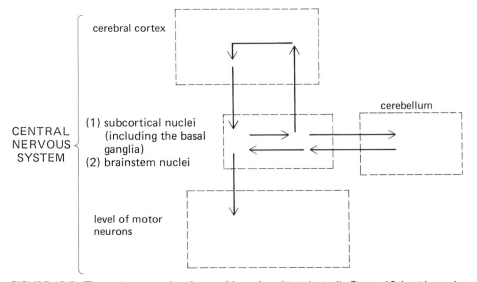

FIGURE 19-9. The multineuronal pathways. Note that this is basically Figure 19-1 without the corticospinal pathway, muscles, and receptors.

Whereas the corticospinal neurons have greater influence over the motorneurons which control muscles that are involved in more discrete movements, particularly those of the fingers and hands, the multineuronal pathways are more involved with posture and the coordination of large groups of muscles, such as the muscles of the trunk and more proximal parts of the limbs. There is a great redundancy of function, however, and just as some fibers of the corticospinal pathway end on interneurons that play important roles in posture, so also do fibers of the multineuronal pathways sometimes end directly on the alpha motor neurons to control discrete muscle movements. Because of this redundancy, loss of function resulting from damage to one system may be compensated for by the remaining systems (although the compensation is generally not complete). Thus, it is wrong to imagine a total separation of function between these two pathways, for the distinctions between them are not at all clear-cut. All normal, coordinated movements, whether automatic or voluntary, require the continual interaction of both pathways.

Cerebral cortex

Many areas of the cortex give rise to the two types of descending pathways described above, but a large number of the fibers come from the posterior part of the frontal lobe, which is therefore called the **motor cortex** (Fig. 19-10). The function of the neurons in the motor cortex varies with location in the cortex. As one starts at the top of the brain and moves down along the side (*A* to *B* in Fig. 19-10), the cortical neurons affecting movements of the toes and feet are at the top of the brain, followed (as one moves laterally down the surface of the brain) by neurons controlling leg, trunk, arm, hand, fingers, neck, and face. The size of each of the individual body parts in Fig. 19-11 is proportional to the amount of cortex devoted to its control; clearly, the cortical areas representing hand and face are the largest. The great number of cortical neurons for innervation of the hand and face is one of the factors responsible for the fine degree of motor control that can be exerted over those parts. The same brain regions that give rise to pathways

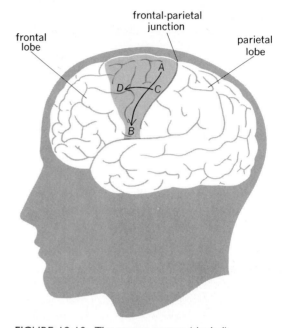

FIGURE 19-10. The motor cortex (shaded).

directly controlling motor neurons of a given muscle also contain neurons that pass to other motor centers in the brain.

Cortical neurons also contribute to different aspects of motor function as one explores from the back of the motor cortex forward (*C* to *D* in Fig. 19-10). Those in the back portion, i.e., closest to the junction of the frontal and parietal lobes, mainly contribute to the execution of specific tasks. Moving anteriorly (forward) in the cortex, this zone of neurons gradually blends into the group whose activity is involved in more complicated motor functions. Thus, damage in this more anterior area (sometimes called the **premotor area**) may be associated with difficulty in changing from one task to another, difficulty in grasping objects in response to visual control, or failure to carry out movements correctly in response to a spoken command. Moreover, this premotor area is an important gateway for relaying processed sensory information from other brain regions (e.g., thalamus and association cortex) to some of the important multineuronal loops.

None of the cortical neurons functions as an

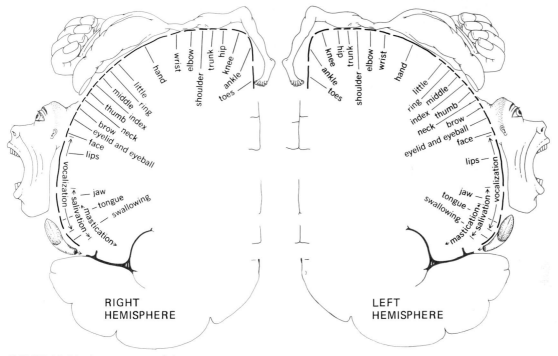

FIGURE 19-11. Arrangement of the motor cortex.

isolated unit. Rather, these neurons are interconnected so that those cortical neurons controlling motor units having related functions fire together. The more delicate the function of any skeletal muscle, the more nearly one-to-one is the relationship between cortical neurons and the muscle's motor units.

As mentioned earlier, cortical neurons governing the activity of a given muscle often receive information from the muscle itself (i.e., from the stretch receptors), the joints affected by the muscle, the muscle's tendons, or even the skin overlying the muscle. Thus, in one sense, one can visualize the cortical cells as integrators of long reflex arcs influenced by the stimulation of peripheral receptors, and, indeed, some modification of movement is explainable in these terms. In addition to these peripheral inputs, which reach the cortex by way of the thalamus, cells of the motor cortex receive input from other regions of the cortex and even from collaterals that loop back from the axons of the corticospinal neurons themselves. The effec-

tiveness of a given set of connections can change both from one instant to the next and over longer periods of time, such as those associated with development and learning. For example, at times—especially during the performance of a movement that has been learned and is, therefore, essentially "preprogrammed"—the inputs to a given cell in the motor cortex may be depressed so that the cell is little influenced by peripheral input.

Throughout this section, we have presented the cerebral cortex as the origin of pathways descending to the motor neurons. However, the neurons of cerebral cortex manifest no spontaneous generation of action potentials. Rather, their firing pattern, like that of almost all neurons, is determined solely by the balance of synaptic activity impinging upon them. In other words, as we mentioned in the earlier description of the overall model of the motor control system, the motor cortex is not the prime initiator of movement; it is simply a tremendously important relay station.

You will have noticed that in our discussion of

inputs to the neurons of the motor cortex, we have not faced the problem of naming the location of the "command" neurons that drive the motor-cortex neurons during voluntary actions. Although we cannot name these regions with any degree of certainty, parts of both the association cortex and cerebellum have been implicated by some research. For example, electrodes placed on the skull to record the electrical activity of the brain during the "decision-making period," pick up over wide areas of association cortex distinctive activity patterns (the so-called **readiness potential**) about 800 ms before the movement begins. Only afterwards, 50 to 60 ms prior to the movement, does sharper electrical activity (the **motor potential**) appear over the motor-cortex region of the hemisphere controlling the involved muscle.

It is also true that if the neurons of motor cortex are stimulated in conscious patients during surgical exposure of the cortex, the patient moves, but not in a purposefully organized way. The movement made depends upon the part of the motor cortex stimulated. The patient is aware of the movement but says, "You made me do that," recognizing that it is not a voluntary movement. In contrast, stimulation of parts of cortex in parietal lobe association areas sometimes causes patients to say that they want to make a certain movement, but the movement does not occur.

Finally, the input to the cortical neurons not only must be directed toward some purpose but must select from a variety of ways of achieving that purpose, since learned motor programs can be executed in many ways with many sets of muscles. For a simple example, a rat trained to press a lever when a signal light flashes, will do so quite consistently, but depending upon the rat's original position in the cage, its movements can be quite varied. If the rat is to the left of the lever, it moves to the right; if it is to the right, it moves to the left. If its paw is on the floor, it raises the paw; if its paw is above the lever, it lowers the paw. The only consistent act is pressing the lever. The movements performed to achieve the purpose are variable and seem almost inconsequential. How a given program is selected is not known.

Note that our system as described also does not take into account the mechanisms by which the "decision" to make a particular movement is reached. What neural events actually occur in the brain to cause one to "decide" to make a certain movement in the first place? Presently we have no insight into the answer to this question.

Subcortical nuclei (including the basal ganglia) and brainstem nuclei

As mentioned above, the multineuronal pathways descending from the cerebral cortex synapse in the basal ganglia, other subcortical nuclei, and several of the brainstem nuclei. Moreover, axons from neurons in the corticospinal pathways have collateral branches which also end in these regions.

Our understanding of the various roles that the subcortical nuclei play in motor behavior is meager, but their most important influence seems to be the feedback effect they exert on the motor cortex via the multineuronal-system loops described earlier. One example of how such a feedback influence might work is as follows: Neurons of the motor cortex fire sustained bursts of action potentials during normal motor activity. But what causes the cortical neurons to fire in this continuous fashion? It has been suggested that neurons in the thalamus, themselves excited by neurons of the motor or premotor cortex, loop back to the cortex and reciprocally excite the cells there (Fig. 19-12). Such a positive-feedback loop could reinforce the activity of neurons of the motor cortex during sustained motor behaviors.

It is also necessary, however, to have inhibitory mechanisms capable of modulating the positive-feedback loop (for example, to stop ongoing activities) and of suppressing unwanted, conflicting behaviors. This function seems to be played by neurons of the basal ganglia, which end in inhibitory synapses upon the thalamic cells (Fig. 19-12). The question now becomes: What systems are responsible for selecting which movements shall be maintained and which suppressed, i.e., which cortical neurons shall be activated and which shall not? It is suggested that the basal ganglia perform this func-

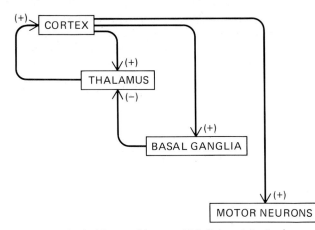

FIGURE 19-12. Neuronal loops which link activity in the cortex, thalamus, and basal ganglia.

tion too. Thus, cells of these nuclei seem to be responsible for selecting and maintaining desired patterns of motor activity while suppressing other antagonistic or unwanted behaviors. The basal ganglia are likely candidates for such a complex task of integration because they are not only anatomically interconnected with the motor cortex and thalamus but also receive input from the premotor and association regions of cortex, the cerebellum, the brainstem nuclei, and even structures of the limbic system.

The theories presented above explain the defects in motor control associated with certain diseases. For example, persons suffering from **Parkinson's disease** tend to maintain ongoing activity and have difficulty establishing new behaviors. Thus, if walking, they tend to keep walking; if seated, they tend to remain seated. One particular subcortical nucleus implicated in this disease is the **substantia nigra** (black substance), which gets its name from the dark pigment in its cells. The neurons of the substantia nigra liberate the transmitter **dopamine** from their axon terminals in the basal ganglia, and dopamine acts to inhibit neurons there. When the substantia-nigra neurons degenerate, as they do in Parkinson's disease, the amount of the dopamine delivered to the postsynaptic cells is reduced; therefore, in turn, the inhibitory activity of the basal ganglia is interfered with, the thalamus is overac-

tive, the positive feedback to the cortex runs unimpeded, and the ability to select specific behaviors is diminished.

The dopamine receptors on the basal-ganglia neurons still respond to dopamine in Parkinson's disease; it is the quantity of transmitter that is deficient. Therefore, L-dopa, which is a precursor of dopamine (Chap. 8), is given to these patients, enters the bloodstream, crosses the blood-brain barrier, and enters neurons where it is converted to dopamine (dopamine itself is not used because it cannot cross the blood-brain barrier). The newly formed dopamine activates the receptors in the basal ganglia and relieves the symptoms of the disease. On the other hand, drugs which block dopamine receptors (sometimes used in the treatment of schizophrenia) have as one of their unwanted side effects symptoms characteristic of Parkinson's disease.

Cerebellum

The cerebellum sits on top of the brainstem, as can be seen in Fig. 8-10. Its most important function is to govern posture and movement, but unfortunately, this function and others of the cerebellum are understood only in a very general sense. One accepted fact is that the cerebellum does influence other regions of the brain responsible for motor activity.

Possibly the cerebellum triggers programs (or subsets of programs) that are used in a given movement. Programs are created on the basis of past experience, i.e., as the result of learning, and they help smooth out the performance of actions by avoiding time-consuming feedback delays. It has been suggested that the programs under cerebellar control are generally small and affect such simple movements as, for example, the program for starting and stopping actions. In fact, the cerebellar "start" and "stop" programs can be considered "subprograms" of more complicated movements.

It has been suggested that the cerebellum smooths actions in a second way, as well. It receives input both from cortex and subcortical centers carrying information about what the muscles *should* be doing and from many afferent systems carrying information about what the muscles *are* doing. If there is a discrepancy between the two, an error signal is sent from the cerebellum to the cortex and subcortical centers, where new commands are initiated to decrease the discrepancy and smooth the action.

The afferent inputs to the cerebellum come from the vestibular system, eyes, ears, skin, muscles, joints, and tendons, i.e., from the major receptors affected by movement. Input from receptors in a single small area of the body end in the same region of the cerebellum as do the inputs from the higher brain centers controlling the motor units in that same area of the body. Thus, information from the muscles, tendons, and skin of the arm arrive at the same area of cerebellar cortex as do the motor commands for "arm" from the cerebral cortex. This permits the cerebellum to compare motor commands with muscle performance.

The symptoms of persons with cerebellar damage are quite clear and generally include the following:

1. They cannot perform movements smoothly. If they reach for an object, their movements are jerky and are accompanied by oscillating, to-and-fro tremors which become more marked as the hand approaches the object.
2. They walk awkwardly with the feet well apart. They have such difficulty maintaining their balance that their gait appears drunken.
3. They cannot start or stop movements quickly or easily, and if, for example, they are asked to rotate the wrist over and back as rapidly as possible, their motions are slow and irregular.
4. They cannot easily combine the movements of several joints into a single, smooth, coordinated motion. To move the arm, they might first move the shoulder, then the elbow, and finally the wrist.

Maintenance of Upright Posture and Balance

The skeleton supporting the body is a system of long bones and a many-jointed spine which cannot stand alone against the forces of gravity. Even when held together with ligaments and covered with flesh, it cannot stand erect without the support given by coordinated muscle activity. This applies not only to the support of the body as a whole but also to the fixation of segments of the body on adjoining segments, e.g., the support of the head on the trunk or the trunk on the legs.

Added to the problem of supporting the body's weight against gravity is that of maintaining equilibrium. A person is a very tall structure balanced on a relatively small base, and the center of gravity (Fig. 19-13) is quite high, being situated just above

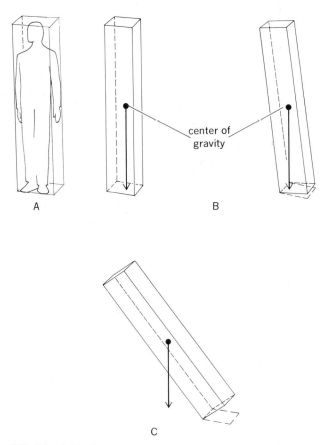

FIGURE 19-13. The center of gravity is the point in a system at which, if a string were attached and pulled upward, all the downward forces due to gravity would be exactly balanced. The center of gravity must remain within the vertical projections of the base if stability is to be maintained. (A) The vertical projections of the base in which the center of gravity must remain for stable posture in human beings. (B) Stable conditions. (C) Unstable conditions.

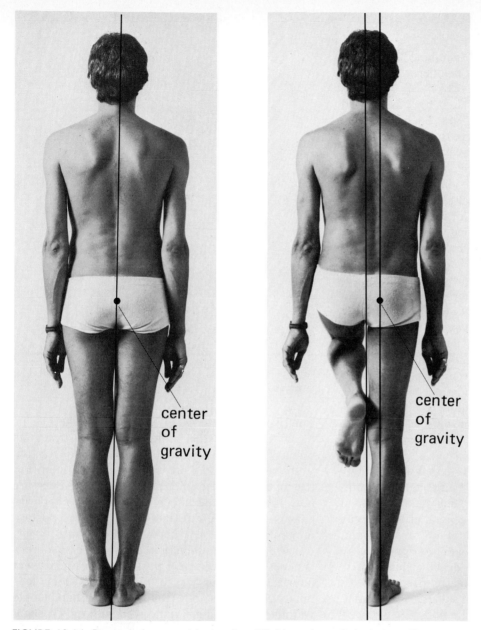

FIGURE 19-14. Postural changes with stepping. (A) Normal standing posture. The line of the center of gravity falls directly between the two feet. (B) As the left foot is raised, the whole body leans toward the right so that the center of gravity shifts over the right foot.

the pelvis. For stability, the center of gravity must be kept within the small area determined by the vertical projections of the base of the body (the feet) that is in contact with the surface upon which the person stands (Fig. 19-13). Yet human beings are almost always in motion, swaying back and forth and from side to side even when attempting to stand still. Moreover, they often operate under conditions of unstable equilibrium and would be toppled easily by physical forces in the environment if their equilibrium were not protected by reflex postural mechanisms.

Such mechanisms do exist, of course. When a normal person raises both arms so that they extend forward, the head and trunk simultaneously move slightly backward; in a similar manner, when a normal subject is asked to bend backward, the knees flex forward. Note the postural changes associated with stepping shown in Fig. 19-14. Such automatic compensatory movements serve to maintain the position of the center of gravity over the person's base of support. One characteristic of postural adjustment is its adaptability, depending upon the way the body was supported initially. Also, there is an element of feedforward control in that if the postural displacement is anticipated, the necessary adjustments are initiated shortly before the movement itself begins.

The maintenance of posture and balance is accomplished by means of complex interacting reflexes, all the components of which we have met previously. The efferent pathways of the reflex arcs are the alpha motor neurons to the skeletal muscles. The afferent pathways come from two basic sources: the eyes and the **proprioceptors (proprioception** is the sense of position of one part of the body relative to another; therefore, the proprioceptors are the receptors from muscles, tendons, joints, and skin, and the vestibular portions of the inner ear).

It is presently not possible to define accurately all the neuronal mechanisms mediating the coordination of the muscles used for maintaining upright posture and balance, but two general principles can be suggested: (1) Rapid postural adjustments, i.e., those occurring within 100 ms or so, appear to be performed automatically by local mechanisms in response to afferent input primarily from receptors within the muscles, joints, and skin. (2) In contrast, slower processes for postural adjustment are performed by central mechanisms that make use of vestibular and visual input as well as information from the muscles, joints, and skin.

The local adjustment mechanisms can only distinguish movements of one part of the body relative to another, and they cannot distinguish movements of the body relative to the external environment, which requires visual and vestibular input as well. Thus, except for the rapid adjustments, the integration of simultaneous input from several receptors is used in determining the appropriate adjustments.

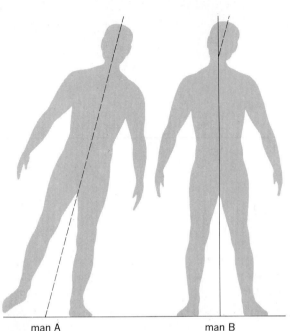

FIGURE 19-15. Interaction of the vestibular, skin, joint, and tendon receptors.

The two figures in Fig. 19-15, who have their heads tilted to exactly the same degree, illustrate this point. The heads are tilted and the inputs from the vestibular apparatus are similar in both men, but the inputs from the joint receptors in the neck are not. Thus, the final integration of the synaptic inputs is quite different, indicating that the posture of man A is unstable whereas that of man B is not.

We have stated that input from several sources is necessary for effective postural adjustments, yet interfering with any one of these inputs alone does not cause a person to topple over. Blind people maintain their balance quite well with only a slight loss of precision, and people whose vestibular mechanisms have been destroyed have very little disability in everyday life as long as their visual system and muscle, joint, and skin receptors are functioning (one such man was even able to ride a motorcycle). The conclusion to be drawn from such examples is that the sensory information received by postural control mechanisms is redundant and that somehow these mechanisms are able to detect when one of the inputs is absent or invalid and shift attention to other inputs that are not. One particular type of information becomes critically important only when the other, redundant inputs have been lost. This redundancy among the sensory inputs allows postural control mechanisms to perform under a wide variety of different environmental conditions.

Walking

The cyclic, alternating movements of walking are controlled by **central pattern generators** within the central nervous system—some in the spinal cord at the level of the motor neurons and (probably) some in the brainstem as well. Although these centers can initiate appropriate motor output in the absence of feedback information from the receptors in muscles, joints, skin, vestibular apparatus, and eyes, such afferent input normally contributes substantially to locomotion by acting on the central pattern generators and on the motor neurons themselves. Such afferent information assures that the output of the central pattern generator is optimally adapted to the environment. For example, as the leg approaches an extended position, information signalling the hip position inhibits further extensor activity and facilitates the flexors. If the limb movement should be slowed by external factors, such as wading through water, the absence of these signals prevents a premature flexion of the limb.

Thus, the pattern generators in the spinal cord at the level of the motor neurons involved in walking are influenced by input from the receptors in the two limbs. They also are affected by input from the multineuronal descending pathways, whose cell bodies, you will recall, are in the brainstem where they receive input from brain regions, such as other brainstem nuclei, the cerebellum, and those structures still unidentified which somehow arrive at the decision to walk. (It is possible that such decisions are made at different places in the brain depending upon the reasons for the movement, e.g., escape, feeding, courtship, or exploratory behavior.) Presumably, then, the spinal cord is under a continual bombardment from the variety of multineuronal descending pathways both during the initiation of activity and during ongoing activity to help adjust the movement to the environment and to the person's goals.

It has been suggested that the spinal central pattern generators are themselves composed of smaller units, each controlling an individual muscle group, such as the ankle extensors. The total output pattern of the limb would be determined by the combined activity of these units, yet the arrangement of individual groups could be altered to allow for such variations in locomotion as walking backward, walking on the knees or crawling, or walking in high-heeled boots.

One attractive hypothesis is that the brain can control stereotyped behaviors such as locomotion simply by activating the appropriate spinal central program generators or their individual units by way of the multineuronal descending pathways. Moreover, it has been suggested that in human beings, the neurons which control stereotyped behaviors lie in a region just below the thalamus (the subthalamus) and specific areas of the upper brainstem. These areas are, however, part of the complicated loops that typically feed into the brainstem cells of the multineuronal pathways, and trying to assign precise functions to them or the many motor regions of the brain with which they connect is uncertain, to say the least.

In summary, the picture that is emerging in the neural control of locomotion is that of an organization of central pattern generators which provide an output to the motor neurons. The pattern of this output may be modified through a complex group of feedback systems that act on the motor neurons, interneurons, and the central pattern generators. Information from the peripheral receptors and information about the neural commands being delivered both reach the cerebellum, which acts, by way of multineuronal descending pathways, on the local motor regions to modify their output. Furthermore, the basic locomotor patterns can be altered to adapt each step to the goals of the person and to her or his environment.

The reader will have noticed our frequent use of such terms as "possibly," "it has been suggested," "hypothesis," etc., which indicate how little is known for certain about locomotion and, indeed, motor behavior in general.

CONSCIOUSNESS AND BEHAVIOR

The term **consciousness** includes two distinct concepts, **states of consciousness** and **conscious experience**. The second concept refers to those things of which a person is aware—thoughts, feelings, perceptions, ideas, dreams, reasoning—during any of the states of consciousness. In contrast, a person's state of consciousness, i.e., whether awake, asleep, drowsy, etc., is defined in two ways: (1) by behavior, covering the spectrum from coma to maximum attentiveness, and (2) by the pattern of brain activity that can be recorded electrically, usually as the electric-potential difference between two points on the scalp. The record is the **electroencephalogram**, or **EEG** (Fig. 20-1). The EEG is such an important tool in identifying the different states of consciousness that we begin with it.

States of Consciousness

Electroencephalogram

The EEG patterns are the result of varying extracellular current flows in the patch of cerebral cortex which underlies the recording electrode, the current flows reflecting varying degrees of activity of the individual cortical neurons. While we usually think of neuronal activity in terms of action potentials, action potentials contribute little to the EEG (except in the unusual circumstance when the activity in a large group of neurons is synchronized). The EEG is largely due to summed postsynaptic potentials, particularly in the large neurons whose processes are perpendicular to the cortical surface. The analysis of the actual waveform of the EEG is very complicated because some of the postsynaptic potentials which cause it are excitatory whereas others are inhibitory, some are near the surface of the cortex close to the recording electrode, others are deeper.

The wave patterns of the EEG are about 100 times smaller than the amplitude of an action potential, and the frequency of the waves may vary from 1 to 30 Hz, the patterns being distinguished from one another by both their frequencies and amplitudes. Changes in EEG patterns are correlated with changes in behavior spanning the entire normal range from alertness to sleep.

The explanation for the wavelike nature, or rhythmicity, of the EEG is not certain. It is currently thought that nuclei in the thalamus contain the rhythm generators which provide for a fluctuating output in nerve fibers leading from the thalamus to the cortex, where they end synaptically on the dendrites of the neurons. The thalamic rhythms may arise either because of an intrinsic pacemaker property of the neurons or because of a feedback network of excitatory and inhibitory neurons.

The EEG is a useful clinical tool because the normal patterns are altered over brain areas that are diseased or damaged. A common neurological disease associated with distinctive, abnormal EEG patterns is **epilepsy**, in which a large collection of neurons discharges in synchrony. This disease is also usually associated with stereotyped changes in behavior, which vary according to the part of the brain that is affected and can include a temporary loss of consciousness.

The waking state

Behaviorally, the waking state is far from homogeneous, comprising the infinite variety of things one can be doing. The prominent EEG wave pattern of an awake, relaxed adult whose eyes are closed is a slow oscillation of 8 to 13 Hz, known as the **alpha rhythm** (Fig. 20-1B), recorded best over the parietal and occipital lobes. The alpha rhythm is associated with decreased levels of attention, and when alpha rhythms are being generated, subjects commonly report that they feel relaxed and happy. A high degree of alpha rhythm is also associated with meditational states. However, people who normally experience high numbers of alpha episodes have not been shown to be psychologically different from others with lower levels, and the relation between brain-wave activity and subjective mood is obscure. People have been trained to increase the amount of alpha brain rhythms by providing a feedback signal such as a tone whenever alpha rhythms appear in the EEG.

When people are attentive to an external stimulus (or are thinking hard about something), the alpha rhythm is replaced by lower, faster oscillations (the **beta rhythm**, 18–20/s, Fig. 20-1A). This transformation is known as **EEG arousal** and is associated with the act of attending to stimuli rather than with the perception itself; for example, if people open their eyes in a completely dark room and try to see, EEG arousal occurs. With decreasing attention to repeated stimuli, the EEG pattern reverts to the alpha rhythm.

Sleep

The EEG pattern changes profoundly in sleep. As a person becomes increasingly drowsy, the alpha rhythm is gradually replaced by slow-wave patterns, and sleep begins. This phase of sleep, called **slow-wave sleep**, is, itself, divided into stages, each having an EEG pattern characterized by progressively slower frequency and higher voltage (Fig. 20-1D to F). The depth of sleep also varies with the different stages, stage 1 (3–7/s) corresponding to light sleep and stage 4 (0.5–2/s) to deep sleep.

Sleep always begins with an initial progression from stage 1 into a period of deep sleep (stage 4). The passage from stage 1 to stage 4 takes 30 to 45 min, and then the process reverses itself, taking the same length of time to return to stage 1.

It is difficult to tell precisely when a person passes from drowsiness into stage 1 and on through the successive stages of slow-wave sleep, for in slow-wave sleep there is still considerable tonus in

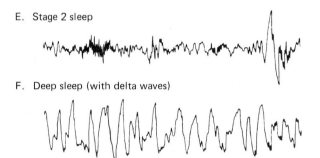

A. Alert (beta rhythm)

B. Relaxed with eyes closed (alpha rhythm)

C. Paradoxical (REM) sleep

D. Stage 1 sleep

E. Stage 2 sleep

F. Deep sleep (with delta waves)

FIGURE 20-1. EEG patterns corresponding to various states of consciousness.

an EEG pattern similar to that of an awake, alert person, this type of sleep is called **paradoxical sleep**.

At the onset of paradoxical sleep, there is an abrupt and complete inhibition of tone in the postural muscles, although periodic episodes of twitching of the facial muscles and limbs and movements of the eyes occur. (Paradoxical sleep is therefore also called rapid-eye-movement or **REM sleep**.) Respiration and heart rate are irregular, and blood pressure may go up or down. When awakened during paradoxical sleep, 80 to 90 percent of the time subjects report that they have been dreaming.

Continuous recordings show that the two states of sleep follow a regular 30- to 90-min cycle, each episode of paradoxical sleep lasting 10 to 15 min (Fig. 20-2). Thus, slow-wave sleep constitutes about 80 percent of the total sleeping time in adults, and paradoxical sleep about 20 percent. The time spent in paradoxical sleep increases toward the end of an undisturbed night.

Although adults spend about one-third of their time sleeping, we know little of the functions served by it. The brain as a whole does not rest during sleep, and there is no generalized inhibition of activity of the cerebral neurons. There is, however, a reorganization of neuronal activity, some individual neurons being less active during sleep than during

the postural muscles and only a small change in cardiovascular and respiratory activity. The sleeper can be awakened fairly easily and, if awakened, rarely reports dreaming. Slow-wave sleep has a characteristic kind of mentation described by subjects as "thoughts" rather than "dreams." The thoughts are more plausible and conceptual; they are more concerned with recent events of everyday life and are more like waking-state thoughts than are true dreams.

Throughout the night there occurs a sequence of light- and deep-sleep episodes. Except for the initial period of light sleep, the light-sleep (stage 1) periods are invariably associated with rapid movements of the eyes behind closed lids and an EEG pattern similar to that of an awake, alert person (Fig. 20-1C). Because the person is asleep but has

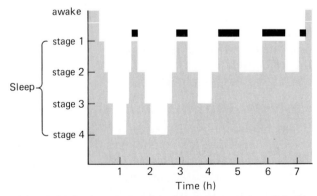

FIGURE 20-2. A typical night of sleep in a young adult (the diagram represents the mean taken from many all-night recordings). The heavy lines indicate periods of paradoxical sleep, which are characterized by low-voltage EEG patterns and rapid eye movements. *(Adapted from Kandel and Schwartz.)*

waking and others showing just the opposite pattern. The total blood flow and oxygen consumption of the brain, signs of its metabolic activity, do not decrease in sleep. It has been suggested that the functional significance of sleep lies not in the short-term recovery but in the relatively long-term chemical and structural changes that the brain must undergo to make learning and memory possible.

Neural substrates of states of consciousness

The prevailing state of consciousness is probably the resultant of the interplay between several neuronal systems. Most are parts of the **reticular formation**. The reticular formation proper lies in the central core of the brainstem in the midst of the neuronal pathways passing between the brain and spinal cord, but neurons in the midline thalamus also function as extensions of this brainstem system. Neurons of the reticular formation have incredibly widespread networks of connections, and they influence and are influenced by virtually all areas of the central nervous system (Fig. 20-3). Neurons with such far-flung connections are perfectly situated to be involved in governing the level of activity of the central nervous system as a whole.

The reticular formation is not homogeneous, and its neurons are sometimes clustered together according to the type of neurotransmitter they release. The three most prominent groups are those that contain norepinephrine, dopamine, and serotonin. Discrete areas of the reticular formation frequently have specific functions, several of which have already been discussed: It helps to coordinate skeletal muscle activity (Chap. 19); it contains the primary cardiovascular and respiratory control centers (Chaps. 11 and 12); and it monitors the huge numbers of messages ascending and descending through the central nervous system (Chaps. 8 and 18). In this chapter we are concerned with its role in determining the states of consciousness.

Neural mechanisms for the sleep-wake cycle. In 1934, it was discovered that after the cerebrum is surgically isolated from the spinal cord and lower three-fourths of the brainstem, the EEG loses the

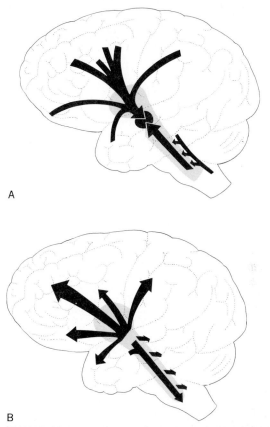

A

B

FIGURE 20-3. (A) Convergence of descending, local, and ascending influences upon the reticular formation (shaded area). (B) Projections from reticular formation to spinal cord, brainstem and cerebellum, and cerebrum. *(Adapted from Livingston.)*

waves typical of an awake animal, indicating that some neural structures in the separated brainstem or spinal cord are essential for arousal and the maintenance of a waking EEG. There are also clusters of neurons in the brainstem necessary for slow-wave and paradoxical sleep (Fig. 20-4).

The areas involved in slow-wave sleep seem to be in the regions of reticular formation that run along the midline of the medulla (the **raphe nuclei**). The cells of these nuclei release the transmitter **serotonin** from their terminals, and one theory postulates that serotonin induces the brain changes that result in sleep. This theory is supported by evidence

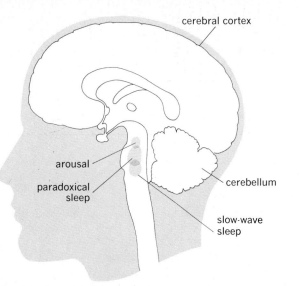

cerebral cortex

arousal

paradoxical
sleep

cerebellum

slow-wave
sleep

FIGURE 20-4. Brainstem structures involved in arousal, para-doxical sleep, and slow-wave sleep. *(Adapted from Jouvet.)*

that introducing serotonin or its precursors into the nervous system causes sleep, whereas blocking serotonin synthesis causes insomnia.

Whereas the neurons of the medullary raphe nuclei are thought to be involved in slow-wave sleep, anatomically and biochemically distinct regions of the brainstem seem to control paradoxical sleep. Some of these cells fire in bursts that are correlated with the rapid eye movements, muscle twitches, and special EEG waves in parts of the brain that are involved in vision and are also associated with paradoxical sleep. Other brainstem neurons fire continually throughout paradoxical sleep, and this activity can be correlated with the EEG pattern of paradoxical sleep and with the loss of tone in postural muscles characteristic of this state. These cells release **acetylcholine** at their axon terminals.

In summary, two opposing systems exist in the brain, one (with several components) a sleep-producing system and the other an arousal system. The two interact in such a way that activity in one suppresses activity in the other, and vice versa. It is the interaction of these two systems which produces the sleep-waking cycles. However, the periodic inhibition of the arousal systems (which would normally cause sleep) can be overridden by input from afferent pathways or other brain centers so that activity in the arousal systems is maintained sufficiently high to keep one awake or to interrupt sleep. In fact, the waking mechanisms seem to be more easily activated than those causing sleep. An example familiar to all parents is that it is much easier to arouse a sleeping child than to get an alert, attentive child to sleep.

In addition to the transmitters mentioned above, there are a number of other sleep-inducing chemical substances which can be found in the blood, urine, cerebrospinal fluid, and brain tissue. One of these causes the large, slow waves of sleep known as the delta waves (Fig. 20-1F), and is called the **delta sleep–inducing peptide (DSIP)**. Another sleep-promoting substance, **substance S**, increases slow-wave sleep when infused into the cerebral ventricles. Despite the fact that we have found substances such as serotonin, DSIP, and substance S which are related to sleep, we do not know the exact relationship of these substances to sleep.

Neural mechanisms for directed attention. The term **directed attention** is taken to mean the avoidance of distraction by irrelevant stimuli while seeking out and focusing on stimuli that are important. Directed attention can be studied in experimental animals by the presentation of a novel stimulus, which causes a change in the EEG pattern from the quiet resting alpha rhythm to the beta rhythm characteristic of arousal. If the novel stimulus happens to be meaningful to the animal, these EEG changes are associated with behavioral changes. The animal stops whatever it is doing and looks around or listens intently, orienting itself toward the stimulus source. This behavior is called the **orienting response**.

Since attention is directed only toward stimuli that are meaningful to the animal, it seems that at some point in the nervous system there must be an estimation of the importance of the incoming sensory information so that attention can be paid to the stimulus or it can be ignored. A decision that the stimulus is irrelevant results in a progressive decrease in response to the stimulus (**habituation**).

For example, when a loud bell is sounded for the first time, it may evoke an orienting response because the animal might be frightened or curious at the novel stimulus. After several ringings, however, the animal makes progressively less response and eventually may ignore the bell altogether. An extraneous stimulus of another modality or the same stimulus at a different intensity restores the original response (**dishabituation**). Habituation is not due to receptor fatigue or adaptation; rather, it involves a depression of synaptic transmission possibly related to a prolonged inactivation of the calcium channels in the presynaptic axon terminals, which results in a decreased calcium influx with depolarization and a decrease in the amount of neurotransmitter released in response to action potentials.

While investigating the neural processes underlying directed attention, researchers have found neurons in the **parietal lobe** which increase their firing rate when the animal notices, looks at, or reaches toward an object that is relevant in some way, e.g., food when the animal is hungry, or drink when it is thirsty. However, if the looking and reaching motions occur spontaneously or if the object is not meaningful, the increase in firing does not occur. It appears that these neurons are part of an apparatus that can associate complex sensory cues with the motivational significance of the object and with the movements associated with looking and reaching.

One brainstem nucleus, the **locus coeruleus**, which projects to the parietal cortex (as well as to many other parts of the central nervous system) is strongly implicated in directed attention. The activity of neurons in this nucleus increases with sensory stimulation, with increased attention, and with the termination of paradoxical sleep. The transmitter released by these neurons (**norepinephrine**) has an inhibitory action, which is thought to decrease the background electrical activity in the brain so that signals from the sensory systems are clearer; i.e., the weak signals are suppressed and the vigorous signals are enhanced, therefore the difference between them is increased. Thus, neurons of the locus coeruleus seem to improve information processing during directed attention.

Conscious Experience

All subjective experiences are popularly attributed to the workings of the mind. This word conjures up the image of a nonneural "me," a phantom interposed between afferent and efferent impulses, with the implication that mind is something more than neuronal activity. The truth of the matter is that physiologists have absolutely no idea of the mechanisms which give rise to conscious experience. Nor are there even any scientifically meaningful hypotheses concerning the problem.

Conscious experiences are difficult to investigate because they can be known only by verbal report. Such studies lack scientific objectivity and must be limited to people. In an attempt to bypass these difficulties scientists have studied the behavioral correlates of mental phenomena in other animals. For example, a rat deprived of water performs certain actions to obtain it. These actions are the behavioral correlates of thirst. But it must be emphasized that we do not know whether the rat consciously experiences thirst; this can only be inferred from the fact that human beings are conscious of thirst under the same conditions.

Thus, one crucial question that cannot be investigated in experimental animals is whether conscious experiences actually influence a behavior. Although intuitively it might seem absurd to question this, the fact is that the answer has important implications for the development of one's concept of humanity. It is possible that conscious experience is an epiphenomenon.

Consider the following sequence of events: The ringing of a telephone reminds a student that he had promised to call his mother. He finishes the page he has been reading and makes the call. What causes him to do so? The epiphenomenon view holds that the conscious awareness accompanies but does not influence the passage of information from afferent to motor pathways. Thus, in this view, behavior occurs automatically in response to a stimulus, memory stores supplying the direct link between afferent and efferent activity. In contrast to this view, the processing of afferent information (the sound of the bell), acting through memory stores, could result in the conscious awareness of his promise, which, in turn, leads to the relevant activity in motor pathways descending from the cortex. There is no way of choosing between these two views at present.

In contrast to this presently unapproachable question, some aspects of conscious experience have yielded, at least in part, to experimentation in human beings. In general, these aspects can be described by two questions: (1) What is the relationship between information arriving over the principal afferent pathways and the conscious experience? (2) Is an anatomically distinct area of the central nervous system involved in conscious experience?

The first question has been approached by studies performed in conscious human subjects whose brains are exposed for neurosurgery. Tiny electrodes lowered into different areas of the thalamus record the activity of single cells. In certain thalamic cells a given somatic stimulus, such as movement of a joint, regularly produces a given response which contains, in coded form, information about the precise location and intensity of the stimulus. The electric response recorded is unchanged regardless of whether the patients are keenly aware of each stimulus or whether their attention is diverted so that they are unaware that they have been stimulated; i.e., the electrical activity is the same whether or not information about the stimulus is incorporated into conscious experience. (The thalamic cells which respond in this manner are in nuclei known to form part of the specific afferent pathway. Thus, information relayed in the specific ascending pathways does not necessarily become part of conscious experience.)

There are other neurons in nonsensory parts of the thalamus which have quite different properties and patterns of activity, which are more closely related to conscious experience. Each of these cells, called **novelty detectors**, responds to input from many parts of the body, but it rapidly ceases responding to repeated stimuli of the same kind as the person's attention to them wanes. The firing pattern is more closely related to the degree to which the person is aware of a given stimulus than to precise information about a certain sensory modality.

Other evidence also suggests that central structures play an important role in conscious experience. Clinical examples of conscious sensory experiences existing in the absence of afferent neural input are not unusual. For example, after a limb has been amputated, the patient sometimes feels as though it were still present. The nonexistent limb, called a phantom limb, can be the "site" of severe pain.

In this context it is interesting to study the conscious experiences of persons undergoing periods of sensory deprivation. Student volunteers lived 24 h a day in as complete isolation as possible—even to the extent that their movements were greatly restricted. External stimuli were almost completely absent, and stimulation of the body surface was relatively constant. At first, the students slept excessively, but soon they began to be disturbed by vivid hallucinations which sometimes became so distorted and intense that the students refused to continue the experiment. The neural bases of these hallucinations are poorly understood, but it has been suggested that the central structures may generate patterns of activity corresponding to those normally elicited by peripheral stimuli when varied sensory input is absent and that conscious experience is not solely dependent upon the senses.

There seems to be an optimal amount of afferent stimulation necessary for the maintenance of the normal, awake consciousness. Levels of stimulation greater or less than this optimal amount can lead to trances, hypnotic states, hallucinations, "highs," or other altered states of consciousness. In fact, alteration of sensory input is commonly used to induce such experiences intentionally. Thus, in reply to our first question, we can say that conscious experience is determined by central structures as well as by peripheral stimuli.

The answer to the second question: Is there a specific brain area in which conscious experience resides? can be gleaned from conscious persons undergoing neurosurgical procedures and from persons who have accident- or disease-inflicted damage to parts of their brains, but the answer is still far from clear. The only available clues have been gained by inference. For example, with evolution comes (we assume) greater complexity of conscious experience. Since the human brain is distinguished anatomically from that of other mammals by a greatly increased volume of cerebral cortex, this is the logical place to look for the seat of conscious experience.

The cortex has been stimulated when patients on the operating table are fully alert and the brain is exposed. If a certain area of association cortex in temporal lobe is stimulated, the subjects may report one of two types of changes in conscious experience. Either they are aware of a sudden change in their interpretation of the present situation, e.g., what they are seeing or hearing suddenly becomes familiar, or strange, or frightening, or coming closer, or going away; or they have a sudden flashback or awareness of an earlier experience. Although they are still aware of where they are, an experience comes to them and repeats itself in the same order and detail as the original experience. It may have been a particular occasion when they were listening to music. If asked to do so, they can hum an accompaniment to the music. If in the past they thought the music beautiful, they think so again. During such electric stimulation, visual or auditory experiences are recalled only if the patient was attentive to them when they originally occurred.

Certain types of conscious experiences have never been produced by such stimulation. No one has ever reported periods when he was trying to make a decision, or solve a problem, or add up a row of figures. It is also interesting that no brain area other than temporal association cortex has been found from which complete memories have been activated. However, one cannot assume that association cortex is the site of the stream of consciousness, since stimulation of these cortical neurons causes the propagation of action potentials to many other parts of the brain.

It is probably necessary to take a different view of the matter, i.e., that consciousness is not localized in a particular part of the brain but is a function of the integrated action of the brain. For example, there is no one point along the ascending pathways and no one particular level of the nervous system below which activity cannot be associated with a conscious sensation and above which it is a recog-

nizable, definable sensory experience. Every synapse along the neural pathways by which information is processed adds an element of meaning and contributes to the conscious experience.

On the other hand, consciousness and, we presume, the accompanying conscious experiences, are inevitably lost when regions deep within the cerebrum or nerve tracts passing from the reticular formation to association cortex are damaged. Although the matter is far from settled, it has been suggested that systems of fibers originating in the brainstem reticular formation and having widespread connections with other areas of the brain somehow determine which of these functional areas is to gain temporary predominance in the on-going stream of the conscious experience. As we mentioned above, the noradrenergic fibers which originate in the locus coeruleus may serve this function.

The concept we want to leave as an answer to the question of where the conscious experience resides is perhaps best presented in the following analogy. In an attempt to say which part of a car is responsible for its controlled movement down a highway, one cannot specify the wheels, or axle, or engine, or gasoline. The final performance of an automobile is achieved only through the coordinated interaction of many components. In a similar way, the conscious experience is the result of the coordinated interaction of many areas of the nervous system. One neuronal system would be incapable of creating a conscious experience without the effective interaction of many others.

Motivation and Emotion

Motivation

Motivation can be defined as those processes responsible for the goal-directed quality of behavior, but it cannot presently be explained in neurophysiological terms. Some motivated behavior is clearly related to homeostasis, i.e., the maintenance of a relatively unchanging internal environment, an example being putting on a sweater when one is cold. In such homeostatic goal-directed behavior specific bodily needs are being satisfied, the word "needs" having a physical or chemical correlate. Thus, in our example the correlate of need is a drop in body temperature, and the correlate of need satisfaction is return of the body temperature to normal. The neurophysiological integration of much homeostatic goal-directed behavior has been discussed earlier (thirst and drinking, Chap. 13; food intake and temperature regulation, Chap. 15; reproduction, Chap. 16).

However, many kinds of behavior, e.g., the selection of a particular sweater on the basis of style, have little if any apparent relation to homeostasis. Much of human behavior fits this latter category and is influenced by a complex set of factors such as habit, learning, experience, and emotions. These factors as well as the actual homeostatically related "needs" are thought to determine the degree of motivation, or drive, behind a particular behavior.

A concept inseparable from motivation is that of **reward** and **punishment**, rewards being things that organisms work for or things which strengthen behavior leading to them, and punishments being the opposite. They are related to motivation in that rewards may be said to satisfy needs. Many psychologists believe that rewards and punishments constitute the incentives for learning. Because virtually all behavior is shaped by learning, reward and punishment become crucial factors in directing behavior. Although some rewards and punishments have conscious correlates, many do not. Accordingly, much of human behavior is influenced by factors (rewards and punishments) of which the individual is unaware.

We have thus far described motivation without regard to its neural correlates. Present knowledge is limited to some of the brain areas (and some of their interconnecting pathways and neurotransmitters) which are important in motivated behavior.

Much of the available information concerning the neural substrates of motivation has been discovered by studying the reinforcing effects of rewards and punishments. Some of the most important of these studies have used **self-stimulation** techniques, in which an unanesthetized experimental animal regulates the rate at which electric stimuli are delivered through electrodes previously im-

planted in discrete brain areas. The animal is placed in a box containing a lever it can press (Fig. 20-5). If no stimulus is delivered to the animal's brain when the bar is pressed, he usually presses it occasionally at random. However, if a stimulus is delivered to the brain as a result of the bar press, a different behavior can result, depending upon the location of the electrodes. If the animal increases his bar-pressing rate above control, the electric stimulus is, by definition, rewarding; if he decreases it, the stimulus is punishing. Thus, the rate of bar-pressing is taken to be a measure of the effectiveness of the reward (or punishment).

Bar-pressing that results in self-stimulation of the sensory and motor systems produces response rates not significantly different from the control rate. Brain stimulation through electrodes implanted in certain areas of the hypothalamus serves as a positive reward. Animals with electrodes in these areas bar-press to stimulate their brains from 500 to 5,000 times per hour! In fact, electric stim-

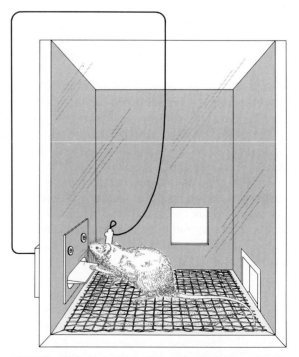

FIGURE 20-5. Apparatus for self-stimulation experiments. *(Adapted from Olds.)*

ulation of some areas of hypothalamus is more re-warding than external rewards; e.g., hungry rats often ignore available food for the sake of electri-cally stimulating their brains. It seems that self-stimulation is triggering those systems that energize certain behaviors, causing them to be performed, in other words, the systems that serve motivation.

High rates of self-stimulation are found in those areas (or some of the neural pathways leading to them) which normally mediate highly motivated be-havior, e.g., feeding, drinking, or sexual behavior. Consistent with this is the fact that the animal's rate of self-stimulation in some areas increases when it is deprived of food; in other areas, it is decreased by castration and restored by the admin-istration of sex hormones. Thus, it appears that neurons controlling homeostatic goal-directed be-havior are themselves intimately involved in the reinforcing effects of reward and punishment.

Although the rewarding sites are more densely packed in the hypothalamus than anywhere else in the brain and animals will bar-press at higher rates when the electrodes are implanted there, self-stim-ulation can be obtained from a large number of sites at different brain levels. Researchers have turned away from looking for specific clusters of neurons whose activation is responsible for motivated be-havior; rather, they prefer to study relatively non-specific systems which affect those parts of the brain involved in such goal-directed behaviors as eating, drinking, and sexual behavior. One of the systems implicated is the **medial forebrain bundle** (Fig. 20-6), a large tract of ascending and descend-ing axons, some long, passing from one end of the brain to the other, and some short, extending only a few millimeters. Fibers in this tract affect virtually every level of the brain, but they have a particularly strong influence on the hypothalamus. Interestingly, neurons of the locus coeruleus, mentioned earlier in the context of directed attention, constitute a significant portion of the fibers in the medial fore-brain bundle.

Chemical mediators for motivation. It is strongly suspected that **norepinephrine** and **dopamine** (both of which belong to the family of catecholamines) are transmitters in the pathways which underlie the brain reward systems and motivation. This state-ment is supported by several kinds of evidence. First, the anatomical sites which give the highest rates of self-stimulation are along catecholamine-mediated pathways. Second, rates of self-stimula-

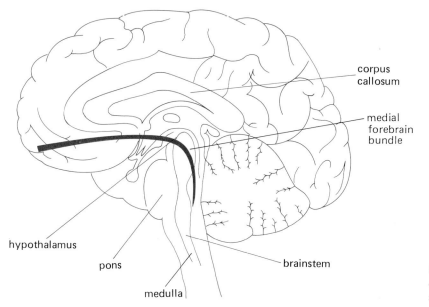

corpus callosum

medial forebrain bundle

hypothalamus

pons

medulla

brainstem

FIGURE 20-6. The medial forebrain bun-dle, a major pathway between the brain-stem and limbic system.

tion by experimental animals are altered by drugs that affect transmission at catecholamine-mediated synapses. For example, drugs (such as amphetamine) which increase synaptic activity in the catecholamine pathways increase self-stimulation rates. Conversely, drugs (such as chlorpromazine, an antipsychotic agent) that lower activity in the catecholamine pathways decrease self-stimulation. However, a recently discovered system of neurons, which is activated by enkephalin, has also been found to support self-stimulation.

The catecholamines and enkephalin are also, as we shall see, implicated in the pathways subserving learning. This is not unexpected since rewards and punishments are believed to constitute the incentives for learning.

Emotion

Related to motivation are the complex phenomena of **emotion**. The emotional tone associated with an action lends strength or intensity to it, and in this way emotion can be seen as a factor contributing to motivation.

There are two ways of looking at emotion. There are **complex behaviors**, such as attack, with emotional expressions, such as laughing or blushing, which are external expressions of an inner feeling; and there are **inner emotions**, which are felt entirely within a person. Scientists are presently trying to understand the operation of the chain of events leading from the perception of an emotionally toned stimulus to the inner emotion experienced, i.e., the feeling of fear, love, anger, joy, anxiety, hope, etc., and to the emotional expressions and complex displays of emotional behavior. Most experiments point to the involvement of the limbic system, which has been studied in experimental animals using either electrode stimulation or surgical destruction of specific areas within it. Of course, there was no way to assess the inner emotional feelings of the animals; instead they were observed for behaviors which are usually associated with emotions in human beings.

Autonomic activity and hormonal changes are particularly evident during complex emotional behavior and even in many emotional expressions; therefore, the hypothalamus (which serves to integrate much of the activity of the autonomic nervous system and the endocrine system with appropriate behaviors and is also part of the limbic system) is one logical place to investigate. Stimulation of certain hypothalamic structures elicits behavior that *seems* to have a strong inner emotional component; yet if the hypothalamus is isolated from other portions of the limbic system, the emotional component of the behavior is lacking.

Scientists investigated other structures of the limbic system while searching for brain structures that account for the missing emotionality. Stimulation of various limbic structures did indeed cause a wide variety of complex emotional behaviors. For example, stimulation of one particular area caused the animal to approach as though expecting a reward. Stimulation of a second area caused the animal to stop the behavior it was performing, as though fearing punishment. Stimulation of a third region caused the animal to arch its back, puff out its tail, hiss, snarl, bare its claws and teeth, flatten its ears, and strike. Simultaneously, its heart rate, blood pressure, respiration, salivation, and concentrations of plasma epinephrine and fatty acids all increased. Clearly, this behavior typified that of an enraged or threatened animal. Moreover, the animal's behavior could be changed from savage to docile simply by electrically stimulating different areas of the limbic system.

Surgical or disease-induced damage to parts of the limbic system also led to a great variety of changes in behavior generally associated with emotion. Destruction of the **amygdala** (a group of nuclei in the tip of the temporal lobe, which we shall encounter again in our discussion of learning, see Fig. 8-12) in some species of experimental animals produced docility in an otherwise savage animal, whereas damage to an area deep within the brain produced vicious rage in a tame animal; the rage caused by this lesion could be counteracted by a lesion in the amygdala. (Some lesioned animals manifest bizarre sexual behavior in which they attempt to mate with animals of other species, and females assume male positions and attempt to mount other animals.)

Limbic areas have been stimulated in awake human beings undergoing neurosurgery. These patients, relaxed and comfortable in the experimental situation, reported vague feelings of fear or anxiety during periods of stimulation to certain areas (even though they were not told when the current was on). Stimulation of other areas induced pleasurable sensations which the subjects found hard to define precisely.

The models that are presently available to explain emotional behavior, emotional expressions, and inner feelings are obviously far from complete. It seems that components of the limbic system (including input from the cerebral cortex) pass information to the hypothalamus. This limbic input conveys information about the meaning of the external stimulus (e.g., whether it is threatening, friendly, sad, etc.), including information gleaned from memory and understanding as well as sensory input. The hypothalamus then integrates the endocrine, autonomic, and even some of the automatic skeleto-motor activities that form appropriate emotional behavior. Moreover, components of the input from other regions of the limbic system serve to inhibit some hypothalamic responses, such as that of the enraged animal described above, so that the hypothalamic response is appropriate for the situation. In addition to its role in sensory integration, memory, and understanding, the cerebral cortex provides the neural mechanisms to direct the skeleto-motor responses to the external event; for example, to approach or avoid a situation. Finally, it seems that the cerebral cortex is necessary for the conscious experience of emotional feelings.

Thus, it seems that stimuli that have emotional meaning trigger activity in the hypothalamus so that the internal state is readied for appropriate activity, e.g., fighting, running, sexual behavior, eating, etc. These autonomic and endocrine alterations occur without conscious control. However, in the complex behavior that accompanies the autonomic and endocrine components of the response, one must interact with the external environment, and structures in the forebrain account for the modulation, direction, understanding, or even inhibition of these behaviors.

Altered States of Consciousness

Schizophrenia

Schizophrenia means simply a fragmentation of mental functioning. Despite extensive research, the causes of schizophrenia remain unclear. It is probably a family of disorders rather than a single disease, but the general symptoms are clear and include altered motor behavior, perceptual distortions, disturbed thinking, altered mood, and abnormal interpersonal behavior. The motor behavior can range from total immobilization (catatonia) to wild, purposeless activity. Perceptual distortions can include hallucinations, particularly auditory ones, such as hearing voices or hearing one's own thoughts out loud. The disturbed thinking can include delusions (illogical thoughts and beliefs that are false or improbable but cannot be changed by contrary evidence or argument), e.g., the belief that one has been chosen for a special mission or is persecuted by others.

Schizophrenia is a fairly common disease, and the chance of becoming schizophrenic is about 1 percent. It can appear gradually during adolescence in an otherwise bright and normal person, or it can occur suddenly after a single precipitating event. Although the cause of schizophrenia is unknown, there is a strong suspicion that there is something biochemically different about people with this disease. Consequently, an amazing variety of tissues have been analyzed for changes, but few differences have been found that relate to schizophrenia in an understandable way.

One of the most promising explanations for the disease, the **dopamine hypothesis**, suggests that there is an abnormality in dopamine metabolism which results either in excessive amounts of this transmitter released at dopamine-mediated synapses (especially in structures of the limbic system) or alternatively, in an increase in the number of dopamine receptors. This hypothesis is not supported by tissue analyses, but it is suggested by the fact that virtually all drugs that alter schizophrenia interfere with transmission of dopamine-mediated synapses. For example, the symptoms are made worse by amphetamine, which causes the release of a number of transmitters, including dopamine, from the axon terminals, and amphetamine psychosis is taken to be a more faithful model of the symptoms of schizophrenia than any other drug-induced psychosis. The drugs most useful in treating schizophrenia (the antipsychotic drugs) are dopamine antagonists which work by blocking dopamine receptors.

Abnormalities in other neurotransmitters and neuromodulators, including serotonin, histamine, acetylcholine, the endorphins, and norepinephrine have been implicated as well. It is possible that these other transmitters modulate (or are modulated by) dopamine. The interlocking of several transmitter systems could, in fact, explain many of the variable results in research on schizophrenia.

Alternative hypotheses concerning the cause of schizophrenia have been proposed. One suggests that potent hallucinogens may be formed from neurotransmitters which are normally in the brain. For example, the hallucinogenic drugs mescaline and psilocin are methylated derivatives of dopamine and serotonin, respectively, and enzymes exist in the brain for this methylation step. Another theory suggests that there are two basic types of the disease, one caused by excess dopamine but the other due to structural deterioration of the brain tissues. Still other theories suggest that the causative agent may be a virus to which genetically predisposed people are susceptible. It is generally accepted that there is a genetic component to the cause of schizophrenia. Whereas 1 percent of the general population will be diagnosed as having the disease, 10 to 16 percent of people having a schizophrenic parent will themselves become schizophrenic.

The affective disorders: Depressions and manias

The affective disorders are those characterized by serious, prolonged disturbances of mood; they include the **depressions**, the **manias**, and swings between these two states, so-called **bipolar affective disorders**. In the depressive disorders, the prominent feature is a pervasive sadness or loss of interest or pleasure, whereas the essential feature of the manias is an elated mood, sometimes with euphoria (i.e., an exaggerated sense of well-being), overconfidence, and irritability. The affective disorders are also accompanied by changes in thought patterns and behaviors. Along with schizophrenia, they form the major psychiatric illnesses today.

Two of the monoamine neurotransmitters, norepinephrine and serotonin, are implicated in the affective disorders. One current hypothesis states that the depressions are due to deficiencies or decreased effectiveness of norepinephrine or serotonin in certain pathways, whereas the manias are caused by functional hyperactivity of these two neurotransmitters.

Much of the evidence supporting the norepinephrine-serotonin hypothesis is circumstantial and controversial; it comes largely from studies of the actions of those drugs used in the affective disorders and from the excretion rates of the various neurotransmitters and their metabolites. The drugs used to combat depressions generally potentiate the effects of the monoamines. Some (the monoamine oxidase inhibitors) inhibit an enzyme responsible for monoamine breakdown, thereby increasing the concentrations of these transmitters in the central nervous system. Others (the tricyclic antidepressants) interfere with the reuptake of norepinephrine and serotonin by presynaptic endings. Thus, both drug types result in an increased concentration of transmitter available to activate receptors on the postsynaptic membrane.

In persons suffering from depression (or going through the depressive phase of a manic-depressive bipolar disorder) there is a decrease in the urinary excretion of the monoamines and their metabolic by-products, suggesting a decreased synaptic release. In contrast, the excretion of these substances increases in manias. However, here one encounters the kind of difficulty met when trying to analyze the cause of a mental disorder: Do the changes reflect disturbances which are the cause of the disease or do they reflect the altered behaviors and mental states resulting from the disease?

Other hypotheses concerning depression suggest that the problem lies in the balance between norepinephrine and serotonin rather than in the absolute quantities of either one, or that the problem lies with the neuromodulators or receptors rather than the neurotransmitters.

Lithium carbonate is the most effective drug currently used in the treatment of the manias; it is highly specific, normalizing mood and slowing down thinking and motor behavior without causing sedation. In addition, it decreases the severity of the swings between mania and depression that occur in the bipolar disorders and, in some cases, is even effective in depression not associated with manias. Lithium decreases the release of norepinephrine from presynaptic terminals and enhances its uptake, actions opposite those of the tricyclic antidepressants. This explanation fits nicely with the hypothesis that norepinephrine is overactive in mania, but it does not explain why lithium is occasionally effective in treatment of the depressions.

Psychoactive drugs, tolerance, and addiction

In the previous sections we mentioned several drugs that are used in an attempt to combat altered states of consciousness. These, as well as other psychoactive drugs, are also used in a deliberate attempt to elevate mood (euphorigens) and produce unusual states of consciousness ranging from meditational states to hallucinations. All these drugs seem to act on neuronal membranes or synaptic mechanisms in one way or another (neurotransmitter release or reuptake, membrane permeability, etc.).

Actually, as we have mentioned, psychoactive drugs are often chemical forms related to neurotransmitters such as serotonin, dopamine (Fig. 20-7), and norepinephrine. Moreover, it is likely that endorphin and the enkephalins, in conjunction with their effects in the pain-inhibiting pathways (Chap. 18), serve as natural euphorigens. As we have seen, several of the naturally occurring euphorigens have been implicated in diseases such as schizophrenia.

In general, there seems to be a tendency for

the transmitter-receptor interaction to stay at the same baseline state and to regain that state if moved from it. Perhaps this balance is upset by some malfunction in disease states such as schizophrenia, depression, and mania. In any case, it may be this tendency to return to a baseline that underlies the dual phenomena of tolerance to and dependence on administered psychoactive drugs.

Tolerance to a drug occurs when increasing doses of that drug are required to achieve effects that initially occurred in response to a smaller dose; i.e., it takes more drug to do the same job. One cause of tolerance may be that the presence of the drug stimulates the manufacture (especially in the liver) of those enzymes that degrade it; as drug concentrations increase so do the concentrations of the degrading enzymes, so that more drug must be administered to produce the same plasma concentrations of the drug and hence the same initial effect.

However, another postulated cause for tolerance has nothing to do with drug degradation but results from the drug's action as a neurotransmitter. The drug's effects add to those of the normally occurring transmitter, thereby producing an increased effect which, by feedback mechanisms, decreases the release of the naturally occurring neurotransmitter. For example, tolerance to opiate drugs such as morphine, which react with some types of the receptors normally activated by enkephalin, may occur in the following way (Fig. 20-8). Enkephalin is normally continually released, at least to some degree, by those neurons which use it as their neurotransmitter, so the enkephalin receptors are always exposed to a certain amount of stimulation. When morphine is present, it binds to receptors not already occupied by enkephalin and thereby increases the analgesic-euphorigenic effects of the enkephalin pathways. As the enkephalin receptors are stimulated more and more strongly, a feedback system (still hypothetical) decreases the firing rate of the enkephalin-releasing neurons, thereby decreasing the synthesis and release of the transmitter and the number of receptors activated by it. The receptors, no longer receiving their usual amount of enkephalin, can therefore accept more

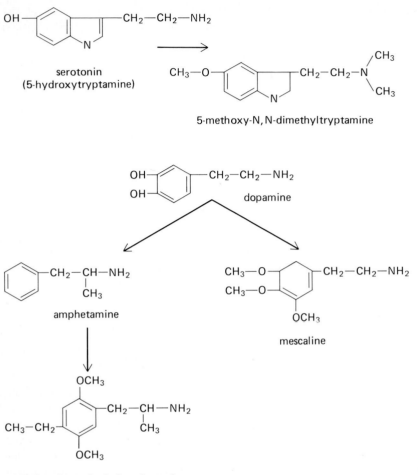

2,5-dimethoxy-4-ethylamphetamine

FIGURE 20-7. Molecular similarities between neurotransmitters (serotonin and dopamine) and some euphorigens (5-methoxy-N,N-dimethyltryptamine, amphetamine, 2, 5-dimethoxy-4-ethylamphetamine, and mescaline). At high doses these euphorigens can cause hallucinations.

morphine. In other words, to regain the previous analgesic-euphorigenic effects of full receptor activation, increased amounts of the drug are required.

This type of reasoning can explain the physical symptoms (for example, disturbed sleep, anxiety, nausea, abdominal cramps, and muscle tremors) associated with cessation of drug use, i.e., **withdrawal**. When the drug is stopped, the receptors receive for some time neither the drug nor the normal neurotransmitter (because the synthesis and release of the neurotransmitter have been inhibited by the previous prolonged drug use).

Addiction. The physical basis for addiction is not known; in fact, the word "addiction," has not even

been defined to everyone's satisfaction. Regardless, the definition finally agreed upon should include the following concepts: (1) Addiction is an atypical response to the exposure to alcohol or certain drugs; (2) it is characterized by the tendency toward progressive increase in consumption of the drug; and (3) it is characterized by a persistent tendency to relapse to use of the substance even after abstinence has been achieved and withdrawal symptoms are no longer in evidence. Thus, because of point (3) above, a person does not have to be using drugs to be an addict.

Alcohol: An example. Medicinally, the central nervous system depressants, such as alcohol, are used

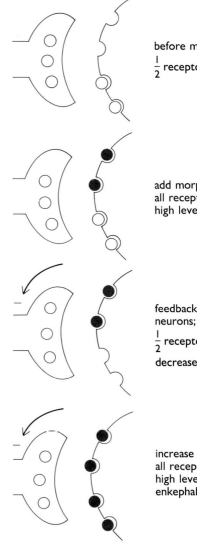

before morphine;
$\frac{1}{2}$ receptors activated

add morphine;
all receptors activated;
high level of analgesic-euphorigenic effect

feedback inhibition of enkephalin-releasing neurons;
$\frac{1}{2}$ receptors activated;
decreased analgesic-euphorigenic effect

increase morphine;
all receptors activated;
high level of analgesic-euphorigenic effect;
enkephalin-releasing neurons inhibited

○ enkephalin

● morphine

⌇ enkephalin receptor

FIGURE 20-8. Hypothesized steps in the production of addiction to morphine.

to produce sedation, sleep, and anesthesia; non-medicinally, they are used for relaxation and relief from anxiety, to produce a mild euphoria, and to promote sleep. With repeated usage they produce tolerance and both the physical and psychological forms of dependence.

Excess ingestion of alcohol causes a variety of central nervous system symptoms, for example, poor coordination, sluggish reflexes, emotional lability, and out-of character behavior (such as belligerant, aggressive behavior in a person typically shy, retiring, and mild-mannered). Sometimes, amnesia for events that happen during the intoxicated period (blackouts) occurs.

The degree of intoxication relates to body size, type of beverage consumed, rapidity of drinking, tolerance, the presence or absence of food in the stomach, and many other factors. Individual response to alcohol consumption varies greatly, but signs of intoxication such as those just described are almost always present at blood levels of 2000 mg/L, and death frequently occurs at concentrations above 5000 mg/L. (A blood alcohol level of 1000 mg/L is the legal upper limit for driving a car.) Unconsciousness usually occurs before the person can drink enough to die, but when death does occur, it is usually because of respiratory depression or vomiting followed by aspiration of the vomitus into the respiratory tract.

Alcohol alters the lipids and proteins of the plasma membrane; this leads to changes in the fluidity of the membrane, which in turn changes the permeability of the membrane to various ions. Furthermore, alcohol inhibits the ATPases in membranes. If alcohol use is continued chronically, these adverse effects are counteracted by changes in the lipid composition of the membranes, which increase their rigidity and make them resistant to the disordering effects of alcohol. This increased rigidity reduces the effect of alcohol on the membrane (a possible explanation for the basis of tolerance to alcohol?), but it also impairs a variety of other membrane functions. The effects of alcohol on various neurotransmitters, particularly the monoamines and opioid peptides because of their role in mood and anxiety, are the focus of extensive research.

Recent research on the basis of alcohol addiction suggests that, in the presence of acetaldehyde (a product of alcohol oxidation), abnormal products are formed during the metabolism of some of the neurotransmitters. They presumably act as "false" neurotransmitters and interfere with the function or metabolism of the normal neurotransmitters in some neural pathways.

The length of the period of intoxication ranges from several hours to 8 to 12 h after alcohol consumption stops. More than 95 percent of the alcohol consumed is metabolized in the liver by oxidation to acetaldehyde and coenzyme A (Chap. 17), and this occurs at a relatively slow rate; for example, it takes 5 to 6 h for an average-sized person to metabolize the alcohol in 120 mL (4 oz) of whisky or 1.2 L (1.25 qt) of beer.

Withdrawal symptoms following cessation of prolonged, heavy drinking are tremors (the "shakes"), nausea and vomiting, a dry mouth (even though the person is not dehydrated), headache and muscular weakness, nightmares, and feelings of anxiety, guilt, and irritability. Symptoms such as hallucinations and seizures can also occur. Withdrawal symptoms disappear in 5 to 7 days.

Coma and brain death

There have been great advances in the last 100 years in the development of techniques of resuscitation; for example, now it is fairly common for artificial respiration to supply gases to the lungs and cardiac massage to restore a nonfunctioning circulation. However, the brain is particularly susceptible to anoxia and it is likely to suffer damage even though resuscitative techniques have restored the functioning of the heart and lungs. Techniques are available to maintain persons in such brain-damaged, comatose states (regardless of their cause), resulting occasionally in a body with adequate cardiovascular and respiratory function but with an inactive nervous system—in other words, a living body which contains a dead brain.

The question "When is a person actually dead?" often has urgent medical, legal, and social consequences. For example, with the advent of organ transplantation and the need for viable tissues, it became imperative to know "Is a person in a coma dead or alive?" **Brain death** is widely accepted by doctors, lawyers, and the general public as the criterion for death, the term being taken to mean "a brain that no longer functions and has no possibility of functioning again." The problem now becomes practical: "How does one know when brain death has occurred? What criteria are valid determinants of brain death?" There is certainly no agreement about these criteria, but the list below gives one set of conservative standards frequently accepted. These criteria are tested only after appropriate therapeutic procedures have been performed, and they must be present for at least 30 min, beginning 6 h after the onset of the suspected brain damage.

1. The person is unconscious and unresponsive, obeying no commands and making no spontaneous sounds or movements.
2. The person is apneic (not breathing). Apnea is considered to have occurred if there is no sign of a spontaneous respiration for a period of 15 min. (Once a patient is on a respirator, it is of course inadvisable to remove him or her for the 15-min test because of the danger of further anoxic damage occurring. Therefore, apnea is diagnosed if the patient's reflexes do not drive respiration at a depth or rhythm different from those of the respirator and if there is no spontaneous attempt to fight the respirator.)
3. There is electrocerebral silence, i.e., a flat EEG with wave amplitudes less than 2 μV. The EEG is the most significant test, since its outcome correlates highly with survival or death; the greater the activity on the EEG recording, the greater the chances of survival. However, this correlation does not hold in those cases when electrocerebral silence follows drug intoxication or periods of severely lowered body temperature.
4. Cortical and brainstem reflexes, such as the pupillary, corneal, cough, swallowing, and gag reflexes are absent; in some cases reflexes mediated by the spinal cord, such as the tendon and stretch reflexes, may be present.
5. Cerebral blood flow is absent for a period of at least 30 min.

Development of the Individual Nervous System

All neurons are present at birth, but during postnatal development there occur many important changes in these neurons: The cells enlarge, the number of dendrites increases, axons elongate, more nerve processes become myelinated, the number of synaptic contacts increases, neurons migrate to new locations, etc. Notice the changes in visual cortex cells in just the first three months of life (Fig. 20-9) and the gradual change in electrical activity (Fig. 20-10). These developmental patterns are reflected in behavioral changes. It is believed that the cortex is relatively nonfunctional at birth, a notion supported by the fact that infants born without a cortex have almost the same behavior and reflexes

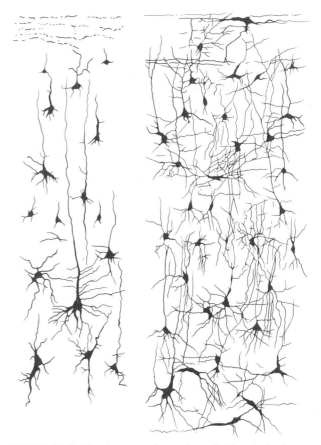

FIGURE 20-9. Visual cortex of a newborn (left) and of a three-month-old child (right). The number of cells is greater in the cortex of the three-month-old because of the migration of neurons into the area. *(Redrawn from J. L. Conel, "The Postnatal Development of the Human Cerebral Cortex," vol. 1. "The Cortex of the Newborn," Harvard University Press, Cambridge, Mass., 1939).*

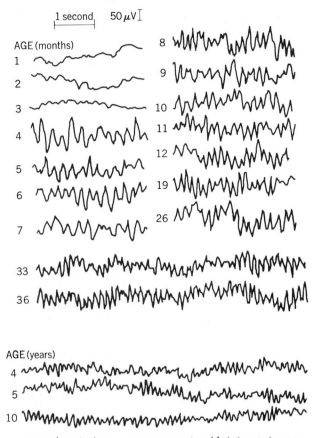

1 second   50 μV

AGE (months)

1

2

3

4

5

6

7

8

9

10

11

12

19

26

33

36

AGE (years)

4

5

10

21

FIGURE 20-10. EEGs from the same individual, showing the onset of the alpha rhythm at four months, the attainment of adult wave frequencies at 10 years of age, and little change thereafter. (Redrawn from D. B. Lindsley, Attention, Consciousness, Sleep, and Wakefulness, in "Handbook of Physiology: Neurophysiology," vol. 3, pp. 1553–1593, American Physiological Society, Washington, D.C., 1960.)

as shown by a normal infant. As cortical function develops, it seems gradually to exert an inhibitory control over the lower (and phylogenetically older) structures.

With cortical development, some of the reflexes whose integrating centers are in subcortical structures come under at least a degree of cortical control. Examples are the rooting and sucking reflexes present in infancy. (The stimulus is a touch on the infant's cheek; in response, the head turns toward the stimulus, the mouth opens, the stimulating object is taken into the mouth, and sucking begins.) The reflex is essential for survival of the young, but it is superseded by other eating behaviors. As the cerebral hemispheres develop and autonomy increases, allowing elaboration of hand feeding, the primitive rooting and sucking reflexes are inhibited. The basic reflex arc does not disappear; it is simply inhibited. For example, while examining a patient with damaged frontal lobes of the brain, a physician standing behind the patient and out of view quietly reached over and touched the patient's cheek. In response, the patient nuzzled the physician's finger until it was in the mouth and even began sucking it; the patient's behavior, released from its normal inhibition because of the brain damage, serves as a perfect illustration of the above point.

During development of the brain and spinal cord, nerve growth and synaptic contact occur with remarkable precision and selectivity. How do the right nerve connections get established in the first place? The best answer presently available is the **chemoaffinity theory**, in which the complicated neural circuits are said to grow and organize themselves according to selective attractions between neurons. It is suggested that these attractions, which are determined by chemical codes under genetic control, involve antigen-antibody types of interactions. Thus, during early stages of their development, neurons acquire individual chemical "identifications" by which they can recognize and distinguish each other. The chemical specificity is precise enough to determine not only the particular postsynaptic cell to which the axon tip will grow, but even the precise area on the postsynaptic cell that will be contacted. (Recall that synapses closer to the postsynaptic cell's initial segment have greater influence over its activity.)

Several other factors also influence the final neural organization: (1) timing of the development of the individual cells (for example, the axons that reach a postsynaptic cell earliest have the greatest chance of forming large numbers of synapses with it); (2) the presence of neighboring neurons and glia (which form structural constraints and guides for the developing nerve processes); and (3) chemical events involving neurotransmitters and enzyme systems in the tip of the developing fiber and adjacent cells.

Of course, adequate nutritive environments are required for normal brain development. The developing brain is extremely sensitive to the effects of malnutrition (i.e., a deficiency of calories, protein, trace metals, or vitamins), and some of the damage produced may be permanent. The level of hormones is also important; for example, decreased thyroid hormone concentrations during development cause a type of mental retardation known as **cretinism**. Although it was formerly believed that the developing fetus ''took what it needed'' nutritionally from the mother and that, in cases of malnutrition after birth, the brain was ''spared'' at the expense of the rest of the body, there is increasing evidence that nutritional deprivation both before and after birth has serious and irreversible effects on the chemical and structural maturation of the brain. Intermittent deprivation affects those cells that are still developing at that time more than cells already mature.

Despite the evidence indicating the selective growth and interconnectivity of nerve cells, the nature-nurture controversy, i.e., how much of development is due to genetically determined patterns and how much to experience and subsequent learning, must also be considered, for in fact, the course of neural development can be altered. However, the period during which this alteration can occur is genetically built into the neuron's developmental timetable. The modifiability of some neurons is limited to their so-called **critical periods** early in development, but in others it persists longer. Thus, some neuronal organizational patterns are highly specified early in development and are unmodifiable thereafter. These periods of maximal structural and functional growth depend upon the availability of proper internal and external environmental conditions for their full development. As an example of such a critical period, in dogs whose eyes were covered to deprive them of sight during the first five weeks of life, the visual systems were anatomically, biochemically, and physiologically retarded; this does not happen if the deprivation occurs beyond the first five weeks of age.

Other neurons remain uncommitted and are modifiable by function and experience until late in development. But in the case of either limited or lengthy periods of modifiability, it can occur only within the constraints imposed by the neurons' genetic code and its experiential history. The genes specify the capacity, whereas the environmental stimuli determine the specific expression and content of the neuronal output.

Learning and Memory

Learning is the increase in the likelihood of a particular response to a stimulus as the consequence of experience. Generally, rewards or punishments, as mentioned earlier, are crucial ingredients of learning, as is contact with and manipulation of the environment. **Memory** is the relatively permanent form of this change in responsiveness, and the postulated neural correlate of memory is called the **memory trace**.

Note that when we speak of learning and memory, we always mention the input of some sort of stimulus (the information to be learned or the memory to be tested) and the output of some sort of behavioral response. Yet it is clear that neural circuitry of the brain has to connect the stimulus and response and that at some point this neural circuitry has to undergo change if learning is to be achieved. Just where in this neural pathway the change (the memory trace) occurs and what form the change takes are the paramount questions in any study of learning, but unfortunately, the answers to these questions are still disappointingly vague. We shall discuss them after a brief description of memory.

Memory

Depending on how long the memory lasts, it is designated as **short-term** (lasting seconds to hours) or **long-term** (lasting days to years). The process of transferring memories from the short- to the long-term form is called **memory consolidation** (Fig. 20-11).

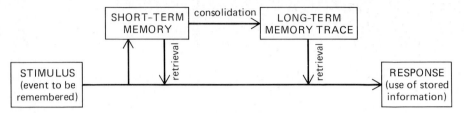

FIGURE 20-11. The events involved in learning and retrieving information.

Short-term memory is a limited-capacity storage process, which serves as the initial depository of information. After the registration of events in short-term memory, they may either pass away or be retained in long-term memory stores, depending upon factors such as attention and motivation. This distribution is not all-or-none, since some fragments of an event may be retained and others lost.

Memory consolidation is a labile process during which the fixation of an experience is susceptible to external interference, such as an attempt to learn conflicting information. The liability lasts for periods of a few seconds to days or even weeks. Not all experiences are consolidated into long-term memories. It is as though, for memory to become fixed, a "fix signal" must follow the event that is to be remembered. Such a signal is probably not specific for a particular event but rather may indicate "whatever just happened, remember it". Thus, the signal for memory consolidation may be likened to the fixation step in photographic developing; the latent image on the film will fade rapidly unless a chemical fixative is applied. The same fixative is used for photographs of any subject and itself contains no specific information. We have no real idea as to the physiological nature of the "fix" signal.

Behavioral investigations and common experience indicate that memories of past events and well-learned behavior patterns normally can have very long life spans. Long-term memories seem to be of two different kinds: procedural and declarative. **Procedural memories** are those involved in remembering *how* to do something, and **declarative memories** are involved in remembering *what* something is or what happened. Persons suffering from severe deficits in declarative memory but having

intact procedural memory can, for example, learn and retain motor skills, but they report no memory of having done so. One case study describes a patient, a pianist, who learned a new piece to accompany a singer at a hospital concert but had no recollection of the event the following morning. Sometimes these two memory types are referred to as verbal (declarative) and nonverbal (procedural).

Not only are memories short- or long-term; they can also be confined to specific modalities. Both points are illustrated in the following example. A man having a severe loss of memory for recent visual information (his memory for auditory and tactile stimuli was normal) was unable to find his way around a newly acquired three-room apartment. This difficulty occurred because he was unable to build up enough visual memories to become spatially oriented. He could, however, find his way around his parents home, where he had lived many years in the past, since the visual memories needed for orientation there were not recent ones.

The location of memory

Something has to change somewhere to account for the altered behavior that occurs with learning, and recent searches for the memory trace have been directed at parts of the limbic system (particularly to the hypothalamus and the specific regions known as the hippocampus and amygdala), the cerebral cortex, and the cerebellum. These searches are carried out not only with experimental animals but also with persons who have had accidental or disease-induced destruction of specific regions of their brains. In fact, it is the clinical studies which have provided the clearest evidence for the existence of verbal versus nonverbal memory and for the clear distinction between short-term and long-term memory.

One difficulty in locating the specific site of the memory trace is that learning in some form or other seems to be a general property of the of the mammalian nervous system and is not limited to any part of it. Nevertheless, it is clear that damage to some locations interferes with the learning of certain tasks. For example, in the man with the loss of recent visual memory, certain neural pathways that connect visual cortex with areas of the temporal lobe were damaged, suggesting that those parts of temporal lobe are necessary for storage of recent visual memories. Certain regions of the cerebellum seem to be necessary for learning discrete, meaningful movements, and the amygdala seems to be necessary for learning to avoid pain. (It is not clear, however, whether the amygdala actually contains the memory trace in these forms of learning or whether, because of its involvement in the emotional (fear) response to pain, it provides the motivation necessary for the learning.) Damage to the temporal lobe and underlying hippocampus results in deficits in humans in declarative but not procedural memories. Perhaps surprisingly, the cerebral cortex does not seem to be necessary for many types of learning except in its role of helping focus the attention of the learner on the task at hand. It is probably essential for memories involving language and complex spatial information.

Another difficulty in localizing memory may be due to the way we remember, generally using many different types of clues, such as "it looks similar to," "it sounds like," "it happened when I was in New York," etc., to help arrive at the desired information. In other words, in searching for a certain memory we often incorporate bits and pieces of memories, each specific to a different sensory modality and, therefore, probably stored in a separate place.

The neural basis of learning and memory

The specific changes occurring during the acquisition of short-term memory are not known for certain, but they appear to involve electrical activity in neurons and brief alterations in synaptic function rather than structural changes. It is known that changes in the amount of transmitter released at specific synapses can persist for weeks without any structural changes taking place. (The mechanisms which alter transmitter release from synaptic terminals were presented in Chap. 8.)

The fact that short-term memory is related to changes in electrical and chemical events rather than structural changes is supported by evidence that conditions which interfere with the electrical activity of the brain (for example, coma, deep anesthesia, electroconvulsive shock, and insufficient blood supply to the brain), also interfere with the retention of recently acquired information. These same states do not usually interfere with previously laid down long-term memory. For example, when a person becomes unconscious from a blow on the head, he or she cannot remember anything that happened for about 30 min before the blow; this is so-called **retrograde amnesia**. As recovery from retrograde amnesia occurs, memories of the most recent events return last and least completely (although islands of memory can return in a discontinuous fashion). The loss of consciousness in no way interferes with memories of experiences that were learned before the period of amnesia.

Several hormones are known to alter the ability of experimental animals (and in some cases, people) to retain memories of recent events. For example, numerous studies have shown that learning and memory are affected by epinephrine and norepinephrine, and certain drugs that affect learning and memory act at least in part through their influences on these two substances.

The opioid peptides enkephalin and endorphin also affect learning and memory. (Although we have encountered enkephalin and endorphin as neurotransmitters, they are released by the pituitary and possibly also the adrenal medulla, and qualify as hormones, too.) They seem to interfere with memory formation particularly when the lesson involves a painful stimulus. Their inhibitory effect on learning may occur in one of two ways, and it is not certain which is correct. They may simply decrease the emotional component (fear, anxiety) of the painful experience associated with the learning situation, thereby decreasing the motivation necessary for learning to occur, or they may interfere with the norepinephrine pathways that facilitate learning. Both enkephalin and enkephalin receptors occur in high concentrations in the parts of the limbic system associated with learning, and it is thought that the opioid synapses occur (presynaptically) on the axon terminals of the norepinephrine-releasing neurons. Increased enkephalin would result in a decreased release of norepinephrine onto the postsynaptic neurons, interfering with memory consolidation.

Adrenocorticotropic hormone (ACTH), oxytocin, and vasopressin (ADH) affect learning and memory, too, but the role they play is even less clear-cut. They may alter memory storage or retrieval directly, or their effect may be on attention or motivation, processes which nevertheless alter the rate of learning.

Several facts relate in general to the hormones mentioned above. One point is that changes in the blood concentrations of most of these substances alter learning and memory, a surprising fact because these substances cannot pass the blood-brain barrier easily. Second, the same hormone may either enhance or impair learning, depending on the nature of the task and the dose of the substance. Third, and perhaps the key factor, is that the opioid peptides, ACTH, vasopressin, and epinephrine are all released from endocrine glands during even mild stress such as that associated with some of the simpler learning tasks studied.

Any physicochemical explanation of learning must be able to account for the following phenomena:

1. Learning can occur very rapidly. In fact, under some situations, learning can occur in one trial.
2. Learning must be translated from an initial form involving action potentials to some permanent form of storage that can survive deep anesthesia, trauma, or electroconvulsive shock, which disrupt the normal patterns of neural conduction in the brain.
3. Information can be retained over long periods of time during which most components of the body have been renewed many times. Learning and memory must therefore reside either in systems that do not turn over rapidly or are self-perpetuating.
4. Information can be retrieved from memory stores after long periods of disuse. The common notion that memory, like muscle, always atrophies with lack of use is largely wrong.
5. Learning opposing responses interferes with the memory of initial responses to the same stimulus.
6. When a specific task is being learned, after an initial period of rapid learning the process seems to slow down and proceed at a rate that offers diminishing returns.
7. Many memories become more vivid and less easily forgotten with time.

Several lines of evidence have shown that long-term memory associated with learning (in contrast to short-term memory) can induce substantial anatomical changes. For example, animals that are exposed to a so-called enriched environment containing toys, social situations, etc., have dramatic changes in the cells in certain regions of their brains (particularly the cerebral cortex, hippocampus, and cerebellum). These changes, as compared to the brains of animals raised in isolation, include an increase in the number of neuroglial cells, greater complexity in the branching patterns of dendrites of nerve cells, increased spines on nerve-cell membranes (these serve as sites for synapses), and changes in the structure of the synapses themselves. (This ability of neural tissue to change because of its past history of activation is known as **plasticity**.) Moreover, when confronted with learning certain specific problems, the animals raised in enriched environments learned faster than those raised in restricted environments, indicating that the structural changes might be associated with performance on learning tasks. Because the early studies of this type were conducted in young animals, the problem of explaining learning in adults remained. However, it has recently been shown that plasticity is present even when the enriched environment begins at middle or old age. Thus, it is clear that the patterns by which neurons are connected are not permanent—as was once supposed.

Structural changes can occur only if there are also chemical changes. Thus, from a practical point of view, researchers are also led to look for chemical processes that underlie the acquisition, storage, and retrieval of information. In fact, drugs that interfere with certain chemical processes in neural tissue also interfere with learning. The substances that interfere most directly with memory are substances known to interfere with protein synthesis, but whether these interfere with the protein synthesis that accompanies structural changes in the neural tissue or whether they interfere with the synthesis of neurotransmitter substances, such as norepinephrine, known to play an essential role in learning, is not known.

Cerebral Dominance and Language

To the casual observer, the two cerebral hemispheres appear to be symmetrical, but they both have functional specializations, the most marked of which is language. Most facets of language—the conceptualization of what one wants to say or write, the neuronal control of the act of speaking or writing, the comprehension of what is spoken or written—are mediated by structures of the left cerebral hemisphere in 97 percent of the population. This is also the dominant hemisphere for fine motor control in the large majority of people, and so most people are right-handed (recall that the left hemisphere controls the movements of the right side of the body). Control of both handedness and language by the left hemisphere suggests that this hemisphere is specialized for controlling successive changes in the precise positioning of muscle groups. Moreover, there is evidence that the perception of events with rapidly changing temporal patterns, such as the perception of speech sounds, is neurally related to the production of movements with rapidly changing temporal patterns. However, the correlation between handedness and language function is not perfect, and for many left-handed people, language

functions are controlled by the left hemisphere (in the remaining left-handed people, language is represented bilaterally or in the right hemisphere).

Just as the left hemisphere is dominant for language in most people, the right hemisphere appears to be dominant for many nonlanguage processes, particularly those requiring spatial ability. The right hemisphere also predominately processes sounds such as musical patterns. Memories are asymmetrical, too; verbal memories are more apt to be associated with the left hemisphere and nonverbal memories, e.g., visual patterns, with the right. It may be that even the emotional responses of the two hemispheres are different, some investigators claiming that when electroconvulsive shock is administered in the treatment of depression, better effects are obtained when the electrodes are placed over the right hemisphere.

Different areas of left cerebral cortex are related to specific aspects of language. Areas in the frontal lobe near motor cortex are involved in the articulation of speech, whereas areas in the parietal and temporal lobe are involved in the sensory functions and in language interpretation, auditory cortex is involved in listening to spoken language, and visual cortex is involved in reading. The language area in the frontal lobe is known as **Broca's area** and lies adjacent to the area of motor cortex which controls the muscles of the face, tongue, jaw, and throat (Fig. 20-12A); the temporal-lobe language area is known as **Wernicke's area**, and it is connected to Broca's area by a thick band of association fibers. The currently accepted model of the formulation of language states that the underlying structure of what phrase is to be said arises in Wernicke's area and is then transferred via the association fibers to Broca's area, where it evokes the precise program for the motor commands necessary to cause the words to be spoken. The program is passed on to the adjacent face area of the motor cortex, which activates the appropriate muscles.

Wernicke's area not only plays a part in speaking but also has a major role in comprehension of the spoken word and in reading and writing. When a word is heard, the sound is initially received in

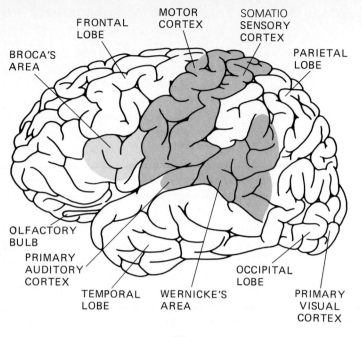

(A)

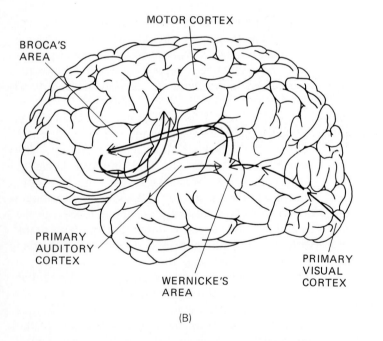

(B)

FIGURE 20-12. (A) The cerebral cortex showing regions involved in language. (B) Pathways indicating parts of brain involved in speaking a word that was heard (color) and speaking a written word (black). *(Adapted from Geschwind.)*

auditory cortex, which is in the temporal lobe adjacent to Wernicke's area, but the signals must pass through Wernicke's area if they are to be understood as a spoken message (Fig. 20-12B). Moreover, when a word is read, the visual signals, after passing through visual cortex, are transmitted through a region of posterior parietal lobe, which in turn elicits the auditory form of the written message in Wernicke's area.

Such cortical specialization is demonstrated by

the **aphasias**, specific language defects not due to mental retardation or paralysis of the muscles involved in speech. Damage to Wernicke's area results in aphasias that are related to conceptualization—the persons cannot understand spoken or written language even though their hearing and vision are unimpaired. In contrast, damage to Broca's area results in expressive aphasias—the persons are unable to carry out the coordinated respiratory and oral movements necessary for language, even though they can move the lips and tongue; they understand spoken language and know exactly what they want to say but are unable to speak. Expressive aphasias are often associated with an inability to write.

The anatomical and physiological asymmetries in the two hemispheres are present at birth. This fundamental asymmetry is more flexible in early years than in later life. For example, accidental damage to the left hemisphere of children under the age of two causes no impediment of future language development, and adequate (though perhaps not equal) language develops in the intact right hemisphere. Even if the left hemisphere is traumatized after the onset of language, language is reestablished in the right hemisphere after transient periods of loss. The prognosis becomes rapidly worse as the age at which the damage occurs increases, so that after the early teens, language is interfered with permanently. The dramatic change in the teens in the possibility of establishing language (or a second language) is possibly related to the fact that the brain attains its final structural, biochemical, and functional maturity at that time. Apparently, with maturation of the brain, language functions are irrevocably assigned, and the utilization of language propensities of the right hemisphere is no longer possible. Interestingly, if the left-hemisphere damage occurs early enough in the child's development so that language functions can be transferred to the right hemisphere, defects occur in those functions in which the right hemisphere normally plays a role; i.e., language develops at the expense of other right-hemisphere functions.

Conclusion

Until recently it was commonly thought that the control of most complex behavior, such as thinking, remembering, learning, etc., was handled almost exclusively by the cerebral cortex. Actually, damage to cortical areas outside of motor and sensory areas produces behavioral results that are subtle rather than obvious (the one exception, language, is highly sensitive to cortical damage), and stimulation of the cortex causes little change in the orientation or level of excitement of the animal.

In general, it is best to consider that particular behavioral functions are not controlled exclusively by any one area of the nervous system but that the control is shared or influenced by structures in other areas. Cortex and the subcortical regions—particularly limbic and reticular systems—form a highly interconnected system in which many parts contribute to the expression of a particular behavioral performance. Moreover, the nervous system is so abundantly interconnected that it is difficult to know where any particular subsystem begins or ends.

In the early seventeenth century Descartes taught that all things in nature, including human beings, are machines. The brain's mode of operation was compared to that of a clock. When computers became widely used, the brain was compared to a computer. A more recent analogy compares the brain to a hologram, a photographic process which records specially processed light waves themselves, rather than the image of an object. These widely divergent analogies only emphasize how little we know of how the brain really functions.

APPENDIX I:
ENGLISH AND METRIC UNITS

	English	Metric
Length	1 foot = 0.305 meter 1 inch = 2.54 centimeters	1 meter = 39.37 inches 1 centimeter (cm) = 1/100 meter 1 millimeter (mm) = 1/1000 meter 1 micrometer (μm) = 1/1000 millimeter 1 nanometer (nm) = 1/1000 micrometer *(1 angstrom (Å) = 1/10 nanometer)
Mass	**1 pound = 433.59 grams 1 ounce = 27.1 grams	1 kilogram (kg) = 1,000 grams = 2.2 pounds 1 gram (g) = 0.037 ounce 1 milligram (mg) = 1/1000 gram 1 microgram (μg) = 1/1000 milligram 1 nanogram (ng) = 1/1000 microgram 1 picogram (pg) = 1/1000 nanogram
Volume	1 gallon = 3.785 liters 1 quart = 0.946 liter	1 liter = 1,000 cubic centimeters = 0.264 gallon 1 liter = 1.057 quarts 1 milliliter (ml) = 1/1000 liter 1 microliter (μl) = 1/1000 milliliter

* The angstrom unit of length is not a true metric unit, but has been included because of its frequent use, until recently, in the measurement of molecular dimensions.
** A pound is actually a unit of force, not mass. The correct unit of mass in the English system is the slug, while the newton is the metric unit of force. When we write 1 kg = 2.2 pounds, this means that one kilogram of *mass* will have a *weight* under standard conditions of gravity at the Earth's surface of 2.2 pounds *force*.

APPENDIX II:
ESSENTIAL NUTRIENTS

Essential nutrient	RDA[a] for healthy adult male, mg	Dietary sources	Major body functions	Deficiency	Excess
Amino acids		From proteins:	Precursors of structural protein, enzymes and coenzymes, antibodies, hormones, metabolically active compounds, neurotransmitters, and porphyrins. Certain amino acids have specific functions:	Deficient protein intake leads to development of kwashiorkor and, coupled with low energy intake, to marasmus	Excess protein intake possibly aggravates or potentiates chronic disease states
Aromatic:		Good sources:			
Phenylalanine	1,100	Legumes, grains,			
Tyrosine	1,100	dairy products,			
		meat, fish			
Basic:		Adequate sources:			
Arginine[b]	0	Rice, corn,			
Lysine	800	wheat			
Histidine[c]	0	Poor sources:			
		Cassava, sweet	(a) Tyrosine is a precursor of epinephrine and other catecholamines and thyroxine		
Branched chain:		potato			
Isoleucine	700				
Leucine	1,000				
Valine	800				
Sulfur-containing:			(b) Arginine is a precursor of polyamines and urea		
Methionine	1,100				
Cystine[d]	1,100		(c) Methionine is required for methyl group metabolism		
Other:					
Tryptophan	250				
Threonine	500		(d) Tryptophan is a precursor of serotonin		
Fatty acids		Vegetable fats (corn, cottonseed, soy oils); wheat germ; vegetable shortenings	Involved in cell membrane structure and function; precursors of prostaglandins (regulation of gastric function, release of hormones, smooth-muscle activity, etc)	Poor growth; skin lesions	Not known
Arachidonic	6,000				
Linoleic	6,000				
Linolenic	6,000				

Appendix 2. Essential Nutrients *(Continued)*

Essential nutrient	RDA[a] for healthy adult male, mg	Dietary sources	Major body functions	Deficiency	Excess
Water-soluble Vitamins:					
Vitamin B_1 (thiamine)	1.4	Pork, organ meat, whole grains, legumes	Coenzyme (thiamine pyrophosphate) in reactions involving the removal of carbon dioxide	Beriberi (peripheral nerve changes, edema, heart failure)	None reported
Vitamin B_2 (riboflavin)	1.6	Widely distributed in foods	Constituent of two flavin nucleotide coenzymes involved in energy metabolism (FAD and FMN)	Reddened lips, cracks at corner of mouth (cheilosis), lesions of eye	None reported
Nicacin	18	Liver, lean meats, grains, legumes (can be formed from tryptophan)	Constituent of two coenzymes involved in oxidation-reduction reactions (NAD and NADP)	Pellagra (skin and gastrointestinal lesions; nervous, mental disorders)	Flushing, burning and tingling around neck, face, and hands
Vitamin B_6 (pyridoxine)	2.2	Meats, vegetables, wholegrain cereals	Coenzyme (pyridoxal phosphate) involved in amino acid metabolism	Irritability, convulsions, muscular twitching, dermatitis near eyes, kidney stones	None reported
Pantothenic acid	4–7	Widely distributed in foods	Constituent of coenzyme A, which plays a central role in energy metabolism	Fatigue, sleep disturbances, impaired coordination, nausea (rare in man)	None reported
Folacin	0.4	Legumes, green vegetables, whole-wheat products	Coenzyme (reduced form) involved in transfer of single-carbon units in nucleic acid and amino acid metabolism	Anemia, gastrointestinal disturbances, diarrhea, red tongue	None reported
Vitamin B_{12}	.003	Muscle meats, eggs, dairy products (not present in plant foods)	Coenzyme involved in transfer of single-carbon units in nucleic acid metabolism	Pernicious anemia, neurological disorders	None reported
Biotin	0.1–0.2	Legumes, vegetables, meats	Coenzyme required for fat synthesis, amino acid metabolism, and glycogen formation	Fatigue, depression, nausea, dermatitis, muscular pains	None reported

Appendix 2. Essential Nutrients *(Continued)*

Essential nutrient	RDA[a] for healthy adult male, mg	Dietary sources	Major body functions	Deficiency	Excess
Choline	Not established; usual diet provides 500–900	All foods containing phospholipids (egg yolk, liver, grains, legumes)	Constituent of phospholipids; precursor of neurotransmitter acetylcholine	Not reported in man	None reported
Vitamin C (ascorbic acid)	60	Citrus fruits, tomatoes, green peppers, salad greens	Maintains intercellular matrix of cartilage, bone and dentine; important in collagen synthesis.	Scurvy (degeneration of skin, teeth, blood vessels; epithelial hemorrhages)	Relatively nontoxic; possibility of kidney stones
Fat-soluble vitamins					
Vitamin A (retinol)	1	Provitamin A (beta-carotene) widely distributed in green vegetables; retinol present in milk, butter, cheese, fortified margarine	Constituent of rhodopsin (visual pigment); maintenance of epithelial tissues; role in mucopolysaccharide synthesis	Xerophthalmia (keratinization of ocular tissue), night blindness, permanent blindness	Headache, vomiting, peeling of skin, anorexia, swelling of long bones
Vitamin D[e] (ergocalciferol)	0.005	Cod-liver oil, eggs, dairy products, fortified milk, and margarine	Promotes growth and mineralization of bones; increases absorption of calcium	Rickets (bone deformities) in children. Osteomalacia in adults	Vomiting, diarrhea, loss of weight, kidney damage
Vitamin E (tocopherol)	10	Seeds, green leafy vegetables, margarine, shortenings	Functions as an antioxidant to prevent cell-membrane damage	Possibly anemia	Relatively nontoxic
Vitamin K[f] (phylloquinone)	0.07–0.14	Green leafy vegetables; small amount in cereals, fruits and meats	Important in blood clotting (involved in formation of clotting factors)	Hemorrhage	Relatively nontoxic; synthetic forms at high doses may cause jaundice
Minerals					
Calcium	800	Milk, cheese, dark-green vegetables, dried legumes	Bone and tooth formation, blood clotting, nerve transmission	Stunted growth, rickets, osteoporosis, convulsions	Kidney stones
Phosphorus	880	Milk, cheese, meat, poultry, grains	Bone and tooth formation, acid-base balance	Weakness, demineralization of bone, loss of calcium	Erosion of jaw (fossy jaw)

Appendix 2. Essential Nutrients *(Continued)*

Essential nutrient	RDA[a] for healthy adult male, mg	Dietary sources	Major body functions	Deficiency	Excess
Sulfur	(Provided by sulfur amino acids)	Sulfur amino acids (methionine and cystine) in dietary proteins	Constituent of active tissue compounds, cartilage, and tendon	Related to intake and deficiency of sulfur amino acids	Excess sulfur amino acid intake leads to poor growth
Potassium	2,500	Meats, milk, many fruits	Acid-base balance, body water balance, nerve function	Muscular weakness, paralysis	Muscular weakness, death
Chloride	2,000	Common salt	Formation of gastric juice; acid base balance	Muscle cramps, mental apathy, reduced appetite	Vomiting
Sodium	2,500	Common salt	Acid base balance, body water balance, nerve function	Muscle cramps, mental apathy, reduced appetite	High blood pressure
Magnesium	350	Whole grains, green leafy vegetables	Activates enzymes; involved in protein synthesis	Growth failure, behavioral disturbances, weakness, spasms	Diarrhea
Iron	10	Eggs, lean meats, legumes, whole grains, green leafy vegetables	Constituent of hemoglobin and enzymes involved in energy metabolism	Iron-deficiency anemia (weakness, reduced resistance to infection)	Siderosis, cirrhosis of liver
Fluoride	2	Drinking water, tea, seafood	May be important in maintenance of bone structure	Higher frequency of tooth decay	Mottling of teeth, increased bone density, neurological disturbances
Zinc	15	Widely distributed in foods	Constituent of enzymes involved in digestion	Growth failure, small sex glands	Fever, nausea, vomiting, diarrhea
Copper	2	Meats, drinking water	Constituent of enzymes associated with iron metabolism	Anemia, bone changes (rare in man)	Rare metabolic condition (Wilson's disease)
Silicon Vanadium Tin Nickel	Not established	Widely distributed in foods	Function unknown (essential for animals)	Not reported in man	Industrial exposures: Silicon—silicosis, Vanadium—lung irritation, Tin—vomiting, Nickel—acute, pneumonitis

Appendix 2. Essential Nutrients *(Continued)*

Essential nutrient	RDA[a] for healthy adult male, mg	Dietary sources	Major body functions	Deficiency	Excess
Selenium	Not established (Diet provides 0.05–0.1 per day)	Seafood, meat, grains	Functions in close association with vitamin E	Anemia (rare)	Gastrointestinal disorders, lung irritation
Manganese	Not established (Diet provides 6–8 per day)	Widely distributed in foods	Constituent of enzymes involved in fat synthesis	In animals: poor growth; disturbances of nervous system; reproductive abnormalities	Generalized disease of nervous system
Iodine	0.15	Marine fish and shellfish, dairy products, many vegetables	Constituent of thyroid hormones	Goiter (enlarged thyroid)	Very high intakes depress thyroid activity
Molybdenum	Not established (Diet provides 0.4 per day)	Legumes, cereals, organ meats	Constituent of some enzymes	Not reported in man	Inhibition of enzymes
Chromium	Not established (Diet provides 0.05–10.12 per day)	Fats, vegetable oils, meats	Involved in glucose and energy metabolism	Impaired ability to metabolize glucose	Occupational exposures: skin and kidney damage
Cobalt	Same as vitamin B_{12}	Same as vitamin B_{12}	Constituent of vitamin B_{12}	Not reported in man	Industrial exposure: dermatitis and diseases of red blood cells
Water	1.5 L/day	Solid foods, liquids, drinking water	Transport of nutrients, temperature regulation, participates in metabolic reactions	Thirst, dehydration	Headaches, nausea, edema, high blood pressure

[a] RDA: Recommended Daily Allowance.
[b] Nonessential if sufficient dietary phenylalanine.
[c] Unnecessary in adults in short-term studies but probably essential for normal growth of children.
[d] Nonessential if sufficient dietary methionine.
[e] If adequate sunlight is available, no dietary intake is required since skin synthesizes this "vitamin," which is properly termed a "hormone" (see Chap. 18).
[f] Synthesized by intestinal bacteria; dietary intake normally unnecessary.
SOURCE: N.S. Scrimshaw and V. R. Young, "The Requirements of Human Nutrition," *Sci. Amer.,* **235**(3):50, Sept. 1976.

GLOSSARY

A band. one of the transverse bands making up the repeating striated pattern of cardiac and skeletal muscle; located in the middle of a sarcomere

A cells. glucagon-secreting cells of the islets of Langerhans of the pancreas; also known as alpha cells

abscess. microbes, leukocytes, and liquified tissue debris walled off by fibroblasts and collagen

absolute refractory period. the time during which an excitable membrane is unable to generate an action potential in response to any stimulus

absorption. movement of materials across a layer of epithelial cells from a body cavity or compartment toward the blood

absorptive state. the period during which nutrients are entering the bloodstream from the gastrointestinal tract

accessory reproductive organs. the system of ducts through which the sperm or ova are transported and the glands emptying into these ducts (in the female the breasts are usually included in this category)

acclimatization. an adaptive change in the functioning of an already-existing genetically based physiological system induced by prolonged exposure to an environmental stress with no alteration in the person's genetic endowment

accommodation. adjustment of the eye for viewing various distances by change of the shape of the lens

acetone. a ketone body produced from acetyl CoA during prolonged fasting or untreated severe diabetes mellitus

acetyl coenzyme A (acetyl CoA). a metabolic intermediate which transfers acetyl groups to the Krebs cycle and to various synthetic pathways

acetyl group. CH_3CO^-

acetylcholine (ACh). a neurotransmitter

acetylcholinesterase. an enzyme which breaks down acetylcholine into acetic acid and choline

ACh. *see* acetylcholine

acid. any compound capable of releasing a hydrogen ion; a solution having a hydrogen-ion concentration greater than that of pure water (i.e., a pH less than 7)

acidity. refers to the hydrogen-ion concentration of a solution; the higher the hydrogen-ion concentration, the greater the acidity

acidosis. any situation in which the hydrogen-ion concentration of arterial blood is elevated; *see also* metabolic acidosis, respiratory acidosis

acquired immune deficiency syndrome (AIDS). a disease characterized by profound inability to resist many infections; the major deficit is a lack of helper T cells

acrosome. a vesicle containing digestive enzymes and located at the head of a sperm

ACTH. *see* adrenocorticotropic hormone

actin. the globular contractile protein to which the myosin cross bridge binds; located in the thin filaments of muscle

action potential. the electric signal propagated over long distances by nerve and muscle cells; characterized by an all-or-none reversal of the membrane potential in which the inside of the cell temporarily becomes positive relative to the outside

activation energy. the energy necessary to disrupt existing chemical bonds or form new ones during a chemical reaction

active hyperemia. the increased blood flow through a tissue resulting from an increase in its metabolic activity; functional hyperemia

active immunity. resistance to reinfection acquired as the result of contact with microorganisms, their toxins, or other antigenic material

active site. the region of an enzyme to which the ligand (substrate) binds

active state. the maximal internal tension which can be generated at any moment as a result of cross-bridge activity in a muscle fiber

active transport. an energy-requiring carrier-mediated transport system in which molecules can be moved across a membrane against an electrochemical gradient

acute-phase proteins. a group of proteins released by the liver during infection

adaptation. a decrease in the frequency of action potentials fired by a neuron despite a stimulus of constant magnitude

adenine. one of the purine bases of DNA and RNA

adenosine diphosphate (ADP). the two-phosphate product of ATP breakdown

adenosine monophosphate (AMP). the monophosphate derivative of ATP; one of the nucleotides in DNA and RNA

adenosine triphosphate (ATP). the major molecule which accepts energy from the breakdown of fuel molecules during its formation from ADP and P_i; donates this energy to cell functions during its breakdown to ADP and P_i

adenylate cyclase. the enzyme which catalyzes the transformation of ATP to cyclic AMP

ADH. *see* antidiuretic hormone

adipocytes. cells specialized for the synthesis and storage of triacylglycerol (neutral fat); fat cells

adipose tissue. a tissue composed largely of fat-storing cells

adolescence. the period of transition from childhood to adulthood in all respects, not just sexual ones

ADP. *see* adenosine diphosphate

adrenal cortex. the endocrine gland which forms the outer shell of each adrenal gland; secretes mainly cortisol, aldosterone, and androgens

adrenal glands. a pair of endocrine glands located just above each kidney; each gland consists of two regions: an outer *adrenal cortex* and an inner *adrenal medulla*

adrenal medulla. the endocrine gland which forms the inner core of each adrenal gland; secretes mainly the hormone epinephrine

adrenaline. the British name for epinephrine

adrenergic. pertaining to norepinephrine or epinephrine; a compound which acts like norepinephrine or epinephrine

adrenocorticotropic hormone (ACTH, corticotropin). a polypeptide hormone secreted by the anterior pituitary; stimulates the adrenal cortex to secrete cortisol

aerobic. in the presence of oxygen

affective disorders. the manias and depressions; diseases characterized by serious, prolonged disturbances of mood as well as changes in thought patterns and behavior

afferent arterioles. vessels in the kidney which convey blood from the arteries to the glomeruli

afferent neuron. a neuron whose cell body lies outside the brain or spinal cord and which carries information to the central nervous system from the receptors at its peripheral endings

afferent pathway. the component of a reflex arc which transmits information from a receptor to an integrating center

affinity. the strength with which a ligand binds to its binding site

afterbirth. the placenta and associated membranes expelled from the uterus after delivery of an infant

aggregation factor. a protein which forms cross links between binding sites on the plasma membranes of specific types of cells, holding them together to form tissues

agonist. a foreign molecule which binds to a receptor site and triggers the cell's response as if the true chemical messenger had combined with the receptor

albumin. the most abundant proteins found in blood plasma

aldosterone. a mineralocorticoid steroid hormone secreted by the adrenal cortex; regulates electrolyte balance

alkaline. having a hydrogen-ion concentration lower than that of pure water (i.e., having a pH greater than 7)

alkalosis. any situation in which the hydrogen-ion concentration of arterial blood is reduced; *see also* metabolic alkalosis, respiratory alkalosis

all or none. an event which occurs either maximally or not at all

allergy. an acquired, specific immune reactivity to environmental antigens such as dust, pollens, and food constituents, involving IgE antibodies

allosteric enzyme. an enzyme whose activity can be altered by modulator molecules binding at a site other than the substrate binding site

allosteric modulation. control of the properties of a protein binding site by modulator molecules which bind to regions of the protein other than the binding site

alpha-adrenergic receptor. a type of receptor for epinephrine and norepinephrine which, by its response to certain drugs, can be distinguished from a beta-adrenergic receptor

alpha helix. a coiled conformation of a polypeptide chain found in many proteins

alpha motor neuron. a large motor neuron which innervates skeletal muscle fibers

alpha rhythm. the prominent 8- to 13-Hz oscillation in the electroencephalograms of awake, relaxed adults with their eyes closed

alternate complement pathway. a sequence for complement activation which is not antibody dependent; bypasses the first few steps in the classical pathway

alveolar dead space. the volume of inspired air which reaches pulmonary alveoli but cannot undergo gas exchange with the blood

alveolar pressure. the air pressure within the pulmonary alveoli

alveolar ventilation. the volume of atmospheric air entering the pulmonary alveoli each minute

alveolus. a small cavity; the thin-walled, air-filled "outpocketing" from the terminal air passageways in the lungs; the cluster of cells at the end of a duct in secretory glands

amine hormone. a hormone derived from the amino acid tyrosine; includes thyroid hormone, epinephrine, and norepinephrine

amino acids. a molecule containing an amino group, a carboxyl group, and a side chain all attached to a single carbon atom; a molecular subunit of protein

amino group. $-NH_2$

aminoacyl-tRNA synthetase. one of 20 different enzymes, each of which catalyzes the covalent linkage of one type of amino acid to its particular type of tRNA during protein synthesis

aminopeptidases. enzymes located in the intestinal epithelial membrane; they break the peptide bond of the amino acid at the end of a polypeptide that has a free amino group

ammonia. NH_3; produced during the breakdown of amino acids; converted in the liver to urea

amnesia. loss of memory; *see also* retrograde amnesia

AMP. *see* adenosine monophosphate

amphetamine. a drug which increases transmission at catecholamine-mediated synapses in the brain

amphipathic molecule. a molecule containing polar or ionized groups at one end and nonpolar groups at the other

amylase. an enzyme which breaks down starch, producing disaccharides and small polysaccharides·

anabolic reactions. synthetic chemical reactions; reactions which put small molecular fragments together to form larger molecules

anaerobic. in the absence of oxygen

analgesia. removal of pain

anatomic dead space. the space within the airways of the respiratory tract whose walls do not permit gas exchange with the blood

androgen. any chemical with actions similar to those of testosterone

anemia. a reduction in total blood hemoglobin

anemic hypoxia. hypoxia in which the arterial P_{O_2} is normal but the total oxygen content of the blood is reduced

angina pectoris. chest pain associated with inadequate blood flow to the heart muscle

angiotensin. angiotensin II

angiotensin I. a peptide generated in the blood by the action of renin on angiotensinogen

angiotensin II. an octapeptide hormone formed by the enzymatic removal of two amino acids from angiotensin I; stimulates the secretion of aldosterone from the adrenal cortex and the contraction of vascular smooth muscle

angiotensinogen. the plasma protein precursor for angiotensin I

"angry" macrophages. macrophages whose activity has been facilitated by chemicals released from T cells

anion. a negatively charged ion, e.g., Cl^-, SO_4^{2-}

anoxia. the absence of oxygen

antagonist. (muscle) a muscle whose action opposes the intended movement; e.g., the finger flexors are antagonists to the finger extensors; (drug) a molecule which competes for a binding site with a chemical messenger normally present in the body and binds to the receptor but does not trigger the cell's response

anterior pituitary. the anterior portion of the pituitary gland; synthesizes, stores, and releases various hormones including ACTH, GH, TSH, prolactin, FSH, and LH

antibody. a protein secreted by plasma cells and capable of combining with the specific antigen which stimulated its production; *see also* "blocking" antibody

antibody-mediated immunity. *see* humoral immunity

anticodon. the three-nucleotide sequence in tRNA that is able to base-pair with a complementary codon in mRNA during protein synthesis

antidiuretic hormone (ADH). a peptide hormone synthesized in the hypothalamus and released from the posterior pituitary; increases the permeability of the kidney's late distal tubules and collecting ducts to water; also called vasopressin

antigen. a foreign molecule, usually a protein, which stimulates a specific immune response against itself when introduced into the body

antihistamine. a chemical which blocks the action of histamine

antrum. (gastric) the lower portion of the stomach, i.e, the region closest to the pyloric sphincter; (ovarian) the fluid-filled cavity in a maturing ovarian follicle

aorta. the largest artery in the body; carries blood from the left ventricle of the heart to the thorax and abdomen

aortic-body chemoreceptor. chemoreceptor located near the arch of the aorta; sensitive to blood P_{O_2}, P_{CO_2}, and hydrogen-ion concentration

aortic valve. the valve between the left ventricle of the heart and the aorta

aphasia. a specific language deficit not due to mental retardation or muscular weakness; deficiency can be in understanding or formulating oral or written language

apnea. cessation of respiration

apocrine secretion. secretion which involves the breaking-away of the entire top of a gland cell, which contains numerous secretory vesicles

appendix. a small fingerlike projection from the cecum in the large intestine

arachidonic acid. the fatty acid which is the major precursor of the prostaglandins

arrhythmia. any variation from the normal heartbeat rhythm

arterial baroreceptors. nerve endings sensitive to stretch or distortion produced by changes in blood pressure in the blood vessels (carotid sinus and arch of the aorta) in whose walls they lie

arteriolar resistance. the resistance to blood flow offered by the arterioles

arterioles. a blood vessel located between an artery and a capillary, surrounded by several layers of smooth muscle; the primary site at which vascular resistance is located

artery. a thick-walled, elastic vessel which carries blood away from the heart

ascending limb. (of Henle's loop) the portion of the renal nephron leading to the distal convoluted tubule

aspartate. formed from the dissociation of the amino acid aspartic acid; a neurotransmitter at some excitatory synapses

asthma. a disease characterized by severe airway constriction and by plugging of the airways with mucus

astigmatism. a defect in vision due to irregularities in the shape of the cornea

atherosclerosis. a disease characterized by a thickening of the arterial wall with abnormal smooth muscle cells, deposits of cholesterol, and connective tissue; results in narrowing of the vessel lumen

atmospheric pressure. the air pressure of the external environment (760 mmHg at sea level); barometric pressure

atom. the smallest unit of matter which has unique chemical characteristics; consists of a complex arrangement of negative electrons moving around a positively charged nucleus; has no net charge; combines with other atoms to form molecules

atomic nucleus. the dense region, consisting of protons and neutrons, at the center of an atom; contains virtually all the mass of the atom

atomic number. the number of protons in the nucleus of an atom; each type of atom has a different atomic number

atomic weight. a relative value which indicates how many times heavier a given atom is than a hydrogen atom; approximately equal to the number of protons and neutrons

ATP. *see* adenosine triphosphate

ATPase. an enzyme which breaks down ATP to ADP and inorganic phosphate; *see* Na,K-ATPase, Ca-ATPase, myosin-ATPase

atrioventricular (AV) node. a region at the base of the right atrium near the interventricular septum, containing specialized cardiac muscle cells through which electrical activity must pass to reach the ventricles from the atria

atrioventricular (AV) valve. a valve between an atrium and a ventricle of the heart; the AV valve on the right side of the heart is the *tricuspid valve*, and that on the left is the *mitral valve*

atrium. a chamber of the heart which receives blood from the veins and passes it on to the ventricle on the same side of the heart

auditory. pertaining to the sense of hearing

auditory cortex. the region of cerebral cortex which receives the nerve fibers from the auditory pathways

autocrine. a chemical which is secreted by a cell into the extracellular fluid surrounding the cell, and which acts upon the very cell that secreted it

autoimmune disease. a disease produced by antibodies or T cells acting against the body's own cells, which results in damage or alteration of cell function

automaticity. capable of self-excitation

autonomic nervous system. a component of the efferent division of the peripheral nervous system; innervates cardiac muscle, smooth muscle, and glands; consists of sympathetic and parasympathetic subdivisions

autoregulation. the ability of individual organs to alter their vascular resistance and, thereby, to control (self-regulate) their blood flow independent of external neural and hormonal influences

autosomal chromosome. *see* chromosome

AV node. *see* atrioventricular node

AV valve. *see* atrioventricular valve

axon. a single process extending from the cell body of a neuron; propagates action potentials away from the cell body; a nerve fiber

axon terminals. the network of fine branches at the end of the axon, each branch ending at a synaptic or neuroeffector junction

axon transport. a process involving intracellular filamentous structures by which materials are moved from one end of an axon to the other end

B cells. (immune system) lymphocytes which, upon activation, proliferate and differentiate into antibody-secreting plasma cells; (endocrine system) insulin-secreting cells in the islets of Langerhans of the pancreas (also known as beta cells)

B lymphocytes. *see* B cells

bacteria. unicellular organisms which have an outer coating, the cell wall, in addition to a plasma membrane; do not have a nucleus or other membrane-bound organelles

balance concept. the statement that if the quantity of any substance in the body is to be maintained at a constant level over a period of time, the total amounts ingested and produced must equal the total amounts excreted and consumed

barometric pressure. *see* atmospheric pressure

baroreceptor. a receptor sensitive to pressure and to rate of change in pressure

basal. resting level

basal ganglia. several nuclei deep in the cerebral hemispheres which code and relay information associated with the control of body movements; specifically, the caudate nucleus, globus pallidus, and putamen

basal metabolic rate (BMR). the metabolic rate when a subject is at mental and physical rest (but not sleeping), at comfortable temperature, and fasted for at least 12 h; the metabolic cost of living

base. any molecule which can combine with a hydrogen ion; a molecular ring of carbon and nitrogen atoms which, with a phosphate group and a sugar, constitutes a nucleotide; *see also* purine bases, pyrimidine bases

basement membrane. a thin proteinaceous layer of extracellular material associated with the plasma membranes of many cells, especially epithelial cells

basic electrical rhythm. the spontaneous depolarization-repolarization cycles of pacemaker cells in the longitudinal smooth muscle layer of the stomach and intestines, which coordinate the repetitive muscular activity of the gastrointestinal tract

basilar membrane. the membrane which separates the cochlear duct and scala tympani in the inner ear; supports the organ of Corti

basophil. a polymorphonuclear granulocytic leukocyte whose granules stain with basic dyes

beta-adrenergic receptor. a type of receptor for epinephrine and norepinephrine which, by its responses to certain drugs, can be distinguished from an alpha-adrenergic receptor

beta rhythm. low, fast oscillations in the EEG pattern of alert, awake adults who are paying attention to (or thinking hard about) something

bicarbonate. HCO_3^-

bile. fluid secreted by the liver; contains bile salts, cholesterol, lecithin (a phospholipid), bile pigments and other end products of organic metabolism, and certain trace metals

bile canaliculi. small ducts adjacent to the liver cells; bile is secreted into them

bile duct. the duct which carries bile from the gallbladder and the liver to the small intestine

bile pigments. colored substances, derived from the breakdown of the heme group of hemoglobin, secreted in bile

bile salts. a group of steroid molecules secreted by the liver in the bile; promote the solubilization and digestion of fat in the small intestine

bilirubin. a yellow substance resulting from the breakdown of heme; excreted in bile as a bile pigment

binding site. a region on a protein which has chemical groups which are able to interact with a specific ligand, binding it to the surface of the protein

"biological clock." cells which function on a rhythmical basis in the absence of apparent external stimulation

biological rhythm. rhythmical changes in some aspect of the body's functioning

biotransformation. alteration of foreign molecules by an organism's metabolic pathways

bipolar affective disorder. an affective disorder characterized by swings in mood between mania and depression; formerly called manic depression

bipolar cells. the neurons in the retina which are postsynaptic to the rods and cones

blastocyst. an early stage of embryonic development consisting of a ball of developing cells surrounding a central cavity

"blocking" antibody. an antibody whose production is induced by cancer cells or tissue transplants and which blocks the killing of those cells by cytotoxic T cells

blood-brain barrier. a group of anatomical barriers and transport systems which closely controls the kinds of substances which enter the extracellular space of the brain from the blood and the rate at which they enter

blood sugar. glucose

blood-testis barrier. a barrier, formed mainly by the Sertoli cells and the tight junctions between them, which limits the movements of chemicals between the blood and the lumen of the seminiferous tubules

blood types. classification of the blood determined by the presence of antigens of the A, B, O, or AB types on the plasma membranes of erythrocytes and by the presence in the plasma of the anti-A or anti-B antibodies (or both or neither)

BMR. *see* basal metabolic rate

bone marrow. a highly vascular, cellular substance in the central cavity of some bones; site of synthesis of erythrocytes, leukocytes, and platelets

Bowman's capsule. the blind sac at the beginning of the tubular component of a kidney nephron

bradykinin. a peptide formed by the action of the enzyme kallikrein on a protein precursor; a vasodilator, increases capillary permeability, and probably stimulates pain receptors

brainstem. a subdivision of the brain consisting of the medulla, pons, and midbrain; located between the spinal cord and cerebrum

bronchiole. a small division of a bronchus (lung airway)

bronchus. a large-diameter air passage which enters the substance of the lung; located between the trachea and bronchiole

buffering. the reversible binding of hydrogen ions by various compounds, e.g., bicarbonate; these reversible reactions tend to minimize changes in acidity of a buffered solution when acid is added or removed

bulk flow. movement of fluids or gases from a region of higher pressure to one of lower pressure

Ca^{2+}. calcium ion

Ca-ATPase. a primary active-transport carrier protein in the plasma membrane (where it actively transports calcium out of the cell) and in the membranes of several cell organelles

calcitonin. a peptide hormone secreted by the parafollicular cells of the thyroid; involved in calcium regulation

calmodulin. an intracellular protein which binds calcium and mediates many of calcium's second-messenger functions

calorie (cal). the unit in which heat energy is measured; the amount of heat needed to raise the temperature of 1 g of water 1°C; one large calorie (Cal), used to represent the caloric equivalent of various nutrients, equals 1000 calories

cAMP. *see* cyclic AMP

cancer. an uncontrolled growth of cells

capacitation. final maturation of sperm in the female

capillary. smallest of the blood vessels; water, nutrients, dissolved gases, and other molecules move between the blood and extracellular fluid across capillary walls

carbamino compounds. a compound resulting from the combination of carbon dioxide with protein amino groups, particularly hemoglobin

carbohydrate. a substance composed of atoms of carbon, hydrogen, and oxygen according to the general formula $C_n(H_2O)_n$, where n is any whole number; e.g., glucose is $C_6H_{12}O_6$

carbon dioxide. CO_2; the major waste product of metabolism

carbon monoxide. CO; a gas which reacts with the same iron binding sites on hemoglobin as does oxygen but with a much greater affinity than oxygen; decreases the oxygen-carrying capacity of blood

carbonic acid. H_2CO_3; an acid formed from H_2O and CO_2

carbonic anhydrase. the enzyme which catalyzes the reaction by which CO_2 and H_2O combine to form carbonic acid (H_2CO_3)

carboxyl group. —COOH

carboxypeptidase. an enzyme secreted into the small intestine by the exocrine pancreas as a precursor, pro-carboxypeptidase; breaks the peptide bond of the amino acid located at the end of a protein and having a free carboxyl group

carcinogen. any of a number of agents, such as radiation, viruses, and certain chemicals, which can induce the cancerous transformation of cells

cardiac. pertaining to the heart

cardiac cycle. one contraction-relaxation episode of the heart

cardiac output. the volume of blood pumped by each ventricle per minute (not the total output pumped by both ventricles)

cardiac muscle. the muscle of the heart

cardiovascular center. a group of neurons in the brainstem medulla which serves as a major integrating center for reflexes affecting the heart and blood vessels

cardiovascular system. the heart and blood vessels

carotid. pertaining to the two major arteries (the carotid arteries) in the neck which convey blood to the head

carotid-body chemoreceptors. chemoreceptors located near the main branching of the carotid artery; sensitive to blood P_{O_2}, P_{CO_2}, and hydrogen-ion concentration

carotid sinus. a dilation of the internal carotid artery just above the main branching of the carotid; location of the carotid baroreceptors

carrier. integral membrane proteins capable of combining with specific molecules, enabling them to pass through the membrane

catabolic reaction. a degradative chemical reaction which results in the fragmentation of a molecule into smaller and smaller parts

catalyst. a substance which accelerates chemical reactions but does not itself undergo any net chemical change during the reaction

catechol-O-methyltransferase. an enzyme which degrades the catecholamine neurotransmitters

catecholamine. the neurotransmitters dopamine, epinephrine, and norepinephrine, all of which have similar chemical structures

cation. an ion having a net positive charge

CCK. *see* cholecystokinin

cecum. the dilated pouch at the beginning of the large intestine into which open the ileum, colon, and appendix

cell. the basic structural and functional unit of living things, i.e., the smallest structural unit into which an organism can be divided and still retain the characteristics we associate with life

cell differentiation. *see* differentiation

cell-mediated immunity. a type of specific immune response mediated largely by cytotoxic T lymphocytes

cell organelles. membrane-bound compartments, non-membranous particles, and filamentous structures which perform specialized functions and constitute the internal structure of cells

cellulose. a straight-chain polysaccharide composed of thousands of glucose subunits; found in plant cells

center of gravity. the point in a body at which the mass of the body is in perfect balance; if the body were suspended from a string attached to this point, there would be no movement

central chemoreceptors. receptors in the brainstem medulla which respond to changes in the H^+ concentration of the brain extracellular fluid

central nervous system (CNS). brain plus spinal cord

central thermoreceptors. temperature receptors in the hypothalamus, spinal cord, abdominal organs, and other internal locations

central venous pool. the blood in the large veins of the thorax and right atrium

central venous pressure. the blood pressure in the large veins of the thorax and right atrium

centrioles. two small bodies composed of nine fused sets of microtubules located in the cell cytoplasm; participate in nuclear and cell division

cephalic phase. (of gastrointestinal control) initiation of the neural and hormonal reflexes regulating gastrointestinal functions by stimulation of the receptors in the head (cephalic receptors)—sight, smell, taste, and chewing

cerebellum. a subdivision of the brain which lies behind the forebrain and above the brainstem and deals with control of muscle movements

cerebral cortex. the 3-mm-thick cellular layer covering the cerebrum

cerebrospinal fluid (CSF). the fluid which fills the cerebral ventricles and the subarachnoid space surrounding the brain and spinal cord

cervix. the lower portion of the uterus; connects the lumen of the uterus with that of the vagina

CG. *see* chorionic gonadotropin

cGMP. *see* cyclic GMP

channel. (membrane) a small passage formed by integral membrane proteins through which certain small-diameter molecules and ions can diffuse across a membrane

chemical bonds. the interactions between the electric forces of subatomic particles which hold atoms together in a molecule

chemical element. matter composed of identical types of atoms; e.g., carbon atoms form one type of chemical element, oxygen atoms another

chemical equilibrum. a situation in which the rates of the forward and reverse components of a chemical reaction are equal, and thus no net change in the concentrations of the reactants or products occurs

chemical specificity. the property of a protein binding site such that only certain ligands (molecules, ions) can bind with it; the fewer the kinds of ligands capable of binding to that site, the greater is its specificity

chemical synapse. a synapse at which chemical mediators (neurotransmitters) released in response to an action potential in one neuron diffuse across an extracellular gap to influence the activity of a second neuron

chemoreceptors. afferent nerve endings (or cells associated with them) which are sensitive to concentrations of certain chemicals

chemotaxin. a general name given to any chemical mediator which stimulates chemotaxis of neutrophils or other phagocytes

chemotaxis. the orientation and movement of cells in a specific direction in response to a chemical stimulus

chief cells. gastric gland cells which secrete pepsinogen, precursor of the enzyme pepsin

cholecystokinin (CCK). a peptide hormone secreted by cells in the upper small intestine; regulates several gastrointestinal activities including motility of and secretion by the stomach, gallbladder contraction, and enzyme secretion by the pancreas; formerly called pancreozymin

cholesterol. a steroid molecule which is the precursor of the steroid hormones and bile salts and a component of plasma membranes

cholinergic. pertaining to acetylcholine; a compound which acts like acetylcholine

cholinesterase. *see* acetylcholinesterase

chorionic gonadotropin (CG). a protein hormone secreted by the trophoblastic cells of the blastocyst and placenta; maintains the secretory activity of the corpus luteum during the first three months of pregnancy

choroid plexus. a highly vascular, epithelial structure lining portions of the cerebral ventricles; responsible for the formation of much of the cerebrospinal fluid

chromatid. one of the two bodies resulting from the duplication of a chromosome during mitosis or meiosis

chromatin. the combination of DNA and nuclear proteins which is found in chromosomes

chromophore. the retinal component of a photopigment molecule

chromosome. a highly coiled, condensed form of chromatin which is formed in the cell nucleus during the process of mitosis and meiosis.

chylomicron. a small (0.1- to 3.5-μm diameter) lipid droplet, consisting of triacylglycerol, phospholipid, cholesterol, free fatty acids, and proteins, which is released from the intestinal epithelial cells and enters the lacteals during fat absorption

chyme. the solution of partially digested food in the lumen of the stomach and intestines

chymotrypsins. a family of enzymes secreted by the exocrine pancreas as precursors, the chymotrypsinogens; break certain peptide bonds in proteins and polypeptides

cilia. hairlike projections from the surface of specialized epithelial cells; sweep back and forth in a synchronized way to propel material along the cell surface

circadian. recurring approximately every 24 h

circular muscle layer. the layer of smooth muscle surrounding the submucosal layer in the stomach and intestinal walls, and whose muscle fibers are oriented circumferentially around these organs

citric acid. a six-carbon molecule which is an intermediate in the Krebs cycle

citric acid cycle. *see* Krebs cycle

classical complement pathway. an antibody-dependent system for activating complement

clearance. the volume of plasma from which a particular substance has been completely removed

climacteric. the entire phase (beginning with menstrual irregularity in the female) involving physical and emotional changes as sexual maturity gives way to cessation of reproductive function

clitoris. a small body of erectile tissue in the female external genitalia; homologus to the penis

clone. a group of genetically identical organisms

CNS. *see* central nervous system

CO$_2$. carbon dioxide

coactivation. almost simultaneous stimulation of alpha and gamma motor neurons which maintains tension on the spindle stretch receptors during skeletal muscle fiber shortening

cochlea. the inner ear; the fluid-filled spiral-shaped compartment which contains the organ of Corti

cochlear duct. a fluid-filled membranous bag which extends almost the entire length of the inner ear, dividing it into compartments; contains the organ of Corti

code word. the three-nucleotide sequence in DNA which signifies a given amino acid

codon. the sequence of three nucleotide bases in mRNA which corresponds to a given code word in DNA

coenzyme. nonprotein organic molecule which functions as a cofactor; generally serves as a carrier molecule which transfers atoms (such as hydrogen) or small molecular fragments from one reaction to another; is not consumed in the reaction in which it participates and can be reused until it is degraded

coenzyme A. the coenzyme which combines with acetyl groups and transfers them from one reaction to another

cofactor. nonprotein substance (e.g., metal ion) which binds to specific regions of an enzyme and thereby either maintains the shape of the active site or participates directly in the binding of substrate to the active site; is not consumed in the reaction in which it participates and can be reused until it is degraded

collagen. a fibrous protein which has great strength; functions as a structural element in the interstitium of various types of connective tissues, e.g., tendons and ligaments

collateral ventilation. movement of air between alveoli through pores in the walls separating the alveoli

collecting duct. the portion of the tubular component of a kidney nephron situated between the distal tubule and the renal pelvis; ducts from several nephrons combine in each collecting duct

colloid. a large molecule to which the capillaries are relatively impermeable, e.g., a plasma protein

colon. the large intestine, specifically that part extending from the cecum to the rectum

color blindness. a defect in color vision due to the absence of one or even two of the three cone photopigments or to a defect in the color-processing neurons in the visual pathway

colostrum. the thin, milky fluid secreted by the mother's mammary glands for a few days after birth prior to the onset of true milk production

coma. a deep, prolonged unconsciousness not fulfilling all the criteria for brain death; e.g., there may be spontaneous movement and some intact reflexes

"command" neuron. neurons or groups of neurons whose activity initiates the series of neural events resulting in a voluntary action

compensatory growth. a type of regeneration present in many organs after tissue damage

competition. the effect that one ligand has upon the binding of another ligand when both are able to bind to the same site

complement. a group of plasma proteins which, upon activation, kill microbes directly and facilitate every step of the inflammatory process, including phagocytosis; the classical complement pathway is triggered by antigen-antibody complexes, whereas the alternate pathway operates independently of antibody

compliance. *see* lung compliance

concentration. the amount of material present per unit of volume of solution; expressed as mass/volume, e.g., g/L or mmols/L

concentration gradient. a gradation in concentration which occurs between two regions having different concentrations

concentric contraction. the contraction by that skeletal muscle of an antagonistic pair of muscles which undergoes the stronger contraction and shortens; *see also* eccentric contraction

conducting portion. those air passages having walls too thick to allow the exchange of gas between blood and air

conducting system. a network of cardiac muscle fibers specialized to conduct electrical activity to different areas of the heart

conduction. the exchange of heat by transfer of thermal energy from atom to atom or molecule to molecule

conductor. material having a low resistance to current flow

cone. one of the two major receptor types for photic energy in the retina; gives rise to color vision

conformation. the three-dimensional shape of a protein

congestive heart failure. a set of signs and symptoms associated with a decreased contractility of the heart, engorgement of the heart, veins, and capillaries with blood, and the presence of edema

connective-tissue cells. cells specialized for the formation and secretion of various types of extracellular elements which serve to connect, anchor, and support the structures of the body

conscious experience. the things of which a person is aware; the thoughts, feelings, perceptions, ideas, and reasoning during any of the states of consciousness

consciousness. *see* conscious experience, state of consciousness

contractility. the force of contraction which is independent of fiber length

contraction. the turning on of the tension-generating process in a muscle

contraction time. the time interval between a muscle action potential and the development of peak twitch tension or shortening by a muscle

control system. a collection of interconnected components which functions to keep a physical or chemical parameter within a predetermined range of values

convection. the process by which air (or water) next to a warm body is heated, moves away, and is replaced by colder air, which, in turn, follows the same cycle

convergence. (neuronal) synapses from many presynaptic neurons acting upon a single postsynaptic neuron; (of eyes) the turning of the eyes inward, i.e., toward the nose, to view near objects

converting enzyme. the enzyme which catalyzes the removal of two amino acids from angiotensin I to form angiotensin II

core temperature. the temperature of the inner body, as opposed to that of the skin

cornea. the transparent structure covering the front of the eye; forms part of the optical system of the eye and helps focus an image of the object on the retina

corneal reflex. closure of the eyelid in response to irritation of the cornea

cornification. layering of flat (squamous) epithelial cells

coronary arteries. the arteries which carry blood to the heart muscle

corpus callosum. a large, wide band of myelinated nerve fibers which connects the two cerebral hemispheres; one of the brain commissures

corpus luteum. an ovarian structure formed from the follicle after ovulation; secretes estrogen and progesterone

cortex. *see* adrenal cortex, cerebral cortex

cortical association areas. regions of the parietal and frontal lobes of the cerebral cortex which receive input from the different sensory modalities, memory stores, etc., and perform further perceptual processing

corticospinal pathway. a descending motor pathway having its nerve cell bodies in the cerebral cortex; the axons pass without synapsing to the region of the motor neurons; the pyramidal tract

corticotropin releasing hormone (CRH). the hypothalamic hormone which stimulates ACTH (corticotropin) secretion by the anterior pituitary

cortisol. a glucocorticoid steroid hormone secreted by the adrenal cortex; regulates various aspects of organic metabolism and has many other actions

cotransport. a form of secondary active transport in which net movement of the actively transported substance is in the same direction as the ''downhill'' movement of the molecule or ion supplying the energy for the ''uphill'' movement of the transported substance

countercurrent multiplier system. the mechanism associated with the loops of Henle which creates in the medulla of the kidney a region having a high interstitial-fluid osmolarity

countertransport. a form of secondary active transport in which net movement of the actively transported substance is in a direction opposite to the ''downhill'' movement of the molecule or ion supplying the energy for the ''uphill'' movement of the transported substance

covalent bond. a chemical bond between two atoms in a molecule in which each atom shares one of its valence electrons with the other

covalent modulation. alteration of the shape and, therefore, the function of a protein by the attachment of various chemical groups to the side chains of certain amino acids of the protein

CP. *see* creatine phosphate

cranial nerve. one of the 24 peripheral nerves (12 pairs) which join the brainstem or forebrain

creatine phosphate. a molecule which transfers phosphate and energy to ADP to generate ATP

creatinine. a waste product derived from muscle creatine

CRH. *see* corticotropin releasing hormone

critical period. a time during the developmental history when a system is most readily influenced by both favorable and adverse environmental factors

cross bridge. a projection extending from a thick filament in muscle; a portion of the myosin molecule capable of exerting force on the thin filament and causing it to slide past the thick one

crossed-extensor reflex. increased activation of the extensor muscles on the side of the body opposite a limb flexion

crossing over. a process in which segments of maternal and paternal chromosomes exchange with each other during chromosomal pairing before the first meiotic division

crystalloid. a low-molecular-weight solute, e.g. Na^+, glucose, or urea

CSF. *see* cerebrospinal fluid

cumulus oophorous. the granulosa cells surrounding the ovum as it projects into the antrum of the ovarian follicle

current. the movement of electric charge; in biological systems, this is achieved by the movement of ions

cutaneous. pertaining to the skin

cyclic AMP (cAMP). cyclic $3',5'$-adenosine monophosphate; a cyclic nucleotide which serves as a second messenger for many nonsteroid hormones, neurotransmitters, and paracrines

cyclic AMP–dependent protein kinases. a group of enzymes activated by cyclic AMP which transfer a phosphate group from ATP to specific proteins, thereby altering the activity of those proteins

cyclic GMP (cGMP). cyclic $3',5'$-guanosine monophosphate; a cyclic nucleotide which acts as second messenger in some cells, possibly in opposition to cyclic AMP

cytochrome system. a chain of enzymes which couples energy (released from the transfer of hydrogen from coenzyme hydrogen carriers to oxygen to form water) to the formation of ATP (from ADP and inorganic phosphate); the enzymes are located in the inner membranes of mitochondria and are associated with oxidative phosphorylation

cytokinesis. a stage of cell division during which the cytoplasm divides to form two new cells

cytoplasm. the region of the cell outside of the nucleus

cytosine. one of the pyrimidine bases in DNA and RNA

cytoskeleton. the rodlike filaments of various sizes in the cytoplasm of most cells; associated with cell shape and movements

cytosol. the watery medium inside cells but outside of cell organelles

cytotoxic T cells. the class of T lymphocytes which, upon activation by specific antigen, directly attack the cell bearing that type of antigen

D cells. somatostatin-secreting cells of the islets of Langerhans of the pancreas

dark adaptation. the improvement in the sensitivity of vision after one is in the dark for some time

dead space. *see* anatomic dead space, alveolar dead space, total dead space

deamination. the removal of an amino ($—NH_2$) group from a molecule such as an amino acid

decremental. decreasing in amplitude

defecation. expulsion of feces from the rectum

degranulation. the fusion of the phagosome and lysosome membranes

dehydroepiandrosterone. an androgen hormone secreted by the adrenal cortex

dendrite. a highly branched process of a neuron which functions mainly to receive synaptic input from other neurons

denervation atrophy. a decrease in the size of muscle fibers whose nerve supply is destroyed

deoxyhemoglobin (Hb). hemoglobin not combined with oxygen; reduced hemoglobin

deoxyribonucleic acid (DNA). the nucleic acid which stores and transmits genetic information; consists of a double strand (double helix) of nucleotide subunits which contain the sugar deoxyribose

depolarize. to change the value of the membrane potential toward zero, so that the inside of the cell becomes less negative

depression. (as an affective disorder) a disturbance in mood characterized by a pervasive sadness or loss of interest or pleasure

descending limb. (of Henle's loop) the segment of a renal nephron into which the proximal tubule drains

desmosome. a type of cell junction whose function is to hold cells together; consists of two opposed plasma membranes separated by a 20-nm extracellular space

diabetes. *see* diabetes mellitus

diabetes insipidus. a disease resulting from defective ADH control of urine concentration; marked by a great thirst and the excretion of a large volume of dilute urine

diabetes mellitus. a disease in which the hormonal control of plasma glucose is defective because of a deficiency of insulin or decreased target-cell response to insulin

diaphragm. the dome-shaped sheet of skeletal muscle which separates the abdominal and thoracic cavities; the principal muscle of respiration

diastole. the period of the cardiac cycle when the ventricles are not contracting

diastolic pressure. the minimum blood pressure during the cardiac cycle; i.e., the pressure just prior to ventricular ejection

diencephalon. that portion of the anterior part of the brain which lies beneath the cerebral hemispheres and contains the thalamus and hypothalamus; considered by some anatomists to be an extension of the brainstem

differentiation. the process by which cells acquire specialized structural and functional properties during growth and development

diffusion. the movement of molecules from one location to another because of random thermal molecular motion; net diffusion always occurs from a region of higher concentration to a region of lower concentration

diffusion equilibrium. the state during which the diffusion fluxes in opposite directions are equal (i.e., the net flux = 0)

diffusion potential. a voltage created by the separation of charge which results from the movement of charged particles

digestion. the process of breaking down large particles and high-molecular-weight substances into small molecules

dihydrotestosterone. a steroid formed by the enzyme-mediated alteration of testosterone; the active form of testosterone in certain of its target cells

2,3-diphosphoglycerate (DPG). a substance produced by erythrocytes during glycolysis; binds reversibly with hemoglobin, causing it to change conformation and release oxygen

disaccharide. a carbohydrate molecule composed of two monosaccharides; e.g., the disaccharide sucrose (table sugar) is composed of glucose and fructose

disinhibition. the removal of inhibition from a neuron, thereby allowing its activity to increase

distal tubule. the portion of the tubular component of a kidney nephron situated between Henle's loop and the collecting duct

disuse atrophy. a decrease in the size of muscle fibers which are not used for a long period of time

diuretic. any substance which inhibits fluid reabsorption in the renal tubule and thereby causes an increase in the volume of urine excreted

diurnal. daily; occurring in a 24-h cycle

divergence. (neuronal) the process in which a single presynaptic neuron synapses upon and thereby influences the activity of many postsynaptic neurons; (of eyes) the movement in which the eyes are turned outward to view distant objects

DNA. *see* deoxyribonucleic acid

DNA-dependent RNA polymerase. *see* transcriptase

DNA polymerase. the enzyme which, during replication of DNA, joins together nucleotides base-paired with the nucleotides of one strand of DNA to form a new strand of DNA

dominant hemisphere. the cerebral hemisphere which controls the hand used most frequently for intricate tasks, e.g., the left hemisphere in a right-handed person

dopa. dihydroxyphenylalanine; an intermediary in the catecholamine synthetic pathway

dopamine. a catecholamine neurotransmitter; a precursor of epinephrine and norepinephrine

dorsal root. a group of afferent nerve fibers which enters the left and right sides of the spinal cord from that region which faces the back of the body

double bond. two covalent chemical bonds formed between the same two atoms, symbolized by =

double helix. a molecular conformation in which two strands of molecules are coiled around each other; the conformation of DNA

down-regulation. a decrease in the total number of target-cell receptors for a given messenger in response to a chronic high extracellular concentration of the messenger

DPG. *see* 2,3-diphosphoglycerate

dual innervation. innervation of an organ or gland by both sympathetic and parasympathetic nerve fibers

ductus deferens. the male reproductive ducts between the epididymis of each testis and the ducts of the seminal vesicles; the vas deferens

duodenum. first portion of the small intestine; that portion of the intestine between the stomach and jejunum

eardrum. *see* tympanic membrane

eccentric contraction. the contraction of that muscle in an antagonistic pair of skeletal muscles which undergoes the weaker contraction and actually lengthens; *see also* concentric contraction

ECG. *see* electrocardiogram

ectopic focus. a region of the heart (other than the SA node) which assumes the role of the cardiac pacemaker

edema. accumulation of excess fluid in the interstitial space

EDV. *see* end-diastolic volume

EEG. *see* electroencephalogram

EEG arousal. the transformation of the EEG pattern from alpha rhythm to the lower, faster oscillations of beta rhythm; associated with increased levels of attention

effector. a cell or cell collection whose change in activity constitutes the response in a reflex or local homeostatic response; often muscle and gland cells, but almost all cells can be effectors

efferent arteriole. a vessel which conveys blood from a renal glomerulus to a set of peritubular capillaries

efferent neuron. a neuron which carries information away from the central nervous system to muscle cells, gland cells, or postganglionic neurons

efferent pathway. that component of a reflex arc which transmits information from the integrating center to the effector

ejaculation. a reflex consisting of contraction of the genital ducts and skeletal muscle at the base of the penis with resulting expulsion of semen from the penis

ejaculatory duct. the continuation of the ductus deferens after it is joined by the duct from the seminal vesicle; joins the urethra within the prostate gland

EKG. *see* electrocardiogram

electric force. that force which causes the movement of charged particles toward regions having an opposite charge and away from regions having a similar charge

electric potential. (or electric potential difference) *see* potential

electrical synapse. a synapse at which local currents resulting from action potentials in one neuron flow through gap junctions joining two neurons, to influence the activity of the second neuron

electrocardiogram (ECG, EKG). a recording at the surface of the skin of the electric currents generated by action potentials of cardiac muscle cells

electrochemical gradient. the force determining the magnitude of the net movement of charged particles; a combination of the electrical gradient (as determined by the voltage difference between two points) and the chemical gradient (as determined by the concentration difference between the same two points)

electrode. a probe to which electric charges can be added (or from which they can be removed) to cause changes in the electric current flow to a recorder

electroencephalogram (EEG). a recording of the electric potential difference between two points on the scalp (or between one point on the scalp and a distant electrode)

electrogenic pump. an active transport system which directly separates electric charge, thereby producing a potential difference

electrolyte. a substance which dissociates into ions when in solution

electromagnetic radiation. radiation composed of waves with electric and magnetic components

electron. a subatomic particle which carries one negative charge and revolves in an orbit around the nucleus of an atom

embolus. a foreign body, such as a fragment of a blood clot or an air bubble, circulating in the blood

embryo. a developing organism during the early stages of development (in human beings, the first two months of intrauterine life)

emission. movement of the contents of the male genital ducts into the urethra prior to ejaculation

emulsification. a fat-solubilizing process in which large lipid droplets are broken into smaller droplets

end-diastolic volume (EDV). the amount of blood in a ventricle of the heart just prior to systole

end-plate potential (EPP). a depolarization of the motor end plate of skeletal muscle fibers in response to acetylcholine; initiates an action potential in the adjacent skeletal muscle plasma membrane

end-product inhibition. inhibition of a metabolic pathway by the action of the final product upon an allosteric site on some enzyme (usually the rate-limiting enzyme) in the pathway

end-systolic volume (ESV). the amount of blood remaining in the ventricle after ejection

endocrine gland. a group of cells which secrete hormones into the extracellular space around the gland cells from which the hormones diffuse into the bloodstream; a ductless gland

endocrine system. all the hormone-secreting glands of the body

endocytosis. the process in which the plasma membrane of a cell invaginates and the invaginations become pinched off forming small, intracellular, membrane-bound vesicles which enclose a volume of material; *see also* phagocytosis

endogenous pyrogen (EP). the protein secreted by monocytes and macrophages, and which acts in the brain to cause fever; probably the same molecule as LEM and IL-1

endolymph. the fluid which fills the membranous sacs of the vestibular apparatus and the cochlear duct of the inner ear

endometrium. glandular epithelium lining the uterine cavity

endoplasmic reticulum. a cell organelle located in the cytoplasm and consisting of an interconnected network of membrane-bound branched tubules and flattened sacs; two types are distinguished: *granular*, which has ribosomes attached to its membranes, and *agranular*, which is smooth-surfaced

β-endorphin. a morphine-like peptide; one of the endorphins; cosecreted with ACTH from the anterior pituitary

endorphins. a group of peptides; some are neurotransmitters at the synapses activated by opiate drugs, while some probably act as paracrines and hormones; "endogenous morphines"

endothelium. the thin layer of cells which lines the cavities of the heart and blood vessels

energy. defined in dynamic terms as the ability to produce change; measured by the amount of work performed during a given change; the total energy content of any object consists of two components: *potential energy* and *kinetic energy*

enkephalin. a peptide which functions as a neurotransmitter at synapses activated by opiate drugs

enterohepatic circulation. a "recycling" pathway for bile salts (and other substances) by reabsorption from the intestines, passage to the liver (via the hepatic portal vein), and passage back to the intestines (via the bile duct)

enterokinase. an enzyme embedded in the luminal plasma membrane of the intestinal epithelial cells; activates pancreatic trypsinogen

enzymatic activity. the rate at which an enzyme converts substrate to product

enzyme. a protein which accelerates specific chemical reactions but does not itself undergo any net change during the reaction; a biochemical catalyst

enzyme repression. inhibition of enzyme synthesis in the presence of product molecules synthesized by that enzyme

eosinophil. polymorphonuclear granulocytic leukocyte whose granules take up the red dye eosin; involved in allergic responses and destruction of parasitic worms

epididymis. a portion of the duct system of the male reproductive system located between the seminiferous tubules and the ductus deferens

epiglottis. a thin, triangular flap of cartilage which folds down and back, covering the trachea, during swallowing

epilepsy. neuronal activity which starts at a cluster of abnormal neurons and spreads to large numbers of cells in adjacent neural tissue, may give rise to convulsions

epinephrine. a hormone secreted by the adrenal medulla and involved mainly in the regulation of organic metabolism; a catecholamine neurotransmitter; also called adrenaline

epiphenomenon. an event which accompanies some action but is not causally related to it

epiphyseal plate. an actively proliferating zone of cartilage near the ends of bones; the region of bone growth

epithelial transport. the movement of molecules from one extracellular compartment across epithelial cells into a second extracellular compartment

epithelium. tissue which covers all body surfaces and lines all body cavities; most glands are formed from epithelial tissue

EPP. *see* end-plate potential

EPSP. *see* excitatory postsynaptic potential

equilibrium. a situation in which no net change occurs within a system and no energy input is required to maintain this state

equilibrium potential. the voltage difference at which there is zero net flux of an ion between two compartments; the voltage gradient produces a flux of that ion equal in magnitude but opposite in direction to the ion flux produced by the concentration gradient

erectile tissue. spongy vascular tissue in the penis and clitoris which becomes engorged with blood and causes the tissue to become erect

erection. a vascular phenomenon in which the penis or clitoris becomes stiff and erect

error signal. the difference between the level of a regulated variable in a control system and the operating or set point for that variable

erythrocyte. a red blood cell

erythropoiesis. formation of erythrocytes

erythropoietin. a hormone secreted mainly by cells of the kidney; stimulates red-blood-cell production

esophagus. the portion of the digestive tract which connects the mouth (pharynx) and stomach

essential amino acids. the amino acids which cannot be formed by the body at all (or at a rate adequate to meet metabolic requirements) and, therefore, must be obtained from the diet

essential nutrients. substances which are required for normal or optimal body function but are synthesized by the body either not at all or in amounts inadequate to prevent disease

estradiol. a steroid hormone of the estrogen family; the major estrogen secreted by the ovaries

estriol. a steroid hormone of the estrogen family; the major estrogen secreted by the placenta during pregnancy

estrogen. a group of steroid hormones which have effects similar to estradiol on the female reproductive tract

estrone. a steroid hormone of the estrogen family

ESV. *see* end-systolic volume

euphorigen. a drug which elevates mood

excitability. the ability to produce action potentials

excitation-contraction coupling. the mechanisms in muscle fibers linking depolarization of the plasma membrane with generation of force by the cross bridges

excitatory postsynaptic potential (EPSP). a depolarizing graded potential which arises in the postsynaptic neuron in response to the activation of excitatory synaptic endings upon it

excitatory synapse. a synapse which, when activated, either increases the likelihood that the membrane potential of the postsynaptic neuron will reach threshold and undergo action potentials or increases the firing frequency of existing action potentials

excretion. the appearance of a substance in the urine, feces, or other material lost from the body

exocrine gland. a group of epithelial cells specialized for secretion and having ducts which lead to a specific compartment or surface, into or onto which the products are secreted

exocytosis. the process in which the membrane of an intracellular vesicle fuses with the plasma membrane, the vesicle opens, and the vesicle contents are liberated into the extracellular fluid

exon. a region of a gene in DNA which contains the code words for a portion of the specific amino acid sequences in a protein

expiration. the bulk flow of air out of the lungs to the atmosphere

expiratory reserve volume. the volume of air which can be exhaled by maximal active contraction of the expiratory muscles at the end of a normal expiration

extension. the straightening of a joint

extensor muscle. a muscle whose activity causes straightening of a joint

external anal sphincter. a band of skeletal muscle around the lower end of the rectum; generally under voluntary control

external environment. the environment which immediately surrounds the external surfaces of an organism

external genitalia. (female) the mons pubis, labia majora and minora, clitoris, vestibule of the vagina, and vestibular glands; (male) the penis and scrotum

external urethral sphincter. a ring of skeletal muscle which surrounds the lower end of the urethra, preventing micturition when contracted

external work. movement of external objects by the contraction of skeletal muscles

extracellular fluid. literally, fluid "outside the cells"; the internal environment in which the cells live; comprises the interstitial fluid and plasma

extrafusal fibers. *see* skeletomotor fibers

extrapyramidal system. *see* multineuronal pathways

extrasystole. a heartbeat which occurs before the normal time in the cardiac cycle

extrinsic clotting pathway. the formation of fibrin clots by a pathway using tissue thromboplastin

facilitated diffusion. a carrier-mediated transport system which moves molecules from high to low concentration across a membrane; energy is not coupled to the system, and equilibrium is reached when the concentrations on the two sides of the membrane become equal

facilitation. the general depolarization of a nerve cell membrane potential when excitatory synaptic input to the cell exceeds inhibitory input

factor XII. *see* Hageman factor

factor XIII. an enzyme in plasma which catalyzes cross-link formation between fibrin molecules to strengthen a blood clot

farsighted. pertaining to the defect in vision which occurs because the eyeball is too short for the lens, so that near objects are focused behind the retina

fat mobilization. the increased breakdown of triacylglycerols and release of glycerol and fatty acids into the blood

fat-soluble vitamins. vitamins which are soluble in nonpolar solvents and insoluble in water; vitamins A, D, E, and K

fatty acid. a chain of carbon atoms with a carboxyl group at one end through which it can be linked to glycerol in the formation of triacylglycerols (neutral fat); *see also* polyunsaturated fatty acid, saturated fatty acid, unsaturated fatty acid

feces. the material expelled from the large intestine during defecation, consisting primarily of water, bacteria, and ingested matter which failed to be digested and absorbed in the intestines

feedback. a characteristic of control systems in which the output response influences the input to the control system; *see also* negative feedback, positive feedback

feedforward. an aspect of some control systems which allows the system to anticipate changes of stimuli to its receptor

ferritin. an iron-binding protein which is the storage form for iron

fertilization. union of a sperm with an ovum

fetus. an organism during the later stages of development prior to birth; in human beings, the period from the second month of gestation until birth

fever. an increased body temperature due to the setting of the "thermostat" of the temperature-regulating mechanisms at a higher-than-normal level

fibrillation. extremely rapid contractions of cardiac muscle in an unsynchronized, repetitive way which prevents effective pumping of blood

fibrin. a protein fragment resulting from the enzymatic cleavage of fibrinogen and having the ability to bind to other similar fragments, i.e., to polymerize, and to turn blood into a solid gel (clot)

fibrinogen. the plasma protein which is the precursor of fibrin

fight-or-flight response. overall activation of the sympathetic nervous system in the response to stress

filtration. movement of essentially protein-free plasma across the walls of a capillary out of its lumen as a result of a pressure gradient across the capillary wall

"first" messenger. the extracellular hormone, paracrine agent, or neurotransmitter which combines with a plasma-membrane or cytosol receptor

flatus. intestinal gas

flexion. the bending of a joint

flexion reflex. flexion of those joints which withdraw an injured part away from a painful stimulus

flexor muscle. a muscle whose activity causes bending of a joint

fluid mosaic model. molecular structure of cell membranes consists of proteins embedded in a bimolecular layer of phospholipids; the phospholipid layer has the physical properties of a fluid, allowing the membrane proteins to move laterally within the lipid layer

flux. the amount of material crossing a surface in a unit of time; *see also* net flux, unidirectional flux

folic acid. a vitamin of the B-complex group; essential for the formation of nucleotides

follicle. *see* ovarian follicle, primary follicle

follicle cell. *see* granulosa cells

follicle-stimulating hormone (FSH). a peptide hormone secreted by the anterior pituitary in both males and females; one of the gonadotropins

follicular phase. that portion of the menstrual cycle during which a single follicle and ovum develop to full maturity prior to ovulation

forebrain. a subdivision of the brain consisting of the right and left cerebral hemispheres (the cerebrum) and the diencephalon

fovea. the area near the center of the retina in which cones are most concentrated and which, therefore, gives rise to the most acute vision

frontal lobe. a region of the anterior cerebral cortex where motor cortex, Broca's speech center, and some association cortex are located

fructose. a five-carbon sugar found in the disaccharide sucrose (table sugar)

FSH. *see* follicle-stimulating hormone

functional hyperemia. *see* active hyperemia

functional unit. a subunit of an organ; all the subunits have similar structural and functional properties

GABA. *see* gamma-aminobutyric acid

galactose. a six-carbon monosaccharide found in lactose (milk sugar)

gallbladder. a small sac underneath the liver; concentrates bile and stores it between meals; contraction of the gallbladder ejects bile into the bile duct which carries it to the small intestine

gallstone. a precipitate of cholesterol (and occasionally other substances as well) in the gallbladder or bile duct

gametes. the germ cells or reproductive cells; the sperm cells in a male and the ova in a female

gametogenesis. the production of gametes

gamma-aminobutyric acid (GABA). a neurotransmitter at some inhibitory synapses

gamma globulin. immunoglobulin G (IgG), the most abundant class of plasma antibodies

gamma motor neuron. a small motor neuron which controls spindle (intrafusal) muscle fibers

ganglion. (pl. ganglia) a cluster of neuronal cell bodies; generally restricted to neuronal clusters located outside the central nervous system

ganglion cells. neurons in the retina which are postsynaptic to the bipolar cells; their axons form the optic nerve

gap junction. a type of cell junction allowing ions and small molecules to flow between the cytoplasms of adjacent cells via small channels or tubes

gastric. pertaining to the stomach

gastric phase. (of gastrointestinal control) the initiation of neural and hormonal reflexes controlling gastrointestinal function by stimulation of the wall of the stomach

gastrin. a peptide hormone secreted by the antral region of the stomach; stimulates gastric acid secretion

gastroilial reflex. a reflex increase in contractile activity of the ileum during the periods of gastric emptying

gastrointestinal system. gastrointestinal tract plus the salivary glands, liver, gallbladder, and pancreas

gastrointestinal tract. mouth, pharynx, esophagus, stomach, and small and large intestines

gene. a unit of hereditary information; the portion of a DNA molecule which contains, coded in its nucleotide sequence, the information required to determine the amino acid sequence of a single polypeptide chain

germ cells. cells which give rise to the male and female gametes (the sperm and ova)

GFR. *see* glomerular filtration rate

GH. *see* growth hormone

gland. a group of epithelial cells specialized for the function of secretion; *see also* endocrine gland, exocrine gland

glial cell. a nonneuronal cell type in the brain and spinal cord; helps regulate the extracellular environment of the central nervous system; neuroglial cell

globin. the polypeptide chains of a hemoglobin molecule

globulin. one of the types of protein found in blood plasma

glomerular filtration. the movement of an essentially protein-free plasma through the capillaries of the renal glomerulus into Bowman's capsule

glomerular filtration rate (GFR). the volume of fluid filtered through the renal glomerular capillaries into Bowman's capsule per unit time

glomerulus. a capillary tuft which forms the vascular component at the beginning of a kidney nephron and the balloonlike hollow capsule (Bowman's capsule) into which the capillary tuft protrudes

glottis. the opening between the vocal cords

glucagon. a peptide hormone secreted by the A cells of the islets of Langerhans of the pancreas; its action on several target cells leads to a rise in plasma glucose

glucocorticoid. one of several steroid hormones produced by the adrenal cortex having major effects on glucose metabolism, e.g., cortisol

gluconeogenesis. synthesis by the liver of glucose from pyruvate, lactate, glycerol, or amino acids; new formation of glucose

glucose. the major monosaccharide in the body; a six-carbon sugar, $C_6H_{12}O_6$

glucose sparing. the switch from glucose to fat utilization by most cells during the postabsorptive state

glutamic acid. an amino acid; a neurotransmitter at some excitatory synapses

glycerol. a three-carbon carbohydrate molecule; forms the backbone of triacylglycerol (neutral fat)

glycine. an amino acid; a neurotransmitter at some inhibitory synapses

glycogen. a highly branched polysaccharide composed of thousands of glucose subunits; the major form of carbohydrate storage in the body

glycogenolysis. the breakdown of glycogen

glycolipid. amphipathic lipid molecules with a covalently bound carbohydrate group

glycolysis. a metabolic pathway which breaks down glucose to form two molecules of pyruvic acid (in the presence of oxygen) or two molecules of lactic acid (in the absence of oxygen)

glycolytic fast fibers. muscle fibers which combine a high myosin-ATPase activity with a high glycolytic capacity

glycoprotein. protein with a covalently linked carbohydrate group

GnRH. *see* gonadotropin-releasing hormone

Golgi apparatus. a cell organelle consisting of membranes and vesicles, usually located near the nucleus; processes newly synthesized proteins for secretion and distribution to cell organelles

Golgi tendon organs. mechanoreceptor endings of afferent nerve fibers; wrapped around collagen bundles in tendons

gonad. gamete-producing reproductive organ, i.e., the testes in the male and the ovaries in the female

gonadotropic hormones. hormones secreted by the anterior pituitary which control the function of the gonads; FSH and LH

gonadotropin-releasing hormone (GnRH). the hypothalamic hormone which controls LH and FSH secretion by the anterior pituitary in both males and females

graded potential. a change in membrane potential which can be varied in magnitude (not all-or-none) in either a depolarizing or hyperpolarizing direction

granuloma. a mass consisting of numerous layers of phagocytic-type cells (the central one of which contains a microbe or potentially harmful nonmicrobial substance) enclosed in a fibrous capsule, which isolates the substance from healthy tissues.

granulosa cells. cells which surround the ovum in the ovarian follicle

gray matter. that portion of the brain and spinal cord which appears gray in unstained specimens and consists mainly of cell bodies and the unmyelinated portions of nerve fibers

ground substance. the material which surrounds cells and fibers in connective tissue; can be watery or gel-like; contains mucopolysaccharides

growth hormone (GH). a peptide hormone secreted by the anterior pituitary; stimulates the release of somatomedins; stimulates body growth by means of its actions on carbohydrate and protein metabolism; somatotropin or somatotropic hormone (STH)

growth hormone release–inhibiting hormone. *see* somatostatin

growth hormone releasing hormone (GRH). the hypothalamic hormone which stimulates the secretion of growth hormone by the anterior pituitary

GTP. *see* guanosinc triphosphate

guanine. one of the purine bases of DNA and RNA

guanosine triphosphate (GTP). a high-energy molecule similar to ATP except that it contains the base guanine rather than adenine

H-Y antigen. the protein which serves as the link between the presence of the Y chromosome and testis differentiation

H zone. one of the transverse bands making up the repeating, striated pattern of cardiac and skeletal muscles; a light region that bisects the A band of striated muscle

habituation. a reversible decrease in the strength of a response upon repeatedly administered stimulation

Hageman factor. factor XII; the initial factor in the sequence of reactions which results in blood clotting

hair cells. mechanoreceptors in the organ of Corti and vestibular apparatus

haploid. having a chromosome number equal to that of the gametes, i.e., half the diploid number found in most somatic cells

hapten. a small molecule which does not itself elicit an immune response but can attach to an existing antigen and thereby acquire the ability to trigger an immune response

Hb. *see* deoxyhemoglobin

HbO$_2$. *see* oxyhemoglobin

H$_2$CO$_3$. carbonic acid

HDL. high-density lipoprotein; a lipid-protein aggregate having a low proportion of lipid

heart attack. damage to or death of cardiac muscle, i.e., a myocardial infarction

heart murmur. the abnormal heart sounds caused by the turbulent flow of blood through narrowed or leaky valves or through a hole in the interventricular or interatrial septum

heart rate. the number of contractions of the heart per minute

heart sounds. the noises which result from vibrations caused by closure of the atrioventricular valves (first heart sound) and pulmonary and aortic valves (second heart sound)

heartburn. pain which seems to occur in the region of the heart but is due to stimulation of pain receptors in the esophageal wall by the acid refluxed from the stomach

heat exhaustion. a state of collapse due to hypotension brought about both by depletion of plasma volume secondary to sweating and by extreme dilation of skin blood vessels; the thermoregulatory centers are still functioning

heat stroke. a positive-feedback situation in which a heat gain greater than heat loss causes body temperature to become so high that the brain thermoregulatory centers do not function, allowing temperature to rise even higher

Heimlich maneuver. used to remove an obstruction causing choking; a forceful elevation of the diaphragm which is produced by a rescuer's fist against a choking victim's abdomen causing a sudden sharp increase in alveolar pressure, which expels the obstructing material

"helper" T cells. the class of T cells which enhances antibody production and cytotoxic T-cell function

hematocrit. the percentage of the total blood volume occupied by blood cells

heme. the iron-containing organic molecule bound to each of the four polypeptide chains of a hemoglobin molecule; the iron-containing portion of a cytochrome

hemoglobin. a red-colored protein, located in erythrocytes; transports most of the oxygen in the blood

hemorrhage. bleeding

hemostasis. stopping the loss of blood from a damaged vessel

Henle's loop. the hairpinlike segment of the tubular component of a kidney nephron situated between the proximal and distal tubules

heparin. an anticlotting agent found in various cells and tissues; an anticoagulant drug

hepatic. pertaining to the liver

hepatic portal vein. the vein which conveys blood from the capillary beds in the intestines (as well as portions of the stomach and pancreas) to the capillary beds in the liver

hertz (Hz). cycles per second; the measure used for wave frequencies

hippocampus. a portion of the limbic system in the brain associated with learning and emotions

histamine. an inflammatory mediator secreted mainly by mast cells; a neurotransmitter of the monoamine class

histocompatibility antigens. antigenic protein molecules on the surface of nucleated cells

histone. one of a group of positively charged proteins associated with DNA to form chromatin

histotoxic hypoxia. hypoxia in which the quantity of oxygen reaching the tissues is normal, but the cell is unable to utilize the oxygen because a toxic agent has interfered with its metabolic machinery

holocrine secretion. a type of secretion which involves the complete disintegration of a secretory cell

homeostasis. the relatively stable physical and chemical composition of the internal environment (extracellular fluid) of the body which results from the actions of regulatory systems

homeostatic system. a control system which consists of a collection of interconnected components and functions to keep a physical or chemical parameter of the internal environment relatively constant

homeotherm. an animal capable of maintaining its body temperature within very narrow limits

hormone. a chemical messenger synthesized by a specific endocrine gland in response to certain stimuli and secreted into the blood, which carries it to the other cells in the body where its actions are exerted

HR. *see* heart rate

humoral immunity. a type of specific immune response in which circulating antibodies play a central role

hydrochloric acid. HCl; a strong acid secreted into the stomach lumen by gland cells lining the stomach

hydrogen bond. a weak chemical bond between two molecules (or parts of the same molecule) in which a negative region of one polarized substance is electrostatically attracted to a positively polarized hydrogen atom in the other

hydrogen ion. H^+; a single free proton; the concentration of hydrogen ions in a solution determines the acidity of the solution

hydrogen peroxide. H_2O_2; a chemical produced by a phagosome and highly destructive to the macromolecules within it

hydrolysis. the breaking of a chemical bond with the addition of the elements of water (—H and —OH) to the products formed; a hydrolytic reaction

hydrostatic pressure. the pressure exerted by a fluid

hydroxyl group. —OH

hyperemia. increased blood flow; *see also* active hyperemia, reactive hyperemia

hyperglycemia. plasma glucose concentration increased above normal levels

hyperpolarize. to change the transmembrane potential so the inside of the cell becomes more negative than its resting state

hypersensitivity. an acquired reactivity to an antigen which can result in bodily damage upon subsequent exposure to that particular antigen; i.e., allergies

hypertension. chronically increased arterial blood pressure

hypertonic solution. a solution containing a higher concentration of effectively membrane-impermeable solute particles than cells contain

hypertrophy. an increase in size; enlargement of a tissue or organ which results from an increase in cell size rather than in cell number

hyperventilation. ventilation greater than that needed to maintain normal plasma P_{CO_2}

hypocalcemic tetany. skeletal muscle spasms due to a low extracellular calcium concentration

hypoglycemia. low blood sugar (glucose) concentration

hypothalamic releasing hormones. hormones released from hypothalamic neurons into the hypothalamo-pituitary portal vessels to control the release of the anterior pituitary hormones

hypothalamus. a brain region below the thalamus; responsible for the integration of many basic behavioral patterns which involve correlation of neural and endocrine function, especially those concerned with regulation of the internal environment

hypotonic solution. a solution containing a lower concentration of effectively membrane-impermeable solute particles than cells contain

hypoventilation. ventilation insufficient to maintain normal plasma P_{CO_2}

hypoxia. a deficiency (but not the complete absence) of oxygen at the tissue level; *see also* hypoxic hypoxia, anemic hypoxia, ischemic hypoxia, histotoxic hypoxia

hypoxic hypoxia. hypoxia due to decreased arterial P_{O_2}

Hz. *see* hertz

I band. one of the transverse bands making up the repeating striated pattern of cardiac and skeletal muscle; located between the A bands of adjacent sarcomeres and bisected by the Z line

ICSH. *see* luteinizing hormone

IgA. the class of antibodies secreted by the lining of the gastrointestinal, respiratory, and urinary tracts

IgD. a class of antibodies whose function is unknown

IgE. the class of antibodies which mediates immediate hypersensitivity ("allergy")

IgG. gamma globulin; the most abundant class of plasma antibodies

IgM. a class of antibodies which, along with IgG, provides the bulk of specific humoral immunity against bacteria and viruses

ileocecal sphincter. a ring of smooth muscle separating the ileum from the cecum, i.e., the small and large intestines

ileogastric reflex. a reflex decrease in gastric motility in response to distension of the ileum

ileum. the final, longest segment of the small intestine

immune enhancement. the protection of a foreign cell from destruction by the combination of antibodies with the foreign antigens, which prevents T cells from combining with the foreign antigens

immune responses. *see* nonspecific immune responses, specific immune responses

immune surveillance. the recognition and destruction of abnormal or mutant cell types which arise within the body

immunity. the physiological mechanisms which allow the body to recognize materials as foreign or abnormal and to neutralize or eliminate them; the mechanisms which maintain the uniqueness of "self"

immunoglobulin (Ig). a synonym for antibody; the five classes are IgG, IgA, IgD, IgM, and IgE

implantation. the event during which the fertilized ovum adheres to the uterine wall and becomes embedded in it

inducer. the molecule which, when combined with the repressor protein, inactivates the repressor thereby freeing the operator gene from inhibition and initiating synthesis of the proteins coded by the operon

induction. stimulation of enzyme synthesis by action of the enzyme substrate on the genetic mechanisms controlling enzyme synthesis

infarction. an area of dead tissue resulting from local ischemia due to diminished circulation supplying the area; an infarct

inferior vena cava. the large vein which carries blood from the lower half of the body to the right atrium of the heart

inflammation. the local response to injury characterized by swelling, pain, and increased temperature and redness in the region of the injury

inhibin. a protein hormone secreted by the Sertoli cells of the seminiferous tubules and by ovarian follicles; inhibits FSH secretion

inhibitory postsynaptic potential (IPSP). a hyperpolarizing graded potential which arises in the postsynaptic neuron in response to the activation of inhibitory synaptic endings upon it

inhibitory synapse. a synapse which, when activated, decreases the likelihood that the membrane potential of the postsynaptic neuron will reach threshold and fire an action potential (or decreases the firing frequency of existing action potentials)

initial segment. the first portion of the axon of a neuron plus the part of the cell body where the axon is joined

inner ear. the cochlea; contains the organ of Corti

inorganic. pertaining to substances which do not contain carbon, e.g., Na^+, Cl^-, and K^+ are inorganic ions

inorganic phosphate. $—HPO_4^{2-}$

inosine. a nucleotide base present in tRNA; forms base pairs with cytosine

insensible loss. the water loss of which a person is unaware; i.e., the loss by evaporation from the cells of the skin and the lining of the respiratory passages

inspiration. the bulk flow of air from the atmosphere into the lungs

inspiratory muscles. those muscles whose contraction contributes to inspiration; the diaphragm is the major inspiratory muscle

inspiratory neuron. a neuron whose cell body is in the brainstem medulla and which fires in synchrony with inspiration and ceases firing during expiration

inspiratory reserve volume. the maximal volume of air which can be inspired over and above the resting tidal volume

insulin. a peptide hormone secreted by the B cells of the islets of Langerhans of the pancreas; stimulates the movement of glucose and amino acids into most cells and stimulates the synthesis of protein, fat, and glycogen

insulin-like growth factors. a group of peptides which have growth-promoting effects; includes the somatomedins

integral proteins. membrane proteins which are embedded in the phospholipid bilayer of the membrane; may span the entire membrane or be located at only one side

integrator. a cell or collection of cells whose output is determined by many often-conflicting bits of incoming information; an integrating center

intention tremor. a tremor which occurs only when the person is trying to perform a movement, not when the person is at rest; associated with cerebellar disease

intercalated disks. the thickened ends of cardiac muscle cells where each cell is connected to adjacent ones; contains desmosomes and gap junctions

intercostal muscles. skeletal muscles which lie between the ribs; contraction results in movement of the rib cage during inspiration or expiration

interferon. a family of proteins which nonspecifically inhibit viral replication and also stimulate the activity of cytotoxic T cells, NK cells, and macrophages

interleukin 1 (IL-1). the protein secreted by macrophages which stimulates activated B cells to proliferate and the helper T cells to secrete interleukin 2; probably the same molecule as LEM and EP

interleukin 2 (IL-2). a protein secreted by helper T cells which causes activated T cells to proliferate; its secretion is stimulated by interleukin 1

internal anal sphincter. a ring of smooth muscle around the lower end of the rectum

internal energy. the energy locked in the intramolecular bonds of organic molecules; when liberated, appears as heat or can be used to perform work

internal environment. the extracellular fluid; includes the interstitial fluid and blood plasma

internal exchanges. changes in the composition of the extracellular fluid due to exchanges with cells and bone

internal genitalia. (female) the ovaries, uterine tubes, uterus, and vagina

internal urethral sphincter. part of the smooth muscle wall of the urinary bladder which, by its contraction and relaxation, opens and closes the outlet from the bladder to the urethra

internal work. energy-requiring activities (work) in the body, e.g., cardiac contraction, membrane active transport, synthetic chemical reactions, and the muscular movement associated with respiration

interneuron. a neuron whose cell body and axon lie entirely within the central nervous system

interphase. the period of the cell-division cycle between the end of one division and the beginning of the next

interstitial cell stimulating hormone (ICSH). *see* luteinizing hormone

interstitial cells. (of the testis) testosterone-secreting cells which lie between the seminiferous tubules in the testes; Leydig cells

interstitial fluid. the extracellular fluid surrounding the cells of a tissue; plasma, which surrounds blood cells, is not considered interstitial fluid

interstitium. interstitial space; the space between tissue cells

interventricular septum. the partition in the heart separating the cavities of the right and left ventricles

intestinal phase. (of gastrointestinal control) the initiation of neural and hormonal reflexes controlling gastrointestinal functions by stimulation of the walls of the gastrointestinal tract

intestino-intestinal reflex. a reflex cessation of contractile activity of the intestines in response to various stimuli in the intestine

intraabdominal pressure. the pressure within the abdominal cavity

intracellular fluid. the fluid in cells; comprises the cytosol as well as the fluid in the nucleus and other cell organelles

intrafusal fibers. *see* spindle fibers

intrapleural fluid. the thin film of fluid in the thoracic cavity between the pleura lining the inner wall of the thoracic cage and the pleura covering the lungs

intrapleural pressure. the pressure within the pleural space generated by the tendency of the lung and chest wall to pull away from each other; also called intrathoracic pressure

intrathoracic pressure. *see* intrapleural pressure

intrinsic clotting pathway. the intravascular sequence of fibrin formation initiated by Hageman factor

intrinsic factor. a glycoprotein secreted by the epithelial lining of the stomach and necessary for the absorption of vitamin B_{12} in the ileum

intron. a region of noncoding base sequences in a gene

ion. any atom or small molecule containing an unequal number of electrons and protons and, therefore, carrying a net positive or negative electric charge, e.g., Cl^-, Na^+, and NH_4^+

ionic bond. a strong electrical attraction between two oppositely charged ions

ionic hypothesis. an explanation of the voltage changes during an action potential in terms of membrane permeability changes and the movement of sodium and potassium ions across the membrane

ionization. the process of removing electrons from or adding them to an atom or small molecule to form an ion

IPSP. *see* inhibitory postsynaptic potential

iris. the ringlike structure which surrounds the pupil of the eye; consists of pigmented epithelium and muscle, which by contraction adjusts the size of the pupil

ischemia. a reduced blood supply to an organ or tissue

ischemic hypoxia. hypoxia in which the basic defect is too little blood flow to the tissues to deliver adequate amounts of oxygen

islets of Langerhans. clusters of endocrine cells in the pancreas; distinct islet cells secrete the hormones insulin, glucagon, somatostatin, and pancreatic polypeptide

isometric contraction. contraction of muscle under conditions in which it develops tension but does not change length

isotonic contraction. contraction of a muscle under conditions in which a load on the muscle remains constant but the muscle length changes (i.e., shortens)

isotonic solution. a solution containing the same number of effectively nonpenetrating solute particles as cells contain

isovolumetric ventricular contraction. the early phase of systole when both the atrioventricular and aortic valves are closed

isovolumetric ventricular relaxation. the early phase of diastole when both the atrioventricular and aortic valves are closed

jaundice. a condition in which the skin is yellowish due to buildup in the blood and tissues of bilirubin as a result of the failure of bilirubin excretion by the liver into the bile

jejunum. the middle segment of the small intestine

K_p. *see* membrane permeability constant

KCl. potassium chloride

keto acid. a molecule containing a carbonyl group (—CO—) and a carboxyl group (—COOH)

ketone bodies. certain products of fatty acid metabolism, i.e., acetoacetic acid, acetone, and β-hydroxybutyric acid, which accumulate in the blood during starvation and in untreated severe diabetes mellitus

killer (K) cells. a class of cells (probably lymphocytes) which are brought into action by the presence of antibody bound to antigen on the surface of a foreign cell and which kill the cell directly without prior phagocytosis

kinesthesia. a sense of movement derived from the position and movement of joints

kinetic energy. the energy associated with an object because of its motion

kininogen. the protein from which the kinins are generated

kinins. peptides split from kininogen in inflamed areas; facilitate the vascular changes associated with inflammation; may also stimulate pain receptors

Krebs cycle. a metabolic pathway located in the mitochondrion which utilizes fragments derived from the breakdown of carbohydrate, protein, and fat and produces carbon dioxide, hydrogen, and small amounts of ATP; the hydrogen is passed on to the oxidative phosphorylation pathway

L-dopa. a precursor of dopamine; administered to patients with Parkinson's disease

lacrimal glands. the glands which secrete tears

lactase. the enzyme which breaks down lactose (milk sugar) into glucose and galactose in the small intestine

lactate. the ionized form of lactic acid

lactation. the production and secretion of milk by the mammary gland

lacteal. a blind-ended lymph vessel in the center of each intestinal villus

lactic acid. a 3-carbon molecule formed by the glycolytic pathway in the absence of oxygen

lactoferrin. a protein secreted by neutrophils; tightly binds iron

lactose. milk sugar; a disaccharide composed of glucose and galactose

larynx. the part of the air passageway between the pharynx and trachea; contains the vocal cords

latent period. the period (several ms) between initiation of the action potential in a muscle fiber and the beginning of mechanical activity

lateral inhibition. a method of refining sensory information in afferent neural pathways whereby the fibers inhibit each other, the most active causing the greatest inhibition of adjacent fibers

lateral sac. the enlarged region at the end of each segment of sarcoplasmic reticulum adjacent to the transverse tubule

law of conservation of energy. the axiom that energy is neither created nor destroyed during any chemical or physical process; the first law of thermodynamics

law of mass action. the maxim according to which an increase in the concentration of a reactant of a chemical reaction causes the reaction to proceed in the direction of product formation; the opposite occurs for a decrease in concentration of a reactant

LDL. low-density lipoprotein; a protein-lipid aggregate which is the major cholesterol carrier in plasma

lecithin. a phospholipid

lens. the adjustable part of the optical system of the eye which helps focus an image of an object on the retina

leukocyte. a white blood cell

leukocyte endogenous mediator (LEM). a protein secreted by monocytes and macrophages; induces decreases in plasma zinc and iron, release of acute-phase proteins, and stimulation of granulocyte production; probably the same molecule as EP and IL-1

leukotrienes. a family of compounds related to the pros-taglandins

LH. *see* luteinizing hormone

lidocaine. a local anesthetic that works by preventing the increase in membrane premeability to sodium ions required for the generation of action potentials

ligand. any molecule or ion which binds to the surface of a protein by noncovalent bonds

ligase. the enzyme which forms recombinant DNA by linking the fragmented ends of two DNA molecules previously split by a restriction enzyme

limbic system. an interconnected group of brain structures within the cerebrum involved with emotions and learning

lipase. a pancreatic enzyme, secreted into the small intestine, which catalyzes the splitting of triacylglycerol to produce free fatty acids and monoglycerides

lipid. any of several types of molecules composed primarily of carbon and hydrogen atoms; characterized by insolubility in water; includes triacylglycerols, phospholipids, and steroids

lipoproteins. aggregates of lipid molecules which are partially coated by protein, the amount of which varies; involved in the transport of lipids in the blood

β-lipotropin. peptide in the endorphin family; precursor of the morphine-like substance β-endorphin

local current flow. the movement of positive ions from a membrane region with a high positive charge through the cytoplasm or extracellular fluid toward a membrane region of more negative charge, and the simultaneous movement of negative ions in the opposite direction; leads to depolarization or hyperpolarization of adjacent segments of a membrane

local potential. a small, graded potential difference between two points; conducted decrementally, has no threshold and no refractory period; e.g., EPSPs and IPSPs

local respose. a biological response initiated by a change in the external or internal environment, i.e., a stimulus, which acts on cells in the immediate vicinity of the stimulus (without the involvement of nerves or hormones)

local spinal reflex. a reflex whose afferents, efferents, and integrating center are located within only a few segments of the spinal cord

locus coeruleus. a brainstem nucleus which projects to many parts of the brain and is implicated in directed attention

long reflex. a type of neural reflex involving the gastrointestinal tract in which the integrating center lies within the central nervous system

long-term memory. a memory which has a relatively long life span

longitudinal muscle layer. the smooth muscle layer surrounding the circular muscle layer in the stomach and intestinal walls and whose fibers are oriented longitudinally along these organs

loop of Henle. *see* Henle's loop

lower esophageal sphincter. smooth muscles of the last 4 cm of the esophagus which can act as a sphincter, closing off the opening of the esophagus into the stomach

LSD. a potent hallucinogenic drug which induces a state resembling schizophrenia

lumen. the space within a hollow tube or organ

lung compliance. the change in lung volume caused by a given change in pressure difference across the lung wall; i.e., the greater the lung compliance, the more stretchable the lung wall

luteal phase. the last half of the menstrual cycle during which the corpus luteum is the active ovarian structure

luteinizing hormone (LH). a peptide hormone secreted by the anterior pituitary; one of the gonadotropins; sometimes called interstitial cell–stimulating hormone (ICSH) in the male; in the female, the sudden increase in LH release during the middle of the menstrual cycle initiates ovulation

lymph. fluid in the lymphatic vessels; derives from interstitial fluid

lymph nodes. small organs, containing lymphocytes, located along the course of the lymph vessels; a site of lymphocyte formation and storage

lymphatics. the system of vessels which conveys lymph from the tissues to the blood

lymphocyte. a type of leukocyte; responsible for specific immune defenses; categorized mainly as B cells and T cells

lymphoid tissue. lymph nodes, spleen, thymus, tonsils, and aggregates of lymphoid follicles such as those lining the gastrointestinal tract

lymphokines. a general term denoting all nonantibody chemical messengers secreted by lymphocytes

lysosome. a spherical or oval cell organelle surrounded by a single membrane; contains digestive enzymes which break down bacteria and large molecules that have entered a cell by endocytosis; can digest damaged components of the cell itself

M line. one of the transverse bands making up the repeating striated pattern of cardiac and skeletal muscle; a thin dark band in the middle of the sarcomere

macrophage. a cell type which functions as a phagocyte in many tissues, processes and presents antigen to the lymphocytes, and secretes chemicals involved in inflammation, proliferation of activated lymphocytes, and total body responses to infection or injury

major histocompatibility complex (MHC). a group of genes which codes for many proteins important for immune function, including the HLA, or histocompatibility antigens

mammary glands. the milk-secreting glands in the breasts

mass movement. contraction of large segments of ascending and transverse colon; propels fecal material into the rectum

mast cell. a connective-tissue cell similar to a basophil except that it does not circulate in the blood; local injury releases histamine and other chemicals from mast cells

matrix. the extracellular substance surrounding connective-tissue cells; has some structural organization, such as connective-tissue fibers, as opposed to being a simple ionic solution

maximal tubular capacity (T_m). the maximal rate of mediated transport of a substance across the wall of the kidney nephrons

mean arterial pressure. the average blood pressure during the cardiac cycle; approximately equal to diastolic pressure plus one-third of the pulse pressure

mechanoreceptor. a sensory receptor which responds preferentially to mechanical stimuli such as bending, twisting, or compressing

median eminence. the region at the base of the hypothalamus at which arteries end in capillary tufts; it is into these capillaries that the hypothalamic releasing and inhibiting hormones are secreted

mediated transport. movement of molecules across membranes mediated by binding to protein carrier molecules in the membrane; characterized by specificity, competition, and saturation; includes facilitated diffusion and active transport

megakaryocyte. a large bone-marrow cell which gives rise to platelets

meiosis. a process of nuclear division leading to the formation of the gametes (sperm and ova); the final cells receive only half of the chromosomes present in the original cell

membrane. a structural barrier composed of phospholipids and proteins; any of the cell membranes, e.g., endoplasmic reticulum, mitochondria, plasma membrane, etc., which provides a selective barrier to the movement of molecules and ions and provides a structural framework to which enzymes, fibers, and a variety of ligands can bind

membrane attack complex (MAC). the group of complement proteins which imbed in the surface of a microbe and kill it

membrane permeability constant (k_p). a proportionality constant defining the rate at which a substance diffuses through a membrane given the value of the concentration gradient across the membrane; the larger the value of k_p, the more permeable the membrane

membrane potential. (or transmembrane potential) the voltage difference between the inside and outside of the cell due to the separation of charge across the membrane

memory cells. B and T lymphocytes which are produced during an initial infection and which respond rapidly during a subsequent exposure to the same antigen

memory consolidation. the processes by which memory is transferred from its short-term to its long-term form

memory trace. the postulated neural substrate of memory

menarche. the onset, at puberty, of menstrual cycling

menopause. the cessation of menstrual cycling at approximately age 50

menstrual cycle. the cyclic rise and fall in female reproductive hormones and processes; has a period of approximately 28 days

menstrual fluid. blood mixed with debris from the disintegrating endometrium

menstruation. the flow of menstrual fluid from the uterus; also called the menstrual period

MES. *see* microsomal enzyme system

mescaline. a hallucinogenic drug which is a methylated derivative of dopamine

mesentery. a sheet of connective tissue which connects the outer surface of the stomach and intestines to the abdominal wall

messenger RNA (mRNA). the form of ribonucleic acid which transfers the genetic information corresponding to one gene (or, at most, a few genes) from DNA, in the nucleus, to the ribosomes in the cytoplasm where proteins are synthesized

metabolic acidosis. an acidosis due to a metabolic production of acids other than carbon dioxide, e.g., the production of lactic acid during severe exercise or of ketones in diabetes mellitus

metabolic alkalosis. an alkalosis resulting from the removal of hydrogen ions from the body by mechanisms other than the respiratory removal of carbon dioxide, e.g., the loss of hydrogen from the stomach with persistent vomiting

metabolic cost of living. *see* basal metabolic rate

metabolic pathway. the sequence of enzyme-mediated chemical reactions by which molecules are synthesized and broken down in cells

metabolic rate. the total energy expenditure of the body per unit time

metabolism. all the chemical reactions that occur within a living organism

metarterioles. blood vessels which connect arterioles and venules directly

methyl group. —CH_3

Mg^{2+}. magnesium ion

micelle. a soluble cluster of amphipathic molecules in which the polar regions of the molecules line the surface of the micelle and the nonpolar regions are oriented toward the center of the micelle; formed from fatty acids, 2-monoglycerides, and bile salts during the digestion of fat in the small intestine

microbes. minute organisms, including bacteria, protozoans, and fungi

microcirculation. circulation in the arterioles, capillaries, and venules

microfilaments. rodlike filaments in the cytoplasm of most cells; can be rapidly assembled and disassembled to allow a cell to change its shape

microsomal enzyme system (MES). enzymes, found in the smooth endoplasmic reticulum of liver cells, which transform molecules into more polar, less lipid-soluble substances; formerly called "drug-metabolizing enzymes"

microtubules. tubular filaments in the cytoplasm, which provide internal support for cells; can be rapidly assembled and disassembled to allow a cell to change its shape and to produce the movements of organelles within the cell

microvilli. small fingerlike projections from the surface of the epithelial cells, which greatly increase the absorptive surface area of a cell; a characteristic of the epithelium lining the small intestine and kidney nephrons

micturition. urination

middle-ear cavity. the air-filled space deep within the temporal bone which contains the three ossicles (ear bones) which conduct the sound waves from the tympanic membrane to the cochlea

migrating motility complex. peristaltic waves which sweep over segments of the intestine after most of a meal has been absorbed

milk intolerance. the inability to digest milk sugar (lactose) because of lack of the intestinal enzyme lactase; leads to the accumulation of large amounts of gas and fluid in the large intestine, which causes pain and diarrhea

milk let-down. the process by which milk is moved from the alveoli of the mammary gland into the ducts, from which it can be sucked

milliliter (mL). a unit of volume equal to 0.001 L

millimole. a concentration equal to 0.001 mole

millivolt (mV). an electrical potential equal to 0.001 V

mineral. a nonorganic (i.e., without carbon) substance; usually a constituent of the earth's crust; the major minerals in the body are calcium, phosphorus, potassium, sulfur, sodium, chloride, and magnesium

mineralocorticoids. steroid hormones produced by the adrenal cortex which have their major effects on sodium and potassium balance; aldosterone is the major mineralocorticoid

mitochondria. rod-shaped or oval organelles in cell cytoplasm which produce most of the ATP used by the cells; site of Krebs cycle and oxidative-phosphorylation enzymes

mitosis. the stage of cell division during which the nucleus divides; the process in cell division in which DNA is duplicated and an identical set of chromosomes is passed to each daughter cell

mitral valve. the valve between the left atrium and left ventricle of the heart

modality. the quality or kind of stimulus; e.g., touch and hearing are different modalities

modulator. a ligand which, by acting at the protein's regulatory site, alters the properties of other binding sites on the protein and thus regulates the functional activity of the protein

molarity. a unit of concentration; the number of moles of solute per liter of solution

mole. (also mol) a unit which indicates the number of molecules of a substance present; the number of moles = weight in grams/molecular weight; one mole of any substance contains the same number of molecules as one mole of any other substance (the number is 6×10^{23}—Avogadro's number)

molecular weight. a number equal to the sum of the atomic weights of all the atoms in a molecule

molecule. a chemical substance formed by linking atoms together

monoamine. the class of neurotransmitters having a basic structure $R—NH_2$, where R is the remainder of the molecule; by convention, excludes the peptides and amino acids

monoamine oxidase. an enzyme which changes the catecholamine neurotransmitters and serotonin into inactive substances

monoamine oxidase inhibitor. a drug which interferes with the action of monoamine oxidase, thereby increasing the concentration of catecholamine neurotransmitters available

monocyte. a type of leukocyte; leaves the bloodstream and is transformed into a macrophage

mononuclear phagocyte system. monocytes and macrophages

monosaccharide. a carbohydrate consisting of a single sugar molecule, which generally contains five or six carbon atoms

monosynaptic reflex. a reflex in which the afferent limb of the reflex arc (the afferent neuron) directly activates the efferent limb (the motor neuron); the stretch reflex is the only known monosynaptic reflex

motor. having to do with muscles and movement

motor cortex. a strip of cerebral cortex along the posterior border of the frontal lobe; gives rise to many (but certainly not all) of the axons descending in the corticospinal and multineuronal pathways to the motor neurons

motor end plate. the specialized region of a muscle-cell plasma membrane which lies directly under the axon terminal of a neuron; analogous to the subsynaptic membrane of postsynaptic neurons

motor neuron. an efferent neuron which innervates skeletal muscle fibers

motor potential. electrical activity which can be recorded over motor cortex about 50 to 60 ms before a movement begins

motor unit. a motor neuron plus the muscle fibers it innervates

mRNA. *see* messenger RNA

mucin. a protein which, when mixed with water, forms mucus

mucopolysaccharide. a group of polysaccharides present in the extracellular matrix of connective tissues

mucosa. the three layers of the stomach and intestinal wall nearest the lumen, i.e., the epithelium, lamina propria, and muscularis mucosa

mucus. the highly viscous solution secreted by the mucous gland cells

Müllerian duct. a part of the embryo which, in a female, develops into the ducts of the reproductive system, but in a male, degenerates

Müllerian inhibiting substance (MIS). a protein secreted by the fetal testes which causes the Müllerian ducts to degenerate

multineuronal pathways. the descending motor pathways which are made up of chains of neurons synapsing in the basal ganglia, several other subcortical nuclei, and the brainstem; only the final neuron of the chain reaches the region of the motor neurons; also called the multisynaptic pathways and the extrapyramidal system

multiunit smooth muscle. smooth muscle which exhibits little, if any, propagation of electrical activity from fiber to fiber and whose contractile activity is closely coupled to its neural input

muscarinic receptor. an acetylcholine receptor which can be stimulated by the mushroom poison muscarine; these receptors are located on smooth muscle, cardiac muscle, some neurons of the central nervous system, and gland cells

muscle. a number of muscle fibers bound together by connective tissue

muscle cells. cells specialized for the production of mechanical forces; may be attached to bone or skin or surround hollow tubes or cavities; muscle fibers

muscle fatigue. a decrease in the mechanical response of muscle with prolonged stimulation

muscle fiber. a muscle cell

muscle spindle. a capsule-enclosed arrangement of afferent nerve-fiber endings sensitive to stretch; a stretch receptor in skeletal muscle

muscle tension. the force exerted by a contracting muscle upon an object

muscle tone. the small amount of tension produced by skeletal muscles under resting conditions; in smooth muscle, tone may be maintained by spontaneous electrical activity within the muscle itself

mutation. any alteration in the base sequence of DNA which alters the genetic information stored there

mV. *see* millivolt

myasthenia gravis. a neuromuscular disease associated with fatigue and skeletal muscle weakness; due to decreased numbers of ACh receptors at the motor end plate

myelin. insulating material covering the axons of many neurons; consists of multiple layers of myelin-forming cell plasma membranes which are wrapped concentrically around the axon

myenteric plexus. a network of nerve cells lying between the circular and longitudinal muscle layers in the esophagus, stomach, and intestinal walls

myoblast. the embryological cell which gives rise to muscle fibers

myocardium. cardiac muscle, which forms the walls of the heart

myoepithelial cells. specialized contractile cells surrounding the alveoli of mammary glands

myofibrils. longitudinal bundles of thick and thin contractile filaments arranged in a repeating sarcomere pattern; located in the cytoplasm of striated muscles

myogenic. originating within a muscle

myoglobin. a protein in muscle fibers, similar to hemoglobin, which binds oxygen

myometrium. uterine smooth muscle

myosin. the contractile protein which forms the thick filaments of muscle

myosin-ATPase. an enzymatic site on the globular head of myosin which catalyzes the breakdown of ATP to ADP and inorganic phosphate, with the release of chemical energy which is used to produce the force of muscle contraction

NaCl. sodium chloride

NAD. *see* nicotinamide adenine dinucleotide

Na,K-ATPase. a primary active–transport carrier protein having ATPase properties; located in all plasma membranes; splits ATP and releases energy which is used to actively transport sodium out of the cell and potassium into the cell

natriuretic hormone. a hormone which inhibits the reabsorption of sodium in the renal nephron

natural antibodies. antibodies to the erythrocyte antigens not present in a person's body; these antibodies are present without prior exposure to the antigen

natural killer (NK) cells. a class of cells (probably lymphocytes) which bind relatively nonspecifically to cells bearing foreign antigens and kill them directly; no prior exposure to the antigen is required

nearsighted. pertaining to the defect in vision which occurs because the eyeball is too long for the lens, so that images of distant objects are focused in front of the retina

negative balance. the situation in which the loss of a substance from the body is greater than the gain, and the total amount of that substance in the body decreases; also used for physical parameters, such as body temperature and energy

negative feedback. the situation in which the response of a reflex arc opposes the stimulus input to that reflex

nerve growth factor. a peptide which stimulates the growth and differentiation of neurons of the sympathetic nervous system

net flux. the amount of material transferred between two regions, equal to the difference between two opposing unidirectional fluxes

neutrophil. a polymorphonuclear granulocytic leukocyte whose granules show preference for neither eosin nor basic dyes; functions as a phagocyte and also releases chemicals involved in inflammation

neutrophil exudation. the amoebalike movement of neutrophils from the lumen of capillaries out into the extracellular space of tissues

NH$_3$. ammonia

NH$_4^+$. ammonium ion

nicotinamide adenine dinucleotide (NAD). a coenzyme which transfers hydrogen from one reaction to another

nicotinic receptors. those acetylcholine receptors which respond to the drug nicotine; primarily, the receptors at the motor end plate and the receptors in the ganglia of the autonomic nervous system

night vision. the only vision possible under conditions of low illumination, i.e., rod vision

nociceptor. a sensory receptor whose stimulation causes the sensation of pain

nodes of Ranvier. spaces between adjacent myelin-forming cells along a myelinated axon where the axonal plasma membrane is exposed to the extracellular fluid

nonpolar molecule. a molecule which contains predominantly neutral chemical bonds (in which electrons are shared equally between atoms); has no polar or ionized groups

nonspecific immune responses. responses which do not require previous exposure to a particular foreign material but nonselectively protect against it without having to recognize its specific identity

nonspecific pathway. a chain of synaptically connected neurons in the brain or spinal cord, all of which are activated by sensory units of several different modalities

nonspecific thalamic nuclei. nuclei in the cerebrum whose axons project to widely dispersed areas of cerebral cortex

norepinephrine. a catecholamine neurotransmitter released at most sympathetic postganglionic endings, from the adrenal medulla, and in many regions of the central nervous system

novocaine. a local anesthetic that works by preventing the increased membrane permeability to sodium ions required to produce an action potential in the axon of a neuron

noxious. poisonous; harmful to health

nuclear envelope. the double membrane surrounding the nucleus of a cell

nuclear pores. openings in the nuclear envelope through which molecular messengers, such as mRNA, pass from the nucleus into the cytoplasm

nucleic acid. a polymer of nucleotides in which the phosphate of one of the repeating nucleotide subunits is linked to the sugar of the adjacent one; functions in the storage and transmission of genetic information; includes DNA and RNA

nucleolus. a prominent, densely staining nuclear structure consisting of granular and filamentous elements; contains the portions of DNA which code for rRNA and is the site of ribosome assembly

nucleoprotein. a complex of nucleic acid with positively charged proteins

nucleotide. the molecular subunit of nucleic acids; consists of a purine or pyrimidine base, a sugar, and phosphoric acid

nucleus. (cell) a large spherical or oval membrane-bound cell organelle present in most cells; contains most of the cell's DNA and some of its RNA; (neural) a cluster of neuronal cell bodies in the CNS

occipital cortex. the posterior region of the cerebral cortex where the primary visual cortex is located

off **response.** an increase in cell activity upon removal of the stimulus; a decrease in the background activity of a cell in response to stimulation

Ohm's law. the relationship between current (*I*), voltage (*V*), and resistance (*R*) such that $I = V/R$

olfactory. pertaining to the sense of smell

olfactory mucosa. the regions of mucous membrane in the upper part of the nasal cavity containing the receptors for the sense of smell

olfactory nerve. cranial nerve I; consists of the axons whose peripheral terminals contain the chemoreceptors for the sense of smell

oligodendroglia. a type of glial cell; responsible for the generation of myelin around nerve fibers in the central nervous system

oligomeric protein. a protein consisting of more than one polypeptide chain

on **response.** the response of a neuron or receptor to the onset of stimulation

oncogene. several genes which appear to be associated with cancer; malignant transformation of tissue appears to be associated with either the turning-on of these genes or an increase in the rate at which they form their specific proteins

oogonia. primitive ova which, upon mitotic division before birth, give rise to the primary oocytes

operating point. the steady-state value maintained by a homeostatic system; the set point

operator gene. the initial gene of an operon; controls the transcription of the genes in the operon

operon. a collection of genes which can be induced or repressed as a unit

opponent color cell. a neuron whose activity is increased by input from cones of one color sensitivity but decreased by cones having a different sensitivity

opsin. the protein component of a photopigment molecule; the rods and three types of cones have different opsins

opsonin. a general name given to any chemical mediator which promotes phagocytosis

optic nerve. cranial nerve II; consists of the axons of the retinal ganglion cells

optimal length *(l₀).* the length at which a muscle fiber develops maximal tension

organ. a collection of tissues joined in a structural unit to serve a common function; e.g., the heart is an organ consisting of muscle, nerve, connective, and epithelial tissues which together serve as a pump for blood

organ of Corti. the collection of structures in the inner ear capable of transducing the energy of waves in a fluid medium into the electrochemical energy of action potentials in the auditory nerve

organ system. a collection of organs which together serve an overall function; e.g., the kidneys, ureters, urinary bladder, and urethra are the organs which constitute the urinary system

organelles. *see* cell organelles

organic. pertaining to substances containing the element carbon

organic acid. an acid molecule containing carbon atoms, e.g. carbonic and lactic acids

orienting response. an animal's behavior in response to the presentation of a novel stimulus, i.e., the animal stops what it is doing and looks around, listening intently and turning toward the stimulus source

osmolarity. the total solute concentration of a solution; one osmole equals one mole of solute particles; a measure of water concentration in that the higher the osmolarity of a solution, the lower its water concentration

osmoreceptor. a neural receptor which responds to changes in the osmolarity of the surrounding fluid

osmosis. the net diffusion of water from regions of high to low water concentration across a membrane; the water always moves from the regions of low solute concentration (high water concentration) to the region of high solute concentration (low water concentration)

osmotic pressure. the pressure that must be applied to a solution on one side of a membrane to prevent the osmotic flow of water into the solution from a source of pure water on the opposite side of the membrane

osteoblast. the type of cell responsible for laying down the protein matrix of bone

otoliths. calcium carbonate "stones" in the gelatinous mass of the utricle and saccule of the vestibular apparatus

oval window. the membrane-covered opening separating the middle-ear cavity from the scala vestibuli of the inner ear

ovarian follicle. the ovum and its encasing follicular, granulosa, and theca cells at any stage of its development prior to ovulation

ovary. the gonad in the female

ovulation. the release of an ovum, surrounded by its zona pellucida and cumulus, from the ovary

ovum. (pl. ova) the female gamete

oxidative. using oxygen

oxidative deamination. a reaction in which an amino group is removed from an amino acid as a molecule of ammonia (NH_3) and replaced by an oxygen atom to form a keto acid

oxidative fast fibers. muscle fibers which combine a high oxidative capacity with a high myosin-ATPase activity

oxidative phosphorylation. the process by which energy derived from the reaction between hydrogen and oxygen (to form water) is transferred to ATP during its formation from ADP and inorganic phosphate; occurs in the mitochondria

oxidative slow fibers. muscle fibers which combine a high oxidative capacity with a low myosin-ATPase activity

oxyhemoglobin. HbO_2; hemoglobin combined with oxygen

oxyntic cells. *see* parietal cells

oxytocin. a peptide hormone of nine amino acids synthesized in the hypothalamus and released from the posterior pituitary; stimulates the mammary glands to release milk and the uterus to contract

P wave. the component of the electrocardiogram reflecting depolarization of the atria

pacemaker potential. the spontaneous depolarization to threshold potential of the plasma membrane of some nerve and muscle cells

pacinian corpuscle. a mechanoreceptor whose structure is specialized to respond to vibrating stimuli

palpitation. a rapid, often irregular beating of the heart which can be felt by the individual

pancreas. a gland in the abdomen near the stomach and connected by a duct to the small intestine; contains both endocrine gland cells, which secrete the hormones insulin, glucagon and somatostatin into the bloodstream, and exocrine gland cells, which secrete digestive enzymes and bicarbonate into the intestine

pancreatic duct. the large duct which carries the exocrine secretion of the pancreas into the duodenum

pancreatic lipase. an enzyme secreted by the exocrine pancreas; acts on triacylglycerols to form 2-monoglycerides and free fatty acids

pancreozymin. *see* cholecystokinin

paracrine. a chemical which, when released, exerts its effects on tissues located near the site of its secretion; by convention, excludes neurotransmitters

paradoxical sleep. the state of sleep associated with small, rapid EEG oscillations (indistinguishable from those of EEG arousal) and with complete loss of tone in the postural muscles, irregular respiration and heart rate, and dreaming

parasympathetic nervous system. that portion of the autonomic nervous system whose preganglionic fibers leave the central nervous system at the brainstem and the sacral portion of the spinal cord; most of its postganglionic fibers release the neurotransmitter acetylcholine

parasympathomimetic. a chemical which produces effects similar to those of the parasympathetic nervous system

parathormone. *see* parathyroid hormone

parathyroid glands. the four parathyroid-hormone–secreting glands on the back of the thyroid gland

parathyroid hormone (PTH). a peptide hormone secreted by the parathyroid glands; regulates calcium and phosphate concentrations of the internal environment; parathormone

parietal cells. gastric gland cells which secrete hydrochloric acid and intrinsic factor; oxyntic cells

parietal lobe. the region of the cerebral cortex where sensory cortex and some association cortex are located

Parkinson's disease. a disease characterized by tremor, rigidity, and a delay in the initiation of movement; due to the degeneration of the dopamine-liberating axons from the substantia nigra

parotid. one of the three pairs of salivary glands

partial pressure. the pressure exerted by one molecular species of gas; e.g., P_{O_2}, the partial pressure of oxygen, is that part of the total gas pressure that is due to oxygen molecules; a measure of the concentration of a gas in a mixture of gases

parturition. birth; delivery of an infant

passive immunity. a resistance to infection resulting from the direct transfer of antibodies or sensitized T cells from one person (or animal) to another

P_{CO_2}. the partial pressure of carbon dioxide

pelvic diaphragm. a sheet of skeletal muscles which constitutes the floor of the pelvis and helps support the abdominal and pelvic viscera

pepsin. an enzyme formed in the stomach from the precursor pepsinogen; breaks protein down to peptide fragments

pepsinogen. the inactive precursor of the enzyme pepsin; secreted by the chief cells of the gastric mucosa

peptide. a short polypeptide chain having less than 50 amino acids

peptide bond. the polar covalent chemical bond

$$(-NH-\overset{\overset{\displaystyle O}{\|}}{C}-)$$ joining two amino acids; formed by a reaction between the amino group of one amino acid and the carboxyl group of a second amino acid; forms the backbone of proteins

perception. the conscious experience of objects and events of the external world which we acquire from the neural processing of sensory information

pericardium. the fibrous connective-tissue sac which surrounds the heart

peripheral chemoreceptors. the carotid and aortic bodies which respond to changes in blood P_{O_2}, P_{CO_2}, and hydrogen-ion concentration

peripheral nervous system. the nerve fibers extending from the brain and spinal cord

peripheral proteins. membrane proteins which are water soluble and can be removed without disrupting the lipid bilayer; thus, they are bound to the membrane surface

peripheral thermoreceptor. cold and warm receptors in the skin and certain mucous membranes

peristaltic waves. a progressive wave of muscle contraction which proceeds along the wall of a tube, compressing the lumen of the tube and causing the contents of the tube to move

peritubular capillaries. capillary network closely associated with the tubular components of the kidney nephrons

permissiveness. the situation whereby small quantities of one hormone are required in order for a second hormone to exert its full effects

pernicious anemia. a disease in which the erythrocytes fail to proliferate and mature because of a deficiency of vitamin B_{12}

peroxisome. a cell organelle similar to a lysosome but of a different chemical composition; can destroy harmful products of chemical reactions

pH. the negative logarithm to the base 10 of the hydrogen-ion concentration of a solution; a measure of the acidity of a solution: the pH decreases as the acidity increases

phagocytosis. a form of endocytosis in which large particles, such as bacteria, are taken into the cell (along with small amounts of extracellular fluid)

phagosome. the membrane-enclosed pouch formed when a phagocyte engulfs a microbe or other particulate matter

pharmacological effects. the effects produced by much larger amounts of a hormone in the body than are normally present

pharynx. throat; a passage common to the routes followed by food and air

phasic. intermittent, as opposed to tonic

phosphodiesterase. the enzyme which catalyzes the breakdown of cyclic AMP to AMP

phospholipid. a subclass of lipid molecules similar to triacylglycerol except that a phosphate group ($—PO_4—$) and a small polar or ionized nitrogen-containing molecule are attached to the third hydroxyl group of glycerol; a major component of cell membranes

phosphoprotein phosphatase. an enzyme which removes phosphate from a protein and returns the protein to its original shape

phosphoric acid. an acid generated during the catabolism of phosphorus-containing compounds; dissociates to form phosphate and hydrogen ions

phosphorylation. addition of a phosphate group to an organic molecule

photopigment. a light-sensitive molecule altered by the absorption of photic energy of certain wavelengths; consists of an opsin bound to a chromophore

photosynthesis. the chemical process by which plants are able to capture the energy in sunlight, with which they form glucose and oxygen using carbon dioxide and water as substrates

pitch. the degree of how high or low a sound is perceived; related to the frequency of the sound wave

pituitary. an endocrine gland which lies in a pocket of bone just below the hypothalamus

pituitary gonadotropins. *see* gonadotropic hormones

placenta. a combination of interlocking fetal and maternal tissues in the uterus; serves as the organ of molecular exchange between fetal and maternal circulations

placental lactogen. a hormone which is produced by the placenta and has effects, particularly in the mother, similar to growth hormone

plasma. the fluid, noncellular portion of the blood; a component of the extracellular fluid compartment

plasma cells. cells which are derived from activated B lymphocytes and which secrete antibodies

plasma membrane. the cell membrane which forms the outer surface of a cell and separates its contents from the surrounding extracellular fluid

plasma protein. the albumins, globulins, and fibrinogen present in the blood plasma

plasmin. a proteolytic enzyme able to decompose fibrin and, thereby, to dissolve blood clots

plasminogen. the inactive precursor of the enzyme plasmin

plasticity. the ability of neural tissue to change its responsiveness to stimulation because of its past history of activation

platelet. a cell fragment present in the blood; plays several roles in blood clotting

pleura. a thin sheet of cells that is firmly attached to the entire interior of the thoracic cage and, folding back upon itself, is attached to the surface of the lungs; forms two completely enclosed sacs within the thoracic cage

pneumothorax. air in the intrapleural space

Po_2. the partial pressure of oxygen

polar body. the small cell resulting from the unequal distribution of cytoplasm during the division of the primary and secondary oocytes

polar bonds. *see* polarized covalent bond

polar molecule. substance containing a number of polar bonds; the region of the molecule to which the electrons are drawn carries a negative charge and the region from which they are drawn carries a positive charge; soluble in water

polarized. having two electric poles, one negative and one positive

polarized covalent bond. a covalent chemical bond in which two electrons are shared unequally between two atoms, the atom to which the electrons are drawn carrying a negative charge and the other atom carrying a positive charge

polymer. a very large molecule formed by linking together many smaller similar subunits

polymorphonuclear granulocyte. subclass of leukocytes; there are three types: neutrophils, eosinophils, and basophils

polypeptide. polymer consisting of amino acid subunits joined in sequence by peptide bonds; peptides and proteins

polyribosome. one strand of mRNA to which are attached a number of ribosomes

polysaccharide. a large carbohydrate molecule formed by linking many monosaccharide subunits together; for example, glycogen and starch

polysynaptic reflex. a reflex employing one or more interneurons in its reflex arc

polyunsaturated fatty acid. a fatty acid which contains more than one double bond

pool. the body's readily available quantity of a particular substance; frequently identical to the quantity present in the extracellular fluid

portal vessels. blood vessels which link two separate capillary networks

positive balance. occurs when the gain of a substance by the body is greater than the loss of that substance, and the total amount of that substance in the body increases; also used for temperature and energy; *see also* negative balance

positive feedback. an aspect of control systems in which an initial disturbance in a system sets off a train of events which increases the disturbance even further; an unstable situation because the output of the system will rapidly and progressively increase in magnitude; *see also* negative feedback

postabsorptive state. the period during which nutrients are not being absorbed by the gastrointestinal tract and energy must be supplied by the body's endogenous stores

posterior pituitary. that portion of the pituitary from which oxytocin and vasopressin are released

postganglionic neuron. a neuron of the autonomic nervous system whose cell body lies in a ganglion and whose axon terminals form neuroeffector junctions; conducts impulses away from a ganglion toward the periphery

postsynaptic neuron. a neuron which conducts information away from a synapse

postsynaptic potential. a local potential which arises in the postsynaptic neuron in response to the activation of synapses upon it; *see also* excitatory postsynaptic potential, inhibitory postsynaptic potential

posttranslational processing. changes in the structure of a polypeptide chain following its synthesis on a ribosome

postural reflexes. those reflexes which maintain or restore upright, stable posture

potential. (or potential difference) the voltage difference between two points

potential energy. the energy associated with an object because of its position or internal structure; chemical energy is a form of potential energy

potentiation. the phenomenon in which the presence of one agent enhances the response to a second agent such that the final response is greater than the sum of the two individual responses

precapillary sphincter. a ring of smooth muscle around a "true" capillary at the point at which it exits from a thoroughfare channel or arteriole

preganglionic neuron. a neuron of the autonomic nervous system whose cell body lies in the central nervous system and whose axon terminals lie within a ganglion; conducts impulses from the central nervous system toward a ganglion

pregnenolone. a precursor of all the adrenal steroids

preprogram. a sequence of neuronal activity which results in a movement independent of afferent input which may occur during the course of the movement

presbyopia. the impairment in vision resulting from stiffening of the lens; makes accommodation difficult

"pressure" autoregulation. the ability of individual organs and tissues to alter their vascular resistance so as to maintain a relatively constant blood flow in the face of changing blood pressure

presynaptic facilitation. *see* presynaptic inhibition

presynaptic inhibition. a relation between two neurons in which the axon terminal of one neuron ends on the synaptic knob of a second neuron; action potentials in the first neuron alter the release of neurotransmitter from the second neuron, which results in a decrease in the magnitude of the postsynaptic potential produced in a third neuron; the opposite is true for presynaptic facilitation

presynaptic neuron. a neuron which conducts action potentials toward a synapse

primary active transport. active transport in which chemical energy is transferred directly from ATP to carrier proteins

primary follicle. any cell cluster in the ovaries which consists of one ovum surrounded by a single layer of follicular (granulosa) cells

primary hypertension. hypertension of unknown cause

primary oocyte. a female germ cell which undergoes the first meiotic division to form a secondary oocyte (and a polar body)

primary spermatocyte. a male germ cell derived from the spermatogonia; undergoes meiotic division to form two secondary spermatocytes

primary visual cortex. the first part of the visual cortex to be activated by the visual pathways

product. the molecules formed in an enzyme-catalyzed chemical reaction

progesterone. a steroid hormone secreted primarily by the corpus luteum and placenta; stimulates secretion by the uterine glands, inhibits contraction of uterine smooth muscle, and stimulates breast growth

program. a related sequence of neural activity preliminary to a motor act

prohormone. a peptide precursor molecule from which are cleaved one (or several) active hormones

projection neuron. a neuron having a long axon which connects it with distant parts of the nervous system or with muscles or glands

prolactin. a peptide hormone secreted by the anterior pituitary; stimulates milk secretion by the mammary glands

prolactin release–inhibiting hormone (PIH). the hypothalamic hormone which inhibits prolactin secretion by the anterior pituitary; identified as dopamine

prolactin-releasing hormone (PRH). one or more hypothalamic hormones which stimulate prolactin release from the anterior pituitary

proliferative phase. that stage of the menstrual cycle between menstruation and ovulation during which the endometrium repairs itself and grows

promoter. a specific sequence of nucleotides at the beginning of a gene; determines which of the two DNA chains is transcribed to form mRNA

pro-opiomelanocortin. a large protein molecule which serves as the precursor for ACTH, the endorphins, and several other hormones

proprioception. the sense of the position of the body in space

prostacyclin. PGI_2; a compound related to the prostaglandins; inhibits platelet aggregation in blood clotting

prostaglandins. a group of unsaturated modified fatty acids which function mainly as paracrines and autocrines; synthesized by many cells of the body

prostate gland. a large gland encircling the urethra in the male; secretes fluid into the urethra

protease. an enzyme which breaks the peptide bonds between amino acids in polypeptide chains

protein. a large polymer consisting of one or more sequences of amino acid subunits joined by peptide bonds

protein binding site. *see* binding site

protein kinase. a group of enzymes, each of which catalyzes the addition of a phosphate group to specific proteins

prothrombin. the inactive precursor to the enzyme thrombin; produced by the liver and normally present in the plasma

protomer. each of the single polypeptide chains in an oligomeric protein

proton. a positively charged subatomic particle in the nucleus of atoms

proximal tubule. the first tubular component of a nephron after Bowman's capsule

psychological fatigue. the mental factors which cause individuals to stop muscle activity even though their muscles are still able to contract

PTH. *see* parathyroid hormone

puberty. the attainment of sexual maturity in the sense that conception becomes possible; as commonly used, it refers to the 3 to 5 years of sexual development which culminates in the attainment of sexual maturity

pulmonary. pertaining to the lungs

pulmonary circulation. the circulation through the lungs; the portion of the cardiovascular system between the pulmonary trunk artery (as it leaves the right ventricle) and the pulmonary veins (as they enter the left atrium)

pulmonary edema. accumulation of fluid (derived from the blood) in the lung interstitium and air sacs

pulmonary stretch receptor. an afferent nerve ending located in the airway smooth muscle and activated by lung inflation

pulmonary surfactant. *see* surfactant

pulmonary trunk. the large artery which carries blood from the right ventricle of the heart to the lungs

pulmonary valve. the valve between the right ventricle of the heart and the pulmonary trunk

pulmonary ventilation. *see* total pulmonary ventilation

pulse pressure. the difference between systolic and diastolic arterial blood pressures

pupil. the opening in the iris of the eye through which light passes to reach the retina

pupillary reflex. contraction of the pupil on exposure of the retina to light

purine bases. double-ring, nitrogen-containing subunits of nucleotides; adenine and guanine

Purkinje fibers. specialized myocardial cells which constitute part of the conducting system of the heart; convey the excitation from the bundle branches to the ventricular muscle

pyloric sphincter. a ring of smooth muscle and connective tissue between the terminal antrum of the stomach and the first segment of the small intestine

pyramidal tract. *see* corticospinal pathway

pyrimidine bases. single-ring, nitrogen-containing subunits of nucleotides; cytosine, thymine, and uracil

pyrogen. *see* endogenous pyrogen

pyruvate. the anion formed when pyruvic acid dissociates, i.e., loses a hydrogen ion

pyruvic acid. the three-carbon keto-acid intermediate in the glycolytic pathway which, in the absence of oxygen, forms lactic acid or, in the presence of oxygen, enters the Krebs cycle

QRS complex. the component of the electrocardiogram corresponding to the depolarization of the ventricles

rate-limiting enzyme. the enzyme in a metabolic pathway most easily saturated with substrate; the step involving the rate-limiting enzyme determines the rate of flow of substrate through the entire metabolic pathway

rate-limiting reaction. the slowest reaction in a sequence of reactions; determines the rate at which the final product can be formed; catalyzed by the rate-limiting enzyme

reactant. a molecule which enters a chemical reaction; in enzyme-catalyzed reactions, the reactant is called the substrate

reactive hyperemia. a transient increase in blood flow following the release of occlusion of the blood supply to an organ

readiness potential. electrical activity which can be recorded over wide areas of association cortex during the "decision-making period," about 0.8 s before a movement begins

receptive field. (of neuron) the area of the body which, if stimulated, results in activity in that neuron

receptive relaxation. the reflex decrease in smooth muscle tension in the walls of hollow organs in response to distension

receptor. (in sensory systems) the specialized peripheral ending of an afferent neuron (or a separate cell intimately associated with it) which detects changes in some aspect of the environment; (in chemical communication) a specific protein binding site in either the plasma membrane or cytosol of a target cell with which a chemical mediator combines to exert its effects

receptor-operated channels. channels in the plasma membrane which are opened (or closed) following binding of a messenger to a specific receptor

receptor potential. a graded potential which arises in the ending of an afferent neuron (or in a specialized cell intimately associated with it) in response to stimulation of the receptor; also called a generator potential

reciprocal innervation. inhibition of motor neurons activating those muscles whose contraction would oppose the intended movement; e.g., inhibition of flexor motor neurons during extensor motor neuron activation

recombinant DNA. DNA formed by joining portions of two DNA molecules previously fragmented by a restriction enzyme

recommended daily allowance (RDA). the daily intake of nutrients considered sufficient to prevent nutritional inadequacy in most healthy persons

recruitment. the activation of additional cells in response to a stimulus of increased magnitude

rectum. the short segment of the large intestine between the sigmoid colon and anus

red muscle fibers. muscle fibers having high oxidative capacity and large amounts of myoglobin

reduced hemoglobin (Hb). *see* deoxyhemoglobin

reflex. a biological control system linking a stimulus with a response and mediated by a reflex arc composed of neural, hormonal, or neural and hormonal elements; in the most narrow sense, an involuntary, unpremeditated, unlearned response to a stimulus

reflex arc. the components which mediate a reflex; usually, but not always, includes a receptor, afferent pathway, integrating center, efferent pathway, and effector

reflex response. the final change brought about by the action of a stimulus upon a reflex arc; effector response

refractory period. the time during which an excitable membrane does not respond to a stimulus whose magnitude is normally sufficient to trigger an action potential; *see also* absolute refractory period, relative refractory period

regulatory site. the site on an allosteric protein at which interaction with a ligand (the modulator) alters the properties of the functional binding sites

relative refractory period. the time during which an excitable membrane will produce an action potential only in response to stimuli of greater strength than the usual threshold strength

relaxation time. the time interval from the peak tension developed by a muscle until the tension has decreased to zero

releasing hormones. hormones which are secreted by the neurons of the hypothalamus and control the release of hormones by the anterior pituitary

REM sleep. *see* paradoxical sleep

renal. pertaining to the kidneys

renal blood flow. the volume of blood delivered to the kidneys per unit of time

renal medulla. the inner region of the kidney

renal pelvis. a large central cavity at the base of each kidney; receives urine from the collecting ducts and empties it into the ureter

renin. an enzyme secreted by the kidneys; catalyzes the splitting off of angiotensin I from angiotensinogen in the blood

repolarize. to return the value of the transmembrane potential to its resting level

repression. inhibition of enzyme synthesis in the presence of product molecules synthesized by that enzyme (or the enzyme chain of which it is a part); product molecules inhibit enzyme synthesis by effects upon the transcription of genes which code for the protein structure of the enzymes

repressor gene. a portion of DNA which contains the genetic instructions for synthesis of a repressor protein

repressor protein. a protein molecule which can bind to an operator gene and inhibit the transcription of the gene into mRNA

residual volume. the volume of air remaining in the lungs after maximal expiration

resistance. the hindrance to movement through a particular substance or tube, e.g., the hindrance to the movement of electric charge through extracellular fluid, or the movement of blood through blood vessels

respiration. (cellular) the consumption of molecular oxygen during metabolism; (respiratory system) the exchange of gas between the cells of an organism and the external environment

respiratory acidosis. an acidosis resulting from failure of the lungs to eliminate carbon dioxide as fast as it is produced

respiratory alkalosis. an alkalosis resulting from the elimination of carbon dioxide from the lungs faster than it is produced

respiratory distress syndrome. a disease afflicting premature infants in whom the surfactant-producing cells are too immature to function adequately

respiratory "pump". the effect of the changing intrathoracic and intraabdominal pressures associated with respiration on the venous return of blood to the heart

respiratory quotient (RQ). the ratio of CO_2 produced to O_2 consumed during metabolism

respiratory rate. the number of breaths per minute

respiratory system. the structures involved in the exchange of gases between the blood and the external environment, i.e., the lungs, the series of airways leading to the lungs, and the chest structures responsible for movement of air in and out of the lungs

resting membrane potential. (*also* resting potential) the voltage difference between the inside and outside of a cell in the absence of excitatory or inhibitory stimulation

resting tremor. a tremor present in persons who are completely at rest; occurs with basal ganglia disease

restriction nuclease. a bacterial enzyme which splits DNA into a number of fragments, acting at different loci in the two strands of the polymer and at points containing a particular sequence of nucleotides

retching. the events that normally accompany vomiting, i.e., deep inspiration, closure of the glottis, elevation of the soft palate, and contraction of abdominal and thoracic muscles, without the actual expulsion of the stomach's contents

reticular activating system. a collection of neurons within the brainstem reticular formation and its thalamic extension whose activity is concerned with alertness and direction of attention to selected events

reticular formation. an extensive network of finely branched neurons extending through the core of the brainstem and linked functionally to a similar group of cells in the midline of the thalamus, receives and interprets information from many afferent pathways as well as from many other regions of the brain

retina. a thin layer of neural tissue containing the receptors for vision and lining the back of the eyeball

retinal. a form of vitamin A; forms the chromophore component of a photopigment

retrograde amnesia. loss of memory for the events during a time immediately preceding a memory-disturbing trauma such as a blow to the head

Rh factor. a group of antigens which may (Rh+) or may not (Rh−) be present on the plasma membranes of erythrocytes

rhodopsin. the photopigment in rods; a light-absorbing glycoprotein

rhythm method. a contraceptive technique in which couples refrain from sexual intercourse near the time of ovulation

ribonucleic acid (RNA). the single-stranded acid involved in the transcription of genetic information and its translation into protein synthesis; its nucleotides contain the sugar ribose; exists in three forms: messenger RNA, ribosomal RNA, and transfer RNA

ribosomal RNA (rRNA). a polymer of ribose-containing nucleotides which is synthesized in the nucleolus, where it is used in the assembly of a ribosome, which then moves to the cytoplasm

ribosome. a cytoplasmic particle which mediates the linking together of amino acids to form proteins: attached to the membranes of the endoplasmic reticulum or suspended in the cytoplasm as a free particle

rickets. a disease in which new bone matrix is inadequately calcified due to a deficiency of 1,25-dihydroxy-vitamin D_3

rigor mortis. death rigor: a stiffness of skeletal muscles resulting from a loss of ATP

RNA. *see* ribonucleic acid

RNA polymerase. an enzyme which joins together the appropriate nucleotides by base pairing to form a molecule of RNA

RNA processing. process during which enzymes remove the noncoding (intron) sequences from a newly formed molecule of RNA

rod. one of the two receptor types for photic energy; contains the photopigment rhodopsin

round window. the membrane-covered opening separating the scala tympani of the inner ear from the middle-ear cavity

RQ. *see* respiratory quotient

rRNA. *see* ribosomal RNA

SA node. *see* sinoatrial node

saccade. a very fast, short, jerking movement of the eyeball

saliva. the secretory product of the salivary glands; a watery solution of various salts and proteins, including the mucins and amylase (ptyalin)

sarcomere. the repeating structural unit of a muscle myofibril; the region of a myofibril extending between two adjacent Z lines

sarcoplasmic reticulum. the endoplasmic reticulum in muscle fibers; forms sleevelike structures around myofibrils in striated muscle; site of storage and release of calcium ions

saturated fatty acid. a fatty acid whose carbon atoms are all linked by single covalent bonds

saturation. the degree to which binding sites are occupied by ligand; if all are occupied, the binding sites are fully saturated, if half are occupied, the saturation is 50 percent

scala tympani. the fluid-filled inner-ear compartment which receives sound waves from the basilar membrane and transmits them to the round window

scala vestibuli. the fluid-filled inner-ear compartment which receives sound waves from the oval window and transmits them to the basilar membrane and cochlear duct

schizophrenia. a disease (or a family of diseases) characterized by altered motor behavior, distortions in perception, disturbed thinking, altered mood, and abnormal interpersonal behavior

Schwann cell. a nonneural cell which surrounds all peripheral nerve fibers; forms the myelin sheath around a myelinated axon in the peripheral nervous system

scrotum. the sac which contains the testes and epididymides

SDA. *see* specific dynamic action

sebaceous glands. glands in the skin which secrete a fatty substance

SEC. *see* series elastic component

second messenger. an intracellular substance (e.g., cyclic AMP and Ca^{2+}) which increases in the cytosol as a result of combination of an extracellular chemical messenger (i.e., the "first" messenger) with a receptor in the plasma membrane or cytosol; alters some aspect of the cell's function and thereby mediates the cell's response to the first messenger

secondary active transport. active transport in which the energy released during the transmembrane movement of one substance (e.g., sodium) from higher to lower concentration is transferred to the simultaneous movement of another solute from lower to higher concentration

secondary peristalsis. esophageal peristaltic waves not immediately preceded by the pharyngeal phase of a swallow

secondary sexual characteristics. the many external differences (hair, body contours, etc.) between male and female which are not directly involved in reproduction

secondary spermatocyte. a male germ cell derived from a primary spermatocyte

secretin. a peptide hormone secreted by cells in the upper part of the small intestine; stimulates the pancreas to secrete bicarbonate into the small intestine

secretion. the elaboration and release of organic molecules, ions, and water by cells in response to specific stimuli

secretory phase. that stage of the menstrual cycle following ovulation and formation of the corpus luteum during which a secretory type of endometrium develops in the uterus

secretory vesicles. membrane-bound vesicles produced by the Golgi apparatus; contain proteins to be secreted by the cell

segmentation. a series of stationary, rhythmical contractions and relaxations of rings of intestinal smooth muscle which mixes the luminal contents of the intestine

semen. the sperm-containing fluid of the male ejaculate

seminal vesicles. a pair of large glands in males which secrete fluid into the two ductus deferens

seminiferous tubule. a tubule in the testis that contains the cells which differentiate into sperm cells

semipermeable membrane. a membrane permeable to some substances but not to others

sensor. first component of a control system; detects changes in some aspect of the environment; a receptor

sensory deprivation. an experimental situation affording isolation as complete as possible from sensory stimulation, even including restriction of movements

sensory information. afferent information which has a conscious correlate

sensory unit. a single afferent neuron plus all the receptors it innervates

series elastic component (SEC). the elastic structures associated with a muscle fiber through which the force generated by the cross bridges is transmitted to the load on a muscle

serosa. a connective-tissue layer surrounding the outer surface of the stomach and intestinal walls

serotonin. a monoamine neurotransmitter; also functions as a paracrine agent in blood platelets and specialized cells of the digestive tract; 5-hydroxytryptamine, 5-HT

Sertoli cell. a type of cell intimately associated with the developing germ cells in the seminiferous tubules; creates a "blood-testis barrier," secretes fluid into the lumen of the seminiferous tubules, and mediates hormonal effects on the tubules

serum. blood plasma from which the fibrinogen has been removed as a result of clotting

sex chromatin. a readily detected nuclear mass not usually found in cells of males; consists of a condensed X chromosome

sex chromosomes. the two chromosomes, X and Y, which determine the genetic sex of an individual, genetic males having one X and one Y, and genetic females having two X chromosomes

sex determination. the genetic basis of an individual's sex, XY determining male and XX, female

sex differentiation. the development of male or female reproductive organs as determined by the individual's genetic make-up

sex hormones. estrogen, progesterone, and testosterone and related hormones

short reflex. a neural reflex in which the afferent and efferent pathways and integrating center are within the plexuses of the gastrointestinal tract

short-term memory. a limited-capacity storage process, possibly lasting only a few seconds, serving as the initial depository of information from which the memory may be transferred to long-term storage or forgotten

sickle-cell anemia. a genetic disease in which one of the 287 amino acids in the polypeptide chains of the hemoglobin molecule is wrong and, at the low oxygen concentrations existing in many capillaries, causes the erythrocytes to assume sickle shapes or other bizarre forms that block the capillaries

sigmoid colon. the S-shaped terminal portion of the colon

single-unit smooth muscle. smooth muscle in which the fibers are joined by gap junctions allowing electrical activity to be conducted from cell to cell, i.e., the muscle responds to stimulation as a single unit

sinoatrial (SA) node. a region in the right atrium of the heart containing specialized cardiac muscle cells which depolarize spontaneously at a rate faster than that of other spontaneously depolarizing cells of the heart; the pacemaker of the heart which determines the rate at which the heart beats

skeletal muscle. a striated muscle attached to skin or bones and responsible for facial expression and movements of the skeleton; controlled by the somatic nervous system

skeletal muscle "pump". the pumping effect of contracting skeletal muscles on the blood flow through vessels underlying them

skeletomotor fibers. the primary fibers of skeletal muscle, as opposed to the modified fibers within the muscle spindle

sleep. *see* paradoxical sleep, slow-wave sleep

sleep centers. clusters of neurons in the brainstem whose activity periodically opposes the tonic activity of the reticular activating system and induces the cycling of slow-wave and paradoxical sleep

sliding-filament mechanism. the process of muscle contraction in which shortening occurs as a result of the sliding of thick and thin filaments past each other

slow viruses. viruses which replicate very slowly and may even become associated with the cell's own DNA and be passed along to the daughter cells upon cell division, altering the cells so they are no longer recognized as "self"

slow-wave potential. a slow oscillation of the membrane potential of some smooth muscle cells with cycle times of seconds or even minutes

slow-wave sleep. the state of sleep associated with large, slow EEG waves, considerable tonus in the postural muscles, and little change in cardiovascular and respiratory activity

smooth muscle. surrounds hollow organs and tubes such as the stomach, intestinal tract, and urinary bladder; not striated; controlled by the autonomic nervous sys-

tem, hormones, and paracrines; *see also* single-unit smooth muscle, multiunit smooth muscle

sodium inactivation. the closing of Na channels in the plasma membrane at the peak of an action potential

soft palate. the nonbony region at the posterior part of the roof of the mouth

solute. molecules or ions dissolved in a liquid

solution. a liquid (solvent) containing dissolved molecules or ions (solutes)

solvent. the liquid in which molecules or ions are dissolved

somatic. pertaining to the body; related to the framework or outer walls of the body, including skin, skeletal muscle, tendons, and joints

somatic chromosomes. the 22 pairs of nonsex chromosomes

somatic nervous system. a component of the efferent division of the peripheral nervous system distinguished from the autonomic nervous system; innervates skeletal muscle

somatic receptor. a neural receptor which responds to mechanical stimulation of the skin or hairs and underlying tissues, to rotation or bending of joints, to temperature changes, or possibly to some chemical changes

somatomedin. a group of growth-promoting peptides released mainly from the liver in response to growth hormone

somatosensory cortex. a strip of cerebral cortex in the parietal lobe in which nerve fibers transmitting somatic sensory modalities (e.g., touch, temperature, etc.) synapse

somatostatin. the hypothalamic hormone which inhibits growth hormone and TSH secretion by the anterior pituitary; a possible neurotransmitter; found also in the stomach and D cells of the pancreatic islets of Langerhans; decreases the rate at which a meal is digested and absorbed

somatotropin. *see* growth hormone

sound wave. a disturbance in the air resulting from variations in densities of air molecules from regions of high density (compression) to low density (rarefaction)

spatial summation. the effects of simultaneous stimuli (or synaptic inputs) to different points on a neuron are added together to produce a potential change greater than that caused by a single stimulus (or synaptic input)

specific dynamic action (SDA). the increase in metabolic rate induced by eating

specific immune responses. responses of lymphocytes and cells derived from them which depend upon prior exposure to a specific foreign material, recognition of it upon subsequent exposure, and reaction to it; *see also* cell-mediated immunity, humoral immunity

specific pathway. a chain of synaptically connected neurons in the brain or spinal cord, all of which are activated by sensory units of the same modality, e.g., a pathway conveying information about only touch or only temperature

specific thalamic nuclei. nuclei in the cerebrum whose axons project to localized, clearly defined regions of cerebral cortex

specificity. selectivity; the ability of a protein binding site to react only with one type (or a limited number of structurally related types) of molecules

sperm. reproductive cells in the male; the male gamete; spermatozoa

spermatid. an immature sperm

spermatogenesis. the manufacture of sperm

spermatogonia. the undifferentiated germ cell which gives rise ultimately to sperm

spermatozoa. *see* sperm

sphincter of Oddi. a ring of smooth muscle surrounding the bile duct at its point of entrance into the duodenum

sphygmomanometer. a device consisting of an inflatable cuff and pressure gauge for measuring arterial blood pressure

spinal nerve. one of the 86 (43 pairs) of peripheral nerves which join the spinal cord

spindle apparatus. a cell structure which appears during nuclear division and consists of a number of microtubules (spindle fibers) passing from centrioles to chromosomes

spindle fibers. the modified skeletal muscle fibers within the muscle spindle; intrafusal fibers; *also* the microtubules associated with the spindle apparatus during nuclear division

spleen. the largest of the lymphoid organs; located in the left part of the abdominal cavity between the stomach and diaphragm

split brain. a condition resulting from cutting of the nervefiber bundles connecting the two cerebral hemispheres, which prevents communication between the two halves of the cerebrum

stapes. the third of the three middle-ear bones; situated so that its footplate rests in the oval window and transmits sound waves to the scala vestibuli of the inner ear

starch. a moderately branched polysaccharide chain composed of thousands of glucose subunits; found in plant cells

Starling's law of the heart. the law that, within limits, an increase in the end-diastolic volume of the heart, i.e., increasing the length of the muscle fibers, increases the force of cardiac contraction

stastis. cessation of blood flow in a vessel

state of consciousness. the degree of mental alertness; e.g., sleep and wakefulness are two states of consciousness

steady state. a situation in which no net change occurs in a system, however, a continual energy input to the system is required to maintain this state; *see also* equilibrium

sternum. breastbone

steroid. a subclass of lipid molecules; the molecule consists of a skeleton of four interconnected carbon rings to which a few polar groups may be attached

STH. *see* growth hormone

stimulus. a change in the environment detectable by a receptor

stress. any environmental change that must be adapted to if health and life are to be maintained; any event that elicits increased cortisol secretion

stretch receptor. afferent nerve endings that are depolarized by stretching, e.g., the nerve endings in the muscle spindle

stretch reflex. a monosynaptic reflex in which stretch of a muscle causes contraction of that muscle; mediated by muscle spindles

striated muscles. muscle which demonstrates a characteristic transverse banding pattern when viewed with a microscope; skeletal and cardiac muscle

stroke. damage to a portion of the brain caused by stoppage of the blood supply to that region because of occlusion or rupture of a cerebral vessel

stroke volume. the volume of blood ejected by a ventricle during one beat of the heart

strong acid. an acid whose molecules ionize completely to form hydrogen ions and corresponding anions when dissolved in water

structural gene. a gene which contains the coded information which specifies the amino acid sequence of a specific protein

subatomic particles. the units of matter which together make up an atom; e.g., electrons, protons, and neutrons are subatomic particles

subcortical nuclei. clusters of neuronal cells buried deep in the substance of the cerebrum; includes the basal ganglia

sublingual glands. one of the three pairs of salivary glands

submandibular glands. one of the three pairs of salivary glands; formerly called submaxillary glands

submucosa. a connective-tissue layer underlying the mucosa in the walls of the stomach and intestines

submucous plexus. a network of nerve cells in the submucosal layer of the esophagus, stomach, and intestinal walls

subset program. the sequence of neural activity which controls a component of a complete motor action

substance P. a neurotransmitter at the synaptic endings of the afferent neurons in the pain pathway as well as other sites

substrate. the reactant in an enzyme-mediated reaction

substrate phosphorylation. occurs in the cytoplasm when a phosphate group is transferred directly from a metabolic intermediate to ADP to form ATP

subsynaptic membrane. that part of a postsynaptic neuron's plasma membrane which lies directly under the synaptic knob

subthreshold potential. any depolarization less than (i.e., farther from zero) the threshold potential

subthreshold stimulus. any stimulus capable of depolarizing the membrane but not by enough to reach threshold

sucrose. a disaccharide carbohydrate composed of glucose and fructose; table sugar

sulfate. —SO_4^{2-}

sulfhydryl group. —SH

sulfuric acid. an acid generated during the catabolism of sulfur-containing compounds; dissociates to form sulfate and hydrogen ions

summation. the increase in muscle tension or shortening which occurs in response to rapid, repetitive stimulation as compared to the response to a single twitch

superior vena cava. the large vein which carries blood from the upper half of the body to the right atrium of the heart

supersensitivity. the increased responsiveness of a target cell to a given messenger, developed in response to chronic very low extracellular concentrations of the messenger

suppressor T cells. the class of T lymphocytes which inhibits antibody production and cytotoxic T-cell function

suprathreshold stimulus. any agent capable of depolarizing the membrane beyond its threshold potential (i.e., closer to zero)

surface tension. the unequal attractive forces between water molecules at a surface which result in a net force which acts to reduce the surface area

surfactant. a phospholipoprotein complex produced by the type II cells of the pulmonary alveoli; markedly reduces the cohesive force of the water molecules lining the alveoli, i.e., reduces the surface tension

sympathetic nervous system. that portion of the autonomic nervous system whose preganglionic fibers leave the central nervous system at the thoracic and lumbar portions of the spinal cord; most of the postganglionic neurons of the sympathetic nervous system release the neurotransmitter norepinephrine

sympathetic trunk. one of the paired chains of interconnected sympathetic ganglia which lie on either side of the vertebral column

sympathomimetic. a chemical which produces effects similar to those of the sympathetic nervous system

synapse. an anatomically specialized junction between two neurons where the electrical activity in one neuron influences the excitability of the second; *see also* chemical synapse, electrical synapse, excitatory synapse, inhibitory synapse

synaptic cleft. the narrow extracelluar space separating the pre- and postsynaptic neurons at a chemical synapse

synaptic knob. the slight swelling at the end of each axon terminal; synaptic terminal

synergistic muscle. a muscle whose action aids the intended motion

synthetic reactions. *see* anabolic reactions

systemic circulation. the circulation through all organs except the lungs

systole. the period when the ventricles of the heart are contracting

systolic pressure. the maximum arterial blood pressure reached during the cardiac cycle

T_m. *see* maximal tubular capacity

T_3. *see* triiodothyronine

T_4. *see* thyroxine

T cells (T lymphocytes). lymphocytes derived from precursors that at one time were in the thymus; *see also* cytotoxic T cells, helper T cells, suppressor T cells

T lymphocytes. *see* T cells

t tubule. *see* transverse tubule

T wave. the component of the electrocardiogram corresponding to repolarization of the ventricles

taste buds. the sense organs for taste; contain chemoreceptors and supporting cells; found on the tongue, roof of the mouth, pharynx, and larynx

tectorial membrane. a structure in the organ of Corti which is in contact with the hairs on the receptor cells

teleology. the explanation of events in terms of the ultimate purpose being served by them

temporal lobe. the region of cerebral cortex where primary auditory cortex and one of the speech centers are located

temporal summation. the potential produced as two or more stimuli (or synaptic inputs) occurring at different times are added together to produce a potential change greater than that caused by a single stimulus (or synaptic input)

tendon. a bundle of collagen fibers which connects muscle to bone and transmits the contractile force of the muscle to the bone

tension. *see* muscle tension.

termination code word. a three-nucleotide sequence in DNA which signifies the end of a gene

testes. the gonads in the male (sing: testis)

testosterone. a steroid hormone produced in the interstitial cells of the testes; the major male sex hormone; maintains the growth and development of the reproductive organs and is responsible for the secondary sexual characteristics of males

tetanus. a maintained mechanical response of a muscle to high-frequency stimulation

TH. *see* thyroid hormone

thalamus. a subdivision of the diencephalon; a way station and integrating center for sensory input on its way to cerebral cortex; also contains motor nuclei and a central core which functions as an extension of the brainstem reticular formation

theca. a cell layer formed from follicle cells and specialized ovarian connective-tissue cells; surrounds the granulosa cells of the ovarian follicle

thermogenesis. heat generation

thermoreceptor. a sensory receptor which responds preferentially to temperature (and changes in temperature) particularly in a low (cold receptors) or high (warm receptors) range

thick filaments. the 12- to 18-nm filaments in muscle cells; consist of myosin molecules

thin filaments. the 5- to 8-nm filaments in muscle cells; consist of actin, troponin, and tropomyosin molecules

thoracic cage. wall of the thoracic cavity; chest wall

thoracic cavity. chest cavity

thoroughfare channel. a capillary which connects an arteriole and venule directly and from which branch the "true" capillaries

threshold. (or threshold potential) the membrane potential to which an excitable membrane must be depolarized in order to initiate an action potential

threshold stimulus. any stimulus capable of depolarizing the membrane to threshold

thrombin. the enzyme which catalyzes the conversion of fibrinogen into fibrin

thrombosis. the formation of a clot in a blood vessel

thromboxane. a substance closely related to the prostaglandins; synthesized from arachidonic acid and other essential fatty acids

thromboxane A_2. a thromboxane which is a stimulant of platelet aggregation in blood clotting

thymine. one of the pyrimidine bases; the one nucleotide base present in DNA but not in RNA

thymosin. a group of hormones secreted by the thymus

thymus. a lymphoid organ lying in the upper part of the chest; site of lymphocyte formation and thymosin secretion

thyroglobulin. a large protein to which thyroid hormones bind; the storage form of the thyroid hormones in the thyroid

thyroid. a paired endocrine gland located in the neck; secretes thyroid hormone and calcitonin

thyroid hormone (TH). the collective term for the amine hormones released from the thyroid, i.e., thyroxine (T_4) and triiodothyronine (T_3); increases the metabolic rate of most cells in the body

thyroid-stimulating hormone (TSH). a glycoprotein hormone secreted by the anterior pituitary; induces secretion of thyroid hormone; also called thyrotropin

thyrotropin. *see* thyroid-stimulating hormone

thyrotropin-releasing hormone (TRH). the hypothalamic hormone which stimulates TSH (thyrotropin) secretion by the anterior pituitary

thyroxine (T$_4$). tetraiodothyronine; an iodine-containing amine hormone secreted by the thyroid

tidal volume. the volume of air entering or leaving the lungs during a single breath during any state of respiratory activity

tight junction. a type of cell junction found in epithelial tissues; extends around the circumference of the cell; restricts the diffusion of molecules across an epithelial layer by way of the extracellular space between cells

tissue. an aggregate of differentiated cells of a similar type united in the performance of a particular function; also commonly used to denote the general cellular fabric of a given organ, e.g., kidney tissue, brain tissue

tissue thromboplastin. an extravascular enzyme capable of initiating the formation of fibrin clots

tolerance. the state in which increasing doses of a drug are required to achieve effects that initially occurred in response to a smaller dose; i.e., it takes more drug to do the same job

tonic. a continuous activity, as opposed to a phasic one

total dead space. alveolar dead space plus anatomic dead space

total energy expenditure. the sum (per unit time) of the energy expended doing external work plus the energy represented in heat production by the body plus any energy stored

total peripheral resistance (TPR). the total resistance to the flow of blood in all the vessels in the systemic vascular tree from the very first portion of the aorta to the very last portion of the venae cavae

total pulmonary ventilation. the volume of air that passes into or out of the respiratory system per minute; equal to the tidal volume times the number of breaths per minute; the minute volume

toxemia of pregnancy. a disease occurring in pregnant women and associated with retention of abnormally great amounts of fluid, the appearance of protein in the urine, hypertension, and possibly convulsions

toxicology. the study of substances which are harmful to the body

TPR. *see* total peripheral resistance

trace elements. a mineral present in the body in extremely small quantities

trachea. the single airway which connects the larynx with the bronchi which enter the right and left lungs

tract. a large bundle of myelinated nerve fibers in the central nervous system

transamination. a reaction in which an amino group (—NH$_2$) is transferred from an amino acid to a keto acid, the keto acid thus becoming an amino acid

transcriptase. DNA-dependent RNA polymerase, the enzyme which joins the nucleotides of mRNA during its formation

transcription. formation of mRNA containing in the linear sequence of its nucleotides the genetic information contained in the DNA gene from which the mRNA is transcribed; the first stage in the transfer of genetic information from DNA to protein synthesis

transduction. the process by which stimulus energy is transformed into an electrical response

transepithelial transport. *see* epithelial transport

transfer RNA (tRNA). a polymer of ribose-containing nucleotides; different tRNAs combine with different amino acids and with an mRNA specific for that amino acid, thus arranging the amino acids in the appropriate sequence to form the protein specified by the codons in mRNA

transferrin. an iron-binding protein which is the carrier molecule for iron in the plasma

translation. assembly of the proper amino acids in the correct order during the formation of a protein according to the genetic instructions carried in mRNA; occurs on the ribosomes of a cell

transmural pressure. the difference between the pressure exerted on one side of the wall of a structure and the pressure exerted on the other side

transverse tubule (t tubule). a tubule extending from the plasma membrane of a striated muscle fiber into the fiber interior, passing between the opposed lateral sacs of adjacent sarcoplasmic-reticulum segments; conducts the muscle action potential into the muscle fiber

TRH. *see* thyrotropin-releasing hormone

triacylglycerol. a subclass of lipid molecules; neutral fat; composed of glycerol and three fatty acids, formerly called triglyceride

tricarboxylic acid cycle. *see* Krebs cycle

tricuspid valve. the valve between the right atrium and right ventricle of the heart

tricyclic antidepressants. drugs which interfere with the reuptake of norepinephrine and serotonin by presynaptic endings, thereby increasing the amount of neurotransmitter present in the synaptic cleft

"trigger zone". that low-threshold·region of a neuron at which action potentials are initiated; sometimes occurs at the initial segment of an axon

triglyceride. *see* triacylglycerol

triiodothyronine (T_3). an iodine-containing amine hormone secreted by the thyroid

tRNA. *see* transfer RNA

trophoblast. the outer layer of cells of the blastocyst; gives rise to the placental tissues

tropomyosin. a regulatory protein capable of reversibly covering the binding sites on actin, thereby preventing their combination with myosin; associated with the thin filaments in muscle fibers

troponin. a regulatory protein bound to the actin and tropomyosin molecules of the thin filaments of striated muscle; site of calcium binding which initiates contractile activity

"true" capillary. a capillary across whose walls material exchange occurs

trypsin. an enzyme secreted into the small intestine by the exocrine pancreas as the precursor trypsinogen; breaks certain peptide bonds in proteins and polypeptides

trypsinogen. the inactive precursor of the enzyme trypsin; secreted by the exocrine pancreas

tryptophan. an essential amino acid; precursor of serotonin

TSH. *see* thyroid-stimulating hormone

tubular reabsorption. the transfer of materials from the lumen of a kidney tubule to the peritubular capillaries

tubular secretion. the transfer of materials from the peritubular capillaries to the lumen of the kidney tubules

turnover rate. the time it takes to replace all the molecules of a specific type with newly synthesized molecules

twitch. the mechanical response of a muscle to a single action potential

tympanic membrane. the membrane stretched across the end of the ear canal

Type I diabetes. diabetes in which the hormone insulin is completely or almost completely absent; insulin-dependent diabetes

Type II diabetes. diabetes in which the hormone insulin is present at near normal or even above normal levels, but the tissues do not respond normally to it; insulin-independent diabetes

tyrosine. an amino acid; a precursor of the catecholamine neurotransmitters and the thyroid hormones

tyrosine hydroxylase. the enzyme which catalyzes the formation of dopa from tyrosine in the catecholamine synthetic pathway; the rate-limiting enzyme in that pathway

ulcer. an erosion, or sore, as in the stomach or intestinal wall

ultrafiltrate. a fluid which is essentially protein-free; formed from plasma as it is forced through capillary walls by a pressure gradient

umbilical vessels. the arteries and veins transporting blood between the fetus and placenta

unidirectional flux. the amount of material crossing a surface in one direction in a unit of time

unsaturated fatty acid. a fatty acid containing one or more double bonds

up regulation. an increase in the total number of target-cell receptors for a given messenger in response to a chronic low extracellular concentration of the messenger; *see* supersensitivity

upper esophageal sphincter. a ring of skeletal muscle surrounding the esophagus just below the pharynx which, when contracted, closes the entrance to the esophagus

uracil. one of the pyrimidine bases; present in RNA but not DNA

urea. NH_2—$\overset{\overset{\textstyle O}{\parallel}}{C}$—$NH_2$: synthesized in the liver by linking two molecules of ammonia with carbon dioxide; the major nitrogenous waste product of protein breakdown

uremia. a general term for the symptoms of profound kidney malfunction; literally, urine in the blood

ureter. a tubule which connects the renal pelvis to the urinary bladder

urethra. the tube which conveys urine from the urinary bladder to the outside of the body

uric acid. a waste product derived from nucleic acids

urinary bladder. a thick-walled sac composed of smooth muscle; stores urine temporarily prior to urination

uterus. womb; a hollow organ in the pelvic region of females; the fertilized ovum implants in the uterine wall and develops within the uterus during pregnancy

v. *see* volt

vagus nerve. cranial nerve X; the major parasympathetic nerve

van der Walls force. an attracting force between nonpolar regions of molecules; due to the weak electromagnetic field which surrounds atoms

varicose vein. an irregularly swollen superficial vein of the lower extremities

varicosity. a swollen region of an axon containing neurotransmitter-filled vesicles along the terminal portions of the axons of autonomic nerve fibers; analogous to the synaptic knob

vas deferens. the ductus deferens

vasectomy. the cutting and tying off of the ductus ("vas") deferens, which results in sterilization of the male without loss of testosterone secretion by the testes

vasoconstriction. a decrease in the diameter of blood vessels as a result of vascular smooth muscle contraction

vasodilation. an increase in the diameter of blood vessels as a result of vascular smooth muscle relaxation

vasopressin. *see* antidiuretic hormone

vein. a wide, thin-walled vessel which carries blood back to the heart from the venules

vena cava. one of two large veins which return systemic blood to the heart

venous return. the volume of blood flowing to each ventricle of the heart from the veins per minute (not the total returned to both ventricles)

ventilation. the bulk-flow exchange of air between the atmosphere and the alveoli

ventral root. a group of efferent fibers which leaves the left and right side of that part of the spinal cord that faces the front of the body

ventricle. a cavity, e.g., a ventricle of the heart or the brain

venule. a small, thin-walled vessel which carries blood from a capillary network to a vein

vesicle. a small, membrane-bound intracellular organelle

vestibular nuclei. neuron clusters in the brainstem which receive primary afferent neurons from the vestibular apparatus

vestibular receptors. hair cells in the gelatinous mass in the semicircular canals, utricle, and saccule

vestibular system. a sense organ (often called the sense organ of balance) deep within the temporal bone of the skull; consists of the three semicircular canals, utricle, and saccule

villi. projections from the highly folded surface of the small intestine; covered with a single layer of epithelial cells

virus. an "organism" that is essentially a nucleic acid core surrounded by a protein coat and lacking both the enzyme machinery for energy production and the ribosomes for protein synthesis; thus, it cannot survive and reproduce except inside other cells whose biochemical apparatus it makes use of; *see also* slow viruses

viscera. the organs in the thoracic and abdominal cavities

viscosity. the property of a fluid, caused by frictional interactions between its molecules, that makes it resist flow

vital capacity. the maximal amount of air that can be moved into the lungs following a maximal expiration; the sum of the inspiratory-reserve, normal tidal, and expiratory-reserve volumes

vitalism. the view that life processes occur because of the presence of a "life force" rather than by physicochemical processes alone

vitamin. an organic molecule which must be present in trace amounts to maintain normal health and growth of an organism but is not manufactured by the organism's metabolic pathways; classified as water-soluble (vitamins C and the B complex) and fat-soluble (vitamins A, D, E, and K)

VLDL. very low density lipoprotein; a lipid-protein aggregate having a high proportion of fat

vocal cords. two strong bands of elastic tissue stretched across the opening of the larynx and caused to vibrate by the movement of air past them, providing the tones of speech

volt (V). the unit of measurement of the electric potential between two points

voltage. a measure of the potential of separated electric charges to do work; the amount of work done by an electric charge when moving from one point in a system to another; a measure of the electric force between two points

voltage-sensitive ion channel. an ion channel in cell membranes which is opened or closed by changes in the membrane potential

vomiting center. the neurons in the brainstem medulla which coordinate the complex reflex of vomiting

vulva. the mons pubis, labia majora and minora, clitoris, vestibule of the vagina, and vestibular glands

wavelength. the distance between two wave peaks in an oscillating medium

weak acid. an acid whose molecules do not completely ionize to form hydrogen ions when dissolved in water

white matter. that portion of the brain and spinal cord which appears white in unstained specimens; contains primarily myelinated nerve fibers

white muscle fibers. muscle fibers lacking appreciable amounts of myoglobin

withdrawal. physical symptoms (usually unpleasant) associated with cessation of drug use

withdrawal reflex. *see* flexion reflex

Wolffian duct. a part of the embryonic duct system which, in a male, remains and develops into the ducts of the reproductive system, but in a female, degenerates

work. a measure of the amount of energy required to produce a physical displacement of matter; formally defined as the product of force (F) times distance (X), $W = FX$; see also external work.

X chromosome. *see* sex chromosomes

Xylocaine. *see* lidocaine

Y chromosome. *see* sex chromosomes

Z line. the band running across the myofibril at each end of a striated muscle sarcomere; anchors one end of the thin filaments

zona pellucida. a thick, clear band separating the ovum from the surrounding granulosa cells

zymogen granule. membrane-bound cytoplasmic vesicle containing inactive precursor enzymes prior to their secretion from a cell; the protein granules secreted by the exocrine portion of the pancreas

REFERENCES
FOR FIGURE ADAPTATIONS

Adolph, E. F.: "Physiology of Man in the Desert," Interscience, New York, 1947.

Anthony, C. P., and N. J. Kolthoff: "Textbook of Anatomy and Physiology," 8th ed., Mosby, St. Louis, 1971.

Bernhard, S.: "The Structure and Function of Enzymes," Benjamin, New York, 1968.

Bloom, W., and D. W. Fawcett: "A Textbook of Histology," 9th ed., Saunders, Philadelphia, 1968.

Boyd, I. A., and T. D. Roberts: *J. Physiol.,* **122**:38 (1953).

Brobeck, J. R., J. Tepperman, and C. N. H. Long: *Yale J. Biol. Med.,* **15**:831 (1943).

Chaffee, E. E., and I. M. Lytle: "Basic Physiology and Anatomy," 4th ed., Lippincott, Philadelphia, 1980.

Chapman, C. B., and J. H. Mitchell: *Sci. Am.,* May 1965.

Comroe, J. H.: "Physiology of Respiration," Year Book, Chicago, 1965.

Daughaday, W. H., and D. M. Kipnis: *Recent Prog. Hormone Res.,* **22**:49 (1966).

Dubois, E. F.: "Fever and the Regulation of Body Temperature," Charles C Thomas, Springfield, Ill. 1948.

Eccles, J. C.: "The Understanding of the Brain," McGraw-Hill, New York, 1973.

Fclig, P., and J. Wahren: *N. Eng. J. Med.,* **293**:1078 (1975).

Ganong, W. F.: "Review of Medical Physiology," 4th ed., Lange, Los Altos, Calif., 1969.

Geschwind, N.: *Sci. Am.,* September 1979.

Goldberg, N. D.,: *Hosp. Prac.,* May 1974.

Gordon, A. M., A. F. Huxley, and F. J. Julian: *J. Physiol.* **184**:170 (1966).

Gregory, R. L.: "Eye and Brain: The Psychology of Seeing," McGraw-Hill, New York, 1966.

Guillemin, R., and R. Burgus: The Hormones of the Hypothalamus, *Sci Am.,* October 1972.

Guyton, A. C.: "Functions of the Human Body," 3d ed., Saunders, Philadelphia, 1969.

Hoffman, B. F., and P. E. Cranefield: "Electrophysiology of the Heart," McGraw-Hill, New York, 1960.

Hubel, D. H., and T. N. Wiesel: *J. Physiol.,* **154**:572 (1960); **160**:106 (1962).

Hume, D. M.: Organ Transplants and Immunity, *Hosp. Prac.,* 1968.

Ito, S.: *J. Cell Biol.,* **16**:541 (1963).

Jouvet, M.: *Sci. Am.,* February 1967.

Kandel, E. R., and J. H. Schwartz: "Principles of Neural Science," Elsevier/North-Holland, New York, 1981.

Kappas, A., and A. P. Alvares: *Sci. Am.,* June 1975.

Kiang, N. Y. S.: *in* D. B. Tower (ed.), "The Nervous System," vol. 3, Raven Press, New York, 1975.

Kim, D. O., and C. E. Molnar: *in* D. B. Tower (ed.), "The Nervous System," vol. 3, Raven Press, New York, 1975.

Kruger, D. T., and J. B. Martin: *N. Eng. J. Med.,* **304**:876, 1981.

Kuffler, S. W., and J. G. Nicholls: "From Neuron to Brain," Sinauer Associates, Sunderland, Mass., 1976.

Lambersten, C. J.: *in* P. Bard (ed.), "Medical Physiology," 11th ed., Mosby, St. Louis, 1961.

Landauer, T. K.: "Readings in Physiological Psychology," McGraw-Hill, New York, 1967.

Lehninger, A. L.: "Biochemistry," Worth Publishers, Inc., New York, 1970.

Lentz, T. L.: "Cell Fine Structure," Saunders, Philadelphia, 1971.

Little, R. C.: "Physiology of the Heart and Circulation," 2d ed., Year Book, Chicago, 1981.

Livingston, R. B.: *in* G. C. Quarton, T. Melnechuk, and F. O. Schmitt (eds.), "The Neurosciences: A Study Program," Rockefeller University Press, New York, 1967.

Loewenstein, W. R.: *Sci. Am.,* August 1960.

McEwen, B. S.: *Sci. Am.,* July 1976.

McNaught, A. B., and R. Callender: "Illustrated Physiology," Williams & Wilkins, Baltimore, 1963.

Merton, P. A.: *Sci. Am.,* May 1972.

Mollon, J. D.: *Ann. Rev. Psych.,* **33**:41 (1982).

Nathanson, I. T., L. E. Towne, and J. C. Aub: *Endocrinology,* **28**:851 (1941).

Olds, J.: *Sci. Am.,* October 1956.

O'Brien, D. F.: *Science,* **218**:961 (1982).

Pedersen-Biergaard, K., and M. Tonnesen: *Acta Endocrinol.,* **1**:38 (1948); *Acta Med. Scand.,* **131** (suppl. 213):284 (1948).

Pitts, R. F.: "Physiology of the Kidney and Body Fluids," 2d ed., Year Book, Chicago, 1968.

Poritsky, R.: *J. Comp. Neurol.,* **135**:423 (1969).

Rasmussen, A. T.: "Outlines of Neuroanatomy," 2d ed., Wm. C. Brown, Dubuque, Ia., 1943.

Routtenberg, A.: *Sci. Am.,* November 1978.

Rushmer, R. F.: "Cardiovascular Dynamics," 2d ed., Saunders, Philadelphia, 1961.

Santen, R., and C. Bardin: *J. Clin. Invest.,* **52**:2617, 1973.

Singer, S., and H. R. Hilgard: "The Biology of People," Freeman, San Francisco, 1978.

Singer, S. J., and Garth L. Nicholson: *Science,* **175**:720 (1972).

Smith, H. W.: "The Kidney," Oxford University Press, New York, 1951.

Steinberger, E., and W. O. Nelson: *Endocrinology,* **56**:429 (1955).

Tepperman, J.: "Metabolic and Endocrine Physiology," 2d ed., Year Book, Chicago, 1968.

von Bekesy, G.: *Sci. Am.,* August 1957.

Wang, C. C., and M. I. Grossman: *Am. J. Physiol.,* **164**:527 (1951).

Warwick, R., and P. L. Williams: Gray's Anatomy, 35th British ed., Saunders, Philadelphia, 1973.

Wersall, J., L. Gleisner, and P. G. Lundquist: *in* A. V. S. de Reuck and J. Knight (eds.), "Myostatic, Kinesthetic, and Vestibular Mechanisms," Ciba Foundation Symposium, Little, Brown, Boston, 1967.

Woodburne, R. T.: "Essentials of Human Anatomy," 3d ed., Oxford University Press, New York, 1965.

Zeifach, B. W.: *Sci. Am.,* January 1959.

SUGGESTED READING

Like all scientists, physiologists report the results of their experiments in scientific journals. Approximately 1100 such journals in the life sciences publish 250,000 papers each year. Every physiologist must be familiar with the ever-increasing number of articles in his or her own specific field of research.

Another type of scientific writing is the review article, a summary and synthesis of the relevant research reports on a specific subject. Such reviews, which are published in many different journals, are extremely useful. Also, the American Physiological Society has a massive program of publishing comprehensive reviews in almost all fields of physiology. These reviews are gathered into a series of ongoing volumes known collectively as the "Handbook of Physiology." Another important series is the *Annual Review of Physiology,* which publishes yearly reviews of the most recent research articles in specific fields and provides the most rapid means of keeping abreast of developments in physiology.

Another type of review which is of great value to the nonexpert is the *Scientific American* article. Written in a manner usually completely intelligible to the layperson, these articles, in addition to reviewing a topic, often present actual experimental data so that the reader can obtain some insight into how the experiments were done and the data interpreted. Many articles published in *Scientific American* can be obtained individually as inexpensive offprints from W. H. Freeman and Company, San Francisco, CA, 94104, which also publishes collections of *Scientific American* articles in a specific area, e.g., psychobiology. Two other magazines which frequently publish excellent reviews of physiology for the nonexpert are *New England Journal of Medicine* and *Hospital Practice.*

Another level of scientific writing is the monograph, an extended review of a subject area broader than the review articles described above. Monographs are usually in book form and generally attempt, not to cover all the relevant literature on the topic, but to analyze and synthesize the most important data and interpretations. At the last level is the textbook, which deals with a much larger field than the monograph or review article; its depth of coverage of any topic is accordingly much less complete and is determined by the audience to which it is aimed.

We have restricted our suggested readings to textbooks and monographs. Their bibliographies will serve as a further entry into the scientific literature. We list first a group of books which cover large areas of physiology.

BOOKS COVERING WIDE AREAS OF PHYSIOLOGY

Cell Physiology

Alberts, Bruce, Dennis Bray, Julian Lewis, Martin Raff, Keith Roberts, and James D. Watson: "Molecular Biology of the Cell," Garland Publishing, New York, 1983.

DeRobertis, E. D. P., and E. M. F. DeRobertis: "Essentials of Cell and Molecular Biology," Saunders, Philadelphia, 1981.

Flickinger, Charles J., Jay C. Brown, Howard C. Kutchai, and James W. Ogilvie: "Medical Cell Biology," Saunders, Philadelphia, 1979.

Giese, Arthur C.: "Cell Physiology," 5th ed., Saunders, Philadelphia, 1979.

Wolte, Stephen L.: "Biology of the Cell," 2d ed., Wadsworth, Belmont, Calif., 1981.

Organ-System Physiology

Berne, R. M., and M. N. Levy: "Physiology," Mosby, St. Louis, 1983.

Ganong, W. F.: "Review of Medical Physiology," 11th ed., Lange, Los Altos, Calif. 1983.

Guyton, A. C.: "Textbook of Medical Physiology," 6th ed., Saunders, Philadelphia, 1981.

Mountcastle, V. B. (ed.): "Medical Physiology," 14th ed., Mosby, St. Louis, 1980.

Selkurt, E. E. (ed): "Physiology," 5th ed., Little, Brown, Boston, 1984.

Anatomy

Basmajian, J. V.: "Grant's Method of Anatomy," 9th ed., Williams & Wilkins, Baltimore, 1975.

Goss, C. M.: "Gray's Anatomy of the Human Body," 29th American ed., Lea & Febiger, Philadelphia, 1973.

Hamilton, W. J., and N. W. Mossman: "Human Embryology," 4th ed., Williams & Wilkins, Baltimore, 1972.

Hollinshead, W. H.: "Textbook of Anatomy," 3d ed., Harper & Row, New York, 1974.

Leeson, T. S., and C. R. Leeson: "Human Structure," Saunders, Philadelphia, 1972.

Luciano, D. S., A. J. Vander, and J. H. Sherman: "Human Anatomy and Physiology," 2d ed., McGraw-Hill, New York, 1983.

Woodburne, R. T.: "Essentials of Human Anatomy," 7th ed., Oxford University Press, New York, 1983.

Pathophysiology

Frohlich, Edward D. (ed.): "Pathophysiology," Lippincott, Philadelphia, 1983.

Porth, Carol: "Pathophysiology," Lippincott, Philadelphia, 1982.

Sodeman, William A., and William A. Sodeman, Jr.: "Pathologic Physiology: Mechanisms of Disease," 6th ed., Saunders, Philadelphia, 1979.

SUGGESTIONS FOR INDIVIDUAL CHAPTERS

Chapter 1

Bernard, C.: "An Introduction to the Study of Experimental Medicine," Dover, New York, 1957 (paperback).

Brooks, C. McC., and P. F. Cranefield (eds.): "The Historical Development of Physiological Thought," Hafner, New York, 1959.

Butterfield, H.: "The Origins of Modern Science," Macmillan, New York, 1961 (paperback).

"The Excitement and Fascination of Science: A Collection of Autobiographical and Philosophical Essays," Annual Reviews, Palo Alto, Calif., 1966.

Fulton, J. F., and L. C. Wilson (eds.): "Selected Readings in the History of Physiology," 2d ed., Charles C Thomas, Springfield, Il., 1966.

Harvey, W.: "On the Motion of the Heart and Blood in Animals," Gateway, Henry Regnery, Chicago, 1962 (paperback).

Leake, Chauncey: "Some Founders of Physiology," American Physiological Society, Washington, D.C., 1961.

Chapter 2

Berns, Michael W.: "Cells," 2d ed., Holt, Rinehart, and Winston, New York, 1983.

Bloom, W., and D. W. Fawcett: "A Textbook of Histology," 10th ed., Saunders, Philadelphia, 1975.

Fawcett, D. W.: "The Cell," 2d ed., Saunders, Philadelphia, 1981.

Swanson, C. P., and P. L. Webster: "The Cell," 4th ed., Prentice-Hall, Englewood Cliffs, N.J., 1977.

Weiss, Leon, and Lambert Lansing (eds): "Histology: Cell and Tissue Biology," 5th ed., Elsevier, New York, 1983.

Chapter 3

Creighton, Thomas E.: "Proteins," Freeman, New York, 1984.

Lehninger, Albert L.: "Principles of Biochemistry," Worth, New York, 1982.

Smith, Emil L., Robert L. Hill, I. Robert Lehman, Robert J. Lefkowitz, Philip Handler, and Abraham White: "Principles of Biochemistry: General Aspects," 7th ed., McGraw-Hill, New York, 1983.

Chapter 4

Alberts, Bruce, Dennis Bray, Julian Lewis, Martin Raff, Keith Roberts, and James D. Watson: "Molecular Biology of the Cell," Garland Publishing, New York, 1983.

Creighton, Thomas E.: "Proteins," Freeman, New York, 1984.

Lloyd, David, Robert K. Poole, and Steven W. Edwards: "The Cell Division Cycle," Academic Press, New York, 1982.

Ruddon, Raymond W.: "Cancer Biology," Oxford University Press, New York, 1981.

Szekely, M.: "From DNA to Protein: The Transfer of Genetic Information," Macmillan, New York, 1980.

Watson, James D.: "Molecular Biology of the Gene," 3d ed., Benjamin, New York, 1976.

Watson, James D., John Tooze, and David T. Kurtz: "Recombinant DNA: A Short Course," Freeman, New York, 1983.

Chapter 5

Baker, Jeffrey J. W., and Garland E. Allen: "Matter, Energy, and Life," 4th ed., Addison-Wesley, Reading, Mass., 1982.

Burton, B. T.: "Human Nutrition," 3d ed., McGraw-Hill, New York, 1975.

Fersht, A.: "Enzyme Structure and Mechanism," Freeman, New York, 1977.

Lehninger, Albert L.: "Principles of Biochemistry," Worth, New York, 1982.

Lehninger, Albert L.: "Bioenergetics: The Molecular Basis of Biological Energy Tranformations," 2d ed., Benjamin, New York, 1971.

Newsholme, E. A., and A. R. Leech: "Biochemistry for Medical Science," Wiley, New York, 1983.

Smith, Emil, Robert L. Hill, I. Robert Lehman, Robert J. Lefkowitz, Philip Handler, and Abraham White: "Principles of Biochemistry: General Aspects," 7th ed., McGraw-Hill, New York, 1983.

Smith, Emil, Robert L. Hill, I. Robert Lehman, Robert J. Lefkowitz, Philip Handler, and Abraham White: "Principles of Biochemistry: Mammalian Biochemistry," 7th ed., McGraw-Hill, New York, 1983.

Stryer, Lubert: "Biochemistry," 2d ed., Freeman, New York, 1981.

Chapter 6

Alberts, Bruce, Dennis Bray, Julian Lewis, Martin Raff, Keith Roberts, and James D. Watson: "Molecular Biology of the Cell," Garland Publishing, New York, 1983.

Andreoli, T. E., J. F. Hoffman, and D. D. Fanestil (eds): "Membrane Physiology," Plenum, New York, 1978.

Harrison, Roger, and George G. Lunt: "Biological Membranes," 2d ed., Halsted, New York, 1980.

Quinn, Peter J.: "The Molecular Biology of Cell Membranes," University Park Press, Baltimore, 1976.

Saier, Milton H., Jr., and Charles D. Stiles: "Molecular Dynamics in Biological Membranes," Springer-Verlag, New York, 1976.

Weissman, Gerald, and Robert Claiborne (eds.): "Cell Membranes: Biochemistry, Cell Biology, and Pathology," Hospital Practice Publishing, New York, 1976.

Chapter 7

Bernard, C.: "An Introduction to the Study of Experimental Medicine," Dover, New York, 1957 (paperback).

Cannon, W. B.: "The Wisdom of the Body," Norton, New York, 1939 (paperback).

Langley, L. L. (ed.): "Homeostasis: Origin of the Concept," Dowden, Hutchinson, and Ross, Stroudsburg, Pa., 1973.

Slonin, N. B. (ed.): "Environmental Physiology," Mosby, St. Louis, 1974.

Tepperman, T.: "Metabolic and Endocrine Physiology," 4th ed., Year Book, Chicago, 1980.

Williams, R. W. (ed.): "Textbook of Endocrinology," 6th ed., Saunders, Philadelphia, 1981.

Chapter 8

Angerrine, J. B., Jr., and C. W. Cotman: "Principles of Neuroanatomy," Oxford University Press, New York, 1981.

Kandel, E. R., and J. H. Schwartz: "Principles of Neural Science," Elsevier/North-Holland, New York, 1981.

Kuffler, S. W., J. G. Nicholls, and A. Robert Martin: "From Neuron to Brain," 2d ed., Sinauer Associates, Sunderland, Mass., 1984.

Matzke, H. A., and F. M. Foltz: "Synopsis of Neuroanatomy," 4th ed., Oxford University Press, 1983.

McGeer, P. L., J. C. Eccles, and E. G. McGeer: "Molecular Neurobiology of the Mammalian Brain," Plenum, New York, 1978.

Ottoson, D.: "Physiology of the Nervous System," Oxford University Press, New York, 1983.

Shepherd, G. M: "Neurobiology," Oxford University Press, New York, 1983.

Chapter 9

Frigley, M. J., and W. G. Luttge: "Human Endocrinology: An Interactive Text," Elsevier, New York, 1982.

Tepperman, J.: "Metabolic and Endocrine Physiology," 4th ed., Year Book, Chicago, 1980.

Williams, R. W. (ed.): "Textbook of Endocrinology," 6th ed., Saunders, Philadelphia, 1981.

Chapter 10

Bourne, G. H. (ed.): "The Structure and Function of Muscle," 2d ed., Academic Press, New York, 1972 (vol. I), 1973 (vols. II and III), 1974 (vol. IV).

Bulbring, Edith, Alison F. Brading, Allan W. Jones, and Tadao Tomita (eds.): "Smooth Muscle," University of Texas Press, Austin, 1981.

Carlson, Francis D., and Douglas R. Wilkie: "Muscle Physiology," Prentice-Hall, Englewood Cliffs, N.J., 1974.

Hoyle, C.: "Muscles and Their Neural Control," Wiley-Interscience, New York, 1983.

Keynes, R. D., and D. J. Aidley: "Nerve and Muscle," Cambridge University Press, Cambridge, 1981.

Needham, Dorothy M.: "Machine Carnis: The Biochemistry of Muscular Contraction in Its Historical Development," Cambridge University Press, New York, 1971.

Peachey, Lee D., and Richard H. Adrian (eds.): "Handbook of Physiology: Section 10, Skeletal Muscle," American Physiological Society, Bethesda, MD, 1983.

Shephard, Roy J.: "The Physiology and Biochemistry of Exercise," Praeger, New York, 1981.

Squire, John: "The Structural Basis of Muscle Contraction," Plenum, New York, 1981.

Twarog, Betty M., Rhea J. C. Levine, and Maynard M. Dewey (eds.): "Basic Biology of Muscles: A Comparative Approach," Raven Press, New York, 1982.

Chapter 11

Berne, R. M., and M. N. Levy: "Cardiovascular Physiology," 4th ed., Mosby, St. Louis, 1981.

Heller, L. J., and D. E. Mohrman: "Cardiovascular Physiology," McGraw-Hill, New York, 1981.

Honig, C. R.: "Modem Cardiovascular Physiology," Little, Brown, Boston, 1981.

Little, R. C.: "Physiology of the Heart and Circulation," 2d ed., Year Book, Chicago, 1981.

Smith, J. J., and J. P. Kampine: "Circulatory Physiology: The Essentials," Williams & Wilkins, Baltimore, 1980.

Chapter 12

Davenport, H. W.: "The ABC of Acid-Base Chemistry," 7th ed., University of Chicago Press, Chicago, 1978.

Levitzky, M. G.: "Pulmonary Physiology," McGraw-Hill, New York, 1982.

Mines, A. H.: "Respiratory Physiology," Raven Press, New York, 1981.

Slonim, N. B., and L. H. Hamilton: "Respiratory Physiology," 4th ed., Mosby, St. Louis, 1981.

West, J. B.: "Respiratory Physiology: The Essentials," 2d ed., Williams & Wilkins, Baltimore, 1979.

Chapter 13

Davenport, H. W.: "The ABC of Acid-Base Chemistry," 7th ed., University of Chicago Press, Chicago, 1978.

Marsh, D. T.: "Renal Physiology," Raven Press, New York, 1983.

Rose, B. D.: "Clinical Physiology of Acid-Base and Electrolyte Disorders," McGraw-Hill, New York, 1984.

Smith, H. W.: "From Fish to Philosopher," Anchor Books, Doubleday, Garden City, N.Y., 1961 (paperback).

Sullivan, L. P., and J. J. Grantham: "Physiology of the Kidney," 2d ed., Lea and Febiger, Philadelphia, 1982.

Valtin, H.: "Renal Function," 2d ed., Little, Brown, Boston, 1983.

Vander, A. J.: "Renal Physiology," 3d ed., McGraw-Hill, New York, 1985.

Chapter 14

Brooks, Frank P. (ed.): "Gastrointestinal Pathophysiology," 2d ed., Oxford University Press, New York, 1978.

Davenport, Horace W.: "A Digest of Digestion," 2d ed., Year Book, Chicago, 1978.

Davenport, Horace W.: "Physiology of the Digestive Tract," 5th ed., Year Book, Chicago, 1982.

Grossman, M., V. Speranza, N. Basso, and E. Lezoche (eds.): "Gastrointestinal Hormones and the Pathology of the Digestive System," Plenum, New York, 1978.

Johnson, L. R. (ed.): "Gastrointestinal Physiology," 2d ed., Mosby, St. Louis, 1981.

Johnson, Leonard R., James Christensen, Morton I. Grossman, Eugene D. Jacobson, and Stanley G. Schultz (eds.): "Physiology of the Gastrointestinal Tract," Raven Press, New York, 1981 (two volumes).

Chapter 15

Adolph, E. F.: "Physiology of Man in the Desert," Interscience, New York, 1947.

Horvath, S. M., and M. K. Yousef (ed.): "Environmental Physiology: Aging, Heat, and Altitude," Elsevier, New York, 1981.

Kluger, M. J.: "Fever: Its Biology, Evolution, and Function," Princeton University Press, Princeton, 1979.

Tepperman, J.: "Metabolic and Endocrine Physiology," 4th ed., Year Book, Chicago, 1980.

Vander, A. J.: "Nutrition, Stress, and Toxic Chemicals: An Approach to Environmental Health Controversies" University of Michigan Press, Ann Arbor, 1981.

Williams, R. W. (ed.): "Textbook of Endocrinology," 6th ed., Saunders, Philadelphia, 1981.

Chapter 16

Johnson, M., and B. Everett: "Essential Reproduction," Blackwell Scientific Publications, Oxford, 1980.

Money, J., and A. E. Ehrhardt: "Man and Woman, Boy and Girl: The Differentiation and Dimorphism of Gender Identity from Conception to Maturity," Johns Hopkins University Press, Baltimore, 1973 (paperback).

Patten, B. M., and B. M. Carlson: "Foundations of Embryology," 4th ed., McGraw-Hill, New York, 1979.

Tepperman, J.: "Metabolic and Endocrine Physiology," 4th ed., Year Book, Chicago, 1980.

Williams, R. W. (ed.): "Textbook of Endocrinology," 6th ed., Saunders, Philadelphia, 1981.

Chapter 17

Barrett, J. T.: "Textbook of Immunology," 4th ed., Mosby, St. Louis, 1983.

Dubos, R.: "Man Adapting," 2d ed., Yale University Press, New Haven, 1980 (paperback).

Vander, A. J. (ed.): "Human Physiology and the Environment in Health and Disease," Freeman, San Francisco, 1976.

Vander, A. J.: "Nutrition, Stress, and Toxic Chemicals: An Approach to Environmental Health Controversies," University of Michigan Press, Ann Arbor, 1981.

Weiner, H.: "Psychobiology and Human Disease," Elsevier, New York, 1977.

Chapter 18

Kandel, E. R., and J. H. Schwartz: "Principles of Neural Science," Elsevier/North-Holland, New York, 1981.

Kuffler, S. W., J. G. Nicholls, and A. Robert Martin: "From Neuron to Brain," 2d ed., Sinauer Associates, Sunderland, Mass., 1984.

Nathan, P.: "The Nervous System," 2d ed., Oxford University Press, New York, 1983.

Ottoson, D.: "Physiology of the Nervous System," Oxford University Press, New York, 1983.

Shepherd, G. M.: "Neurobiology," Oxford University Press, New York, 1983.

Chapter 19

Kandel, E. R., and J. H. Schwartz: "Principles of Neural Science," Elsevier/North-Holland, New York, 1981.

Kuffler, S. W., J. G. Nicholls, and A. Robert Martin: "From Neuron to Brain," 2d ed., Sinauer Associates, Sunderland, Mass., 1984.

Ottoson, D.: "Physiology of the Nervous System," Oxford University Press, New York, 1983.

Shepherd, G. M.: "Neurobiology," Oxford University Press, New York, 1983.

Chapter 20

Kandel, E. R., and J. H. Schwartz: "Principles of Neural Science," Elsevier/North-Holland, New York, 1981.

Kuffler, S. W., J. G. Nicholls, and A. Robert Martin: "From Neuron to Brain," 2d ed., Sinauer Associates, Sunderland, Mass., 1984.

McGeer, P. L., J. C. Eccles, and E. G. McGeer: "Molecular Neurobiology of the Mammalian Brain," Plenum Press, New York, 1978.

Ottoson, D.: "Physiology of the Nervous System," Oxford University Press, New York, 1983.

Shepherd, G. M.: "Neurobiology," Oxford University Press, New York, 1983.

INDEX